Proceedings of the 41st
INDUSTRIAL WASTE CONFERENCE
May 13, 14, 15, 1986

Proceedings of the 41st
INDUSTRIAL WASTE CONFERENCE
May 13, 14, 15, 1986

Purdue University
West Lafayette, Indiana

The Purdue Industrial Waste Conference is under
the direction of:

The School of Civil Engineering
The Division of Conference and Continuation
Services

in cooperation with the

Indiana Department of Environmental Management
Indiana Department of Natural Resources
Indiana Environmental Management Board
Indiana Section of American Society of Civil
Engineers
Indiana Section of the American Water Works
Association
Indiana State Board of Health
Indiana Stream Pollution Control Board
Indiana Water Pollution Control Association

CRC Press
Taylor & Francis Group
Boca Raton London New York

CRC Press is an imprint of the
Taylor & Francis Group, an informa business

First published 1987 by CRC Press
Taylor & Francis Group
6000 Broken Sound Parkway NW, Suite 300
Boca Raton, FL 33487-2742

Reissued 2018 by CRC Press

© 1987 by LEWIS PUBLISHERS, INC.
CRC Press is an imprint of Taylor & Francis Group, an Informa business

No claim to original U.S. Government works

Publisher's Note
The publisher has gone to great lengths to ensure the quality of this reprint but points out that some imperfections in the original copies may be apparent.

Disclaimer
The publisher has made every effort to trace copyright holders and welcomes correspondence from those they have been unable to contact.

ISBN 13: 978-1-315-89028-9 (hbk)
ISBN 13: 978-1-351-06938-0 (ebk)

Visit the Taylor & Francis Web site at http://www.taylorandfrancis.com and the
CRC Press Web site at http://www.crcpress.com

PREFACE

The 41st Industrial Waste Conference was sponsored by the School of Civil Engineering of Purdue University.

Ninety-six technical papers were presented during the three days of the conference. Five papers were presented but not submitted for publication.

The papers are divided into the following 14 sections: Pretreatment Programs and Systems, Physical/Biological Systems, Hazardous/Toxic Wastes, Pulp and Paper Mill Wastes, Dairy Wastes, Plating Wastes, Oilfield and Gas Pipeline Wastes, Food Wastes, Coal, Coke, and Power Plant Wastes, Dye Wastes, Landfill Leachate, Metal Wastes, Laws, Regulations, and Training, and Miscellaneous.

As indicated in the table of contents, numerous papers at this year's conference dealt with the topics of PHYSICAL/BIOLOGICAL SYSTEMS (Section 2) and HAZARDOUS/TOXIC WASTES (Section 3).

The conference keynote speaker was Dr. Betsy Ancker-Johnson, Vice President, Environmental Activities Staff, General Motors Technical Center, Warren, Michigan. Dr. Ancker-Johnson emphasized that pollution control must take a multi-media approach, even though laws and regulations are primarily single-media based; e.g., Clean Water Act, Clean Air Act, etc. She presented the General Motors approach in this regard by citing several examples of "at-source" control on manufacturing processes and improvement of material use efficiency to minimize waste generation.

A duo, called "Echo," from Lafayette, Indiana, entertained the audience with various musical selections at the conference banquet. At the luncheon on Thursday, Harry L. Stout, Professor of Foreign Languages and Literature at Purdue University, made a presentation entitled: "Sherlock Holmes Decoded."

At the end of this proceedings is a comprehensive 10-year index (current and previous volumes). This index is cross-referenced by author and various subjects for the utility of the reader.

The efforts of all who gave freely of their time toward the success of the 41st Purdue Industrial Waste Conference are gratefully acknowledged.

JOHN M. BELL, Editor
Associate Professor of Environmental Engineering
School of Civil Engineering
Purdue University

CONTENTS

Section 1.   Pretreatment Programs and Systems

Section 2.   Physical/Biological Systems

Section 13. Laws, Regulations, and Training

Section 14. Miscellaneous

Section One
PRETREATMENT PROGRAMS AND SYSTEMS

1 IMPLEMENTATION OF AN INDUSTRIAL WASTEWATER PRETREATMENT PROGRAM

Paul C. Martyn, Supervising Civil Engineer

Jay G. Kremer, Head, Industrial Waste Section
County Sanitation Districts of Los Angeles County
Whittier, California 90607

BACKGROUND

The Sanitation Districts of Los Angeles County were established in 1925 under authority granted by the State of California, County Sanitation District Act of 1923. This act is currently embodied in Sections 4700 through 4894 of the California Health and Safety Code. The statutes associated with this act permitted the creation of sanitation districts to treat and dispose of wastewater on a regional basis, generally according to logical drainage areas, irrespective of city boundaries. As part of the authority granted by this act, sanitation districts had the ability to extend treatment services to areas even outside their boundaries by developing separate sewer service contracts.

There are now 27 separate sanitation districts within the Sanitation Districts of Los Angeles County which provide sewerage service to most of the Los Angeles County area outside the City of Los Angeles. In the central service area, 15 sanitation districts operate as the Joint Outfall System to service over 9,000 industrial companies and about four million people. Within the Districts, 75 cities provide the local collector sewer system while the Districts operate the large trunk sewers and wastewater treatment plants. In the Joint Outfall System, five inland tertiary activated sludge treatment plants provide about 150 mgd of treatment capacity. Treated wastewater from these plants is used for many reclamation purposes such as irrigation, industrial water supply and recharge of the underground domestic water supply. The Joint Water Pollution Control Plant (JWPCP) in Carson currently treats about 360 mgd of wastewater by advanced primary and partial secondary treatment with the effluent discharged into the Pacific Ocean. Secondary treatment is provided for about 200 mgd at the JWPCP.

Certain sections of the Joint Outfall System service area contain significant industrial development. Included in this group of industrial dischargers are some 20 petroleum refineries, several major secondary fiber paper mills, food processing industries, chemical manufacturing plants, and 520 electroplating companies, as defined by Federal categorical regulations. Inherent in such a large industrial base is the need to control toxic discharges and high strength wastes. A failure to provide such control may result in wastewater qualities which are incompatible with conventional municipal treatment processes.

Although the Districts have had a defined industrial waste control program since about 1952, a comprehensive "modern" program was first established in 1972. At this time it became evident that such control was needed to prevent major secondary treatment plant upsets and to protect the quality of reclaimed water. Concurrently the developing environmental movement gave impetus to this step.

OBJECTIVES OF AN INDUSTRIAL WASTE PRETREATMENT PROGRAM

At the Districts the objectives of an industrial waste program include the following:

1. Protect Districts' sewer and treatment plant capital facilities against damage from industrial wastes.
2. Prevent adverse effects on Districts' employees or Districts' wastewater conveyance, treatment or disposal processes.
3. Prevent harm to private property or adverse effects on public health.

4. Prevent public nuisances such as obnoxious odors.
5. Recover costs of industrial wastewater treatment and disposal from companies discharging the wastewater.
6. Protect the environment from toxic pollutants discharged by industry and comply with NPDES discharge limits and sludge quality standards for Districts' treatment plants.
7. Comply with Federal laws regulating industrial pollutant discharges to the sewerage system.

The United States EPA "General Pretreatment Regulations for Existing and New Sources" (40 CFR Part 403) of January 28, 1981, state that the intent of the regulations is to:

1. Prevent interference with treatment plant operations.
2. Prevent pass-through of pollutants in violation of limits of the NPDES discharge permit for the publicly owned treatment works (POTW).
3. Prevent municipal sludge contamination.

The intent of the national EPA pretreatment regulations should be the intent of all local industrial waste pretreatment programs. However, the inflexibility and the inherent limitations of a national program require that POTWs evaluate those special circumstances and waste discharge conditions that pertain to their operations. It may be necessary to implement more stringent effluent limitations on industrial dischargers or to provide specific discharge requirements on industries not regulated by the EPA. The EPA has recognized this situation and has specified that specific local limitations must be developed if pollutants discharged by a nondomestic source pass through, interfere or create special problems at a POTW.

NEED FOR INDUSTRIAL WASTE PRETREATMENT PROGRAMS

Many POTWs have been slow in implementing effective industrial waste pretreatment programs. In an assessment of past performance, a 1981 report by JRB Associates [1] which reviewed (in part) the control of industrial dischargers by POTWs, found that the aforementioned pretreatment goals were not being consistently achieved. About 75% of the POTWs had reported recurrent operation and maintenance or process upset problems attributed to industrial waste discharges. Over 60% of the POTWs evaluated did not monitor their influent wastewater for toxic heavy metals or toxic organic chemicals. Moreover, about 70% of the POTWs had no industrial waste monitoring program.

The need for an effective industrial waste pretreatment program is also mandated by the Clean Water Act. As specified in the 40 CFR 403 regulations of January 28, 1981, POTWs with flows in excess of 5 mgd or which serve significant industrial dischargers are required to have an approved pretreatment program.

ELEMENTS OF INDUSTRIAL WASTE REGULATORY PROGRAM

The Districts' program and the Federal pretreatment regulations contain essentially the same elements. These components to a pretreatment program are considered basic to an effective regulation of industrial discharges:

1. The POTW should have a clearly established legal authority to implement regulatory controls. Such authority is generally articulated in a wastewater ordinance.
2. Industrial waste discharge permits should be issued to condition and control the release of pollutants.
3. Effluent limitations must be developed to control toxic and problem pollutants. These limitations should complement national pretreatment regulations.
4. There must be a random program of monitoring and inspection of industrial waste discharges.
5. Industrial waste regulatory personnel should be adequately informed and knowledgeable of regulations and wastewater treatment processes.
6. Self-monitoring reports should be required of all major industrial companies.
7. Violations of waste discharge regulations should be enforced through civil and/or criminal actions.
8. The POTW should have the ability to halt or prevent the discharge of toxic pollutants to the sewerage system through industrial waste discharge permit suspension or revocation.
9. A means must exist to adequately fund the industrial waste regulatory program. This includes the collection of industrial waste treatment and disposal costs from companies in proportion to their burden on the sewerage system.

Table I. Industrial Waste Engineers Areas of Work Concentration

Area of Concentration	Federal Standard Industrial Classification Numbers (SIC Nos.)[a]
Petroleum Production, Refining and Associated Industries	1311 to 1389, 2911 to 2999, 4922 to 4925, 5171 to 5172, 5541,
Chemical Industry	2812 to 2899, 3011 to 3079, 5161, 7342, 7395,
Metal Finishing Toxic Wastes, Transportation Industries	3111 to 3199, 3211 to 3299, 3411 to 3499, 3511 to 3599, 3612 to 3699, 3711 to 3799, 3811 to 3873, 4011 to 4013, 4111 to 4172, 4212 to 4231, 4911, 4931 to 4939, 5085, 5511 to 5521, 7391, 7397, 7512 to 7539, 7531 to 7549,
Fiber Industries (paper, textiles, etc.) and Primary and Secondary Metals Industries	2211 to 2299, 2311 to 2399, 2411 to 2499, 2511 to 2599, 2611 to 2661, 2711 to 2795, 3312 to 3399, 7211 to 7220, 7332,
Food Industries, Hospitals	0111 to 0191, 0211 to 0291, 0711 to 0783, 2011 to 2099, 5141 to 5149, 5411 to 5499, 5812 to 5813, 7512 to 7539, 8062 to 8069, 8071 to 8072, 8081,
Industrial Wastewater Monitoring,[b]	Includes larger volume wastewater dischargers from all SIC No. categories.

Notes: [a]For specific identification of these categories see *Standard Industrial Classification Manual*, 1972, U.S. Government Printing Office.
[b]Includes engineering aspects of monitoring for wastewater quantity and quality and supervision of Monitoring Crews.

ORGANIZATION OF THE DISTRICTS' INDUSTRIAL WASTE PROGRAM

At the Districts, the Industrial Waste Section is divided into four subsections. These subsections include the Permit Processing, Field Engineering, Industrial Waste Engineering and Surcharge Processing groups; these groups cover all of the major elements of the industrial waste program. Laboratory analyses required by the program are provided by the Districts' extensive Water Quality Laboratory Section.

It is believed appropriate for industrial waste functions at a large sewerage agency to be centralized for effective industrial waste control purposes. Some sewerage agencies have fragmented industrial waste functions such as placing inspection services under the operations section, permit processing under the engineering section, and surcharge collection under the accounting section. As these functions are all interrelated, it is believed most efficient to have them centralized.

Currently six project engineers work in the Industrial Waste Engineering subsection. These engineers are assigned specific areas of industrial waste technology and work concentration areas (Table I). As shown in Table I, engineers concentrate on the industrial waste problems from specific SIC number groupings of industrial companies. The engineers develop in-depth knowledge of industrial processes that create pollutants and the pretreatment methods used to control the discharge of these pollutants. In addition, most of the industrial waste engineers employed to resolve specific discipline industrial waste problems have graduate level training in environmental engineering with either MS or PhD degrees. Such expertise is believed necessary to properly address the multiple industrial waste problems found in a large metropolitan area.

An example of the work of these industrial waste engineers is the elimination of small quantities of polychlorinated biphenyl (PCB) discharges from the Districts' sewerage system [2]. Districts' industrial waste regulations prohibit the discharge of any measurable quantities of "total identifiable chlorinated hydrocarbons" (TICH) which includes DDT, selected pesticides, and PCBs. A large paper mill was found to be discharging moderately large amounts of PCBs to the sewer. The industrial waste engineer, working cooperatively with the company, discovered that the PCBs originated with carbonless carbon paper office forms which were purchased as waste paper secondary fiber by the paper mill. By eliminating this source of secondary fiber, the PCB discharge problem was solved.

The industrial waste engineers are also responsible for coordinating the effort to implement those Federal categorical pretreatment regulations within their assigned area of expertise. This would include the review and comment on draft pretreatment standards, the preparation of summary documents outlining final regulations for affected industries, and the drafting of baseline monitoring report forms. Moreover, they are also responsible for insuring that these regulations are incorporated into waste discharge permits and that the mandatory 90-day compliance and on-going self-monitoring reports are submitted on time.

INDUSTRIAL WASTE PERMIT PROGRAM

Key to the development of an effective industrial waste permit program is the availability of information on the discharger. This should include the following:

- The location of the industrial discharger should be identified and the points of connection to the sewer system should be detailed.
- The quantity and chemical quality of the wastewater discharge should be provided.
- The industrial processes generating wastewater and a general description of the discharge should be available.
- Any unusual discharge characteristics need to be identified; these would include the potential for slug dumps and seasonal discharges.
- Any hazardous or toxic materials stored in meaningful quantities on-site should be noted.
- The pretreatment needs for industry need to be identified.
- The person at the company responsible for the wastewater discharge needs to be noted.

In large metropolitan areas where the above-mentioned information has not been obtained, its acquisition may be difficult and expensive. An industrial waste survey, as recommended by EPA, provides a means to develop this information. As much of this information is not static, it is believed that an industrial waste permit program, which requires updating as major wastewater discharge characteristics change, provides an effective way to continuously obtain such information. Alternately, permit expiration dates can be used; however, this procedure may result in needless updating while allowing major discharge modifications to remain unknown or not recognized in intervening periods.

The Districts' Wastewater Ordinance requires that all companies discharging significant quantities of wastewater to the sewer apply for an Industrial Wastewater Discharge Permit for each sewer outlet. Companies such as restaurants, laundromats, small dry cleaners, and most small service businesses are not included in the Districts' current permit program. All new industrial companies must obtain a permit before their wastewater can be accepted for treatment. The fairly rigorous permit review which follows, provided by a group of plan evaluation engineers, is believed essential in that potential waste discharge problems can be corrected before they occur. Industries which have the potential of discharging significant quantities of hazardous materials and do not have adequate treatment facilities have been prevented from connecting to the sewer until the noted deficiencies have been corrected.

Currently, permit information and resulting data are partially computerized. With the existing system, permit listing and sorting functions according to: permit number sequence, alphabetical order of the companies' names, SIC number, District number, location, permit classification or flow rate are possible. All industrial effluent monitoring data is planned to be computerized in the future. At present Districts'-obtained data but not industrial self-monitoring is on a computer system. The present computer system has the ability to organize complex information and generate sophisticated statistical reports and studies. These can be used by the Districts as the basis for management decisions regarding operations, design, economic, and regulatory requirements.

In the permit review process, careful attention is given to the use of toxic or hazardous materials in the industrial process and to the methods for disposal of such materials. A Districts' spill containment program requires that companies having tanks of toxic materials contain or berm the tanks such that a tank rupture will not cause a spill of such materials to the sewer. Flammable materials, cyanides, acids, heavy metals in solution, and toxic solvents must be located in separate spill containment areas such that the rupture of the largest single tank will be contained in the bermed or diked area. For example, at metal plating companies, any tank containing over 10 pounds of a heavy metal or cyanide in solution must be located in a spill containment area with cyanide tanks contained separately from any acid tanks. This program has significantly decreased the incidence of treatment plant upsets due to spills of heavy metal plating and other toxic wastes.

Since July 1981, the Sanitation Districts have required all companies with over 100,000 gpd of wastewater flow and with potential flammable waste problems, such as petroleum refineries, petroleum production facilities, and chemical manufacturing plants, to install and properly operate a combustible gas detection meter. Currently, 25 industrial dischargers (22 petroleum associated facilities and 3 chemical manufacturing plants) are subject to this requirement. The Districts require that these meters continuously record the explosivity of the vapor emanating from the industrial wastewater discharge and sound an alarm when the limit reaches 20% of the lower explosive limit (LEL). Through use of these meters, the Districts hope to prevent explosions in the sewers caused by flammable industrial waste.

Self-monitoring by the industrial dischargers is required by the permit program. All dischargers of significant quantities of toxic wastes are to monitor four times per year, while EPA categorical dischargers are required to provide such monitoring at least twice per year and major nontoxic dischargers at least once per year. These data are reported on a standard Districts' form which requires a certification statement to be signed by a company representative and requires that the laboratory conducting the analyses, as well as the monitoring location, be identified. In that the major intent of such monitoring is to keep the industry informed of their discharge compliance status, as a general rule, enforcement actions are not initiated against self-reported effluent violations.

As part of the Districts' permit and surcharge program, all companies having wastewater flows over 50,000 gpd are required to provide a full-time indicating, recording and totalizing flow meter. These flow meters are not required to be of any specific type but are required to be well engineered and properly designed. The difficulty found in simple verification of the flow in closed-channel meters has resulted in the Districts requiring open-channel meters wherever possible. Included with these flow meters are contact closure devices which are capable of activating automatic sampling equipment so that flow-proportioned samples may be obtained. All meters are required to be calibrated at regular frequencies with those meters having mechanical parts requiring more frequent calibration. Waste flows established by these flow meters are used to determine industrial waste surcharge payments and discharge masses for toxic materials. The largest number of flow meters is of the Palmer-Bowlus flume type.

ON-SITE INSPECTION OF INDUSTRIAL COMPANIES

Inspection of industrial companies is necessary to ensure compliance with industrial waste regulations. This field inspection program includes visiting industrial companies to investigate whether their discharges are in compliance with effluent standards, tracing toxic wastewater constituents in the sewer to locate industrial sources causing treatment plant upsets, and having the inspectors act as field representatives for the Districts. In dealing with industrial dischargers, it has been the Districts' philosophy to provide assistance to companies in their pollution control efforts whenever practicable rather than immediately instituting extensive enforcement and legal actions to coerce companies into compliance.

The present inspection group consists of eleven inspectors operating in three teams under the leadership of one supervising inspector. Each of these teams covers a specific area and each team has a senior inspector as a leader. This procedure allows for more than one inspector to be familiar with a given inspection area. As such, there is a greater flexibility in handling emergency conditions, both in off-hour situations and in those instances which require a concentrated work effort. Coordination between team members is accommodated by each inspection team having a brief morning meeting where current industrial waste, safety, and other job-related subjects are discussed. Once each week, all teams meet at the Districts' main office where complex industrial waste problems and major program goals are discussed.

The industrial waste inspectors have continuous access to a large body of information. The industrial waste discharge permits, blueprints of the companies' sewer connections, and correspondence with the main office have been reproduced on microfiche. While two sets of these microfiche are kept in the main office, a third set is distributed among the industrial waste inspectors according to their jurisdictions. Portable viewers allow for immediate recall of this information.

Compliance with discharge requirements is verified by the industrial waste inspectors through several means. In addition to visual observations of discharge conditions, field test kits are used to approximate concentrations of certain toxic materials and to establish levels of pH, sulfide, and ionized cyanide where change can be appreciable in transporting the sample to the laboratory. Cali-

Table II. Industrial Wastewater Effluent Limitations
(Phase I Limits)

Parameter	Maximum Discharge Concentration, (mg/L)
Arsenic	3
Cadmium	15
Chromium	10
Copper	15
Lead	40
Mercury	2
Nickel	12
Silver	5
Zinc	25
Cyanide (total)	10
Total Identifiable Chlorinated Hydrocarbon	Essentially None

brated explosimeters are used to analyze gas conditions in the sewer. Grab samples of industrial waste discharges are randomly taken to determine a company's compliance with the Districts' effluent limits. The industrial waste inspectors also check the adequacy of hazardous waste manifests; a company's failure to properly document anticipated waste loads may indicate that this material is being improperly sewered.

LOCAL WASTEWATER EFFLUENT REGULATIONS

It is considered appropriate that industrial effluent regulations be established to reflect local needs. At the Districts the effluent limits shown in Table II were established to meet the treatment plant discharge limits of the California Ocean Plan of 1972 and the need to eliminate upsets at the Districts' activated sludge treatment plants due to toxic conditions. These effluent limits are applicable to all industrial dischargers and are enforceable on grab samples. The ability to enforce effluent limits on grab samples is believed to complement the Federal pretreatment program limits which emphasize compliance as determined by 24-hour composite samples.

As described by Eason et al. [3], the Districts made a calculation of preliminary effluent limits using the formula below:

$$L = PR [Lp - La (1 - S)]$$

where
$$Lp = Le/(1 - E)$$

(1)

The terms used in these equations are defined below:

1. L is the calculated maximum concentration allowable in an industrial discharge.
2. P is the assumed ratio of maximum to average concentration in industrial wastewater containing a given pollutant.
3. R is the ratio of wastewater discharged containing a given pollutant to the total wastewater flow.
4. Lp is the calculated permissible influent concentration at the Districts' treatment plants necessary to meet the Ocean Plan discharge limit for a specified POTW removal efficiency.
5. Le is the existing Ocean Plan effluent limit, either maximum or average.
6. E is the anticipated fraction removed at the treatment plant for a given pollutant. Removal efficiencies used assume full secondary biological treatment.
7. La is the existing influent concentration before source control.
8. S is the fraction discharged by controllable industrial sources.

In addition to the criteria listed above, more stringent limitations were applied if additional protection was needed to prevent the contamination of treatment plant sludges or if a lower limit could be reasonably achieved by metal plating dischargers practicing good housekeeping procedures.

MONITORING OF INDUSTRIAL DISCHARGERS

In addition to the grab samples taken by the industrial waste inspectors, the Districts employ several industrial waste monitoring crews to take composite samples. Chain-of-custody procedures are followed on all compliance samples to insure the integrity of the sample results used in enforcement actions. These procedures are used when sampling EPA categorical dischargers. If a company is thought to be intentionally discharging pollutants to the sewer, these monitoring crews are used to provide surveillance sampling of the discharger in which bracket monitoring techniques are used in the adjacent sewer system.

The composite samples are taken with unattended automatic equipment. In sampling on-site at industrial dischargers, the equipment is normally serviced every day and the sample brought to the laboratory. The samplers currently used are of the ISCO brand. All samplers are iced in the sampler base for preservation and transported to the laboratory in an ice chest. Most monitoring of industrial dischargers involves one receptacle to contain all composite samples. However, if equipment tampering or an unusual discharge is expected, 24 individual sample bottles can be used to collect hourly samples which can then be analyzed separately. The presence of a partially filled or empty sample bottle in a series of bottles may indicate that equipment tampering occurred. Evidence tape can be used to assist in securing the sampling equipment.

When taking industrial waste samples, care must be taken to obtain a sample accurately reflecting the wastewater going to the sewer. Monitoring crew personnel try whenever possible to place the sampler intake tube in a moving wastewater stream where turbulence is a maximum. Isokinetic sampling is ideally sought; however, debris accumulation is commonly encountered when the sample tube is pointed in the direction of flow. Transport velocities must be sufficient to carry representative solids loadings to the sampler; in this regard the Districts have found that sample strainers often work to under-represent wastewater strengths. A preferred sampling location is immediately downstream of a Palmer-Bowlus or Parshall flume, downstream of a turbulent weir discharge, or in a sewer with rapid turbulent flow. Care must be taken when sampling in sample boxes, sumps, or clarifiers due to solids buildup which may occur with such flow impoundments.

ENFORCEMENT OF INDUSTRIAL WASTE REGULATIONS

The effectiveness of an industrial waste regulatory program is at least partly dependent upon the effectiveness of the enforcement program. The Districts have a patterned enforcement program which, while somewhat deliberate, leads to resolution of industrial waste problems. The Districts' program uses a three-level enforcement notice program. For nonemergency violations, an initial industrial violation will generally result in an oral warning followed by a written Information Notice citation if left uncorrected. On the Information Notice a correction time is specified, usually 30 days. If uncorrected, a Violation Notice is issued, again with a correction deadline. A Final Notice of Violation is issued if the violation still remains uncorrected. At each enforcement notice level a short letter is sent to a management level person in the company stating the problem, requesting a resolution, and stating the possible actions if the violation is left uncorrected. At the Final Notice of Violation level, the letter to the company requires that company personnel attend a compliance meeting at the Districts' office. In the compliance meeting, Districts and company personnel confer with the purpose of obtaining a technically adequate and economical solution to the industrial waste violation. A follow-up letter to the company after this meeting states the conditions agreed upon and states a timetable for their implementation.

Failure to eliminate the violations at this level will result in the Districts taking the matter to the Los Angeles County District Attorney for a compliance conference. At this conference, compliance measures are established and a timetable set for their implementation. Failure to meet these requirements will result in the District Attorney filing a misdemeanor court action against the company. However, if the Districts believe that an intentional and flagrant violation of effluent limits occurred, the matter can be directly referred to the District Attorney for prosecution.

Historically, the penalties obtained in the misdemeanor actions had not in general been commensurate with the seriousness of the offense. The typical penalty given was a small fine of several hundred dollars and a probationary period of usually two years. This situation has now begun to change. With a judiciary more in tune with environmental issues and aggressive action being pursued by the District Attorney's office, fines have now begun to exceed $100,000 with jail times required for those determined to be criminally responsible.

COLLECTION OF INDUSTRIAL WASTE TREATMENT COSTS

The Districts have collected from major industrial dischargers conveyance, treatment, and disposal costs since the 1972–73 fiscal year. On an annual basis the Districts' prevailing capital, and operation and maintenance costs, are allocated to the parameters flow, COD, and total suspended solids; these costs are then divided by the mass loading of each of these constituents in the Districts' sewerage system. To these unit rates are added the cost of operating the industrial waste program, which is again allocated to the parameters flow, COD, and total suspended solids and then divided by the industrial mass loading of each of these constituents in the Districts' sewerage system. These two unit rates are combined to establish annual charges for flow, COD, and total suspended solids. In addition, a peak flow charge exists for major dischargers which encourages an efficient use of sewerage facilities. This peak flow charge assesses a fee for discharges that exceed the average daily flow rate and which occur between the hours of 8:00 a.m. and 10:00 p.m.

The industrial waste revenue program is based on a self-monitoring concept where dischargers are responsible for determining effluent flow rates, and the larger industrial waste dischargers are also responsible for determining wastewater strengths. This practice requires that the Districts also monitor those dischargers that are covered by such a program; extensive monitoring occurs whenever major disputes are discovered. As part of this process, audits of the discharge statements are routinely conducted.

At present, the industrial waste revenue program is divided into a Short Form Program of one to six million gallons per year (MGY) using a charge for wastewater flow only, a User Charge program for hospital users based on annual occupancy levels and a Long Form program (over six MGY) where charges are based upon monitored values for total flow, peak flow, COD, and suspended solids. Revenue obtained is about 90% from the Long Form and about 10% from the Short Form and User Charge companies. Of the 1,500 companies on the program, about 500 are on the Long Form program. Revenue for the 1984–85 fiscal year totaled $15.5 million.

COST OF INDUSTRIAL WASTE REGULATIONS

The Districts' expenditures for the Industrial Waste Section totaled $2.7 million for the 1984–85 fiscal year. The individual components of these expenditures by category were as follows:

Permit Processing	$ 505,000
Sampling and Flow Measurements	360,000
Inspection	824,000
Industrial Waste Engineering	530,000
Surcharge Processing	191,000
Industrial Waste Laboratory Analyses	270,000
Legal Assistance	20,000
	$2,700,000

SUMMARY

An effective industrial waste control program is necessary for any POTW receiving significant amounts of nondomestic wastes. Of particular importance in these programs should be the consideration of local limits to augment those discharge standards established by the Federal categorical pretreatment program. Such limits should then be diligently enforced. A failure to develop local limits or a failure to ensure that there is adequate compliance with them by industrial dischargers may result in a wastewater that is incompatible with existing municipal treatment processes. Unnecessary discharge of pollutants to the environment is likely to ensue.

In a large municipal sewerage system, the proper regulation of industrial waste is a complex matter. With the knowledge required to understand the many unique industrial processes which generate wastewater and the procedures available to treat these discharges, it is believed essential to have a competent, technically-oriented staff available to coordinate this regulation. It is believed most effective when this staff addresses the problems of developing industrial waste discharge permits, reviewing and resolving discharge problems and implementing a revenue program in a centralized and coordinated manner.

REFERENCES

1. "Assessment of the Impacts of Industrial Discharges on Publicly Owned Treatment Works," a report for EPA, prepared by JRB Associates, McLean, VA, (November 1981).
2. Rhee, Choong Hee, Leslie D. Rose, and Jay G. Kremer, "Control of Polychlorinated Biphenyls in the Sanitation Districts of Los Angeles County," presented at the ASCE Conference in San Francisco, (1979).
3. Eason, John E., Jay G. Kremer, and Franklin D. Dryden, "Progress in the Sanitation Districts' Industrial Waste Source Control Program," presented at CWPCA Conference, (1977).

2 INDUSTRIAL WASTE PRETREATMENT USING A CELROBIC ANAEROBIC REACTOR

Donald F. DeAngelis, Environmental Process Supervisor

Louis Giokas, Senior Environmental Engineer
Badger Engineers, Inc.
Cambridge, Massachusetts 02142

James M. Nelson, Assistant City Engineer

David L. Welsh, Wastewater Plant Superintendent
City of Beloit
Beloit, Wisconsin 53511

INTRODUCTION

The purpose of this chapter is to present a case history for the Celrobic anaerobic pretreatment plant located in Beloit, Wisconsin.

A Celrobic anaerobic wastewater pretreatment plant has been constructed and is pretreating industrial wastewater prior to discharge into the municipal wastewater collection system. The pretreatment plant is an excellent example of cooperative effort by private industry and a municipality to solve a wastewater handling problem.

A new enzyme production facility was planned for construction by Enzyme Bio-Systems Ltd. in the Beloit, Wisconsin Industrial Park. The increased organic wastewater load from the enzyme plant would have utilized all the reserve capacity of the existing municipal wastewater treatment plant. To alleviate this problem, the decision was made by Enzyme Bio-Systems Ltd. and the City of Beloit to pretreat the enzyme plant wastewater along with all the other wastewater from the industrial park. The pretreatment plant greatly reduces the organic load to the municipal wastewater treatment plant (POTW), thereby reserving plant capacity for further growth in the greater Beloit area.

The case history presented herein defines the need for wastewater pretreatment, explains the development of the design basis, summarizes wastewater characteristics and laboratory treatability tests, describes the Celrobic anaerobic process, discusses engineering, procurement, construction and start-up of the facility, and summarizes operating results.

BACKGROUND

During 1983, Enzyme Bio-Systems Ltd., an enzyme producer and subsidiary of CPC International executed a site selection survey for a Midwestern site. The proposed enzyme plant was to manufacture enzymes used for the processing of various carbohydrates. The enzyme products from the plant were to be produced by fermentation followed by recovery and purification.

As in most fermentation processes, the disposal of fermentation liquor after the desired products have been recovered, is an issue which must be addressed due to the high strength of the wastewater. After an extensive review of various plant sites, Enzyme Bio-Systems Ltd. selected a site at the Beloit Industrial Park in Beloit, Wisconsin adjacent to Interstate 90.

The industrial park consists of a diversified cross-section of commercial/industrial users as shown in Table I. All the users discharge process and/or sanitary wastewater into the municipal collection system for treatment in the City's treatment plant. The municipal primary and secondary plant treats wastewater from the industrial park as well as servicing the rest of the City's residential, commercial, and industrial users. The municipal plant has a maximum capacity of 9.5 mgd and 27,000 lb/d of BOD.

Since the City POTW was operating near its maximum capacity, a moratorium on the addition of new wastewater sources was in affect.

Table I. Beloit Industrial Park Sewer (Commercial Users)

Commercial User	Type of Industry
Broaster	Oven Manufacturer
Bud Weiser	Car Dealer
Dillion Inn	Motel
Enzyme Bio-Systems	Enzyme Manufacturer
Freeman Shoe	Warehouse & Outlet
Frito-Lay	Corn & Potato Processing
Hormel	Food Canning
Louis Building	Machine Tools Manufacturers
D & M Manufacturers	Metal Fabricators
McDonalds	Restaurant
Plantation Motel	Motel
Wendys	Restaurant
Yates American	Machine Tools Manufacturers

The discharge of wastewater from the proposed enzyme plant would have overloaded the existing City plant. It was apparent that pretreatment of the enzyme plant wastewater was necessary, if Enzyme Bio-Systems was to locate in Beloit. Enzyme Bio-Systems decided to enter into an agreement with the City of Beloit to pretreat the enzyme plant wastewater along with all other wastewater from the industrial park. This agreement provides financial and operational benefits to both parties. Since it was expected that the combined enzyme plant and industrial park wastewater BOD would be 3370 mg/L and that methane produced from the pretreatment plant could be used in the enzyme plant, anaerobic technology was selected for pretreatment.

PROBLEM DEFINITION

Prior to construction of the enzyme plant, the major wastewater sources were Frito-Lay and Hormel. For the most part, the minor sources discharged sanitary wastes. Flows and characteristics of the industrial park wastewater were developed from limited historical records. Schedule pressure for the new enzyme plant precluded a thorough wastewater source survey which would have provided an understanding of the operation of the individual sources and would have developed data which accurately described the variations in flow and composition for each source. The wastewater flows and characteristics for the enzyme plant were expected values since the enzyme plant was under construction and no wastewater existed. The expected values were developed by Enzyme Bio-Systems by considering the number of process units in their new facility, production rates, and water usage. In addition, Enzyme Bio-Systems had previously performed pilot studies for their new enzyme processes and used this information to assist in the characterization of their wastewater.

Due to the impacts on the downstream municipal wastewater plant, it was imperative that the pretreatment plant be on-line and operational before start up of the enzyme plant, which was scheduled for March 1985. In March 1984, Enzyme Bio-Systems contracted Badger Engineers to engineer, procure, and construct a Celrobic anaerobic pretreatment plant.

DESIGN BASIS DEVELOPMENT

The first step in designing any treatment system is to develop a solid design basis in order to optimize selection and sizing of equipment. A thorough understanding of the influent wastewater characteristics is mandatory, including identification of major constituents and flow rates, as well as their daily fluctuations.

Due to the enzyme plant project time constraints, a long term sampling program of the existing industrial park wastewater, and an extensive pilot plant program were not feasible. Since performance guarantees were required, it was decided that confirmatory laboratory batch treatability tests would be performed during the design phase. The performance guarantees included percent BOD removal, methane production, and utility consumption.

A wastewater sample from the existing industrial park interceptor and a synthesized sample of the enzyme plant wastewater were combined in appropriate portions to perform batch treatability tests and wastewater characterization studies. The main objective of these tests was to determine the level to which the combined industrial park wastewater was anaerobically treatable, if toxic or inhibitory components were present, and if so, to identify which of the wastewater sources were toxic or non-

Table II. Characterization of Combined Enzyme Bio-Systems/Industrial Wastewater[a]

Total COD	5,300 mg/L
Soluble COD	4,000 mg/L
BOD_5	2,600 mg/L
BOD_{10}	3,100 mg/L
BOD_{15}	4,500 mg/L
BOD_{20}	4,500 mg/L
Total Kjeldahl Nitrogen	170 mg/L
Oil and Grease	1,500 mg/L
Ammonia Nitrogen	38 mg/L
Phosphorus	99 mg/L
Sodium	300 mg/L
Potassium	170 mg/L
Total Calcium	250 mg/L
Soluble Calcium	250 mg/L
Magnesium	68 mg/L
Nickel	0.05 mg/L
Iron	10 mg/L
Zinc	0.52 mg/L
Cobalt	Less than 0.05 mg/L
Manganese	0.50 mg/L
Chlorides	544 mg/L
Sulfate	120 mg/L
Sulfite	Less than 1.0 mg/L
Sulfide	2.4 mg/L
Nitrate	0.075 mg/L
Nitrite	0.071 mg/L
VFA	560 mg/L
TSS	2,160 mg/L
VSS	2,080 mg/L

[a]All values are totals unless otherwise specified.

biodegradable. Other objectives included obtaining better definition of the characteristics of wastewater.

Results of the wastewater characterization for the combined wastewater are presented in Table II.

Four sets of anaerobic batch treatability tests were performed to determine levels of biodegradability and methane production. Four sets of tests were performed to identify the impacts of: 1) treating only the industrial park wastewater; 2) treating only the Enzyme Bio-Systems wastewater; 3) treating the total combined wastewater; and 4) treating the soluble portion of the combined wastewater. The anaerobic batch tests were carried out in syringes where seed sludge, nutrient solution, and the test wastewater were combined and incubated at 37 C for 35 days. Figure 1 results from batch test data and shows the rate of methane formation as a percent versus time. As shown in Figure 1, most of the sample wastewater was biodegraded after 15–20 days.

Table III summarizes the results of the batch tests. Column 1 indicates the amount of COD from the wastewater sample added to the test syringe. Column 2 is a calculation of the COD removed based on the methane produced (i.e., Kinetic COD), and Column 3 compares methane produced as a percentage of COD added. The results of these tests were:

- The industrial park wastewater sample Test 1 was essentially completely anaerobically degradable, including the insoluble COD.
- The wastewater sample from the enzyme plant Test 2 was anaerobically treatable with kinetic COD efficiencies of about 75% at the concentrations expected in the full scale Celrobic plant.
- The combined wastewater sample Test 3 and Test 4 were similarly treatable with kinetic COD efficiencies of about 75%.

After determining the combined industrial park and enzyme plant wastewater was anaerobically treatable, a final design basis was developed. The design basis was developed using results of the wastewater characterization and anaerobic batch treatability tests. In addition, reactor sizing was

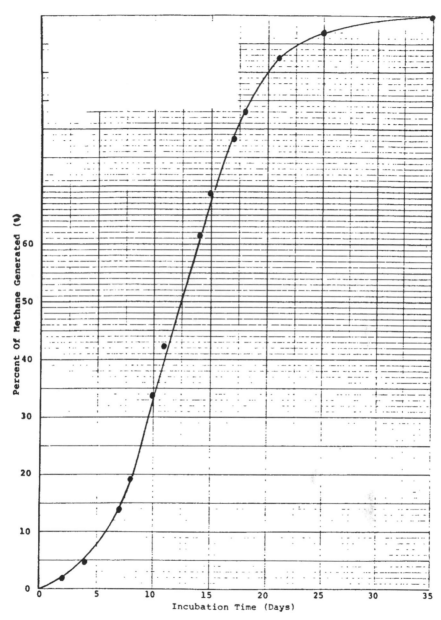

Figure 1.   Batch test results.

based on experience gained from the existing Celrobic anaerobic reactors. The design basis takes into account the expected initial wastewater flow and loadings as well as future anticipated conditions.

The plant is designed to pretreat a combined wastewater flow of 802,000 gallons per day. A summary of the influent and effluent design parameters for the initial and future conditions are given in Table IV. The influent design parameters are based on data provided by Enzyme Bio-Systems Ltd. and the City of Beloit. The major difference in the initial conditions versus the future conditions is total flow. The initial flow conditions are based on the actual current (1985) wastewater flows from

Table III. Batch Treatability Tests — Summary

			(1)	(2)	(3)
			COD (mg)		
Test #	Description	Dilution	Total Added	Kinetic Less Blank	Corrected COD, % Kinetic
1A	Industrial Waste	None	144	144	100
1B	Only	0.30:1	43	45	105
2A	Enzyme Waste	1:4	81	62	75
2B	Only	1:10	32	14	45
3A	Unfiltered	None	186	207	110
3B	Combined	1.2:5	74	55	75
3C	Waste	1:5	37	25	70
4A	Filtered	None	140	92	65
4B	Combined Waste	1.2:5	56	39	70

the industrial park plus the anticipated flow from the new enzyme plant. The future flow conditions allow for increases in existing industrial park flows plus expansion of the enzyme plant.

BOD concentrations are as received from Enzyme Bio-Systems Ltd.; COD concentrations were developed from BOD data using a BOD/COD ratio of 0.68. The effluent design parameters are expected values.

Other major design assumptions developed with Enzyme Bio-Systems and the City are as follows:

- Feed sulfate levels would not present a toxicity problem.
- Sufficient buffering capacity would be available in the wastewater.

Table IV. Design Parameters

INFLUENT

	Initial		Future	
Flow (Daily Average)	502,000 gpd (350 gpm)		802,000 gpd (560 gpm)	
Parameter	mg/L	lb/day	mg/L	lb/day
BOD	3,370	14,110	3,830	25,620
COD	4,960	20,756	5,630	37,660
TSS	2,395	10,025	2,280	15,230
Oil and Grease	170	715	300	2,000
pH	6.5-7.5			
Average Temperature	68°F			
Temperature Range	60°F–75°F			
BOD/COD Ratio	0.68			

EFFLUENT

	Initial		Future	
Flow (Daily Average)	502,000 gpd (350 gpm)		802,000 gpd (560 gpm)	
Parameter	mg/L	lb/day	mg/L	lb/day
BOD	675	2,820	765	5,120
COD	990	4,150	1,125	7,530
TSS	400	1,675	400	2,675
Oil and Grease	60	250	75	500
pH	6.5-7.5			
Average Temperature	70°F			

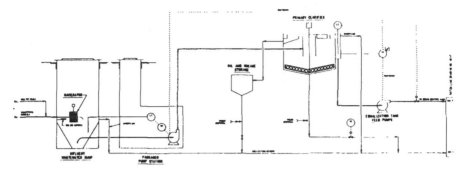

Figure 2.   Process flow diagram of pretreatment plant.

- Macro and micro nutrients would be available in the wastewater.
- No toxic or inhibitory substances would be present in the wastewater.
- Feed calcium levels would not be high enough to adversely affect the reactor performance.
- It was assumed that there were no chemical components which would adversely affect the formation of methane.
- It was assumed no stormwater would be handled.

PROCESS DESCRIPTION

The wastewater treatment system is an upflow, fixed film, random packed, anaerobic biological system designed to pretreat the combined industrial wastewater prior to discharge into the Beloit POTW. The primary objective of the Celrobic reactor is to remove and convert COD-BOD into usable fuel gas. The pretreatment plant consists of a macerator, influent pump station, primary clarifier, equalization tank, and Celrobic anaerobic reactor. The offgas handling system includes a blower to compress offgas for use in the enzyme plant boilers, and an offgas flare to dispose of the gas during boiler shutdown or emergencies. Process flow diagrams of the pretreatment plant are presented in Figures 2 and 3.

As shown in Figure 2, wastewater flows by gravity into the influent wastewater sump from two separate underground sources; the industrial park interceptor and an independent discharge, from the enzyme plant. The combined flow passes through a macerator in order to reduce larger solids prior to pumping. In addition, a bypass overflow with a coarse bar rack is provided for periods when the macerator requires servicing. After maceration, the combined wastewater discharges into the influent

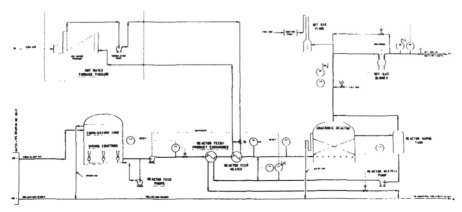

Figure 3.   Process flow diagram of pretreatment plant.

wastewater sump. The lift station sump pumps are the variable flow type and are used to eliminate on-off flow surges to the primary clarifier. The lift station pumps are housed in the packaged underground pump station.

The combined wastewater enters the primary clarifier for removal of settleable solids as well as oil and grease. Solids which settle in the primary clarifier are periodically discharged into the underground waste treatment bypass line which discharges directly into the City interceptor. Floating oil and grease are skimmed from the primary clarifier and discharged into an oil and grease storage tank. Oil and grease are periodically trucked offsite for disposal although they can be alternately discharged back into the City interceptor. Clarified effluent from the clarifier is pumped to the equalization tank using the equalization tank feed pumps.

The combined wastewater flow is pumped into the equalization tank in order to minimize fluctuations in feed flow and composition. As shown in Figure 3, equalization tank contents are mixed by circulating wastewater through three equalization tank eductors. Wastewater is pumped from the equalization tank, based on flow control, via the reactor feed pumps through the reactor feed/product heat exchanger. The object of this heat exchanger is to recover heat from the anaerobic reactor effluent. Wastewater feed from the reactor feed/product heat exchanger then passes through a reactor feed heater where hot water is used to control the reactor temperature. Hot water is supplied by a hot water furnace package.

Heated raw wastewater is then mixed with reactor recycle. By recycling the reactor effluent, two critical objectives are promoted: reactor stability and good hydraulic efficiency. It was assumed that the raw waste contains adequate alkaline material to buffer the water in the reactor. This alkaline material helps maintain the appropriate pH of the system by buffering short-chain fatty acids, carbon dioxide, and other acids formed in the reactor.

Although cellular synthesis is relatively small in most anaerobic processes, appropriate quantities of nutrients are required for synthesis of cellular material. For wastes deficient in nitrogen and phosphorus containing compounds, phosphoric acid and ammonia are generally the nutrients of choice. These nutrients are mixed with the buffered waste stream. In this case, analysis of the waste streams indicated sufficient phosphorus and nitrogen were available in the industrial park wastewater.

Wastewater which has been heated, flows into the anaerobic reactor. The anaerobic reactor utilizes an upflow, fixed film, random-packed design. Microorganisms that decompose the waste attach themselves to the packing. This attachment mechanism ensures that the solids retention time in the reactor is considerably longer than the hydraulic retention time. This relationship of solids retention time to hydraulic retention time is a key to a successful high-rate anaerobic process.

The nature and quantity of attached microorganisms, the dilution of fresh feed with recycled reactor effluent, and the process control system provide a high system tolerance to shock loads and toxic materials. The entrapment and occlusion of the biomass in the random packing enhances operating stability more than any other single factor. The occlusion phenomena, which differentiates a random packed reactor from a reactor with oriented packing, and from suspended growth type reactors, shields a large portion of the reactor biological population from toxicants by limiting diffusion of the toxicant into the biomass. Once the toxicant has been removed from the reactor, the occluded biomass which has not been affected by the toxic conditions, rapidly compensates for the surface coating of biomass which has been killed via diffusion of the toxic material.

A portion of the microbial population contained in the reactor, hydrolytically convert larger organic molecules into volatile fatty acids. Other segments of the microbial population convert the fatty acids into methane and carbon dioxide. During these acid- and gas-forming reactions, a relatively small fraction of the waste substrate is utilized for cell synthesis and cellular requirements.

In order to realize high quality effluents from a single-stage anaerobic reactor, proper mixing must be accomplished in conjunction with favorable biokinetics. This mixing in the anaerobic filter is accomplished by the upflow design, liquid recycling, and gas bubble formation. As a result of the upflow design, floating waste components do not accumulate in the top of the reactor. To prevent the loss of good mixing characteristics in the reactor through over-accumulation of microorganisms, biomass control is utilized. Biomass control is an on-line method of managing the biomass content of the reactor.

The treated effluent overflows from the anaerobic reactor into a reactor surge tank and is pumped out using a reactor recycle pump. Part of this flow is recycled to the filter and part passes through reactor feed-product exchanger prior to discharge in the City interceptor.

The gas formed in the anaerobic reactor, composed of methane, hydrogen sulfide, carbon dioxide, and water is compressed by the offgas blower for use in the boilers in the enzyme plant. An offgas flare is also included so the offgas can be safely burned when the boiler is not in use, in the event a power failure or other emergency.

PROJECT EXECUTION

Enzyme Bio-Systems Ltd. began evaluating wastewater pretreatment options for the proposed plant during the last quarter of 1983, and the first quarter of 1984. Since the completion of the new enzyme plant was to occur during the first quarter of 1985, a short schedule in addition to a competitive cost was important for the pretreatment project. To accomplish these objectives, Enzyme Bio-Systems Ltd. solicited lump sum, turnkey competitive bids.

The project was awarded to Badger Engineers, Inc., a sublicensor of Celrobic technology during March of 1984. The project was completed in 10-1/2 months. A chronology of project milestones appears in Table V.

PLANT OPERATION

Startup

Construction of the pretreatment plant was completed on February 7, 1985, and plant commissioning began on February 18, 1985. Commissioning and startup activities were performed for the next two weeks. In order to facilitate a quick startup, the anaerobic reactor was seeded with anaerobic sludge from the primary and secondary sludge digesters located at the City's municipal plant. Prior to adding seed sludge, the anaerobic reactor was filled with City water and heated to 90°F using the hot water furnace and the reactor feed heater. Approximately 50,000 gallons of seed sludge were transferred via tank truck by the City operators and pumped into the reactor. Feed wastewater was not introduced immediately into the reactor. The contents of the reactor were mixed for a period of two days by the recycle pump and maintained at 90°F. During this time, the primary clarifier, equalization tank, and offgas flare were commissioned, and wastewater allowed to fill the equalization tank.

Initially, wastewater feed to the reactor was started at a low hydraulic and organic loading rate in order not to wash out the seed sludge or organically overload the reactor. The initial flow rate was set at 150 gpm with a corresponding organic loading of 0.1 lb COD/ft3/day. Over the next six weeks, feed flow rate was incrementally increased to 280 gpm with organic loading up to 0.225lb COD/ft3/day. At this time, the reactor was treating all of the available wastewater. Wastewater flow and concentrations were below the design basis because the Enzyme Plant was not at full production.

Operation

The pretreatment plant is located approximately five miles from the City's wastewater treatment plant. City operators from the municipal plant make daily trips to the pretreatment plant for routine operation and maintenance. Typically, the tasks performed include:

Task	Frequency
• Take wastewater samples for analysis in City's laboratory	Once Per Day
• Record all instrument readings	Once Per Day
• Check water seals in flare and reactor overflow	Once Per Shift
• Backwash or clean recycle pump suction strainer	Once Per Shift
• Drain condensate from all offgas piping and instrument air compressor	Once Per Day
• Clean pH probes and check pH with portable pH meter	Once Per Shift
• Clean clarifier scum box and check oil and grease flow to sewer	As Necessary
• Perform routine equipment maintenance	As Necessary
• Fill out operation log book and note any changes or problems	Once Per Shift

The City's municipal plant operates on three 8-hour shifts, and the operators visit the pretreatment plant once per shift. A total of three to four hours per day is required at the site for normal operation. At times, it is necessary to spend additional time at the plant to clean heat exchangers, maintain pumps, clean out tanks, and for general maintenance.

OPERATING DATA SUMMARY

A summary of the major influent wastewater characteristics for the time period between March 1985 and March 1986 is presented in Table VI. The summary shows a comparison of the design basis

Table V. Project Milestones

Milestone	Date
Award & Start Engineering	March — 1984
Batch Treatability Test Completion	May — 1984
Major Equipment Purchased	June — 1984
General Site Preparation	July — 1984
Erection of Major Vessels	October — 1984
Packing Loaded in Reactor	November — 1984
Field Check of P & I Diagrams	January — 1985
Operator Training	February — 1985
Mechanical Completion	Feb. 7 — 1985
Reactor Seeded	Feb. 22 — 1985
Feed Introduced	Feb. 25 — 1985

and initial operation (March to May 1985) and more recent operation (December 1985 to March 1986).

The waste characteristics during the period of initial operation show that the wastewater BOD and COD concentrations, while highly variable, were lower than the design basis. As mentioned previously, the enzyme plant was not at full capacity during this period and a combination of lower flow and lower organic concentrations limited the ability to increase the reactor organic loading to design rates.

The more recent characteristics of the wastewater show that the BOD and COD concentrations are higher than design and extremely variable. Industrial park contributors have been made aware of this problem, and efforts have been successfully made recently to reduce the organic loading by the individual sources. The pretreatment plant has been able to process all of the wastewater, even at the higher organic concentrations because the design was based on future expansion at the industrial park, i.e., higher flowrates than were expected for 1986.

Both periods of operation shown in Table VI show much higher oil and grease concentrations than expected. This has created problems and requires more operator attention than expected for the primary clarifier and the oil and grease storage system. The City of Beloit and the industrial park contributors are presently involved in determining if the oil and grease levels can be reduced at their source.

Table VII presents a summary of the operating performance of the Celrobic anaerobic reactor. The time period shown (February and March 1986) was chosen because it is a period of high organic loadings and represents the current operation of the reactor, one year after startup. Table VII compares actual reactor performance with the expected performance used during design.

As shown in Figure 4, the organic loading has been highly variable during February and March 1986. This is caused by high variations in influent COD concentrations. Although the reactor loading

Table VI. Influent Wastewater Characterization

Parameter	Design Basis		Initial Operation (Mar. 1985 thru May 1985)		Present Operation Dec. 1985 thru Mar. 1986	
	Initial	Future	Average	Range	Average	Range
Total COD, mg/L	7,720	8,250	4,600	1,800–8,900	9,650	3,500–32,000
Soluble COD, mg/L	4,960	5,630	1,870	800–6,200	6,740	1,200–27,000
Soluble BOD$_5$, mg/L	3,370	3,830	2,050	860–3,150	5,730	1,500–20,100
Oil and Grease mg/L	170	300	1,350	300–3,900	1,710	600–4,370
TSS, mg/L	2,400	2,280	2,410	600–6,400	2,520	940–6,950
Total Kjeldahl Nitrogen, mg/L	170	170	90	60–110	NA*	NA*
Phosphorus, mg/L	100	100	10	4–20	28	13–73
Sulfate, mg/L	120	120	45	30–65	64	24–130
BOD/COD Ratio	0.68	0.68	.41	.22–.62	.62	.36–.90
pH	6.5–7.5	6.5–7.5	9.2	5.9–11.8	8.2	6.1–12.4

* not available.

Table VII. Celrobic Reactor Performance

	Design Basis	Actual Operation	
		Average	Range
Organic Loading, lbCOD/ft3/D	0.625	.44	0.2–0.83
BOD Removal Efficiency (%)	75	75	50–83
COD Removal Efficiency (%)	75	68	45–85
TSS (mg/L)	400	460	130–1,200
Oil and Grease (mg/L)	75	570	110–1,500
pH	6.5–7.5	6.9	6.8–7.2

has not been stable, the BOD removal efficiencies have averaged 75% and have been as high as 83%. Figure 5 shows the daily BOD removal efficiencies during February and March 1986.

As can be seen in Table VII, the COD removal efficiencies have been lower than expected. This is because the current BOD/COD ratio is 10% lower than the expected ratio indicating that the COD portion of the wastewater is not as biologically degradable as predicted. The actual reactor effluent pH and TSS levels are very close to the expected values. The actual oil and grease concentration in the reactor effluent is 7 to 8 times higher than expected. This is because the oil and grease concentration in the feed to the clarifier is 15 times higher than originally designed.

OPERATING ISSUES

After one year of operation, four major operating issues were identified as impacting the efficient operation of the pretreatment plant.

pH Excursions

Four days after initially seeding the reactor, the influent pH dropped approximately to 2.0, which caused a reactor pH decrease to 5.2. After neutralizing the reactor with sodium hydroxide, additional seed sludge was added, and the reactor put back on line. During the design basis development, it was decided not to install a buffer addition system due to the high pH and alkalinity of the influent wastewater. In order to resolve the pH excursion problem, three changes were made: 1) a pH monitor-

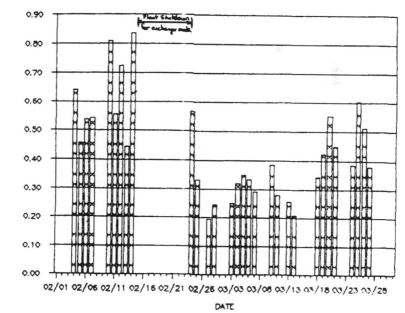

Figure 4.   Reactor organic loading (2/86–3/86).

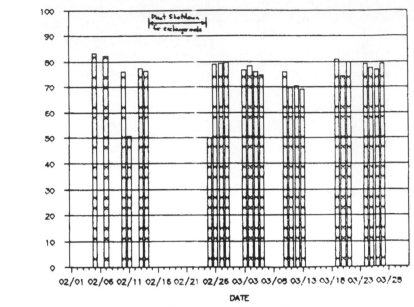

Figure 5.   Reactor BOD removal efficiency (2/86-3/86).

ing and reactor feed shutdown system was installed in July 1985; 2) an upstream neutralization system, not part of the pretreatment plant, was modified to prevent shock loads of sulfuric acid in November 1985; and 3) a sodium hydroxide storage tank and addition system were added to the pretreatment plant.

To date, the sodium hydroxide addition system has not been used since modification to the upstream neutralization system has prevented further pH excursions.

Wastewater Variability

As shown in Table VI, wastewater flows, COD, BOD, TSS, and oil/grease concentrations vary considerably when compared with the pretreatment plant design basis. Organic loading changes as high as 50% per day are experienced with only minor transient changes in performance following the large changes in loading.

Odors

Although odors have not impacted performance of the pretreatment plant, odors are an annoyance. One of the major sources of odors was from the reactor pressure/vacuum relief valve. The cause was identified as "high" pressure surges as offgas from the reactor flowed through the flare water seal. This was resolved by lowering the flare water seal level from 9" H_2O to 4" H_2O in order to lower reactor operating pressure. Other sources of odors included unsealed manhole connections, vents, etc. These sources have been eliminated to prevent odors in the future.

Heat Exchanger

Under design conditions, the wastewater feed temperature is 60°F, the operating temperature of the reactor is 98°F, and the pretreatment plant effluent is 63°F. In order to achieve this level of heat conservation, a feed-effluent heat exchanger was required. This relatively close "approach temperature" of 3°F was selected for two reasons: 1) at $5 per million BTU's, a one degree rise in the effluent temperature is worth approximately $12,000 per year in operating costs; and 2) the solids contained in the waste samples settled very rapidly, and with a clarifier preceding the heat exchanger, fouling of the heat exchanger with solids was not predicted.

A plate and frame feed-effluent heat exchanger was installed primarily because of its cost effectiveness and because it is easier to clean than other types of heat exchangers. Operating experience has indicated that the feed side of the heat exchanger fouls. The solids which foul the heat exchanger are small (1mm x 1mm) grease coated potato skins which, due to their neutral buoyancy, pass through the clarifier and foul the feed side of the heat exchanger. The result is that over a few weeks period, as the differential pressure across the heat exchanger increases, bypassing of the feed side of the heat exchanger is required to maintain feed flow, and the temperature of the reactor falls, which results in lower reactor conversions, until the heat exchanger is disassembled, cleaned, and returned to service. Testing is underway to resolve this operating problem.

SUMMARY AND CONCLUSIONS

Combined wastewater from the Beloit Industrial Park is anaerobically treatable.

Seeding and startup of the Celrobic reactor was accomplished relatively rapidly using municipal anaerobic digester supernatant.

The Celrobic plant has been in operation since February 1985 and has:

- Removed as much as 34,000 lbCOD/D.
- Achieved 75 to 80% BOD removals at loadings of 0.2 to .65 lbBOD/ft3/D and 0.2 to 0.8 lbCOD/ft3/D.

Operation of the Celrobic system has been remarkably stable even though:

- The actual wastewater feed is considerably different than the design basis.
- Extremes in feed pH have occurred.
- Rapid variations in wastewater characteristics and reactor loading have occurred.

The pretreatment plant has achieved the above results with considerably less operator attention than a comparable aerobic treatment system.

To minimize the potential for odor associated with anaerobic treatment, a great deal of attention must be paid to mechanical details.

Considerable care should be taken to define wastewater characteristics when selecting heat exchangers for wastewaters containing suspended solids.

The pretreatment plant has successfully performed in terms of BOD removal, methane production, and consumption of utilities despite extreme variations in wastewater feed and minor mechanical problems.

3 MATERIAL SUBSTITUTION LOWERS INDUSTRIAL WASTE TREATMENT COSTS

Gail E. Montgomery, Project Engineer

Bruce W. Long, Project Engineer
Black and Veatch Engineers-Architects
Kansas City, Missouri 64114

INTRODUCTION

The United States Environmental Protection Agency (EPA) has been moving toward the development and implementation of industrial pretreatment standards for both existing and new point sources since the mid-1970's. The responsibility of imposing the pretreatment standards, in most cases, has been assigned to municipalities following the development and adoption of a local industrial sewer use ordinance. Pollutant discharge limitations of the local ordinances vary from location to location, and since the municipality must comply with its NPDES permit and ensure that the sludge from its wastewater treatment plant is suitable for handling and disposal (including possible reuse), the local ordinances are often more restrictive than the federally mandated standards.

Pretreatment standards have been promulgated on two different criteria, only one of which applies to a given industrial category or subcategory. The first places a concentration limit on individual pollutant parameters with no explicit limit on the volume of the process wastewater discharged. The second limits the daily mass of a pollutant discharged to the sewer and bases the mass limitation on some unit of the particular industry's production process, such as the surface area of parts cleaned or plated or per unit mass of a particularly significant raw material used.

Industries can achieve compliance with mass-based standards by a combination of approaches, such as reducing the volume of wastewater discharged and/or reducing the concentration of pollutants by appropriate treatment. Mass-based standards encourage use of water conservation measures, for as the volume of wastewater discharged decreases, the concentration of pollutants allowable under the EPA pretreatment standards increases. At some point, however, the allowable pollutant concentration may become limited by the local sewer use ordinance.

An example of the impact of flow reduction measures on allowable pollutant discharge concentrations is illustrated by the data in Table 1.

These data apply to a midwestern manufacturing facility that will be required to meet pretreatment standards for zinc, nickel, manganese, and mercury. The Table I data illustrate that reduction of the plant's wastewater volume from 20,000 gallons per day (gpd) to 5,000 gpd would permit a four-fold

Table I. EPA Allowable Monthly Average Discharge Concentrations vs. Daily Flow

Constituent, mg/L	Flow, gpd				
	20,000	15,000	10,000	5,000	2,000
Cyanide	0.08	0.11	0.16	0.32	0.80
Chromium	0.12	0.16	0.24	0.48	1.20
Zinc	0.41	0.55	0.82	1.64	4.10[b]
Nickel	0.84	1.12	1.68	3.36	8.40[c]
Mercury	0.07[a]	0.09[a]	0.14[a]	0.28[a]	0.70[a]
Silver	0.11	0.15	0.22	0.44[d]	1.10[d]
Manganese	0.19	0.25	0.38	0.76	1.90

[a]Limited by city ordinance standard of 0.02 mg/L.
[b]Limited by city ordinance standard of 2.6 mg/L.
[c]Limited by city ordinance standard of 4.0 mg/L.
[d]Limited by city ordinance standard of 0.40 mg/L.

Table II. Wastewater Characteristics
(Based on 24 Hour Flow Composited Samples)

Parameter	Concentration, mg/L		Limit
	Average	Range	
Oil and Grease	160	80–360	100
TSS	470	120–760	300
Zinc	31	26–48	0.4
Nickel	2.2	1.5–35	0.8
Manganese	3.9	0.3–5.0	0.2
Mercury	0.018	0.013–0.021	0.02
VOC	1.1	0.5–1.5	0

increase in the allowable concentrations of zinc, nickel, and manganese but would have no impact on the permitted level of mercury since the city standards apply.

LABORATORY INVESTIGATIONS

A sampling program was undertaken to determine the raw wastewater characteristics and to ascertain treatment requirements. Time-and flow- composited 24 hour samples were collected on days representing different plant production cycle requirements. The problem wastewater characteristics are presented in Table II. It was concluded that treatment requirements would be focused on zinc, nickel, and manganese since VOC contamination could be eliminated through source control.

A two-pronged approach was undertaken to arrive at the most economical but effective and reliable treatment system. The first effort was to reduce the volume of water used and ultimately discharged. Plant engineering and operating personnel successfully reduced the daily volume of wastewater discharged by more than 50% primarily through the reduction of excess flow and usage rates of process water, recycling, and the implementation of multiple-use cascading water quality flow patterns. The conservation efforts are an ongoing commitment, and plant personnel continue to look for points where less water can be used or where water of lower quality would be satisfactory.

The second effort was to determine the most cost-effective treatment processes to reduce the concentrations of zinc, nickel, and manganese to their respective discharge limits of 0.4, 0.8, and 0.2 mg/L. A review of the literature and our previous designs concluded that alkaline precipitation was the most likely candidate technology.

A treatment pH of 9.5–10.0 was selected for testing based upon the solubility curves for the metals of concern as shown on Figure 1. Laboratory tests on a simulated plant wastewater confirmed the effectiveness of the proposed treatment process, alkaline precipitation followed by filtration, producing a filtrate quality very close to the theoretical values. Laboratory tests were then performed on a grab sample of the actual plant wastewater. The analytical results from the tests are presented in Table III.

These data showed zinc to be in solution at levels several orders of magnitude higher than the theoretical value and three orders of magnitude higher than expected for actual treatment plant operation. The first reaction was that the laboratory must have made an error in the analysis or in calculating the concentration. However, a check on quality control samples run with the test samples confirmed the accuracy of the results. The second reaction was that there must be something in the wastewater that had been overlooked – a substance or substances that could effectively shield zinc from alkaline precipitation.

Discussions with plant personnel revealed that the soap used in certain product washers contained approximately 30 grams per liter of ethylene diaminetetracetic acid (EDTA), a very strong chelating agent. Washwaters containing EDTA constituted more than 20% of the total plant process wastewater. In addition, the soap was used for general cleanup throughout the plant. To further compound the problem, the plant uses a high molecular weight polymer to bind zinc into the final product. This polymer is activated at elevated pH and effectively binds divalent cations.

A soluble zinc versus pH diagram was prepared using a flow-composited 24-hour sample of the plant's combined process effluent. The solubility curve derived from the test results is presented on Figure 2. Of particular interest are two facts: 1) the pH at which minimum zinc solubility was obtained was 12, approximately 3 pH units higher than the minimum solubility in a water solution of soluble zinc salt, and 2) the lowest measured soluble zinc concentration was 5 mg/L, significantly higher than

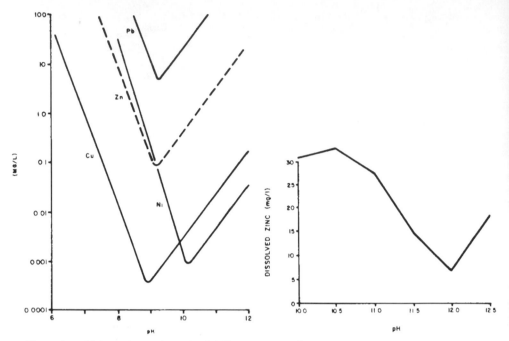

Figure 1.   pH dependence of metal solubility.

Figure 2.   Zinc solubility curve for plant wastewater.

the allowable discharge limit of 0.4 mg/L. Pure zinc solutions were prepared and varying concentrations of EDTA or the polymer were added. The results confirmed effective complexing by both materials. It was recommended that plant chemists investigate new soaps that could be substituted for the EDTA-containing soap which would render zinc more accessible to precipitation. Because the EDTA soap had been used successfully for many years, and because production personnel would be reluctant to accept change, it was decided to plan on continued use of the EDTA-bearing soap and to look for an effective process to treat the wastewater. Working in close cooperation with the soap and polymer manufacturers, numerous chemical treatment steps were tried. Measures as extreme as acidification to pH 1, followed by raising the pH to 9.5, did little to release the bound zinc. Attempts to substitute calcium and other divalent metals for the complexed zinc were of no avail. Precipitation by other anions, including carbonate, sulfide, and phosphate, was equally ineffective.

Other treatment processes, including ultrafiltration and ion exchange, were tested without success. The ineffectiveness of cationic exchange resins suggested the possibility that an anionic exchange resin might succeed in attracting and binding active anionic sites on the polymer. This attempt also failed.

The next treatment procedure tested was aimed at removal of the organic materials with attendant removal of complexed metals. Activated carbon was found to effectively remove the metal complexes, lowering zinc concentrations in the treated effluent to 0.1 mg/L or less. As expected, this process was not as effective in removing uncomplexed zinc.

Table III. Laboratory Test Results for Filtered Samples

Metal/pH	Concentration (mg/L)				
	10	11	12	13.5	Limit
Zinc	128	85	182	90	0.4
Nickel	0.50	0.48	0.22	0.34	0.8
Manganese	0.04	0.06	0.03	0.03	0.2
Mercury	3.6	3.0	2.3	2.5	0.02

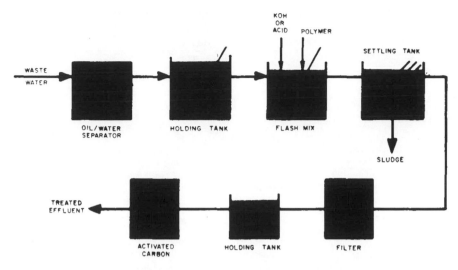

Figure 3.   Schematic of treatment process.

The last process tested was a two-step treatment: wastewater samples were first treated at pH 9.5 to 10.0 to precipitate uncomplexed zinc and nickel. The wastewater was then settled, the supernatant filtered, and the filtrate treated by activated carbon adsorption. Zinc concentrations less than 0.4 mg/L were consistently obtained, at varying ratios of free and complexed zinc. The activated carbon also provided the additional benefit of removing mercury and trace VOC's.

Based upon the favorable results of bench-scale tests, the treatment process schematic shown in Figure 3 was developed. After removal of free oils in an oil/water separator, a day's production of wastewater is stored for batch treatment on the following day. The pH of the pretreated wastewater is adjusted to between 9.5 and 10.0 (this often requires addition of acid rather than alkali), and the metal hydroxide precipitate is separated in a settling tank. The sludge is thickened and dewatered using a plate and frame filter press. Settling tank supernatant is filtered and stored in a holding tank where pH is adjusted as required to achieve maximum adsorption on granular activated carbon as the final treatment step. The quality of the carbon column effluent is high enough for reuse in several plant operations.

Spent carbon is withdrawn and returned to the supplier for regeneration. Tests indicated no loss of regeneration capability as a result of the treatment.

Although this treatment process would effectively treat the wastewater to below discharge limits, it was again recommended that a search be initiated to look for alternate soaps. The basis of concern was the very high rate of carbon utilization associated with removal of the EDTA-zinc complexes. Laboratory scale studies indicated a carbon utilization rate on the order of 3 to 10 gallons of process water treated per pound of carbon exhausted, which is two to three orders of magnitude higher than carbon exhaustion rates for more typical carbon wastewater treatment systems.

PILOT PLANT TESTING

In order to ascertain that the results obtained in the bench scale tests were representative of the plant wastewater under actual operating conditions, and were typical of results to be expected from day-to-day operation of the treatment plant, a pilot plant testing program was designed and implemented.

Objectives of the pilot plant program were:

- To verify that the proposed treatment scheme would meet the effluent requirements.
- To evaluate the operation of the pilot plant over an extended period since the wastewater characteristics varied considerably from day to day.
- To determine the carbon exhaustion rate.
- To evaluate the cost-effectiveness of different types of activated carbon.

Table IV. Zinc Concentrations Following Treatment

	Dissolved Zinc mg/L	Total Zinc mg/L
Raw Wastewater	22	34
Plate Settler Effluent	21	22
Dual Media Filter Effluent	21	21
Carbon Column #1 Effluent	—	0.2
Carbon Column #2 Effluent	—	0.1
Carbon Column #3 Effluent	—	0.1
Discharge Limit (at 20,000 gpd flow)	—	0.4

The pilot plant was designed as a batch process to be operated at 1 - 2 gallons per minute. It contained the following major components:

- 600 gallon raw wastewater holding tank to hold a 1-2 day sample of wastewater.
- Coagulation tank for addition of polymer and slow mixing.
- Plate settler for solids settling and removal of the sludge.
- Dual media filter to remove particulate material and sludge carryover.
- Three activated carbon columns in series for the removal of metals and organics.
- Plate and frame filter press for sludge dewatering.

Adjustment of pH to approximately 10 was accomplished by addition of sulfuric acid or potassium hydroxide in the raw wastewater tank. Intermediate holding basins were provided for additional pH adjustment ahead of the dual media filter and the carbon columns.

Before starting treatment of each batch, raw wastewater samples were collected before and after pH adjustment. During operation, hourly samples were taken of the effluent from the plate settler, the dual media filter, and each of the three carbon columns. The hourly samples were composited daily and analyzed for total and dissolved zinc, nickel, and manganese. During the carbon exhaustion rate testing, the hourly samples from each of the three carbon columns were also analyzed.

Initial testing of the pilot plant was intended to establish a baseline for the effectiveness of each treatment component and to determine if the overall process could consistently meet the discharge criteria. For these tests, the carbon columns were filled with a total of 110 pounds of activated carbon, resulting in an empty bed contact time of approximately 30 minutes. Approximately 500–600 gallons of wastewater were treated over a two-day period at a rate of one gallon per minute.

Findings from the initial testing indicated wide day-to-day variations in the influent suspended solids and metals concentrations. Problems were also encountered as a result of inadequate removal of oil and grease from the raw wastewater tank. Most of the suspended and particulate solids were removed in the plate settler, resulting in a clear effluent. However, the analyses showed that the dissolved metals were not effectively removed through the pH adjustment, coagulation-settling, and filtration treatment steps. At times, as much as 90% of the dissolved metals in the raw wastewater were still present following sand filtration. These dissolved metals were effectively removed in the activated carbon columns, bringing the total metal concentrations in the final effluent well below the discharge standards. Results of a typical run showing zinc concentrations following each treatment process are shown in Table IV.

Following the initial testing and verification of treatment capability, the study moved to the next phase which included testing for carbon exhaustion rates. This included placing a measured amount of granular activated carbon in each carbon column, operating the plant as described above, and collecting hourly samples of the effluent from each carbon column for metals analysis. Results of these analyses were expected to show low initial values, increasing with time as available adsorption sites were filled until no further removal was achieved. The point at which the effluent concentration equals the influent concentration was designated as the exhaustion point, and the carbon exhaustion rate was calculated as the number of gallons of wastewater treated per pound of carbon used. A hypothetical case showing the expected exhaustion curves for three columns in series is shown on Figure 4.

Results from actual tests generally followed the expected pattern, as shown on Figure 5. However, since the pilot plant has operated only during the daytime shift and was shut down overnight, some breaks appear in the data. In addition, some inconsistencies in the effluent concentrations can be seen

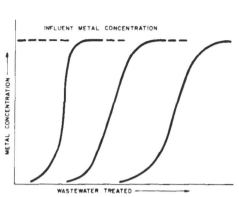

Figure 4. Expected carbon exhaustion curve.

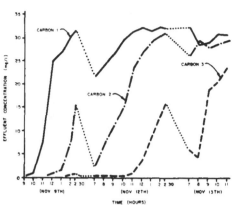

Figure 5. Measured carbon exhaustion curve.

at the end of one day's operation and at the start of the following day. This suggests that contact time in the carbon columns was a factor; the long contact period overnight may have allowed the complexed metals to find additional adsorption sites. However, the recovery effect of increased contact time decreased as additional wastewater was treated.

The exhaustion points were determined and exhaustion rates were computed to be approximately 22 gallons treated per pound of carbon exhausted. This rate was considerably better than the range of 3 to 10 gallons per pound obtained from the bench scale tests. However, it still represented a very high carbon use rate which would result in high annual carbon replacement costs for wastewater treatment. This projected high annual operating cost, $75,000–$100,000, was of considerable concern to the client so other alternatives were considered including substitution of materials in the manufacturing process and cleanup operations and recycling of treated wastewater before carbon treatment.

One of the alternatives proposed previously was use of a different soap in the product washers and for cleanup activities. Plant personnel had experimented with a substitute soap that does not contain EDTA but had found some disadvantages including:

- Requirement of a deionized water supply to work effectively.
- Slightly less effective in cleaning than the soap currently in use.
- Slight odor problem.
- Different color.
- Objections of work force regarding any change.

However, because of the economic impact, it was decided to collect pilot plant data on the substitute soap to determine its effect on treatment and carbon utilization.

One pilot plant run was made with wastewater containing the substitute soap. The results showed a substantial decrease in the dissolved metals concentrations following pH adjustment. Results of the testing and comparison with previous runs using the original soap are shown in Table V.

Carbon exhaustion tests were run, but no exhaustion rate could be computed since the carbon did not reach exhaustion during the test. After treatment of 530 gallons, the zinc concentration in the

Table V. Zinc Concentrations in Raw Wastewater Tank Following pH Adjustment

	Total Zinc mg/L	Dissolved Zinc mg/L
Original Soap		
Batch #4	35	25
Batch #6	32	32
Substitute Soap		
Batch #7	27.5	1.9

effluent from the first carbon tower was only 1.5 mg/L; zinc concentrations in the effluent from the second and third carbon columns were less than 0.1 mg/L. From these data and from the relative concentrations of dissolved zinc applied to the carbon columns, it is evident that the amount of carbon required to treat this wastewater would be an order of magnitude less than for the EDTA-containing wastewater.

Since the plant was considering installation of a deionized water system for a portion of the plant for other reasons, the potential cost savings of using the substitute soap provided the impetus to install the system. With the new system, the substitute soap provides effective cleaning while reducing the wastewater treatment plant cost.

Results of further tests indicated that use of the soap, coupled with flow reduction and recycling, could eliminate the need for activated carbon treatment for metals removal since the total mass in the effluent would be less than the discharge limit. However, at the client's request, the carbon treatment system was left in for mercury and VOC removal and to protect against exceeding the metals discharge limits.

CONCLUSIONS

This case study has shown that technology is available to effectively treat difficult waste streams. However, the overall cost of treatment must also be considered. When treatment costs are high, material substitution may afford an effective means of achieving treatment goals at reduced cost. Factors to be considered for material substitution include the effect on the production process, acceptability to plant personnel, and cost.

4   A Case History of the Design, Construction and Startup of a System for Biological Treatment of Wastewater From a New Ethanol Production Plant

W. Vince Lord, Jr., Environmental Supervisor
Tennol Energy Company
Jasper, Tennessee 33347

J. G. Walters, Principal Engineer
Infilco Degremont, Inc.
Richmond, Virginia 23229

James E. Smith, Jr., Project Engineer
Harbert International, Inc.
Birmingham, Alabama 35201

INTRODUCTION

The Tennol Energy Company constructed and started a new ethanol production plant near Jasper, Tennessee, which was completed in late 1985. The plant first produced alcohol in December, 1985, from the same feedstock as used for centuries by the inhabitants of Tennessee for making their famous sourmash bourbon, namely, corn. Because of its location in a very pristine valley and a small community of 4,000, Tennol was required to construct a pretreatment plant to reduce 8,000 mg/L BOD by 97% before discharge to the city of Jasper wastewater treatment system.

Annual alcohol production is 25 million gallons per year with the byproducts of CO_2 being sold to nearby gas distributors, and the distillers dried grains with solubles (DDGS) being marketed as animal feed.

The alcohol production facilities as engineered by Lummus Crest, Inc., Bloomfield, New Jersey, were based on technology from Buckau-Wolf, a Germany-based company. The design and selection of equipment for the wastewater treatment plant was the responsibility of Infilco Degremont, Inc., Richmond, Virginia, with construction by Southland Power Constructors, Birmingham, Alabama. Buckau-Wolf technology of thick mashes creates a smaller volume of wastewater but results in a higher concentration of BOD_5. The waste loading of 0.3 lb BOD/bushel is slightly lower than that found in other ethanol plants using a corn feedstock.

This case history will present, in detail, the evaluation of various systems and the rationale for the selection of a methanization system for the first-stage treatment followed by a low F/M (food to mass) activated-sludge system. This paper will also discuss system design, startup experiences, actual versus estimated wastewater characteristics, and operating results available to the date of the paper presentation.

TREATMENT REQUIREMENTS

Tennol's plant is officially designated a pretreatment plant since the waste after treatment is discharged to the City of Jasper's collection system and wastewater treatment facility. The Jasper system of aerated lagoons has a design capacity of 0.78 mgd of domestic sewage at 250 mg/L BOD. Prior to Tennol's discharges, the Jasper plant received an average of 0.2 mgd. Jasper's plant, therefore, had capacity to accept the pretreated wastewater from Tennol provided the BOD and suspended solids did not exceed 250 mg/L.

BASIS FOR DESIGN

Since there is not another plant operating with the exact technology that is used in the production of alcohol by Tennol, there was no opportunity to adequately characterize the waste or conduct pilot

Table I. Design Parameters[a]

Source	GPD	BOD$_5$ mg/L	BOD$_5$ lb/day	COD lb/day	TSS mg/L
Process Water	125,000	8,900	9,300	18,600	3,000
Floor Washing	17,500	1,000	150	300	500
Rain Work — Off	21,000	300	50	100	100
Totals	163,500		9,500	19,000	

[a]The sanitary waste and boiler and cooling water blowdowns will bypass the treatment plant and be discharged directly to the city sewer without pretreatment. The alcohol production facilities will operate 24-hours/day, 7-days/week. Wastewater from the above sources will be on a daily basis except for the rainwater. The effluent from the wastewater treatment plant will discharge to the city sewer and will average 250 mg/L or less of suspended solids and BOD$_5$.

plant studies; therefore, the design was based on educated assumptions. The design of the wastewater system was based on the design criteria given in Table I.

Figure 1 shows the flow diagram of the collection systems from the various parts of the production facilities and their routing to the Tennol Wastewater Treatment Plant.

Due to the high concentration of BOD in the wastewater, consideration was given to systems that included the use of trickling filters, activated sludge and methanization systems as well as combinations of these. The system selected as most suitable for the pretreatment of this wastewater stream was the use of a first stage methanization system followed by a low F/M activated sludge system for the second stage treatment. The use of either aerated lagoons or activated sludge systems was ruled out due to high energy costs associated with the treatment of the high strength waste. Trickling filters were considered but were removed from consideration because of lack of experience and inefficiency as a first-stage treatment in the distilling industry.

The use of methanization in a contact anaerobic digester has several benefits besides being an efficient method of treatment. There is less sludge for disposal with an anaerobic sludge system, and the methane generated is of fuel grade quality.

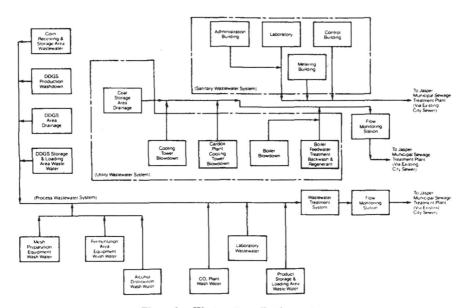

Figure 1.   Wastewater collection system.

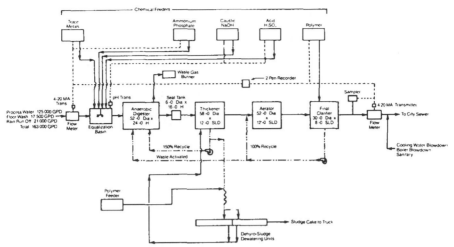

Figure 2.   Wastewater treatment system.

SYSTEM DESIGN

The plant consists essentially of four stages which are:

- Equalization
- Anaerobic Contact Digestion
- Activated Sludge
- Sludge Dewatering

The wastewater progresses through the system as shown in Figure 2. The raw wastewater is pumped at random rates from sumps located throughout the plant into the pretreatment equalization basin for neutralization and nutrient addition. The waste is then pumped at a constant rate into the anaerobic fermenter (digester). The digester effluent flows by gravity through a degassifier, through the thickener, and into the activated sludge system. The sludge from the activated sludge final clarifier is returned to the aeration tank. Thickened anaerobic sludge from the bottom of the thickener is recycled to the digester to maintain a high concentration of methanization microorganisms. Waste (excess) sludge is sent to the DeHydro[R] sludge concentration system. The effluent from the final clarifier flows by gravity into the Jasper sewer collection line after metering and sampling.

Equalization System

The purpose of the equalization basin (Figure 3) is to accumulate the production facilities' random flows which may be as high as 1100 gpm for 10 minute intervals. The raw wastewater is also conditioned and delivered to the digester by pumping at a constant hydraulic rate. The pH is controlled by the automatic addition of acid/caustic as required. Nutrient feeders have been furnished and are paced from the inlet flowmeter as required. A mechanical mixer keeps the contents of the tank homogeneous. The pH of the equalization basin is constantly monitored by a pH element and transmitter which is displayed on pH controller at the wastewater control panel. The pH controller provides an on/off control to activate both the caustic feed pumps and acid feed pumps to maintain a pH of 5.5 to 7.5.

Trouble warnings are received by a remote annunciator located in the production facilities' control room. The system includes high flow and high level alarms, motor control breaker trip alarms, and methane flare not-lit alarms.

Anaerobic Contact Digestion System

The anaerobic contact digestion system is shown in Figure 4 and consists of a reactor with homogeneous suspended bacteria. In order to maintain the required concentration of bacteria, anaerobic sludge is returned from the clarifier thickener, much as would be the case in an activated sludge

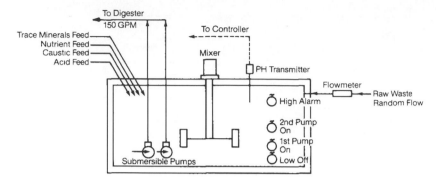

Figure 3.   Equalization system.

system. This contact process differs from fixed cultures in that the bacteria are kept in a homogeneous state in a fluid culture as opposed to growing on supports such as plastic, porous materials, sand, etc.

The suspension is kept homogeneous in the digester by means of gas mixing. Part of the 120,000 cu ft/day of methane gas which is produced by the digester fermentation is recycled through a circular row of gas pipes located in the center of the digester by means of a compressor. The gaslift effect provides very efficient mixing of the digester. The anaerobic mixed liquor leaving the digester still contains tiny gas bubbles which are trapped in the bacteria and sludge. To improve the settleability of the sludge, it is necessary to remove the gas by using a vacuum chamber. The removed gas is returned to the digester for compression.

The digester is sized on the basis of 0.38 lb COD/cu ft/day resulting in a digester volume of approximately 50,000 cu ft. The flow from the digestion tank is by gravity through the degasification system and into the clarifier-thickener, and the thickened sludge is recycled back to the digester to maintain a 15,000 mg/L suspended solids concentration. The recycled flow to the digester is approximately 150% of the raw waste flow into the digester. The thickener was designed on the basis of a suspended solids loading of 16 lb/sq ft/day.

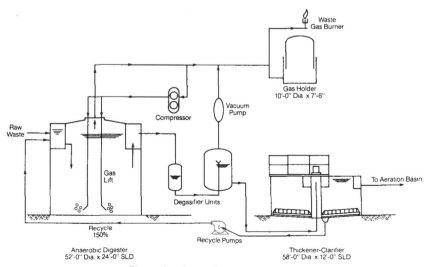

Figure 4.   Anaerobic contact system.

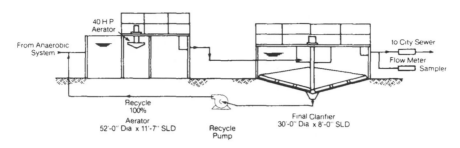

Figure 5.   Activated sludge system.

Activated Sludge System

This is a typical activated sludge system as shown in Figure 5 with the basic components being an aeration tank, followed by a clarifier with the sludge being returned from the clarifier to the aeration basin and the waste sludge returned to the anaerobic digester.

The activated sludge system was designed to treat 950 lb/day of BOD_5, assuming 90% removal of BOD in the anaerobic digester. The oxygen requirement was calculated on the basis of 1.3 lb of oxygen per lb of BOD_5 removed and a residual 3.0 mg/L dissolved oxygen in the aeration basin. The theoretical calculations for a surface aerator indicated a requirement of 32-mhp. A 40-hp Vortair[R] Aerator was furnished in order to give flexibility and to provide high residual dissolved oxygen to combat the growth of filamentous organisms.

The 30'-0" diameter final clarifier is conservatively designed at approximately 230 gpd/sq ft based on a flow of 163,500 gpd. The recycled sludge is designed to operate at 100% of the flow into the aeration basin.

The clarified effluent flows into the Jasper municipal collection system. The effluent line from the clarifier has a composite sampler and a flowmeter which paces the sampler. A polymer feeder may be used prior to the final clarifier if required in the case of upsets to help limit the suspended solids discharged to the City of Jasper.

Sludge Dewatering System

A dual-bed DeHydro[R] Vacuum-Assisted Sludge Dewatering System (Figure 6) was selected because of its capability of dewatering a wide variety of sludges to a truckable condition. The operation is a batch-volume reduction process which involves the compaction and compression of a sludge through a controlled filtration process. Since no two sludges have the same constituents and act differently, the sludges for treatment will require specific polymers and polymer concentrations, pumping rates, vacuum times, etc. The degree of conditioning may be varied to allow for flexibility in handling the different types of sludges and dewatering rates as well as considerations of operating costs.

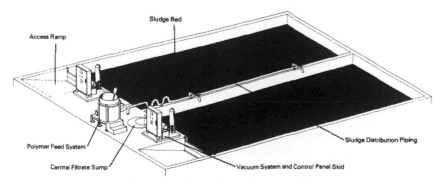

Figure 6.   Sludge dewatering system.

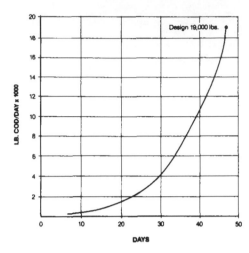

Figure 7.   COD loading rate during digester startup.

The sludge is distributed across the dewatering mats with the velocity and pumping rates being controlled to allow for: 1) effective mixing of the polymer and sludge in the serpentine mixer, 2) a sludge velocity which prevents breakdown or emulsion of the chemically conditioned particles, and 3) a rate of flow which distributes the sludge evenly over the dewatering mat without blinding the surfaces of the mat and allowing even distribution of the sludge across the bed. The distribution process should take about one hour.

The sludge is first allowed to dewater by gravity, and later a vacuum is applied to the bed to help draw the trapped water out of the compacted sludge layer.

STARTUP PROCEDURE

The anaerobic treatment of a complex organic material is considered to be a two-stage process even though it may occur in a single tank. Complex materials such as fats, proteins, and carbohydrates are hydrolyzed, fermented, and biologically converted into simple organic acids. It is in the second stage where methane fermentation occurs and waste stabilization is accomplished. In this stage, the organic acids are converted by special groups of methane bacteria which turn the organics into a gaseous product of CO_2 and methane. The acid formers tend to multiply and grow much more quickly than do the methane formers and; therefore, when starting up the system, it is necessary to carefully increase the organic loads to the digester over a period of time.

After hydrotesting, approximately 25% of the water was drained from the tank and seed sludge was added. The first day after seeding, waste was added with a COD of approximately 5% by weight of the sludge weight. The COD loading was increased by 10% per day. During this procedure it was necessary to observe the volatile acids versus alkalinity ratio and if the ratio increased, then caustic was added and/or the rate of feed decreased to the digester. Figure 7 shows a theoretical curve that was prepared as a guide for the loading procedures to enable the digester to receive the design loadings.

The ideal procedure would be to have obtained seed sludge from an anaerobic digester treating an identical waste. Since this was not practical, sludge was obtained from a municipal wastewater treatment system. In early November 1985, approximately 44 cu yd of anaerobic sludge was added to the digester. The sludge had been treated with polymer and centrifuged to a dry solids concentration of approximately 15%. The sludge proved somewhat difficult to transfer into the digester; however, this was solved by putting it into the clarifier thickener, diluting it with water, and then pumping it with the recycle pumps to the digester.

The charge of sludge amounted to approximately 12,000 lb and the first day 600 lb of COD was added. Concentrated corn syrup and cornmeal were used as a feed source for the digester. The cornmeal was more readily available, and for subsequent dosings of the digester this was predominantly used even though the corn syrup might have been more effective due to not having to hydrolyze

Table II. Analytical Data (1/15/86 to 4/24/86)

	Average BOD, mg/L	Average COD, mg/L	Average Sus. Solids, mg/L	Average Flow, gpd
Influent, grab	4,470	11,570	3,490	—
Digester Effluent	3,070	9,740	6,520	—
Aeration Basin	1,260	5,680	4,680	—
Plant Effluent	105	490	205	66,080

the solids. Laboratory analysis showed the cornmeal to have a COD of 1.4 gram per gram and the syrup of 1.6 gram COD per gram syrup.

Logs were kept of the addition of the corn syrup and cornmeal into the digester in order to know the loading of COD's that were applied. Since the laboratory facilities at the plant were not complete, outside laboratory services were obtained to run BOD_5, COD, volatile acids, alkalinity, etc. to monitor the operation. At this time, the influent to the digester was minimal and consequently there was little, if any, flow going through the digester into the activated sludge system.

The original design of the plant did not include a heat exchanger since it was believed that the temperature of the water reaching the plant would be sufficient to maintain a temperature in the digester of 90–95°F. Since the plant was started up in the colder months of the year and the water in the tank was ambient temperature, it was necessary to procure a temporary heat exchanger to bring the contents of the digester to the designed temperature level. A permanent heat exchanger has been installed since it now appears that the temperature of the wastewater influent will be variable and is insufficient to maintain a proper temperature in the digester.

It was decided around the first of December to seed the activated sludge system with some sludge from the Jasper treatment system. Cornmeal and corn syrup were also added to the aeration basin to provide a food source for the bacteria. Startup of the production facilities began on a limited basis in mid-December resulting in the first flow to the pretreatment plant.

The flows from the production facilities were very erratic due to startup problems resulting in corn solids, caustic and stillage being received at the pretreatment plant. The original design of the plant did not include a composite sampler on the influent to the plant, and, therefore, it was virtually impossible to know what was being received by the plant. Grab samples were taken, but it was difficult to make any calculations as to loadings.

When the wastewater treatment plant started to receive a waste stream on a daily basis from the production facilities that was relatively high in COD and BOD_5, the food to the activated sludge plant was sufficient to cause the mixed liquor solids to increase fairly rapidly.

As normally can be expected in the startup of a new plant, there were a few mechanical and hydraulic problems. Several lines were sized wrong, valves were located at the improper locations, and strainers on a well water system for pump packing plugged daily.

PLANT OPERATION

Difficulties have been experienced in plant operation, particularly with the highly variable loadings to the plant. There have been times when unfermented corn has come to the plant and, typically, the settleable solids are much higher than expected.

pH fluctuations, particularly on the alkaline side, have been much higher than expected, and it has been necessary at times to manually add caustic or acid to the waste since the acid/caustic feed system was not sufficient to keep pace with these changes. Larger pumps and day-tanks are being fabricated to replace drum handling of acid and caustic.

The corn solids have been high as a result of the plant's fermenter cleaning system. Modifications are being made to eliminate the high suspended solids from reaching the pretreatment plant.

The effluent has been able to meet the 250 mg/L BOD_5 and suspended solids limits on a regular basis. The effluent, digester, and aeration data are summarized in Table II.

Though the alcohol plant is not scheduled to reach full production until the fall of 1986, the design and operation of the pretreatment plant should allow for existing pretreatment standards to be met. The plant's waste sludge and effluent have already met EPA toxicity and priority pollutant requirements, respectively.

Section Two
PHYSICAL/BIOLOGICAL SYSTEMS

5 FLOC-LOADING BIOSORPTION CRITERIA FOR THE TREATMENT OF CARBOHYDRATE WASTEWATERS

Mervyn C. Goronszy, Executive Technical Manager
Transfield, Inc.
Irvine, California 92715

W. Wesley Eckenfelder, Professor
Environmental and Water Resources Engineering
Vanderbilt University
Nashville, Tennessee 37235

INTRODUCTION

Biosorption is an essential mechanism in the application of biomass selectivity principles for the control of filamentous sludge bulking. A number of systems incorporating biological selector concepts have been reported over recent years showing variable success.

This paper provides information on biosorption relative to floc-loading and biomass degradable fraction for a number of readily degradable carbohydrate wastewaters. Procedures which enable optimum biosorption, regeneration of biomass biosorptive capacity together with degradation kinetics, overall oxygen requirements, and sludge yield are described. Parameters which favour the selection of a non-filmentous biomass are presented.

BIOSORPTION PHENOMENA

Sorption of soluble organic compounds is a fundamental mechanism in the metabolism of substrate by microorganisms. This phenomenon was originally observed by Porges et al. [1] and Eckenfelder [2] but little was done with application to process design and operation until recently. Sorption can be defined as that phenomenon by which a rapid removal of soluble organics occurs upon contact of the substrate with activated sludge. This reaction is very rapid occurring in a matter of minutes. There is a limit to the mass of organics which can initially be removed by a unit mass of sludge. This limiting removal is in fundamental terms a function of the storage capacity of the biomass and the rate of biological oxidation of the substrate. A maximum removal by biosorption of 0.65 mg COD/mg VSS from skim milk was found by Porges et al. [1]. This removal was stored in the cell as glycogen which was metabolized over a three-hour period.

The contact stabilization process was developed to take advantage of the adsorptive and biosorptive properties of activated sludge [3]. This process achieves 90–95% BOD and suspended solids removal with only 15–30 minutes mixing of raw sewage with a well-stabilized sludge.

The removal of BOD in the activated sludge process in mediated by two principal mechanisms. The first is a physical removal by which the particulates, both organics and inerts, are enmeshed and/or adsorbed by the sludge matrix. Solubilization of adsorbent particulate organics takes place by enzymatic reactions which contribute to the net soluble organic fraction which is then available for subsequent biological degradation. Removal of soluble organics is a biological phenomenon.

Porges et al. [1], Quirk [4], Eckenfelder [2,5], Goronszy et al. [6], Chang [7], Hager [8], and Flippen et al. [9] report observations of rapid, soluble organic uptake by activated sludge with a variety of substrates.

Biosorptive transport mechanisms are especially significant in activated sludge systems which are configured with plug-flow hydraulics or which incorporate a short retention contacting region in order to generate similar sorptive transport conditions which occur in plug-flow systems. A high level of initial absorption can be evident in those systems where a high substrate to biomass availability exists, a situation which under certain waste specific and loading conditions can favour the generation of a floc-forming nonfilamentous biomass over filamentous forms [10].

Factors which should be considered as important in the sorptive transfer process include the fraction of soluble and colloidal substrate that is constituent in the wastewater, the degradable fraction of biomass taking place in the reaction, the availability of that fraction of biomass in the system and its ability to participate in the biosorptive and degradation processes. The degradable fraction of biomass, which is related to sludge age as

$$X_d = \frac{X'_d}{1 + X'_n b\theta_c} \tag{1}$$

controls the yield of solids and the net oxygen consumed. The biodegradable fraction is assumed to quantitatively relate to the active or viable fraction of biomass.

It has been observed that significant biosorption occurs in contact periods as low as two minutes, and depending on the mass ratio of substrate to microorganism population, it may continue for as long as 30 minutes. The sorbed organics are subsequently oxidized and synthesized. Upon contact of sludge with substrate, the specific oxygen uptake rate rapidly rises to a maximum and then follows normal kinetics until the residual soluble organics are removed without further storage. Eckenfelder [2] has shown a relationship between concentration of substrate and biomass (10 minutes contact time),

$$\frac{S_i}{S_o} = e^{-K_i X_v/S_o} \tag{2}$$

where

$$X_v/S_o = (\text{FLOC-LOAD})^{-1} \tag{3}$$

Continued removal of substrate occurs in accordance with a first order reaction,

$$\frac{S}{S_i} = e^{-k_p X_d X_v t} \tag{4}$$

In those cases where measurement is in terms of COD or TOC and where a fraction of the influent concentration is non-degradable and hence not responsive to enzymatic sorptive transport the equation may be written,

$$\frac{S - S_n}{S_i - S_n} = e^{-k'_p X_d X_v t} \tag{5}$$

It could also be expected that biosorption is functional on concentration of organics, the active or degradable fraction of biomass and the biological load driving force ratio or floc-load as,

$$\frac{S_i}{S_o} = e^{-k_i X_d X_v/S_o} \tag{6}$$

It has also been reported that the method of generating the sludge has a profound influence on its sorption capabilities [11]. Under batch feeding conditions the microorganisms grow alternatively at high and low substrate concentrations. It has been observed that the substrate uptake rate in batch fed cultures is higher than those fed continuously.

In a competing system of a reactive and a less reactive species, batch feeding will result in the selection of the reactive species because of its higher "overcapacity" for substrate uptake. Van den Eynde defined "overcapacity" as a situation whereby cells are maintained in a state in which they are always ready to take up substrates suddenly introduced to the culture. The reactive species will be favoured because of its larger overcapacity allowing it to take up the largest amount of substrate which will ultimately result in a higher growth rate leading to its predominance in the mixed population. The same logic can be applied to activated sludge systems. Activated sludge systems are mixed cultures and with a continuous supply of substrates, the system selects for cells with a relative high degree of energetic coupling and a low reactivity. If filamentous microorganisms belong to this group, sludge bulking can be explained. With batch feeding, or as in a plug-flow system, species with the highest reactivity will be selected, capable of taking up the substrates rapidly during the exogenous phase. Species producing reserve substances are favoured over species producing break down metabolites. The faster uptake rate and an efficient metabolism of reserves will result in a faster overall growth rate and will lead to predominance in the activated sludge. This hypothesis equates to the fact

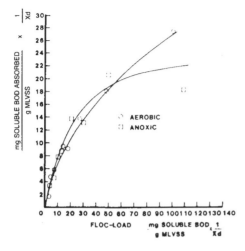

Figure 1. Aerobic-anoxic absorption for primary sewage. $BOD_{tot} = 120$ mg/L; $BOD_{sol} = 80$ mg/L; 2 minutes contact [6].

that sludges grown in batch or plug-flow systems, show a high substrate uptake rate and an increased capacity for reserve polymer generation.

Houtmeyers [12] has also reported that floc-forming bacteria have a higher biosorptive removal capacity than filamentous organisms.

It is not necessary to maintain aerobic conditions during the initial contacting period for sorption to occur. Essentially the same degree of sorption, for similar floc-load conditions, has been observed under anoxic and aerobic conditions (Figure 1). The principal applications of the sorption phenomena to process design and operation has been in the control of filamentous sludge bulking. By rapidly removing the degradable organics, which serve as food source for the filaments, their growth is restricted.

In this respect it has been observed [13] that prevention and control of bulking sludge, by means of a high substrate to biomass reaction zone, will only be effective if a large part of the available substrate is actually taken up by the biomass. Various floc-loading conditions have been described in the literature which purport to be effective in providing a competitive advantage for floc-forming organisms. Table I summarizes floc-load versus substrate removal data taken from a number of sources [14].

For a given sludge, the removal of soluble organics is related to floc-load; as the floc-load increases, the fraction removed decreases depending upon the characteristics of the substrate. Filamentous organisms can also be generated at high floc-loads. In order to regenerate the biosorptive capacity of a sludge, it is necessary to oxidize accumulated substrate and intermediate storage products. If filamentous growth is to be minimized, it is necessary to have a knowledge of floc-load, biosorption and regeneration time criteria which are conducive to the generation of a well- settling activated sludge.

Grau and Chudoba [15] have shown that sludge settling properties can be related to the regeneration time of a previously sorbed sludge together with the proportion of substrate that needs to be oxidized to fully restore the absorption capacity.

SUMMARY OF BASIC PRINCIPLES FOR CONTROL OF FILAMENTOUS BULKING

Activated sludge bulking has been associated with a number of organisms and is a condition which can be averted by specific design and operational variables. In setting up a protocol for the control of filamentous bulking the following should be considered [16].

- Filamentous microorganisms are present in most activated sludge. Exceptions are complex organics, i.e., chemical industry wastes, in which the filaments cannot utilize the carbon source. Bulking problems appear when the filaments overgrow the floc-formers in wastewater containing readily degradable wastes.
- Overgrowing of filamentous microorganisms in activated sludge is affected by:
  - Composition of treated wastewater with respect to soluble colloidal fractions of organics, lipids, long chain fatty acids, etc.

Table 1.   Summary of Floc-Load, Sorption Data [14]

DOMESTIC — 10 mins. contact								
Floc-load (COD based)	55	52	63					
% Organic removed	56	36	27					
F/M	0.1	0.2	0.4					
DOMESTIC — 1 min. contact								
Floc-load (COD based)	275	222	163	109	82	55	27	
% Organics removed	35	38	33	54	59	78	79	
DOMESTIC — 2 mins. contact, sludge age 13 days								
Influent BOD	160	120	166					
Floc-load (BOD based)	13	34	40					
% Organics removed	47	34	37					
SYNTHETIC — influent COD 180 mg/L, F/M 0.08								
Floc-load (BOD based)	200	166	162	141	20	17	16	15	13
% Organics removed	25	35	47	47	81	93	92	83	85
SYNTHETIC — F/M 0.08								
Influent COD	180	170	240	240	170			
Floc-load (COD based)	15	16	20	49	75			
% Organics removed	83	92	81	25	47			

- Actual concentration of dissolved oxygen in the aeration tank
- Concentration of soluble substrate under which microorganisms grow
- Parameters relating to the process such as organic loading, sludge age, active fraction of biomass.
- Nitrogen and phosphorus deficiency.
- The microorganism which initially accumulates most of the substrate, through sorptive transport, will be dominant in the system provided the regeneration period for exhaustion of all of the accumulated substrate is sufficient.
- The biosorptive capacity of all microorganisms, in aeration systems without separate sludge regeneration, will only be fully restored if an endogenous reaction condition exists.
- Filamentous microorganisms in activated sludge are generally believed to be slow-growers, having low values of maximum substrate removal, saturation constant, and accumulation capacity.
- Floc-forming microorganisms in activated sludge are generally specified as being fast growers with high values of maximum substrate removal, saturation constant, and accumulation capacity.
- Faster growing starvation susceptible filamentous microorganisms will be removed during extended periods of endogenous metabolism.
- The relative net growth of filamentous microorganisms can be controlled by varying substrate concentration and the relativity of exogenous and endogenous reaction conditions.

SPECIFIC OXYGEN UPTAKE RATE

In can be reasoned that because only the active biomass uses oxygen during the course of substrate utilization and maintenance activity, the measurement of oxygen uptake rate should reflect the extent of microbial activity of that sludge. It appears that oxygen uptake rate of an activated sludge is a function of the substrate concentration to a point, except under substrate saturation conditions. It would also appear that the maximum oxygen uptake rate has a fixed value and thus can be used to represent the microbial activity of an activated sludge under a fixed set of reaction conditions. It might be contemplated that specific oxygen uptake rate should relate to the degradable fraction of the biomass.

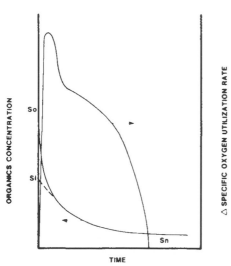

Figure 2.   Schematic of absorption phenomena.

In the absence of techniques being available to easily measure the active fraction of biomass and associated enzyme assisted sorption phenomena, a simple measure of the response of a biomass to absorption effects can be obtained using the response in specific oxygen utilization rate of a biomass following a pulse load stimulus. Quantitative initial absorption can be determined by extrapolation of the substrate degradation kinetic relationship to zero-time in order to determine a theoretical initial substrate concentration which fits the kinetic model. Subtraction of this value from a calculated initial concentration affords the determination of absorbed substrate which can be expressed as a percentage removal of soluble substrate or as a mass ratio relative to the mass of volatile or degradable biosolids initially present in the biosorption reaction. Figure 2 schematically depicts the response time mechanisms.

This procedure can be used to obtain a relative measurement of the absorptive capacity or potential of a biomass, the premise being that a high degree of absorptivity from a given reference state for a given substrate with a defined soluble component is accompanied by a marked response in specific oxygen uptake rate and a lesser level of absorptivity is indicated by a lesser response in specific oxygen uptake rate.

EXPERIMENTAL

In order to determine the magnitude of sorptive transport, it is necessary to generate an activated sludge which has been acclimated to the substrate under consideration. For continuous systems, it is necessary to run the reactors for a period equal to three sludge ages in order to attain biological steady state conditions. For batch systems acclimation will generally be reached in about three weeks. Batch acclimation has the advantage of generally being able to produce a sludge that has a relatively low population of filamentous microorganisms. Batch acclimation, depending upon the organic loading and aeration period, has the added advantage, because of the feed starve mode of operation, of providing a single sludge which has a wide range of sorptive preconditioning.

In order to determine sorptive-regeneration parameters, it is necessary to first subject the acclimated sludge to various loading conditions in order to determine the floc-load soluble substrate removal relationship. In principle an aliquot of sludge is taken on which the oxygen utilization rate (OUR), MLVSS, and soluble organic (TOC or COD or BOD) concentrations are determined. A suitable volume of the sludge is removed from the reactor and allowed to settle with supernatant liquid being removed to leave 1/2 of the original volume. This sludge is then aerated, sampled as before, and split into several volumes to which an equal volume of substrate is added, the concentration of which is varied in order to cover the range of floc- loadings that is being studied. A total volume of 1 liter for each floc-load is sufficient. Aeration via a fish tank diffuser stone will also adequately mix each reactor. Sampling at time intervals of 2, 10, 20, 30, and 60 minutes, with analyses for MLVSS, OUR, and soluble organics will provide the data base for quantitative assessment of floc-load and biosorption.

Table II.   Composition of Concentrated Nutrient Broth

Component (in 4 Litres)	Weight (Grams)
Nutrient Broth	403
Glucose	145
Yeast extract	35
KH_2PO_4	19
NaOH	3.6
$NaHCO_3$	270
$MgSO_4$	39
$CaCl_2$	24
$FeCl_3$	4.2
$MnSO_4$	0.05

COD of stock solution approximately 160,000 mg/L

Given that the floc-load is selected by the above procedure, regeneration- accumulation data is found as follows. Similarly determine the initial MLVSS, OUR and soluble organic concentration of the test sludge. Concentrate and decant as before. Add an equal volume of substrate to the concentrated sludge (aerated) of approximate concentration to yield the target floc-load and continue aeration for 20 minutes, during which time samples are taken at intervals of 2, 10, and 20 minutes for MLVSS, OUR, and soluble organics. After 20 minutes aeration, the sludge is either settled or centrifuged in order to remove 50% of the volume as supernatant. The concentrated sludge is re-aerated in order to regenerate its biosorption capacity. A volume of the original sludge is kept under aeration and sampled at appropriate intervals for MLVSS, OUR, and soluble organics. The concentrated sludge is sampled at say hourly intervals for the same parameters. Aliquots are taken from the regenerator and recontacted with substrate to target at the initial floc-load condition; each contactor is aerated and sampled at 2, 10, and 20 minute intervals for MLVSS, OUR, and soluble organics.

Sludge from three different units, acclimated at different organic loadings, were each subjected to the same floc-load conditions and the same basic parameters measured. The degradable fraction of each sludge was determined through aerobic digestion for 28 days. A nutrient broth solution made up of concentrate, diluted with tap water, was used as substrate (Table II).

ANALYTICAL PROCEDURES

Laboratory tests were performed in accordance with *Standard Methods for the Examination of Water and Wastewater* [17] unless specified otherwise. Soluble organic concentration as chemical oxygen demand (COD) was determined by the semi-microtube method [18]. COD samples were preserved using concentrated sulfuric acid to a pH < 2. Samples were filtered through a 0.45 mm membrane filter. Soluble organic concentration as total organic carbon (TOC) was determined using a Beckman Model 915 TOC analyzer. All TOC samples were filtered through glass fiber filters followed by 0.45 mm membrane filters. Samples were acidified with concentrated hydrochloric acid to a pH < 2 and stripped with nitrogen gas to remove any inorganic carbon prior to analysis. Oxygen uptake rates (OUR) were determined as per *Standard Methods* and converted to specific oxygen uptake rate (SOUR).

RESULTS AND DISCUSSION

Batch experiments at floc loadings of 870, 660, 345, 170, 80, and 44 mg COD/g VSS (degradable fraction 0.53) were performed to determine the specific oxygen uptake rate response profiles relative to the sludge biosorption capacity. The oxygen uptake rate response profile is shown in Figure 3 which reveals the maximum value in each case varies with the magnitude of the floc- load. Typical data for a high floc-load condition is shown in Figure 4 where accompanying data on substrate removal and accumulation of volatile solids over the course of the run are also shown. These data can typically provide data on the net oxygen coefficient for substrate degradation (including the endogenous component), VSS yield, and degradation kinetics for the substrate. The rate of degradation of volatile solids required for the determination of X_d can also be used to provide an estimate of the endogenous decay coefficient which can be used to back calculate the endogenous oxygen requirement.

Tables III, IV, V, and VI summarize percent substrate removal at various floc-loadings for contact times of 2, 10, 20, and 30 minutes using nutrient broth, dextrose and peptone-dextrose substrates. The

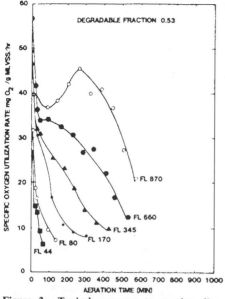

Figure 3. Typical response to varying floc-load, constant X_d [20].

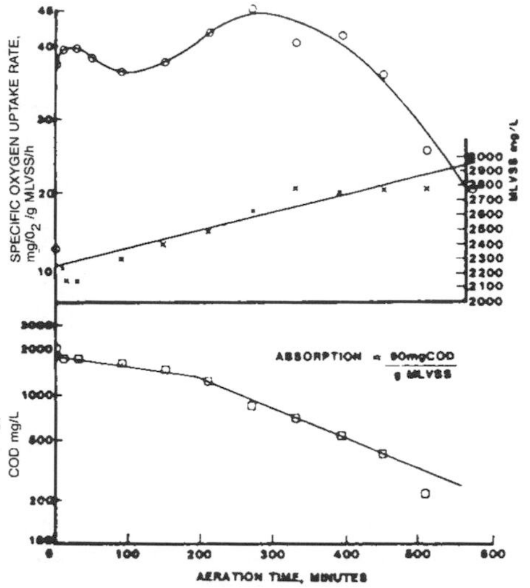

Figure 4. Typical reactor response to high floc-load, 870 mg COD/gm MLVSS [20].

Table III. Percent Organics Removal (as TOC) for a Nutrient Broth Substrate [19]

Contact Time	Floc Loading (mg COD/g VSS)				
(min)	70	132	257	473	938
2	40	67	19	2	—
10	48	74	42	26	9
20	53	56	57	39	18
30	68	56	65	51	30

Table IV. Percent Organics Removal (as TOC) for a Dextrose Substrate [19]

Contact Time	Floc Loading (mg COD/g VSS)				
(min)	53	111	209	394	693
2	—	12	44	23	19
10	75	63	44	29	35
20	68	88	48	35	39
30	82	90	72	59	31

Table V. Percent Organics Removal (as TOC) for a Dextrose Substrate [19]

Contact Time	Floc Loading (mg COD/g VSS)				
(min)	49	97	176	361	619
2	52	29	22	—	67
10	74	51	39	20	30
20	83	82	58	27	16
30	66	85	62	49	28

Table VI.   Percent Organics Removal (as TOC) for a Peptone-Dextrose Substrate [19]

Contact Time (min)	Floc Loading (mg COD/g VSS)				
	41	61	97	180	346
2	35	19	10	18	28
10	22	60	56	41	35
20	35	55	65	64	—
30	32	47	70	73	66

sludge was acclimated to nutrient broth substrate at an F/M loading of 0.4. Those data when plotted as the inverse of floc-load versus percent organics remaining produce a straight line, the slope of which gives a value for K_i, the kinetic removal rate for that biosorption time period. Initial removal rates determined for a contact time of 10 minutes for nutrient broth (floc-load range 40–470 mg/g VSS) and dextrose (floc-load range 50–700 mg/g VSS) were 0.154 and 0.082, respectively, (K' is dimensionless) (Figures 5 and 6).

It has been postulated that initial biosorption is related to the active fraction of biomass. This is quantitatively shown in Figure 7 where the oxygen uptake rate response of three sludges having a different degradable fraction were stimulated by the same floc-load condition. The accompanying plot of substrate absorbed at theoretical zero time is shown in Figure 8. A plot of ΔSOUR versus log X_d similarly yields a straight line. This suggests that there is a direct relationship between the mass of organics that can be absorbed initially by a biomass and its associated maximum rate of oxygen uptake of the form,

$$X_d = e^{KU/X_d X_v} \qquad (7)$$

The results of experiments devised to determine the reaeration time for a sludge order to regenerate its biosorption capacity are shown in Figures 9 and 10. The regeneration time for a floc-load condition can be simply obtained from the SOUR versus time plot, being the time taken to restore the SOUR to its initial value (Figures 3 and 9). The time at which the SOUR peaks should give the time at which maximum or saturation absorption conditions, for that sludge, are reached such as in Figure 9 where that time corresponds to 10 minutes. The maximum SOUR absorptive response of 55 mg O_2/g

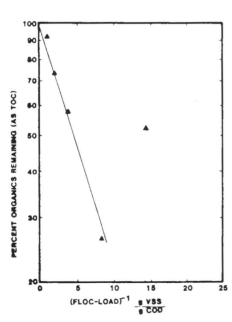

Figure 5. Percent organics remaining versus floc-load^{-1} for nutrient broth substrate at a 10-minute contact time [19].

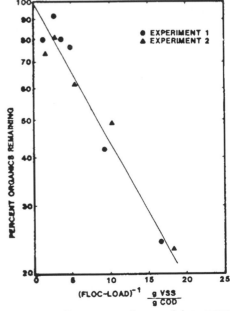

Figure 6. Percent organics remaining versus floc-load^{-1} for dextrose substrate at a 10-minute contact time [19].

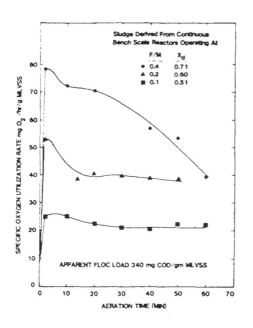

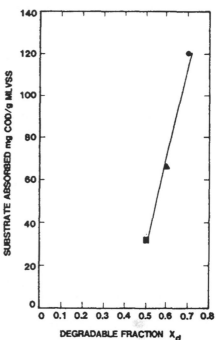

Figure 7. Specific oxygen utilization rate response versus X_d at same floc-load [20].

Figure 8. Initial absorption versus degradable fraction [20].

MLVSS/L was restored and maintained after 60 minutes of aeration. Alternatively optimum regeneration time can be determined from re-contacting experiments (Figure 10) where the percent organics removal for various contact time is plotted against aeration time. An 8-minute biosorption rate of 0.037 can be calculated for these data while the regeneration time is between 3 and 4 hours.

It has been suggested [15] that in order to prevent activated sludge filamentous growth, it is necessary to remove a certain fraction of substrate by biosorptive means. In doing so, failure to adequately regenerate the biosorptive capacity of the biomass will quickly lead to the generation of filamentous growth. This data base shows the level of soluble organics removal that can be achieved at various floc-load conditions. This sort of data is necessary in order to be able to design activated

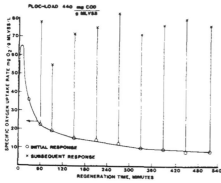

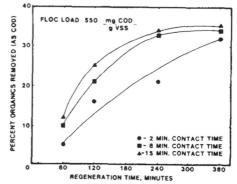

Figure 9. Oxygen utilization response to initial and subsequent contacting.

Figure 10. Substrate removal versus regeneration time at various contact times [19].

sludge systems that can be operated without bulking. There is evidence to suggest that sludge bulking may result from the treatment of similarly readily degradable carbohydrate substrates where the initial floc-loading conditions exceed 100 mg COD/g MLVSS [13]. This hypothesis has been borne out by studies in which failure to remove most of the degradable substrate by biosorption led to filamentous bulking in a subsequent complete-mix reactor stage.

CONCLUSIONS

These studies have shown that with proper selection of initial floc-loading conditions, substantial removals of soluble organics in a readily degradable wastewater can be achieved. Other studies have shown that this rapid removal of organics can be used to prevent or minimize the occurrence of sludge bulking conditions in the activated sludge process.

The principles and procedures outlined in this paper can be used as a basis for the determination of treatability data necessary for configuring and sizing full-scale activated sludge facilities. Batch sorption and related treatability studies enable fundamental process design parameters to be established without hindrance from sludge bulking conditions which are frequently generated in bench scale units where readily degradable carbohydrate wastes are concerned.

In applying biosorption principles to the design of activated sludge facilities, it should be remembered that the settling characteristics of a sludge will be functional on a number of factors which include reactor configuration, dissolved oxygen concentration, organic loading rate, mixed liquor suspended solids concentration, and the nature of the wastewater. Reactor configurations which promote alternating high floc-loading with rapid transport of soluble organics into the biomass, combined with extended periods of endogenous metabolism will result in the growth of a nonfilamentous activated sludge culture. Such feast-famine growth conditions will completely eliminate bulking at low organic loadings.

The period of aeration allocated to restoring the sorption potential of a biomass is in turn an important parameter in design. While slow growing filaments can be eliminated due to their low rates of substrate assimilation at high floc- load conditions, fast growing filaments need to be eliminated as a result of a loss in viability during an extended period of endogenous metabolism following the metabolism of intermediate substrate storage products.

NOMENCLATURE

X_v	— volatile suspended solids concentration, mg/L
X_d	— degradable fraction, dimensionless
X'_d	— degradable fraction upon generation, dimensionless
X'_n	— non-degradable fraction upon generation, dimensionless
b	— endogenous rate coefficient, days^{-1}
θ_c	— sludge age, days
S_o	— influent organics concentration, mg/L
S_i	— organics concentration after absorption at zero time, mg/L
S_n	— non-degradable concentration of organics, mg/L
S	— organics concentration during reaction, mg/L
K_i	— absorptive kinetic rate constant based on X_v, days^{-1}
k_p	— specific first order rate coefficient based on X_d, days^{-1}
k'_p	— modified specific first order rate coefficient based on X_d and accounting for non-degradable organics, days^{-1}
K	— constant based on oxygen uptake rate
U°	— maximum oxygen uptake rate response to load stimulus, mg/L/h

REFERENCES

1. Porges, N. Jasewicz, L., and Hoover, S. R., "Biochemical Oxidation of Dairy Wastes. VII Purification, Oxidation, Synthesis and Storage," *Proc. 10th Industrial Waste Conference*, Purdue University (1955).
2. Eckenfelder, W. W., and O'Connor, D. J., *Biological Waste Treatment,* Pergamon Press, London, England (1961).
3. Ulrich, A. H., and Smith, M. W., "The Biosorption Process of Sewage and Waste Treatment," *Sewage and Industrial Wastes*, 23, 10, 1248 (1951).
4. Quirk, T. P., "Design Data for Biological Treatment of Combined Wastes," *Sewage and Industrial Wastes*, 31, 1288 (1959).

5. Eckenfelder, W. W., *Principles of Water Quality Management,* CBI Publishing Company, Inc., Boston, Mass (1980).
6. Goronszy, M. C., Barnes, D., and Irvine, R. L., "Intermittent Biological Waste Treatment Systems — Process Considerations," *Water,* A.I.Ch.E. Symp Series, 77, 129-136 (1980).
7. Chang, C. N., "Biosorption and Reaction Kinetics in the Activated Sludge Process," *Ph.D. Dissertation,* Vanderbilt University (1982).
8. Hager, B. G., "Biosorption Mechanisms and Their Applications in Achieving Improved Reactor Capacity," *Ph.D. Dissertation,* Vanderbilt University (1984).
9. Flippen, H. T., Eckenfelder, W. W., and Goronszy, M. C., "Control of Activated Sludge Bulking in a Carbohydrate Wastewater Using a Biosorption Contactor," *Proceedings 39th Industrial Waste Conference,* Purdue University (1984).
10. Chudoba, J., et al., "Control of Activated Sludge Filamentous Bulking Part I and Part II," *Water Res.* (1973). 1163-1182 and 1389-1406, and Part II, *Water Res.,* 8, 231-237 (1974).
11. Eynde, Van den, E., Geerts, J., Maes, B., and Verachtert, H., "Influence of the Feeding Pattern on the Glucose Metabolism of Arthrobacter sp. and Sphaerotilus natans, Growing in Chemostat Culture, Simulating Activated Sludge Bulking," *Europ. J. Appl., Microbiol. Biotechnol.* 17, 35-43 (1983).
12. Houtmeyers, J., "Relations Between Substrate Feeding Patterns and Development of Filamentous Bacteria in Activated Sludge Process," *Agricultural,* 26, 1-135 (1978).
13. Eikelboom, D. H., "Bulking of Activated Sludge," Ellis Harwood Publishers, Chichester, England. Eds. Chambers, B., and Thomlinson, E. J.
14. Goronszy, M. C., and Eckenfelder, W. W., "A Design Approach for Plug-Flow Activated Sludge Plants Treating Municipal Wastewaters," *59th Annual WPCF Conference,* Kansas City (1985).
15. Grau, *et al., Bulking of Activated Sludge,* Ellis Harwood Publishers, Chichester, England. Eds. Chambers, B., and Thomlinson, E. J. (1982).
16. Grau, J., "Control of Activated Sludge Filamentous Bulking — VI. Formulation of Boric Principles," *Water Res.,* 19, 8, 1017 (1985).
17. American Public Health Association, *Standard Methods for the Examination of Water and Wastewater,* New York, NY, 16th Ed. (1984)
18. Himebaugh, R. R., and Smith, M. J., "Semi-micro Tube Method for Chemical Oxygen Demand," *Analytical Chemistry,* 51, 7, 1085 (1979).
19. "An Activated Sludge Modification to Eliminate Bulking, Minimize Volume and Enhance Nutrient Removal," Unpublished Report, AWARE Inc., (1985).
20. Goronszy, M. C., Eckenfelder, W. W., and Cevallos, J., "Sludge Bulking Control for Highly Degradable Waste Waters Using the Cyclically Activated Sludge System," *Proc. 40th Industrial Waste Conference,* Purdue University (1985).

6 EFFECT OF FEED CONCENTRATION ON FORMATION AND UTILIZATION OF MICROBIAL PRODUCTS DURING AEROBIC TREATMENT

A. F. Gaudy, Jr., Professor

Y. K. Chen, Graduate Student (Deceased)
Department of Civil Engineering
University of Delaware
Newark, Delaware 19716

T. S. Manickam, Sanitary Engineer
N.Y. State Health Department
Albany, New York 12237

P. C. D'Adamo, Environmental Engineer
Tatman and Lee, Consulting Engineers
Wilmington, Delaware 19801

M. P. Reddy, Environmental Engineer
Orange County Public Utility Company
Orlando, Florida 32801

INTRODUCTION

More than four decades have passed since Monod [1] published his empirical equation (Equation 1) relating specific growth rate, μ, to substrate concentration for noninhibitory carbon sources.

$$\mu = \mu_{max}\left(\frac{S}{K_s + S}\right) \tag{1}$$

It has been over three decades since Monod [2] and Novick and Szilard [3] independently published the equations and concepts which have come to be known as the theory of continuous culture (see Equation 2 for substrate concentration).

$$S = \frac{K_s (D + k_d)}{\mu_{max} - (D + k_d)} \tag{2}$$

These equations were elaborated upon and extended for cell recycle systems by Herbert and his co-workers in the 1950's and 1960's [4,5] (see Equation 3 for substrate). Originally Equations 2 and 3 did not contain the term

$$S = \frac{K_s [D(1 + \alpha - \alpha c) + k_d]}{\mu_{max} - [D(1 + \alpha - \alpha c) + k_d]} \tag{3}$$

for cell decay, k_d. In later years it was found that a better prediction of biomass output, X_w, resulted if this term were included.

These concepts have been of interest to both basic and applied researchers, especially to environmental engineers interested in biological treatment of organic wastewaters. Studies have been conducted to test the applicability of the theory of continuous culture to heterogeneous microbial populations of sewage origin growing on simple sugars in once-through chemostats. Results of studies in our laboratories showed that Equations 1 and 2 were applicable to such populations [6,7]. In these continuous flow studies substrate was measured both as COD and as carbohydrate, and it was observed that the effluent COD was, except at high dilution rates approaching wash-out, very low, although approximately 10 times higher than effluent carbohydrate. Thus, there was evidence that metabolic products were elaborated by the cells either as products of cell degradation or as products of substrate degradation, e.g., metabolic intermediates. It was apparent that one could not predict effluent soluble substrate precisely, but the theory was without doubt shown to be applicable to microbial systems other than the pure cultures for which it had been originally devised, and the amount of organic matter in the effluent was shown to be controlled by specific growth rate, μ or μ_n.

A subsequent study by Ramanathan and Gaudy [8] was designed to test the applicability of the equations of Herbert for cell recycle systems (see Equation 3) to heterogeneous populations using sedimentation for cell separation. It was found that the concentration factor c (c = X_R/X) was not a practical means of control for such systems, and model equations using X_R rather than c as one of the means of controlling specific growth rate were suggested as a useful modification for design and operation of activated sludge processes. Later this control strategy was successfully tested in laboratory pilot plant studies, and the equations were modified to include cell decay [9,10]. Lawrence and McCarty [11] also developed a model which was in general consonance with the theory of continuous culture. Whereas the models of McKinney [12] and of Eckenfelder [13] were developed more or less independently of the theory of continuous culture, they nonetheless contain some elements of the concepts originally laid down by the basic biological researchers cited previously. Gaudy and Kincannon have compared these four models conceptually and computationally [14]. The major difference between the predictive equations of Gaudy and co-workers and others in the pollution control field is that the former equations are derived by writing a materials balance around the reactor only, rather than the entire reactor-clarifier system. Also, the equations differ from those of Herbert because the return sludge concentration, X_R, rather than the ratio X_R/X is employed as a system descriptor and control parameter. The equations of Gaudy and co-workers are derived elsewhere [15]; the equation for S is given below because it is germane to portions of the following discussion.

$$S = \frac{-b \pm \sqrt{b^2 - 4ac}}{2a} \tag{4}$$

$$a = \frac{\mu_{max}}{1 + \alpha} - D - \frac{k_d}{1 + \alpha}$$

$$b = D(S_i - K_s) - \frac{\mu_{max}}{1 + \alpha}\left(S_i + \frac{\alpha X_R}{Y_t}\right) + \frac{k_d}{1 + \alpha}(S_i - K_s)$$

$$c = K_s D S_i + \frac{K_s k_d}{1 + \alpha} S_i$$

Examination of Equations 2 and 3 shows that, according to the theory of continuous culture, the effluent substrate concentration is independent of the influent or feed substrate concentration. In recent years this aspect of the fundamental theory has been challenged by workers in the pollution control field because several, including ourselves, have observed that when using COD as a measure of substrate, the soluble residual COD generally tends to increase somewhat with increased feed COD even when a continuous culture system is run at a constant net specific growth rate; for example, compare effluent COD values obtained at S_i = 1,000 mg/L COD and effluent COD values obtained when S_i was 3,000 mg/L COD in Table I. It should be pointed out that examination of Equation 4 shows that S has some dependence on S_i in the model of Gaudy and co-workers. However, this does not come about because the equation is inconsistent with the "theory" of continuous culture. It is recalled that S_i affects biomass concentration, X, and, since X_R is an independent control variable in this model, an increase in X due to an increase in S_i will also increase μ, which in turn predicts an increase in S. A more detailed discussion of this aspect may be found elsewhere [14,16]. However, in agreement with the findings of other investigators, Manickam and Gaudy [16] have shown that there is an increase in S for increased S_i beyond that predicted using Equation 4. The increase was found to be rather small at relatively low values of μ_n (or high sludge age, θ_c) .

To accommodate the experimental observation of increased S with increased S_i, the model of Eckenfelder has been modified to include the direct dependence on S_i [17]. Grau et al. have also proposed a model in which S is dependent on S_i [18]. Benefield and Randall [19] endorsed the approach of Grau and they questioned the usefulness of μ_n or θ_c as a major strategy for controlling effluent quality at treatment plants. Grady and Williams have devised an empirical equation based upon their experimental data [20]. Chen and Hashimoto have presented a modification of the equation of Contois to relate S_i to S [21]. Baskir and Hansford have adapted the product formation model developed by Luedeking and Piret for lactic acid fermentation [22]. Daigger and Grady analyzed data collected previously in their laboratories and formulated a relationship for residual total soluble organic material from continuous culture reactors which attributes the soluble residual organic matter

Table I. Comparison of Feed and Effluent Concentrations for Heterogeneous Microbial Populations Growing on Glucose (at Several Specific Growth Rates) in a Chemostat

$D = \mu_n = 1/\bar{t}$ hr^{-1}	$S_i{}^a = 1000$ [6]		$S_i{}^a = 3000$ [8]	
	S^b	$S_c{}^c$	S^b	$S_c{}^c$
1/24	48	5	221	43
1/18	54	4	87	9
1/12	45	4	112	11
1/6	69	9	120	17
1/4	92	8	–	–
1/3	237	214	113	17
1/2	874	698	224	31
1/1.5	913	764	1569	1225
1/1			2745	2535

$^a S_i$ = Influent COD, mg/L.
$^b S$ = Effluent COD, mg/L.
$^c S_c$ = Effluent carbohydrate, mg/L.

largely to formation of three types of microbial products [23]. These are: 1) products formed during fairly rapid degradation of exogenous carbon source, 2) cell products associated with growth which may leak during replication, and 3) products more closely associated with cell degradation. All three processes are assigned unique first order rate constants, and the contribution of each is proportional to the amount of cells present and in turn to the feed concentration S_i. This approach is an attempt to conceptualize and quantify the production and elaboration of soluble cell products which exert a Δ COD or BOD but are not original feed substrate. These concepts fit qualitatively the experimental observations of Chudoba et al. who observed a U-shaped plot of effluent soluble COD and BOD versus specific loading rate (relatable to specific growth rate) [24]. Benefield and Randall have suggested that the product model of Daigger and Grady has somewhat limited significance because of the fact that most of the effluent BOD or degradable COD in actual activated sludge effluents does not arise from soluble organics but from unsettled particulate organic matter [25]. However, it would seem to us that this fact may increase rather than decrease the potential significance of increased soluble organics simply because soluble product concentrations may be eventually predictable and controllable whereas currently there is no really quantitative approach to predicting suspended and colloidal organic concentrations in clarifier effluents. Thus, it would seem advisable to reserve as much as possible of the allowable discharge of oxygen-demanding organics for suspended materials by seeking ways to make the leakage of soluble degradable COD as small as possible. It seems to us that the subject could have considerable significance depending upon the magnitude of the leakage and the subsequent fate of the plant effluent. If there is a considerable increase in S for increased S_i at a given specific growth rate(or sludge age or specific substrate removal rate), then the occurrences may have considerable engineering significance. On the other hand, if the increase is small, it may be neglected and we can continue to apply models in consonance with the theory of continuous culture and accept the roughness of the approximation as due largely to the colligative measures we employ to assess the strengths of influents and effluents. In the work of Manickam and Gaudy cited earlier [16], the increase in S above that predicted by the effect of S_i on μ was small, amounting to only 10–15 mg/ L COD per 1,000 mg/L COD in the feed, and only a portion of this exerted a biochemical oxygen demand. This was felt to be too small an increase to warrant changing or adjusting the models, at least at relatively slow specific growth rates typical of activated sludge processes. On the other hand Grady and co-workers [20,26],using once-through systems at higher specific growth rates, observed higher values, approximately 40 mg/L COD per 1,000 mg/L COD in the feed. These data were the main source of the experimental results employed by Daigger and Grady in formulating the total product model cited previously.

It is recognized [23,27] that more base-line data regarding the phenomenon is needed and the work reported below is aimed at expanding the pool of available information on the effect of S_i on soluble residual organic matter in reactor effluents. There are many unanswered questions regarding production, accumulation and biodegradability of soluble organic matter produced during growth and decay of biomass, and this work is intended to provide insights into the extent of the phenomenon and the nature and biochemical behavior of "residual COD" from biomass grown on several carbon sources.

Table II. Mineral Composition of Synthetic Waste per 500 mg/L Glucose, Sorbitol or Phenol COD

Constituents	Amounts[a]		
	b	c	d
Ammonium Sulfate, $(NH_4)_2SO_4$	250	300	500
Magnesium Sulfate, $MgSO_4 \cdot 7H_2O$	50	50	100
Ferric Chloride, $FeCl_3$	0.25	0.25	1.0
Manganous Sulfate, $MnSO_4 \cdot H_2O$	5	5	10
Calcium Chloride, $CaCl_2$	3.75	3.75	30
Sodium Chloride, NaCl	—	—	40
1 Molar Phosphate Buffer Solution pH 7.0	5	50	—
2 Molar Phosphate Buffer Solution pH 8.0	—	—	10
Tap Water	50	50	20

a = All units = mg/L.
b = Glucose or sorbitol, all dilution rates.
c = Phenol, D = 0.03125 hr^{-1}.
d = Phenol, D = 0.018 hr^{-1}.

MATERIALS AND METHODS

Continuous Growth Studies in Once-Through Chemostats

Two-liter once-through chemostats were run at several dilution rates with various levels of carbon sources, either glucose, sorbitol or phenol. The composition of the feed is given in Table II. The units were vigorously aerated and checked frequently for complete mixing conditions; the dissolved oxygen was maintained above 6 mg/L and the temperature of the aerating liquid was 22 ± 2°C. Chemostats were run for a period of 2 to 4 weeks at each dilution rate. Daily samples were taken for soluble COD. Some COD's were run in accord with methods outlined in Standard Methods for the Examination of Water and Wastewater [28], and others were run using commercially available procedures [29]. Biomass was measured as dry suspended solids (103–105°C, 2 hours) retained on a membrane filter, pore size 0.45 μm. In some runs soluble BOD's were run frequently using seed organisms from the chemostat effluent [30]. Samples were also taken for analysis of the specific carbon source, either glucose [31], sorbitol [31], or phenol [32] and for carbohydrate [31].

Pilot Plant Studies on Municipal Sewage

During a long-term study [33] designed to assess the effects of certain priority pollutants on municipal treatment plants, it was necessary to run a control activated sludge unit. The unit which was of the "internal cell recycle" type was run at θ_c = 5 days and at a temperature of 22 ± 2°C. The municipal sewage used was found to be rather weak and the pilot plant feeds were supplemented with 50 mg/L glucose. During the study the soluble COD in the feed ranged from 125 mg/L to 374 mg/L. Soluble COD in the pilot plant effluent was determined frequently and these data provided an opportunity to determine whether there was any noticeable effect on S over the observed range of S_i values. The study was conducted over a period of 18 months.

Batch Studies to Assess Residual COD

Batch studies were run at several values of S_o using either glucose or phenol to assess the amount of residual COD and its fate. The experimental protocol will be described as the results are presented.

RESULTS

Continuous Growth Studies in Once-Through Chemostats

GLUCOSE. Figure 1 shows the effect of increased S_i on S measured as COD, BOD_5 and carbohydrate when the system was held at a dilution rate of 0.031 hours^{-1} ($\bar{t}$ = 32 hours, θ_c = 1.33 days). Over the range of S_i from 500 to 5,000 mg/L carbon source, the effluent COD increased from approximately 30 to 80 mg/L. The mode of increase was not linear but appeared to approach,

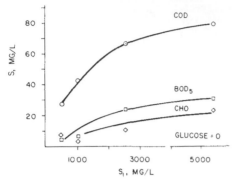

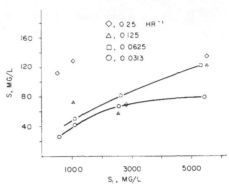

Figure 1. Relationship between S_i and S at a net specific growth rate, $\mu_n = 0.0313$ hr^{-1}, for heterogeneous microbial populations growing on glucose in a once-through reactor.

Figure 2. Relationship between influent and effluent COD at several net specific growth rates for heterogeneous microbial populations growing on glucose in a once-through reactor.

asymptotically, a maximum value as S_i was increased. No glucose was detected in the filtrate at any influent glucose level. Both total carbohydrate and BOD$_5$ increased with increasing S_i. Comparison of the BOD$_5$ and COD concentrations provides some indication that a somewhat higher fraction of the residual COD consisted of readily available organic substrate at the two higher levels of S_i than at the lower feed levels.

In Figure 2 the effect of S_i (glucose) on S (measured as COD) is compared for several values of D. For D = 0.0313 and 0.0625 hours^{-1}, there was a consistent increase in S with increased S_i but at the higher D values, 0.125 and 0.25 hours^{-1}, the increasing trend is not so clearly indicated.

SORBITOL. Figure 3 shows results using sorbitol as carbon source for a dilution rate of 0.125 hours^{-1}. The results are similar to those for glucose except that the absolute values of residual S's are higher than for glucose due to the faster specific growth rate employed. There was some evidence for the presence of an increasing concentration of the original exogenous substrate at the two highest feeding concentrations. However, the test for sorbitol is a general test for sugar alcohols and, thus, does not have the same degree of specificity as does the test for glucose. The relationship is drawn as a curvilinear one as in Figure 1, but there is less definite indication that the data can be so represented than there was for glucose.

PHENOL. The results of studies run at dilution rates of 0.0313 and 0.018 hours^{-1} using phenol as substrate (see Figure 4) seem best fitted to a straight line relationship between S_i and S. There is a slight increase in the rate of increase at the faster growth rate, 30/1,000 compared to 20/1,000 mg S COD/mg S_i COD. Analyses for phenol were performed but phenol was not present in any of the samples taken. Analyses for carbohydrate in the effluent were not run.

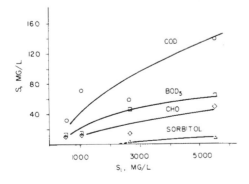

Figure 3. Relationship between S_i and S at a net specific growth rate, $\mu_n = 0.125$ hr^{-1}, for heterogeneous microbial populations growing on sorbitol in a once-through reactor.

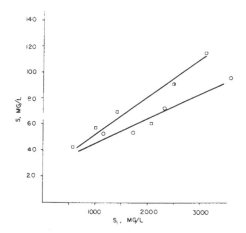

Figure 4. Relationship between S_i and S at net specific growth rates of 0.0313 hr^{-1} and 0.018 hr^{-1} for heterogeneous microbial populations growing on phenol in a once-through reactor.

Pilot Plant Studies on Municipal Sewage

In general municipal sewage is characterized by soluble COD values much lower than those employed in the previously described studies. Thus, in assessing the practical ramifications of the relationship between S_i and S to the treatment of municipal sewage, it was useful to compare the effect of increased S_i on S for a system growing essentially on municipal sewage at a constant growth rate. The study cited in the Materials and Methods section afforded such an opportunity. A bench scale activated sludge unit was operated at θ_c = 5 days for approximately 1.5 years. During this time, the influent soluble COD ranged from 125 to 374 mg/L. In order to determine whether there was a tendency for soluble effluent COD to increase for higher values of influent soluble COD, the recorded influent soluble COD's were arranged into groups with a range of 50 mg/L. All S values corresponding to a set of S_i values were averaged and these are shown in Table III. It is seen that there was no indication of increased S with increased S_i.

Batch Studies to Assess Residual COD

In the foregoing studies, it could not be determined whether the residual organic matter arose from intermediate products of active metabolism of the exogenous substrate or from products of cell lysis and autodigestion. Indeed, it is reasonable to expect that the growth rate and perhaps the biomass concentration itself may play a role in determining the source of the residual. For example, at low growth rates and high biomass concentrations one might expect that the residual COD arises largely due to autodigestion of MLSS. Regardless of the metabolic source, some of the residual is metabolically available to microorganisms as evidenced by the fact that it exerts a BOD.

When one does batch experiments at increasing values of S_o the same general trend occurs as in continuous reactors, i.e., there is an increase in residual substrate, S, with increased S_o. However, such an increase may not represent a truly significant amount of substrate. In an experiment using cells acclimated to phenol, 8 batch reactors were seeded identically with 950 mg/L biomass and fed phenol at COD concentrations ranging from 150 to 3,900 mg/L. The experiment was run for 120 hours, but for all these systems the COD removal had essentially terminated in 60 hours. Table IV compares S_o values with corresponding measured values of S after 60 hours of aeration. It can be seen

Table III. Effect of Feed COD (S_i) of Municipal Sewage (Supplemented with 50 mg/L Glucose) on Effluent Soluble COD (S) for Treatment by Activated Sludge at θ_c = 5 Days

S_i Range	n	S_i Mean	S Mean	% REM
125–174	5	157	46	71
175–224	33	211	35	83
225–274	105	251	37	85
275–324	80	298	39	87
325–374	24	346	37	89

Table IV.   Comparison of Feed (Phenol) and Residual COD in Batch Experiments with an Initial Biomass Concentration of 950 mg/L

$S_o{}^a$	S^b
150	38
450	45
900	48
1625	32
1850	55
2650	65
3150	70
3900	75

$^a S_o$ = Initial substrate phenol COD concentration, mg/L.
$^b S$ = Final (residual) COD, mg/L (measured COD after 60 hours of aeration).

that for a 25-fold increase in S_o there was approximately a 2-fold increase in S; that is, S increased with S_o but the increase was not really significant. There were, of course, more cells in the system as S_o increased, and this may have accounted for the higher degree of "cleanup" of the soluble COD. On the other hand, there were more cells to undergo autodigestion and, thus, more opportunity to produce soluble cell debris. Thus, it seems very important to study the metabolic availability and rate of metabolic utilization of these products as well as their production.

A similar experiment was run using glucose as carbon source. Two reactors were seeded with 1,160 mg/L of acclimated biomass. To one was added 1,000 mg/L COD and to the other 10,000 mg/L COD. These systems were aerated for the next 5 days and samples removed for analysis for COD and glucose. Occasionally samples were taken for measurement of biomass concentration. The results are shown in Figure 5. At the low S_o level ($S_o/X_o \cong 1$) the COD was removed very rapidly. A residual of 70 mg/L was recorded at 10 hours. The residual remained at this concentration until the termination of the experiment at 120 hours. At 24 hours, biomass concentration was 1,750 mg/L.

For the system fed 10,000 mg/L glucose COD, glucose disappeared within 24 hours, whereas the COD was essentially removed at 72 hours. From hour 60 to 72 it decreased from 1,116 to 304 mg/L. Over the next 48 hours there was a further slow decrease. At 120 hours, 268 mg/L soluble residual COD was registered. Thus, S rose approximately fourfold for a tenfold increase in S_o. This is a somewhat more significant increase than for the phenol batch study previously presented, and it may have come about due to the proportionally lower initial seed level in the high S_o system wherein the ratio S_o/X_o was 10 compared to 4 for the phenol experiments. Also the rate of substrate removal was greater in the glucose experiment, and faster growth may enhance opportunities for release of metabolic intermediates.

It is important to note that the COD decreased slightly during the last 48 hours for the system run with an initial S_o of 10,000 mg/L. Biomass concentration was determined occasionally, and the highest value recorded was 5,740 mg/L at 72 hours. At 120 hours, biomass concentration was 4,990 mg/L; thus, in addition to slow removal of soluble COD there is evidence of auto-digestion of the biomass during this time. We had noted this type of occurrence in preliminary studies and had determined in this experiment to ascertain how much of the residual COD was actually available as carbon source to the cells. One could not make such an assessment directly from the results of the batch study because it is not known whether the small observed decrease in soluble COD is due to very slow metabolism of portions of the 304 mg/L present at 72 hours or whether the COD was being metabolized more rapidly but being replenished or replaced by release of COD from the autodigesting biomass. In order to determine if a significant portion of the soluble COD at 72 hours was capable of serving as carbon source, a sample of mixed liquor was filtered at 72 hours and seeded with a small portion of cells taken from the unit at that time. This new system was aerated and samples withdrawn for measurement of soluble COD remaining. The initial cell concentration was approximately 30 mg/L and the initial soluble COD was 304 mg/L. The COD remaining is plotted in the insert graph (see triangles) on Figure 5. For comparison the soluble COD in the original unit (see circles) is also plotted as are the COD values in the reactor which had been fed 1,000 mg/L COD. It is clear from the auxiliary experiment at high F/M (10/1) that nearly 50% of the residual COD was readily metabolizable whereas in the system from which it was taken (at lower F/M) it appeared that only 10% was biologically removable organic material. Thus, it is clear that models suggested to predict microbial product concentration in effluents should also include rate constants for product utilization as well. The product model recently proposed by Rittman et al. is addressed to this point [34].

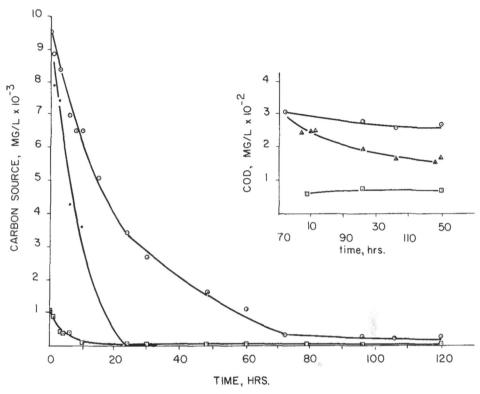

Figure 5.   Course of removal of substrate and resultant residual COD at feed levels, S_o, of 1,000 and 10,000 mg/L Glucose COD; Θ, Soluble COD for system fed 10,000 mg/L COD; •, Glucose for system fed 10,000 mg/L COD; $\square$, Soluble COD for system fed 1,000 mg/L COD. Insert: Removal of 72 hour residual COD in reseeding experiment, $\triangle$, compared to residual COD in original experimental system.

DISCUSSION AND CONCLUSIONS

It seems clear from these results as well as our previously reported work and the work of others that the residual COD increases for increased influent COD in the feed. Whether this occurrence is of real significance to design and operation of all biological treatment facilities is a question which is not entirely answered, but it clearly involves the magnitude of the occurrence as well as one's choice of method for measurement of plant performance. For the most part treatment plants are installed to remove putrescible organic material, not COD. Thus, not only the magnitude of increase in S for increased S_i, but the aerobic metabolic fate of the increased S as well as the rate at which it exerts its biochemical oxygen demand are of significance. It was apparent from the results for glucose and sorbitol that both the BOD and COD increased for increased S_i. In all cases, the seed for the BOD determination was obtained from the pilot plant population, i.e., the one which produced the residual. Somewhat less than 50% of the residual COD registered as BOD_5. The results of the batch study indicated that approximately 50% of the 300 mg/L residual COD was utilized as a carbon source for growth. Thus, it can be stated that over the short term about half of the residual COD may be "readily" utilized as carbon and energy source when it is known that the residual consists wholly of microbial product, i.e., none of the residual COD is composed of compounds of the original feed COD. When this is not known, as for example with whole municipal or industrial wastes, this situation is somewhat clouded and such determinations become more difficult to make. But, 50% appears to be a reasonable approximation for the near-term biological availability of residual COD. From other work it is known that over the long term more than 90% of the residual COD is subject to biological removal provided long-term contact is provided [35]. Thus, when the residual COD is released to the aqueous environment, e.g., a receiving body of water, it is possible that 90% or more can be removed by aerobic biomass. Whether such release has a deleterious effect on the dissolved oxygen resource is clearly a matter of both the rate of expression of its O_2 demand and the overall amount of material released. It is clear, however, that even at the relatively fast net specific growth rates (low θ_c's) employed in these studies, the residual COD did not represent that of the original waste; it is in essence a new waste produced by the treatment process. Fortunately, its rate of biochemical availability appears to be rather slow, thus tending to lessen the stress on the limited reaeration capacity of most receiving streams.

Two facts seem worthy of emphasis. It is known that the increase in S for increased S_i is less at lower growth rates than it is at faster growth rates [16]. Also, it is well known that S is lower at lower net growth rates for a given S_i. These facts argue for employing low net specific growth rates (high θ_c's) in design and operation. It is also known that the specific growth rate at which a process is run exerts a significant effect on the gross chemical composition of the residual COD. For.example, it has been found that as the operating growth rate was decreased in an activated sludge pilot plant, a lower fraction of the filterable COD consisted of high molecular weight (presumably less easily metabolized) components [36]. Thus, increased starvation afforded by slower growth rate either enhanced the development of biochemical and/or ecological conditions under which the more slowly and less readily degradable high molecular weight microbial products could be metabolized or otherwise hampered production and accumulation of such compounds. In either case, slowing down the growth rate of the activated sludge system decreased both the total amount of soluble COD and the proportion of high molecular weight (> 100,000) components it contained.

In general, the results obtained in this study as well as those we have previously reported lead us to conclude that the rise in effluent COD per 1,000 mg/L rise in degradable influent COD is not of sufficient magnitude to warrant a change in the currently employed design and operational models or procedures, especially since approximately only half of this residual COD can be expected to be readily available as carbon and energy sources and may in most cases be expected to be metabolized at a slower rate than the original biodegradable feed COD. Furthermore, the studies with municipal sewage did not indicate a pattern of increase in S for increased S_i. Since it has taken more than 30 years to transfer physically demonstrable physiological-kinetic growth theory from systems well-defined with respect to substrate and cells to serviceable models for design and operation of heterogeneous systems, it would seem a mistake to suggest replacing it in design practice largely for the sake of accommodating such a gross measurement of substrate as COD. It would seem more productive to seek improvement in the parameters by which we measure S_i and S.

However, it should be emphasized that the phenomenon is of considerable scientific interest and can be of significance in design for cases where very high strength wastes are being treated. It is also a phenomenon of some consequence to chemical treatment prior to discharge to a receiving stream, e.g., chlorination and ozonation [37]. Also prediction of the amount of residual product and its chemical composition can be of great importance when treated effluent is to be prepared for immediate reuse. That is, metabolic products may have a serious effect on subsequent treatment processes preparatory to direct reuse.

In conclusion, although the increase of S with S_i can be of importance to subsequent chemical-physical processes to which the effluent may be subjected, it does not appear to be of sufficient magnitude to warrant modification of current design and operational models. Furthermore, all of the findings augur well for recommending the use of slow growth rate (high cell age) processes. Much more work on the biological fate of the product COD in the natural environment is needed.

ACKNOWLEDGMENT

Portions of the experimental work were conducted at Oklahoma State University with support from USEPA. Other portions of the experimental work were conducted at the University of Delaware with support from a UNIDEL Foundation Grant. Data analysis and manuscript preparation were supported under a grant from the Delaware Water Research Center, USDI.

REFERENCES

1. Monod, J., "Recherches sur la Croissance des Cultures Bacteriennes," Hermann et Sie, Paris (1942).
2. Monod, J., "La Technique de Culture Continue. Theorie et Applications," *Ann. Inst. Pasteur*, 79, 390–410 (1950).
3. Novick, A., and Szilard, L., "Description of the Chemostat," *Science*, 112, 715–716 (1950).

4. Herbert, D., Elsworth, R., and Telling, R. C., "The Continuous Culture of Bacteria—A Theoretical and Experimental Study," *J. Gen. Microbiol.*, 14, 601–622 (1956).
5. Herbert, D., "Theoretical Analysis of Continuous Culture Systems," Soc. of Chem. Ind., *Monograph No. 12*, 21–53, London (1960).
6. Gaudy, A. F., Jr., Ramanathan, M., and Rao, B. S., "Kinectic Behavior of Heterogeneous Populations in Completely Mixed Reactors," *Biotech. and Bioeng.*, 9, 387–411 (1967).
7. Peil, K. M., and Gaudy, A. F., Jr., "Kinetic Constants for Aerobic Growth of Microbial Populations Selected with Various Single Compounds and with Municipal Wastes as Substrates," *Appl. Microbiol.*, 21, 253–256 (1971).
8. Ramanathan, M., and Gaudy, A. F., Jr., "Effect of High Substrate Concentration and Cell Feedback on Kinetic Behavior of Heterogeneous Populations in Completely Mixed Systems," *Biotech. and Bioeng.*, 11, 207–237 (1969).
9. Srinivasaraghaven, R., and Gaudy, A. F., Jr., "Operational Performance of an Activated Sludge Process with Constant Sludge Feedback," *Jour. Water Poll. Contr. Fed.*, 47, 1946–1960 (1975).
10. Gaudy, A. F., Jr., Srinivasaraghaven, R., and Saleh, M., "Conceptual Model for Activated Sludge Processes," *Jour. Env. Engr. Div. A.S.C.E.*, 103, 71–84 (1977).
11. Lawrence, H. W., and McCarty, P. L., "Unified Basis for Biological Treatment Design and Operation," *Jour. San. Engr. Div., A.S.C.E.*, 757–778 (1970).
12. McKinney, R. E., "Mathematics of Complete-Mixing Activated Sludge," *Jour. San. Engr. Div., A.S.C.E.*, 88, 87–113 (1962).
13. Eckenfelder, W. W., Jr., "Theory and Practice of Activated Sludge Process Modifications," *Water and Sewage Works*, 108, 145–150 (1961).
14. Gaudy, A. F., Jr., and Kincannon, D. F., "Comparison of Design models for Activated Sludge," *Water and Sewage Works*, 123, 66–70 (1977).
15. Gaudy. A. F., Jr., and Gaudy, E. T., *Microbiology for Environmental Scientists and Engineers*, McGraw-Hill Book Co., Inc., New york (1980).
16. Manickam, T. S., and Gaudy, A. F., Jr., "Studies on the Relationship Between Feed COD and Effluent COD During Treatment by Activated Sludge," *Proc., 34th Annual Ind. Waste Conf.*, Purdue University pp. 854–867 (May 1979).
17. Adams, C. E., Eckenfelder, W. W., and Hovious, J. C., "A Kinectic Model for Design of Completely-Mixed Activated Sludge Treating Variable-Strength Industrial Wastewaters," *Water Research*, 9, 37–42 (1975).
18. Grau, P., Dohanyos, M., and Chudoba, J., "Kinetics of Multicomponent Substrate Removal by Activated Sludge," *Water Research*, 9, 637–642 (1975).
19. Benefield, L. D., and Randall, C. W., "Evaluation of a Comprehensive Kinetic Model for the Activated Sludge Process," *Jour. Water Poll. Contr. Fed.*, 49, 1636–1641 (1977).
20. Grady, C. P. L., Jr., and Williams, D. R., "Effects of Influent Substrate Concentration on the Kinetics of Natural Microbial Populations in Continuous Culture," *Water Research*, 9, 171–180 (1975).
21. Chen, Y. R., and Hashimoto, A. G., "Substrate Utilization Kinetic Model for Biological Treatment Processes," *Biotech. and Bioeng.*, 22, 2081–2095 (1980).
22. Baskir, C. I., and Hansford, G. S., "Product Formation in the Continuous Culture of Microbial Populations Grown on Carbonhydrates," *Biotechnol. and Bioeng.*, 22, 1857–1875 (1980).
23. Daigger, G. T., and Grady, C. P. L., Jr., "A Model for the Bio-Oxidation Process Based on Product Formation Concepts," *Water Research*, 11, 1049–1057 (1977).
24. Chudoba, J., Miroslav, P., and Emmerova, H., "Residual Organic Matter in Activated Sludge Process Effluents III. Degradation of Amino Acids and Phenols Under Continuous Conditions," *Scientific Papers—Inst. of Chemical Technology*, Prague F13 (1968).
25. Benefield, L. D., and Randall, C. W., *Biological Process Design for Wastewater Treatment*, Prentice-Hell, Inc., Englewood Cliffs, NJ (1980).
26. Grady, C. P. L., Jr., Harlow, L. J., and Riesing, R. R., "Effects of Growth Rate and Influent Substrate Concentrations on Effluent Quality from Chemostats Containing Bacteria in Pure and Mixed Culture," *Biotech. and Bioeng.*, 14, 391–410 (1972).
27. Vandevenne, L., and Eckenfelder, W. W., "A Comparison of Models for Completely Mixed Activated Sludge Treatment Design and Operation," *Water Research*, 14, 561–566 (1980).
28. A. P. H. A., *Standard Methods for the Examination of Water and Wastewater*, 14th Ed., Method 508, Washington, D.C. (1975).
29. Hach Chemical Company, *Hach Water Analysis Handbook*, Loveland, CO, 2-175-179 (1980).
30. A. P. H. A., *Standard Methods for the Examination of Water and Wastewater*, 14th Ed., Method 507, Washington, D.C. (1975).
31. Ramanathan, M., Gaudy, A. F., Jr., and Cook, E. E., "Selected Analytical Methods for

Research in Water Pollution Control," *Manual M-2*, Center for Water Research in Engineering, Bioenvironmental Eng., Oklahoma State University, Stillwater, OK (1968).
32. A. P. H. A., *Standard Methods for the Examination of Water and Wastewater*, 14th Ed., Method 510c, Washington, D.C., (1975).
33. Gaudy, A. F., Jr., Kincannon, D. F., and Manickam, T. S., "Treatment Compatibility of Municipal Waste and Biologically Hazardous Industrial Compounds," NTSI, PB83-105536, *EPA-600/2-82-075a ORD Report*, U.S. Dept. of Commerce (1982).
34. Rittman, B. E., Bae, W., Namkung, E., and Lu, C. J., "A Critical Evaluation of Microbial Product Formation in Biological Systems," prepared for presentation at I.A.W.P.R.C. Conference, Sao Paulo, Brazil (1986).
35. Gaudy, A. F., Jr., and Blachly, T. R., "A Study of the Biodegradability of Residual COD," *Jour. Water Poll. Control Fed.*, 57, 332–338 (1985).
36. Saunders, F. M., and Dick, R., "Effect of Mean-Cell Residence Time on Organic Composition of Activated Sludge Effluents," *Jour. Water Poll. Control Fed.*, 53, 201–215 (1981).
37. Watt, R. D., Kersch, E. J., and Grady, C. P. L., "Characteristics of Activated Sludge Effluents Before and After Ozonation," *Jour. Water Poll. Control Fed.*, 57, 157–166 (1985).

7 IRON AND MANGANESE REMOVAL USING A ROTATING BIOLOGICAL CONTACTOR

Ellen K. Russell, Environmental Scientist
CH2M HILL
Milwaukee, Wisconsin 53203

James E. Alleman, Associate Professor
School of Civil Engineering
Purdue University
West Lafayette, Indiana 47906

INTRODUCTION

This investigation describes the removal of two heavy metals, iron (Fe) and manganese (Mn), within a rotating biological contactor (RBC) wastewater treatment system. The effects of these metals on the RBC treatment process, in addition to their accumulation within the biofilm are studied. Rotating biological contactor biofilms can become coated with metals that cause a decreased transfer of nutrients, waste products, and gases across the biofilm surface [1]. Iron and manganese, two readily oxidized and deposited metals, can accumulate and may cause a decline in treatment efficiency. This could be accompanied by an increase in the weight load on the shaft and biofilm attachment media. In light of past and present RBC failures [2], this latter aspect becomes an important consideration of this study.

LITERATURE REVIEW

Both iron and manganese are ubiquitous throughout nature, constituting roughly 5% and 0.1% of the earth's lithosphere, respectively [3]. Their occurrence has been frequently documented in systems as diverse as a wastewater treatment plant and a vertebrate's circulatory system [4,5]. Each metal has two common oxidation states. In the reduced +2 form, iron is referred to as ferrous iron, and in the oxidized +3 state it is ferric iron. Similarly, manganese is referred to as the manganous ion (Mn +2) and the manganic ion (Mn +4).

The ability of iron and manganese to undergo reversible oxidation and reduction reactions represents the focal point of all discussions concerning their roles and fates within aquatic systems (including an RBC system). Their solubility is influenced by the pH, oxidation-reduction potential, organic and inorganic constituents, and biological components (macro and microorganisms) of their environment.

Both metals are also efficient scavengers of heavy metals [6]. Practical application of this ability has been made in water and wastewater treatment for the removal of other heavy metals, phosphorus compounds, and organic contaminants [7]. Iron and manganese are critical micronutrients for various organisms including bacteria, fungi, protozoa, metazoa, and algae. In particular, certain species of bacteria are classified by their association with iron and manganese [8]. These organisms include *Thiobacillus, Leptothrix, Sphaerotilus, Gallionella, Siderocapsa, Metallogenium, Pedomicrobium,* and *Pseudomonas.*

Typically, environmental engineers are concerned with the occurrence of iron and manganese as pollutants or nuisance compounds in ground and surface drinking water supplies [9]. These compounds may also be present as pollutants in stormwater runoff and domestic, commercial, and industrial wastewaters [10]. Ironically, manganese and iron are often added directly to the same treatment processes that have been designed to remove them.

For example, in wastewater treatment plants, compounds such as ferric chloride, ferrous chloride, and ferrous sulfate are used for phosphorus precipitation [11]. Their addition is normally to the aeration basins, and phosphorus compounds are subsequently removed during secondary clarification. Manganese in the permanganate form is used in the water treatment process for oxidation of

59

Table I. Influent Feed Composition

Component	Concentration in Influent (mg/L)
COD (measured)	160.0
BOD$_5$ (measured)	150.0
NH$_4$-N (measured)	<1
NO$_3$-N (measured)	0.3
Total N (calculated)	12.0
Cl⁻ (measured)	74.0
Total P (calculated)	2.92
Total P (measured)	2.00
Mg (calculated)	0.12
TSB (7.025 g/L in carboy)	150.0
Sucrose (2.6 g/L in carboy)	56.0
pH	7.32

organics. In the process, a manganese containing sludge is generated. Iron sludges are also generated during phosphorus removal.

In municipal biofilm plants, iron salts should be added immediately before the final clarifiers to avoid staining and coating of the biofilms. In a study conducted by the U.S. EPA [11], 52% of all municipal plants surveyed (both biofilm and activated sludge) used some type of iron salt for phosphorus removal.

Because of the beneficial associations that iron and manganese have with the water and wastewater treatment processes, it is also of interest to ascertain what detrimental effects these metals could possibly be having on the treatment processes. To limit the field of investigation, it was decided to study a rotating biological contactor (RBC) wastewater treatment system as a candidate for exposure to iron and manganese. The following describes the materials and methods used to determine the effects of these metals.

MATERIALS AND METHODS

Depending upon the geographical area and the amount and type of industrial input, iron and manganese may be present in varying concentrations in the influent wastewater. Typical generators of iron and manganese containing waste streams include the iron and manganese ore mining industry, iron and steel industry, and various chemical industries. Pretreatment regulations governing the discharge of either metal have been established to prevent burdensome quantities of these metals from entering wastewater treatment plants [12].

In this instance, both metals were supplied jointly to the influent of a bench-scale RBC unit in soluble concentrations ranging from 0 to 30 mg/L. These concentrations were considered representative of those encountered in a typical municipal wastewater treatment plant receiving some industrial input [13,14]. The metals were supplied across a 5-month period.

The RBC reactor was designed to simulate a full-scale system. It had a liquid volume of 0.036 m³ and a hydraulic retention time of roughly 6 hours. The average flow through the RBC was 140 L/day, which traveled in a fashion approximating plug-flow. The total disc surface area amounted to approximately 3.3 m² which translated to a hydraulic loading of 0.04 m²/m³/day. Similar loading rates have been used by previous RBC investigators and manufacturers [15,16,17].

The wastewater supplied was a mixture of Trypticase Soy Broth, sucrose, and tapwater (see Table I). The soluble BOD$_5$ loading to the system was 0.0054 kg/m² disc surface area/day. Figure 1 shows the pumping arrangement used to supply the reactor.

A total of 40, 22.9 cm diameter plexiglass discs were rotated at 22 rpm. The total disc surface area was 3.28 m². The reactor was divided into 4 stages with 10 discs per stage. Discs were designed to be easily removed for biofilm sampling (see Figure 2). Sampling areas (6.54 cm² surface area) had been previously assigned. Discs were kept approximately 40% submerged at all times. The shaft consisted of a solid stainless steel rod (1.27-cm diameter) that would not corrode or break.

Both metals were added according to the scheme supplied in Table II. Concentrations of each were increased by set increments at the beginning of each operation month. This continued for 5 months. Previous to any metal addition, the unit was operated for 7 months during which a "pseudo" stable

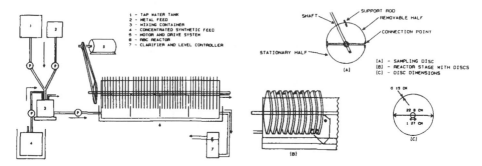

Figure 1.   Schematic of RBC system.   **Figure 2.   Schematic of RBC sampling disc.**

biofilm was allowed to develop and background metals were monitored. By "pseudo" stable it is meant that the biofilm was not subjected to shocks of any sort and was allowed to develop a mature population of organisms.

Throughout the unit's operation, data were collected in the form of chemical oxygen demand (COD) removal, solids production, pH, temperature, dissolved oxygen (DO), amounts of suspended, soluble, and biofilm associated iron and manganese, and biofilm mass measurements.

RESULTS AND DISCUSSION

Results indicated that a mean soluble COD removal efficiency of 86% was obtained throughout the period of reactor operation (see Figure 3). Seventy percent of the soluble COD was removed within the first stage of the reactor. The slight decline in treatment efficiency experienced towards the end of operation could not be attributed entirely to the metals, since temperatures decreased somewhat during this period of operation (i.e., winter months).

Mean soluble iron and manganese removals were 98% and 86%, respectively, over the entire period of operation. Manganese removal declined sharply to 21% following the second week of a continuous addition of 15 mg/L iron and 15 mg/L manganese (see Figure 4). It is likely that manganese, a more soluble metal than iron, remained in solution longer and was not deposited as quickly as was iron.

Analysis of the metals within the biofilm at various points along the shaft of the RBC showed definite trends. Figures 5 through 16 are representations of the metal accumulation that occurred. The iron/manganese dosage (in mg/L) is given in each figure title. Essentially, iron was found in significantly (alpha = 0.05) higher concentrations than manganese in the biofilm on the discs in the first two stages of the reactor. Manganese was found in the biofilm on the discs in stages two and three in amounts statistically greater than the amounts of iron present. The explanation for this behavior is that iron has a lower solubility than manganese, and therefore oxidizes and deposits more rapidly than does manganese.

Metals also accumulated to the largest extent in the center regions of the discs. Peripheral biofilm accumulated less metal, probably because of the faster rotational velocities existing at the periphery of

Table II. Metal Additions

Metal Addition Phase	Exposure Dates	Influent			Amount Reagent for Concentrated Feed (g/L)	
		Conc. Fe^{+2} (mg/L)	Conc. Mn^{+2} (mg/L)	Text Abbrev.	$FeCl_2$	$MnCl_2$
1	1/31/84-9/12/84	0	0	[0/0]	0	0
2	9/12/84-10/11/84	5	5	[5/5]	1.59	1.59
3	10/11/84-11/8/84	5	7.5	[5/7.5]	1.59	2.39
4	11/8/84-12/7/84	7.5	7.5	[7.5/7.5]	2.39	2.39
5	12/7/84-1/7/85	15	15	[15/15]	4.78	4.78
6	1/7/85-2/2/85	30	30	[30/30]	9.55	9.55

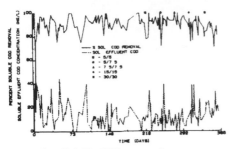

Figure 3.   Soluble COD removal.

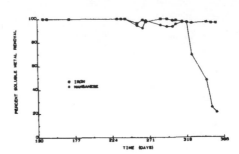

Figure 4.   Iron and manganese removal.

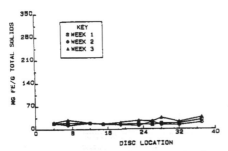

Figure 5.   Iron accumulation per total solids
(0/0).

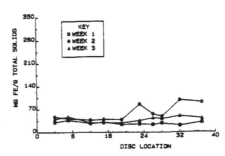

Figure 6.   Iron accumulation per total solids
(5/5).

the disc as opposed to those in the center. Figures 17 and 18 show the appearance of the biofilm on discs in the first and third stages of the reactor, respectively.

The biofilm on the discs with large amounts of accumulated iron appeared amorphous and organisms were obscured by the iron that had been deposited. In contrast, the appearance of the biofilm on a disc that had high amounts of manganese accumulated (see Figure 18) appeared flaky and mica-like. Both photographs attest to the metal accumulation within the RBC biofilm. It is remarkable that despite the apparent blockage of biofilm surface exchange sites, the soluble COD treatment efficiency did not decline more than it did.

An increase in total dry biofilm weight of 309 g was attributed to the addition of total iron and manganese amounts of 263 and 274 g, respectively. Of these 309 g, 50 g were due to iron and 83 g to manganese. The overall weight gain minus the iron and manganese weight contribution was 176 g. Metal accumulation within the biofilm reached maximum levels of 34% and 78% on a weight of metal

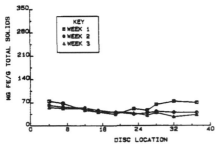

Figure 7.   Iron accumulation per total solids
(5/7.5).

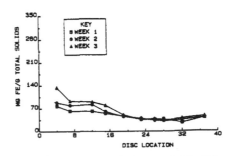

Figure 8.   Iron accumulation per total solids
(7.5/7.5).

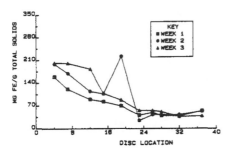

Figure 9.   Iron accumulation per total solids (15/15).

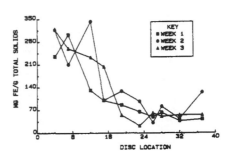

Figure 10.   Iron accumulation per total solids (30/30).

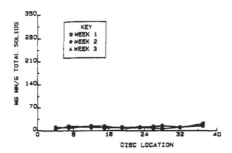

Figure 11.   Manganese accumulation per total solids (0/0).

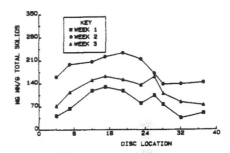

Figure 12.   Manganese accumulation per total solids (5/5).

to weight of dry total biofilm solids basis for iron and manganese, respectively. Figure 19 depicts the total accumulation of the two metals according to location within the RBC shaft.

When the above metal accumulation results are extrapolated to a full-scale, 150,000 ft^2 RBC shaft, the potential for weight increase may be as high as 7.5 tons [18]. This type of load increase on the RBC shaft could prove detrimental and cause increased fatigue. When considering past shaft, bearing, and media failures (see Table III) this situation becomes important.

Not only did the metals themselves account for the weight increase, but it is postulated that iron and manganese along with their coprecipitates (in the form of carbonates and hydroxides) may have physically entrapped and cemented biomass to such an extent that normal sloughing was hindered causing an additional biofilm weight increase. This is supported by solids production data which show a steady decline in total suspended solids leaving the reactor as metal addition increased. Finally, the growth found on RBC discs was microscopically observed to contain organisms, both filamentous

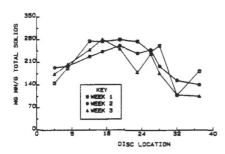

Figure 13.   Manganese accumulation per total solids (5/7.5).

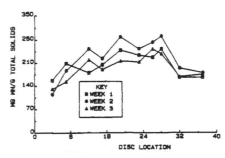

Figure 14.   Manganese accumulation per total solids (7.5/7.5).

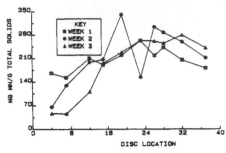
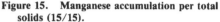
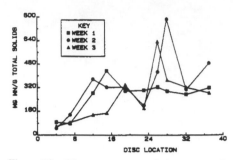

Figure 15.   Manganese accumulation per total
solids (15/15).

Figure 16.   Manganese accumulation per total
solids (30/30).

and capsule-like, that have a natural tendency to deposit oxidized forms of metals, particularly iron and manganese, upon their surfaces.

CONCLUSIONS

In conclusion, results indicate that the RBC treatment process was efficient in removing the metal concentrations applied with very little simultaneous decline in soluble COD removal. However, the weight gain experienced by the RBC unit resulting from deposition of the metals within the biomass could prove detrimental to normal operation of a full-scale unit. It would be prudent to avoid upstream addition of iron compounds in RBC treatment plants for the above reason alone. This includes measures to prevent recirculation of iron compounds as well as their direct upstream applica-

Figure 17.   Scanning electron micrograph of
iron deposition on biofilm. Black bar
within box represents 100 μ.

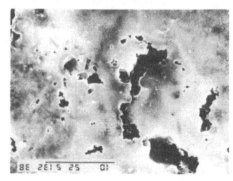

Figure 18.   Scanning electron micrograph of
manganese deposition. Black bar within box
represents 10 μ.

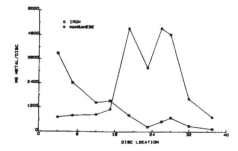

Figure 19.   Total metal mass accumulated in
biofilm.

Table III. Summary of Shaft-Related Failures, [2]

Plant	Number of Shafts	RBC System Startup Year	Broken Shafts	Bearing Failures	Media Support Failures
Cheboygan, MI	8	1978	3	0	0
Cleves, OH	6	1977	0	0	0
Edgewater, NJ	4	1973	0	1	0
Gladstone, MI	6	1974	1	0	0
Gloucester, NJ	4	1974	1	2	0
Hamilton, NJ	48	1979	6	0	0
Hartford, MI	2	1978	0	0	0
Ionia, MI	12	1978	12	0	0
North Huntington, PA	4	1975	3	2	0
Rhinelander, WI	10	1977	6	0	0
Selden, NY	12	1974	0	8	0
Thermopolis, WY	2	1978	0	0	0
Voorhees, NJ	6	1976	3	0	0
Wappinger Falls, NY	2	1978	0	1	4
Washington, NJ	3	1974	0	0	3
Winchester, KY	24	1977	6	2	0

tion. It may also be that special compensations should be made in the design and operation of an RBC unit that unavoidably must be exposed to heavy metals.

REFERENCES

1. Atkinson, B. and E. L. Swilley, "The Effect of Ferric Chloride on the Removal Efficiency of a Biological Film Reactor," *Water Resources*, 1:687-693 (1967).
2. Brenner, R. C., "Design Information for Rotating Biological Contactors," *U.S. EPA Report 430/9-84-008*, Cincinnati, Ohio (September 1984).
3. Maynard, J. B., *Geochemistry of Sedimentary Ore Deposits*, Springer-Verlág: New York (1983).
4. Wetzel, R. G., *Limnology*, W. B. Saunders Company: Philadelphia (1975).
5. Berdow, R., ed. *Merck Manual*, Merck, Sharp, and Dohme Research Laboratories: Rahuay (1982).
6. Stumm, W., and G. F. Lee, "The Chemistry of Aqueous Iron," *Schweizensche Zcitshrift Fur Hydrologie*, 22(1):295-319 (1960).
7. Singh, S. K., and V. Subramanian. "Hydrous Fe and Mn Oxides— Scavengers of Heavy Metals in the Aquatic Environment," *CRC Critical Rev. in Environmental Control*, 14(1):33-90 (1984).
8. Ghiorse, W. C., "Biology of Iron and Manganese Depositing Bacteria," *Ann. Rev. Microbiol*, 38:515-550 (1984).
9. Breland, E. D., and L. R. Robinson, *Iron and Manganese Removal from Low Alkalinity Groundwater*, Mississippi State University:Mississippi (1967).
10. Patterson, J. W., and P. S. Kodukula, "Metal Distribution in Activated Sludge Systems," *J. Water Pollution Control Fed.*, 56(5):432-441 (1984).
11. Schmidt, C. J., "Review of Techniques for Treatment and Disposal of Phosphorus—Laden Chemical Sludges," *U.S. EPA Report 600/2-79-983*, Cincinnati, Ohio (August 1979).
12. Code of Federal Regulations, Federal Register, Title 40: Protection of the Environment, Parts 420, 421, 424, 434, and 440 (1984).
13. Oliver, B. C., and E. G. Cosgrove, "The Efficiency of Heavy Metal Removal by a Conventional Activated Sludge Treatment Plant," *Water Resources*, 8:869-874 (1974).
14. Cheng, M. H., J. W. Patterson, and R. A. Minear, "Heavy Metals Uptake by Activated Sludge," *J. Water Pollution Control Fed.*, 47(2):362-376 (1975).
15. Poon, C. P., Y. L. Chao, and W. J. Mikucki, "Factors Controlling Rotating Biological Contactor Performance," *J. Water Pollution Control Fed.*, 51(3):601-611 (1979).
16. Kinner, N., D. L. Balkwill, and P. L. Bishop, "The Microbiology of Rotating Biological Contactor Films," *Proc. lst Inter. Conf. on Fixed-Film BiologicalProcesses*, Kings Island, Ohio, pp. 184-209, (1982).

17. Friedman, A. A., and L. E. Robbins, "Effect of Disc Rotational Speed on Biological Contactor Efficiency," *J. Water Pollution Control Fed.*, 51(11):2678–2690 (1979).
18. Bowman, M. D., and J. T. Gaunt, "Fatigue Behavior of Rotating Biological Contactor Shafts," *Technical Report*, prepared for U.S. EPA, Cincinnati, Ohio (April 1982).

8 WASTEWATER TREATMENT BY A BIOLOGICAL-PHYSICOCHEMICAL TWO-STAGE PROCESS SYSTEM

Milos Krofta, President
Krofta Engineering Corporation
Lenox, Massachusetts 01240

Lawrence K. Wang, Director
Lenox Institute for Research, Inc.
Lenox, Massachusetts 01240

PROJECT DESCRIPTION

The effectiveness of a Krofta Sandfloat Sedifloat clarifier (Type SASF-5) on treating the secondary effluent at the Norwalk Wastewater Treatment Plant, Ohio, has been investigated.

Norwalk is located in northern Ohio, approximately 50 miles southwest of Cleveland and supports a population of 14,500 people. The existing Norwalk Wastewater Treatment Plant consists of one primary sedimentation, one aeration basin, one primary trickling filter, two secondary trickling filters and one chlorination- sedimentation basin. The plant treats combined domestic sewage and food-processing waste.

At present the Norwalk Wastewater Treatment Plant discharges a secondary effluent containing an average of 100.4 mg/L of total suspended solids (TSS), 7.64 mg/L of phosphate, and 40.57 mg/L of 5-day biochemical oxygen demand (BOD).

The primary objective of this investigation was to further treat the secondary effluent by a tertiary Sandfloat Sedifloat clarifier in order to meet the US. Environmental Protection Agency established tertiary effluent standards:

Total suspended solids: 10 mg/L
Biochemical oxygen demand, 5-day: 10 mg/L
Phosphate: 1 mg/L

PROCESS CONCEPT

Many multiple stage wastewater treatment systems have been developed based on the following hypotheses:

a. Wastewater generally contains both biodegradable and non-biodegradable dissolved organic pollutants. If a biological process system is used first for removal of biodegradable organics (i.e., BOD), then a supplemental physicochemical process system consisting of mixing, flocculation and sedimentation can be used to remove the remaining non-biodegradable organics hoping that most of the non-biodegradable organics can be flocculated.

b. Wastewater generally contains both flocculatable and non-flocculatable dissolved organics. If a physicochemical process system is used first for removal of flocculatable organics, then a biological process system (i.e., activated sludge, trickling filter, lagoon, biological tower, rotating biological contactor, etc.) can be used to remove the remaining non-flocculatable organics hoping that most of the non-flocculatable organics are biodegradable.

c. Non-biodegradable and non-flocculatable dissolved organics can usually be removed by granular activated carbon, ion exchange and/or membrane processes.

Several important chemical process facts have been overlooked by environmental engineers and scientists for many decades until now. Some residual non-flocculatable dissolved organics in the clarified effluent of the first-stage physicochemical system (consisting of flocculation and clarification) can be made flocculatable in a second-stage physicochemical system (also consisting of floccula-tion and clarification) when adequate chemicals are used as the flocculating agents and/or flotation aid.

Another important fact that dissolved air flotation can replace sedimentation for clarification has also been overlooked.

Many refractory surfactants, oil and grease, floating algae, volatile organics, etc., cannot be removed by a conventional flocculation and sedimentation system but can be removed by innovative flocculation and flotation. In other words, the non-flocculatable organics can be floatable.

Finally sand filtration has been a proven technology for efficient tertiary waste treatment. Incorpo-ration of filtration into the waste treatment plant for final polishing will ensure that the governmental effluent standards on TSS can be met. Total COD effluent and BOD will decrease when TSS decreases.

The following sections introduce the state-of-the-art multiple stage (mainly two-stage) process systems as well as the newly developed two-stage biological-Sandfloat Sedifloat process system.

Conventional Biological-Physicochemical Two-Stage System

BRW Textiles operates a knit-dyeing and finishing mill in Bangor, PA, USA. Wastewater from the mill is cooled via heat exchanger, equalized, and then treated, first, biologically and second, chemi-cally. The raw wastes enter the treatment plant at an average rate of 0.72 mgd. Ammonia supplemen-tation is provided when necessary, and the wastes are equalized in a 1-million gallon basin. No aeration is provided, but hydrogen peroxide is added in the mill to help maintain an adequate dissolved oxygen level. The first-stage activated-sludge process is used for biological treatment. Twelve hours' aeration with an MLSS concentration of 2,500–3,500 mg/L results in 95% BOD_5 removal. Chemical antifoam-ing agents have had to be added to alleviate foaming problems in the aeration basin. Presently, solids are centrifuged and disposed of via landfill. Raw waste is unusually high in color and high in temperature. Heat exchangers remove 20°F, and holding for equalization results in additional cool-ing. The temperature in the aeration basin averages 80°F while raw waste temperature averages 123°F. The biological process produces little color removal.

As a result, alum is applied at 300 mg/L with 5 mg/L of anionic polymer in a second-stage physicochemical system. After rapid mixing, flocculation, and settling, 75% color removal was achieved with respect to the raw waste.

Biological-Sandfloat Sedifloat Two-Stage System

The newly developed two-stage system involves the use of a Sandfloat Sedifloat clarifier (or a Sandfloat clarifier) in the second stage for upgrading of an existing biological system (activated sludge, trickling filter, biological tower, lagoon, rotating biological contactor, etc.) which is now considered as the first stage treatment.

The treatment efficiency of this new system will be much better than that of the conventional Biological-Physicochemical Two-Stage System because: 1) the Sandfloat Sedifloat has a built-in tertiary filtration which produces tertiary treatment results; 2) the Sandfloat Sedifloat has a built-in flocculation and dissolved air flotation clarifier which can remove floatable pollutants which may be nonbiodegradable and/or non flocculatable; and 3) the Sandfloat Sedifloat is much cheaper and uses less land space in comparison with a conventional flocculation and sedimentation system.

Additional technical data concerning tertiary wastewater treatment and multiple-stage process systems can be found elsewhere [1,2].

DESCRIPTION AND OPERATION OF SANDFLOAT SEDIFLOAT

Figure 1 shows the bird's view of the Sandfloat Sedifloat clarifier (Type SASF-5) which was used as the second-stage tertiary physicochemical treatment unit at Norwalk Wastewater Treatment Plant, Ohio.

The dimension of the clarifier's outside tank is 5 feet (1.5 meters) diameter by 7 feet (2.1 meters) high.

Referring to Figure 1, the Sandfloat Sedifloat consists of the components as noted in Table I.

In normal operation of a Sandfloat Sedifloat, the pressure pump (#10) takes clarifier water from the main tank above the sandbeds and feeds the air dissolving tube (#11), with compressed air entering

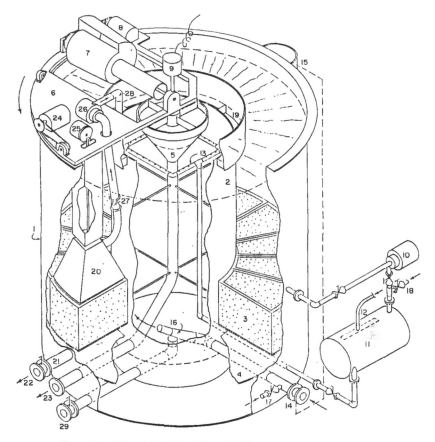

Figure 1.   Schematic of Sandfloat Sedifloat clarifier (type SASF-5).

through the panel (#12). The aerated water enters the flocculation tank via a pipeline and discharges through the distribution pipes (#13) at the top of the flocculation tank just prior to the inlet overflow into the main flotation tank (#1). The chemically treated Sandfloat Sedifloat influent wastewater enters the flocculation tank through the inlet regulating valve (#14) and through the nozzle (#16) in a jet motion causing the slow mixing which furthers the flocculation process. The level control (#15) is used for setting the desired level in the main flotation tank and keeps the level constant. The fine bubbles generated by the air dissolving tube are released from solution through the distribution pipes (#13) and mixed with solid particles (flocs) at the top of the flocculation tank (#2). The air bubbles attach to the solids and overflow into the main flotation tank via the deflector ring (#19). The solids are thereby floated to the surface of the Sandfloat Sedifloat where the sludge blanket forms. The floated sludge is removed from the surface of the water with the spiral scoop (#7) and discharged to the sludge handling system via the central sludge well (#5). The scoop is mounted on the carriage (#6) which rotates on the upper rim of the main tank and the sandbed backwashing pump (#26) and auxiliary backwashing equipment. Sludge flow can be regulated by adjusting the water level in the main tank or by adjusting the scoop speed with the scoop variable speed motor controller (#8). Power is supplied to the carriage components through the electrical rotary contact (#9).

   Water is now clarified by flotation in the main tank, passes down through the sand bed sections for final polishing and enters a common clearwell. The finished water goes through the pipeline (#21) and this flow is regulated by the clarified water discharge pump (#22). The residual flocs which are not floated out are trapped on the surface of the sand (#3) where a fine layer of solids begins to form. There are 23 sand bed sections in the Sandfloat Sedifloat which are individually backwashed. As the filter progressively plugs, the filtration rate decreases and the section becomes ready for backwashing.

Table I. Sandfloat Sedifloat Components

Item #	Description
1	Outside tank, 5 ft. diameter by 7 ft. high
2	Inside flocculation tank
3	Sandbed assembly with screen
4	Tank bottom
5	Sludge collection funnel
6	Moveable carriage assembly
7	Spiral scoop
8	Scoop variable speed drive
9	Electrical rotary contact
10	Pressure pump
11	Air dissolving tube (type 60)
12	Compressed air addition point on dissolving tube
13	Aerated water distribution pipes
14	Raw water inlet regulating valve
15	Tank level control sensor
16	Raw water inlet jet nozzle in flocculation tank
17	Alum chemical addition point
18	Polyelectrolyte addition point
19	Deflector ring into main flotation tank
20	Backwash hood assembly
21	Clarified water pipeline
22	Clarified water flow regulating valve
23	Floated sludge discharge pipe
24	Main carriage drive
25	Motor to lift backwash hood assembly
26	Backwash suction pump
27	Check valve on suction pipe to prevent backflow
28	Dirty backwash water discharge pipe to recycle water back to flocculation tank for reprocessing
29	Drain line

Each section has a corresponding "Stop Tab" near the rim of the main tank which activates a limit switch on the carriage. The filters are set up for backwashing at a predetermined time interval by skipping a number of sections before stopping to backwash. This number is set with a counter mounted on the control panel. Skipping a greater number of sections before backwashing increases the run time for each filter section. By using a prime number of filter sections (#23), a given section will be ready for backwashing only after all the other sections have been washed. In this way, the backwashing is uniformly extended over the complete filtering time. A backwash hood (#20) with a size corresponding to the size of one sand section is hung on the carriage. The backwash pump suction line opens into this hood. When the carriage stops, a motor (#25) lowers the hood and presses it against the top rim of the sand section. When the hood stops the wash pump automatically starts and pulls water from the clearwell up though the sand, lifting and fluidizing the sand. The wash pump run time is adjusted with a timer to provide a complete washing of the sand in each section. The dirty backwash water is discharged through the pipeline (#28) back to the flocculation tank for reprocessing. When the wash pump stops, a delay timer allows the sand to settle before the hood is raised again. A check valve (#27) stops any return flow of backwash water to the hood. The hood is raised and the carriage then goes to the next section for backwashing.

PILOT PLANT DEMONSTRATION

Secondary effluent from Norwalk Wastewater Treatment Plant (pretreated by primary sedimentation, aeration, and trickling filter or also chlorination-sedimentation) was continuously pumped to the Sandfloat Sedifloat (Type SASF-5) by means of a submersible sump pump at a rate of 30 gpm. Inorganic chemicals, such as aluminum sulfate, were added approximately 40 feet ahead of the Sandfloat Sedifloat to allow for thorough mixing before entering the central flocculating compartment. Organic anionic polymer, such as Nalco 7769 or equivalent, was added in line just prior to the Sandfloat Sedifloat inlet compartment. The chemically treated wastewater then entered the central

Table II. Treatment Data of Norwalk Ohio Waste Water Treatment Project

Date 1985	Sampling Time	pH (unit) In	Out	TSS[a] (mg/L) In	Out	Phosphates[a] (mg/L) In	Out	BOD[a] (mg/L) In	Out	Alum[c] Dosage (mg/L)
5/9	2:00 PM	7.9	5.9							
5/9	4:00 PM	7.25	6.4	104	11	8.8	0.8	54	5	128
5/9	6:00 PM	7.3	6.25							
5/10	9:00 AM	7.3	6.85							
5/10	11:00 AM	7.25	6.70	136	6	9.2	0.4	51	3	318
5/10	1:00 PM	7.45	6.55							
5/10	4:00 PM									105
5/13	12:45 PM	7.20	6.40							
5/13	2:30 PM	7.20	6.65	122	12	9.6	1.1	NA	NA	173
5/13	4:30 PM	7.35	6.35							
5/14	11:10 AM	7.05	6.35							
5/14	12:30 PM	7.15	6.25	106	<1	–	0.26	45	<2	154
5/14	2:00 PM	7.20	6.25							
5/14	4:15 PM	7.15	6.55							
5/15	10:45 AM	6.9	6.35							
5/15	3:15 PM	6.85	6.45	72	4	6.65	0.32	43	3	132
5/15	4:45 PM	7.0	6.5							
5/16[b]	9:30 AM	7.15	6.65							
5/16[b]	11:30 AM	7.2	6.55	28	9	7.0	0.69	20	4	120
5/16[b]	2:00 PM	7.25	6.60							
5/16[b]	4:00 PM	7.10	6.50							
5/17[b]	Composite	7.31	7.13	132	7	4.5	0.23	32	4	120
5/20	Composite	NA	NA	103	8	7.75	0.58	39	4	120

[a]TSS, Phosphate, and BOD tests done on daily composites.
[b]Testing done on 5/16/85 and 5/17/85 was performed on chlorination-sedimentation effluent. All other tests performed on secondary trickling filter effluent.
[c]0.25 mg/L of anionic polymer Nalco 7796 was dosed in conjunction with alum throughout the entire project.

flocculating compartment with slow mixing action where flocs were produced. The flocculated wastewater was then floated by dissolved air flotation and finally polished by sand filtration. Both the influent to and effluent from the 30-gpm pilot Sandfloat Sedifloat unit were taken by Norwalk Wastewater Treatment Plant technicians for analysis by the Norwalk City Laboratory. The chemical dosages and floated sludge were determined by the Lenox Institute for Research staff.

From April 29 to May 3 and May 6 to May 8, 1985, the unit was set up and pretested. The actual testing period was after May 9, 1985. Experimental results generated in the period May 9–20, 1985, are presented in Table II.

Wastewater samples were taken at least three times a day during the testing period. Total suspended solids (TSS), phosphate, and 5- day biochemical oxygen demand (BOD) tests were all done on daily composites as indicated in Table I. While only alum dosages were recorded in the last column of Table I, 0.25 mg/L of anionic polymer Nalco 7796 was dosed in conjunction with alum throughout the entire project. Experimental results are summarized below:

Sandfloat Sedifloat Influent:

TSS	range	28–136	mg/L
	average	100.4	mg/L
Phosphate	range	4.5–9.6	mg/L
	average	7.64	mg/L
BOD	range	20–54	mg/L
	average	40.57	mg/L

Sandfloat Sedifloat Effluent:

TSS	range	< 1–12	mg/L
	average	7.25	mg/L
	% removal	92.8	
Phosphate	range	0.23–1.1	mg/L
	average	0.54	mg/L
	% removal	92.9	
BOD	range	< 2–5	mg/L
	average	3.57	mg/L
	% removal	91.2	

It can be seen that on the average the Sandfloat Sedifloat (Type SASF-5) met the U.S. Environmental Protection Agency's tertiary effluent standards (TSS = 10 mg/L; BOD = 10 mg/L; P = 1 mg/L) in the testing period May 9–20, 1985. Considering the final testing period (May 16–20, 1985) when both chemical dosages (120 mg/L of alum and 0.25 mg/L of anionic polymer Nalco 7796) and mechanical operational conditions were optimized, the Sandfloat Sedifloat effluent met the Federal government's effluent standards all the time.

The average sludge consistency of all floated sludge was measured to be 3.2% solids, which was excellent.

The average removal efficiencies of Sandfloat Sedifloat in the testing periods were: 92.8% for TSS removal, 92.9% for P removal, and 91.2% for BOD removal.

Based on the optimized chemical dosages (i.e 120 mg/L alum and 0.25 mg/L polymer) and assuming the alum cost is US $0.08 per pound and polymer cost is US $1.08 per pound, the chemical costs on an annual basis are determined to be US $0.2463 per 1000 US gallons, which is very reasonable.

In summation, wastewater treatment by a biological-physicochemical process system has been demonstrated to be both technically and economically feasible. The most promising physicochemical process unit is the newly developed Sandfloat Sedifloat clarifier, Type SASF. An existing secondary biological wastewater treatment plant can be easily upgraded to a tertiary wastewater treatment plant by addition of a Sandfloat Sedifloat (or equivalent) in series. The capital cost of a Sandfloat Sedifloat clarifier is very low because of its short detention time (15 minutes).

REFERENCES

1. Krofta, M., and L. K. Wang, "Tertiary Treatment of Secondary Effluent by Dissolved Air Flotation and Filtration," *Civil Engineering for Practicing and Design Engineers*, Vol. 3, P. 253–272, NTIS-PB83-171165 (1984).
2. Wang, L. K., and B. C. Wu, "Development of Two-Stage Physical-Chemical Process System for Treatment of Pulp Mill Wastewater," Lenox Institute for Research Inc., *Technical Report No. LIR/05- 84/2*, 65 p., (May 1984).

9 DESIGN CONSIDERATIONS FOR CYCLICALLY OPERATED ACTIVATED SLUDGE SYSTEMS TREATING DOMESTIC WASTEWATERS

Mervyn C. Goronszy, Executive Technical Manager
Transfield, Inc.
Irvine, California 92715

INTRODUCTION

Variable volume or cyclically operated activated sludge technology has emerged as a viable alternative to the use of constant volume, constantly aerated conventional activated sludge methodology. The development and application of the variable volume approach has been well documented in the literature [1,2,3,4,5,6]. A wide array of wastewaters including municipal, food processing and many industrially derived wastes are amenable to treatment in a variable volume facility [7,8] for which there are a number of proprietary and well-developed systems. While previously considered as an appropriate technology for small treatment applications, the need for cost effective and high quality treatment performance has seen the adoption of this simple and reliable methodology for much larger installations. Capital cost effectiveness over conventional systems of around 30% has been demonstrated for variable volume municipal wastewater systems to a size of around 15 mgd. Similar savings in both capital and operational costs can be realized in many industrial wastewater treatment applications.

The emergence of this treatment methodology, apart from the cost savings, stems primarily from process advantages that can be realized. A major advantage derives from the configuration and aeration sequencing whereby initial floc-loading conditions are generated which maximize biosorptive transport of soluble substrate. These conditions can be used to favour the generation of floc-forming microorganisms over most filamentous forms [9]. This principle is used to good advantage in the treatment of wastewaters which have traditionally been associated with filamentous bulking. Long sludge age systems operating at low temperatures treating domestic wastes and systems treating essentially carbohydrate wastes having a high soluble organic content are typical situations where bulking sludge is frequently encountered.

System design requires the determination of a basin volume in which a desired degree of treatment can be achieved, a mass and rate of application of oxygen, the generation of biomass, the rate at which surface liquors can be withdrawn and a method for optimizing the use of the basin volume for hydraulic load equalization. A design procedure based on material balances, active biomass, floc-loading parameters, biosorptive capacity with related biosorptive regeneration criteria, and biomass zone settling velocity is presented.

CYCLIC OPERATIONAL PRINCIPLES

All systems use a common basin in which to carry out the biodegradation reactions together with the separation of biomass and liquid in order to produce a treated effluent. Biodegradation is achieved during an aeration sequence which is followed by a non-aeration sequence which is necessary to enable the same basin to function as a solids-liquid separation unit. Time sequences are therefore used to accomplish in a single basin what is spatially achieved in separate basins in conventional constant volume activated sludge systems. Time sequencing treatment provides an operational flexibility which is not available in conventional systems. The basic cycle consists of three discrete periods which accomplish AERATION, SETTLEMENT, DECANTATION (effluent removal through surface skimming).

During the period of a cycle, the liquid volume within the basin increases from a set minimum operating bottom water level in response to a varying influent flow rate. Aeration and mixing ceases at a predetermined period of the cycle to allow the biomass to flocculate and settle under quiescent conditions. After a specific settling period, the treated effluent is removed as supernatant returning the liquid level in the basin to the minimum operating bottom water level after which the cycle is

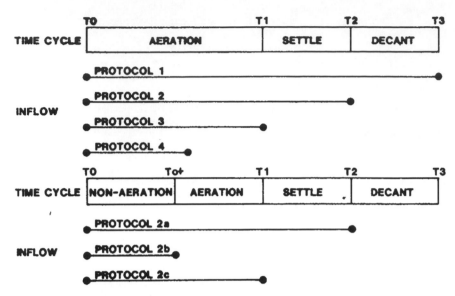

Figure 1.   Schematic of various feeding protocols for variable volume reactor systems.

repeated. Solids are wasted as required to maintain the biomass at manageable concentrations. The volume of solids liquor that is transferred in order to maintain a balanced system is minimal by comparison to conventional constant volume systems which necessarily waste from the recycle flow. The most important feature of cyclic activated sludge systems is the absence of separate solids-liquid separation and associated facilities which are necessary for the conventional constant volume system in order to replenish the reactor with biomass that is necessarily removed to effect its separation from the treated effluent. Various operational protocols can be used with variable volume reactor systems depending upon the waste type, the design objective, and number of vessels to be used in the overall system (Figure 1).

Protocols 1, 2, 2a can be used in essentially single vessel facilities. Interruption of inflow during the decant sequence requires a small holding compartment within the vessel, from which the stored waste flow is transferred to the main reactor configuration. This single vessel approach is usually limited to small systems, treating a low strength waste of less than 100,000 gpd or alternatively a high strength low volume industrial waste. For redundancy requirements it is usual to provide a minimum of two basins in which case protocols 1, 2, 2a are accommodated as a continuous inflow to the system. Flow diversion is used for Protocols 2, 2a in order to prevent inflow to a basin during the decant sequence. Protocols 2b, 3, 3a, 4 are usually accommodated in more than two basins. Protocol 4 is particularly suited to the treatment of waste streams which are known to exhibit toxic effects on the biomass. The feeding pattern permits inflow and biodegradation to take place until a predetermined upper concentration limit (less than the toxic concentration) is reached after which biodegradation (aeration) continues until effluent discharge criteria are reached.. Two basin system designs in which the settle and decant sequences of one basin equate to the aeration sequence of the second basin and vice versa provide for optimum sizing and use of aeration units. Performance data for a two basin system operating on a protocol 1 feed arrangement are shown in Table 1.

SOLIDS-LIQUID SEPARATION

The generation of a sludge which exhibits good settlement properties is a most important consideration in the design and operation of all activated sludge systems. This is particularly so for cyclically operated systems as the zone settling velocity is a deterministic factor regarding the hydraulic capacity of a facility. Settling, compaction and separation properties of activated sludge are related to the relative numbers of filamentous and floc-forming microorganisms in the total floc. Excessive growth of filamentous microorganisms is correlatable with bulking sludge conditions [10]. A minimal presence of filamentous organisms is also deleterious as it leads to a weak flocculating sludge which results

Table I. 2 Basin Variable Volume Reactor Performance – Feeding Protocol 1
Tullahoma, Tennessee Facility

Date April 86	Flow mgd	S_o mg/L	X_o mg/L	MLVSS(1) mg/L	MLSS(1) mg/L	MLVSS(2) mg/L	MLSS(2) mg/L	ΔX lb/d	S_e mg/L	X_e mg/L	$N_{ae}(1)$ mg/L	$N_{ae}(2)$ mg/L
1	1.882	430	264	3,040	4,610	3,720	5,600	2,550	50	7	—	17.0
2	1.743	390	204	—	4,720	—	5,330	12,030	24	9	—	—
3	1.660	500	176	—	4,380	—	5,120	2,700	30	—	—	—
4	1.442	307	160	3,190	4,770	3,900	5,780	2,630	27	11	5.8	12.0
5	1.362	00	164	3,290	4,470	3,910	5,830	3,045	—	13	6.0	14.0
6	1.290	—	152	3,550	5,270	3,670	5,490	3,980	—	10	<1	15.0
7	1.806	—	288	3,160	4,805	—	5,510	3,190	—	9	—	10.0
8	2.040	—	352	—	4,750	—	5,210	2,530	—	18	—	—
9	1.533	350	208	—	4,610	—	4,980	2,440	10	8	—	—
10	1.400	520	408	—	—	—	—	—	8	13	<1	7.0
11	1.313	—	350	—	5,080	—	5,490	2,560	—	28	—	—
12	1.181	—	96	—	—	—	—	—	—	5	<1	9.0
13	1.043	440	96	3,400	5,140	3,830	5,730	1,650	11	8	<1	<1
14	1.436	—	252	3,270	4,990	4,060	6,110	3,550	—	7	<1	6.3
15	1.229	550	440	—	5,020	—	5,940	2,870	19	18	—	—
16	1.316	580	324	—	5,390	—	6,020	7,130	13	12	—	—
17	1.216	—	876	—	—	—	—	—	—	15	—	3.8
18	1.035	380	224	—	—	3,840	5,810	2,880	10	20	<1	5.2
19	1.022	—	195	3,420	5,170	—	—	3,200	—	6	1.6	4.8
20	1.464	280	164	—	—	—	—	—	5	19	<1	5.1
21	1.519	280	288	—	5,010	—	5,480	4,020	6	14	—	—
22	1.324	318	256	—	4,650	—	5,530	4,440	10	12	<1	2.6
23	1.237	—	—	3,360	5,150	3,980	5,990	5,010	—	—	<1	<1
24	1.221	302	288	—	4,890	—	5,610	3,835	12	12	—	12.0
25	1.100	365	470	—	4,730	—	5,420	—	12	22	—	—
26	0.900	—	—	—	4,630	—	5,120	—	—	—	<1	3.5
27	0.830	338	320	3,310	5,090	4,040	5,190	3,560	8	17	—	—
28	1.230	415	448	—	4,860	—	5,220	4,670	10	15	—	5.5
29	1.191	500	416	—	5,230	—	5,730	3,330	9	4	—	6.2
30	1.164	380	424	—	5,240	—	5,920	4,820	9	19	—	4.3

in effluent turbidity and hence an elevated content of organics. A number of factors either singly or in combination are known to promote poor settling sludges which in many instances are species specific (Table II). Mass transfer limitations and availability of soluble substrate and/or dissolved oxygen via diffusional transport mechanisms into the biological floc greatly influence the type of sludge that is generated [13]. It has been observed that where there is a low availability of soluble organics, the growth rate of specific filamentous organisms exceeds that of flocculating organisms with the result that filamentous growth predominates. Because of diffusional resistances associated with net particle size, the driving force for organics to penetrate the floc is insufficient which can result in favoured

Table II. Dominant Filament Types Indicative of Activated Sludge Operational Problems [11,12]

Suggested Causative Conditions	Indicative Filament Types
low F/M	M. parvicella, Nocardia sp. H. hydrossis, 0041, 0675, 0092, 0581, 0961, 0803
low DO	1701, S. natans; Possibly 021N and Thiothrix sp.
presence of sulfides	Thiothrix sp., Beggiatoa sp., possibly 021N
low pH	fungi
nutrient deficiency (N and/or P)	Thiothrix sp., possibly 021N

filamentous growth. The provision of an initial high substrate concentration profile, whereby there is a rapid transport of soluble organics into the cellular biomass, enables conditions which favour the excessive growth of filaments to be circumvented. If the biomass has a low potential for sorptive removal of soluble organics, then provision of a high substrate profile alone will not in itself be effective. Further justification for the competitive growth phenomena relates to the fact that in a bulking sludge, extended filaments, and hence by inference, filamentous biomass is generally only of the order of 1% of total sludge solids [13].

A flocculating biomass will therefore proliferate where high sorptive transport of soluble organics prevail. This is achieved in variable volume reactor systems by feeding and operation as a simple batch system or by proper configuration of inlet feed sections to provide for appropriate floc-loading conditions. In essence, it is necessary to provide a similar reaction environment as that which prevails in plug-flow configured facilities if low organic loading bulking sludge causative conditions are to be avoided.

The parameter floc-loading is used as a measure of the substrate driving force applied to the biomass which is in effect a timeless organic loading parameter. Aerobic conditions are not essential for the occurrence of sorption. Sorptive transport is functional on the fractional absorptive capacity of the biomass which is principally related to its degradable fraction and rate of metabolism of previously absorbed organics. Figure 2 shows typical data for sorptive removal of total and soluble BOD for a domestic wastewater with a sludge having a degradable fraction of 0.69 for both aerobic and anoxic conditions. In conducting the experiments, it was observed that soluble substrate (or BOD) was released from the biomass held under anoxic maintenance conditions (Figure 3). The released organics were resorbed together with the measured fraction of added soluble organics during the two minutes of aerobic contacting used to determine fractional sorptive transport for the various floc-loadings. The net removal of soluble and total organics for both aerobic and anoxic biomass is essentially identical. Sorptive transport of soluble organics is therefore unaffected by the duration of the air-off sequence in cyclically operated systems.

Sludge settlement characteristics, as measured by SVI, were found to be markedly affected by the air-on to air-off ratio [4,14], and while it was found that equal sequences in a cycle, in the treatment of domestic wastes, maximized nitrification-denitrification performance, it also ensured the best sludge settleability. In these full-scale plant studies, a feeding protocol over the full cycle period was practiced which provided the floc-loading environment sufficient to enable the floc-formers a competitive growth advantage. Introduction of an anoxic mixing sequence into the air-off sequence in order to promote more efficient biological denitrification resulted in a rapid growth of a filamentous biomass due to a reduced peak substrate to biomass concentration in each cycle. The condition was reversed by eliminating the anoxic mixing sequence. Cycles of both four and six hours behaved similarly with an equivalent anoxic influent sequence of two and three hours, respectively. The substrate loading (floc-loading) conditions that are necessary to prevent sludge bulking are waste and sludge specific [15].

Bench scale studies using essentially protocols 3 and 3a found that unstable settling behavior occurred repeatedly when the basin functioned as a complete mix reactor in which the substrate to biomass concentration ratio was continuously low. Stable settling characteristics resulted when the feeding protocol was altered to provide an initial high substrate to biomass concentration ratio [16].

Feeding protocols and the method of obtaining a sufficiently high initial substrate to biomass profile has a profound influence on the volume, number of basins, and interconnected pumping facilities which need to be provided in practice. Minimum requirements in all respects are obtained from a feed protocol which has a return flow of sludge from the aeration basin mixing with the influent waste flow, the total flow then passing to an inlet section which is in continuous fluid communication with the main aeration basin. While protocol 1 has proved very effective where carbonaceous treatment only is to be considered, influent bypassing under certain temperature conditions has been shown to cause a reduction in nitrogen removal. This situation is circumvented by interruption of inflow during the decant sequence. Schematic arrangement for a two basin system is shown in Figure 4. Waste strength and sludge condition dictates the number of inlet sections. A variable volume reactor with three reactor compartments in continuous fluid communication has provided a means of reducing a waste strength of 2300 mg/L to less than 20 mg/L BOD without sludge bulking [17].

MUNICIPAL WASTEWATER TREATMENT

Most variable volume systems are sized to operate without primary settling facilities. Process design is based on sludge age considerations and includes nitrification and denitrification criteria.

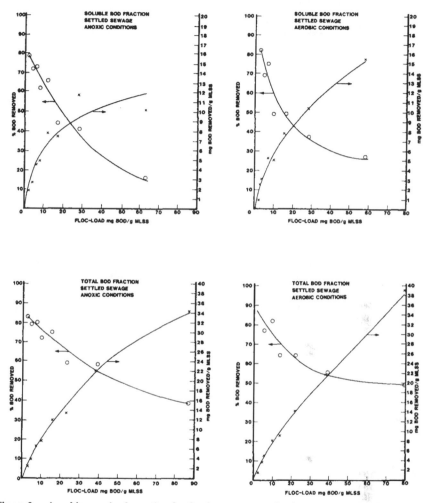

Figure 2.   Aerobic-anoxic absorption for 2-minute contact of primary sewage. BOD_{tot} = 120 mg/L and BOD_{sol} = 80 mg/L.

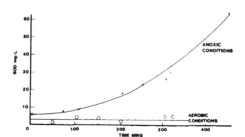

Figure 3.   Biomass BOD release during anoxic conditions.

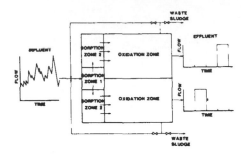

Figure 4. Schematic of two-basin, cyclic activated sludge system.

Carbonaceous Input

Influent organic load is composed of a soluble colloidal fraction of the influent BOD plus a component which is derived from the solubilization of influent degradable solids [18,19].

$$BOD(equiv) = f_sS_o + f_vf_df_xX_i \tag{1}$$

$$Net\ BOD_r = f_sS_o + f_vf_df_xX_i - S_e \tag{2}$$

Solids Balance

The development of models for constant volume systems using degradable fraction and active biomass has been reported [20]. These are equally applicable to variable volume systems for which a mass balance on volatile solids yields,

$$\Delta X_v = a[f_sS_o + f_vf_df_xX_i - S_e] + f_v(1 - f_x)X_i + f_v(1 - f_d)f_xX_i - bf_BX_dX_vt \tag{3}$$

Where fraction of biological solids is;

$$f_B = \frac{a[f_sS_o + f_vf_df_xX_i - S_e]}{a[f_sS_o + f_vf_df_xX_i - S_e] + f_v(1 - f_x)X_i + f_v(1 - f_d)f_xX_i} \tag{4}$$

The degradable fraction is

$$X_d = \frac{x'_d}{1 + x'_nb\theta_c} \tag{5}$$

Volatile primary solids degradation is assumed to be an exponential decay function

$$(1 - fd) = e^{-Kp\theta c} \tag{6}$$

where θ_c is the aerobic sludge age.

Net solids generation is

$$\Delta X = (1 - f_v)X_i + 0.11[a(f_sS_o + f_vf_df_xX_i - S_e)] + \Delta X_v \tag{7}$$

The relationship for sludge age is

$$\theta_c = \frac{- \{Y - bX_vt(f_BX'_d + X'_n)\}}{2YbX'_n} + \frac{\sqrt{\{Y - bX_vt(f_BX'_d + X'_n)\}^2 + 4X_vtbX'_nY}}{2YbX'_n} \tag{8}$$

where

$$Y = a[f_xS_o + f_vf_df_xX_i - S_e] + f_v(1 - f_x)X_i + f_v(1 - f_d)f_xX_i \tag{9}$$

Data used to determine degradability constants for primary solids were generated by Barth [21]. Mixtures of activated sludge and primary solids were aerated for 60 days in order to establish values of K_p and K_b (Figure 5). Exponential decay functions were assumed for the degradation reactions. Degradation of activated sludge biosolids alone is described by

$$X_{vb}(t) = X_{vb}(o)e^{-K_bt} \tag{10}$$

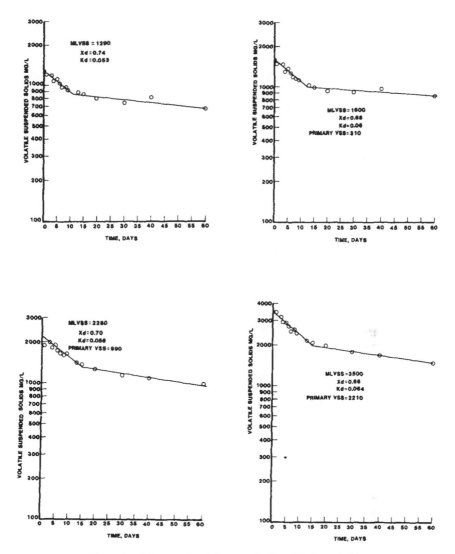

Figure 5. Biomass plus primary volatile solids degradation.

For activated sludge volatile biosolids plus primary volatile solids is

$$X_{vb}(t) = [X_{vb}(o) + 0.8 \times 0.6(X_{vp} + X_{vp}e^{-K_pt})]e^{-K_bt} + X_{vp}e^{-K_pt} \qquad (11)$$

Subtraction of the two equations yields a function for degradable volatile solids,

$$X_v(t) = 0.48(X_{vp} + X_{vp}e^{-K_pt})e^{-K_bt} + X_{vp}e^{-K_pt} \qquad (12)$$

The data shown yields values for endogenous degradation and solubilization of primary volatiles of $0.053d^{-1}$ and $0.16\ d^{-1}$, respectively.

NITROGENOUS INPUT

Input nitrogenous load is composed of a number of constituents, the relative proportions of which markedly affect the design, particularly at cold temperatures. The various fractions in the influent are expressed as

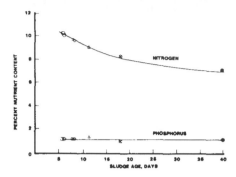

Figure 6.   Nutrient content in biomass.

$$N_{ti} = N_{ai} + N_{ui} + N_{pi} + N_{oi}$$

$$N_{ai} = f_{na}N_{ti} \text{ and } N_{ui} = f_{nu}N_{ti}$$

(13)

Excluding nitrates, the various effluent components are similarly expressed,

$$N_{te} = N_{ae} + N_{ue} + N_{pe} + N_{oe}$$

(14)

The nitrogen balance must consider the total influent nitrogen incorporated in the net sludge generated per day and the effluent TKN (both unbiodegradable and particulate associated). Nitrogen incorporated into biomass is a function of sludge age and can be estimated from Figure 6 [22]. In raw domestic sewage, the ammonia fraction is approximately 0.7, the soluble non- degradable organic fraction is approximately 0.03, and the effluent refractory soluble organic nitrogen is typically 1 to 1.5 mg/L.

NITRIFICATION

Ammonia nitrogen is biologically oxidized to nitrate nitrogen consuming oxygen (4.33 mg O_2/ mg NH_3–N). Nitrification rate for nitrifying bacteria has been estimated as 1.04 mgNH_3-N/mg nitrifier VSS/d (20°C) with a cell yield of 0.15 mg VSS/mg NH_3-N[23]. The critical sludge age is given by

$$\theta_C \geq 2.13 \, e^{0.098 \, (15 - T)}$$

(15)

where θ_C is the aerobic sludge age. System design which incorporates volumes and time periods of non-aeration require θ_C to be adjusted accordingly. The minimum sludge age can also be calculated from

$$\theta_{CN} = \frac{1}{\mu_N(1 - f_{xt}) - b_{nT}}$$

(16)

Values of μ_N are variously quoted at 0.33 to 0.65d^{-1}. For conservatism select the lower value unless actual data are available. Plant design and operation at unaerated mass fractions of 0.5 require a doubling of the critical sludge age parameter. If the unaerated sludge mass fraction exceeds 0.7, the mass of sludge generated can be expected to markedly increase due to a reduction in solubilization of the adsorbed and enmeshed influent volatile solids fraction. In single sludge systems where both carbonaceous and nitrogenous oxidation occur, the actual nitrification rate depends on the fraction of nitrifying organisms in the sludge. Temperature, pH, and dissolved oxygen concentrations also affect the nitrification rate. The fraction of nitrifiers and nitrification rate can be estimated as follows [24].

$$f_n = \frac{0.15 \, N_{ar}}{0.15 \, N_{ar} + aSr}$$

(17)

$$K_n = 1.04 \, f_n X_v \, 1.05^{(T - 20)}$$

(18)

In the absence of a positive dissolved oxygen concentration, provided the input rate of oxygen meets the oxygen demand rate, nitrification will proceed but at a rate less than the maximum. A maximum rate will generally be achieved when the bulk dissolved oxygen concentration reaches 2 mg/L. Data from full scale plants [25] shows net nitrification rates, after accounting for nitrogen

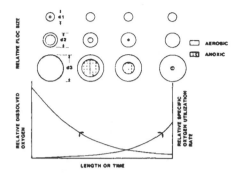

Figure 7.   Schematic of oxygen penetration into
biological floc.

incorporated into cell biomass, to range between 1.2–1.5 mg NH_3-N/g MLVSS/hr (20°C). For a
nitrifying biomass fraction of 3.5%, expressed in terms of degradable mass fraction of 0.7, the rate
becomes 1.7–2.1 mg NH_3-N/g MLVSS/hr.

DENITRIFICATION

Cyclic systems by virtue of their mode of operation offer a simple and efficient methodology for
combined or single sludge nitrification and denitrification. Internal wastewater carbonaceous organic
compounds are used as the hydrogen acceptor in the denitrification sequence. Nitrate is used as an
oxygen substitute thus reducing the energy requirements for aerobic degradation. Denitrification can
take place under both aerobic and anaerobic conditions. Biological floc, depending on the bulk
dissolved oxygen concentration, biomass specific oxygen utilization rate and net equivalent particle
size, can have a substantial anoxic fraction (Figure 7). The anoxic floc fraction is frequently greater
than 50% in the inlet end of a conventional plug-flow system, where the oxygen demand is at a
maximum, and also for significant periods of the aeration sequence in cyclic systems. The oxygen and
alkalinity recovery during denitrification is 2.8 mg O_2/mg NO_3-N and 3.57 mg $CaCO_3$/mg NO_3-N,
respectively. This is accompanied by a removal of BOD_5 of approximately 2 mg/mg NO_3-N and a
release of ammonia nitrogen of 0.02 mg NH_3-N/mg NO_3-N. The carbon-nitrogen ratio (as BOD:N)
required for the reaction to proceed efficiently varies between 3:1 to 6:1 depending on the carbon state
[26,27]. Denitrification in cyclic systems is generally through endogenous nitrate respiration where
carbon is derived from the sludge biomass and from carbon compounds adsorbed by the biomass.
Denitrification rates of 1.2 mg N/g MLSS/h and 0.2 mg N/g MLSS/L (20°C) have been reported for
aerobic and quasi-aerobic conditions, respectively, using domestic wastewater as the carbon source.

Full-scale plant operational data essentially using influent protocols 1, 2 and 2a gave rates for
endogenous denitrification of 0.9 ± 0.2 mg NO_3-N/g MLVSS/h and 0.15 NO_3-N/g MLVSS/h (20°C)
for anoxic and quasi-anoxic conditions, respectively [25,28]. The system sludge age was 25 days,
which when expressed as an aerobic sludge age of 12.2 days enables a value of X_d of 0.70 to be derived
($b = 0.06$ d^{-1}). These rates then become 1.3 ± 0.3 and 0.2 mg NO_3-N/g active MLVSS/h, respectively. For a cyclic system with co-current nitrification-denitrification, the rate of endogenous denitrification can be estimated from

$$K_{dn} = 1.3\ X_d X_v\ 1.09^{(T - 20)} \tag{19}$$

The aerobic denitrification rate is dependent on the anoxic sludge portion of the biological floc and
the availability of carbon substrate. Higher rates of denitrification may be achieved in feeding and
operational protocols whereby advantage can be taken of other than endogenous and adsorbed
carbon sources.

OXYGEN REQUIREMENTS

Net oxygen requirements are calculated from

$$O_2/day = a'BOD_r + 1.4\ bf_b X_d X_v t\ (1.04)^{T-20} + 4.33\ NOD - 2.8\ DEN \tag{20}$$

PHOSPHORUS REMOVAL

The cyclic operation of variable volume reactor systems provides a means whereby phosphorus can be removed biologically. Biological phosphorus removal requires an anaerobic stage or period followed by an aerobic period. The biomass must reach an oxidation reduction potential of less than -150 mV to insure proper reaction conditions. The anaerobic stage is thought to play the role of a fermentation zone where acidogenic bacteria transform organic matter into volatile fatty acids which are then used by the phosphate removing bacteria and stored in the form of poly-B-hydroxybutyrate. The energy required for this storage would be supplied by the hydrolysis of the poly-phosphate reserves, which explains the release of phosphorus during the anaerobic phase. Cyclic aeration sequences promote volatile fatty acid generation in the settled sludge. The removal of nitrate from within the sludge mass has been generally thought to be a prerequisite for phosphorus release as is a readily degradable carbon source. More recent research has shown that acetate, propionate and formate each induce effective phosphate release even in the presence of nitrate which suggests that the phosphate release phenomenon is primarily dependent on the nature of the feed rather than the anaerobic state as such. The notion of simple dosing of volatile fatty acid compounds (such as acetate) during the air-off sequence in cyclic systems appears to offer a positive and controllable means of promoting effective biological phosphorus removal. Present understanding indicates biological phosphorus removal occurs as a result of specific organisms, namely *Acinetobacter*, which release phosphorus under anaerobic conditions in the presence of acetate, and subsequently rapidly uptake phosphorus under aerobic conditions. There is a general consensus that an anaerobic environment is required, together with a minimum COD/TKN ratio in the wastewater and a minimum anaerobic retention time in order to achieve phosphorus release. Indications are that the propensity to achieve excess phosphorus removal is a function only of the magnitude of readily biodegradable COD during anaerobiosis above 25 mg COD/L together with the magnitude of the anaerobic sludge mass fraction [29].

Protocol 1 in full scale cyclic activated sludge systems has been associated with total phosphorus removals of 50 to 75% for equal air-on air-off ratios in 4 and 6 hour cycles. Other modes of operation have been studied in bench scale studies, again for equal air-on air-off ratios, using 8 hour cycles where biological phosphorus removal has proved to be very effective [30]. It is a simple matter to augment biological removal with chemical precipitants to insure minimum effluent phosphorus concentrations where required. Protocols by which maximum biological phosphorus removal are achieved may in some cases prove to be disadvantageous in terms of selecting for good sludge settling.

SURFACE SKIMMING-EFFLUENT REMOVAL

The removal of surface elements from a variable volume reactor requires that the induced horizontal or local fluid velocities caused by the withdrawal of liquid does not cause a breakdown of the sludge cohesive forces at the sludge interface. The net rate at which liquid can be removed in order to return the liquid level to the designated operating bottom water level is a function of the relative position of the sludge blanket interface. All but protocol 1 provide a perfectly quiescent condition for the effluent removal operation. Very effective surface liquor removal is provided by variable level decanting devices [5] which have a rate and position adjustability in the event that biomass settleability suffers an upset. A fixed-level decanting device does not have this flexibility and, if used, would reduce the operational versatility and flexibility of the overall system.

Studies in a model tank of 12 ft. depth have been conducted [31] in which surface skimming rates were established. Rates of removal equivalent to a vertical depth rate of 0.12, 0.26, and 0.52 ins./sec. showed that a liquid level above the sludge blanket of approximately 4", 7", and 30" was required in order to prevent solids entrainment at these rates. Mean horizontal velocities based on an initial free liquid end area above the sludge blanket were 0.016, 0.032, and 0.048 ft./sec. Apparent horizontal velocities at the onset of solids entrainment, similarly based on free liquid end area above the sludge blanket, were 0.16, 0.15, and 0.11 ft./sec.

Typical sludge settling data using a one-litre measuring cylinder and taken from a number of cyclic activated sludge systems over an extended period are summarized in Figure 8.

It should be noted that full scale systems exhibit a lower apparent sludge volume index due to the absence of "wall" and "arching" effects. Effective surface skimming has been obtained in full scale facilities in which the surface level depletion rate is of the order of 1.2 in./min.

Facilities operating on protocol 1 have, with certain configurations and cycles, shown the existence of short circuiting. This has been established using Rhodamine WT and ammonia nitrogen as tracers. Up to 30% of influent flow during settle and decant for a 4 1/2 hour cycle duration has been shown to

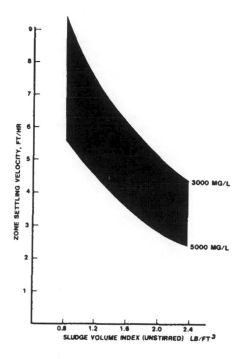

Figure 8.   Sludge settlement characteristics for
cyclic activated sludge systems.

be removed with the effluent, the fraction depending on the temperature differential between influent and basin contents.

CONCLUSIONS

The various aspects which need to be considered in the process design of cyclically operated activated sludge systems have been presented. A rational design approach considering component removal of particulate and soluble degradable organics has been presented. The method makes use of several years of operational data taken from a number of full-scale facilities. Cyclically operated systems can be designed for almost complete ammonia removal through appropriate selection of operational parameters.To assure a high ammonia removal the systems aerobic SRT must exceed a critical value which is temperature dependent. Interruption of inflow during decantation prevents degradation of effluent quality due to short-circuiting effects. Cycles in which the air-on to off ratio is equal give optimal nitrification-denitrification performance.

For those cyclic systems which operate at a batch process (feeding protocols 2b, 2c, 3, and 4) a sludge of good settleability will only be generated if a threshold loading condition is exceeded which in concept qualitatively relates to the loading conditions at the inlet of a "plug-flow" system. Batch feeding and a cyclic operation will not necessarily mitigate against the growth of filamentous organisms. A level of feeding less than the "threshold load" will result in growth conditions which are similar to those that exist in a continuous feed complete-mix system. In oth ̄ words a definite "threshold load" must be exceeded in order to set up conditions which provide for a continuous more rapid growth of floc-forming organisms over filamentous forms. The notion ˙herefore presented that there is a "threshold load" condition which must be exceeded which is waste and sludge specific for which a sufficient level of sorptive transport of soluble organics must take place in order to produce the net result that leads to a growth advantage for floc-forming microorganisms. Operation near this "threshold load" will result in the generation of a sludge having a relatively low zone settling velocity (1–2 ft/hr) because of the comparatively high number of filament clumps that will exist under these circumstances. A rapid shift in population to favour filamentous growth can accompany relatively minor reductions in loading.

Selector principles can be used to configure cyclic systems so that "threshold load" criteria are exceeded. Augmentation of the cyclic mode of operation with a cyclic step feed load pattern within a cycle, together with a low volume flow of biomass directed to contact the feed, provides a simple and very effective means for exceedance of the "threshold load" condition.

A similar design approach can be used for industrial wastewater treatment facilities. It is normal to conduct bench scale treatability studies in such cases in order to obtain the relevant process criteria.

EXAMPLE-DOMESTIC WASTEWATER

$Q_a = 9$ mgd	$Q_p = 13$ mgd, $Q_w = 20.7$ mgd
$S_o = 200$ mg/L	$S_e = 20$ mg/L
$X_i = 200$ mg/L	$X_e = 20$ mg/L
$N_{ae} \leq 2$ mg/L	$T = 10°C$
$f_s = 0.6$	$f_sS_o = 0.60 \times 200 = 120$ mg/L
$f_x = 0.7$	$f_v = 0.7$
$f_vX_i = 140$ mg/L	$f_xf_vX_i = 100$ mg/L
$N_t = 35$ mg/L	$N_a = 25$ mg/L
$f_{xt} = 0.5$	$b_{n10°C} = 0.027$
$a = 0.6$	$K_b = 0.06$
$K_p = 0.15$	

Provide design for a 4 basin system operating on 4-hour cycles with 2 hours aeration in each cycle. First estimate of sludge age,

$$\theta_{CN} = \frac{1}{\mu_N(1 - f_{xt}) - bn10°} = 7.2 \text{ days}$$

$(1 - f_d) = e^{-0.15(7.2)}$

$f_d = 0.66$

$\Delta X_v = 0.6[0.60 \times 200 - 10 + 0.66 \times 0.7 \times 200 \times 0.7] + 0.3 \times 0.7 \times 200 + 0.7 \times 0.34 \times 0.7 \times 200$

$= 104.8 + 42 + 33.3$

$= 180$ mg/L

$f_B = \frac{105}{180} = 0.58$

$X_d = \frac{0.8}{1 + 0.2 \times 0.041 \times 7.2} = 0.76$

Select $X_v = 3500$ mg/L

Endogenous
Solids $= 0.041 \times 0.58 \times 0.76 \times 3500 \times 0.5$

$= 32$ mg/L

$\Delta X_v = 180 - 32 = 148$ mg/L

$\Delta X = 220$ mg/L

$\theta_c = \frac{X_vt}{\Delta X_v} = \frac{3500 \times 0.5}{148} = 11.8$ days

Repeat above:

$f_d = 0.84$

$\Delta X_v = 175$ mg/L

$f_B = 0.66$

$X_d = 0.73$

Endogenous
Solids $= 35$ mg/L

$\Delta X_v = 138$ mg/L

θ_c = 12.7 days (checks with 11.8 days)

ΔX = 211 mg/L

N incorporated with sludge growth = 0.09 x 0.66 x 175

$$= 10.4 \text{ mg/L}$$

N_u = 1.5; N_{pe} = 20 x 0.09 = 1.8

N_{ui} = 0.03 x 35 = 1.05 mg/L

N_{pi} = 0.1 x 0.03 x 200 x 0.05 = 0.3 mg/L

N_{oi} = 35 − (25 + 1.05 + 0.3)

= 8.65 mg/L

ΔN_t = 33.65 − (10.4 + 1.5 + 1.8)

= 20 mg/L

f_n = $\dfrac{0.15 \times 20}{0.15 \times 20 + 0.6 \times 190}$ = 0.025

K_n = 56.7 mg/L/d

= 0.68 mg NH_3-N/g MLVSS/h   (at 10°C)

ΔN to be oxidized = 1403 lb/d

X_v = $\dfrac{1403 \times 2 \times 10^6 \times 10^{-3}}{24 \times 0.68}$ = 171936 lb

$Q_a S_o$ = 14969 lb

$\dfrac{Q_a S_o}{X_v}$ = 0.087

X_v/basin = 42,984 lb

Q_a/basin = 12,533 ft³/h

Q_p/basin = 18,104 ft³/h (peak 2 hour)

Inflow/basin/4-hr cycle = 57653 ft³

= 431244 gal

Q_w/basin = 28827 ft³/h

= 215625 gal/h

X/basin = $\dfrac{42984}{0.76}$ = 56558 lb

Size for TWL − BWL = 4 ft

V_{BWL} = 56558 x 2.4 + 3 x 14413

= 178978 ft³

V_{TWL} = 236630 ft³

Basin dimension = 80 ft x 174 ft x 16.5 ft (TWL)

Using endogenous carbon sources only,

K_{dn} = 1.3 x 0.72 x 3,500 x 1.09^{10-20}

= 1.38 mg/L/h

= 0.40 mg/g/h

Endogenous denitrification = 825 lb/d

Oxic denitrification = $\dfrac{0.2}{1.3}$ x 825

= 127 lb/d

NO_3 − N denitrified = 952 lb/d

$$= 70\% \text{ of total oxidized}$$

$$O_2 = 0.6\text{x}187\text{x}9\text{x}8.34 + 1.4\text{x}0.06\text{x}0.66\text{x}0.73\text{x}171936\text{x}1.04^{10-20} + 4.33\text{x}1403 - 2.8\text{x}952$$

$$= 8422 + 4700 + 6075 - 2666$$

$$= 16531 \text{ lb/d}$$

Figure 9 shows four alternative idealized flow configurations for this facility. Size each basin to accord with floc-load criteria for selection of floc-forming biomass. Size blowers to function with continuous delivery to provide 345 lb process oxygen per hour with a configuration which permits alternate cycling to each of the two basin pairs.

NOMENCLATURE

a   — cell yield coefficient, mg VSS/mg BOD, COD, TOC

a′   — oxygen utilization coefficient, mg O_2/mg BOD, COD, TOC

b   — endogenous rate coefficient, days^{-1}

b_T — $b_{20} \times 1.04^{T-20}$

b′   — oxygen utilization coefficient for endogenous respiration, days^{-1}

b_n — specific endogenous mass loss rate for *Nitrosomonas*, days^{-1} = 0.04 at 20°C

e   — subscript, effluent

b_{nT} — $b_{n20} \times 1.04^{T-20}$

f_b — fraction biological volatile solids

f_d — fraction of f_x degraded

f_x — fraction degradable primary volatile solids

f_v — fraction of volatile suspended solids

f_s — fraction of influent BOD, COD which is soluble colloidal

f_{na} — fraction influent TKN as ammonia

f_{nu} — fraction influent TKN unbiodegradable soluble organic nitrogen

f_{xt} — unaerated mass fraction

f_n — fraction nitrifers

i   — subscript, influent

K_b — endogenous decay rate activated sludge volatiles, days^{-1}

K_p — decay rate primary volatile solids, days^{-1}

K_n — nitrification rate, mg/L/d

K_{dn} — denitrification rate, mg/L/d

N_t — TKN, mg/L

N_a — ammonia nitrogen, mg/L

N_u — soluble unbiodegradable organic nitrogen, mg/L

N_p — particulate unbiodegradable organic nitrogen, mg/L

N_o — biodegradable organic nitrogen, mg/L

N_{ar} — ammonia nitrogen removed, mg/L

S_o — influent organic concentration, mg/L

S_e — effluent organic concentration, mg/L

S_r — substrate removal across reactor, mg/L

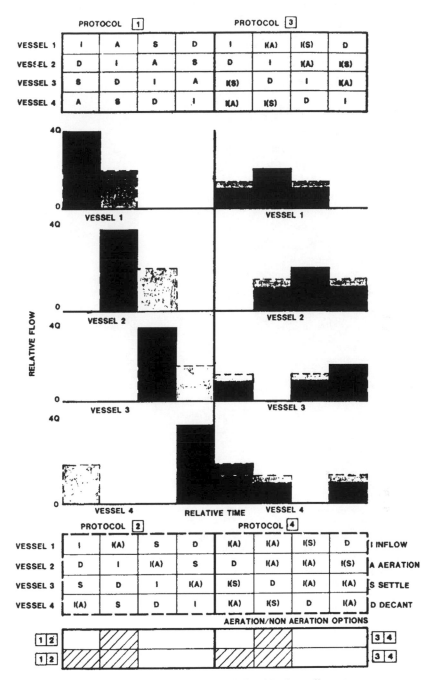

Figure 9.   Alternative feed protocols for 4-basin, cyclic system.

S_i — soluble organics remaining after initial absorption, mg/L

SVI   — sludge volume index, ft³/lb

t   — hydraulic retention time, days

t   — time

T   — temperature °C

X_v — mixed liquor suspended solids, mg/L

X_a — active fraction, dimensionless

X_d — degradable fraction, dimensionless

X'_d — degradable fraction upon generation, dimensionless

X'_n — non-degradable fraction upon generation, dimensionless

ΔX_v — excess biological volatile solids, mg/L

X_{vb} — volatile biosolids, mg/L

X_{vp} — volatile primary solids, mg/L

X_i — influent suspended solids, mg/l

θ_C — sludge age, days

θ_{CN} — sludge age for nitrification, days

μ_N — maximum nitrifier specific growth rate at 20°C and pH of 7.2

$\mu_N(T)$ — $\mu_{N20} \times 1.123^{T-20}$

NOD   — nitrification oxygen demand

DEN   — denitrification oxygen equivalent

Q_a — average dry weather flow, mgd

Q_p — peak dry weather flow, mgd

Q_w — wet weather flow, mgd

REFERENCES

1. Ardern, E., and Lockett, W. T., "Experiments on the Oxidation of Sewage Without the Aid of Filters," *J. Soc. Chemical Ind.*, 33, 523 (1914).
2. Ardern, E., and Lockett, W. T., "The Oxidation of Sewage Without the Aid of Filters, Part III," *J. Soc. Chemical Ind.*, 34, 937 (1915).
3. Pasveer, A., "A Case of Filamentous Activated Sludge," *J. Wat. Pollut. Control Fed.*, 41, 1340 (1969).
4. Goronszy, M. C., "A Study of the Intermittent Extended Aeration Process Including Nitrification and Denitrification," Unpublished report (1975).
5. Goronszy, M. C., "Intermittent Operation of the Extended Aeration Process for Small Systems," *J. Wat. Pollut. Control Fed.*, 51, 2, 274 (1979).
6. Irvine, R. L., et al., "Municipal Applications of Sequencing Batch Treatment at Culver, Indiana," *J. Wat. Pollut. Control. Fed.*, 55, 484 (1983).
7. Goronszy, M. C., and Barnes, D., "Continuous Single Vessel Activated Sludge Treatment of Dairy Wastes," *Water*, AIChE Symp. Series, 76, 271-277 (1979).
8. Staszak, C. N., et al., "Full-scale Sequencing Batch Reactor Use in a Commercial Hazardous Waste Facility," *Proc 39th Industrial Waste Conference*, Purdue University (1984).
9. Chudoba, J., Grau, P., and Ottova, V., "Control of Activated Sludge Filamentous Bulking- II; Selection of Microorganisms by Means of a Selector," *Water Res.*, 7, 1389–1406 (1973).
10. Sezgin, M., et al., "A Unified Theory of Filamentous Activated Sludge Bulking," *J. Wat. Pollut. Control Fed.*, 50, 363 (1973).
11. Eikelboom, D. H., "Filamentous Organisms Observed in Activated Sludge," *Water Res.*, 9, 365 (1975).
12. Richard, M. G., Hao, D. S., and Jenkins, D., "Growth Kinetics of *Sphaerotilus* Species and Their Significance in Activated Sludge," *55th Annual WPCF Conference*, St. Louis, Missouri (1982).
13. Lau, A. O., et al., "The Competitive Growth of Floc-forming and Filamentous Bacteria: A Model for Activated Sludge Bulking," *J. Wat. Pollut. Control Fed.*, 56, 52 (1984).

14. Goronszy, M. C., and Barnes, D., "Sequentially Operated Systems for Bulking Sludge Control," *Process Biochem.*, 15, 7, 42 (1980).
15. Flippin, T. H., Eckenfelder, W. W., and Goronszy, M. C., "Control of Sludge Bulking in a Carbohydrate Wastewater Using a Biosorption Contactor," *Proc. 39th Industrial Waste Conference*, Purdue University (1984).
16. Silverstein, J., and Schroeder, E. D. "Control of Activated Sludge Settling in an SBR," *A.S.C.E. Env. Engg. Conference*, Colorado, 238 (1981).
17. Goronszy, M. C., Eckenfleder, W. W., and Cevallos, J., "Sludge Bulking Control for Highly Degradable Wastewaters Using the Cyclic Activated Sludge System," *Proc. 40th Industrial Waste Conference*, Purdue University (1985).
18. Guyer, W., "The Effect of Particulate Organic Material on Activated Sludge Yield and Oxygen Requirement," *Prog. Wat. Tech.*, 12, 79–95 (1980).
19. Chudoba, J., and Tucek, F., "Production Degradation and Composition of Activated Sludge in Aeration Systems Without Memory Sedimentation," *J. Wat. Pollut. Control Fed.*, 57, 3, 201 (1985).
20. Goronszy, M. C., and Eckenfelder, W. W., "A Design Approach for Plug-flow Activated Sludge Plant Treating Municipal Wastewater," *58th Annual WPCF Conference*, New Orleans (1984).
21. Barth, J., "The Effect of Primary Solids on the Activated Sludge Process," *MS Thesis*, Vanderbilt University (1985).
22. Gaddapati, P., *MS Thesis*, Vanderbilt University (1984).
23. Wong-Chong, G. M., and Loehr, R. C., "The Kinetics of Microbial Nitrification," *Water Res.*, 9, 1099 (1975).
24. Jarrett, P. T., Eckenfelder, W. W., and Goronszy, M. C., "The Optimization of Nitrification and Denitrification in Plug-flow Activated Sludge Systems Breaking Municipal Wastewaters," *A.S.C.E. Env. Engg. Conference*, Cincinnati, July 8–10 (1986).
25. Goronszy, M. C., "Intermittently Aerated Activated Sludge Systems," *Principles of Wastewater Treatment and Design*, Queensland University Continuing Education Series, Ed. P. Greenfield, 244–305 (1981).
26. Christensen, M. H., "Denitrification of Sewage by Alternating Process Operation," *Prog. in Wat. Tech.*, 7, 2, 339–347, Pergaman Press (1975).
27. Balakrishnan, S., and Eckenfelder, W. W., "Nitrogen Relationships in Biological Treatment Processes – III, Denitrification in the Modified Activated Sludge Process," *Wat. Res.*, 3, 177–188 (1969).
28. Goronszy, M. C., and Irvine, R. L., "Nitrification Denitrification in Intermittently Aerated Activated Sludge Systems and Batch Systems," *Proceedings EPA International Seminar on Control of Nutrients in Municipal Waterwaste Effluents*, San Diego, California, September 9–11 (1980).
29. Ekama, G. A., et al., "Biological Excess Phosphorus Removal," *Theory Design and Operation of Nutrient Removal Activated Sludge Processes*, South Africa Water Research Commission (1984).
30. Manning, J. F., and Irvine, R. L., "The Biological Removal of Phosphorus in a Sequencing Batch Reactor," *J. Wat. Pollut. Control Fed*, 57, 1, 87 (1985).
31. Goronszy, M. C., "Effluent Withdrawal Studies for the Prototype 4000-person EA Unit to be Constructed at Bathurst Wastewater Treatment Works," Unpublished report (1975).

10 BREWERY WASTEWATER TREATMENT BY CONTACT OXIDATION PROCESS

Chang-Wu Huang, Associate Professor
Environmental Engineering Department
Wuhan Institute of Urban Construction
Wuhan, China

Yung-Tse Hung, Professor
Civil Engineering Department
Cleveland State University
Cleveland, Ohio 44115

INTRODUCTION

The brewery plants produce a large quantity of wastewaters, which contain high concentration of organic pollutants, low concentration of nutrients, and have large variation in these parameters[1]. Proper treatment of brewery wastewater is necessary in order to minimize the detrimental effect on the environment. Since the wastewaters contain high concentration of biodegradable organics, the most commonly used treatment method has been the biological treatment process[1,2].

Beers are beverages of low alcoholic content of 2–7%. The basic raw materials for manufacture of beer are cereals of which barley is the most important material. Types of cereals used include barley, flaked rice, corn, oats, and wheat, with rice and millet used in China. Hops are also added to produce a more or less bitter taste and to control the subsequent fermentation process. Brewing sugars, syrups such as corn sugar or glucose, and yeast are also added during beer making.

The flow diagrams for beer brewing are shown in Figures 1 and 2. The beer brewing is carried out in two stages: 1) malting of barley and 2) brewing of beer from malt. In the malting process, barley is removed from storage and placed in a tank and steeped with water to bleach out the color and allowed to germinate while air and water are introduced to stimulate growth of enzymes. During the growth period, oxygen is utilized and carbon dioxide is released while the enzyme diastase is produced. Diastase is the biological catalyst that converts the dissolved starch to the disaccharide maltose and then to monosaccharide glucose by the enzyme maltase. The monosaccharide glucose is then fermented to beer by yeast in stage 2. In stage 2, the brewing of beer, as shown in Figure 2, can be divided into the following steps:

1. Mashing of coarse ground malt with water.
2. Converting insoluble starch into liquefied starch, and the soluble malt starch into dextrin and malt sugars in the pressure cookers.
3. Mixing the resulting boiling cooker mash with the rest of the malt in the malt tun, raising the temperature to 168°C to prepare the brewer's wort, and cooled.

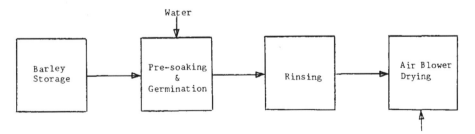

Figure 1. Flow diagram for molting of barley.

90

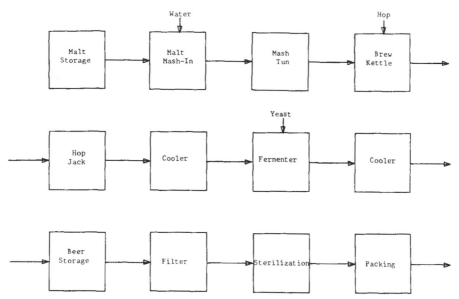

Figure 2.   Flow diagram for brewing of beer from malt.

4. Adding yeast for alcoholic fermentation of the wort.
5. Separating the yeast and the malt residues from the final product beer by filtration.
6. Barrelling and bottling the filtered beer.

Wastewaters from the malting process contain water soluble substances of barley leached to the soaking water including pentose, sucrose, glucose, cellucose, protein, and minerals. A 0.5 to 1.5% of barley on a weight basis will end up in the wastewaters. Two thirds of the pollutants are organic while the remaining are inorganic salts including potassium and calcium silicate, sulfate, and phosphate.

Brewing processing wastewaters consist of wastewaters from brew kettle, fermenter, storage, and packing operations. The main pollutants in the brewing processing wastewaters are spent enzymes and proteins. Settling can remove and recover a portion of these insoluble organic pollutants.

Soft packing media contact oxidation process is an innovative treatment process recently developed in China. The process is an aerated attached growth biofilm process. The soft packing media is made of synthetic fiber, rayon, with specific gravity of 1.02, a unit weight of 3 kg/m³, and a void ratio of larger than 0.99 (Table I). It is a low cost material developed in China. The details of this soft packing media are shown in Figure 3. The arrangement of media in the oxidation tank is depicted in Figure 4. Table II listed the design parameters for the soft packing media. The diameter of fiber is 0.07 mm while the diameter of fiber bundle is 120 mm. There are 81,000 fibers per fiber bundle and 71 fiber bundles per center rope. The distance between center ropes is 120 mm while the distance between fiber bundles is 60 mm. The specific surface area of media is 2472 m²/m³ of tank volume. This provides a large surface area for bacteria growth and attachment. Air diffusers are located at the bottom of the contact oxidation tank to provide air for mixing the tank contents and for oxygen requirement for

Table I. Soft Packing Media Characteristics

Parameters	Values
Material	Rayon fiber
Specific weight	1.02
Unit weight (kg/m³)	3
Void ratio	≥0.99
Aldehyde content	≥25%
Anti-pulling strength (g/fiber)	6.8−7.1
Maximum elongation ratio	≤12%

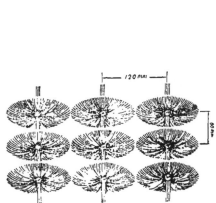
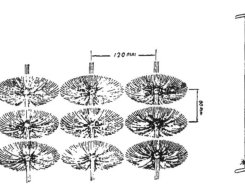

Figure 3.   Soft packing media.

Figure 4.   Soft packing media in oxidation tank.

aerobic biological oxidation. Since a high bio-mass concentration can be maintained in the soft packing media contact oxidation tank, high treatment efficiency can be achieved even at high organic loading conditions.

The objectives of this study were to determine the feasibility of using soft packing media contact oxidation process in treating high strength brewery wastewaters and to develop the design parameters for the soft packing media contact oxidation process.

MATERIALS AND METHODS

Wastewaters

Table III shows the BOD_5 values for various wastewater streams in the Qingdao Brewery plant. These include fermenter rinse water, hop rinse water, packing spray water and barley rinse water of first rinse to the fifth rinse operations. The fermenter rinse water and the hop rinse water had a very high BOD of up to 12,000 mg/L. The fermenter rinse water contained settleable spent yeast, protein and beer residues, and soluble organics. Hop rinse water contained settleable spent hop residues, protein, and soluble organics. For barley rinse waters, the first rinse water had the highest BOD of up to 1050 mg/L, while the fifth rinse had the lowest BOD of 150 mg/L as to be expected. The wastewaters exhibited a tea brown color and had high suspended solids concentration. It contained difficult to settle colloidal organic pollutants. The combined wastewaters from various wastewater streams had an average BOD of 300 to 500 mg/L.

With respect to waste characteristics and wastewater flow rate, the barley rinse waters are the most difficult to be treated and consequently were selected as feed wastewaters in the pilot plant study.

Table II. Design Data for Soft Packing Media

Parameters	Values
Distance between center ropes (mm)	120
Distance between fiber bundles (mm)	60
Diameter of fiber bundle (mm)	120
Number of fiber bundles per center rope	71
Number of fiber per fiber bundle	81,000
Diameter of fiber (mm)	0.07
Specific surface area (m^2 / m^3 tank volume)	2,472
Weight of media per unit volume of tank (kg/m^3)	2.5–3
Unit cost of media ($/$m^3$)	17

Table III. Brewery Wastewater Characteristics

Waste Stream	BOD$_5$ (mg/L)	Characteristics
Various rinse waters in fermentation process	1,200–12,000	Settleable spent enzymes Proteins Brewery residues Soluble organics
Hop rinse waters	4,800–12,000	Settleable spent hop residues Proteins Soluble organics
Packing spray water	57–1,000	Beer Dirt
Malting process wastewaters		Tea brown color High TSS
1st rinse water	653–1,050	Colloidal organics
2nd rinse water	330–520	
3rd rinse water	240–260	
4th rinse water	240–300	
5th rinse water	150–280	

Pilot Plant

The process flow diagram for the pilot plant for barley rinse wastewater treatment is shown in Figure 5. Raw wastewaters first underwent primary settling to remove easily settleable materials. Settled wastewaters were then pumped to an equalization tank. Coagulant was added to the inlet to the first contact oxidation tank to enhance the removal of color and organic pollutants. The bio-oxidation of organic substances took place in the first and the second contact oxidation tanks, which used the soft packing media contact oxidation process for biological treatment. Wastewaters were settled in the final settling tank for the separation of biological solids from the treated effluent, which was discharged to the receiving waters. The design flow rate was 0.5 to 1 m³/hr with an actual maximum flow of 1 to 1.5 m³/hr. The volumetric loading rate was 2 to 3 kg BOD removed / m³ media / day with a maximum value of 4 to 8 kg BOD removed / m³ media / day.

In the first oxidation tank, wastewaters flowed downward while the air flow was upward in a counter current flow mode. In the second tank the upward wastewater flow and the air flow were concurrent flow. The depth of water in the oxidation tank was 2.5 m. The air pressure required at the bottom of tank was 0.3 kg/cm². Since the air compressor provided an air pressure of 6 kg/cm², the pressure would require reduction to 0.3 kg/cm² at the point of release at the bottom of the oxidation tank. Table IV shows the dimensions of various treatment units in the pilot plant. There are three primary settling tanks in series, each with 5.67 m³ in volume. The volume of the first and second oxidation tanks were 0.99, and 0.95 m³, respectively. The final settling tank and the clear water tank were each 5.5 m³ in volume.

Acclimation and Startup of Pilot Plant

Initial seed for acclimation was obtained from raw wastewaters and from natural environment. For initial acclimation, the raw wastewaters, seed and nutrient salts were added to the oxidation tank and aerated with batch mode operation. After 24 hours of aeration, tank contents were allowed to settle under quiescent conditions. The top half of the tank contents was drained and filled with raw

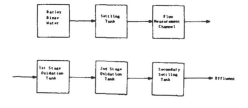

Figure 5. Flow diagram of soft packing media contact oxidation process for treatment of brewery wastewater.

Table IV. Dimensions of Treatment Units

Treatment Unit	Length (mm)	Width (mm)	Effective Height (mm)	Number	Volume (m³)
Primary settling tank	2,900	1,600	1,200	3a	16.7
Flow measuring channel	1,800	500	600	1	0.54
1st oxidation tank	630	630	2,500	1	0.99
2nd oxidation tank	630	630	2,400	1	0.95
Secondary settling tank	2,900	1,600	1,200	1	5.5
Clear water tank	2,900	1,600	1,200	1	5.5

aIn series.

wastewaters and nutrient salts. The contents were aerated on a batch basis. After 12 hours aeration, decanting of top half of tank contents and refilling with raw wastewaters and nutrient salts were repeated. After batch aeration for a total period of 48 hours, continuous reactor operation mode was used instead. The flow rate was 0.3 to 0.5 m³/hr. After 24 hours of continuous reactor treatment, biofilm was formed and increased the apparent volume of soft packing media in the oxidation tank. Microscopic examination revealed the presence of microorganisms on the biofilm. The COD (chemical oxygen demand) removal efficiency was 30 to 50%. After 10 days of continuous mode operation, the knots of fiber connections had expanded to 10 mm spheres, which was filled with gray yellow biofilm. The COD removal efficiency increased to higher than 70%, and the pilot plant had reached normal operating condition.

Operation of Pilot Plant

Tables V and VI showed the operating conditions for the pilot plant. The air to wastewater flow ratio based on volume basis was 27 to 1. The hydraulic detention time was 1.9 hours. The effluent DO(dissolved oxygen) level was 2-4 mg/L for the first oxidation tank and was 4-6 mg/L for the second oxidation tank. The effluent pH was 6 to 7.

Table V. Operating Parameters for Pilot Plant Study

Parameters	Values
Water temperature (°C)	≥ 10
Influent flow (m³/hr)	0.5-1 (design)
	1-1.5 (actual)
BOD loading rate	
(kg BOD_5 removed/m³ media/day)	2-3 (design)
	4-8 (actual)
MLSS (mg/L)a	
1st oxidation tank	4,650
2nd oxidation tank	1,100
pH	7-7.5

aMLVSS = 0.8 MLSS.

Table VI. Normal Operating Conditions of Pilot Plant

Parameters	Values
Influent flow (m³/hr)	1
Air flow / Influent flow ratio (volume/volume)	27 : 1
Hydraulic detention time in oxidation tank (hr)	1.9
pH	6 − 7
Effluent dissolved oxygen concentration (mg/L)	
1st oxidation tank	2 − 4
2nd oxidation tank	4 − 6

Table VII. Summary of Performance Data of Pilot Plant Study

Parameters	Values
Influent BOD (mg/L)	489
Effluent BOD (mg/L)	39
% BOD removal	92
Influent COD (mg/L)	413
Effluent COD (mg/L)	109
% COD removal	73.6
1st oxidation tank effluent DO (mg/L)	2–4
2nd oxidation tank effluent DO (mg/L)	4–6
Air to water flow ratio (volume/volume)	27 : 1
Excess sludge production (kg / kg BOD_5 removed)	0.55
BOD loading to oxidation tank (kg BOD_5 removed / m³ media / day)	4–8

Wastewater Sampling and Analysis

Water samples were taken from influents and effluents from first and second oxidation tanks for determination of COD, TSS (total suspended solids), BOD (biochemical demand), temperature, pH and dissolved oxygen. Influent flow rate and air flow rate were also measured. Testing procedures were in accordance with *Standard Methods*[3].

RESULTS AND DISCUSSIONS

Pilot Plant Study

Table VII presents the experimental results for the 18 months pilot plant study for brewery wastewater treatment. Influent had 489 mg/L BOD and 413 mg/L COD and was reduced to 39 mg/L BOD and 109 mg/L COD after biological treatment using soft packing media contact oxidation process. The removal efficiency was 92% for BOD removal and 73.6% for COD removal. The air flow to wastewater flow ratio on a volume basis was 27 : 1, which was sufficient to satisfy the oxygen requirement for bio-oxidation and to maintain a 2−4 mg/L DO in the first oxidation tank and a 4−6 mg/L DO in the second oxidation tank effluents. The design parameters developed in this study are listed in Table VIII. The influent BOD was higher than the influent COD possibly due to the nitrogenous demand (nitrification) of brewery wastewaters exerted in the 5-day BOD test.

Coagulation Study

Coagulation study was conducted to improve the COD removal efficiency in the pilot plant study. Coagulant was added to the influent to the first oxidation tank. Four coagulants were investigated, which included ferric chloride ($FeCl_3$), ferrous sulfate ($FeSO_4$ 7 H_2O), alum ($Al_2(SO_4)_3$, and polyaluminum chloride or chlorhydrol ($Al_2(OH)_5Cl$), as shown in Table IX. At a coagulant dosage of 60 mg/L, polyaluminum chloride was found to be the best coagulant and was used in the pilot plant study. Table X presents the treatment efficiency in terms of COD and BOD removal for the combined contact oxidation and coagulation processes. Water samples were taken from the influent to the first oxidation tank and from the effluent from the final settling tank. Polyaluminum chloride was added to the influent to the first oxidation tank at dosages of 50 to 240 mg/L in order to determine the best

Table VIII. Design Parameters for Full Scale Plant

Parameter	Value
Influent BOD_5 (mg/L)	200–600
Bio-oxidation effluent BOD_5 (mg/L)	20–40
Wastewater temperature (°C)	≥ 10
Air to water ratio (volume/volume)	27 : 1
D. O. level	
1st oxidation tank (mg/L)	2–4
2nd oxidation tank (mg/L)	4–6
BOD_5 loading (kg BOD_5 removed / m³ media / day)	4–8
$Al_2(OH)_5Cl$ dosage (mg/L)	60

Table IX. Coagulants Used in Pilot Plant Study

Coagulant	pH	Dosage (mg/L)
$FeCl_3$	4−11	60
$FeSO_4$ 7 H_2O	8−11	60
$Al_2(SO_4)_3$	6−8.5	60
$Al_2(OH)_5Cl$	6−8.5	60

dosage for the COD and BOD removal. BOD removal efficiency was about 98% in the dosage range of 50 to 240 mg/L tested. The dosage of polyaluminum chloride in the range tested did not have any effect on percent BOD removal. The COD removal efficiency was from 77 to 96% for the range of dosage tested. The best COD removal efficiency of 95% was obtained at a dosage of 80 mg/L. With dosages higher than 240 mg/L, effluent became tea brown in color due to the color present in the coagulant. With coagulant less than 50 mg/L, effluent was turbid. It was recommended that a dosage of 60 mg/L be used which could achieve 94% COD and 98% BOD removal using the combined soft packing media contact oxidation and coagulation treatment processes.

With a coagulant dosage of 60 mg/L, and a chlorine dosage of 3.75 g/ton of water for chlorination of final effluents, the COD removal efficiency was increased 17% while the BOD removal efficiency was increased 3% when compared to the treatment process without coagulation and chlorination.

Soft Packing Media Contact Oxidation Process

In treating wastewaters at high organic loading rates, the suspended growth aerobic treatment system tends to exhibit sludge bulking problem with poor settled sludge resulting in an effluent with high suspended solids. The inclusion of soft packing media in the contact oxidation process is intended to eliminate the sludge bulking problems associated with treatment of organic wastewaters at high organic loading rates. In the soft packing media contact oxidation process, the biofilm is attached to the surface of packing media, which could maintain a high bio-mass concentration without sludge bulking problem associated with the suspended growth treatment systems. The filamentous bacteria, attached to the packing media, grow at a fast rate in the biological oxidation of high concentration of sugars, organic acids, and alcohols present in the brewery wastewaters with the utilization of organic carbon contained in these organic compounds in the cell synthesis of filamentous bacteria. It appears that soft packing media contact oxidation process is an excellent treatment method for brewery wastewaters and food processing wastewaters where filamentous bacteria growth would render activated sludge process incapable of meeting effluent quality standard regarding suspended solids. Since the soft packing media is very flexible and would move in all directions in accordance to the mixing current, the rope with fiber bundles would make wave-like motions depending upon the agitation or mixing in the tank provided by the diffused air. The violent agitation of water flow would keep the continual renewal of biofilm on the soft packing media. The old biofilm sloughed off is replaced by the new biofilm, which is a very thin film and maintains an active biomass. This can handle the high organic loading rate encountered in the brewery wastewater treatment. After 18 months pilot plant study, there was no clogging for the soft packing media in the contact oxidation tank.

Table X. Summary of Performance Data with Addition of $Al_2(OH)_5Cl$ to 1st Oxidation Tank

Dosage (mg/L)	Infl COD (mg/L)	Effl COD (mg/L)	% COD removal	Infl BOD (mg/L)	Effl BOD (mg/L)	% BOD removal
240	368	44	88	598	9.2	98
160	312	48	85	438	9.7	98
100	268	29	90	369	6.3	98
100	192	8	95	233	3.8	98
80	224	4	96	−	−	−
60	250	54	78	249	21.2	92
60	275	16	94	420	9.2	98
50	362	24	93	524	10.1	98
50	441	102	77	−	−	−

Table XI. Comparisons of Characteristics of Various Packing Media

Type of Media	Weight (kg/m³)	Specific Surface Area (m²/m³)	Void Space (%)	Cost ($/m³)
Paper Honeycomb	19	159	98.8	58.6
Cloth Honeycomb	31	158	98.8	108
Soft Media	3	2,472	79.9	16.7

Soft Packing Media

In the past the packing media used in China for the contact oxidation process consisted mainly of paper or cloth, with special surface treatment, of honeycomb configuration. Table XI shows the comparisons among paper, cloth and the soft media made of rayon synthetic fibers. The soft packing has the lowest density of 3 kg/m³ compared to 19 kg/m³ for paper media and 31 kg/m³ for cloth media. The soft packing media also has the largest specific surface area of 2472 m²/m³, which is about 16 times the value for cloth or paper media. The cost of soft packing media is also very attractive at $16.6/m³, which is about one third the cost of paper media and about one sixth the cost of cloth media. The other advantage of using soft packing media is the long life of the media material under wastewater treatment conditions compared to either paper or cloth media, which tend to deteriorate in a relatively short period of operation.

Temperature Effect on Soft Packing Media Contact Oxidation Process Performance

During the pilot study, the air temperature dropped below 4°C in December resulting in the formation of a thin layer of ice on the surface of the tanks. The treatment efficiency was decreased in this period of time. In the subsequent months the recovery in the treatment plant performance was very slow and incomplete. It appears that the high organic loading rate used in the normal operation of a treatment plant should be reduced in the winter, the period of low air temperature, in order to maintain a satisfactory treatment efficiency. The other solution is to maintain a water temperature higher than 10°C by minimizing the heat loss from the wastewater piping especially from the overhead piping by proper insulation. This measure will ensure the proper water temperature of higher than 10°C even in the winter. The temperature in the packing wastewater and fermentation rinse wastewater was higher than 20°C. After mixing of these two wastewater streams with other streams in the brewery plant, the overall temperature in the combined wastewater would be exceeding 15°C.

Biological Observation

A biological indicator has been used in the evaluation of wastewater treatment performance. This normally involves the determination of types and number of organisms present in the wastewater treatment systems.

In the soft packing media contact oxidation process, the presence of a large number of attached and mobile protozoa following the formation of the biofilm would indicate an active biomass has been established in the biofilm. The protozoa would graze on the biofilm and use bacteria as their food source. Rotifers, ciliates, and insects also use organisms present in the biofilm as food source. This helps the renewal of biofilm on a continuous basis. The presence of these predators in the soft packing media contact oxidation process would ensure an excellent treatment condition present in the treatment system.

Biological Sludges

The sloughing off of the old biofilm together with the non-biodegradable suspended solids present in the raw wastewater forms the excess sludge in the soft packing media contact oxidation process. In a 30 minutes settling test, the settled sludge volume was 30% of the original volume.

A high concentration of biomass was maintained in the soft packing media contact oxidation process especially in the first oxidation tank. The biomass concentration was computed based on the sum of the biomass present in the biofilm and the biomass contained in the sloughing off biosludge. The first oxidation tank had a very high biomass concentration of about 4650 mg/L while in the second oxidation tank it was 1100 mg/L. The wet sludge ball on the soft packing media was 126.8 g and 45.75 g in the first and second oxidation tank, respectively, with a corresponding water content of

96.6 and 97.7%. The MLVSS (mixed liquor volatile suspended solids) to MLSS (mixed liquor suspended solids) ratio was about 0.80. The excess sludge production rate was 0.55 kg sludge produced/kg BOD removed.

Correlation Between Parameters

Based on the experimental results, the following correlations were obtained.

    a. Influent BOD_5 vs influent COD
      $BOD_5 = 65.54 + 0.93\ COD$
      $COD = 0.90\ BOD_5 - 8.06$
         Correlation coefficient $= 0.916$

    b. Influent BOD_5 vs influent OC (oxygen consumption)*
      $BOD_5 = 25.52 + 2.42\ OC$
      $OC = 55.94 + 0.23\ BOD_5$
         Correlation coefficient $= 0.745$
         *OC = oxygen consumption using potassium permanganate $KMnO_4$

CONCLUSIONS

Soft packing media contact oxidation process could effectively remove organic pollutants from brewery wastewaters with a BOD removal efficiency of 92% and a COD removal efficiency of 74%.

There was no clogging in the soft packing media during the 18 months operation of the pilot plant.

Soft packing media made of rayon synthetic fibers provides a large surface area for biofilm attachment and is capable of treating brewery wastewater at a high organic loading rate.

Soft packing media showed no sign of deterioration in the 18 months of pilot plant operation.

Presence of protozoa, rotifers, and ciliates on or near the biofilm serves as a biological indicator for excellent treatment performance and presence of active biofilm.

Polyaluminum chloride addition to the first contact oxidation tank as coagulant enhanced the removal of BOD and COD from brewery wastewater. This combined soft packing media contact oxidation coagulation process together with chlorination increased COD removal efficiency by 17% and BOD removal efficiency by 3%. The polyaluminum dosage was 60 mg/L and the effective chlorine dosage was 3.75 g/ton of water.

REFERENCES

1. Mahmud, Z., "Bottle Washing and Brewery Wastewater Treatment," *Proc. of 34th Purdue Industrial Waste Conf.*, pp. 375-384 (1979).
2. Le Clair, B. P., "Performance Monitoring Program – Molson's Brewery Deep Shaft Treatment System," *Proc. of 39th Purdue Industrial Waste Conf.*, pp. 257-268 (1984).
3. American Public Health Association, *Standard Methods for the Examination of Water and Wastewater*, 15th edition, American Public Health Association, Washington, D.C. (1980).

11 SLUDGE QUALITY INDEX—EVALUATION OF SLUDGE SETTLEABILITY

Richard Fox, Project Engineer
Angelbeck Environmental Engineers, Inc.
Toledo, Ohio 43606

Donald Angelbeck, Associate Professor
Department of Civil Engineering
University of Toledo
Toledo, Ohio 43606

INTRODUCTION

Major advances in process control of biological activated sludge operation have been made possible through the continued development of solids flux analysis. Widespread availability of microcomputers and technical software have only recently opened the potential benefits of solids flux analysis for daily operational control at municipal and industrial wastewater treatment facilities. Solids flux analysis involves measuring sludge settleability which leads to a sludge settling constraint. This information can be used to estimate operational levels of mixed liquor concentration within the aeration tanks and underflow concentration of the final clarifier for different combinations of influent and recycle flows.

The variable SQI, or Sludge Quality Index, has been devised to quantify sludge settleability as it is measured and used in solids flux analysis. This paper will report the field data which supports the theoretical development and definition of the Sludge Quality Index. Two traditional solids flux models along with an empirical SVI model will be investigated to obtain solids flux curves and the associated Sludge Quality Indexes.

The Sludge Quality Index, a single quantitative measure of sludge settleability, is needed for two reasons. First, this quantity can be directly compared with other process variables to develop relationships that can be used to determine operational corrections to the activated sludge system. Secondly, the measured SQI quantity along with SQI trending over the recent past can be used in algorithms to calculate control adjustments of flow splits within a step-feed activated sludge process.

DEFINITION OF THE SLUDGE QUALITY INDEX

The Sludge Quality Index can be calculated from any batch solids flux curve. Initial settling curves are obtained by conducting settleability tests (Settleometer Tests) on several concentrations of a waste activated sludge. This data can be used to construct the solids flux curve either graphically or with mathematical modeling. Traditional plots of batch solids flux settling capacity versus concentration produce curves which vary in both shape and location depending upon overall sludge settling characteristics. Models used to construct these curves normally involve two variables, neither of which singly or adequately represent changes in the overall sludge settleability. Furthermore, a general index for sludge settleability should not be dependent on any one model for a solids flux curve.

The definition for the Sludge Quality Index is relatively simple once a solids flux settling curve has been obtained. Place a recycle operating line, with a slope equal to a reference recycle flow rate divided by the area of the clarifier, tangent to the solids flux curve. The value of the concentration intercept of this line is defined as the SQI (Figure 1). Physically, the SQI is the minimum underflow sludge concentration that could be expected from operation using a reference recycle pumping rate or the maximum recycle pumping rate. We have chosen a standard recycle flux of 700 gpd/ft² to define SQI. If SQI data are compared within a specific plant, then the SQI defined by the maximum recycle flux would be a more realistic variable which describes the minimum recycle sludge concentration at a given time. However, if sludges between different plants are to be compared, the SQI defined by a reference recycle flux must be used since each plant may have different recycle pumping capacities.

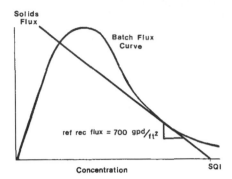

Figure 1.   Graphical definition of SQI.

Calculating an SQI provides a single quantitative variable which describes the current sludge settling character by combining the effects of the shape and position of the solids flux settling curve. The result is a sludge index which may prove more accurate and fundamental than the much maligned but difficult to replace Sludge Volume Index. If the necessary investment of time and money for laboratory measurement of sludge settleability and computer assisted data manipulation has already been made, use of the SQI is a logical extension from the current use of solids flux analysis in determining optimal recycle strategies. While the use of SQI values is invariably restricted to those who are performing advanced methods of process control with solids flux analysis, it may also add to the development of such methods and toward wider acceptance of such techniques.

SOLIDS FLUX MODELS

The time requirements for either data manipulation or the corresponding use of the solids flux curves necessitates the use of a microcomputer. This is especially true if the sludge settling curves are to be updated at least twice a week. Computer calculations require a mathematical model for solids flux settling capacity as a function of concentration. Two highly accepted models were developed by Dick and Young [1] and Vesilind [2] and will be presented in detail below.

The major issue of contention in solids flux analysis is not which of these models to choose; rather, it is whether the methods used to create the batch settling curve accurately reflect how sludge is actually settling in the field. Researchers have repeatedly shown that the height, diameter, and stirring mechanism within settleometers drastically effect the shape and location of the solids flux curve[3,4]. Researchers have used large settleometers to reduce these effects but these are not practical or available for widespread use[1,5,6]. Keinath has proposed a procedure which uses measurements of multiple operating states of the full scale system to provide the information needed to model the solids flux curve[3]. Daigger and Roper have recently presented work which uses empirical SVI correlations with settleometer tests to obtain future solids flux models with SVI measurements alone[6]. Our experience in developing a microcomputer system which models activated sludge process operation has yielded another approach[7,8]. Laboratory settleometer tests can be performed with simple and inexpensive, commercially available settleometer equipment but must then be "scaled" to actual plant flow conditions by a computer algorithm. The results of scaling solids flux models will be shown later as the two solids flux models are compared. Our research in this area has just begun so the information presented is limited and studies are continuing.

Batch solids flux is the product of the settling velocity as a function of sludge concentration and the sludge concentration. Settleometer tests provide data pairs of unhindered settling velocity and concentration for several dilutions of return activated sludge, RAS, over a wide range of concentrations. Two methods of relating this data have led to two traditional models.

Log-log Model

The first method involves a linear regression of the log of velocity versus the log of concentration. The resulting models for velocity and solids flux were first reported by Dick and Young[1].

$$V = a \cdot C^{-n} \tag{1}$$

$$G = a \cdot C^{1-n} \tag{2}$$

```
100 CLS:REM FIND SQI OF LOG-LOG CURVE
110 PRINT:PRINT:PRINT
120 PRINT "Input BO AND B1 from Plot of LOG(V) ";
130 PRINT "in ft/day vs Conc. in #/ft3"
140 PRINT:PRINT TAB(20);:INPUT "BO ---------> ";BO
150 PRINT:PRINT TAB(20);:INPUT "B1 ---------> ";B1
160 INPUT "Ref Recycle Flux  gpd/ft2 -----> ";VRMAX
170 VRMAX=-1*VRMAX/7.48:REM IN FT/D
180 DEF FN G(X)= EXP(BO)*X^(B1+1)
190 CT=(VRMAX/EXP(BO)/(B1+1))^(1/B1)
200 PRINT:PRINT:PRINT TAB(20);
210 SQILOG=(VRMAX*CT-(FN G(CT)))/VRMAX:REM #/FT3
220 PRINT "SQI Log-log  = ";SQILOG*16025;"  mg/l"
230 END
```

Figure 2.   Basic program to calculate SQI for log-log model.

where:

   a = experimentally determined constant

   n = experimentally determined constant

   C = concentration of solids

   V = settling velocity of sludge

   G = solids flux

To retain the notation that is used in the original regression analysis, Dick's model can be rewritten as:

$$\log(V) = B_0 + B_1 \cdot \log(C) \tag{3}$$

$$V = e^{B_0} \cdot C^{B_1} \tag{4}$$

$$G = e^{B_0} \cdot C^{B_1 + 1} \tag{5}$$

where:

   B_0 = intercept of the linear regression on the log-log plot of initial settling velocity and concentration

   B_1 = slope of the linear equation (negative value)

Calculating the SQI for the log-log model is relatively easy. The slope of the recycle operating line is the reference recycle flow divided by the area of the clarifier. The concentration where this line is tangent to the batch flux curves can be calculated by setting this recycle flux equal to the derivative of Equation 5. The SQI can be calculated using this point and the slope of the recycle flux. A short computer program is listed in Figure 2 that calculates SQI values for a log-log model.

Semi-log Model

The second method involves a linear regression of the log of velocity directly versus the concentration. The resulting models for velocity and solids flux were first reported by Vesilind[2].

$$V = V_0 \cdot e^{-K'C} \tag{6}$$

$$G = C \cdot V_0 \cdot e^{-K'C} \tag{7}$$

where:

   V_0 = unhindered settling velocity of individual sludge particles

   K' = experimentally determined constant

To retain the notation that is used in the original regression analysis, Vesilind's semilog model can be rewritten as:

$$\log(V) = A_0 + A_1 \cdot C \tag{8}$$

$$V = e^{A_0} \cdot e^{A_1 C} \tag{9}$$

$$G = C \, e^{A_0} \cdot e^{A_1 C} \tag{10}$$

where:

   A_0 = intercept of the linear regression on the semilog plot of initial settling velocity and concentration

   A_1 = slope of the linear equation (negative value)

Calculating the SQI for the semilog model is more involved than the log-log model. The slope of the recycle operating line remains the same. The concentration where this line is tangent to the batch flux curve can be found using a Newton-Raphson iterative routine. This procedure is complicated by the fact that the semilog batch flux curve has an inflection point. There are two possible points of tangency with the same slope of recycle flux. The desired point of tangency must lie to the right of the inflection point. In addition, the inflection point also represents the point of tangency for the highest possible recycle hydraulic flux. If the reference recycle hydraulic flux used to calculate the SQI is greater than this maximum possible hydraulic flux, then no SQI value can be calculated. Therefore, the definition of the SQI for the semilog model must be modified. If possible, the SQI is the underflow concentration resulting from operation at the reference recycle hydraulic flux; otherwise, the SQI is the underflow concentration resulting from operation at the maximum recycle flux allowable for the semilog model, that is the slope of the tangent at the inflection point. The SQI can be calculated as before using the correct tangency point and the corresponding slope of the recycle flux. A short computer program is listed in Figure 3 that calculates SQI values for a semilog model.

Results

Settleometer Test results from Bowling Green, Ohio that have been modeled with both of these methods are shown in Table I. Both models achieve a high coefficient of determination (i.e., R^2). The two models are compared graphically in Figure 4 in relation to the measured laboratory solids flux data points. The semilog model produces a more traditional solids flux curve. The model starts with zero solids flux at zero concentration, increases to a maximum and then decreases as concentration continues to increase, passing through an inflection point as the curvature changes. The graph also points out an inherent weakness of the semilog model. The semilog curve fails to pass near the measured solids flux point for the lowest concentration; consequently, the semilog curve is flattened and spread out. The resulting effect of the right side of the semilog curve is shown in Table I where the maximum possible recycle flux (i.e., determined by the inflection point) ranges between 98 and 316 gpd/ft². These values are considerably less than the normal operating recycle fluxes at the Bowling Green Facility. Our standard 700 gpd/ft² reference recycle flux could not be used in the SQI calculation for the semilog model. Therefore, SQI values based on a 125 gpd/ft² recycle flux are given for both models. The log-log model shown in Figure 4 produces a solids flux curve which goes to infinity as the concentration approaches zero. Fortunately, solids flux analysis is not affected by solids flux modeling at low concentrations. Therefore, the best solids flux model is one that best models solids flux settling along the right hand side of the solids flux curve.

Scaling of Laboratory Data

We have found that direct use of laboratory settling curves from these simple settleometer tests has consistently failed to predict actual operational conditions for MLSS and underflow concentrations within the activated sludge system. We have used 1-liter and 2-liter settleometers as well as 3 ft deep and 6 in. diameter settling columns with and without slow (0.25 to 2 rpm) stirring speeds. Batch solids flux curves generated by any of these laboratory tests most often have not accurately described full scale clarifier performance. However, manageable routine settleometer tests with minimal side wall effects can be performed using 2-liter settleometers provided that the test procedure related to mixing, etc. is consistent. The discrepancy between laboratory measurement and actual operating conditions can be corrected by shifting laboratory curves to a more "true" operating position. Scaling either one or both of the laboratory determined constants for either solids flux model could bring this about.

Determining the location of the improved batch flux curve requires the use of the pivot point or state point concept for activated sludge operation[5,7,9]. During non-transient conditions, a mass balance of solids between the aeration tank and clarifier determines a unique pivot point and corresponding MLSS and underflow concentrations for any given combination of influent and recycle flows. This pivot point concept, shown in Figure 5, is the basis for all later calculations involved in solids flux analysis of activated sludge operation. The resulting relationship between influent flow, recycle flow, concentrations, and the location of the solids flux curve can be used in an iterative procedure to shift the laboratory curve. System measurements of the influent flow, recycle flow, and the underflow concentration at the time of the laboratory settling test during non-transient system operation are used to modify the laboratory solids flux curve to one that reflects full-scale operation. A sample of the results of the scaling process is shown in Figure 6 for both models.

The approach used in scaling the models was an attempt to change the location of laboratory curves without changing their shape. This has proved relatively simple for both models because it involved

```
100 CLS:REM FIND SQI OF SEMILOG CURVE
110 PRINT:PRINT:PRINT
120 PRINT "Input AO AND A1 from Plot of LOG(V) ";
130 PRINT "in m/hr vs Conc. in g/l"
140 PRINT:PRINT TAB(20);:INPUT "AO ---------> ";AO
150 PRINT:PRINT TAB(20);:INPUT "A1 ---------> ";A1
160 PRINT:INPUT "Ref Recycle Flux  gpd/ft2 -----> ";VRMAX
170 DEF FN G(X) = X*EXP(AO)*EXP(A1*X):REM G()=FLUX
180 DEF FN V(X) = EXP(AO)*EXP(A1*X)*(1+A1*X)
190 REM FIND INFLEXION PT
200 DEF FN F(X) = A1*EXP(AO)*EXP(A1*X)*(2+X*A1)
210 DEF FN D(X) = A1^2*EXP(AO)*EXP(A1*X)*(3+X*A1)
220 XR=-.1:ES=-.01:CI=0
230 GOSUB 1000
240 CI=XR:REM GM/L
250 VI=FN V(CI)
260 REM FIND SQI
270 VRMAX=-1*VRMAX/7.48/3.28/24:REM  M/HR
280 IF ABS(VRMAX)>ABS(VI) THEN VRLIM=VI ELSE VRLIM=VRMAX
290 DEF FN F(X) = EXP(AO)*EXP(A1*X)*(1+X*A1) - VRLIM
300 DEF FN D(X) = A1*EXP(AO)*EXP(A1*X)*(2+X*A1)
310 XR=10
320 GOSUB 1000
330 CT=XR:REM GM/L
340 SQISEMI=(VRLIM*CT-(FN G(CT)))/VRLIM:REM GM/L
350 PRINT:PRINT:PRINT TAB(20);
360 PRINT "SQI Semilog  = ";SQISEMI*1000;"  mg/l"
370 PRINT:PRINT:PRINT TAB(20);
380 PRINT "MAX REC FLUX  = ";VI*-1*7.48*3.28*24
390 END
1000 REM Newton-Raphson Root Finder Subroutine
1010 XN=XR - (FN F(XR)/FN D(XR))
1020 IF XN=0 THEN 1070
1030 EA=ABS((XN-XR)/XN)*100
1040 IF EA<=ES THEN 1999
1050 XR=XN
1060 GOTO 1010
1070 PRINT "ROOT NOT REACHED"
1999 RETURN
```

Figure 3.   Basic program to calculate SQI for semilog model.

changing only one of the two parameters in each case. Interestingly, this single parameter was A_1 for the semilog model and B_0 for the log-log model. While this approach has provided good results, there may be better combinations of parameter adjustment in the scaling process.

Successful shifting of laboratory solids flux curves and solids flux analysis in general is determined by whether predicted operating levels are accurate. Accuracy data for MLSS and underflow concentrations are presented in Table II for a municipal activated sludge system which has been generated with a shifted log-log model. The measured values were obtained with centrifuge readings for time savings and converted to concentrations with a correlation routine.

Alternative Method — Using SVI to Obtain Solids Flux Curve

Daigger and Roper have presented a method by which the solids flux curve can be modeled with the input of a simple daily SVI measurement[6]. A semilog model was used with a constant value for A_0. Preliminary investigations with solids flux data from another municipal facility showed that Roper's

Table I. Model Results from Settleometer Tests[a]

| | Semilog Model | | | | | Log-log Model | | | | |
Date	A_0[b]	A_1[b]	R^2	SQI[c] (125)	Max[d] Rec	B_0[e]	B_1[e]	R^2	SQI[c] (700)	SQI[c] (125)
02–03	0.6168	−0.4082	.9916	10490	148	0.2905	−2.1756	.9842	4530	10010
02–07	0.4082	−0.3330	.9878	*12010	120	0.7118	−1.9884	.9815	4680	11120
02–10	1.1864	−0.4214	.9891	12130	261	0.6558	−2.2247	.9783	5570	12080
02–13	1.0290	−0.4656	.9464	10510	223	0.3382	−2.1831	.9914	4660	10270
02–19	0.2095	−0.3206	.9547	*12480	98	0.6055	−1.9728	.9594	4360	10450
02–21	0.7653	−0.3874	.9331	11650	171	0.5835	−2.1517	.9835	5090	11330
02–24	1.3699	−0.4470	.9670	11990	314	0.4420	−2.5233	.9958	6180	12240
02–27	1.3772	−0.4946	.9755	10850	316	0.0993	−2.6016	.9811	5660	10980

*SQI based on Maximum possible recycle flux.
[a]Based on 1986 Settleometer Test data from Bowling Green, Ohio.
[b]Units for A_0 and A_1 are from plotting velocity (m/hr) versus conc. (g/L).
[c]SQI given in terms of mg/L for ( ) gpd/ft² reference recycle flux.
[d]Maximum recycle flux in terms of gpd/ft².
[e]units for B_0 and B_1 are from plotting velocity (ft/day) versus conc. (lb/ft³).

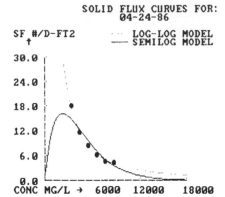

Figure 4. Graphical comparison of two models.

model produced solids flux curves which were not located near either original or adjusted log-log or semilog modeled curves. Therefore, the SVI correlation seems to be dependent on the general sludge settling characteristics for a particular plant and the correlation procedures would need to be repeated at each individual facility. Trial and error adjustments to Roper's empirical relationships did provide

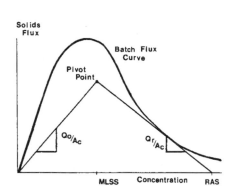

Figure 5. Solids flux pivot point concept.

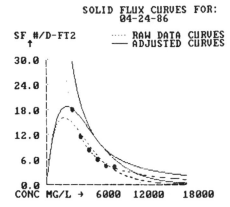

Figure 6. Graphical adjustment of batch flux curves.

Table II. Accuracy of Solids Flux Prediction Using a Log-log Model

Date	MLSS % Accuracy	RAS % Accuracy
2-19[a]	87.9	98.7
2-20	71.2	86.5
2-21[a]	90.2	90.6
2-22	93.5	96.1
2-23	86.6	94.2
2-24[a]	83.6	87.9
2-25	92.0	91.8
2-26	89.2	93.1
2-27[a]	86.7	99.6
2-28	78.9	82.6

[a]Settleometer Updated
Note: 1986 Second Shift Data from Bowling Green, Ohio.

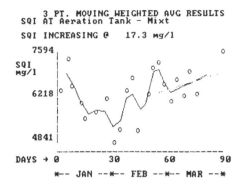

Figure 7.   Moving weighted average trending of SQI data.

realistic solids flux curves for this particular facility. Comparing several solids flux curves generated with this modified SVI method with shifted log-log modeled curves showed that the relative changes in location of the solids flux curves were reflected by both of these models. While the SVI method shows promise, research should be directed toward the following areas: 1) our data have shown that the A_0 varies greatly from day to day analysis; 2) did the use of large settleometers produce data which accurately reflects full-scale sludge settling?; 3) is the SVI an adequate control measurement since its magnitude is dependent on MLSS concentration and not on sludge settleability alone?; 4) will the empirical relationship always be constant?; and 5) do the resulting solids flux curves produce reliable concentration estimates during solids flux analysis?

SQI RELATIONSHIPS

Sludge settling characteristics are best represented by solids flux settling curves which are obtained by directly measuring sludge settling at different concentrations. The SQI variable, obtained from the solids flux curve, is a single quantity which may best represent sludge settleability. Relative changes in SQI reflect measured changes in sludge settleability. The SQI variable can be plotted alone and trended to indicate how sludge settleability changes with time(Figure 7). The SQI quantity can also be compared to other process variables.

SQI and SVI

The magnitude of SQI changes oppositely of SVI values (Figure 8) as would be expected since SVI^{-1} represents the average sludge concentration within the sludge blanket during the batch settling test. Furthermore, the range of the SVI domain is a function of sludge settleability as well as MLSS concentration. For example, an SVI of 200 and a MLSS of 2000 mg/L has much different settling character than a sludge with an SVI of 200 and a MLSS concentration of 4000 mg/L. The SQI, however, is obtained using solids flux data which incorporates both settling character (i.e., settling velocity information) as well as concentration information and produces a value which is indicative of sludge settling performance under limiting hydraulic constraints. Relating both indexes directly, therefore, shows no definitive correlation between these two variables(Figure 9).

STEP-FEED MANAGEMENT USING SQI

A step-feed aeration system provides an additional control variable to the activated sludge process. Splitting the waste flow among the aeration tank bays changes operation along a continuum from plug flow to contact stabilization. Usually, step-feed adjustments are made to improve sludge settleability. To develop a meaningful control algorithm for a step-feed process, a reliable method is needed to evaluate sludge settling characteristics. The SQI quantity calculated with each settleometer update can be used to decide when and how large of flow split adjustment is needed to increase sludge settleability. Flow split decisions are important because there is a natural tradeoff between increasing sludge settleability and decreasing treatment effectiveness[10].

The step-feed computer algorithm developed by the authors retrieves up to 90 past SQI data and performs a trend analysis to indicate whether the settling characteristics of the sludge are increasing or decreasing. Assessment is made using weighted moving average trend analysis techniques on past SQI

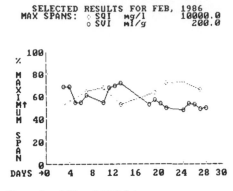

Figure 8.   SQI and SVI data.

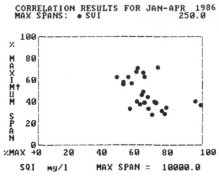

Figure 9.   Correlation graphing: SVI versus SQI.

data. The most recent six SQI trended values are linearized. The maximum and minimum SQI defined by a 95% confidence interval using the normal distribution are calculated. Deterioration or improvement of sludge settleability is made through a point system based upon: 1) the slope of the linearization; 2) comparison of the current SQI with the SQI predicted from the moving average linearization; and 3) comparison of the current SQI with the 95% maximums and minimums. If the point system indicates settleability is decreasing, an appropriate shift of the hydraulic pattern away from plug flow mode is recommended. Conversely, if sludge settleability is increasing, then a hydraulic pattern is recommended which approaches the plug flow mode thus increasing treatment performance[7].

SUMMARY

Sludge settleability plays such a prominent role in activated sludge operation, yet it has been difficult to quantify as a process control variable. The Sludge Quality Index is a way to reduce the complete solids flux settling curve into a single quantity representing sludge settleability. This new index is not a primary end in itself. It most probably will only be used by those who will use solids flux techniques. Nevertheless, with today's technology, the SQI concept in conjunction with solids flux techniques provides a technically sound methodology to assess sludge settling performance. This additional step forward from the traditional solids flux analysis provides a sludge index that can be used to compare changes in sludge settleability with other process control variables, provides a parameter that can be used for step-feed adjustment and/or coagulant addition, and also provides an absolute measure of sludge settleability, a parameter which has been absent from activated sludge technology over the years.

The choice of which type of model to use for the solids flux relationship depends on preference since both the log-log model and the semilog model produce adequate results. While the traditional shape of the semilog is appealing, its use complicates both solids flux analysis calculations and the definition of the SQI. Initial data from the Bowling Green Facility indicates that the semilog batch settling curves are too flat or spread out and thereby give unrealistically low recycle constraints in solids flux analysis. Therefore, until further studies are made comparing these two solids flux models, the log-log model is preferable to these authors.

The major issue that needs resolving is whether the solids flux curves, obtained through either laboratory measurement or a SVI relationship, actually reflect field operation. We have set forth one solution to the discrepancy between laboratory solids flux curves and actual full scale operation. Scaling of the solids flux model or shifting the initial solids flux curve may be the best method to assure meaningful results in solids flux analysis regardless of the method of obtaining the initial solids flux curve. Studies currently underway are continuing using both log-log and semilog models to determine the best shifting procedures.

Our work with solids flux analysis and the subsequent defining of the Sludge Quality Index have taken place within the development of an activated sludge operational control system of microcomputer programs. These programs which include the data manipulations of settleometer test updates, optimal recycle and wastage calculation, and step-feed management are incorporated within a plant-wide program system. In fact, most of the figures present herein are examples of some of the graphics capabilities available to the users.

ACKNOWLEDGEMENTS

The authors would like to thank all the personnel at the Bowling Green (Ohio) Water Pollution Control Facility and the Fremont (Ohio) Water Pollution Control Center who have used the CONTROL software programs, provided analytical assistance, and freely given their support and suggestions.

REFERENCES

1. Dick, R. I., and K. W. Young, "Analysis of Thickening Performance of Final Settling Tanks," *Proceedings of the 27th Purdue Industrial Waste Conference*, Purdue University, W. Lafayette, Indiana (1972).
2. Vesilind, P. A., "The Influence of Stirring in the Thickening of Biological Sludge," *J. Water Pollut. Control Fed.*, 41(5):Part 2, R197–R214 (1969).
3. Keinath, T. M., "Operational Dynamics and Control of Secondary Clarifiers," *J. Water Pollut. Control Fed.*, 57(7):770–776 (1985).
4. Vesilind, P. A., *Treatment and Disposal of Wastewater Sludges*, Ann Arbor, Michigan, Ann Arbor Science Publishers, Inc. (1975).
5. Keinath, T. M., M. D. Ryckman, C. H. Dana, and D. A. Hofer, "A Unified Approach to the Design and Operation of the Activated Sludge System," *Proceedings of the 31st Purdue Industrial Waste Conference*, Purdue University, W. Lafayette, Indiana (1976).
6. Daigger, G. T., R. E. Roper, Jr., "The Relationship between SVI and Activated Sludge Settling Characteristics," *J. Water Pollut. Control Fed.*, 57(8):859–866 (1985).
7. Angelbeck, D. I., R. J. Fox, "The Use of Microcomputers to Optimally Control Activated Sludge Process Operation," presented at The 3rd National Conference on Microcomputers in Civil Engineering, Orlando, Fl. (Nov. 1985).
8. Fox, R. J., "The Use of a Microcomputer for Operational Control of the Activated Sludge Process," *M. S. Thesis*, University of Toledo (1985).
9. Dick, R. I., "Activated Sludge Final Settling Tank Analysis," *Seventh Symposium on Wastewater Treatment, Montreal, Canada (1984).*
10. West, A. W., *Operational Control Procedures for the Activated Sludge Process*, Parts 1, 2, 3A, 4, Appendix, 1976 Summary Update and 1978 Summary Update, USEPA (1973–1978).

12 SUSPENDED SOLIDS REMOVAL FROM THE EFFLUENT OF A FIXED-FILM ANAEROBIC REACTOR

Karel Kapoun, Senior Environmental Engineer
Badger Engineers, Inc.
Cambridge, Massachusetts 02142

Frederic C. Blanc, Professor

James C. O'Shaughnessy, Professor
Department of Civil Engineering
Northeastern University
Boston, Massachusetts 02115

INTRODUCTION

The purpose of this investigation was to evaluate Dissolved Air Flotation as a method for removal of suspended solids from the effluent stream of a randomly packed fixed-film anaerobic reactor (Celrobic). The test work reported here include 5 months operation with and without addition of organic flocculent. All the experimental work was performed at the Celanese Chemical Company plant located in Pampa, Texas, an integrated chemical production facility that includes one of the world's largest acetic acid plants.

Feed to the Celrobic reactor is the process wastewater from the entire facility. The organic contaminants are primarily volatile fatty acids and, to a lesser degree, lower alcohols, ketones, esters, and aldehydes.

The Celrobic reactor effluent contains some soluble organic contaminants which are refractory to the bacterial degradation and suspended solids which are predominantly anaerobic bacteria washed out from the reactor packing with some inorganic solids.

Because of its ability to remove both settleable and floatable solids, the Dissolved Air Flotation (DAF) system was selected for pilot plant testing.

BACKGROUND OF DISSOLVED AIR FLOTATION TESTS

The DAF unit used as a feed the effluent stream from the full size randomly packed anaerobic reactor (Celrobic System) [1].

The Celrobic reactor has a volume of approximately 1.5 million gallons. The reactor feed rate during the testing period fluctuated between 400 gpm and 600 gpm while the COD removal efficiency was maintained at approximately 80%. The average COD concentration in the reactor feed has been reported as 16.7 grams/liter. Net sludge production in the Celrobic reactor has been determined to be 0.038 lb/lb of COD removed. This represents daily sludge production of between 2,440 and 3,660 lb/day. Since the Celrobic reactor effluent contains approximately 400 mg/L TSS, the daily sludge removal due to the suspended solids in the reactor effluent is less than the solids production. This leaves some of the biomass (suspended solids) to accumulate in the reactor. The biomass is normally attached to the randomly packed media and must be periodically removed.

The pilot DAF system (13.5 ft² rectangular unit) was manufactured by Komline-Sanderson [2]. The DAF aeration system is based on partial recycle of the treated effluent via a pressurized and aerated retention tank. The float removal is effected through a timer-operated, countercurrent float skimmer. In addition the DAF pilot plant was equipped with a flocculation tank with a "picket-fence," variable-speed flocculator and chemical dosing equipment.

DAF TESTING PROCEDURES

Initially, the DAF unit was tested without chemicals while the Celrobic reactor was operated in the normal steady-state condition producing low TSS effluent. The later part of the testing included high

TSS feed conditions to the DAF unit, normally associated with the removal of biomass from the Celrobic reactor. Testing during this period included operation of the DAF unit both with and without chemical treatment.

The actual testing consisted of DAF unit operation at a constant feed flow rate with the recycle rate being changed at 8 hour intervals. The testing was then repeated at the different feed flow rates to simulate hydraulic loading between 0.5 gpm/ft² and 2.0 gpm/ft². Each different flow rate included four different recycle rates (0.5/1, 1/1, 1.5/1, and 2/1). This testing procedure also simulated variations in solids loading and air-to-solids ratio. The additional random variable was feed TSS.

The feed into the DAF unit, the clarified effluent, and the float were analyzed for TSS and VSS. The feed and effluent were also analyzed for total COD. In addition, turbidity determination, particle size distribution, and settleable solids tests were performed on selected samples from the DAF effluent.

Prior to chemical treatment operation, a series of Jar tests were run to determine the most efficient coagulation/flocculation chemicals.

The measured data for all test runs were statistically analyzed using the MINITAB computer program [3]. Since the emphasis of the pilot plant testing was on finding a method for removal of suspended solids from the anaerobic reactor effluent, the statistical analysis concentrated on evaluating relationships between effluent TSS and the following variables:

Feed TSS
Hydraulic loading
Solids loading
Air-to-solids ratio

The computer output evaluation and the best fit and/or lack of fit relationship are discussed along with the pertinent process comments in their respective sections.

DAF TESTING WITHOUT FLOCCULATING CHEMICALS

Testing of DAF During Low Feed TSS Conditions

During this phase of the testing, hydraulic loading varied between extremes of 0.5 gpm/ft² and 2.2 gpm/ft² with most of the data points between 1.0 and 2.0 gpm/ft². Suspended solids in the DAF feed fluctuated between 160 mg/L and 780 mg/L with a mean of 409 mg/L. The volatile suspended solids for the same period ranged from 120 mg/L to 420 mg/L with a mean of 303 mg/L. The corresponding VSS/TSS ratio was calculated to be 0.73 or 73%.

Suspended solids in the DAF overflow (effluent) fluctuated between 83 mg/L and 347 mg/L with a mean of 197 mg/L. Corresponding volatile solids were determined to be between 76 mg/L and 227 mg/L. The VSS mean was calculated to be 154 mg/L. The VSS/TSS ratio for the effluent stream was calculated to be 0.81 or 81%. Suspended solids concentration in the float from the DAF unit is directly related to operation of the scraper mechanism. Because of the freezing weather which occurred during the tests, and the possibility of float freezing in the uninsulated float discharge pipe, operation of the scraper mechanism was not optimized. However, during this phase of testing, the average float solids concentration was determined to be 3.2% dry suspended solids.

Effluent TSS versus Feed TSS. The best fit was obtained through a linear regression. The relationship between the effluent TSS and feed TSS is graphically presented in Figure 1. The standard deviation with respect to effluent TSS (37 mg/L) has been used to plot the outside lines in Figure 1, therefore, indicating an expected range of values of effluent TSS concentration for the various feed TSS concentrations. For an average feed TSS of 409 mg/L, the corresponding effluent TSS will fluctuate between 153 mg/L and 227 mg/L with the expected value of 190 mg/L. The relatively high values of the effluent TSS are attributed to the large quantities of suspended solids (approximately 70% of TSS) which were smaller than 8 microns in diameter, very light and for all practical purposes non-floatable and non-settleable. Settling tests were also conducted on the DAF effluent and the results have shown that no settleable solids were observed in the DAF effluent. However, all samples contained relatively high residual turbidity (between 80 and 100 NTU), therefore, indicating a presence of large quantities of suspended solids.

Effluent TSS versus Hydraulic Loading. The statistical analysis indicated that there is no significant relationship between the effluent TSS and the hydraulic loading in the range of operation investigated (0.5–2.0 gpm/ft²). The lack of correlation in the above data can be explained by the assumption that the DAF has been operated in the generally acceptable range of hydraulic loading,

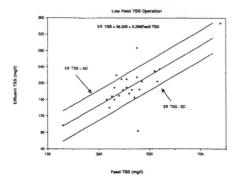

Figure 1. DAF effluent TSS versus DAF feed TSS.

and within this range there is very little influence on the effluent quality. This is probably due to the character of the particles and the particle size distribution discussed previously. It is assumed that the DAF will remove all settleable suspended solids while operating within acceptable hydraulic loading range. However, the further lowering of hydraulic loading will not improve the effluent quality below that of removing all settleable solids.

Effluent TSS versus Solids Loading. No statistically significant relationship between effluent TSS and solids loading has been found. This is probably due to insufficient sample size and the operation of the DAF in the low range of the solids loading. During the normal operation of the Celrobic reactor, the reactor effluent (or DAF feed) contains on average 400 mg/L of suspended solids. This figure, coupled with hydraulic loading of 2.0 gpm/ft² represents a solids loading of approximately 0.4 lb/ft²-hr. Since the bulk of the sample points is in the solids loading region below 0.5 lb/ft²-hr, the influence of solids loading on the effluent TSS cannot be statistically determined.

Effluent TSS versus Air-to-Solids Ratio. The best statistical relationship has been obtained from the second degree polynomial regression. The relationship between the effluent TSS and Air-to-Solids Ratio is graphically presented in Figure 2. The upswing in the regression curve may be explained as follows:

> Air-to-Solids ratio is a function of the feed suspended solids concentration, recycle ratio, and the saturator operating conditions/efficiency. Since the saturator operating conditions and the efficiency remained constant during the testing period, the increase in A/S ratio can be achieved only by increasing the recycle ratio. The increased recycle ratio places an additional hydraulic load on the DAF unit, possibly resulting in lower effluent quality.

COD Reduction. During the later part of this testing period, COD in the feed and the effluent was monitored and compared with suspended solids removal. It has been found that during the COD/TSS removal monitoring, the average suspended solids removal was 49.2% while the average COD removal was determined to be 9.0%. The filtration of Celrobic effluent samples using analytical membrane filters (0.45 microns size) showed an average reduction in COD between unfiltered and

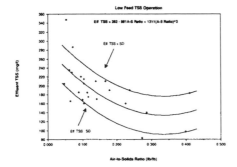

Figure 2. DAF effluent TSS versus A-S ratio.

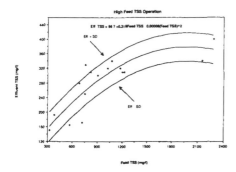

Figure 3.   DAF effluent TSS versus DAF feed TSS.

filtered samples to be 17.3%. Therefore, it is apparent that COD is a constant portion of TSS, and the more solids that are removed, the more COD in the waste stream is reduced.

Testing of DAF During High Feed TSS Conditions

During this period the suspended solids concentration in the DAF influent fluctuated between 325 mg/L and 2300 mg/L, and the hydraulic loading was kept between 1.52 gpm/ft^2 and 1.85 gpm/ft^2 with an average of 1.79 gpm/ft^2.

Measured data obtained during this test period were again statistically analyzed, and the results are discussed below.

Effluent TSS versus Feed TSS. Relationship between the effluent TSS and feed TSS is graphically presented in Figure 3. The suspended solids concentration in the DAF effluent was higher than that observed during the previous testing period. The TSS fluctuated between 150 mg/L and 400 mg/L with a mean of 279 mg/L. Corresponding volatile solids were measured as between 145 and 370 mg/L with a mean of 260 mg/L.

The higher suspended solids in the effluent was due to the increased TSS in the DAF feed stream. The mean feed TSS concentration for this part of investigation has been calculated as 776 mg/L. This figure, however, includes some low TSS values normally associated with a steady-state operation of the Celrobic reactor.

Effluent TSS versus Solids Loading. The high suspended solids feed to the DAF unit resulted in an increase in the solids loading. The solids loading fluctuated between 0.3 lb/ft^2/hr and 2.1 lb/ft^2/hr. This wide operating range and sufficient data points provided the basis for evaluation of the relationship between effluent TSS and solids loading. The measured relationship between the effluent TSS and the solids loading is shown in Figure 4.

Effluent TSS versus Air-to-Solids Ratio. No statistically significant relationship between effluent TSS and air-to-solids ratio has been found for the data points covered in this phase of the experiments.

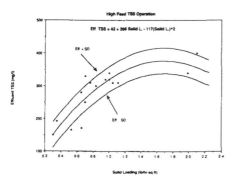

Figure 4.   DAF effluent TSS versus solids loading.

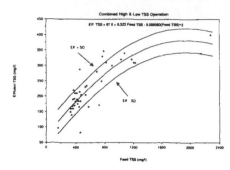

Figure 5. DAF effluent TSS versus DAF feed TSS.

COMBINED HIGH AND LOW SUSPENDED SOLIDS OPERATION

Statistical evaluation of combined data (high and low TSS in the DAF feed) has shown that statistically significant relationships exist between the DAF effluent TSS and the following operating parameters:

Feed TSS
Solids loading
Air-to-Solids ratio

The relationship between effluent TSS and hydraulic loading showed a high degree of randomness. In order to evaluate hydraulic loading, an indirect calculation method was used.

Since solids loading is function of feed TSS and hydraulic loading, knowledge of feed TSS and *solids loading each correlated to effluent TSS is sufficient to calculate the corresponding* hydraulic loading. Examination of test data (Figures 5 through 7) yielded the following operating parameters:

For the feed TSS concentration of 400 mg/L (low TSS conditions):

Effluent TSS $-$ 180 mg/L
Solids load $-$ 0.30 lb/ft^2/hr
Hydraulic loading $-$ 1.5 gpm/ft^2
Air-to-solids ratio $-$ 0.14 lb/lb

For the feed TSS concentration of 1050 mg/L (high TSS conditions):

Effluent TSS $-$ 310 mg/L
Solids load $-$ 0.95 lb/ft^2/hr
Hydraulic loading $-$ 1.8 gpm/ft^2
Air-to-solids ratio $-$ 0.01 lb/lb

Based on the above data, the hydraulic loading for the DAF specification should not be higher than 1.5 gpm/ft^2.

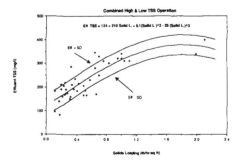

Figure 6. DAF effluent TSS versus solids loading.

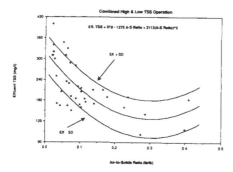

Figure 7. DAF effluent TSS versus A-S ratio.

DAF TESTING WITH CHEMICAL PRE-TREATMENT

During the period of high TSS feed conditions, the DAF pilot plant was tested employing chemical pre-treatment. Based on Jar Testing results, a strongly cationic, medium molecular weight polymer was selected.

Test runs were conducted with a constant feed rate (1.74 gpm/ft²) and constant recycle (1/1). The main variables were feed composition (DAF feed composition changed between 920 and 1960 mg/L TSS) and chemical dosage.

The Jar Tests results, using the same polymer solution as used in the chemical feed, indicated that 100 mg/L of polymer produced the best supernatant and the best settling flocs. Based on this observation, the DAF unit was initially operated with 100 mg/L of polymer. The effluent TSS concentration fluctuated between 16 mg/L and 148 mg/L with an average of 55 mg/L. After lowering polymer dosage to 65 mg/L, the quality of effluent deteriorated only slightly. The average TSS for this period was calculated as 64 mg/L.

Further lowering of polymer dosage to 30 mg/L, however, resulted in a dramatic decrease in the effluent quality. The average TSS concentration increased to 168 mg/L. The visual observation and the settleable solids test revealed that even at under-dosed conditions, the 150 mg/L of TSS represented 8 ml/L of settleable solids and 210 mg/L TSS yielded 12 ml/L of settleable solids. In comparison, the effluent from non-chemical treatment (200 mg/L TSS) did not contain any settleable solids.

COD Reduction

During the non-chemical operation of DAF unit, the highest recorded COD reduction was 13.8%. During the chemical treatment phase of DAF operation, the lowest recorded COD reduction was 25.9%. The average COD reduction for the polymer treatment operation was reported as 38.1%. This figure corresponds with observed COD reduction during Jar Tests. Since the reduction in COD due to the suspended solids removal was higher than during previous non-chemical operation of the DAF unit, it is assumed that the chemical flocculation/coagulation removes some of the soluble COD or COD associated with particles smaller than 0.45 microns.

CONCLUSIONS

Besides the development of design criteria for the DAF, several important conclusions can be drawn from the testing which are pertinent to the design and operation of the DAF. Even during periods when the concentration of TSS in the DAF feed is less than 400 mg/L, the concentration of TSS in the effluent will not fall significantly below 200 mg/L. This is because there exists a background concentration of TSS that is not removable by the DAF without the use of chemicals.

Based on the test results, the expected DAF effluent with chemical pre-treatment will be as follows:

During the low TSS feed conditions – 50 mg/L
During the high TSS feed conditions – 100 mg/L

In addition to over 90% of suspended solids removal, approximately 35% of the total COD can be removed together with the TSS. Because of the additional benefit of COD removal, the chemical treatment could be used as an emergency backup during the organic overload conditions. This arrangement will keep the operating cost at the minimum levels while providing a reasonable protection for the unit operations located downstream of the anaerobic reactor/DAF system.

REFERENCES

1. The Badger Company, Inc., "A Celrobic System Case History—The Celanese Pampa Plant," The Badger Company Publication, Sales Department (May 1983).
2. Komline-Sanderson, Peapack, New Jersey, *Instruction Manual Pilot Scale Dissolved Air Flotation System DAF-15* (July 1984).
3. Pennsylvania State University, Statistic Department, *MINITAB-Computer Program*, Release 82.1.

13 TREATMENT OF COMBINED MUNICIPAL/PACKING HOUSE WASTEWATER USING AN INNOVATIVE CONTINUOUSLY FED—INTERMITTENTLY OPERATED ACTIVATED SLUDGE PROCESS: A DESIGN RATIONALE

Kevin S. Young, Vice-President
J. R. Wauford & Company
Consulting Engineers, Inc.
Jackson, Tennessee 38305

INTRODUCTION

This paper presents the procedure and the rationale behind the procedure used to design a continuously—fed, intermittently aerated, settled and discharged activated sludge facility to treat a combined municipal and packing house wastewater. Technical criteria for process selection included: 1) Capability to hydraulically accommodate peak/base flow ratios of up to 9:1; 2) Capability to consistently meet year-round secondary effluent standards summarized in Table I with seasonal winter/summer ratios for influent organic and hydraulic loadings of 1.35:1 and 1.50:1, respectively; and 3) Capability to accommodate slug loadings of relatively inert suspended solids.

Preliminary evaluations of one conventional and two innovative biological treatment processes were conducted in early 1983. Comparisons of the conventional oxidation ditch, interchannel clarifier variation of the oxidation ditch and the continuously-fed, intermittently operated activated sludge system were made based on process performance capabilities, capital costs and life cycle costs. Neither the interchannel clarifier variation of the oxidation ditch nor the continuously-fed, intermittently-operated, activated sludge system had been operated on a full-scale basis in this country in early 1983. These two innovative processes were evaluated using foreign full scale and domestic pilot plant data. The comparison of these three technologies indicated that the continuously-fed, intermittently-operated, activated sludge system could meet all of the technical criteria for process selection and could provide the lowest life cycle cost.

WASTEWATER CHARACTERIZATION

The 1980 U.S. Census reported Union City's population as 10,436. With the exception of the packing house, Union City has a broad base of clean industry. The packing house kills and processes approximately 700 cattle and 3,000 hogs per day. Pretreatment is provided prior to discharge into the municipal sewer system by a grease removal unit and a 4.5 acre anaerobic lagoon with an average depth of nine feet. Discharge from the anaerobic lagoon is monitored by the City for flow, BOD_5 and suspended solids three times per week.

Analysis of a temporal plot of several year's daily flow, BOD_5, and suspended solids data for the combined municipal/packing house waste stream indicated a distinct difference in values of these wastewater parameters between winter and summer. Statistical plots of values of each parameter for the individual domestic waste stream and packing house waste stream for .the periods May through November and December through April indicated distinct seasonal differences in flow and organic and suspended solids loadings. Median values for existing flow and organic and suspended solids loadings are summarized in Table II.

Table I. Effluent Standards (Union City, Tennessee)

BOD_5	30 mg/L
Suspended Solids	30 mg/L
Ammonia Nitrogen	15 mg/L
Dissolved Oxygen	1.0 mg/L

Table II. Existing Median Seasonal Wastewater Characteristics (Union City, Tennessee)

	Winter (December–April)		Summer (May–November)	
	Municipal	Packing House	Municipal	Packing House
BOD$_5$ (Lb/Day)	1925	770	1145	300
Suspended Solids (Lb/Day)	4100 (Combined)		2650 (Combined)	
Flow (Mgd)	3.05	0.69	1.71	0.69

The increased municipal flow during winter months is indicative of the high groundwater infiltration experienced by most municipal sewer systems in the region. The increased packing house BOD$_5$ loading during winter months can be attributed to the decreased efficiency of the anaerobic lagoon due to colder temperatures. The increase in municipal BOD$_5$ loading is unexplained.

Design flows and waste loadings were determined using population projections and industrial production projections and are listed in Table III.

PROCESS DESCRIPTION

The continuously-fed, intermittently-operated activated sludge system selected for this project was an ICEAS$_{tm}$ process. The ICEAS$_{tm}$ process was developed in Australia and can operate in single or multiple tank configurations.

The ICEAS$_{tm}$ process is one of four different types of periodic processes currently identified in the literature[1]. A single reactor vessel is used for aeration and sedimentation. Flow into the reactor is continuous and aeration, sedimentation and decantation of supernatant all take place within time oriented cycles. Because the ICEAS$_{tm}$ process allows flow into the reactor vessel during sedimentation, and decantation of supernatant, a transverse baffle must be used to insure that disturbance of the sludge blanket is kept to a minimum and short circuiting of untreated influent does not occur during decantation. The transverse baffle extends from tank bottom to above top water level. Openings for wastewater flow are located along the tank bottom.

The volume isolated at the inlet of the reactor vessel by this baffle wall is referred to as the pre-react zone or "selector." In addition to the hydraulic benefits provided by the transverse baffle wall, the concept of "floc-loading" or instantaneous food to mass ratio (F/M)$_i$ is nurtured in this isolated inlet volume. During sedimentation and decantation, the higher strength influent is contacted with the sludge blanket at the bottom of the baffle wall. This type of contacting has been shown by several investigators to promote the growth of a non-bulking sludge[2] and dampen the variability of diurnal specific oxygen uptake rates[3].

Initial conversations with the state wastewater regulatory agency, the Tennessee Department of Health and Environment, indicated that an intermittent discharge might require an effluent equalization basin prior to discharge to the receiving stream. In order to eliminate the need for an effluent equalization basin, a four tank configuration was devised with cycle times for aeration, sedimentation, and decantation selected to provide a near constant discharge from the process.

The basic cycle selected was three hours consisting of 90 minutes for aeration, 45 minutes for settling and 45 minutes for decanting. The basic or "normal" operating cycles for the four tank

Table III. Design Flows and Waste Loads (Union City, Tennessee)

Parameter	Value
Summer Base Flow	4.03 MGD
Winter Base Flow	6.03 MGD
Seasonal Sustained Flow	9.03 MGD
Instantaneous Peak Flow	15.7 MGD
Summer BOD$_5$ Loading	4,600 LBS/DAY
Winter BOD$_5$ Loading	6,200 LBS/DAY
Summer TKN Loading	2,353 LBS/DAY
Winter TKN Loading	2,000 LBS/DAY
Summer Suspended Solids Loading	6,672 LBS/DAY
Winter Suspended Solids Loading	10,008 LBS/DAY

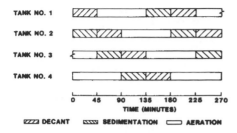

Figure 1. **Basic operating cycles for four tank ICEAS process.**

configuration are graphically depicted in Figure 1. A second cycle designated as a "storm" cycle to hydraulically accommodate the instantaneous peak design flow is required. The "storm" cycle selected was two hours consisting of 60 minutes for aeration, 30 minutes for settling, and 30 minutes for decanting.

The cycles depicted in Figure 1 indicate that a constant hydraulic discharge can be achieved by this tank and cycle configuration.

In practice, continuous discharge is not achieved if a constant rate decant device is used since diurnal variations in influent flow rate do not allow the liquid level in the tanks to reach maximum design decant liquid level on each cycle.

Because aeration in the basins is intermittent, the aeration and mixing equipment selected had to be capable of operating intermittently without clogging. At the time a design decision was required, most full scale experience with intermittently-operated aeration and mixing systems in the United States, Australia, and Japan was with the jet aeration system. Jet aeration systems involve the mixing of air and recirculated mixed liquor through nearly congruent nozzle discharges. Only the amount of air required to supply oxygen for biochemical reactions is supplied, and the remaining energy required for mixing is supplied by the recirculated mixed liquor. Since the design decision to use jet aeration was made, several full-scale intermittent activated sludge systems have become operational in the United States and Canada using coarse bubble diffused aeration systems, reportedly without major problems[4].

The experience base for selection of a decant mechanism was quite limited at the time a design decision was required. The gutter type weir decanter, positively actuated in a vertical direction by a hydraulic cylinder was the only successfully tested decant mechanism located during research for design purposes.The gutter type decanter consists of a stiffened gutter fabricated from a pipe section or from heavy gauge sheet metal in sections of up to 15 to 20 feet in length. The gutter is drained through its bottom at several points by attached manifold pipes which, in turn, are attached and drain into a header pipe. The header pipe serves as the rotating pivot point from which the gutter moves downward in an arc during decant. As the gutter is lowered into the liquid, only the front lip is allowed to become submerged and acts as a sharp crested weir. A hinged baffle is mounted in front of the gutter, which pushes scum away from the gutter's point of entry as it is lowered and prevents floatables from entering the gutter during decant. The header pipe is connected to outlet structures on one or both ends by a mechanical seal which allows for rotation.

REACTOR SIZING

Reactor sizing was determined based on F/M values, sludge settling characteristics, decanter capacities, and hydraulic requirements.

The required reactor volume at bottom water level was determined using the following equation:

$$V = \frac{(So)\ (Q)}{(F/M)\ (X)} \tag{1}$$

where:

F/M = (lb BOD_5)/(day) ÷ lb MLSS (day^{-1}) on a 24 hour basis. If the effective F/M ratio is selected based on aeration time, this value must be multiplied by the ratio (hours of aeration per day)/(24 hours per day) to determine F/M for this equation.

S_o = Influent BOD_5 concentration (mg/L)

Q = Influent Flow Rate (mgd)

X = Concentration of MLSS at Bottom Water Level (mg/L)

V = Reactor Volume (million gallons)

F/M as defined in Equation 1 equates to F/M_i (instantaneous F/M) as defined in the literature[3] if F in the F/M_i expression is the average daily BOD_5 concentration and M in the F/M_i expression is the average daily MLSS concentration.

Without pilot plant data, values for both F/M and X were assumed using the limited information from the literature[3,5]. Values of F/M_i (equivalent to F/M in Equation 1) reported to produce acceptable effluents varied from 0.04[5] to 0.5[3] in conjunction with values of X ranging from 5,000 mg/L[5] to 2,500 mg/L[3].

For design purposes, a conservative value for F/M of 0.1 (0.2 based on aeration time) was selected. Because of seasonal organic loading criteria, varying values for X were selected to produce the same reactor volume for both winter and summer organic loadings. A value of X for winter design conditions of 4,100 mg/L was selected. Reactor volume was calculated using Equation 1 as follows:

$$\text{Bottom Water Level Volume} = \frac{(123 \text{ mg/L } BOD_5) (6,030,000 \text{ gal/day})}{(0.1 \text{ day}^{-1}) (4,100 \text{ mg/L})}$$

$$\text{Bottom Water Level Volume} = 1,809,000 \text{ gallons or } 241,845 \text{ ft}^3$$

The summer value for X was calculated by rearranging Equation 1 as follows:

$$\text{Summer Bottom Water Level MLSS} = \frac{(137 \text{ mg/L } BOD_5) (4,030,000 \text{ gal/day})}{(0.1 \text{ day}^{-1}) (1,809,000 \text{ gallons})}$$

$$\text{Summer Bottom Water Level MLSS} = 3,052 \text{ mg/L, say } 3,050 \text{ mg/L}$$

This summer value for X is within the range of successful experience reported. in the literature[3,5].

The minimum bottom water level (BWL) depth was calculated based on assumed sludge settling characteristics. For determination of minimum BWL depth, a maximum MLSS value of 5,000 mg/L at BWL was assumed. Literature values for Sludge Volume Index (SVI) for pilot and full scale facilities with similar MLSS and cyclic operating parameters ranged from 44 ml/g to 150 ml/g[5]. A conservative value of 180 ml/g at 5,000 mg/L MLSS was used for design purposes which equates to 2.4 ft^3/lb.

The volume required by a settled sludge with a SVI of 180 ml/g and a bottom water level MLSS value of 5,000 mg/L may be calculated as follows:

$$\text{Settled Volume} = (5,000 \text{ mg/L}) (1.809 \text{ mg}) (8.34) \frac{2.4 \text{ ft}^3}{\text{lb}}$$

$$\text{Settled Volume} = 181,045 \text{ ft}^3$$

The percentage of the total bottom water level volume occupied by settled sludge under the assumed conditions is calculated as follows:

$$\text{Percent Volume Occupied By Settled Sludge} = \frac{181,045 \text{ ft}^3 \times 100}{241,845 \text{ ft}^3}$$

$$= 74.86\%, \text{ Say } 75\%$$

Assuming a two foot separation between bottom water level and the top of the sludge blanket under the assumed design conditions, the minimum bottom water level depth may be calculated as follows:

Table IV. Maximum Active Depths For Various Flow Conditions

Flow Condition	Cycle Orientation	Max. Active Depth (ft)
Summer Base	Normal	1.67
Winter Base	Normal	2.50
Seasonal Peak	Normal	3.74
Instantaneous Peak	Storm	4.34

$$\text{Minimum BWL Depth} = \frac{2 \text{ Feet}}{1 - 0.75}$$

$$\text{Minimum Bottom Water Level Depth} = 8 \text{ Feet}$$

Reactor surface area required may now be calculated as follows:

$$\text{Area Required Per Tank} = \frac{241,845 \text{ ft}^3 / 4}{8 \text{ ft}}$$

$$= 7,556 \text{ ft}^2$$

The active depth from bottom water level to top water level (TWL) for any flow condition and any cycle orientation can be calculated:

$$\text{Max. Depth} = \frac{(Q)(a + s)}{A} + BWL \, \text{- - -} \tag{2}$$

where:

$$Q = \text{Influent Flow to a Tank (cfm)}$$
$$a = \text{Aeration Time (minutes)}$$
$$s = \text{Settling Time (minutes)}$$
$$A = \text{Surface Area of Tank (ft}^2)$$
$$BWL = \text{Bottom Water Level (ft)}$$

The maximum depth for all design flow criteria as determined by Equation 2 are listed in Table IV.

The maximum active depth of 4.34 feet will occur during "storm" cycle operation at instantaneous peak flow rates. For design purposes, this is rounded up to 5.0 feet; therefore, the vertical weir travel during decant will be based on 5.0 feet per 45 minutes for "normal" cycle and 5.0 feet per 30 minutes for "storm" cycle.

The width of each tank is dependent upon the length of decanter weir required at peak flow rates during "normal" cycle and during "storm" cycle. The recommended capacity of the decant mechanism selected is approximately 18 ft³/ft³/Min.

The maximum decant rate required under "normal" cycle operation will occur during the sustained seasonal design flow rate of 9.03 mgd. The length of decanter weir required per tank under "normal" cycle conditions may be calculated as follows:

$$\text{Max. Decanter Length for "Normal" Cycle} = \left[\frac{210 \text{ ft}^3}{\text{min.}} + \frac{(5.0 \text{ ft})(7,556 \text{ ft}^2)}{45 \text{ min.}} \right] \div \frac{18 \text{ ft}^3}{\text{ft.min.}}$$

$$= 58.31 \text{ ft., Say 60 ft.}$$

The maximum decant rate under "storm" cycle operation must be checked and may be calculated as follows:

Table V. Summary of Reactor Sizing (Union City, Tennessee)

Number of Tanks	:	4
BWL Volume of Each Tank	:	60,552 ft³
BWL Depth	:	8.0 ft
TWL Depth	:	13.0 ft
Freeboard	:	2.0 ft

$$\text{Max. Decant Rate for "Storm" Cycle} = \left[\frac{364 \text{ ft}^3}{\text{min}} + \frac{(5.0 \text{ ft})(7,556 \text{ ft}^2)}{30 \text{ min}}\right] \div 60 \text{ ft}$$

$$= 27.05 \text{ ft}^3/\text{ft.min}$$

This value is within the manufacturer's allowable peak flow limits.

The minimum width required to install 60 feet of the selected decanter is approximately 80 feet. A square tank providing a surface are# of 7,556 ft² has dimensions of 86.93 feet by 86.93 feet.

For design purposes, a square tank 87.0 feet by 87.0 feet was selected. A longitudinal baffle extending from the floor of the tank to above top water level was used to divide each square tank into two effective reactor areas with length to width ratios of 2:1.

A summary of the reactor sizing is presented in Table V.

TANK DETAILS

Each of the four identical reactor tanks is baffled as shown in Figure 2 to inhibit hydraulic short circuiting and to promote biosorption during settling and decanting periods.

OXYGEN REQUIREMENTS

Oxygen requirements for carbonaceous BOD_5 oxidation and nitrification were calculated using a method after Eckenfelder[6]. Oxygen requirements used for design are listed in Table VI.

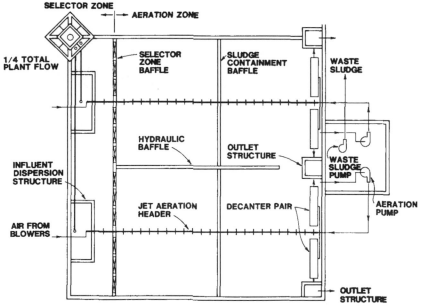

Figure 2.   Tank details.

Table VI. Design Oxygen Requirements

	Winter Conditions	Summer Conditions
Carbonaceous BOD$_5$ Oxidation	4,633	4,888
Nitrification	7,162	8,703
Denitrification (Credit)	(3,184)	(4,975)
Total Oxygen Requirements	8,611	8,616

Table VII. Comparison of Certain Design Parameters for Union City With Extended Aeration and Conventional Activated Sludge Processes

Design Parameter	Union City Criteria	Extended Aeration Criteria [7]	Conventional Activated Sludge Criteria [7]
F/M (day^{-1})	0.1	0.05 to 0.15	0.2 to 0.4
Volumetric Organic Loading (lb BOD$_5$ Applied/ft^3 Reactor/day)	0.019 (Summer) 0.026 (Winter)	0.1 to 0.4	0.3 to 0.6
MLSS (mg/L)	3050 (Summer) 4100 (Winter)	3000 to 6000	1500 to 3000
Hydraulic Detention (hrs.)	13.04 (Summer) 9.46 (Winter)	18 to 36	4 to 8
Clarifier Solids Loadings (lb MLSS/ft^2/hr)	3.33 (Quiescent)	0.2 to 1.0	0.6 to 1.2
Solids Retention Time (days)	11 (Summer) 7.8 (Winter)	20 to 30	5 to 15

Two centrifugal blowers, operating continuously drawing 76 bhp each at standard conditions, are required to meet design oxygen requirements for the total plant.

MIXING REQUIREMENTS

The mixing requirements per tank, in excess of the air flow, were determined by the jet aeration system supplier as 11,700 gpm of mixed liquor recirculation. Two mixed liquor recirculation pumps, both operating for 90 minutes per "normal" cycle drawing 39 bhp and pumping 5850 gpm, are required to meet mixing requirements in each tank. This operational arrangement is equivalent to four pumps operating continuously for the total plant.

IMPACT ON PLANT HYDRAULICS AND OTHER UNIT PROCESSES

The flow characteristics of intermittently operated processes must be given consideration in sizing plant outfall lines. Maximum decant rates must be calculated and used for sizing outfall lines.

If jet aeration systems are used for aeration and mixing, special care must be taken in sizing influent wastewater bar screens and in designing recirculation pump suction inlets.

If centrifugal blowers are used to supply oxygen, special consideration must be given to selecting a blower with appropriate head-discharge characteristics to accommodate variable depths during aeration.

DESIGN SUMMARY

Certain design criteria are presented in Table VII to compare the design of the Union City continuously-fed, intermittently-oeprated actived sludge system to the conventional extended aeration activated sludge system.

REFERENCES

1. Irvine, R. L., et al., "An Organic Loading Study of Full-Scale Sequencing Batch Reactors," *Journal Water Pollution Control Federation*, Vol. 57, No. 8, p. 847 (1985).
2. Goronszy, M. C., et al., "Sludge Bulking Control for Highly Degradable Wastewaters Using the Cyclic Activated Sludge System," *Proceedings of 40th Annual Purdue Industrial Waste Conference* (1985).
3. Goronszy, M. C., et al., "Intermittent Biological Waste Treatment Systems — Process Considerations," *Water — AIChE Symposium Series 209*, Vol. 77, p. 129 (1981).
4. Arora, M. L., et al., "Technology Evaluation of Sequencing Batch Reactors," *Journal Water Pollution Control Federation*, Vol. 57, No. 8, p. 867 (1985).
5. Goronszy, M. C., "Intermittent Operation of the Extended Aeration Process for Small Systems," *Journal Water Pollution Control Federation*, Vol. 51, No. 2, p. 274 (1979).
6. Eckenfelder, W. W., *Water Quality Engineering for Practicing Engineers*, Barnes and Noble, New York, N.Y. (1970).
7. Metcalf & Eddy, Inc., *Wastewater Engineering: Treatment, Disposal, Reuse*, 2nd Edition, McGraw — Hill Book Company, New York, N.Y. (1979).

14 BIOLOGICAL TREATABILITY OF INDUSTRIAL WASTEWATER AND WASTE MACHINE TOOL COOLANTS AT JOHN DEERE DUBUQUE WORKS

Loren Polak, Advanced Engineering Analyst
Environmental Control
Plant Engineering Department
John Deere Dubuque Works
Dubuque, Iowa 52001

INTRODUCTION

John Deere Dubuque Works is a major industrial equipment manufacturing facility that produces a variety of earthmoving heavy construction equipment. In contrast to most manufacturing facilities, which discharge into municipal wastewater treatment systems, the wastewater from John Deere Dubuque Works is treated and discharged directly into a receiving waterway, the Mississippi River.

As a result of the 1972 Federal Water Pollution Control Act Amendments, John Deere Dubuque Works was required to meet more stringent wastewater effluent limitations in 1977, and construction of a new treatment plant was completed that year to meet the new standards. The 1977 treatment plant includes physical/chemical processes (Figure. 1) to treat pollutants such as solids, oils, heavy metals, acids, and alkali. However, it does not include a treatment process to remove soluble organic pollutants. In recent years, as a result of changes in manufacturing process chemicals, the inability to remove soluble organic pollutants has caused the treatment plant to intermittently experience difficulty meeting its biochemical oxygen demand (BOD_5) effluent limitation. Because the treatment plant is a direct discharge facility, the Iowa Department of Water, Air and Waste Management has required that the plant meet a BOD_5 standard of 30 mg/L (30-day average) and 45 mg/L (daily maximum).

BACKGROUND

Investigations were conducted in the early 1980's to determine the BOD_5 contribution from each wastewater category in the hope that the problem could be eliminated through wastewater source management. Where feasible, efforts were made to reduce or eliminate the BOD_5 inputs from the major contributors. The efforts included: 1) monitoring some of the wastewater sources to ensure that BOD_5 producing chemicals were not being wasted unnecessarily; 2) imposing some constraints on the types of process chemicals that could be used by selectively preventing the use of high BOD_5 chemistries; and 3) shipping one waste category (spent machine tool coolant) to an off-site treatment facility, because it represented an extremely high BOD_5 input.

An investigation was also conducted into the use of chemical agents to oxidize the contaminants generating the BOD_5. In fact, one of the oxidizing agents (hydrogen peroxide) has been used on a full-scale basis.

Although the combined efforts have been somewhat effective, they have not been entirely successful in eliminating the BOD_5 problem, and they are beset by some disadvantages. For example, attempting to control BOD_5 by placing constraints on the selection of chemicals has impeded efforts to reduce the costs of some manufacturing operations. Secondly, shipping spent machine tool coolants off-site for treatment has resulted in the disadvantages of both greater environmental liabilities and higher treatment costs. Thirdly, the use of hydrogen peroxide as an oxidizing agent has proven to be only partially successful, and it too has a cost disadvantage because of the ongoing need to purchase the chemical.

It was decided that another solution was needed and that biological treatment should be investigated because it seemed to offer the most economical means of controlling the BOD_5 problem and it offered several other advantages. First, with biological treatment it should be possible to eliminate the constraints on high BOD_5 chemicals. Secondly, it should be possible to reestablish the on-site chemical treatment of spent coolants. Thirdly, it should not be necessary to make the ongoing purchases of

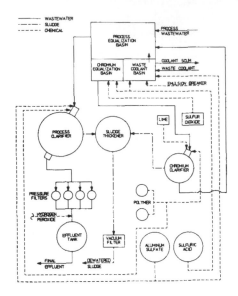

Figure 1. Schematic of industrial wastewater treatment plant.

hydrogen peroxide. Last of all, it should provide a means to consistently meet the BOD$_5$ effluent standard.

To determine the feasibility of biologically treating the industrial wastewater and spent machine tool coolants, a biological treatment study was conducted. It had two major objectives: 1) to determine whether the industrial wastewater and spent machine tool coolants could be successfully treated in a biological system; and 2) if biological treatment was possible, to determine process parameters for the design and operation of a full-scale system. Although other biological treatment options were considered, the activated sludge (aerobic suspended-growth) process was selected for the study. Biological oxidation of chemically flocculated wastewater using the activated sludge process has been shown to be an effective treatment method[1].1 It is also the most familiar to personnel at John Deere Dubuque Works because an existing on-site activated sludge treatment plant is used to treat domestic wastewater.

EXPERIMENTAL METHOD

Bench-scale reactors are commonly used to assess the treatability of wastewater by determining treatment efficiencies and process coefficients under controlled conditions[2,3]. For this study, a small continuous flow bench-scale reactor (Figure 2) was used to simulate the activated sludge process. The reactor .was a 12.5 liter (3.3 gal) plexiglass tank with baffles to separate the aeration chamber from the clarifier and to provide sludge recirculation. The study was conducted for approximately eight months in 1984.

To begin the operation of the reactor, it was filled with mixed liquor from the on-site domestic wastewater activated sludge treatment plant. Oxygen was supplied by conducting compressed air through a porous stone diffuser to generate fine bubbles. The dissolved oxygen concentration in the aeration chamber was determined daily and was controlled at 2.0–4.0 mg/L by adjusting the air flow with a needle valve on the air line. After trying different mixers and mixing techniques, an adjustable magnetic stirring apparatus was chosen. It was placed under the aeration chamber with a large mixer "pill" inside the aeration chamber. This arrangement provided adequate mixing without generating excessive turbulence in the clarifier section of the reactor. Influent was fed to the continuous flow reactor with a variable speed peristaltic pump.

During the early portion of the study, the influent to the reactor was the effluent from the final clarifier of the existing physical/chemical treatment plant. The clarifier effluent was pumped directly to the reactor at selected feed rates.

During the remaining portion of the study, three different spent coolants were investigated. The coolants, representing the types used at the John Deere Dubuque Works, included two synthetic

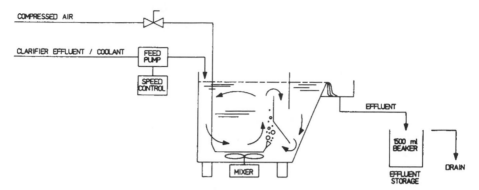

Figure 2. Schematic of continuous-flows, bench-scale reactor.

alkanolamine based coolants, designated as Coolant A and Coolant B. The third coolant, designated as Coolant C, was an emulsified oil coolant named Deersol 20 which is formulated at John Deere Dubuque Works. The BOD_5 and COD values of the spent coolants at their process concentrations are shown in Table I. Each type of spent coolant was individually studied for several weeks.

To prepare a spent coolant for treatment in the reactor, a representative sample of the coolant was first collected from process coolant systems in the factory. In previous coolant investigations it was discovered that coolants obtained from factory process systems had higher values for BOD_5 and COD than did laboratory prepared coolant mixtures of.the same concentrations. Since it was the objective of the study to simulate the actual discharge of spent coolants, the representative coolant samples were obtained from factory centralized machine tool coolant systems. Each coolant was chemically pretreated in similar fashion.

The pretreatment procedure involved first lowering the pH of the spent coolant to approximately 7.0–7.5 with sulfuric acid and adding a cationic emulsion breaker polymer at a predetermined dosage while the contents were being vigorously mixed. Next, a predetermined dosage of aluminum sulfate (alum) was added while the contents continued to be mixed. After a few minutes, the mixing was slowed and stopped to allow the separation of the flocculated phase from the water phase. The pretreatment of the coolant is intended to remove emulsified oil.

Since Coolants A and B are synthetic coolants, the pretreatment procedure had little, if any, impact on their BOD_5 and COD concentrations. Synthetic coolants typically do not contain emulsified oil[4]. However, Coolant C is an oil emulsified coolant, so it was greatly affected by the pretreatment process. Its BOD and COD concentrations were reduced approximately 90–95% by the pretreatment process.

The water phase from the pretreated coolant was then diluted with raw industrial wastewater at specific mixture ratios to simulate the discharge of pretreated spent coolant into the industrial waste-water. The mixtures were then treated with alum and anionic polymer in the same manner that industrial wastewater is normally treated in the physical/chemical treatment plant. The treated mixtures were daily prepared and placed into a 100 liter storage tank. The contents were pumped from the tank to the reactor at a uniform rate throughout the coolant investigation portion of the study. Three different concentrations for each type of spent coolant were fed to the reactor.

Following the development of a suitable biota culture in the reactor, each type of waste was individually treated to determine the systems capabilities, tolerances, and optimum modes of operation. Influent and effluent BOD_5, COD, and suspended solids concentrations were determined twice per week. Influent and effluent ammonia (NH_3-N), nitrate (NO_3-N), and orthophosphate (PO_4)

Table I. Characteristics of Coolant Categories at Process Concentrations

Coolant	Type	Concentration	BOD_5 (mg/L)	COD (mg/L)
A	Synthetic Alkanolamine	5%	16,500	50,000
B	Synthetic Alkanolamine	5%	20,400	58,000
C	Emulsified Oil	5%	12,000	105,000

Table II. Operating Parameters and Performance of Treating the Existing Treatment Plant Clarifier Effluent

Hydraulic Detention (Hrs)		24	16	8
BOD$_5$ (mg/L)	Influent	32	30	35
	Effluent	< 4	< 4	< 4
COD (mg/L)	Influent	300	350	280
	Effluent	100	100	90
NH$_3$-N (mg/L)	Influent	0.73	0.62	0.75
	Effluent	0.10	0.21	0.25
MLSS (mg/L)		2300	2400	2500
MLVSS (mg/L)		1800	1800	1900
MCRT (Days)		59	39	21
F:M (g BOD$_5$/g MLVSS/d)		0.024	0.033	0.073
F:M (g COD/g MLVSS/d)		0.22	0.38	0.58
U (g COD/g MLVSS/d)		0.15	0.270	0.40
BOD$_5$/COD (Influent)		0.11	0.086	0.13
SVI (ml/g)		100–180	70–90	80–90
Y net (g MLVSS/g COD)		0.094	0.11	0.12
pH (aeration chamber)		7.5–8.4	7.5–8.2	7.6–8.6
OUR (mg O$_2$/g MLVSS/h)		1.8	2.8	3.6

concentrations were determined weekly to verify the satisfactory availability of nutrients and the extent of nitrification. Influent flow rate, aeration chamber pH, and aeration chamber dissolved oxygen (D.O.) were determined daily. Mixed liquor suspended solids (MLSS), mixed liquor volatile suspended solids (MLVSS), and sludge settleability rates were determined twice per week. Mixed liquor oxygen uptake rates were also determined once per week.

RESULTS AND DISCUSSION

The bench-scale reactor was operated at three organic loading rates for each of the four waste categories. During the portion of the study when the influent to the reactor was strictly the existing treatment plant's final clarifier effluent, the range of organic loading rates was selected on the basis of flow rate. This is reflected in the three different hydraulic detention times in Table II. During the other portions of the study, when spent pretreated coolants were being investigated, ranges of organic loading rates were selected on the basis of spent coolant substrate concentrations.

Treatment of Clarifier Effluent

The results presented in Table II indicate that the effluent from the existing treatment plant can be successfully treated in an extended aeration mode of an activated sludge wastewater treatment system. The BOD$_5$ was almost totally removed, averaging < 4.0 mg/L, and COD was substantially reduced.Of course, the influent BOD$_5$ was very low compared to typical domestic wastewater influent, but it did average slightly above the allowable 30 mg/L BOD$_5$ treatment plant effluent standard. Furthermore, there were individual days when the influent BOD$_5$ concentration was as high as 130 mg/L and the effluent BOD$_5$ did not exceed 9.0 mg/L.Ammonia removal was also evident, although the influent ammonia concentrations were so low that little nitrification potential existed.

Obviously, the range of feed rates and MCRT's selected for this portion of the study were far more conservative than necessary to meet the BOD$_5$ effluent limitation. But the real objective of this part of the study was to determine whether the existing treatment plant's effluent could be properly treated in an extended aeration activated sludge system. Only after that capability was established was it feasible to add the influence of spent coolants to the reactor.

Other parameters shown in Table II include the food to microorganism (F:M) ratios for COD and BOD$_5$. It is typical for F:M ratios to be low in extended aeration types of activated sludge systems[5]; however, the F:M ratios based on BOD$_5$ were even lower than the 0.05–0.15 range that is considered typical for treating domestic wastewater[2,6].

The low specific utilization rates (COD basis) relative to the F:M ratios indicate the inefficiency of the bacteria in removing the substances that generate the COD. Those COD generating substances

Table III. Operating Parameters and Performance of Treating Coolant A

Pretreated Coolant (mg/L)		500	1000	1500
BOD$_5$ (mg/L)	Influent	190	360	540
	Effluent	5	18	30
COD (mg/L)	Influent	580	1100	1600
	Effluent	130	250	350
NH$_3$-N (mg/L)	Influent	2.1	19	100
	Effluent	4.8	2.2	21
MLSS (mg/L)		5600	5700	6200
MLVSS (mg/L)		4500	4600	5000
MCRT (Days)		44	32	18
F:M (g BOD$_5$/g MLVSS/d)		0.08	0.15	0.21
F:M (g COD/g MLVSS/d)		0.25	0.47	0.63
U (g COD/g MLVSS/d)		0.20	0.36	0.49
BOD$_5$/COD (Influent)		0.33	0.33	0.34
SVI (ml/g)		80–90	60–90	60–80
Y net (g MLVSS/g COD)		0.08	0.13	0.21
pH (aeration chamber)		7.4–8.2	7.1–8.5	6.8–8.2
OUR (mg O$_2$/g MLVSS/h)		5.2	6.8	8.5

represent the chemically oxidizable leftovers from the physical/chemical treatment plant. The bacterial biodegradation difficulty is also reflected in the low influent BOD$_5$/COD ratios (0.086–0.13). Typical domestic wastewater has a BOD$_5$/COD ratio of approximately 0.4–0.8.

The sludge volume index (SVI) results (Table II) demonstrate quite acceptable settleability and compaction. Net yield (Y) of MLVSS per unit removal of COD was low, which reflects the long MCRT's of extended aeration and the difficulty the bacteria had assimilating the COD generating substrate.

The pH in the aeration chamber was within bacterial tolerances for maintaining proper growth. Oxygen uptake rate (OUR) was also presented in Table II to allow comparisons of relative bacterial activity[7].

Treatment of Coolant A

The portion of the study devoted to the biological treatment of spent pretreated coolants had the major objective of determining what concentrations of pretreated coolants could be added to the existing industrial wastewater and be able to obtain an effluent that was well below the 30 mg/L (30-day average) BOD$_5$ limit. Using the term "pretreated" in reference to Coolants A and B is somewhat misleading. They are alkanolamine based synthetic coolants. They do not contain emulsified oil and they reject tramp oil. Therefore, the emulsion splitting pretreatment step served no real purpose in the study other than to maintain consistency for comparative purposes with Coolant C[8].

The results in Table III indicate that spent Coolant A can be added to the industrial wastewater up to a concentration of at least 1000 mg/L and be assured of an effluent BOD$_5$ concentration well below 30 mg/L. It is impressive how drastically BOD$_5$ and COD increased when spent Coolant A was added to the industrial wastewater. Without any coolant added, the industrial wastewater had a BOD$_5$ concentration averaging only 30–35 mg/L and a COD average of 300 mg/L (Table II). But with only 1000 mg/L of spent Coolant A added, the BOD$_5$ increased to an average of 360 mg/L and COD increased to an average of 1100 mg/L. It becomes obvious why a relatively small input of Coolant A into the treatment plant can have a very adverse impact on the plant's effluent BOD$_5$ concentration, especially since the treatment plant has an average production day flow of only 150,000 GPD.

The addition of spent Coolant A to the industrial wastewater also greatly increased the concentration of ammonia (Table II versus Table III). That was expected because Coolant A is an alkanolamine based coolant and the amine can be converted to ammonia. Although there was no evidence of nitrification at the 500 mg/L Coolant A concentration, nitrification did occur at the 1000 mg/L and 1500 mg/L Coolant A concentrations. It is not obvious why nitrification did not occur at the 500 mg/L concentration, nor why it did not occur more completely at the 1500 mg/L concentration.

Table IV. Operating Parameters and Performance of Treating Coolant B

Pretreated Coolant (mg/L)		500	1000	1500
BOD$_5$ (mg/L)	Influent	270	490	610
	Effluent	16	31	34
COD (mg/L)	Influent	760	1300	1500
	Effluent	150	260	240
NH$_3$-N (mg/L)	Influent	13	13	39
	Effluent	1.7	6.6	15
MLSS (mg/L)		5500	6200	5600
MLVSS (mg/L)		4400	5000	4500
MCRT (Days)		49	26	17
F:M (g BOD$_5$/g MLVSS/d)		0.12	0.19	0.27
F:M (g COD/g MLVSS/d)		0.34	0.51	0.66
U (g COD/g MLVSS/d)		0.27	0.41	0.55
BOD$_5$/COD (Influent)		0.36	0.38	0.41
SVI (ml/g)		70–80	50–60	50–70
Y net (g MLVSS/g COD)		0.089	0.14	0.22
pH (aeration chamber)		6.8–8.3	7.3–8.2	7.9–8.3
OUR (mg O$_2$/g MLVSS/h)		7.4	9.3	9.8

Although the existing industrial wastewater treatment plant does not have an ammonia effluent limitation, the concentration of ammonia in the effluent does cause some concern. If the effluent ammonia concentration is appreciable, the presence of nitrifying bacteria can be reflected as higher BOD$_5$ in the effluent BOD$_5$ test. A nitrification inhibitor has not been authorized by the regulatory agency in testing the treatment plant's effluent BOD$_5$ concentration.

The mixed liquor suspended solids (MLSS) concentration range of 5600–6200 mg/L (Table III) is at the high end of what is considered typical for extended aeration (3000–6000 mg/L), and the range of MCRT's selected for the treatment of Coolant A expands beyond the typical 20–30 day range for extended aeration[2]. However, the high concentration of bacteria and long residence times appeared to be necessary to maintain low effluent BOD$_5$ in treating spent Coolant A.

The F:M ratios based on BOD$_5$ (Table III) are within the typical extended aeration range of 0.05–0.15 for treating domestic wastewater, but the F:M ratios based on COD are higher than the more typical range of 0.08–0.25 for the extended aeration treatment of domestic wastewater[2]. If COD can be assumed to approximate ultimate BOD, it follows that the bacteria received relatively high amounts of "food" by extended aeration standards, but the organic substrate was not as easily removed or assimilated as domestic wastewater organic pollutants are. This is supported by the relatively low BOD$_5$/COD ratios of the influent which were approximately 0.33 while typical domestic wastewater, as mentioned earlier, has a BOD$_5$/COD ratio of approximately 0.4–0.8. However, the 0.33 ratio is far higher than the 0.086–0.13 BOD$_5$/COD ratios of the treatment plant's clarifier effluent (Table II), which indicates that Coolant A substrate is more easily biodegraded than the leftovers from the physical/chemical treatment plant. The values for Y net and OUR display a similar pattern of being higher for treating Coolant A than for the treatment of clarifier effluent. The SVI values indicated good sludge compaction and settling characteristics, and the pH ranges were quite acceptable for maintaining proper bacterial growth.

Treatment of Coolant B

Coolant B, like Coolant A, is an alkanolamine based synthetic coolant, although it differs slightly in composition. The results (Table IV) from the treatment of Coolant B show, on the average, that pretreated Coolant B generated higher influent BOD$_5$ and COD values than Coolant A at the same substrate concentrations (Table III). The overall Coolant B effluent BOD$_5$ and COD values also exhibited higher values than for Coolant A. The results (Table IV) further indicate that spent Coolant B can be added to the industrial wastewater up to a concentration of at least 500 mg/L and be assured of an effluent BOD$_5$ concentration well below 30 mg/L. This compares to 1000 mg/L of spent Coolant A that could be added and be assured of adequate BOD$_5$ reduction (Table III).

Treated Coolant B effluent ammonia concentrations (Table IV) averaged in the same general range as the ammonia concentrations following treatment of Coolant A (Table III). Again, there was some

Table V. Operating Parameters and Performance of Treating Coolant C

Pretreated Coolant (mg/L)		5000	7000	9000
BOD$_5$ (mg/L)	Influent	80	110	140
	Effluent	8	14	22
COD (mg/L)	Influent	560	800	1100
	Effluent	160	240	330
NH$_3$–N (mg/L)	Influent	8.6	9.5	11.1
	Effluent	0.4	0.8	1.1
MLSS (mg/L)		4900	4800	4700
MLVSS (mg/L)		3900	3900	3800
MCRT (Days)		40	31	21
F:M (g BOD$_5$/g MLVSS/d)		0.040	0.056	0.073
F:M (g COD/g MLVSS/d)		0.28	0.40	0.57
U (g COD/g MLVSS/d)		0.20	0.28	0.40
BOD$_5$/COD (Influent)		0.14	0.14	0.13
SVI (ml/g)		100–110	90–100	90–110
Y net (g MLVSS/g COD)		0.12	0.14	0.16
pH (aeration chamber)		7.8–8.4	7.6–8.2	7.6–8.1
OUR (mg O$_5$/g MLVSS/h)		2.2	3.0	3.9

evidence of nitrification, but it did not occur to completion as evidenced by the remaining ammonia in the effluent (Table IV).

The other operating and performance values for the treatment of Coolant B followed patterns similar to the values for the treatment of Coolant A. The general differences between the two coolants were: 1) the influent and effluent BOD$_5$ and COD values were higher for Coolant B than Coolant A, which was mentioned previously; and 2) Coolant B demonstrated traits that indicated it was more readily biodegradable than Coolant A.

The biodegradability difference was demonstrated by the U values, which were closer to the F:M ratios (COD basis) for the treatment of Coolant B (Table IV) than they were for the treatment of Coolant A (Table III). That suggests the bacteria were more efficient in removing the Coolant B substrate than they were in removing the Coolant A substrate. Further supporting the theory that Coolant B is more easily biodegraded than Coolant A are the BOD$_5$/COD ratios, which were higher for Coolant B influent than for Coolant A influent at the same concentrations. The Y net and OUR values were also slightly higher for Coolant B than for Coolant # which indicates that spent Coolant B is more easily assimilated by bacteria.

Treatment of Coolant C

Unlike Coolants A and B, Coolant C is an emulsified oil type of coolant. Therefore, the emulsion splitting pretreatment step did serve a real purpose by removing much of the oil, and perhaps some other organic components, in the spent coolant. Prior to the pretreatment of spent Coolant C, its COD concentration was approximately twice as high as for Coolants A and B at the same process concentrations (Table I). However, after chemical treatment of the spent coolants, the COD concentration for Coolant C at the 5000 mg/L concentration (Table V) compared closely with the COD concentration for Coolant A at the 500 mg/L concentration (Table III) and was lower than the COD for Coolant B at the 500 mg/L concentration (Table IV). That illustrates the relative impacts of pretreatment on the three spent coolants and demonstrates the substantial removal of the COD generating substrate in Coolant C.

The concentrations of treated spent Coolant C in the industrial wastewater (5000 mg/L, 7000 mg/L, and 9000 mg/L) (Table V) were selected on the basis of full-scale treatment capabilities. The three concentrations cover the upper limit within which it is practicable to chemically treat spent coolants at. the existing physical/chemical treatment plant. That limitation is based on the volume capacity of the waste coolant basin and the existing flow rate through the treatment plant.

Effluent BOD$_5$ following biological treatment in the bench-scale reactor was below 30 mg/L at the 9000 mg/L pretreated spent Coolant C substrate concentration (Table V). This indicates that an extended aeration activated sludge system could successfully treat pretreated spent Coolant C at the treatment plant's maximum rate of contribution to the industrial wastewater.

Since Coolant C does not contain the alkanolamine compounds found in Coolants A and B, the influent ammonia concentrations were not nearly as high for Coolant C (Table V) as they were for Coolants A and B (Tables III and IV) . There was evidence of nitrification as reflected by the reductions in ammonia (Table V), but the relatively low concentrations of ammonia in the influent did not offer much potential for nitrification.

The MLSS and MLVSS concentrations were held slightly lower in the reactor for the treatment of Coolant C than for the treatment of Coolants A and B. That was necessary to approximate the same range of MCRT's for the treatment of all three coolants. Nevertheless, the F:M ratios were lower for the treatment of spent Coolant C (Table V) than for the treatment of the other two coolants (Table III and IV).

The influent organic substrate, which consists of the organic leftovers following the chemical treatment of Coolant C, seemed to be relatively difficult to biologically remove. This was indicated by the U values, which were lower relative to the F:M ratios for Coolant C than they were for Coolants A and B. This means that the bacteria were less efficient in removing the organic leftovers from the chemical treatment of Coolant C than in removing the Coolant A and B organic substrates. Also, the BOD_5/COD influent ratios were lower for Coolant C than for Coolants A and B which indicates a lower biodegradability rate.

The OUR values for the treatment of Coolant C were also lower than for the treatment of Coolants A and B. That could be the result of lower bacterial assimilation of the organic substrate and/or the lower organic feed rate into the system. The SVI values demonstrated good sludge settling for the treatment of Coolant C (Table V), as was the case for the treatment of all waste categories in the study. The pH ranges were also quite acceptable throughout the study and did not have any impact on differences of biological treatability among the waste categories.

SUMMARY AND CONCLUSIONS

A bench-scale activated sludge wastewater treatment study was conducted to determine the biological treatability of the industrial wastewater and waste machine tool coolants generated at John Deere Dubuque Works. The reactor was operated in the extended aeration mode.

Although caution must be taken in using process parameter values from a bench-scale study and applying them to a full-scale system, they do offer some general guidance in the design and operation of a full-scale system.

A continuous flow activated sludge system can be used to further treat the John Deere Dubuque Works chemically treated industrial wastewater to obtain a low effluent BOD_5 concentration.

If the addition of spent synthetic coolants to the industrial wastewater is controlled, a continuous flow activated sludge system can be used to obtain a low effluent BOD_5 concentration following activated sludge treatment.

Spent pretreated emulsified oil coolant can be added to the industrial wastewater at the limit of the existing treatment plant's coolant treatment capacity and obtain a low effluent BOD_5 concentration following activated sludge treatment.

Spent synthetic coolants displayed higher BOD_5, COD, and ammonia inputs than the pretreated emulsified oil coolant.

Spent synthetic coolant substrates were more readily treatable than the organic substrates remaining in the chemically treated industrial wastewater or in the pretreated emulsified coolant.

Nitrification of ammonia occurred in the biological treatment of all waste categories, but it did not occur to completion for the higher synthetic coolant substrate concentrations.

Documentation of feed rates, operating parameter values, and environmental conditions provided general guidelines for the design and operation of an extended aeration activated sludge system for treating the waste categories.

REFERENCES

1. Rebhun, M., N. Gahlil, and N. Narkis, "Kinetic Studies of Chemical and Biological Treatment for Renovation," *Journal of the Water Pollution Control Federation*, 57 (1985).
2. Metcalf & Eddy, Inc., (Revised by George Tchobanoglous), *Wastewater Engineering: Treatment, Disposal, Reuse*, McGraw-Hill Book Co. (1979).
3. U.S. Environmental Protection Agency (Technology Transfer), *Process Design Manual for Upgrading Existing Wastewater Treatment Plants* (1974).

4. Vahle, H. R., "Synthetic Metalworking Fluids: A Closer Look," *Tooling and Production* (February 1982).
5. ASCI & WPCF, *Wastewater Treatment Plant Design*, Lancaster Press, Inc. (1977).
6. Great Lakes—Upper Mississippi River Board of State Sanitary Engineers, *Recommended Standards for Sewage Works,* Health Education Service, Inc. (1978).
7. Straub, C. P. (Editor), *CRC Critical Reviews in Environmental Control*, CRC Press, Vol. 15, Issue 2 (1985).
8. Napier, S., and K. E. Rich, "Waste Treatability of Aqueous-Based Synthetic Metalworking Fluids," *Journal of the American Society of Lubrication Engineers* (June 1985).

15 VOLATILIZATION OF ORGANICS IN ACTIVATED SLUDGE REACTORS

Don F. Kincannon, Professor
School of Civil Engineering
Oklahoma State University
Stillwater, Oklahoma 74078

Ali Fazel, Vice President
Environmental Engineering Consultants, Inc.
Stillwater, Oklahoma 74075

INTRODUCTION

During the past few years more and more attention has been focused on the treatment of wastewater containing organic priority pollutants. In addition, specific concern has been focused on the volatilization or stripping of organics during biological wastewater treatment. Concern has been expressed over which priority pollutants will be stripped and to what extent. This has caused an interest in the possibility of air pollution and in turn has caused renewed interest in the fate of priority pollutants during biological wastewater treatment. Kincannon and co-workers [1–8] have presented considerable work in the past in regards to the fate of priority pollutants during biological wastewater treatment. This paper presents an extension of this work. The objective of this work was to determine whether or not the addition of powdered activated carbon to the activated sludge reactor would decrease the level of stripping of the priority pollutants from the reactor.

EXPERIMENTAL METHODS

The general experimental plan specified bench scale continuous flow activated sludge reactors to be used to treat a synthetic wastewater containing selected organic priority pollutants. The bench scale activated sludge system is shown in Figure 1. The system consisted of a clear plexiglass internal recycle reactor. The activated sludge reactor had a volume of 2.8 liters. The settling compartment had a volume of 1.7 liters. The reactor and settling compartment both had a plexiglass cover. The wastewater was pumped from a sealed feed tank to the reactor. The effluent from the settling unit flowed by gravity to a collection tank. The off-gas was pulled by a vacuum pump through a 1/8" purge trap which contained six inches of Tenax and four inches of Selica Gel. The influent air and off-gas were measured with air flow meters.

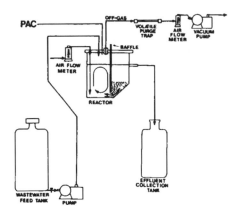

Figure 1.   Experimental reactor.

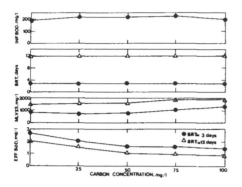

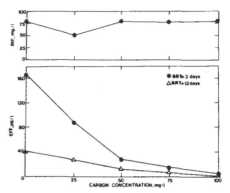

Figure 2. The effect of powdered activated carbon dosage and SRT on the performance of an activated sludge system receiving a wastewater containing toluene.

Figure 3. The effect of powdered activated carbon and SRT on effluent toluene levels.

Activated sludge for initial seeding was obtained from a local municipal activated sludge plant. The wastewater fed to the reactors was a synthetic wastewater containing ethylene glycol, ethyl alcohol, glucose, glutamic acid, acetic acid, phenol, ammonium sulfate, phosphoric acid, salts, and the specific priority pollutant under study. The synthetic wastewater was made up to provide a BOD of approximately 250 mg/L. The priority pollutants studied were benzene, toluene, ethylbenzene, and 1,2-dichloroethane. Two activated sludge reactors were operated side by side. One was operated at a sludge retention time (SRT) of 3 days and the other reactor was operated at a SRT of 12 days. The flow rate was controlled to give a hydraulic retention time of 8 hours.

RESULTS

The objective of this study was to determine the fate of priority pollutants in an activated sludge reactor and to determine whether or not powdered activated carbon would alter this fate. The results of this study are presented based upon the priority pollutants added to the reactor.

Toluene

The two activated sludge reactors were fed the synthetic wastewater which contained 100 mg/L of toluene. One reactor was operated at a SRT of 3 days, and the second reactor was operated at a SRT of 12 days. Both units were subjected to carbon dosages of 0, 25, 50, 75, and 100 mg/L (based upon influent flow rate). At each carbon dosage, the reactors were operated until steady state was achieved with respect to MLVSS and then complete analysis was conducted on the units. Figure 2 shows the performance of the activated sludge systems receiving the wastewater containing toluene. It is seen that all systems produced an effluent low in BOD. The system operated at a SRT of 12 days produced an effluent with a lower BOD than the system operated at a SRT of 3 days. It is also seen that as the carbon dosage was increased from 0 mg/L to 50 mg/L, the effluent BOD decreased; however, as the carbon dosage was incrased above 50 mg/L, no significant decrease in effluent BOD was observed.

Figure 3 shows the concentration of toluene that was observed in the effluent of the various systems. It is seen that the system operated at a SRT of 12 days had a lower effluent level of toluene than the system operated at a SRT of 3 days. It is also seen that as the concentration of powdered activated carbon was increased, the concentration of toluene in the effluent was decreased. In fact, at carbon dosages of 100 mg/L the concentration of toluene in the effluent was barely detectable.

Figure 4 shows the fate of toluene in the activated sludge systems. It is seen that approximately 12-16% of the toluene was stripped from the reactor when no carbon was added to the system. The percent of toluene stripped was reduced to approximately 2% by the addition of 50 mg/L of powdered activated carbon. Additional carbon dosages made very little difference.

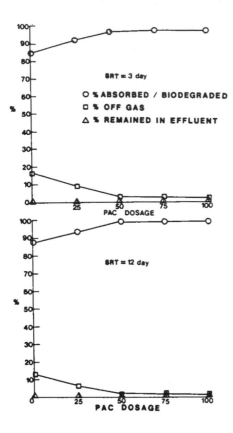

Figure 4. Fate of toluene in activated sludge system.

Ethylbenzene

In this part of the study, 40 mg/L of ethylbenzene was added to the base mix feed and used as the synthetic wastewater. Again, two systems were operated, one at a SRT of 3 days and the other at a SRT of 12 days. Figure 5 shows the performance of the activated sludge systems receiving the ethylbenzene wastewater. It is seen that excellent removal of BOD was achieved. It is also seen that the addition of carbon dosages up to 50 mg/L only slightly improved the effluent BOD levels.

Figure 6 shows the concentration of ethylbenzene in the effluent of the various systems. The ethylbenzene was reduced from 35 mg/L to approximately 275 ug/L in the 3-day system and 140 ug/L in the 12-day system when no carbon was added. As carbon dosages were increased, the concentration of ethylbenzene in the effluent of both systems decreased to almost undetectable levels.

Figure 7 shows the fate of ethylbenzene in the activated sludge systems. Approximately 15% of the ethylbenzene was stripped in the 3-day system whereas only approximately 5% was stripped in the 12-day system. The addition of powdered activated carbon had no affect on the amount of ethylbenzene stripped.

1,2-Dichloroethane

Figure 8 shows the performance of the activated sludge systems receiving a synthetic wastewater containing approximately 150 mg/L of 1,2-dichloroethane. The effluent BOD was very low for both systems with the 12-day SRT system having an effluent BOD slightly less than the 3-day SRT system. With the effluent BOD's being so low, the powdered activated carbon had very little effect on reducing the effluent BOD. The addition of the carbon had a significant effect on the mixed liquor volatile suspended solids. The MLVSS increased to almost 4000 mg/L and 8000 mg/L in the two reactors.

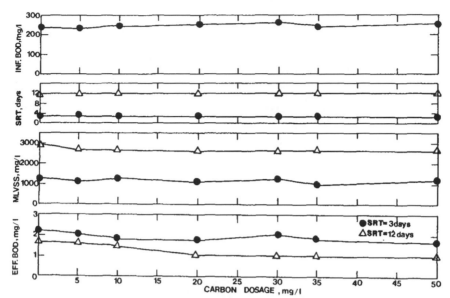

Figure 5. The effect of powdered activated carbon dosage and SRT on the performance of an activated sludge system receiving a wastewater containing ethylbenzene.

Figure 9 shows the effect of the powdered activated carbon on the effluent concentrations of 1,2-dichloroethane. It is seen that the effluent concentration of 1,2-dichloroethane for the systems receiving no carbon was 8 and 9 mg/L. This is significantly higher than that observed for toluene and ethylbenzene. It can also be observed that it required very high concentrations of carbon to reduce the

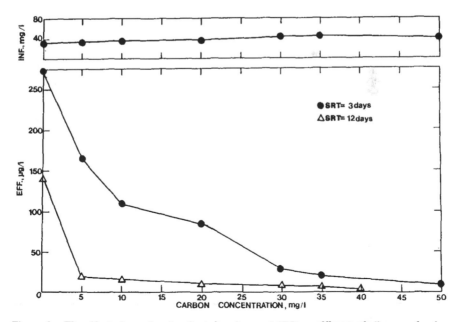

Figure 6. The effect of powdered activated carbon and SRT on effluent ethylbenzene levels.

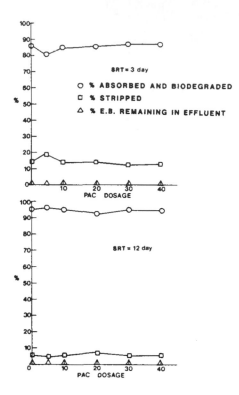

Figure 7.   Fate of ethylbenzene in activated
sludge system.

effluent concentrations any significant amount. A carbon dosage of 2500 mg/L was required to reduce the effluent concentration of 1,2-dichloroethane to approximately 2 mg/L. It is also interesting to note that the SRT at which the reactor was operated had no effect on the effluent level of 1,2-dichloroethane.

Figure 10 shows the fate of 1,2-dichloroethane in the activated sludge reactors. It is seen that approximately 92-96% of the 1,2-dichloroethane was stripped when no activated carbon was added to the reactors. The level of stripping was reduced by the addition of powdered activated carbon; however, very high concentrations of carbon were required to achieve any significant reduction. As the quantity stripped was decreased, the amount absorbed and/or biodegraded was increased. An attempt was made to distinguish between adsorption and biodegradation in these studies; however, this effort was not successful.

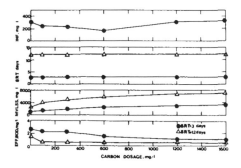

Figure 8.   The effect of powdered activated
carbon dosage and SRT on the performance
of an activated sludge system receiving a
wastewater containing 1,2-dichloroethane.

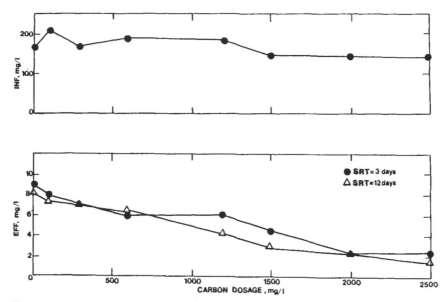

Figure 9.   The effect of powdered activated carbon and SRT on effluent 1,2–dichloroethane
levels.

Combined Pollutant Studies

It was also of interest to study a system that was receiving a number of priority pollutants.
Therefore, systems containing benzene, toluene, ethylbenzene, and 1,2–dichloroethane were studied.
Figure 11 shows the performance of the systems receiving the four priority pollutants. Both systems

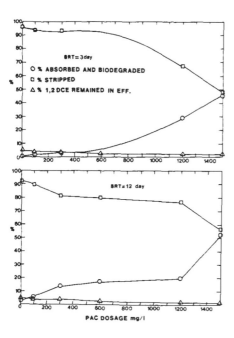

Figure 10.   Fate of 1,2–dichloroethane in an
activated sludge system.

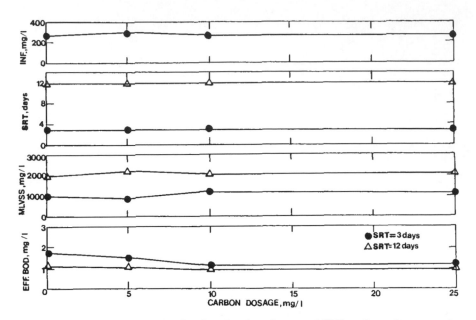

Figure 11.   The effect of powdered activated carbon dosage and SRT on the performance of an activated sludge system receiving a wastewater containing benzene, ethylbenzene, toluene, and 1,2–dichloroethane.

achieved very low effluent BOD levels. The powdered activated carbon dosages had very little effect on the effluent BOD.

Figure 12 shows the concentration of the four pollutants in the effluent of the reactors operated at SRT's of 3 and 12 days. It is seen that for the 12-day SRT system the effluent levels for the four pollutants were at approximately the detection limits. Detectable levels were observed in the effluent of the reactor operated at a SRT of 3 days for benzene, toluene, and ethylbenzene. The addition of powdered activated carbon decreased the effluent concentration for all three of these pollutants. The concentration of 1,2–dichloroethane in the effluent of the 3-day system was approximately the detection limit.

The fate of the four pollutants in the system operated at a SRT of 3 days is shown in Figure 13. Approximately 15% of benzene, ethylbenzene, and toluene was stripped in this system, and the addition of powdered activated carbon had very little, if any, effect on the amount of pollutant stripped. Approximaately 85% of the benzene, ethylbenzene, and toluene was bio- degraded in this system. Approximately 95% of the 1,2–dichloroethane was stripped in this system when no activated carbon was added. The addition of 5 and 10 mg/L of powdered activated carbon decreased the percent stripped to approximately 84%. The decrease in the quantity stripped was found to be adsorbed.

The fate of the pollutants in the system operated at a SRT of 12 days is shown in Figure 14. Without the addition of powdered activated carbon, 12% of the benzene was stripped, and 88% was biode- graded. Fifteen percent of the toluene was stripped, and 85% was biodegraded, and 15% of the ethylbenzene was stripped with 85% being biodegraded. The addition of powdered activated carbon had no effect on the fate of benzene and toluene. It did have a slight effect on ethylbenzene. The addition of 5 and 10 mg/L of carbon resulted in 12% being stripped and 88% being biodegraded/ adsorbed. The removal of 1,2–dichloroethane again was accomplished by stripping, and the addition of powdered activated carbon at dosages of 5 and 10 mg/L had very little effect on the fate of the pollutant. The percentage of 1,2–dichloroethane that was stripped varied from 95 to 99%.

DISCUSSION

In this study, varying dosages of powdered activated carbon were administered to activated sludge systems operating at 3 and 12-day SRT's. The systems received a wastewater comprised of a synthetic

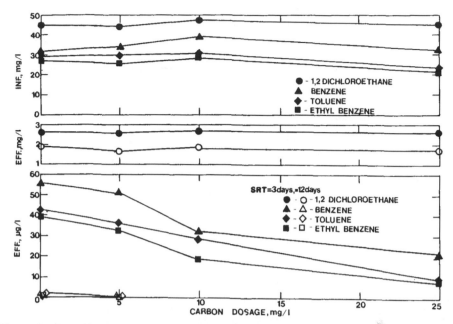

Figure 12.   The effect of powdered activated carbon and SRT on effluent levels of combined system containing benzene, ethylbenzene, toluene, and 1,2–dichloroethane.

improved the effluent quality but also was responsible for reducing off-gas emission for toluene. It was found in the case where 1,2- dichloroethane was administered to the activated sludge, air stripping accounted for virtually all of the removal of this pollutant. The addition of powdered activated carbon slightly improved the effluent quality and greatly reduced the off-gas emission of 1,2–dichloroethane; however, very high dosages of carbon were required to achieve this reduction.

The combined pollutant studies showed similar results. Only slight improvements in the effluent quality were noted when powdered activated carbon was added. This may have been because the systems were achieving an excellent effluent without carbon addition. At these low BOD's, it would be difficult to do much better. Virtually no additional reduction of 1,2–dichloroethane concentrations in the effluent was noted as the powdered activated carbon dosage was increased from 0 to 25 mg/L. However, in all cases the system that was operated at a SRT of 12 days was able to achieve lower effluent concentrations than the system that was operated at a SRT of 3 days.

It must be recognized that the systems used in this study were closed systems. This was done in order to facilitate the capture and measurement of the off-gases. The fact that the systems were closed could probably have had a significant effect on the stripping of benzene, toluene, and ethylbenzene. The closed system caused a concentration of the pollutant to be present in the air space above the liquid level of the reactor. This would decrease the driving force for stripping and, therefore, cause the pollutant to remain in solution longer. This would provide more opportunity for the microorganisms to biodegrade the pollutant. Thus, more of the material would be biodegraded and less stripped in a closed system than in an open system. This may provide a control mechanism for controlling the emissions of volatile pollutants from wastewater treatment plants. In addition, when powdered activated carbon was added to the reactor, very low levels of benzene, toluene, and ethylbenzene were observed in the effluent. Thus, a combination of a closed reactor and the addition of powdered activated carbon may be a very effective control process for volatile pollutants that are subject to biodegradation.

The pollutant 1,2–dichloroethane was not biodegradable, and it was found that the only removal mechanism was stripping. The addition of powdered activated carbon did cause removal by adsorption; however, the amount of the required dosage was extremely high.

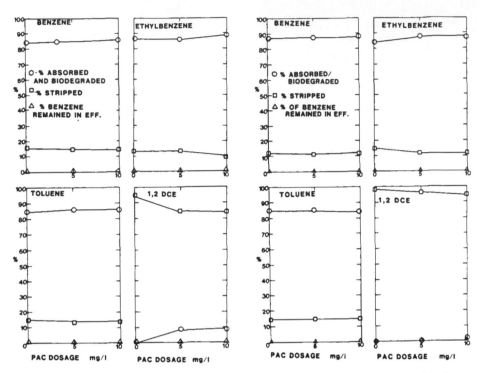

Figure 13.   Fate of pollutants in combined system for SRT = 3 days.

Figure 14.   Fate of pollutants in combined system for SRT = 12 days.

REFERENCES

1. Kincannon, Don F., Enos L. Stover, and Yu-Ping Chung, "Biological Treatment of Organic Compounds Found in Industrial Aqueous Effluents," presented at American Chemical Society National Meeting, Atlanta, GA (March 29–April 3, 1981).
2. Stover, Enos L. and Don F. Kincannon, "Biological Treatability of Specific Organic Compounds Found in Chemical Industry Wastewaters," presented at the 36th Purdue Industrial Waste Conference, Purdue University, West Lafayette, IN, May 12–14, 1981, and *JWPCF*, Vol 55 (January 1983).
3. Kincannon, Don F. and Enos L. Stover, "Fate of Organic Compounds During Biological Treatment," presented at National Conference on Environmental Engineering, ASCE, Atlanta, GA (July 8–10, 1981).
4. Kincannon, Don F., Enos L. Stover, Virgil Nichols, and David Medley, "Removal Mechanisms for Biodegradable and Non-Biodegradable Toxic Priority Pollutants in Industrial Wastewaters," Proceedings Industrial Wastes Session, 54th Annual Water Pollution Control Federation Conference, Detroit, MI, October 4–9, 1981, and *JWPCF*, Vol. 55 (February 1983).
5. Kincannon, Don F. and Enos L. Stover, "Stripping Characteristics of Priority Pollutants During Biological Treatment," presented at Secondary Emissions Session of the 74th Annual American Institute of Chemical Engineers Meeting, New Orleans, LA (November 8–12, 1981).
6. Kincannon, Don F., Anne Weinert, Robin Padorr, and Enos L. Stover, "Predicting Treatability of Multiple Organic Priority Pollutant Wastewaters from Single Pollutant Treatability Studies," *Proceedings of the 37th Purdue Industrial Waste Conference*, pp. 641–650, Purdue University, West Lafayette, IN (May 11–13, 1982).
7. Stover, Enos L., David E. McCartney, Faramarz Dehkordi, and Don F. Kincannon, "Variability Analysis During Biological Treatability of Complex Industrial Wastewaters for Design," *Proceed-*

ings of the 37th Purdue Industrial Waste Conference, p. 773-783, Purdue University, West Lafayette, IN (May 1982).

8. Kincannon, Don F., Enos L. Stover, David E. McCartney, and Faramarz Dehkordi, "Reliable Design of Activated Sludge Systems to Remove Organic Priority Pollutants," Presented at the 55th Annual Conference, Water Pollution Control Federation, St. Louis, MO (October 3-8, 1982).

16 TRANSIENT TEMPERATURE RESPONSE OF PACT® AND ACTIVATED SLUDGE SYSTEMS TO TEMPERATURE CHANGES

Yu-Chia T. Chou, Development Engineer

Gerald J. O'Brien, Research Associate
Jackson Laboratory
E. I. DuPont DeNemours and Company
Deepwater, New Jersey 08023

INTRODUCTION

Du Pont's Chambers Works facility in Deepwater, New Jersey, is the site of one of the largest industrial wastewater treatment plants in the country. The treatment plant, which has a 40 mgd capacity, treats wastes from a myriad of industrial processes and from a diversity of outside industries. An upset condition occurred during a severe cold spell in the winter of 1984, which adversely affected the treatment plant's ability to remove dissolved organics. Remedial actions were taken, including raising the wastewater temperature, but the expected DOC removal was not reached until several weeks later. Laboratory experiments showed that a time lag exists after the temperature is raised before steady state, DOC removal is reached. This paper will review the case history, the laboratory studies, and the corrective action taken.

BACKGROUND

A schematic of the Chambers Works Wastewater Treatment Plant (WWTP) is shown as Figure 1. The primarily acidic wastes from Chambers Works (CW) and from outside businesses (OWB) are mixed together and neutralized with lime. The neutralization adjusts the pH for bacterial decomposition and precipitates heavy metals. Polymer is added to flocculate the suspended solids, which are then concentrated, filtered, and buried in the CW secure landfill. The landfill is double-lined, monitored, and permitted for the deposition of hazardous waste solids.

The overflow from the clarifiers enters three, four-million-gallon aerators. The biodegradable DOC is removed by the bacteria while activated carbon is added to adsorb residual color, non-biodegradable organics, and heavy metals, and to aid settling. Sparged air suspends the solids and supplies oxygen for the metabolic needs of the bacteria. Polymer aids the flocculation of the solids in the secondary clarifiers. A small portion of the MLSS (mixed-liquor, suspended solids) is wasted while most of the MLSS is recirculated to the aerators. The overflow is mixed with non-contact cooling water in a 15-acre settling basin before it is discharged to the river. Water quality is closely monitored for compliance with the NPDES Permit parameters.

CASE HISTORY

The inlet water temperature to the aerators is governed primarily by the Delaware River, which supplies process water to the CW site. January of 1984 was abnormally cold, and the aerator temperature remained below 14°C during the month. On January 24, a surge in the inlet dissolved organic carbon (DOC) concentration occurred and the dissolved oxygen concentration rose, which indicated a toxic shock ("hit") to the bacteria. The DOC removal deteriorated immediately and remained below 50% for two weeks despite a series of remedial actions.

- The carbon dose was doubled to adsorb toxins.
- Outside bacteria from another industrial waste treatment plant was added to the aerators to increase the viable bacteria concentration.
- The wastewater temperature was increased from 14 to 22°C by steam injection to increase the bacterial metabolism.
- Additional residence time was provided by putting the standby third aerator on-line.

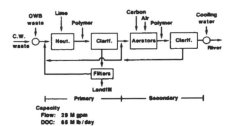

Figure 1.   The process flow sheet of Chambers
Works Wastewater Treatment Plant.

• Nutrients and readily biodegradable organics were added to promote bacterial growth.

The DOC removal steadily increased from 45 to 65% during the eight-day period of the temperature increase, but three weeks were required to reach the steady-state value of 90% (Figure 2). Laboratory experiments were run to determine the effects of temperature and the other remedial actions. Temperature changes were found to be the most important factor affecting DOC and color removal.

TRANSIENT TEMPERATURE RESPONSE—LITERATURE REVIEW

Ludzack et al. [1] has reported that two weeks or more were required for the microbial population to reach characteristic equilibrium and performance after a significant temperature shock. The acclimation time required for different wastewater feeds at 5°C was about five times as long as at 30°C, and the steady state BOD and COD removal efficiencies were 10% lower at 5°C than at 30°C. Benedict and Carlson [2] have monitored the endogenous respiration rate of biomass and effluent COD and TSS after rapid changes in temperature from 19 to 4°C and from 19 to 32°C. The COD removal at 4°C was only slightly lower than at 19°C. However, when the temperature was raised to 32°C, the effluent COD and TSS required in excess of 26 days to reach the new steady state.

Koshy [3], using a synthetic glucose waste, studied the response of activated sludge composition, effluent COD, and biological solids content variations to temperature shocks. Depending on the magnitude and rate of temperature changes, 2 to 7 days were required for the effluent COD and biological solids to reach the new steady states. The effluent COD increased either due to reduced biodegradation of the glucose substrate or the release of metabolic intermediates of the biosludge.

Eden et al. [4] found that it took about six weeks for the nitrilotriacetic acid removal efficiency to reach a new steady state after a temperature rise from 5 to 20°C in continuous activated-sludge systems. Also, a time lag of ten days occurred before any response to the initial temperature change was observed. When the temperature was decreased from 20 to 7.5°C over a four-week period, a four-week time lag was observed before any significant decrease in removal efficiency occurred. Stiff and Rootham [5] fed nonionic surfactants to continuous, activated-sludge units and decreased the temperature from 20 to 12°C. No significant time lag was observed, but steady state was not reached until ten days later.

The literature results suggested that the three-week period to reach a steady state, DOC removal in Chambers Works WWTP was due to the transient response to the temperature increase. Laboratory experiments were carried out to study the transient response to temperature shocks.

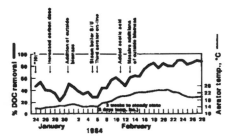

Figure 2.   The transient DOC removal of
Wastewater Treatment Plant in response to
temperature increases during the winter upset
of 1984.

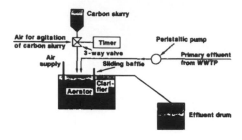

Figure 3.   The schematic diagram of laboratory flow-through aerator.

EQUIPMENT AND METHODS

The six bench-top aerators used in this study were modified Eckenfelder units with internal clarifiers and an aerator volume of 6.5 to 7.5 liters (Figure 3). These units were used to simulate the biological processes occurring in the secondary treatment of our WWTP and were equipped for feeding powdered activated carbon. The wastewater feed was the same stream as was fed to the plant aerators. The feed flow rates were controlled by peristaltic pumps to provide a hydraulic, residence time of eight hours to the Eckenfelder aerators, and the sludge age was kept at about 30 days. Powdered activated carbon at 45 mg/L dosage, based on wastewater flow, was fed to the PACT® units. Daily composite samples of wastewater feed and aerator effluents were analyzed for DOC and color concentrations.

DOC was determined by a Beckman total organic carbon analyzer (Model 915A, Fullerton, CA). Color was measured in terms of APHA units by visual comparison with standards in a Hellige Aqua Tester (Garden City, NY).

There were two continuous aerators operating in the laboratory at ambient temperatures since 1976. The PACT® unit was fed with 45 mg/L activated carbon constantly, and the biomass unit was a conventional activated sludge aerator. They served as control units for system performance comparisons with four test aerators.

One PACT® and one biomass unit were operated at 14°C for three months to assure steady state was reached. The other two were PACT® units, which were operated at ambient temperature. The two units at 14°C were raised in temperature, while the other two test units at ambient temperature were lowered to 10°C.

TRANSIENT RESPONSE TO TEMPERATURE INCREASE

The temperature of the two cold test units was raised from 14 to 21°C at a rate of 0.5°C per hour. The transient responses of DOC removal efficiencies of these two units with their controls at 21°C are shown in Figures 4 and 5 and summarized in Table I. It took about three weeks for the PACT® unit to reach the steady state removal efficiency of the control unit while the biomass unit required six weeks or more to reach the new steady state, which was 4.5% less than that of the control biomass unit. The

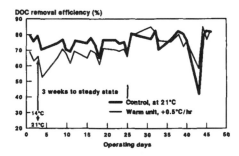

Figure 4.   The transient DOC removal of the PACT® unit in response to a temperature increase from 14 to 21°C.

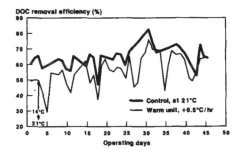

Figure 5.   The transient DOC removal of the biomass unit in response to a temperature increase from 14 to 21°C.

Table 1. Pact® and Activated Sludge
Summary of System Performance to Temperature Shocks

Unit	Temp Change °C	ΔT/Δt (°C/hr)	Days for DOC Removal to Reach Steady State	Days for DOC Removal to Show Deterioration	DOC % Before	DOC % After	Color % Before	Color % After
					Steady State Removal Efficiency			
					DOC		Color	
					Before Temperature Shock	After	Before	After
Activated Sludge	21	0	control	control	68	67	32	25
PACT®	20	0	control	control	80	81	82	61
Activated Sludge	14→21	+0.5	>40	—	55	63	31	25
PACT®	14→21	+0.5	22	—	67	81	70	61
PACT®	21→10	-0.15	no control for comparison	0	76	62	76	60
PACT®	19→9	-1.30	no control for comparison	3	81	65	82	61

removal efficiency of the PACT® unit was consistently 10–15% greater than the activated-sludge unit before and after the temperature shock.

The DOC result with the PACT® unit was almost identical to that observed in the CW WWTP. The six-week time lag observed in the activated-sludge unit was similar to those reported in the literature. The activated carbon, since it is the only difference between the two systems, must be responsible for the superior DOC removal and the shorter response time to temperature shock.

The color removal efficiency of the PACT® unit (Table I) was also much greater than the biomass unit because of the added carbon. The decrease in color removal efficiency of both units after the temperature increase was due to changes in the incoming wastewater composition as evidenced by the corresponding decrease in color removal of the control units.

COMPARISON OF LABORATORY AND PLANT RESULTS

Figure 2 shows that the DOC removal for the CW WWTP increased steadily from 45 to 65% during the eight-day period when the temperature was increased from 14 to 22°C at 1°C per day. However, three weeks were required before the steady state value of 90% removal was reached. The laboratory PACT® unit's temperature was raised from 14 to 21°C in less than a day. In both the PACT® laboratory unit and the CW WWTP, three weeks were required for the DOC removal to reach steady state after similar temperature increases.

TRANSIENT RESPONSE TO TEMPERATURE DECREASE

The temperatures of the two, ambient PACT® units were decreased at different rates. When the temperature was decreased from 21 to 10°C at a slow rate of 0.15°C per hour, the PACT® unit indicated no time lag in deterioration of DOC removal efficiency (Figure 6). However, when the temperature was quickly dropped from 19 to 9°C at a rate of 1.3°C per hour, the DOC removal efficiency of the PACT® system did not deteriorate until three days later (Figure 7). After about two weeks of operation at this cold temperature, this quickly-cooled PACT® unit had a biological solids washout, and the DOC removal efficiency deteriorated even further. The above results suggest that the rate of temperature change determines the time lag in the PACT® system and subsequent bacterial viability.

During this experimental period, the color removal efficiencies were not affected by the temperature shocks. The apparent decrease in color removal is due to the change in incoming waste, as shown by the control units, rather than due to the temperature shock. However, previous experiments in the laboratory did indicate less color removal efficiency at low temperatures. The color removal of the PACT® unit was 35–50% greater than the activated sludge unit before and after the temperature shock.

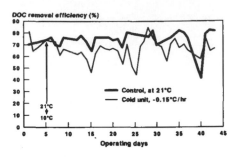
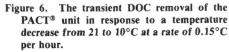

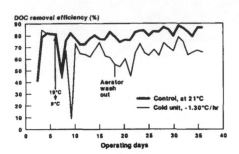

Figure 6. The transient DOC removal of the PACT® unit in response to a temperature decrease from 21 to 10°C at a rate of 0.15°C per hour.

Figure 7. The transient DOC removal of the PACT® unit in response to a temperature decrease from 19 to 9°C at a rate of 1.30°C per hour.

PREVENTIVE MEASURES

Maintaining ~20°C aerator temperature during the winter months is an economically impractical solution for CW because of the amount of steam which is required to heat the large water flow. The chosen approach was to reduce the probability of a "hit" by storing compatible outside wastes greater than 1% in DOC concentration. The storage and slow release of the high organic wastes eliminates surges in the DOC concentration and provides additional dilution of any toxins inadvertently received from the outside waste business. This procedure was successfully used in 1985 and 1986.

FUTURE WORK

Laboratory experiments are in progress to determine whether bacteria with longer sludge ages are more susceptible to a toxic shock after prolonged operation at low temperatures. Mesophile and Psychotroph ratios will be measured before and after the temperature change, and the effect of the rate of temperature change and the role of carbon in altering the time lag will be examined.

SUMMARY

The slow recovery from a toxic "hit" was due to low temperature operation. A time lag of about three weeks was observed in the laboratory and plant for the DOC removal to reach steady state after the temperature was increased from 14 to 21°C. In the conventional activated-sludge system, the time to reach steady state was in excess of six weeks, which was in good agreement with the literature. The added activated carbon in the PACT® unit reduced the transient temperature response time and improved the DOC removal efficiency. When the temperature was decreased from 21 to 10°C, the laboratory PACT® unit's DOC removal deteriorated rapidly within three days. The actions taken to prevent a recurrence were to reduce the probability of a "toxic hit" since temperature control was economically prohibitive.

ACKNOWLEDGEMENT

The authors would like to express their sincere appreciation to Mrs. Lorraine Kite for typing this manuscript.

REFERENCES

1. Ludzack, F. J., R. B. Schaffer and M. B. Ettinger, *Journal WPCF*, Vol 33, No. 2 (1961).
2. Benedict, A. H. and D. A. Carlson, *Journal WPCF*, Vol 45, No. 1 (1973).
3. Koshy, G. T., *Ph.D. dissertation*, Oklahoma State University (1968).
4. Eden, G. E., Culley, G. E., and Rootham, R. C., *Water Research*, V. 6, 877–883 (1972).
5. Stiff, M. J., and Rootham, R. C., *Water Research*, V. 7, 1407–1415 (1973).

17 BIOLOGICAL TREATMENT OF MINING EFFLUENTS

J. P. Maree, Chief Researcher

A. Gerber, Researcher

E. Hill, Assistant
National Institute for Water Research
Council for Scientific and Industrial Research
Pretoria, South Africa 0001

A. R. McLaren, Water Quality Control Officer
Gold Fields of South Africa, Ltd.
Johannesburg, South Africa 2000

INTRODUCTION

Mining effluents are major contributors to mineralization of receiving waters and may prove toxic to men, animals, and plants due to unacceptably high concentrations of heavy metals and cyanide. Its reuse potential is limited due to the presence of saturated calcium sulphate, which causes scaling of pipes and equipment. Mining effluents originate from both underground sources and metallurgical process plants. Its sulphate content results from bacterial oxidation of pyrite, while metallurgical processing adds sulfate in the form of spent sulfuric acid. Sulphate, calcium, heavy metals and cyanide may be removed by well-established demineralization processes, such as reverse osmosis and electrodialysis, but these are costly, hence the need for the development of alternative processes. A promising new process entails the biological reduction of sulphate to sulphide by the bacterium *Desulfovibrio desulfuricans*. Sulphide can, in turn, be converted to elemental sulphur.

Various researchers have studied biological sulphate reduction. Middleton and Lawrence (1) determined the kinetics of microbial sulphate reduction in complete mix reactors using acetic acid as carbon source and observed a sulphate reduction rate of 0.29 g SO_4/L/d (biomass concentration not specified). Cork and Cusanovich (2) developed a continuous purge system, using an inert carrier gas (75% argon and 25% CO_2), to feed sulphide removed from actively growing cultures of *Desulfovibrio desulfuricans* to cultures of *Chlorobium thiosulfatophilum*, for oxidation to sulphur. A sulphate reduction rate of 6.3 g SO_4/L/d was observed in a completely mixed reactor containing 12,600 mg/L lactic acid, at pH 6.5 and temperature 30°C (biomass concentration not specified). Some 91% of the hydrogen sulphide produced was swept off into a sulphide oxidizing chamber containing *Chlorobium*, where 88% was converted to sulphur. This represents an overall sulphur yield of 80%. Hilton et al. (3) applied a similar process, known as the BIOSULFIX process, in which hydrogen sulphide was removed in an external stripping chamber, and observed a sulphate reduction rate of 6.5 g SO_4/L/d. Maree and Strydom (4) studied sulphate reduction in a packed bed reactor and observed reduction rates of the same order of magnitude.

Oversaturated calcium carbonate levels and unutilized carbonaceous material prevent water from being reused directly after anaerobic treatment. Maree (5) showed that these products can be removed successfully by applying hydrogen sulphide stripping, clarification and aerobic treatment. Aerobic treatment reduced the organic content from 1,100 to 300 mg/L COD.

The specific purpose of this study was to study calcium sulphate removal from lead mine effluent in the presence of heavy metals and organics such as xanthates and carbamates, used during mineral flotation. A secondary objective was to find alternative ways of effecting clarification without resorting to the addition of chemicals. Typically, 500 mg/L $FeCl_3$. which adds unwanted salts to the water, is required for the latter purpose.

The lead mine involved is situated at Black Mountain in the North Western Cape. Being situated in an arid region, water is scarce and the daily requirement of 4.5 ML is pumped from the Orange River over a distance of some 58 km. After utilization in the flotation plant, where lead, copper and zinc are separated selectively, a side stream of 3.5 ML/d is bled off and discharged into a nearby vlei to maintain acceptable levels of dissolved salt and organics in the process water. The discharge water

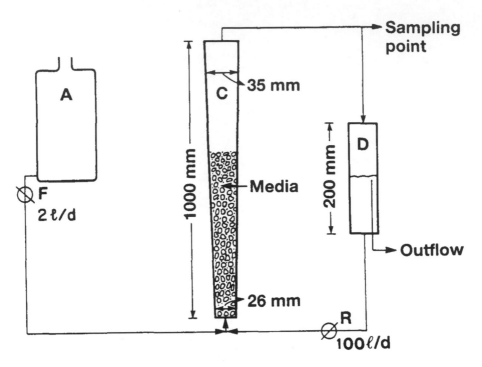

Figure 1.   Schematic diagram of laboratory-scale plant used for biological sulphate reduction.
   A : Feed tank containing mine water and 3 ml/L molasses
   C : Anaerobic packed bed reactor
   D : Distribution chamber
   F : Feed pump
   R : Recirculation pump
   Total liquid volume = 1 L
   Media volume = 0,5 L
   Retention time = 12 h

creates environmental problems due to excessive concentrations of heavy metals, organics, and calcium sulphate.

EXPERIMENTAL

Continuous-flow reactor

The reactor system used in the study is given schematically in Figure 1. The columnar anaerobic packed bed reactor had a tapered bottom to facilitate fluidization of the medium when flushing out excess sludge and heavy metal precipitates. This was done regularly at an upflow velocity of 50 m/h, achieved by increasing the recirculation rate. One reactor volume of fluid was generally sufficient to remove the sludge. Backwash samples were withdrawn at various time intervals and analysed for suspended solids (SS) and volatile suspended solids (VSS) to determine the relationship between sludge age and specific biomass production. The distribution chamber served as reservoir for recirculation liquid and was equipped with an overflow facility to dispose of the effluent. Fluid was recirculated at a rate of 100 L/d (compared with the inflow rate of 1 L/d) between the anaerobic reactor and distribution chamber. A Masterflex peristaltic pump was used for recirculation purposes and a NIWR cylinder pump for feeding. One-half of the anaerobic reactor was filled with dolomitic pebbles with a diameter of 2 to 3 mm. The void ratio of the medium was 50%. Feedstock, which was stored at 4°C, consisted of process water obtained from Black Mountain Lead Mine and with composition as shown in Table I supplemented with 3 mL molasses/L. A biologically active film was established on the stone medium by inoculating the reactor with activated sludge from a laboratory activated sludge plant.

Table I.   Chemical Composition of Untreated and Treated Water

Parameter	Unit	Untreated	Untreated & molasses	Treated
pH		6.7	5.3	7.0
Kjeldahl nitrogen (as N)	(mg/L)	8.6	11.2	10.7
Ammonia nitrogen (as N)	(mg/L)	6.8	3.4	2.8
Nitrate nitrogen (as N)	(mg/L)	1.4	2.2	1.6
Sulphate (as SO_4)	(mg/L)	2050	2050	100
Free sulphide (as SO_4)	(mg/L)	0	0	1400
Total phosphate (as P)	(mg/L)	2.9	2.4	1.9
Orthophosphate (as P)	(mg/L)	1.1	1.3	0.8
Chloride (as Cl)	(mg/L)	536	560	551
Total alkalinity (as $CaCO_3$)	(mg/L)	20	80	1999
COD (as O_2)	(mg/L)	502	3300	1662
Sodium (as Na)	(mg/L)	245	250	10.4
Potassium (as K)	(mg/L)	43	136	137
Calcium (as $CaCO_3$)	(mg/L)	2125	2025	1950
Magnesium (as Mg)	(mg/L)	40	38	38
Silicon (as Si)	(mg/L)	3.7	4.2	7.9
MBAS (as LAS)	(μg/L)	3355	5363	4588
Aluminium (as Al)	(μg/L)	100	144	270
Boron (as B)	(μg/L)	710	2500	2242
Fluoride (as F)	(μg/L)	2117	1863	1246
Arsenic (as As)	(μg/L)	0	0	0
Chromium (as Cr)	(μg/L)	34	32	0
Cadmium (as Cd)	(μg/L)	120	14	9
Cobalt (as Co)	(μg/L)	195	195	127
Copper (as Cu)	(μg/L)	2400	2400	61
Gold (as Au)	(μg/L)	298	266	202
Iron (as Fe)	(μg/L)	82	3700[a]	705
Lead (as Pb)	(μg/L)	2100	1500	170
Mercury (as Hg)	(μg/L)	8	1	1
Manganese (as Mn)	(μg/L)	1623	2200	414
Nickel (as Ni)	(μg/L)	70	100	92
Selenium (as Se)	(μg/L)	0	0	0
Silver (as Ag)	(μg/L)	100	100	0
Zinc (as Zn)	(μg/L)	741	991	111
Cations	(me/L)	58.6	59.3	57.3
Anions	(me/L)	58.8	60.7	57.7

[a]increase in iron concentration is due to its presence in molasses.

Batch Studies

Hydrogen sulphide stripping was studied in batch systems by aerating 20 L volumes of anaerobically treated water, gathered over a period of 20 days, for 10 h.

Aerobic treatment likewise was conducted in batch systems, after supplementing process water with ammonia-N and phosphate-P to concentration levels of 20 and 4 mg/L respectively. Bacterial growth was initiated by adding 200 mL thickened activated sludge (with a suspended solids concentration of 5,500 mg/L) to 20 L of anaerobically treated water. Air was supplied through a diffuser at a rate sufficient to maintain the dissolved oxygen content at 4 mg/L. The sludge was allowed to settle once daily, upon which 600 mL of the aqueous supernatant was replaced with the same volume of stored water (also supplemented with ammonia and phosphate).

Analytical

In the continuous flow reactor samples were taken daily from the distribution chamber and analyzed for sulphate, sulphide, alkalinity, chemical oxygen demand (COD), and pH. Complete chemical analyses were performed on the raw and treated water after steady state operation had been obtained. Manual determinations of sulphate, COD, alkalinity, sulphide and pH were carried out on filtered

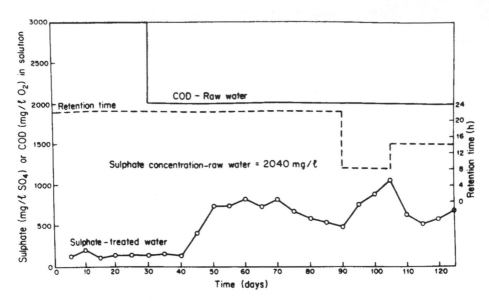

Figure 2.   Temporal variation of the sulphate concentration during biological treatment. Each plotted point on graph for treated water represents the mean value of 5 determinations.

samples according to analytical procedures as described in *Standard Methods* (6). Direct flame atomic absorption and automatic colorimetric techniques were used for the other chemical determinations. The COD value obtained was corrected by subtracting the COD equivalent of the sulphide content of the water (96 mg as sulphate equivalent to 64 mg COD).

RESULTS AND DISCUSSION

Anaerobic treatment

The reactions that occur in the anaerobic reactor can be summarized as follows:

Fermentation. According to Jørgensen (7), sulphate reducing bacteria are capable of oxidizing the main products of bacterial fermentations. Therefore, it is expected that fermentation bacteria, which live symbiotically with sulphate reducing bacteria, biodegrade the sugar components in molasses to products such as lactate and pyruvate, which serve as energy source for the sulphate reducing bacteria. The conversion of sucrose to lactate can be represented by the following reaction (8)

$$C_{12}H_{22}O_{11} + H_2O \rightarrow 4CH_3CHOHCOOH \tag{1}$$

Anaerobic Respiration. The sulphate reducing species, *Desulfovibrio desulfuricans*, utilizes lactate to produce acetate according to the following reaction (9):

$$2CH_3CHOHCOO^- + SO_4^= \rightarrow 2CH_3COO^- + 2HCO_3^- + H_2S \tag{2}$$

Reaction 2 can also be represented as

$$2C + SO_4^= + 2H_2O \rightarrow H_2S + 2HCO_3^- \tag{3}$$

The results in Table I show that 1950 mg/L sulphate (as SO_4) was reduced in the process while only 1,400 mg/L sulphide (as SO_4) was produced. The difference of 595 mg/L sulphide can be ascribed to biological sulphur production, heavy metal precipitation, or stripping.

Figure 2 shows results obtained when the effluent was treated continuously in a packed bed reactor. Sulphate concentration was decreased from 2,050 to less than 200 mg/L when 3 mL/L molasses (equivalent to a COD dosage of 3,000 mg/L) was supplied (days 0–40). As it is not essential for the mine involved to achieve complete sulphate removal, the molasses dosage was decreased from 3 mL/L to 2 mL/L (equivalent to 2,000 mg/L COD) in an effort to achieve optimum utilization of the

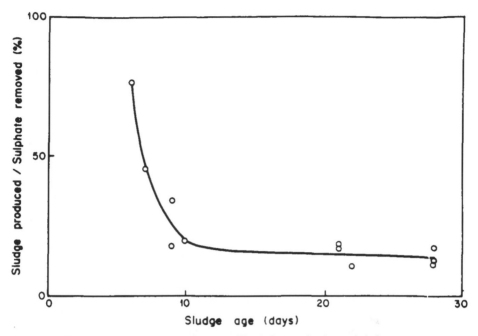

Figure 3.   Relationship between specific sludge production and sludge age.

biodegradable fraction. However, decreasing either the molasses dosage, or the retention time, led to impaired sulphate removal. Decreasing the molasses dosage from 3 to 2 mL/L on day 30 resulted in the effluent sulphate concentration increasing from 130 to 800 mg/L, where it levelled off (days 50–70). During days 70 to 90, better sulphate removal was obtained without changing the amount of molasses supplied, or any other conditions. This is ascribed to bacterial adaptation to the limited molasses supply, which led to fermentation of those components of molasses that were not utilized otherwise. A decrease in the retention time from 21 h to 8 h on day 90 resulted in a gradual increase in the sulphate concentration over the next 15 days, from 500 to 1060 mg/L. By increasing the retention time to 14 h on day 105, effective sulphate removal was restored.

The theoretical amounts of COD required and alkalinity produced during bacterial respiration, both per unit of sulphate removed, are 0.67 and 1.04 respectively (Equation 3). The average *experimental* values were 0.86 and 1.06, respectively. It is observed that the theoretical alkalinity production corresponded well with observed values. The discrepancy pertaining to theoretical and experimental COD demand simply implies that COD is utilized in other reactions beside respiration. The major ones are those involved in fermentation according to reaction 1 and production of bacterial cells.

The specific sludge production as a function of sludge age is given in Figure 3, which indicates that the former decreases rapidly with increasing retention time up to 10 d, after which it stablizes. The sludge production stabilized at 0.10 compared to the volatile solids level of 0.05 (both based on one unit of sulphate removed). The disposal of sludge has a cost disadvantage and from this point of view large-scale plants should be run at sludge ages longer than 10 days to minimize sludge production. However, long sludge ages also imply less active bacteria and strong heavy metal accumulation. The advantage of lesser waste sludge production should be weighed against these possible disadvantages.

Reducing the sulphate content in 5 ML/d of effluent from 2,500 to 200 mg/L, is equivalent to a mass removal of 11.5 tons of sulphate. The relationship between sludge age and sludge production suggests that 1.15 t/d of sludge (100% solids content) would be produced concomitantly.

The observed volatile suspended solids (VSS) represents only 41.0% of the suspended solids (SS) mass. This low value is ascribed to the presence of calcium carbonate in the sludge.

One potential problem associated with excess sludge disposal is that of odour, resulting from H_2S remaining in the sludge. The following remedial measures can be applied in such cases:

Table II.   Quality of Water Before and After Hydrogen Sulphide Stripping (results in mg/L)

Determinand	Before	After
Sulphide (as SO_4)	700	85
Calcium (as $CaCO_3$)	1,708	1,324
Alkalinity (as $CaCO_3$)	910	514
Sulphate (as SO_4)	900	879
pH	7.4	8.0

(i) Treat the sludge with an iron source such as ferric sulphate to precipitate the sulphide as ferric sulphide.

(ii) Treat the sludge biologically in an aerobic process to convert residual sulphide to sulphate.

Heavy metal precipitation. Heavy metals such as copper, lead, manganese, iron, silver, and zinc are effectively removed in the anaerobic stage (Table I). Heavy metal sulphide precipitates accumulated together with the bacterial biomass in the anaerobic reactor and were periodically removed by backwashing along with removal of excess sludge. Elimination of heavy metal poisoning of the biological system, prevention of heavy metal pollution of receiving waters, and recovery of precious metals via their sulphide salts count among the advantages associated with this removal.

Calcium carbonate crystallization. Calcium and alkalinity values of 1360 and 2315 mg/L (as $CaCO_3$) respectively (Table I), indicate that the effluent after primary anaerobic treatment is oversaturated with respect to calcium carbonate. Yet, only some 140 mg/L $CaCO_3$ crystallizes out in this stage. It is thought that this can be ascribed to:

(i) the relatively low pH value of 7.0 after anaerobic treatment,

(ii) the organic components of molasses being converted mainly to acetate, and not to carbon dioxide, which is required as the carbonate source for the reaction to take place, and

(iii) certain organic components of molasses, such as phenol, or fermentation products such as lactic acid, which inhibit calcium carbonate crystallization in line with the action of organics such as polyacrylic acid which has this capability (10).

Removal of end products produced during anaerobic treatment

Water leaving the anaerobic stage was unacceptable for reuse due to residual concentrations of COD, sulphide, and supersaturated calcium carbonate. Therefore, two additional stages were required for the removal of these end products, namely stripping for sulphide, and aerobic treatment for COD and calcium carbonate removal.

Air stripping. During air stripping of the anaerobically treated water, both sulphide and calcium are removed from solution within 3 hours to levels of 85 mg/L (as SO_4) and 1324 mg/L (as $CaCO_3$), respectively (Table II and Figure 4). Hydrogen sulphide is stripped as a result of reduced partial pressure at the liquid-air interface. The pH of the anaerobically treated water is about 7, at which 32% of the total sulphur species in solution is in the $[H_2S(aq)]$ form and amenable to stripping. The crystallization of calcium carbonate lags that of sulphide removal, which would be in accordance with the hypothesis that the former process is inhibited by organics which is stripped off together with sulphide. However, at this stage the effluent is a colloidal suspension with turbidity as high as 215 NTU. During an earlier investigation, Maree (5) determined that 500 mg/L ferric chloride was required for clarification of this water. This option is, however, unacceptable due to the associated high chloride dosage level.

Aerobic treatment. During aerobic treatment, COD and sulphide were removed to concentrations of 294 and 40 mg/L (as COD and SO_4) respectively, while 500 mg/L calcium (as $CaCO_3$) crystallized (Table III). The turbidity also decreased from 200 to 15 NTU. The reduction in the COD content is related to the biodegradation of compounds, such as acetate, produced in the anaerobic stage. These compounds are utilized to supply the energy requirements of living organisms and to create carbonaceous elements suitable for synthesis of new cells. The residual of 287 mg/L represents phenol, which is not readily biodegradable.

The untreated water of the lead mine has a COD value of 500 mg/L due to some or all of the following organics (11):

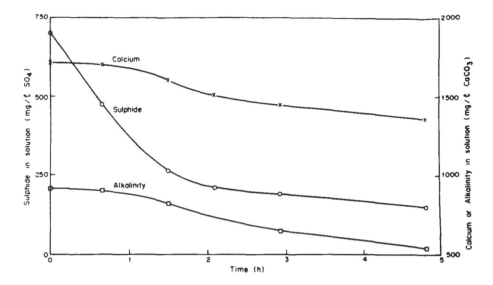

Figure 4.   Influence of air stripping on the removal of sulphide, calcium and alkalinity in solution.

- Isopropyl ethyl thiono-carbamate
- Methyl isobutyl carbinol
- Sodium ethyl xanthate
- Aryl dithiophosphate and
- Dioctyl sodium sulphosuccinamate.

The fact that the COD value of 300 mg/L after anaerobic and aerobic treatment is less than the 500 mg/L of the untreated water shows that the abovementioned organics were partially degraded. However, a similar residual COD value was obtained when water with COD value of only 50 mg/L from another mine was treated. This would indicate that the residual COD content after treatment of both mine waters was probably due to weakly biodegradable components of molasses and not due to organics in the effluents which may interfere with the metallurgical processes.

The decrease in the calcium and alkalinity values indicates that further calcium carbonate crystallization takes place in the aerobic stage. This is due to conversion of soluble organic material to carbon dioxide, which serves as the carbonate source during crystallization. An advantage of this reaction is that the calcium carbonate aids coagulation in the aerobic stage, as indicated by turbidity reduction from 200 to 15 NTU. Therefore, the 500 mg/L ferric chloride that was required for clarification during an earlier investigation could be eliminated.

Table III.   Quality of Water Before and After Aerobic Treatment (results in mg/L)

Determinand	Before	After
pH	8.0	8.4
COD (as O_2)	875	294
Sulphate (as SO_4)	940	1,100
Sulphide (as SO_4)	150	48
Alkalinity (as $CaCO_3$)	800	260
Calcium ($CaCO_3$)	1,652	1,150
Orthophosphate (as P)	4	0.2
Kjeldahl nitrogen (as N)	23	18
Ammonia (as N)	20	15
Nitrate (as N)	0	7.0
Turbidity (NTU)	200	15

The sulphide concentration was reduced from 150 to 40 mg/L (as SO_4) in the aerobic stage. This biological reaction is effected by chemolithotrophic bacteria such as *Ferrobacillus ferrooxidans*, *Thiobacillus ferrooxidans*, and *Thiobacillus thioxidans* (12); Richard and Lundgren (13) or *Thiobacillus thioparus* (14) and serves to limit the occurrence of unacceptably high sulphide concentrations in the final effluent.

General and economic considerations

The practical applicability of a process ultimately depends on its economics. The most costly item of the proposed sulphate reduction process is the organic carbon source, namely molasses. Approximately 2.5 ml (3.4 g) molasses syrup is required to remove 2,480 mg sulphate. Accepting the price of molasses as 8 SA c/kg in South Africa, this cost alone would amount to 27 SA c/m³ to treat water saturated with 2,480 mg/L sulphate. The corresponding molasses cost in California would amount to 0.28 US \$/m³ assuming that molasses is commercially available at 75 US \$ per 2000 lb. By using waste carbon sources such as sewage sludge, if available in sufficient quantities, the treatment cost would be less (4).

If it is mainly heavy metals that need to be removed from mining effluents, this process can serve as a sulphide source by treating a side stream for sulphate reduction. Sulphide can be generated more cheaply than chemicals such as sodium sulphide. The cost related to the carbon source for the biological production of 1 kg sulphide (as SO_4) amounts to RO,ll, while the price of the equivalent amount of sodium sulphide is R1,89 per kg Na_2S (as SO_4) (SA Rl = US 0.50 \$).

CONCLUSIONS

During anaerobic treatment, the sulphur species composition of mining effluent was reduced from the initial 2,050 mg/L sulphate to less than 200 mg/L sulphate (as SO_4), 1,400 mg/L sulphide (as SO_4), elemental sulphur and metal sulphides. Heavy metals were removed while calcium carbonate reached supersaturation levels due to the presence of organics.

The products present in the effluent from the anaerobice stage, namely hydrogen sulphide, calcium carbonate and soluble organic carbon, were successfully removed by H_2S stripping and aerobic treatment. Gaseous hydrogen sulphide content was decreased to 50 mg/L (as SO_4) and the COD to 300 mg/L .

The molasses cost amounts to 28 US c/m³ in California when water with a sulphate content of 2,050 mg/L is treated for complete sulphate removal. The process provides a cheap source of hydrogen sulphide for removal of heavy metals.

It is concluded that this integrated biological process is a viable option for sulphate removal and can be considered for large scale application.

ACKNOWLEDGMENTS

The authors express their thanks to Gold Fields of South Africa for financial support and to the Analytical Department of the NIWR for performing the automated chemical analysis.

REFERENCES

1. Middleton, A. C., and Lawrence, A. W., "Kinetics of Microbial Sulphate Reduction," *J. Wat. Pollut. Control Fed.*, 1659–1670 (1977).
2. Cork, D. J., and Cusanovich, M. A., "Sulphate Decomposition: A Microbiological Process," *Metallurgical Applications of Bacterial Leaching and Related Microbiological Phenomena*, Edited by Murr, L. E., Torma, A. E., and Brierly, J. A. New Mexico, Academic Press, 207–221 (1978).
3. Hilton, B. S., Oleszkiewicz, A., and Oziemblo, Z. J., *Sulfate Reduction as an Alternative to Methane Fermentation of Industrial Wastes*, Canadian Society for Civil Engineering Annual Conference, SASKATOON, SK., 243–261 (1985).
4. Maree, J. P., and Strydom, Wilma F., "Biological Sulphate Removal in an Upflow Packed Bed Reactor," *Water Research*, 19(9) pp. 1101–1106 (1985).
5. Maree, J. P., "A Biological Process for Sulphate Removal from Industrial Effluents," *Water S.A.* (1986).
6. American Public Health Association, *Standard Methods for the Examination of Water and Wastewater*, 16th Edition [1985].

7. Jørgensen, B. B., "Ecology of the Bacteria of the Sulphur Cycle with Special Reference to Anoxic-oxic Interface Environments," *Sulphur Bacteria* London, 543–561 (1982).
8. Lebel, A., Do Nascimento, H. C. G., and Yen, T. F., "Molasses Promoted Biological Sulphur Recovery from High Sulphate Wastes," *Proceedings, 40th Industrial Waste Conference, West Lafayette,* Indiana (1985).
9. Cork, D. J., and Cusanovich, M. A., "Continuous Disposal of Sulphate by a Bacterial Mutualism," *Developments in Industrial Microbiology,* 20, 591–602 (1979).
10. McCartney, E. R., and Alexander, A. E., "The Effect of Additives Upon the Process of Crystallization," *J. Coll. Sci.,* 13, 383–396 (1958).
11. Twidle, T. R., and Engelbrecht, P. C., "Developments in the Flotation of Copper at Black Mountain," *Journal of the South African Institute of Mining and Metallurgy,* 84(6), 164–178 (1984).
12. Razzell, W. E., and Trussell, P. C., "Isolation and Properties of an Iron-Oxidizing *Thiobacillus,*" *J. Bacteriol.,* 85, 595–603 (1963).
13. Richard, F., and Lundgren D. G., "A Comparative Nutritional Study of Three Chemoautotrophic Bacteria: *Ferrobacillus ferrooxidans, Thiobacillus ferrooxidans* and *Thiobacillus thiooxidans,*" *Soil Sci.,* 92, 302–313 (1961).
14. Kelly, D. P., "Biochemistry of the Chemolithotrophic Oxidation of Inorganic Sulphur," *Sulphur Bacteria,* London, 499–528 (1982).

18 ANAEROBIC TREATMENT OF HIGH STRENGTH, HIGH SULFATE WASTES

Barry L. Hilton, Research Assistant

Jan A. Oleszkiewicz, Associate Professor
Department of Civil Engineering
The University of Manitoba
Winnipeg, Manitoba
Canada R3T 2N2

INTRODUCTION

The high rate of accumulation of sulfates constitutes a growing environmental concern. Large quantities of sulfates are generated by the electric power industry from flue gas desulfurization sludges. The pulp and paper industry, pharmaceutical industry, and segments of the food industry produce effluents which are high in sulfates and other species of oxidized sulfur compounds. The application of methanogenic treatment processes to these wastes may be hindered by the resulting reduction of these oxidized sulfur species by sulfate reducing bacteria (SRB) to sulfides. The threshold of inhibition of sulfides, generally acknowledged to be toxic to biomass, in particular the methane producing bacteria (MPB), has been assumed to be 200–300 mg S^{2-}/L [1].

The selectivity of various sulfate reducing bacteria and methane producing bacteria for energy substrates is quite varied. As shown in several recent studies, the spectrum of organics utilized by SRB's is much wider than in the case of methanogens [2–4]. This explains early successes in utilizing SRB's to accelerate anaerobic stabilization of sewage sludges [5,6]. More recently, the mixed-culture work of Middleton and Lawrence [7], DLA Inc. [8], Hilton et al. [9], Oleszkiewicz and Hilton [10,11], and Olthof et al. [12] have indicated substantial advantages of sulfidogenic pathways in application to high sulfate complex industrial waste streams. In the sulfidogenic pathways, organic matter is oxidized using sulfate as an electron acceptor in contrast to the conventional methanogenic pathway where CO_2 acts as an electron acceptor and where the volatile fatty acids are broken down to methane as an end product.

Middleton and Lawrence [7], and Obayashi and Cork [13], using seed cultures obtained from anaerobic digesters, grew SRB's to the exclusion of methanogens, using acetate as the carbon source. In another study [14], the acetate consumption rate was shown to be 15-fold higher for SRB than for MPB's. Lovley et al. [15] found that the SRB's usually outcompete MPB due to lowering the partial pressure of hydrogen below levels that could effectively be utilized by MPB. Based upon those studies, one would expect that in the presence of fermentative bacteria forming short-chain fatty acids, especially lactate and acetate, SRB's would oxidize the lactate and acetate to carbon dioxide and water and would reduce sulfate to sulfide. In the process, one would expect that the methanogenic population either would be outcompeted for substrates or would be inhibited by the generated sulfide.

PURPOSE AND SCOPE

The purpose of this study was to maximize sulfate reduction using a minimal quantity of carbon. It was expected that the lactate formed during acidogenesis would first be incompletely oxidized to acetate and carbon dioxide by incomplete oxidizing SRB's then the remaining acetate would be completely oxidized to carbon dioxide by the complete oxidizing SRB's. In other words, it was expected that the mixed culture would consist of acidogenic anaerobes, incomplete oxidizing SRB's, and complete oxidizing SRB's. Continuous stripping of residual sulfides was operated to eliminate problems of transient inhibition. High sulfate loads, in excess of 1.0 g S^{6+}/L/d were accompanied by high carbon loads in excess of 1.5 g TOC/L/d in order to define the optimum conditions for sulfate reduction in terms of the carbon to sulfur (C/S) ratio in the feed and in terms of maximum organic loading (B_v) and sulfate-sulfur loading (L_s) to the system.

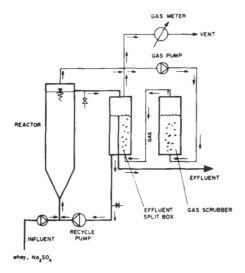

whey, Na₂SO₄

Figure 1. Diagram of Laboratory Scale System.

EQUIPMENT AND PROCEDURES

Two 2.5 L upflow anaerobic sludge bed reactors were operated in parallel (Figure 1) and housed in an environmental chamber operated at 35°C. Recycled product gases were used to purge hydrogen sulfide into a solution of ARI-311C™. The reactors were fed a mixture of acclimated sludge obtained from previous experiments and with anaerobic sludge from the City of Winnipeg North End Treatment Plant anaerobic digesters.

The raw wastewater mixture was composed of powdered whey and sodium sulfate diluted with tap water. The feed was maintained at 3°C and was fed to the reactors by Masterflex™ peristaltic pumps at rates from 2.0–3.25 L/d (HRT = 0.8–1.3 d). The feed concentration was varied from 1.7–6.5 g TOC/L and 0.63–3.0 g S^{6+}/L. Recycle pumps maintained an upflow velocity of 0.5–0.75 m/h in the reactors. Norprene™ pump tubing and stainless steel pump heads were used to minimize pump tube failure.

Influent and effluent TOC values, pH, effluent solids (VSS), influent and effluent sulfates, and gas production, were measured 5–7 times per week. The MLVSS in the reactors were measured once per week. The VFA in the reactors were measured 1–3 times per week. The concentration of sulfides in the reactors were measured 3–7 times per week, both before and after the stripping box.

All routine measurements were according to APHA [16]. Gas volume was measured by Triton WRC Model 181 low flow meters; gas analyses were performed using a Gow-Mac 550 Thermal Conductivity GC with a Poropak Q column. Volatile fatty acids (C2–C5) were analyzed on a Gow-Mac 750 Flame Ionization GC with a Chromasorb 101 column. Total Organic Carbon (TOC) was analyzed on a Dohrmann DC-80 Total Organic Carbon Analyzer. Sulfates were analyzed using a Technicon Autoanalyzer in accordance with the procedure in APHA [16]. Sulfides were analyzed using an Orion specific ion probe with daily standardization.

RESULTS AND DISCUSSION

The kinetics of biomass growth and TOC removal will be described and related to the rate of sulfate reduction. The substrate utilization rate is defined as the mass of substrate removed in an interval of time [17]:

$$r_{su} = \Delta S/\Delta t = \text{mass of sulfate-sulfur reduced/time interval; g/L/d} \qquad (1)$$

$$r_{cu} = \Delta C/\Delta t = \text{mass of substrate (TOC) removed/time interval; g/L/d} \qquad (2)$$

where ΔS and ΔC denote the difference between the influent and effluent concentration of SO_4-S and TOC, respectively.

The specific substrate utilization rate, U, is defined as [17]:

$$U_s = (r_{su}) \, (1/X) = (\Delta S/\Delta t) \, (1/X) = \text{mass of sulfate-sulfur reduced/unit biomass time;} \quad (3)$$
$$(\text{g/g/d}); \, (\text{L/d})$$

or

$$U_c = (r_{cu}) \, (1/X) = (\Delta C/\Delta t) \, (1/X) = \text{mass of substrate (TOC) removed/unit biomass time;} \quad (4)$$
$$(\text{g/g/d}); \, (\text{L/d})$$

Since the biomass yield, Y, is defined as [17]:

$$Y = \Delta X/r_{su} = (\Delta X/\Delta t) \, (\Delta t/\Delta S) = \Delta X/\Delta S = \text{biomass increase/mass of sulfate reduced} \quad (5)$$
$$(\text{g/g})$$

or

$$Y = \Delta X/r_{cu} = (\Delta X/\Delta t) \, (\Delta t/\Delta C) = \Delta X/\Delta C = \text{biomass increase/mass of carbon removed} \quad (6)$$
$$(\text{g/g})$$

and the specific growth rate, μ, is defined as:

$$\mu = (\Delta X/\Delta t) \, (1/X) \quad (\text{g/g/d}); \, (\text{L/d}) \quad (7)$$

then the specific substrate utilization rate may be expressed as [17]:

$$U = \mu/Y_o; \quad (\text{L/d}) \quad (8)$$

where Y_o is the observed yield of biomass.

In a steady-state situation, the net growth rate of biomass is equal to the total growth decreased by the biomass lost due to endogenous energy requirements, defined by the decay coefficient, k_d. This traditionally has been defined as [17]:

$$\mu = Y_t U - k_d \quad (9)$$

The yield of biomass, Y_t, is defined as the "true" yield of biomass, in fact a theoretical maximum yield of biomass from the substrate utilized, assuming that all the substrate is incorporated into new cells, without any losses for energy of maintenance. The relationship between the observed growth yield, Y_o, and true yield, Y_t, is [18]:

$$Y_o = Y_t/(1 + k_d/\mu) \text{ or } 1/Y_o = 1/Y_t + (k_d/\mu) \, (1/Y_t) \quad (10)$$

Prior to analyses, all influent and effluent concentrations were multiplied by Q/V to achieve mass loading values in g/L/d and are written as r_{su} or r_{cu}. Therefore, for the purposes of the following discussion, values denoting concentration have the factor Q/V implicitly included. Sludge concentrations were converted to g/L. Wherever possible, these values were compared with the values of other researchers.

The whey powder used in this experiment was 68% lactose. The anaerobic fermentation of lactose, a disaccharide sugar, results in the formation of monosaccharide sugars, glucose and d-galactose [19]. Depending upon the pH, substrate concentrations, and the particular group of microbes present during the fermentation process, these sugars may be further broken down into lactate, ethanol, acetate, and carbon dioxide through the mechanism of anaerobic glycolysis [18]. Below pH 6.0, lactate production is reduced and acetate production increases whereas at pH = 6–8, lactate production predominates [18]. Dissimilatory sulfate reduction is then performed by two groups of anaerobic sulfate reducing bacteria (SRB). One group, the incomplete oxidizers (ex. *Desulfovibrio* spp., *Desulfotomaculum* spp., *Desulfosarcina* spp.) [4] oxidizes two moles of lactate, pyruvate, α-glycerophosphate, or propionate to two moles of acetate and two moles of carbon dioxide or carbonate in the process of reducing one mole of sulfate to sulfide [4]. The reaction representing the incomplete oxidation of lactate to acetate during the process of dissimilatory reduction of sulfate to sulfide is [4]:

$$2CH_3CHOHCOO^- + SO_4^{2-} \rightarrow 2CH_3COO^- + 2HCO_3^- + H_2S \quad (11)$$
$$\text{lactate} \qquad\qquad\qquad \text{acetate}$$

A second group, complete oxidizing SRB, also called acetoclastic SRB (ex. *Desulfobacter* spp.) [4], oxidizes one mole of acetate to two moles of bicarbonate while reducing one mole of sulfate to sulfide in accordance with the following reaction [4]:

$$CH_3COO^- + SO_4^{2-} \rightarrow 2HCO_3^- + HS^- \quad (12)$$
$$\text{acetate}$$

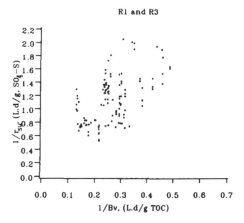

Figure 2. 1/Sulfur Reduced, $1/r_{su}$, L/d/g, versus 1/TOC load applied, $1/B_v$, L/d/g.

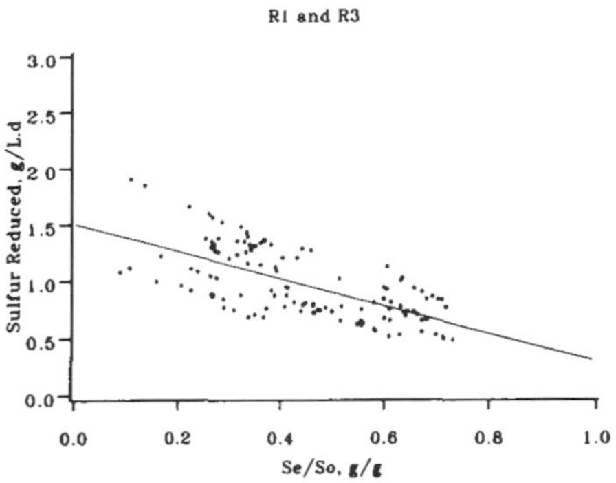

Figure 3. r_{su}, Sulfur Reduced, g/L/d versus oxidized residual sulfur in the effluent, S_e, divided by oxidized sulfur in the influent, S_o, g/g.

In both of the above cases, 2 moles of carbon are oxidized per mole of sulfur reduced for a C/S = 2 (M/M) or C/S = 0.75 (g/g). Any carbon removed in excess of this requirement is either assimilated into cell matter or is converted into other forms such as methane.

Rate of Sulfate Reduction

A reciprocal plot of the rate of sulfate reduction versus TOC load applied to the reactors resulted in the curve shown in Figure 2. The resulting equation is of the form:

$$1/r_{su} = a_1 + a_2/B_v \tag{13}$$

Rearrangement of the terms and substitution for the constants results in an equation of the form:

$$r_{su} = B_v/(0.37 B_v + 2.33) \tag{14}$$

which describes the rate of sulfate-sulfur reduction, in g/L/d, in terms of the TOC load applied to the reactors. The experimental data fit the model for TOC loads of 1.5–6 g TOC/L/d.

The substrate utilization rate, r_{su}, for the rate of sulfate-sulfur reduction can be plotted against S_e/S_o as is shown in Figure 3. Ideally, the intercept of the ordinate, where $(S_e/S_o) = 0$, represents the maximum rate of substrate utilization. The intercept of the abscissa is where $(S_e/S_o) = 1$. The slope of the line, KX, is a constant times the MLVSS in the reactor. If the intercept of the abscissa is $\neq 1$, then there is some uncertainty in the data. For the data reported in this experiment, the maximum rate of reduction, the intercept, is 1.5 g S/L/d which compares with 1.5 g S/L/d reported by Pipes [20] and by Sadana and Morey [6] and is double the 0.7–0.8 g S/L/d range of steady-state values achieved in pilot plant and other studies [12].

If Equation 14 is solved for B_v, one could predict the required TOC load to attain a specified rate of sulfate reduction:

$$B_v = (2.33 r_{su})/(1 - .37 r_{su}); \quad (g/L/d) \tag{15}$$

Substituting $r_{su} = 1.5$ g S/L/d into Equation 15, $B_v = 7.85$ g TOC/L/d which is approximately 21 g COD/L/d which is a high loading rate. If the load were to be reduced to 2.8 g TOC/L/d (7.5 g COD/L/d), the corresponding rate of sulfate-sulfur reduction would be 0.83 g S/L/d. The experimental data showed that for an $r_{su} = 0.8$, the actual B_v was 3.3–4.1 g TOC/L/d. Although the data and calculations indicate that high rates of sulfate-sulfur reduction were possible at high rates of TOC loading, steady-state conditions were more easily maintained at loading rates up to 3.5 g TOC/L/d.

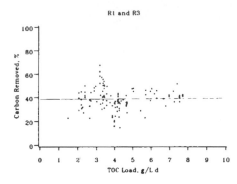

Figure 4. TOC percent removal versus TOC load applied.

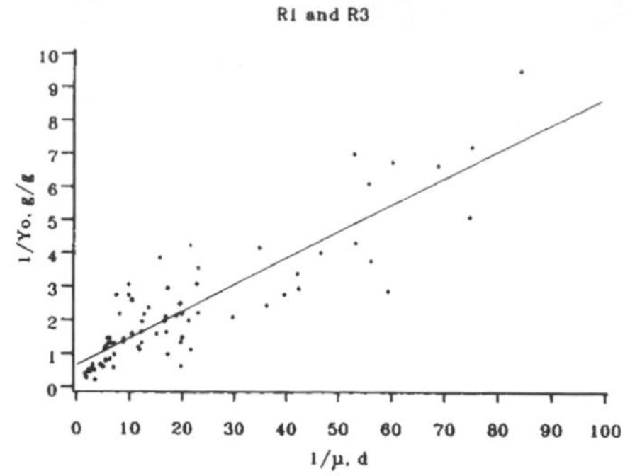

Figure 5. Reciprocal plot, $1/Y_o$, for TOC, versus u^{-1}.

Total Organic Carbon Utilization

A plot of the efficiency of removal versus TOC load (Figure 4) indicated a uniform efficiency of removal of 40% (0.4 g TOC removed/g TOC load/d). This uniform efficiency contrasts to the normal expectations for biological reactors where the removal efficiency decreases with increased load, especially when the reactors are loaded in excess of 4 g TOC/L/d (10 g COD/L/d). The constant ratio of TOC removed to TOC load in this experiment, 40%, regardless of the sulfate reduced, would suggest either that sulfate was not the only electron sink in the reactor or that sulfate reduction was a linear function of the TOC load applied. In the former case, Chartrain and Zeikus [21] have shown that sulfate reducing bacteria, in the absence of sulfate, will use methanogens as the terminal electron acceptor. In the latter situation, sulfate reduction, r_{su}, would have been numerically equal to a constant percentage of the TOC load. The data presented here indicated that r_{su} did not increase as a constant percentage of TOC load but gradually approached an assymptotic value, 1.5 g/L/d. Since the pure culture studies of Domka and Szulczynski [22] indicated a direct correlation between the quantity of sulfate reduced and the quantity of lactate fed and since the microbes in this study appeared to be lactate utilizing incomplete oxidizers, the maximum value for r_{su} may reflect a limitation in the rate of lactate generation in this experiment. If indeed that is the case, then sulfate reduction in these reactors is carbon limited due to a limitation in the quantity of available lactate.

A plot of $1/Y_o$ (TOC) versus $1/\mu$ from this experiment is shown in Figure 5. The intercept is $1/Y_t$ and the slope of the line is both k_d/Y_t and U_c. The specific substrate utilization rate, U_c, is grams of TOC removed per day per gram of biomass in the reactor. For TOC, the calculated values were: $Y_t = 1.3$, $k_d = 0.1$ d^{-1} and $U_c = 0.081$ g/g/d. If one assumes that COD = 2.67 TOC, then $Y_t = 0.45$ in terms of grams VSS per gram COD removed, which compares with that found in aerobic, activated sludge systems rather than anaerobic treatment systems [23]. A comparison of different growth yields from sulfate reduction studies is shown in Table I. The microbial decay coefficient, $k_d = 0.1$ d^{-1}, is high for anaerobic systems where k_d should approach 0.01 but again is similar to values encountered in aerobic activated sludge systems [19]. The comparison to aerobic treatment systems would appear to be justified since the carbon source is oxidized during the process of the dissimilatory reduction of sulfates.

Table I. A Comparison of Different Growth Yields from Sulfate Reduction Studies

Substrate	Yield (g VSS)/(g TOC)	Culture	Source
Acetate	.16	Mixed	[7]
Lactate	.16	*Desulfovibrio* spp.	[23]
Lactate	.88	Mixed	[7]
Propionate	.4	*Desulfobulbus* spp.	[24]
Pyruvate	.27	*Desulfobulbus* spp.	[23]
Formate	.24	*Desulfobulbus* spp.	[23]
Whey	.50	Mixed	This study

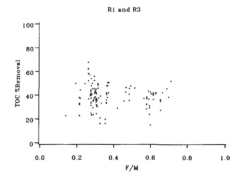

Figure 6. TOC percent removal versus F/M ratios.

A regression analysis was performed on $1/Y_o$ (for sulfate-sulfur) versus $1/\mu$. The slope of the line was the specific substrate utilization rate for sulfate-sulfur, $U_s = 0.08$.

The specific substrate utilization rate, $U_c = 0.081$, which is higher than $U_c = 0.05$ reported by Olthof et al. [12], may reflect the higher loading rates in this study. The ratio of U_c/U_s gave a C/S = 1.0, which is close to the theoretical C/S = 0.75 (Equation 12). The average observed yield for carbon, based upon observed values, was 0.50 g/g with standard deviation = 0.25. The average observed yield for sulfur, based upon observed values, was 0.47 g/g with a standard deviation = 0.22. Calculating the C/S based upon observed yields, C/S = 1.07 (g/g). This ratio is only slightly higher than found by Domka and Szulczynski [22] for growth of *Desulfovibrio vulgaris* on lactate, malate, or α-glycerophosphate.

The Food to Microorganism ratio, F/M, is defined as mass of the substrate fed (g TOC/d) divided by the mass of organisms in the reactor (g VSS).

$$F/M = B_v/X; \quad g\ TOC/g\ VSS/d \tag{16}$$

A plot of TOC removal efficiency versus F/M ($\Delta C/C_o$ versus C_o/Xt) is shown in Figure 6. The trend indicates that at F/M ratios below 0.3, removal efficiency increased. In the study of Olthof et al. [12], the efficiency of COD removal was 60% at an F/M = 0.1 (0.5–2.5 g COD/L/d) and decreased at increasing values of F/M. In that study, 18% of the COD removal was attributable to the production of methane and that 42% of the COD removal was attributable to sulfate reduction. In the work of Oleszkiewicz and Hilton [11], carbon removal efficiencies up to 95% were achieved with loadings up to 2.2 g TOC/L/d (5.8 g COD/L/d) with 52% of that carbon removal being accounted for from the production of methane. Therefore, no more than 43% of the carbon removal in that study was attributable to sulfate reduction. In the study of Maree et al. [25], COD removals of 49–66% were reported. Since methane production was not reported, COD removals attributable to methane production could not be calculated. One could conclude that, based upon the studies of Olthof et al. [12], and Oleszkiewicz and Hilton [11], lower F/M ratios and lower carbon volumetric loadings (B_v) could lead to improved carbon removals as a result of increased carbon removal through the process of methanogenesis. However, it also would appear that the lower loadings would not improve the utilization of carbon in the reduction of sulfate.

Middleton and Lawrence [7] calculated that the maximum specific growth rate was $\mu_m = 0.33\ d^{-1}$ for incomplete oxidizing SRB's and found $\mu_m = 0.13\ d^{-1}$ for complete oxidizing SRB's. In this study, the data did not permit the calculation of μ_m. The μ_m for this study was estimated as being 0.3 d.$^{-1}$ The average specific growth rate, an average of observed values of μ ($\Delta X/X$), was $\mu = 0.12\ d^{-1}$. For a $\mu_m = 0.33$, the Θ_c^m would be 3.0 days which means that the minimum SRT should be 2.1 days. For design purposes, reactors normally are designed for 2–5 times the Θ_c^m which would suggest that, for incomplete oxidizing SRB's, the SRT should be in the range of 6–15 days [17]. It would be expected that an SRT greater than 10 days would provide no advantage if incomplete oxidizing SRB's are to be utilized for sulfate reduction and for carbon removal.

The solids retention time (SRT) for this system ranged from 2 to 75 days as shown in Figure 7. Above an SRT of 3 days, the efficiency of TOC removal and the rate of TOC removal were relatively unaffected by increases in the SRT which would suggest that the SRT for the system exceeded the minimum required SRT, $\Theta_c m$. The fraction of volatile solids in the total effluent solids was 85% in comparison to the 45–50% in the reactor sludge solids. This would suggest that much of the retained sludge in the reactors was old and that it contained mineral precipitates and a concomitant low

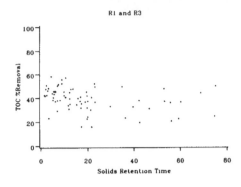

Figure 7.   TOC percent removal versus solids retention time, d^{-1}.

biological activity per unit mass. The higher volatile fraction in the sludge lost from the system would indicate the presence of relatively new biological growth of presumably much higher activity than the retained sludge. The activity of the solids was not tested. The same findings were reported in the study of Maree et al. [25], when the SRT was 10 days or greater, the yield of sludge, g TS/g S reduced remained constant and in the work of Middleton and Lawrence [7], maximum sulfate reduction (acetate as the carbon source) occurred at an SRT of 2 days or greater. In the latter study, the volumetric loadings were less than 5% of the loadings employed in this experiment.

Acetate Utilization

The average acetate concentration in the reactors was 3075 mg/L with a standard deviation = 996 which indicated a low uptake rate of acetate in the reactors. Pure culture studies have shown that some SRB which prefer lactate as a carbon source also metabolize acetate during the process of sulfate reduction [4]. When microorganisms are able to utilize two different carbon sources but utilize one carbon source more easily than another, this process is called diauxie. To test for the presence or absence of diauxie (in this case lactate utilization rather than acetate utilization), the following experiment was performed. The feed pumps were turned off, and the reactor supernatant was tested for TOC and sulfates. The TOC was in excess of 4 g/L and the sulfates were in excess of 0.2 g S/L in the reactor supernatant. During the 14 day period when the feed pumps were turned off, there was no removal of TOC and no sulfate reduction. If diauxie were present, then TOC removal and sulfate reduction should have occurred. Since TOC removal and sulfate reduction did not occur, it was concluded that there was no diauxie.

At the end of this phase of the experiment, the feed pumps and gas recycle pumps were turned off, but the liquid recycle pumps were left running. After a two month break, the reactor supernatnant was tested for TOC, VFA, and sulfates. Analyses showed that 77% of the remaining TOC was removed, 97% of the remaining VFA were removed, and 97% of the remaining sulfates were reduced. This utilization of VFA for sulfate reduction closely corresponded with the work of Middleton and Lawrence [7]. Due to the lag time required to utilize the VFA for sulfate reduction, the additional TOC removal and sulfate reduction were a result of a shift in the microbial population and were not due to diauxie.

In the studies of Olthof et al. [12] and Oleszkiewicz and Hilton [11], carbon removal by the process of methanogenesis augmented carbon removal by sulfidogenesis. In the study reported here, carbon removal by the process of methanogenesis was minimal. As a result, there were high residual concentrations of acetate in the effluent.

The upper limit on the rate of sulfate reduction in these reactors could be the result of several factors: 1) inhibition by metabolic products (toxicity); 2) insufficient carbon; 3) insufficient sulfur; 4) insufficient biomass; or 5) substrate limitation in other nutrients.

If inhibitory compounds were formed between sulfides and the metabolic products of the SRB's, inhibition would have resulted in a buildup of VFA and a gradual decrease in the rate of TOC removal from the system as the TOC load increased. These inhibitory products could be in a form such as $HSCH_2COO^-$ or CH_3COSH. The presence or absence of these compounds in the reactors was not known since testing for these compounds was not performed. The toxicity of these compounds to the microbial biomass in these reactors is not known at this time.

The accumulation of VFA, especially acetate, did occur in these reactors; however, the TOC removal, r_{cu} ($\Delta C/\Delta t$), increased with an increase in the load to the reactors with little or no change in the removal efficiency ($\Delta C/C_o$, %). Therefore, if toxic metabolic products were present, they were inhibiting the acetoclastic MPB's and SRB's and directly or indirectly affected the rate of sulfate reduction by the dominant population of SRB's. The presence or absence of toxicity will be defined in future work.

Between day 50 and day 70, the feed was supplemented with lactate. The rate of sulfate reduction increased stoichiometrically as predicted by the reaction in Equation 4 and acetate accumulated in the effluent of the reactors. Therefore, it was concluded that these reactors utilized incomplete oxidizing, lactate utilizing sulfate reducing microbes. This would suggest that the rate of sulfate reduction is dependent upon the rate of lactate formation during the acidogenic phase of the process and, therefore, is lactate limited which corresponds with the results of the work of Domka and Szulczynski in which sulfate reduction was dependent upon the quantity of lactate in the culture [26].

Excess sulfate-sulfur was always present in the reactors; only rarely was the reactor sulfate-sulfur concentration below 150 mg/L. During periods of high TOC loading and high sulfate loading, the reactor sulfate-sulfur concentration was greater than 500 mg S^{6+}/L. Therefore, the maximum rate of sulfur reduction was not limited by the concentration of sulfur.

The maximum concentration of volatile solids in the reactors was 23 g/L, which coincided with the maximum observed rate of sulfur reduction in the reactors and the greatest TOC load to the reactors. The specific substrate utilization rate was 0.06–0.08 g S^{+6}/g VSS/d. Due to the continuous agitation of the effluent split-boxes for gas purging, solids lost in the effluent were not recaptured. In one study, flocculant sludge bed reactors were shown to lose sludge and refused to produce granular sludge in methanogenic systems; therefore, clarifiers or layers of media were employed to retain the biomass [27]. In the study of Isa et al. [28], it was shown that SRB's (most likely acetoclastic) were lost more rapidly than MPB's in continuous flow reactors. Consequently, one might conclude that improved solids retention could increase the sulfate-sulfur reducing capacity of the system.

Separate 2.5 L continuous flow reactors were connected in series to the two 2.5 L units receiving the whey and sodium sulfate. These reactors collected all the solids as well as the liquid which was lost from the first or primary reactors. The pH of the secondary reactors remained between pH 6.5 and 8.2. When the first reactor was operating at a steady-state with pH > 6.5, the second reactor removed 8% of the TOC to raise the total TOC removal from 35% to 40%. During the same period, the second reactor reduced an additional 24% of the sulfur to raise the total sulfate-sulfur reduced from 70% to 77%. Based upon the latter data, one would conclude that, for this system, additional retention most likely would not significantly improve sulfur reduction nor TOC removal.

The work of Middleton and Lawrence [7] showed that sulfate could be reduced by acetoclastic sulfate reducing bacteria obtained from the supernatant of an anaerobic digester. Based upon that work, it was expected that it would be possible to maintain a co-culture of incomplete and complete oxidizing SRB in one reactor where the incomplete oxidizers would oxidize lactate to acetate and the complete oxidizers would oxidize acetate to carbon dioxide.

In spite of high concentrations of acetate in the reactors, sulfide concentrations less than 80–100 mg/L, and redox potentials below −200 mv, sulfate reduction by acetoclastic SRB was not evident until after feed to the reactors had been discontinued for several weeks. The work of Middleton and Lawrence [7] indicated that, at 32°C, a 10 day SRT was sufficient. However, in that work, the volumetric loading was less than 10% of the loading utilized in the study reported here. To date, the authors are not aware of studies where it has been noted that incomplete and complete oxidizing SRB's have been grown in the same culture to effect sequential carbon utilization during the process of sulfate reduction. The attempt to culture both groups of SRB in a single continuous flow reactor in the present experiment was unsuccessful.

Very little methane production was recorded during the course of this experiment. The low production may have been due to the presence of unknown inhibitory compounds; it may have been the result of occasional brief excursions of sulfide concentrations in excess of 200 mg/L; it may have been the result of the incomplete oxidizing SRB outcompeting the acetoclastic SRB and MPB for hydrogen, or it may have been the result of some other unknown factor. The work of Olthof et al. [12], using a variety of substrates including sewage sludge and a specialty industrial waste, Maree et al. [23], using molasses, and Saw and Anderson [29], using food processing wastes, showed a similar lack of methane production in experiments of several weeks duration.

CONCLUSIONS

The maximum rate of sulfate-sulfur reduction by this system was 1.7 g S^{6+}/L/d; this was achieved at 5.7 g TOC/L/d and at pH 6.8.

The observed growth yield, $Y_o = 0.5$, and "true" growth yield, Y_t were very high when compared with acetoclastic sulfidogenic or methanogenic anaerobic reactors and compared with values normally associated with aerobic systems.

Methanogenesis did not occur under conditions of high sulfate load in spite of low concentrations of sulfides and in spite of sufficient concentrations of acetate (a methanogenic precursor).

Total carbon removal from a sulfidogenic reactor, operating at high TOC loads, using a carbon substrate such as whey which will ferment to lactate, was limited to 40-50% of the carbon being fed to the system.

The data presented here indicated that acetoclastic SRB were not active in the continuous flow reactors.

ACKNOWLEDGEMENTS

This research was sponsored by a grant from the Natural Sciences and Engineering Research Council of Canada. The authors acknowledge the advice of Dr. A. B. Sparling, Department of Civil Engineering, and Dr. G. Blank, Department of Food Sciences, during the course of the research. Thanks go to Mrs. Judy Tingley and Stan Mateja for their technical assistance in performing analyses.

NOMENCLATURE

B_v	— Organic loading, TOC, or COD, (kg/m³/d,g/L/d)
L_s	— Sulfate loading, $S-SO_4^{2-}$ (kg/m³/d, g/L/d)
MLVSS	— Mixed liquor volatile suspended solids in the reactor, g/L
MPB	— Methane Producing Bacteria
μ	— Specific growth rate, d^{-1}
μ_m	— Maximum specific growth rate, d^{-1}
Q	— Flow, L/d
r_{su}	— Substrate utilization rate, $(S_o-S_e(Q/V))$, g/L/d
r_{su}	— Sulfate-sulfur reduction rate, $(S_o-S_e)(Q/V)$, g/L/d
S_e	— Effluent substrate concentration, TOC or SO_4-S, mg/L
S_o	— Influent substrate concentration, TOC or SO_4-S, mg/L
SRB	— Sulfate Reducing Bacteria
Θ_c	— Solids residence time (SRT) numerically equivalent to $1/\mu$, d
Θ_c	— The minimum SRT at which washout occurs.
TOC	— Total Organic Carbon
U_c	— Specific substrate utilization rate, r_{cu}/X, TOC, g/g/L/d
U_s	— Specific substrate utilization rate, r_{su}/X, Sulfate-sulfur, g/g/L/d
V	— Reactor volume, L
VFA	— Volatile Fatty Acids
X	— Reactor concentration of cells, MLVSS, g/L
$\Delta X/\Delta t$	— Growth of cells, VSS, g/L/d
Y_o	— Observed yield ratio of the mass of cells formed to the mass of substrate consumed, $\Delta X/r_{cu}$, g/g
Y_t	— Maximum yield coefficient, the ratio of the mass of cells formed to the mass of substrate consumed, g/g

REFERENCES

1. Kroiss, H., and Plahl-Wabnegg, F., "Sulfide Toxicity with Anaerobic Waste Treatment," Anaerobic Waste Treatment Conference, Amsterdam, 72 (1983).
2. Imhoff, S. D., and Pfening, N., "Isolation and Characterization of a Nicotinic Acid-degrading Sulfate-reducing Bacterium, *Desulfococcus niacini, Archives of Microbiology*, 136, 194 (1983).
3. Laanbroek, H. J., and Pfening, N., "Oxidation of Short-chain Fatty Acids by Sulfate Reducing Bacteria in Freshwater and in Marine Sediments," *Archives of Microbiology*, 128, 330 (1981).
4. Postgate, J. R., *The Sulphate Reducing Bacteria*, Cambridge University Press, Cambridge (1984).
5. Burgess, S. G., and Wood, L. B., "Pilot Plant Studies in Production of Sulfur from Sulfate-enriched Sewage Sludge," *Journal of the Science of Food and Agriculture*, 12, 326 (1961).
6. Sadana, J. C., and Morey, A. V., "Microbial Production of Sulfide from Gypsum," *Journal of Science and Industrial Research*, 21C, 124 (1962).
7. Middleton, A. C., and Lawrence, A. W., "Kinetics of Microbial Sulfate Reduction," *Journal Water Pollution Control Federation*, 49, 1659 (1977).
8. DLA, Inc., "Hydrogen Sulfide Recovery from Calcium Sulfate — Phase I. A Quantitative Feasibility Study," *Final Report*, W. R. Grace Co. (1982).
9. Hilton, B. L., Oleszkiewicz, J. A., and Oziemblo, Z. J., "Sulfate Reduction as an Alternative to Methane Fermentation of Industrial Wastes," *Proceedings Canadian Society of Civil Engineering Annual Conference*, May 27–31, 1985, Saskatoon, Sask., 243 (1985).
10. Oleszkiewicz, J. A., and Hilton, B. L., "Anaerobic Treatment of High Sulfate Wastes," In *New Directions and Research in Waste Treatment and Residuals Management*, June 23–28, 1985, The University of British Columbia, Vancouver, B. C., 864 (1985).
11. Oleszkiewicz, J. A., and Hilton, B. L., "Anaerobic Treatment of High Sulfate Wastes," *Canadian Journal of Civil Engineering*, in press (1986).
12. Olthof, M., Kelly, W. R., Oleszkiewicz, J. A., and Weinreb, H., "Development of Anaerobic Treatment Process for Wastewaters Containing High Sulfates," *Proc. 40th Ind. Waste Conf.*, Purdue Univ. (1986).
13. Obayashi, A. W., and Cork, D. J., "Microbial Sulfur Recovery." *Management of Industrial Pollutants by Anaerobic Processes*, Report EPA-600/2-83-119, National Technical Information Service (1983).
14. Schönheit, P., Moll, J., and Thauer, R. K., "Kinetic Mechanism for the Ability of Sulfate Reducers to Outcompete Methanogens for Acetate," *Archives of Microbiology*, 132, 285 (1982).
15. Lovley, D. R., Dwyer, D. F., and Klug, M. J., "Kinetic Analysis of Competition Between Sulfate Reducers and Methanogens for Hydrogen in Sediments," *Applied and Environmental Microbiology*, 43, 1373 (1982).
16. APHA, *Standard Methods for the Examination of Water and Wastewater*, 16th Edition, American Public Health Association, New York (1985).
17. Benefield, L. D., and Randall, C. W., *Biological Process Design for Wastewater Treatment*, Prentice-Hall, Inc., Englewood Cliffs, NJ (1980).
18. Gaudy, A. F., Jr., and Gaudy, E. T., *Microbiology for Environmental Scientists and Engineers*, McGraw-Hill, New York (1980).
19. Zubay, G., *Biochemistry*, Addison-Wesley, Reading, MA (1983).
20. Pipes, W. O., Jr., "Sludge Digestion by Sulfate Reducing Bacteria," *Proc. 15th Ind. Waste Conf.*, Purdue Univ., 308 (1960).
21. Chartrain, M., and Zeikus, J. G., "Microbial Ecophysiology of Whey Biomethanation: Characterization of Bacterial Trophic Populations and Prevalent Species in Continuous Culture," *Appl. Environ. Microbiol.*, 51, 188 (1986).
22. Domka, F., and Szulczynski, M., "The Effect of Organic Substrate Concentration on Activity for Microbiological Reduction of Sulfates," *Acta Microbiol. Pol.*, 28, 237 (1979).
23. Metcalf and Eddy, Inc., *Wastewater Engineering: Treatment, Disposal, Reuse*, Second edition, McGraw-Hill, New York (1979).
24. Magee, E. L., Jr., Ensley, B. D., and Barton, L. L., "An Assessment of Growth Yields and Energy Coupling in *Desulfovibrio*," *Arch. Microbiol.*, 117, 21 (1978).
25. Maree, J. P., et al., "Biological Treatment of Mining Effluents," *Proc. 41st Ind. Waste Conf.*, Purdue Univ. (in press).
26. Widdell, F., and Pfennig, N., "Studies on Dissimilatory Sulfate-Reducing Bacteria that Decompose Fatty Acids, II. Incomplete Oxidation of Propionate by *Desulfobulbus propionicus* gen. nov., sp. nov," *Arch. Microbiol, 131*, 360 (1982).
27. Oleszkiewicz, J. A., Hall, E. R., and Oziemblo, Z. J., "Performance of Laboratory Anaerobic Hybrid Reactors with Varying Depths of Media," *Environ. Technol. Lett.*, in press (1986).

28. Isa, Z., Grusenmeyer, S., and Verstraete, W., "Sulfate Reduction Relative to Methane Production in High-rate Anaerobic Digestion: Technical Aspects," *Appl. Environ. Microbiol.*, 51, 580 (1986).
29. Saw, C. B., and Anderson, G. K., "Comparison of the Anaerobic Contact and Packed Bed Processes for the Treatment of Edible Oil Wastewaters," *Proceed. 41st Indust. Waste Conf.*, Purdue Univ. (in press).

19 BIOPHYSICAL TREATMENT OF LANDFILL LEACHATE CONTAINING ORGANIC COMPOUNDS

Sheila F. McShane, Senior Engineer

T. E. Pollock, Manager
Industrial Services Department
James M. Montgomery, Consulting Engineers, Inc.
Pasadena, California 91109

Alon Lebel, Project Engineer
BKK Corporation
West Covina, California 91791

Bryan A. Stirrat, Principal
Bryan A. Stirrat and Associates
City of Industry, California 91744

INTRODUCTION

BKK Corporation (BKK) operates the BKK Landfill in West Covina, California, which is located approximately 20 miles east of central Los Angeles. The total area occupied by the landfill is approximately 580 acres. The landfill operations commenced in 1963 when non-hazardous municipal and commercial wastes were accepted. In 1972 the landfill began accepting hazardous waste and that phase of the operation continued until 1984 when it was voluntarily discontinued. BKK has continued dumping non-hazardous wastes on top of the hazardous waste area to complete the contouring of the landfill for adequate site draining and landfill gas control. In 1987 the non-hazardous waste disposal operations will commence in a new canyon on-site away from the previous hazardous waste area. It is expected that disposal operations will continue until 1995. The landfill is presently accepting 7,500 tons per day of non-hazardous wastes.

To mitigate environmental impacts associated with the landfill, BKK operates an extensive leachate/gas collection system, as well as an extensive series of monitoring wells. The purpose of the majority of these facilities is to collect all liquids emanating from the trash disposal area and treat them onsite. Figure 1 presents the liquid management systems within the landfill. Currently, mildly contaminated landfill waters are used for dust control onsite, and the more contaminated leachate wastestreams are subjected to air stripping prior to onsite solidification and redisposal at the landfill. Due to regulatory restrictions at the landfill and the labor intensive nature of these processes, BKK wishes to discontinue the solidification of the leachate and redisposal onsite. It is BKK's intent to implement a program to dewater the landfill. To aid in accomplishing this, a series of horizontal drains and vertical wells extending into the trash prisms have been installed and more are planned for installation in the near future. This dewatering activity is part of BKK's plan to close and maintain that portion of the landfill containing deposits of hazardous waste in accordance with RCRA requirements.

Several disposal options for the leachate were evaluated. The disposal options considered included discharge to a publicly owned treatment works (POTW), discharge to a surface water body or on-site use for landscape irrigation and dust control utilizing the POTW for disposal of excess as needed. The last option is considered the most viable at this time. In order to utilize any of these options, treatment of the leachate would be required. Biological, physical/chemical, and biophysical processes were evaluated to determine their potential suitability for treatment of the leachate from the landfill. A biophysical process which utilizes a suspended growth biological culture supplemented by the addition of powdered activated carbon (PAC) to the aeration basin was recommended as the basic treatment system. This is a proprietary system named PACT (Powdered Activated Carbon Treatment) which is licensed by Zimpro, Inc. of Rothchild, Wisconsin. The biophysical treatment system has been used successfully for the treatment of hazardous waste and in industrial applications with highly variable influents [1,2,3]. Since landfill leachate contains a heterogeneous mixture of biodegradable com-

167

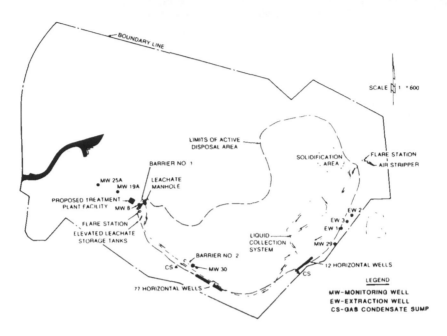

Figure 1.   Liquid management facilities at the BKK sanitary landfill.

pounds, biologically resistant compounds, and potentially toxic materials, the PACT process appeared to be a suitable process for its treatment.

A biophysical system has a lower capital cost, material requirements, and requires less land area than an activated sludge system followed by a carbon contacting system. The combined biological-physical processes also involve treatment synergism with associated higher contaminant removal rates than an activated sludge system—carbon system. The PACT process has been reported to have successfully treated wastes containing priority pollutants. In the presence of PAC, biological cultures showed no adverse reactions to slug doses or continuous feed of priority pollutants [4]. One reason that the PACT process provides more stable treatment than a conventional activated sludge system is that potentially toxic compounds in a wastestream can be adsorbed thereby protecting the biological system. A higher sludge age can be maintained in a PACT system than in an activated sludge system. The higher sludge age results in greater treatment stability. In a PACT system satisfactory sludge settlement can be maintained at the higher sludge ages. The presence of powdered activated carbon increases sludge density and compaction. At higher sludge ages the sludge generated by an activated sludge system tends to form filamentous bacteria which retard sludge settling [5]. It has also been noted that the PACT process has been used to decrease foaming in systems by the removal of surfactants.

Although a review of the available literature indicates that the PACT process has a successful track record of treating a wide variety of toxic industrial wastes and priority pollutants, the process has a limited history of use with landfill leachates such as those found at the BKK Landfill. The lack of historical operating data precluded the definitive determination of the degree of contaminant removal that could be obtained in this type of treatment system. Because of this, an experimental program was implemented to determine the suitability and effectiveness of the PACT process for treatment of BKK's leachate. A description of the design operation and results of the experimental program are presented hereafter.

METHODS AND MATERIALS

Table I presents the design criteria used in the experimental study. Four 55 gallon drums were utilized as reactors in this experiment. The first reactor was operated as an activated sludge system only, with no PAC addition. The second reactor was operated with a mixed liquor PAC concentration

Table I. Design Criteria

Item	Units	Value
Reactor Type	—	Fill and Draw
Reactor Influent	—	Baker Tank Contents
Reactor Volume	gallon	55
Feeding Schedule	—	Twice per day
Total Testing Time	days	30 (startup) + 60
		(3 cell turnovers)
PAC Level Maintained		
Reactor 1	mg/L	0 (control)
Reactor 2	mg/L	5,000
Reactor 3	mg/L	10,000
Reactor 4	mg/L	10,000
Mean Cell Residence Time	day	20
Hydraulic Detention Time	day	3
Volumetric Air Input	scfm/cu. ft.	0.28
Air Flow/Reactor	scfm	1.7

of 5,000 mg/L. The third and fourth reactors were operated as replicate systems with PAC concentrations of 10,000 mg/L. The reactors were operated in a complete-mix, extended aeration, fill and draw (semibatch) mode with a mean cell residence time of 20 days. The experiment systems are shown in Figure 2. The units were serviced ("fed") by wasting sludge from the mixed liquor, then sludge was settled by shutting off the air flow and effluent was wasted as necessary to maintain the desired hydraulic retention time (HRT). Influent leachate, makeup PAC slurry, and nutrients were then added to the systems and aeration was resumed. To prevent potential organic compound emissions from the reactors, off-gasses were piped to the landfill's flares for incineration. An additional tank was used for the storage and maintenance of a biomass supply that could have been used for reseeding purposes in case of an upset in one of the reactors.

Throughout the course of the study, daily analyses were performed to determine the reactors' solids inventory (total suspended solids (TSS), volatile suspended solids (VSS), and PAC concentrations), as well as the influent and effluent soluble chemical oxygen demand (SCOD) values. Soluble biochemical oxygen demand (SBOD), priority pollutants, and other selected organic compounds (acetone, methyl ethyl ketone, and tetrahydrofuran) were measured on a weekly basis. All analyses utilized

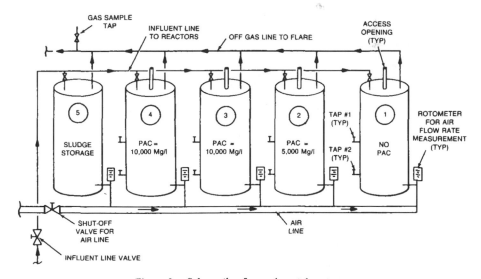

Figure 2.   Schematic of experimental system.

standard EPA analytical procedures. Soluble, rather than conventional, COD and BOD were measured because of the interference caused by the presence of PAC particles in conventional testing.

The choice of PAC concentrations used in the reactors was based on a literature review of reported PACT pilot and full scale operational parameters for the treatment of chemical wastes, contaminated groundwater, and wastewater containing priority pollutants [1,4,6,7]. The PACT type used was a wastewater grade, with a medium surface area and was of a coarse grind to enhance sludge settleability.

RESULTS AND DISCUSSION

Influent Characteristics

The BKK leachate can be characterized as having high dissolved solids, high SCOD and SBOD levels, and relatively low metal concentrations. The leachate contains significant amounts of the volatile compounds acetone, methyl ethyl ketone (MEK), tetrahydrofuran (THF), and 1,2 dichloroethane (DCE). Other priority pollutants are present at relatively low concentrations. A summary of the experimental systems' average influent chemical composition is presented in Table II.

Solids Inventory

Figure 3 presents the total suspended solids concentrations that were maintained in each of the reactors over the course of the experimental study. After the solids buildup during the acclimation period, the TSS generally remained stable in all the systems. It appears that the differences in the TSS levels among the reactors primarily reflect the powdered activated carbon addition.

When PAC is added to a biological system, the determination of the components in the solids phase of the system becomes more difficult than in an activated sludge system. Other than radiotracing, the most accurate method presently available for determination of PAC and biomass concentrations is the nitric acid digestion method [8]. In this technique, nitric acid is used to solubilize the biomass growing on the surface of the carbon while leaving the PAC intact. In practice there are several sources of error associated with this technique. The PAC particles may be solubilized to a certain degree by the nitric acid. The calculation of the biomass concentration involves two volatile component determinations. Since the volatile component determination is not an accurate analytical procedure, the error is compounded in the biomass calculation. Since the nitric acid digestion method was used in this study, biomass determination results obtained fluctuated; however, the values determined generally indicate that similar biosolids generation rates were seen in both the activated sludge and PACT reactors. Published information on PACT systems indicate that biosolids generation rates are generally the same or are slightly lower than those found in activated sludge systems [9].

SCOD and SBOD Removals

The reduction of SCOD in all the experimental reactors was significant. The extent of SCOD reduction is graphically shown through the effluent versus influent plot in Figure 4. The effluent SCOD values for the various reactors are "blown-up" and are shown in Figure 5. The graph indicates that the PACT systems (Reactor Nos. 2, 3, and 4) consistently reduced the SCOD leachate levels to lower concentrations than the activated sludge system (Reactor No. 1). The PACT systems' effluent peaks are less severe than those of activated sludge indicating the PACT treatment systems were more stable. This phenomena was especially apparent during the acclimation phase of the experiment.

Reactor No. 2, which maintained a PAC mixed liquor concentration of 5,000 mg/L, produced an effluent SCOD concentration that was marginally higher and more erratic than that of Reactor Nos. 3 and 4, which contained 10,000 mg/L PAC. The performance difference among the PACT reactors, however, is slight as the effluent variation among them averaged approximately 40 mg/L. Cumulative probability plots assuming a normal data distribution was developed for the SCOD influent and effluent values. This data is presented in Table III.

A summary of influent and effluent SBOD values for the test systems is presented in Table IV. The SBOD removal efficiency was greater in the PACT systems than in the activated sludge system. Reactors Nos. 3 and 4, which contained the higher PAC concentrations, had marginally higher SBOD removals than Reactor No. 2.

Table II. Average Influent Chemical Concentrations to the Experimental Systems

Constituent	Average Concentration (mg/L)
General Parameters	
SCOD	3,237
SBOD	1,603
TSS	527
TDS	6,010
Oil and Grease	71.6
Inorganic Compounds	
Alkalinity (as CaCO$_3$)	1,433
Cadmium	0.018
Copper	0.05
Cyanide	<0.02
Lead	<0.05
Nickel	0.15
Silver	0.02
Zinc	0.11
Total Kjeldahl Nitrogen (TKN)	77.4
Orthophosphate	1.1
Organic Compounds	
Acetone	13.4
Methyl Ethyl Ketone	10.9
Tetrahydrofuran	2.9
1,2 Dichloroethane	2.3
Benzene	0.16
1,1 Dichloroethane	0.16
Methylene Chloride	1.1
Toluene	0.51
1,2 Dichloropropane	1.3
1,1,2 Trichloroethane	0.23
Isophorane	0.063
bis (2-Ethylhexyl) Phthalate	0.14
Naphthalene	0.14
2,4 Dimethylphenol	0.42
Phenol	0.98

SCOD Removal Kinetics

A three day hydraulic retention time was maintained in all the test reactors for the majority of the experimental period. Towards the end of the experiment, however, a special test was performed on Reactor No. 4 to determine the effect of the variation of the HRT on SCOD removal. The HRT was incrementally decreased from 3 days to 1.5 days over approximately 2 weeks. The mean cell residence was maintained constant at 20 days. A kinetic study ws performed at several of the HRT values by periodically withdrawing mixed liquor samples from the reactor after feeding. The samples were

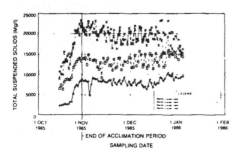

Figure 3.   Total suspended solids
concentrations maintained in the reactors.

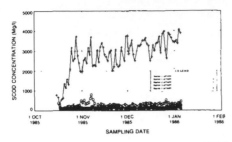

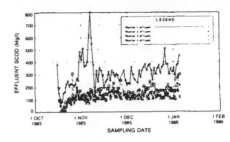

Figure 4. Influent and effluent SCOD concentrations maintained in the reactors.

Figure 5. Effluent SCOD concentrations maintained in the reactors.

Table III. Summary of SCOD Values

	50th Percentile Value (mg/L)	90th Percentile Value (mg/L)
Influent	3,200	3,788
Reactor No. 1 Effluent	314	405
Reactor No. 2 Effluent	164	228
Reactor No. 3 Effluent	122.5	159
Reactor No. 4 Effluent	127.5	172.5

immediately filtered and analyzed for SCOD concentrations. The results of this testing are presented in Figure 6. The results indicate that the SCOD removal rates increased as the initial reactor concentration increased and the HRT decreased. The SCOD concentrations were ultimately reduced to approximately the same effluent concentrations for each of the reactors.

Poor sludge settleability and excessive scum buildup became a problem when the HRT was reduced to 1.5 days. This resulted in a loss of solids from the system. The cause of these problems may have

Table IV. Summary of SBOD Values

	Average Value (mg/L)	Average Removal Efficiency (%)
Influent	1,530	
Reactor No. 1 Effluent	90	94
Reactor No. 2 Effluent	41	97
Reactor No. 3 Effluent	31	98
Reactor No. 4 Effluent	34	98

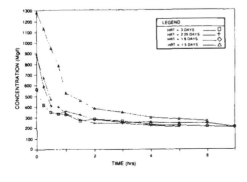

Figure 6. SCOD removal kinetics with varying hydraulic retention times in reactor 4.

Table V. Summary of Acetone, Methyl Ethyl Ketone and Tetrahydrofuran Values

	Acetone		MEK		THF	
	Average Value (mg/L)	Average Removal Efficiency (%)	Average Value (mg/L)	Average Removal Efficiency (%)	Average Value (mg/L)	Average Removal Efficiency (%)
Influent	13.4		10.9		2.9	
Reactor No. 1 Effluent	0.47	96	0.36	97	0.61	79
Reactor No. 2 Effluent	0.04	>99	0.06	99	0.22	92
Reactor No. 3 Effluent	0.03	>99	0.05	99	0.29	90
Reactor No. 4 Effluent	0.17	99	0.12	99	0.36	88

been due to a combination of factors. They include toxicity problems associated with high initial organic loadings, a short term increase in biomass concentrations and/or entrainment of bacterial floc in oil and grease. During this period of the testing, slug loadings of oil and grease appear to have been entering the reactors from some of the leachate storage tanks.

Organic Compound Removals

Effluent waters from the four experimental reactors were periodically sampled and analyzed for volatile organic compounds and base/neutral and acid (BNA) extractable compounds. The acetone, MEK, and THF concentrations in the reactor effluents were determined on a weekly basis. A summary of the results of this testing is presented in Table V. All four reactors effectively removed acetone and MEK. The presence of PAC in Reactor Nos. 2, 3, and 4 resulted in superior average removals of 99% and greater. There was no discernible difference between the performance of the reactors with varying PAC dosages. The removal efficiences of THF were somewhat less than those for the removal of acetone and MEK. Again, however, as was observed for acetone and MEK removals, the presence of PAC increased the removal efficiency of the process but the two-fold increase in the PAC dosage (Reactor Nos. 3 and 4 versus Reactor No. 2) did not result in an apparent increase in removal efficiency.

On three occasions, complete priority pollutant, volatile organic compound scans of the reactor influent and effluents were conducted, and one set of representative results is presented in Table VI. The priority pollutant compound listed in the table includes only those volatile compounds which were present at influent concentrations of 0.1 mg/L or greater. The removal of volatile compounds were, in the majority of cases, to levels at or below detection limits.

Table VI. Reactor Effluent Volatile Priority Pollutants Concentrations and Removal Efficiencies of November 20, 1985[a]

		VOC Concentrations, mg/L				
Date	Compound[b]	Influent	Reactor No. 1 Effluent	Reactor No. 2 Effluent	Reactor No. 3 Effluent	Reactor No. 4 Effluent
11/20/85	Benzene	0.17	<0.25	<0.05	<0.05	<0.05
	1,1–DCE	0.14	<0.25	<0.05	<0.05	<0.05
	1,2–DCE	<0.25	<0.25	<0.05	<0.05	<0.05
	MC	<0.25	<0.25	<0.05	<0.05	<0.05
	Toluene	0.68	0.25	<0.05	<0.05	<0.05
	1,2–DCP	4.7	<0.25	<0.05	<0.05	<0.05
	1,1,2–TCA	0.22	<0.25	<0.05	<0.05	<0.05

[a]Non-detected values are reported as less than the detection limit that prevailed for that particular analysis.
[b]Legend:
   DCE = Dichloroethane
   MC = Methylene Chloride
   DCP = Dichloropropane
   TCA = Trichloroethane

Table VII. Reactor Effluent BNA Extractable Organic Compound Concentrations and Removal Efficiencies[a]

Date	Compound		BNA Extractable Organic Concentrations, mg/L			
		Influent	Reactor No. 1 Effluent	Reactor No. 2 Effluent	Reactor No. 3 Effluent	Reactor No. 4 Effluent
11/20/85	bis (2E)pH[b]	0.27	<0.01	<0.02	<0.02	<0.02
	Isophorane	0.17	<0.005	<0.005	<0.005	<0.005
	Naphthalene	0.05	<0.005	<0.005	<0.005	<0.005
	2,4-DMP[c]	1.1	<0.005	<0.005	<0.005	<0.005
	Phenol	2.3	<0.005	<0.005	<0.005	<0.005

[a]Non-detected values are reported as less than the detection limit that prevailed for that particular analysis.
[b]bis (2-Ethylhexyl) Phthalate.
[c]2,4-Dimethylphenol.

The concentrations of BNA extractable compounds in the reactor influent and effluents are presented in Table VII. Removals of these compounds were achieved to levels below the detection limits for all reactors.

The effluent volatile and BNA extractable compound concentrations were summed and averaged as an estimate of the effluent total toxic organic (TTO) concentrations. The resulting average and maximum influent and effluent TTO levels are summarized in Table VIII. The lowest average and maximum TTO concentrations were achieved with the reactors containing PAC. All four reactors, howver, produced effluents of average TTO levels of less than 1 mg/L.

SLUDGE SETTLEABILITY

Daily sludge volume indices (SVI) were determined for the mixed liquor sludges of the activated sludge reactor and the biophysical PACT reactors. The SVI is a relative measure of the settleability of a biological sludge and is defined as the volume of sludge which settles in a graduated cylinder, in a given period of time, per unit mass of mixed liquor suspended solids. For this study, mixed liquor was settled in a 100 mL graduated cylinder for 30 minutes. The procedure as used herein differs from the usual method specified for the determination of SVI which uses a settling volume of 1,000 mL. The use of the small mixed liquor volume probably resulted in higher SVI values than would be the case if the larger volume were used. The smaller cylinder diameter would tend to increase the side wall effects which could result in sludge bridging and a slower sludge interface settling rate. The results of this test are valuable, however, for making comparisons of the settling characteristics of the sludges.

The daily SVI's of the four reactors are presented graphically in Figure 7. The observed fluctuations and variabilities of the SVI values for a given reactor were the result of a number of factors. First, daily fluctuations occurred, as would be expected, because of the inherent variabilities associated with the sampling of the mixed liquor solids with the conduct of the test and with the required determination of the suspended solids concentrations. Second, the data for all reactors appear to be underlain by a trend of increasing SVI with each successive week of the experimental program. And, third, the data for Reactor Nos. 3 and 4 were influenced by special testing which commenced in the latter part of December. The special testing where the HRT of Reactor No. 4 was varied was discussed earlier.

Two possible causes for the apparent trend of increasing SVI over the duration of the experimental study are an increasing mixed liquor solids concentration and the relatively long MCRT that was employed. Although the SVI, by definition, is independent of sludge solids concentration, concentra-

Table VIII. Total Toxic Organic Compound Summary

	Average Value (mg/L)	Maximum Value (mg/L)
Influent	7.5	16.1
Reactor No. 1 Effluent	<0.85	<1.76
Reactor No. 2 Effluent	<0.36	<0.72
Reactor No. 3 Effluent	<0.37	<0.72
Reactor No. 4 Effluent	<0.60	<1.42

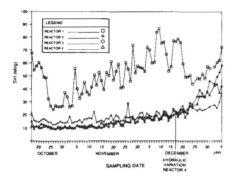

Figure 7. SVI values maintained in the reactors.

tion effects can occur during the test, which could affect the rate of settling. Additionally, relatively high SVI values can be characteristic of activated sludges with long MCRT values in the range of 20 days or more.

The settleability of the PACT sludge, as measured by SVI values, is greater than that of the activated sludge. It is thought that in a PACT system, sludge settleability is enhanced due to the formation of floc composed of PAC particles held together in a bacterial organism net.

Chemical Characteristics of Sludge

The disposal options available for the sludge generated by the proposed treatment system are dependent on the characteristics of that sludge. An extraction procedure (EP) toxicity test was performed on both the activated and PAC sludge that was produced in this experiment. The EP Toxicity test results are used to identify a waste as being hazardous vis-a-vis the criteria of Federal Regulation 40 CFR Part 261 (RCRA). The results of the test, as well as the maximum concentration levels permitted, are presented in Table IX. All of the constituent concentrations for both the activated and sludge test samples are well below the maximum concentrations allowed. The sludges can therefore be considered non-hazardous as defined by the criteria set for EP Toxicity.

SUMMARY AND CONCLUSIONS

The BKK leachate can be characterized as having high dissolved solids, high SCOD and SBOD levels and relatively low metals concentrations. The leachate contains significant amounts of acetone, MEK and THF and 1,2 DCE. Other priority pollutants are present at relatively low concentrations. In the experimental study performed on the leachate, both the PACT and conventional activated sludge systems significantly reduced constituents of concern in the leachate, namely: SCOD, SBOD, acetone, MEK, THF, and other organic priority pollutants.

Although both the activated sludge and PACT systems significantly reduced the concentrations of the contaminants of concern, the PACT system did perform more efficiently and stably. Table X presents a summary of the comparison of treatment effectiveness between the activated sludge and PACT systems. The PACT systems averaged 94 to 96% removal of SCOD as compared to 88% for the activated sludge system. The PACT system also exhibited less severe SCOD effluent peaks than the activated sludge system, especially in the acclimation phase. The SBOD removal achieved by the PACT system averaged 97 to 98%; that of the activated sludge was 94%. The PACT systems had higher removals of acetone, THF, and MEK than the activated sludge system. Both types of experimental systems consistently reduced priority pollutant concentrations to levels below detectable limits. The settleability of PACT sludge, as evidenced by SVI values, was better than that of conventional activated sludge. Generally, the PACT reactors with the higher PAC dosages of 10,000 mg/L in the

Table IX. EP Toxicity Test Results

Constituent	Activated Sludge Concentration[a] (mg/L)	PACT Sludge Concentration[b] (mg/L)	Maximum Concentration[c] (mg/L)
Arsenic	<0.002	<0.002	5.0
Barium	3.8	2.4	100.0
Cadmium	<0.005	<0.005	1.0
Chromium	<0.02	<0.02	5.0
Lead	<0.05	<0.05	5.0
Mercury	<0.0002	<0.0002	0.2
Selenium	0.005	0.003	1.0
Silver	<0.01	<0.01	5.0
Endrin	<0.001	<0.001	0.02
Lindane	<0.001	<0.001	0.4
Methoxyclor	<0.001	<0.001	10.0
Toxaphene	<0.001	<0.001	0.5
2,4-D	<0.001	<0.001	10.0
2,4,5-TP Silvex	<0.001	<0.001	1.0

[a]Sample from Reactor No. 1.
[b]Sample from Reactor No. 3.
[c]Maximum concentration is as defined in 40 CFR 261.

Table X. Comparison of Treatment Effectiveness of Activated Sludge and PACT Systems

Item	Activated Sludge System	PACT Systems[a]
Average SCOD Removal Efficiency (%)	88	94–96
50th Percentile Value–SCOD Effluent Concentration (mg/L)	314	122.5–164
90th Percentile Value–SCOD Effluent Concentration (mg/L)	405	159–228
Average SBOD Removal Efficiency (%)	94	97–98
Average SBOD Effluent Concentration (mg/L)	90	31–41
Average Effluent Acetone Concentration (mg/L)	0.47	0.04–0.17
Average Effluent MEK Concentration (mg/L)	0.36	0.06–0.12
Average Effluent THF Concentration (mg/L)	0.61	0.22–0.36
Average Effluent TTO Concentration (mg/L)	<0.85	<0.36–<0.60
Effluent TOC Concentration (mg/L)	100	38–57
50th Percentile SVI Value (ml/L)	52	14–18

[a]Values reflect the range of performances for the three reactors tested.

mixed liquor performed marginally better than that system with a PAC concentration of 5,000 mg/L. The sludge produced by both the activated sludge and PACT processes was non-hazardous in nature.

The results of this experimental study confirm the suitability and effectiveness of the PACT process for treatment of the BKK leachate. At the present time the design of a full scale PACT treatment system for the treatment of the leachate is underway.

REFERENCES

1. Zadonic, L. A., "Comprehensive Site Cleanup at Bofors Nobel, Inc.," *Proceedings of the Industrial Waste Symposia*, 57th Annual Water Pollution Control Federation Convention (1984).
2. Flynn, B. P., and Stanik, J. G., "Startup of a Powdered Activated Carbon-Activated Sludge Treatment System," *Journal Water Pollution Control Federation*, Volume 51, No. 2, 358 (1979).
3. Adams, Alan D, "Improving Activated Sludge Treatment with Powdered Activated Carbon," *Proceedings 29th Purdue Industrial Waste Conference*, (1972).
4. Cormack, J. W., et al., "A Pilot Study for the Removal of Priority Pollutants by the PACT Process", *Proceedings of 38th Purdue Industrial Waste Conference* (1983).
5. Heath, Harry W., "Bugs and Carbon Make a PACT," *Civil Engineering*, 81–83 (April 1986).

6. Copa, W. M., C. A. Hoffman, and J. A. Meidl, "Powdered Carbon-Activated Sludge Treatment of Mid-State Disposal Site Leachate," Presented at Eighth Annual Madison Waste Conference (1985).
7. Rollins, R. M., Ellis, E. C., and Berndt, C. L., "PACT/Wet Air Regeneration of an Organic Chemical Waste," *Proceedings of 37th Purdue Industrial Waste Conference* (1984).
8. Schultz, J. R., "PACT Process Mechanisms," *Ph.D. Thesis*, Clemson University (1982).
9. Flynn, B. P., "A Model for the Activated Carbon-Activated Sludge Treatment System," *Proceedings of the 30th Purdue Industrial Waste Conference* (1975).

20   COMPARISON OF THE ANAEROBIC CONTACT AND PACKED BED PROCESSES FOR THE TREATMENT OF EDIBLE OIL WASTEWATERS

C. B. Saw, Senior Research Associate

G. K. Anderson, Senior Lecturer

J. A. Sanderson, Research Associate
Department of Civil Engineering
University of Newcastle upon Tyne
Newcastle upon Tyne
United Kingdom NE1 7RU

INTRODUCTION

The process of neutralization, which is carried out during the refining of certain edible oils, produces a soapstock from which fatty acids are recovered by means of acid splitting. Acid splitting is often carried out by the addition of sulfuric acid to the soapstock which causes the free fatty acids to separate from the acidic water phase. Once this separation has occurred, the water phase is drained to a fat separator where residual free fats are recovered. The final acid water represents the effluent from the process and contains predominantly glycerol, fat, and sodium sulfate.

Conventional treatment of acid water involves a combination of chemical, physical, and biological processes with substantial mechanical input required for the dewatering of byproduct sludges. Chemical treatment with lime, followed by sedimentation, has been carried out in order to reduce the concentration of fatty matter to a level which can be tolerated by conventional aerobic biological treatment processes, e.g., activated sludge. Since chemical treatment alone removes only 50% of the COD, there is still a need for a sizable aerobic process to fully treat the effluent. This combination of treatment is known to be effective but involves high capital and operating costs.

Possible Role of Anaerobic Process in the Treatment of Acid Water

Experience from sewage sludge digestion has shown that anaerobic systems have the ability to degrade fatty matter. As such, it was thought that anaerobic biological treatment would be preferred to chemical treatment as a preliminary treatment process for acid water. The benefits sought by adopting the above process were:

- degradation of fats,
- low chemical requirements,
- low sludge production,
- low power requirements, and
- moderate COD removal efficiency (> 70%).

For anaerobic treatment to be economically attractive in terms of minimizing the capital costs, it is also important that such treatment should be possible in one of the new generation of anaerobic reactors, such as, a contact digester, packed-bed digester, upflow sludge blanket digester or the expanded/fluidized bed digester. However, there are no such processes operating in the U.K. on edible oil refining wastewaters, and therefore it was decided, following preliminary laboratory tests, that pilot plant studies should be carried out in order to clearly identify the feasibility of treating acid water by means of an anaerobic process.

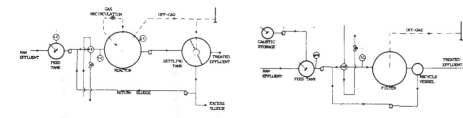

Figure 1. Schematic of anaerobic contact process.   **Figure 2.** Schematic of anaerobic filter.

Description of the Pilot Plants

Due to fears of the possible formation of a fatty scum in upflow processes, two pilot plants, the completely mixed contact process, and the upflow anaerobic filter, were operated in parallel to establish a comparison of the effects of process configuration. Flow diagrams of each process are shown in Figures 1 and 2.

Each plant had a reactor volume of 0.5 m³, and as may be seen from the flow diagrams, raw wastewater, following correction of pH and nutrient addition, was fed continuously, by means of separate feed pumps, to each unit.

In the case of the contact process, the wastewater entered the reactor where it was intimately mixed with the anaerobic biomass. Inflow of the wastewater caused an overflow of the mixed liquor to a sedimentation tank where the micro-organisms separated from the liquor, leaving a relatively clear liquid to be discharged as the treated effluent from the settling tank. In order to maintain a high concentration of micro-organisms in the reactor, the separated biomass in the sedimentation tank was recycled continuously back into the reactor.

The upflow packed bed process, known as the anaerobic filter consisted of a 3 m high, 0.5 m diameter reactor partly filled with 50 mm polypropylene plastic media. The wastewater was pumped into the bottom of the reactor along with a recycled flow of treated effluent in order to give a net upflow velocity of 15–20 m/d. Treated effluent, not requiring any further clarification, was displaced from the top of the reactor into a small recycle vessel from where the flow was divided into the recycled flow and the final effluent. The packed bed within the reactor thus remained fully submerged and the anaerobic bacteria developed primarily at the bottom and to a lesser extent within the interstices at the upper levels of the packed bed.

Both plants were fully insulated and heated in order to maintain a temperature of 35°C ± 1°C. Off-gases from each reactor were monitored for flow rate using dry-type meters and for methane by means of an in-line dedicated analyser.

The Possible Role of Sulfate Reducing Bacteria in Anaerobic Treatment Processes

The presence of sulfate in the wastewater (as shown in Table I) was a major consideration in this study due to the possible adverse effects of hydrogen sulfide which could be produced by the action of sulfate reducing bacteria. Much of the research in this field has been to attempt to identify a threshold

Table I. Characteristics of the Raw Wastewaters

Parameter		Range
COD	(mg/L)	1010–8200
pH		1.5–2.0
SS	(mg/L)	380–1420
TKN	(mg/L)	55–65
PO_4-P	(mg/L)	90–100
TFM	(mg/L)	140–1510
SO_4	(mg/L)	3100–7400
Na	(mg/L)	2940–3000

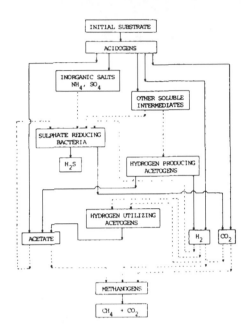

Figure 3.   Schematic of substrate dissimilation in anaerobic biological reactors.

sulfide concentration below which methanogenesis is relatively unaffected. Rudolfs and Amberg [1] concluded that soluble sulfide had a linear effect on methane formation, up to a limit of 165 mg/L, but that additions of sulfide up to 300 mg/L had little effect on the production of volatile acids. Lawrence et al. [2] also observed that methane production falls significantly with increasing sulfide concentration before an increase in volatile acids occurs. They concluded that below 200 mg/L sulfide there are no significant toxic effects, but that above 200 mg/L sulfide, severe toxicity occurs and gas production completely ceases.

A considerable amount of research has been carried out on the relationship between methanogenic bacteria and sulfate reducing bacteria (SRB) which exist in salt water and fresh water sediments [3–6]. The general conclusion is that SRBs inhibit methanogenic bacteria by outcompeting them for hydrogen and acetate. Lovley et al. [7] further concluded that methanogenic bacteria and SRBs can coexist in the presence of sulfate and that the outcome of competition at any time was a function of the rate of hydrogen production, the relative population sizes, and sulfate availability. Thus, methanogenic bacteria and SRBs should coexist in environments where the rate of uptake of hydrogen and acetate by the SRBs is lower than the rate of hydrogen and acetate production.

In addressing the possibility of a toxicity effect of sulfide on methanogenesis, the significance of competition can easily be missed. As shown in the sediment work, methanogenic bacteria and SRBs can coexist. A simple description of the competition which takes place between these two groups is shown schematically in Figure 3. Since both the methanogenic bacteria and SRBs have the ability to utilize acetate and hydrogen, it would seem reasonable to assume that in a carbon limited system, rich in sulfate, the SRBs would dominate and should replace the methanogens as the terminal group responsible for the majority of the COD removal capacity.

A further objective of this study was, therefore, to identify the capability of a SRB-dominated system to provide efficient COD removal from a fat-bearing wastewater.

RESULTS AND DISCUSSION

Tables II, III, IV, V, VI, and VII summarize the performance of the two anaerobic systms as they were operated at organic loading rates from 0.42 to 1.12 kg COD/m³/d and 0.41 to 3.39 kg COD/m³/d for the contact process and anaerobic filter, respectively. The hydraulic retention time ranged from 9.26 to 1.98 days and 8.06 to 0.67 days, respectively, for the contact process and anaerobic filter.

Table II. Average Performance Data for the Anaerobic Contact and Anaerobic Filter Processes (Operating Conditions and TFM-Related Performance)

Process	Period days	Flow L/d	H.R.T. days	O.L.R. kg/COD/m³/d	M.L.S.S. mg/L	Eff. S.S. mg/L	Inf. TFM mg/L	Eff. TFM mg/L	Removal %
Contact	0–26	54	9.26	0.42	7400	1065	582	67	88
Contact	27–54	93	5.38	0.51	10430	245	548	56	90
Contact	55–143	252	1.98	1.12	11360	390	600	61	90
Filter	0–26	62	8.06	0.41	—	608	691	45	93
Filter	27–49	146	3.42	0.67	—	295	282	49	83
Filter	50–95	272	1.84	1.70	—	580	594	70	88
Filter	96–143	471	1.06	1.95	—	335	285	63	78
Filter	201–221	747	0.67	3.39	—	300	—	54	—

Startup and Development of the SRB-Dominated System

Sludge from a heated municipal sludge digester was screened through a size 80 mesh and used as the initial seed material for both reactors. 500 litres of this material containing 20,000 mg/L suspended solids (64% volatile) was added to each system, and the temperature was adjusted to 35°C before application of the raw effluent commenced.

An initial organic loading rate of 0.42 and 0.41 kg COD/m³/d was applied to the contact process (CP) and anaerobic filter (AF), respectively. This relatively low loading was maintained for a period of 25 days as shown in Table II in order to permit gradual acclimatization of the biomass to the industrial wastewater (acid water).

Both methane and hydrogen sulfide were evolved within 24 hours of applying the acid water. The initial methane yield was, however, only 0.23 m³/kg COD applied and fell steadily over the 25 day period to approximately 0.02 m³/kg COD. Gas production fell to negligible levels during this period, at the end of which the gas composition in the headspace of both reactors averaged 75% methane,

Table III. Average Performance Data for the Anaerobic Contact and Anaerobic Filter Processes (Gross COD-Related Performance)

Process	Period days	O. L. R. kg COD/m³/d	Inf. COD mg/L	Eff. COD mg/L	Filtered Eff. COD mg/L	Removal (Whole) %	Removal (Filtered) %
Contact	0–26	0.42	3865	1940	1515	50	61
Contact	27–54	0.51	2734	1250	981	54	64
Contact	55–143	1.12	2213	1741	1324	21	40
Filter	0–26	0.41	3306	1620	1092	51	67
Filter	27–49	0.67	2295	1174	978	49	57
Filter	50–95	1.70	3125	1795	1381	43	56
Filter	96–143	1.95	2068	1564	1153	24	44
Filter	201–221	3.39	2269	1590	1028	30	55

Table IV. Average Performance Data for the Anaerobic Contact and Anaerobic Filter Processes (Sulfate-Related Performance)

Process	Period days	O.L.R. kg COD/m³/d	SO₄ Loading Rate kg SO₄/m³/d	Inf. SO₄ mg/L	Eff. SO₄ mg/L	SO₄ Reduction %
Contact	0–26	0.42	0.49	4574	2093	54
Contact	27–54	0.51	1.16	6258	3526	44
Contact	55–143	1.12	1.83	3631	2067	43
Filter	0–26	0.41	0.50	4000	2290	43
Filter	27–49	0.67	1.22	4164	1710	59
Filter	50–95	1.70	2.44	4482	2544	43
Filter	96–143	1.95	3.25	3454	2423	30

Table V. Carbonaceous Content of Treated Effluent

Process	Period days	COD mg/L	Volatile Acids Total mg/L	Volatile Acids Acetic mg/L	Volatile Acids COD Equiv. mg/L	Fatty Matter TFM mg/L	Fatty Matter COD Equiv. mg/L	Organic Carbon mg/L
Contact	0–26	1515	125	88	150	67	191	—
Contact	27–54	981	125	90	150	56	160	—
Contact	55–143	1324	231	173	277	61	174	179
Filter	0–26	1092	124	88	149	45	128	—
Filter	27–49	978	124	86	149	49	140	—
Filter	50–95	1381	332	250	398	70	200	—
Filter	96–143	1153	260	190	312	63	180	172
Filter	201–221	1028	299	229	348	54	154	—

Table VI. Distribution of COD Components in Treated Effluent

Process	Period days	Total COD (Filtered) mg/L	Carbonaceous COD mg/L	Carbonaceous COD % Total	Sulfide Related COD mg/L	Sulfide Related COD % Total
Contact	0–26	1515	341	23	1174	77
Contact	27–54	981	310	32	671	68
Contact	55–143	1324	451	34	873	66
Filter	0–26	1092	277	25	815	75
Filter	27–49	978	289	30	689	70
Filter	50–95	1381	598	43	783	57
Filter	96–143	1153	492	43	661	57
Filter	201–221	1028	502	49	526	51

at the end of which the gas composition in the headspace of both reactors averaged 75% methane, 20% carbon dioxide and 5% hydrogen sulfide on a volume to volume basis, reflecting the relative solubilities of each gas.

Development of the SRB-dominated system was clearly rapid during this first phase of loading, but methanogenesis never completely stopped.

Removal of Total Fatty Matter (TFM)

As shown in Table II, the TFM removal efficiency ranged from 78 to 93%, with both processes providing a similar performance. Thus, for an average inlet TFM concentration ranging from 282 to 691 mg/L, the treated effluent TFM concentration ranged from 45 to 70 mg/L. Inspection of the inside of the reactors after 143 days of operation confirmed that the fatty matter had been degraded since there was no visible accumulation.

Table VII. Averaged Carbonaceous COD Removal Characteristics of the SRB-Dominated Digesters

Process	Period days	O.L.R. kg COD/m³/d	Carbonaceous COD Utilization Rate kg COD/m³/d	Carbonaceous COD Removal Efficiency %	COD:SO₄ Ratio Applied kg/kg	COD:SO₄ Ratio Utilized kg/kg
Contact	0–26	0.42	0.38	91	0.84	1.42
Contact	27–54	0.51	0.45	89	0.44	0.89
Contact	55–143	1.12	0.89	80	0.61	1.13
Filter	0–26	0.41	0.38	92	0.83	1.77
Filter	27–49	0.67	0.59	87	0.55	0.82
Filter	50–95	1.70	1.37	81	0.70	1.30
Filter	96–143	1.95	1.48	76	0.60	1.53
Filter	201–221	3.39	2.64	78	—	—

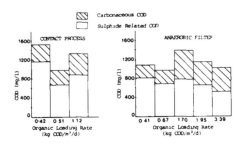

Figure 4.   Concentrations of carbonaceous- and sulfide-related COD contributing to the gross treated effluent COD.

The TFM removal efficiencies achieved by the anaerobic processes, even at relatively low hydraulic retention times, were at least equivalent to those which would be achieved by chemical treatment. The resultant treated effluent TFM concentrations are acceptable for discharge to sewer and would not cause difficulties in a polishing activated sludge plant.

Gross COD Removal Efficiency

In common with many studies on the anaerobic treatment of high sulfate bearing wastewaters, both anaerobic processes demonstrated negligible gas production. COD removal efficiencies, as shown in Table III, appear to be only moderate, ranging from 21 to 54% and 24 to 51% for the CP and AF processes, respectively. However volatile fatty acids concentrations (VFA) in the treated effluent were consistently low and ranged from 124 to 332 mg/L for both processes as shown in Table V. Combined with the low TFM concentrations in the treated effluent, these low levels of VFAs seem inconsistent with the apparently low COD removal efficiencies. As an additional check, the concentration of organic carbon was measured in the effluent from each process in order to determine whether or not the VFAs and TFM represented the sole source of carbon in the effluent. As shown in Table V, the organic carbon from the VFAs and TFM combined, accounted for most of the organic carbon in the treated effluent. This proved that, not only was hydrolysis of the raw effluent proceeding efficiently, but also that the removal of carbonaceous matter was proceeding efficiently. Clearly then, the COD in the treated effluent was largely due to inorganic material, and the obvious source was sulfides formed by the SRB's.

Relative Contribution of Carbonaceous COD and Sulfides to the Treated Effluent COD

Sodium sulfide is oxidized in the COD test, as shown below:

$$Na_2S + 2O_2 \rightarrow Na_2SO_4 \tag{1}$$

Thus, 2 g oxygen is required to oxidize 1 g sulfide. By subtracting the carbonaceous ÇOD (estimated from organic carbon measurements, or from VFA and TFM concentrations) from the measured filtered effluent COD, it is possible to obtain an estimate of the COD exerted by dissolved sulfides. As shown in Table VI and Figure 4, sulfide-related COD accounted for between 66 to 77% of the total COD in the case of the contact process and between 51 to 75% of the COD in the case of the anaerobic filter.

Removal of Carbonaceous COD

Taking into account the effect of the dissolved sulfides on effluent COD, the true COD removal characteristics of the SRB system may be estimated. As may be seen from Table VII and Figure 5, the removal efficiency of carbonaceous COD was very high in both processes with efficiencies ranging from 76 to 92% and carbonaceous COD removal rates ranging from 0.38 to 2.64 kg COD/m3/d for equivalent organic loading rates of 0.41 to 3.39 kg COD/m3/d. The effect of organic loading rates on the substrate utilization rate is shown in Figure 6, from which it is evident that at the lower organic loading rates (< 1.5 kg COD/m3/d) at which the CP was operated, the performances of the CP and the AF were very similar. In addition, at the higher loading rates (3.39 kg COD/m3/d), which the AF was able to accomodate, efficiency of treatment hardly diminished, indicating that the maximum treatment capacity of the system had not been reached.

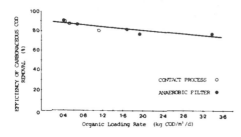

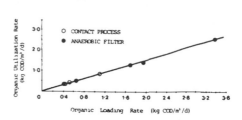

Figure 5.   Effect of organic loading rate on the   Figure 6.   Effect of organic loading rate on the
efficiency of carbonaceous COD removal.   rate of utilization of carbonaceous COD.

COD:Sulfate Utilization Ratio

Sulfate reducing bacteria (SRB) such as *Desulfovibrio desulfuricans* derive energy for synthesis and maintenance from the metabolism of organic matter and use sulfate as their terminal electron acceptor. The reduction of sulfates in this process may be expressed by the following equation [3]:

$$8H^+ + 8e^- + SO_4^= \xrightarrow{\text{Sulfate Reducing Bacteria}} S^= + 4H_2O$$

Thus, the reduction of one mole of sulfate (96 g) corresponds to the oxidation of eight equivalents of organic matter or about 64 g COD. The theoretical COD:SO$_4$ ratio required for the reduction of sulfate is thus 0.67:1.0. However, this excludes the removal of COD for cell growth (synthesis) and the effects of COD removal by other microorganisms such as methanogenic bacteria operating within the same system, which would tend to result in a higher COD:SO$_4$ ratio.

In this study, the applied COD:SO$_4$ ratios ranged from 0.44 to 0.84, but the COD:SO$_4$ utilization ratios ranged from 0.82 to 1.53 as shown in Table VII. The higher utilization ratios observed in the anaerobic filter were believed to be due to greater methanogenic activity which may have been encouraged by the spatial separation of the species throughout the height of the reactor. However, the fact that methane continued to be produced at all in both processes was surprising since the limiting substrate was clearly carbon with sulfate being very definitely in excess as shown in Table IV. Both systems demonstrated higher COD:SO$_4$ ratio in the initial period after startup, reflecting the greater methanogenic activity which occurred during that period when the SRB population was not fully established. Excluding these initial loading phases, the average COD:SO$_4$ utilization ratios were 1.01 kg/kg for the contact process and 1.22 kg/kg for the anaerobic filter.

Effect of Process Configuration on Retention of Biomass

It has been convenient, for the purpose of comparison between the two processes, to use organic loading rate and organic utilization rate as a measure of process performance. Of course the treatment capacity of any biological system is primarily a function of the mass of active bacteria retained within the reactor, which is in turn controlled by the configuration of the reactor.

In the case of the contact process, the concentration of bacteria within the reactor is controlled by the efficiency with which they can be separated from the treated effluent. Where a sedimentation tank is used, as in this case, the efficiency of separation is affected by the rate of settling. The maximum rate of hindered settling was measured in a 50 cm high 3.5 litre cylinder equipped with a 1 rpm stirrer, which was developed in the U.K. by the Water Research Centre [8]. Figure 7 shows the effect of the suspended solids concentration and organic loading rate on the rate of hindered settling.

The data show that, in order to maintain a settling rate of greater than 5 m/d, for organic loading rates above 1.0 kg COD/m³/d, the mixed liquor suspended solids (MLSS) should be maintained at less than 10,000 mg/L. This will result in a limitation on the maximum organic loading rate which may be applied to the contact process. In this study, without deliberate sludge wastage, the completely mixed reactor operated with average MLSS concentration ranging from 7,400 to 11,360 mg/L as shown in Table II. Overall microbial yields calculated were, respectively, 0.303, 0.102, and 0.220 kg SS/kg COD utilized for the three operating periods studied. The average yield from these three figures is 0.21 kg SS/kg COD, which is 15% higher than the average yield of 0.179 kg SS/kg COD reported

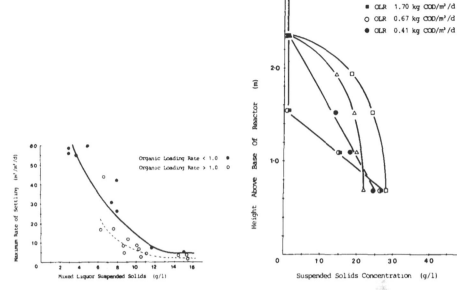

Figure 7.   Effect of MLSS and organic loading rate on maximum rate of settling (contact process).

Figure 8.   Effect of organic loading rate and liquid depth on suspended solids concentration within the upflow packed-bed reactor.

for a wholly methanogenic system [9]. As the loading rate on the contact process is increased, the MLSS concentration will tend to rise above the range of 10,000–11,000 mg/L. Clearly, in order to maintain a MLSS concentration of approximately 10,000 mg/L which would sustain acceptable rates of settling, it would be necessary to deliberately waste sludge from the system. Under such circumstances, with the resultant drop in mean solids retention time, a drop in treatment efficiency would be expected, and this will be tested in further work on the pilot plant.

In the case of the anaerobic filter, no external separation device was found to be necessary, since the biomass remained within the reactor under all conditions tested. Most of the solids within the reactor appeared to be suspended and not attached, and the concentration profile of these suspended solids within the reactor is shown in Figure 8. The effluent suspended solids concentrations, as shown in Table II, ranged from 295 to 608 mg/L and were unaffected by upflow velocities as high as 19.8 m/d, created by applying a high rate of effluent recycle. Information is still being gathered on the activity profile within the reactor, but in general, the VFA concentration was found to range from 200 to 500 mg/L as acetic at the bottom of the reactor, 300 to 800 mg/L at mid-height and 124 to 332 mg/L in the final effluent. This indicates that although most of the effluent treatment takes place in the bottom 1.5 m, there is still significant activity in the top half of the reactor.

Nutrient Requirements

The COD:nitrogen: phosphorus ratio in the raw wastewater was, on average, 100:1.2:2.6. Following the addition of 30 mg/L each of nitrogen and phosphorus (as di-ammonium phosphate) to the raw effluent, the average concentrations of total Kjeldahl nitrogen and phosphate-phosphorus in the treated effluent ranged from 25 to 40 and 65 to 80 mg/L, respectively. The ratio of nutrient utilization was thus approximately 100:1.6:0.9, and the excess measured in the treated effluent at all times indicate that phosphorus supplementation was not necessary and that the need to supplement with nitrogen was only marginal.

pH Adjustment

Acid water is strongly acidic with a pH of less than 2.0. No attempt has yet been made to identify the optimum pH for the SRB-dominated system. In the case of all operational conditions so far studied, the raw wastewater has been neutralized with sodium hydroxide to within a pH range of 6.5 to 7.2. The pH of the reactor contents has reflected this upstream adjustment and has operated within the range of 6.75 to 7.5 with an associated alkalinity of 1500–2500 mg/L as $CaCO_3$.

Downstream Treatment

The anaerobic process, utilizing sulfate reducing bacteria, has demonstrated its capability in this work to remove fatty matter and a high percentage of the carbonaceous COD in a single stage. The process has obvious potential, particularly in the shape of the anaerobic filter, to provide cost effective pretreatment of acid water. However, before that end can be achieved, the COD exerted by dissolved sulfides must be removed. One method, which is currently being investigated as an extension to this work, is the oxidation of the sulfide to sulfur using a pure oxygen system. If successful, then the anaerobic sulfate reducing process could be an attractive alternative to existing treatment technology in providing greater than 80% removal of TFM and COD in a single stage.

CONCLUSION

Both anaerobic processes provided efficient removal of fatty matter without the need for preliminary chemical treatment. Treated effluent TFM concentrations did not exceed 100 mg/L, and in general the removal efficiency was in the range of 80 to 90%.

Both anaerobic processes provided efficient removal of carbonaceous COD with an efficiency ranging from 76 to 92%. However, overall COD removal was only in the range of 21 to 54% due to the oxygen demand exerted by dissolved sulfides in the treated effluent. This work has shown that anaerobic processes operating on sulfate-bearing acid water from edible oil refining result in the development of a microbial system with sulfate reducing bacteria rather than methanogenic bacteria as the main terminal group. The resultant biological sulfate reducing systems provide a high efficiency of carbonaceous COD removal at rates at least comparable with methanogenic systems. Further work is now required to develop a suitable downstream treatment process in order to remove the effect of dissolved sulfides.

The anaerobic filter and the anaerobic contact process provided virtually identical results at loading rates less than 1.5 kg COD/m³/d. However, the anaerobic contact process will probably not be able to operate with the same efficiency as the filter at higher loading rates, since it would be necessary to limit the mass of bacteria within the contact process's reactor by wasting sludge in order to permit efficient clarification of the treated effluent in a sedimentation tank. The anaerobic filter, on the other hand, is capable of retaining a high mass of microorganisms without the need of sedimentation. This process has been able to operate at a hydraulic retention time which is virtually a third of that of the anaerobic contact process and suffered little or no fall in treatment efficiency at average loading rates of up to 3.39 kg COD/m³/d and a hydraulic retention time of 0.67 day. At this stage, therefore, the anaerobic filter with its high efficiency, stability and easy mode of operation appears to be the more preferable process of the two under test.

REFERENCES

1. Rudolfs, W., and Amberg, H. R., "White Water Treatment – II. Effect of Sulfide on Digesters," *Sewage Ind. Waste*, 24, 1278 (1952).
2. Lawrence, A. W., McCarty, P. L., and Guerin, F. J. A., "The Effects of Sulfides on Anaerobic Treatment," *Air and Wat. Pollut. Int. J.*, 10, 207 (1966).
3. Abram, J. W., and Nedwell, D. B., "Inhibition of Methanogenesis by Sulfate Reducing Bacteria Competing for Transferred Hydrogen," *Arch. Microbiol.*, 117, 89 (1978).
4. Abram, J. W., and Nedwell, D. B., "Hydrogen as a Substrate for Methanogenesis and Sulfate Reduction in Anaerobic Saltmarsh Sediments," *Arch. Microbiol.*, 117, 93 (1978).
5. Martens, C. S., and Berner, R. A., "Interstitial Water Chemistry of Anoxic Long Island Sound Sediments. 1. Dissolved Gases," *Limnol. Oceanogr*, 22, 10 (1977).
6. Mountfort, D. O., and Asher, R. A., "Role of Sulfate Reduction Versus Methanogenesis in Terminal Carbon Flow in Polluted Intertidal Sediment of Waimea Inlet, Nelson, New Zealand," *Appl. Environ. Microbiol.*, 42, 252 (1981).

7. Lovley, D. R., Dwyer, D. F., and Klug, M. J., "Kinetic Analysis of Competition Between Sulfate Reducers and Methanogens in Sediments," *Appl. Environ. Microbiol.*, 43, 1373 (1982).
8. White, M. J. D., "Settling of Activated Sludge," *Technical Report TR11*, Water Research Centre (1975).
9. Donnelly, T., "The Kinetics and Mathematical Modelling of an Anaerobic Contact Digester," *Ph.D. Thesis*, University of Newcastle upon Tyne, U.K. (1984).

21 ANAEROBIC TREATMENT OF HIGH NITROGEN, HIGH TDS INDUSTRIAL WASTES

Shahab Shafai, Research Assistant

Jan A. Oleszkiewicz, Associate Professor
Department of Civil Engineering
University of Manitoba
Winnipeg, Manitoba
Canada R3T 2N2

G. D. Hooper, Engineer
MacLaren Engineers, Inc.
Winnipeg, Manitoba
Canada R3L 2T4

INTRODUCTION & OBJECTIVES

Effluent from a pharmaceutical plant was treated anaerobically. The wastewater was from an estrone manufacturing plant and contained significant quantity of spent pregnant mare's urine (PMU). The plant operates during winter months only from October to March. The process block diagram is similar to the one shown in Figure 1. The raw PMU goes through a succession of chemical processes of extraction-evaporation-acidification. The effluent wastewater characteristics are listed in Table I. The spent PMU is very high in total dissolved solids (TDS), nitrogen, and organics. The wastewater from this plant is discharged into the city main sewer and along with the domestic sewage from the city is treated in a combination of an extended aeration activated sludge plant and a lagoon system. The municipal wastewater treatment plant (MWTP) is organically overloaded. As part of the sewage treatment facilities expansion, this study was to determine the feasibility of a separate pretreatment of the spent PMU on site to achieve ammonification and some organics removal by anaerobic means thus facilitating nitrification in the MWTP by lowering the presently high organic loadings. The specific objectives were to examine the minimum dilution required to achieve ammonification, the extent of COD removal in an anaerobic process, and the possibility of inhibition and/or toxicity due to total dissolved solids (TDS) and/or free ammonia. An approximately three month period was allotted for the study, and it was run under quasi-steady-state conditions.

EQUIPMENT & METHODS

Two types of anaerobic reactors were used: continuously fed upflow reactors and three series of batch reactors. Parallel to this study, separate biomethanation potential (BMP) tests were run.

Continuous Flow Studies

In continuous flow studies, three parallel upflow anaerobic reactors were used (Figure 2). The reactors were made of plexiglass. Reactors 1 and 2 were upflow sludge blanket (USB) reactors with conical bottoms. Reactor 3 was an anhybrid reactor of cylindrical shape with its upper 75% of volume filled with plastic rings one inch in diameter. This reactor was based on the original concept introduced earlier by DLA [1].

Each reactor was connected to a split box, installed to equalize pressure between the recycle line and the reactor top. The recycle line allowed blending of the raw and the recycled wastewater just before the influent end of the reactors. Each reactor was equipped with variable speed and recycle pumps. The reactors were placed in a walk-in environmental chamber kept at 35°C. Three separate feed buckets and three separate effluent storage tanks were housed in an adjoining environmental chamber which was maintained at 5°C. Gas was evacuated through a gas lock flask, allowing visual inspection of gas production, and a Triton low flow gas meter. Tygon tubing was used for all gas and liquid lines.

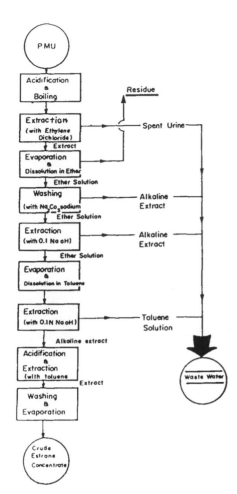

Figure 1.   Schematic of estrone concentrate manufacture from pregnant mare's urine (PMU).

All reactors were loaded with locally available flocculant anaerobic sludge on day 1. On day 39 R1 (total volume 2.46 l) was emptied and was filled with 100% imported granular sludge only. Reactor 2 (2.5 l) and 3 (2.80 l) were supplied with only 30% and 25% (volume basis) of granular sludge per total sludge volume, respectively.

The applied loading rates ranged from 1.5–11.8 kg COD/m³/d. Dilutions used were from 1:22 to 1:5, corresponding to the influent TDS concentrations of 5.2 to 21 g/L and influent COD of 2.8 to 11.5 g/L. Hydraulic residence time (HRT), over the 80 day study period, ranged between 15 and 20 hrs.

Feed was prepared using spent PMU diluted with tap water. Nutrients and microelements were added in form of KH_2PO_4, $FeCl_3$, $MgSO_4$, $NiCl_2$, $CoCl_2$, $ZnCl_2$, and $CuSO_4 \cdot H_2O$. Hydrochloric acid was used for pH adjustment.

Table I. Raw Waste Characteristics

TOC	24.3 g/L
COD	62 g/L
TDS	11.4 %
TKN	9.7 g/L
NH_3-N	3 g/L

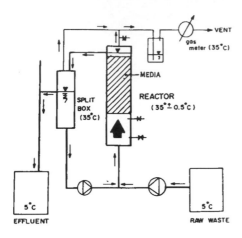

Figure 2.   Schematic of the anhybrid
reactor 3.

Testing and Analyses

All tests were performed according to the 15th Edition of *Standard Methods* [2]. Gas analyses were performed on a Gow-Mac chromatograph equipped with a thermal conductivity detector. Volatile fatty acids (VFA) were analysed on a Gow-Mac gas chromatograph equipped with a flame ionization detector. TOC was determined using a Dohrman-DC-80, using an UV reactor and an infra red detector. TKN and NH_3-N were measured using a Tecator Kjeltec distillation system. COD was measured using a Bausch and Lomb Spectronic 20 spectrophotometer.

Batch Experiments

In batch experiments, three series of anaerobic batch reactors (Figure 3) 450 ml in volume were used over a period of 80 days. Three series of initial Food/Microorganism loadings, 0.25, 0.50, 0.80 kg COD/kg VS/d, were used. The TDS concentration varied from 5 to 35 g/L. The sludge used consisted of a mixture of flocculant and granular sludge. The initial feed included the same nutrients and microelements combination as used in the continuous study. The reactors were equipped with gas volume measurement and gas sampling and pH measurement ports. They were placed in a water bath incubator maintained at 35°C. The sludge in the reactors was mixed manually twice a day. The gas volume was measured once a day. Gas composition in batch experiments was analyzed once every 2 days initially and whenever gas production was noticed at later stages of the study. pH was measured weekly using a Fisher pencil thin gel filled combination electrode.

RESULTS

Continuous Flow Studies

Figure 4 shows the histogram of COD removal (%) and loading for Reactor 2 (USB). At initial load of 5 kg/m³/d, starting from day 18, the COD removal increased initially from 55% to 64% on day 24. A further increase in load to 6.6 kg/m³/d, on day 24, led to a decrease in COD removal to 24% on day 27. The removal continued to drop to a minimum value of 18% on day 33, at which time the load was decreased to 4.0 kg/m³/d. This decrease in load resulted in an increase in COD removal efficiency. COD removal increased, even with a stepwise load increase, to as high as 72%, until day 58 when the loading was increased to 6.6 kg/m³/d, at which time the removal started to drop again. Further increase in load, from day 58 on, resulted in subsequent decrease in COD removal. COD removal efficiencies also behaved in a similar fashion for the other 2 reactors. Conversion of Org-N to NH_3-N (deamination or ammonification) for R2 is also shown in Figure 4. Efficiency of ammonification (percent conversion of N_{org} to NH_3-N) followed a similar course to the efficiency of COD removal with respect to changes in loading. Increasing ammonification was noticed at low loads, and decreasing performance at high loads was evident. All three reactors achieved ammonification of up to 100%.

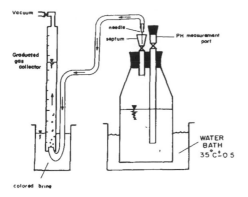

Figure 3.   Schematic of batch reactors.

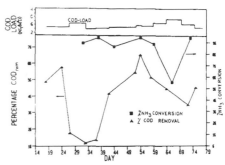

Figure 4.   Percent COD removal and percent nitrogen conversion R2.

Table II compares the best performance for the 3 reactors in terms of mass of removed COD/mass VS/d. Reactor R2 with mainly flocculant sludge was the best performer. At a load of 8.9 kg/m³/d, it achieved a COD removal of 0.15 kg/kg VS/d. R3, also with mainly flocculant sludge, followed with 0.12 kg/kg VS/d COD removal. R1 with 100% granular sludge was the poorest performer with 0.10 kg/kg VS/d removal.

Figure 5 shows the effect of influent TDS concentration on gas production. An increase in TDS beyond a certain threshold resulted in a drop in G.P. Gas production increased until TDS was increased to 21 g/L, corresponding to dilution of 1: 5.4, at which point the gas production dropped drastically. Similar sharp drop in G.P. was noticed at 16.4 and 18.6 g/L TDS for R2 and R3, respectively. Performance of all three reactors deteriorated sharply at approximately 17 g/L TDS. Figure 6 shows a negative correlation between methane production expressed in terms of L CH_4/g $COD_{infl.}$ and TDS for all three reactors and for the batch reactors.

Batch Studies

To further study the effects of TDS and to determine the level of TDS toxicity, anaerobic batch reactors were set up. Figure 7 is a typical cumulative gas production (G.P.) curve obtained in this study. The effect of TDS on the gas production for batch reactors, which was obtained using three series of batch reactors each operating under a different initial food to microorganisms ratio (0.25, 050, 0.80 kg COD/kg VS/d), together with the data from the continuous flow studies are presented in Figure 6.

A concurrent batch study using an unacclimated sludge attempted to show the effects of acclimation on TDS tolerance (Figure 8). As the acclimation period increased from 10 to 40 days, the peak methane production occurred at progressively larger TDS levels (from 6 g/L TDS after 10 days to 13 g/L TDS after 40 days).

DISCUSSION

Effects of Total Dissolved Solids

There is a considerable number of studies in the literature, documenting the inhibitory effect of TDS on microbial activity. Davis et al.[3] and Kincannon et al.[4] found that salinity may inhibit microbial activity. High TDS levels cause bacterial cells to dehydrate because of osmotic pressure and the cells die[5]. Davis et al.]3] showed that when the toxic level of TDS was reached, gas production was severely affected while only a small decrease in bacterial population was recorded. In their study, the TDS was found inhibitory to one group of organisms: the gas producers. In this study, Figure 5 clearly shows a drop in microbial activity (gas production) as TDS levels rise, suggesting inhibition due to TDS. Results from the subsequent batch studies are plotted in Figure 6, confirming this.

Table II. Comparison of the Three Reactors Operating in Unsteady State Conditions

Parameter	Reactor		
	R5	R6	R7
Reactor volume (L)	2.46	2.50	2.80
Mass of sludge (g/L react)			
TS	66.7	52.4	55.1
VS	49.4	27.5	37.1
Best Performance in terms of COD mass removed			
L (kg/m³/d)	9.7	8.9	9.2
F/M (kg/kg/d)	0.20	0.32	0.25
COD_{rem}(kg/m³/d)	4.7	4.1	4.6
Corresponding Ratio (%)	50	51	49
COD_{rem}(kg/kgVS/d)	0.10	0.15	0.12
Best Performance in terms of Ratio of COD_{eff}/COD_{infl}			
L (kg/m³/d)	9.7	5.4	4.3
F/M (kg/kg/d)	0.20	0.20	0.12
Ratio (%)	50	72	71
Corresponding COD_{rem} (kg/m³/d)	4.7	3.9	2.9
COD_{rem} (kg/kgVS/d)	0.10	0.14	0.08

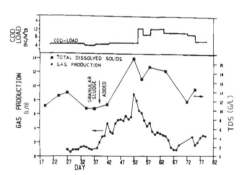

Figure 5.   Gas production and TDS. 100% granular sludge (R1).

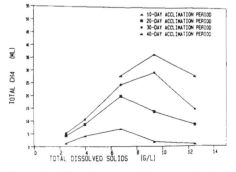

Figure 6.   Gas production from the influent COD load.

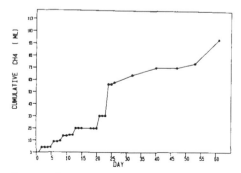

Figure   7.   Total   methane   production: F/M(I) = 0.25; TDS = 8 g/L.

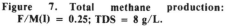

Figure 8.   Cumulative methane production versus TDS.

Davis et al.[3] concluded that at salt level of 1.3% (13 g/L), G.P. was severely affected, 13 g/L being the threshold of toxic inhibition. Using data from the batch study (Figure 6), it can be concluded that TDS levels of 8–10 g/L seem to be inhibitory to the gas producing microorganisms in this study.

DeBaere et al.[5] concluded that adaptation affects tolerance of methanogens to TDS. Abram and Nedwell[6] suggested that methanogenesis is possible at high TDS levels after a period of acclimation and that methane production in marine or salt sediments is documented at 35 g/l NaCl. The curves from Figure 8 with peak CH_4 production occurring at progressively high TDS levels suggest that methanogens with prolonged acclimation will develop halophillic properties and tolerate higher TDS levels.

Effects of Ammonia

As ammonification was found to be virtually complete in all three flow through reactors, it was deemed necessary to investigate the possibility of NH_3-N toxicity. In their paper, studying the effects of ammonia toxicity, both McCarty and McKinney[7] and Zeeman et al.[8] have reported nitrogen concentrations as mg/L of "ammonium ion." In this report this is assumed to mean "total ammonia nitrogen" as defined in *Standard Methods*[2]. McCarty and McKinney[7] found that 1700 mg/L of NH_3-N was the threshold toxicity level for CH_4 production and Zeeman et al.[8] found that there was a discontinuous linear negative correlation between NH_3-N concentration and CH_4 production rate. The threshold level above which methane production was possible only after a prolonged period of acclimitization was estimated at 1700 mg/L NH_3-N. Zeeman et al.[8] also concluded that a buildup of VFA during lag phase indicated that acid formers were less inhibited by NH_3-N than the methanogens. Sathananthan[9] observed that inhibition of total Ammonia-N (NH_3-N) was related primarily to the concentration of free Ammonia $(NH_3)g$—in gaseous form.

DeBaere et al.[5] suggested that free-ammonia should be kept below 80–100 mg/L for optimal performance. Webb et al.[10] concluded that the threshold for free-NH_3 inhibition was above 135 mg/L and below 225 mg/L. McCarty et al.[7] observed that when free-ammonia concentration exceeded 150 mg/L, the units died. In anaerobic reactors, the total Ammonia-N exists in 2 forms, NH_4^+-ion and free-NH_3, according to the following equation:

$$NH_4^+ \rightleftarrows NH_3 + H^+ \qquad (1)$$

At constant temperature, an increase in pH will cause an increase in free-NH_3 concentration or a shift to the right in Equation 1. Keeping pH constant and increasing the temperature will also increase the free-NH_3 concentration. The percentage of total ammonia in form of free-ammonia can be calculated, using the following equation[11]:

$$f = \frac{1}{10^{(pKa - pH)} + 1} \times 100 \qquad (2)$$

where:

f = % of total ammonia in the un-ionized state
pKa = dissociation constant for ammonia
$$= 0.0901821 + \frac{2729.92}{T}$$
T = temperature (°K)

To investigate the effect of free-NH_3 on CH_4 production, the ammonia concentrations used by Zeeman et al.[8] were converted to free NH_3-N, using pH = 7.75 and Temp. = 30°C. A negative correlation was obtained between CH_4 production and free-NH_3 concentration (Figure 9). The free-ammonia levels in both flow through and batch reactors were well below the toxic levels reported in the literature. Maximum NH_3-N concentration was 50 mg/L for flow through reactors and 66 mg/L for batch reactors. However, plotting the average volume of methane produced per reactor versus the free-NH_3 concentration for batch reactors (Figure 10) showed a similar type of a curve as Figure 9, which is plotted using results from the study by Zeeman et al.[8] with gas production dropping with increased free NH_3-N concentration in the liquid phase. The most recent hypothesis on the nature of ammonia toxicity has been extended by Sprott[12], who postulated that ammonium ion itself may act as an inhibitory factor by displacing Mg^{++} ion at plasma membrane surrounding methanogens. The membrane protein that activates methane production from CO_2 inside the cell is adversely affected by this Mg^{++} removal. Since ammonium ions outside the cell exist in an equilibrium with free-NH_3, this free-NH_3 may diffuse through the cell membrane and draw off cell's protons (H^+) upsetting the acidic

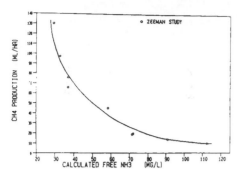

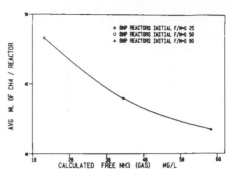

Figure 9. Methane production versus free ammonia.

Figure 10. Average milliliters of methane versus free ammonia.

nature of the inner cell. The cell draws in protons from the outside to maintain internal pH balance, thereby building up positive charge inside the cell. To maintain its internal balance, the cell releases the potassium ion, K^+ and in the process it dies[12].

CONCLUSIONS

Conversion of N_{org} to $N-NH_3$ (ammonification) for both flow-through and batch reactors was easily accomplished at a wide range of loads.

Ammonification process could be monitored by gas production (G.P.): a drop in ammonification was always signaled by decreased G.P.

Process upsets affected both the COD removal and ammonification efficiencies.

Accomodation of target COD loads in excess of 10 kg/m³/d could not be accomplished in the course of this short term study (3 months).

Performance was affected by total dissolved solids levels (TDS). TDS was strongly inhibiting to methanogenesis and ammonification at concentrations over 17 g/L in flow through reactors and at concentrations in excess of 10 g/L in batch reactors.

The minimum recommended dilution for spent PMU wastewater was concluded to be 1: 6.7. This may be improved by prolonged acclimation.

Acclimation time increases the tolerance to TDS and may improve resistance to free ammonia toxicity.

REFERENCES

1. DLA: Anaerobic Treatment and Energy Recovery—International Seminar, DLA Inc. Pittsburgh, PA (November 1981).
2. *Standard Methods for the Examination of Water and Wastewater*, 15th ed., New York: American Public Health Association (1980).
3. Davis, E. M., J. R. Bishop, and R. K. Guthrie, "MIcrobial Behavior in Hypersaline Wastes: Effects on BOD and Degradability", *Proceedings of the 33rd Annual Purdue Industrial Waste Conference*, Indiana (1978).
4. Kincannon, D. F., and A. F. Gaudy, "Some Effects on High Salt Concentrations on Activated Sludge", *J. Water Poll. Control Fed.*, 38(7): 1148–1159 (1966).
5. DeBaere, L. A., M. Devocht, P. Van Assche, and W. Verstraete, "Influence of High NaCl and NH₄Cl Salt Levels on Methanogenic Associations," *Water Res.*, Vol. 18, No. 5, 543–548 (1984).
6. Abram, J. W., D. B. Nedwell, "Inhibition of Methanogenesis by Sulfate-Reducing Bacteria Competing for Transferred Hydrogen," *Arch. Microbiol.*, 117, 89–92 (1978).
7. McCarty, P. L., and R. E. McKinney, "Salt Toxicity in Anaerobic Digestion," *J. Water Pollution Control Federation*, 33, 399-415 (1961).
8. Zeeman, G., W. M. Wiegant, M. E. Koster-Tueffers, and G. Lettinga, "The Influence of the Total Ammonia Concentration on the Thermophilic Digestion of Cow Manure," *Agr. Wastes*, 14, 19–35 (1985).

9. Sathananthan, S. "Ammonia Toxicity in Anaerobic Digesters," *M. S. Dissertation*, Univ. of Newcastle-Upon -Tyne (1981).
10. Webb, A. R., and R. R. Hawkes, "The Anaerobic Digestion of Poultry Manure: Variation of Gas Yield with Influent Concentration And Ammonium-Nitrogen Levels," *Agri. Wastes*, 14, 135–156 (1985).
11. Thurston, R. V., et al. "Aqueous Ammonia Equilibrium Calculations," *Technical Report No. 74-1*, Fisheries Bioassay Laboratory, Montana State University, Bozeman, Montana (1974).
12. Nurski, J. quoting D. Sprott (NRC, Ottawa) "The Third Kingdom," *Science Dimension*, 17–23 (May 1985).

22 PERFORMANCE OF A HYBRID ANAEROBIC PROCESS

George V. Crawford, Project Manager
Gore & Storrie Ltd.
Toronto, Canada M4G 3C2

Gerald H. Teletzke, President
GS Processes Inc.
Scottsdale, Arizona 85258

This paper describes the process development, full scale design and performance of a 2,500 m³/day capacity hybrid anaerobic treatment process located at the Lakeview Water Pollution Control Plant in Mississauga, Ontario. The municipal treatment plant serves a residential population of 350,000 and an industrial fraction that includes a major corn starch processing facility. This hybrid anaerobic process treats the byproduct liquors from a sludge thermal conditioning process at the plant. Decant tanks and vacuum filters are employed to dewater the conditioned sludges. The vacuum filtrate is returned to the decant tanks. The byproduct liquors being treated by the anaerobic process are the overflows from the decant tanks.

Prior to 1975, the plant sludges were anaerobically digested. Thermal conditioning was not employed. The sludge thermal conditioning facility was commissioned in 1975, and until 1985 the byproduct liquors were returned to the step aeration activated sludge plant for treatment. The liquors are high in COD and BOD averaging 14,000 mg/L and 7,000 mg/L, respectively. The byproduct liquors constitute over 20% of the total plant organic load. In 1985, a full scale hybrid anaerobic process was put into operation to treat the byproduct liquors and thereby reduce the plant organic load.

The design of the hybrid anaerobic process followed six years of laboratory and pilot scale process development and study. The studies tested a variety of anaerobic processes and configurations including sludge blanket, fluidized bed, filter and hybrid reactors. Upflow and downflow filters were studied. Random and modular type filter medias were investigated. The most suitable design for thermal conditioning liquor treatment was found to be a unique hybrid configuration combining anaerobic suspended growth and fixed film filter technology. The HYAN process HYbrid ANaerobic design was adopted and implemented.

PROCESS DEVELOPMENT

Sludge thermal conditioning liquors (TCL) are generated at many municipal and industrial waste treatment facilities as a byproduct of sludge dewatering operations. As part of the thermal conditioning process, biological cells in the sludge are lyzed and oxidized. A significant portion of the particulate solids matter is converted to dissolved solids. After the conditioned solids have been separated from the liquid by settling and dewatering, the liquid phase contains high concentrations (10,000 to 20,000 mg/L) of COD. This liquid stream must undergo treatment before it can be discharged to any receiving stream.

Because of the nature of the waste, and because of its obvious potential for anaerobic treatment, the Region of Peel and the Ontario Ministry of the Environment commissioned Gore & Storrie Limited in 1978 to conduct a laboratory scale investigation into the potential for anaerobic treatment of the TCL generated at Lakeview. Subsequent studies and design effort by the firm led to the development of the HYAN anaerobic hybrid reactor system.

Some early research by others on the treatment of sludge thermal conditioning liquors considered the use of the anaerobic filter as proposed by Young and McCarty [1]. Such works considered the treatability of TCL in anaerobic filters but did not explore the longer term effects of solids accumulation, solids inventory control, short circuiting and cleaning.

During 1978 and 1979 Gore & Storrie conducted a laboratory scale treatability study of the Lakeview TCL using five anaerobic filters operating in parallel [2,3]. BOD removals of 70 to 95% and COD removals of 60 to 75% were achieved at various loadings up to 24 kg COD/m³/day. The study calculated load rates in terms of COD load per unit mass of reactor solids per day. The study was not of sufficient duration, however, to determine the effect of high reactor solids concentrations on performance, short circuiting, reactor pluggage and process control.

The Wastewater Technology Centre in Burlington, Ontario, a research facility of Environment Canada, first began its investigation of anaerobic treatment processes in 1980. Gore & Storrie Limited was retained to provide technical assistance to the project and to coordinate the work between Environment Canada and the Ontario Ministry of the Environment. TCL from the Lakeview plant was used as the waste feed material. The study evaluated four pilot scale anaerobic process configurations in parallel: an upflow anaerobic filter reactor, a downflow fixed film columnar reactor, a fluidized bed reactor and a sludge blanket reactor [4]. After one year of operation, the COD removal was 67% by the anaerobic filter, 58% by the columnar reactor and 50% by the fluidized bed reactor, all at loadings of 20 kg COD/m³/day.

Throughout that study, the sludge blanket process could not be successfully loaded above 5 kg COD/m³/day. External settlers and solids return means were required. Hydraulic retention times of at least 49 hours were necessary. A sludge blanket reactor, successfully treating brewery waste and with a fully developed granular sludge bed, was subsequently fed TCL. This reactor suffered a loss of performance and again the granular sludge characteristics could not be maintained on TCL feed. Failure of the sludge blanket process was attributed to the poor settling characteristics of the biomass produced and the frequent loss of solids to the reactor effluent. Similar studies at Green Bay Wisconsin were also unable to establish a granular sludge having good settling characteristics when treating TCL. The required hydraulic retention time in Green Bay averaged 58 hours [5].

From 1981 until 1983, Gore & Storrie operated twelve laboratory scale anaerobic filter reactors in a three phase study to investigate ways to reduce the time required for startup. Lakeview TCL was used as the feed waste. Various startup techniques were investigated and developed [6]. Some of these techniques, including the use of a target reactor effluent acid level of 2,000 mg/L as a feed control parameter, are now commonly used in the field. The effects of dynamic load variations and pH control on anaerobic filter reactors were also investigated.

New information on filter solids accumulation and its effect on short circuiting and reactor performance was developed in the same study (7). After one year of operation at high COD loadings the removal performance in several reactors decreased from near 70% to between 26 and 42%. Removal of only 3 to 5 kg COD/m³/day could be achieved. Reactor liquid volumes had been reduced to between 8 and 19% because of solids accumulation. The reduced performance of the fixed film filter reactors was attributed to the reduced hydraulic retention time (HRT) caused by the reduced reactor volumes. Much of the accumulated solids mass was the result of the precipitation and deposition of inert solids material within the filter reactor.

The dissolved solids concentration of TCL is high. The shift in pH from the feed waste at pH 4.5 to the reactor pH of 7.0 causes the precipitation of dissolved inert solids. The addition of alkalinity for pH control was found to aggravate and increase the precipitation and undesirable accumulation of these inert solids.

The study calculated the rate of COD load per unit mass of volatile suspended solids but concluded such rates were not significant for anaerobic filter reactors. Because of the high degree of deposition of inert solids, much of the fixed film biomass solids were coated and not in contact with the passing waste stream. The calculation of such load rates for the non-attached biomass solids was not part of the study.

In January of 1983 a hybrid anaerobic reactor configuration was constructed at laboratory scale. That reactor was operated in parallel with an anaerobic filter reactor. Each received Lakeview TCL as feed [8]. After three years of continuous operation at high loadings, removals of 70% of soluble COD and 88% of soluble BOD continue to be achieved in the hybrid reactor. Reactor loadings are above 6.3 kg COD/m³/day at a total reactor HRT of 38 hours.

HYBRID ANAEROBIC REACTOR DESIGN

Various forms of hybrid anaerobic reactor designs have been proposed [9–14]. Most have considered an upflow configuration having a filter above a sludge blanket reactor. Some designs have included the recirculation of effluent to the lower feed zone [9]. Others propose the withdrawal of gas from below the filter zone but above the sludge blanket zone [10]. Floating filter media beds have been used as gas-liquids-solids separators. Glass beads have been proposed as a filter medium, noting that no fixed film growth will occur in the filter zone [11]. The filter zone of such a hybrid, however, consequently does not contribute to the reactor organic removal performance.

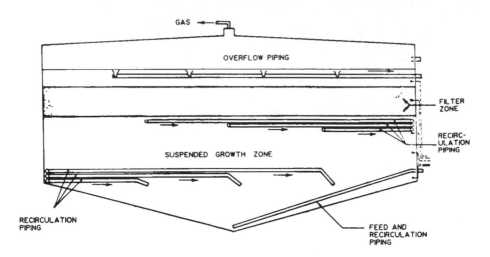

Figure 1.   Conceptual schematic of the HYAN reactor.

Most hybrids have been based on sludge blanket technology in the lower zone. Such hybrids, however, do not address the problem of excess solids accumulation in the filter zone. To improve contact and feed distribution in the lower zone, effluent recycle has often been employed. This, however, increases the effective rise rate velocity in the filter zone. Excess solids accumulation in the filter and high effluent solids loss will result as evidenced from several anaerobic filter studies that considered the use of recycle [7].

Removal of the gas below the filter zone is a novel idea, reducing turbulence within the filter zone. This configuration, however, is equivalent to the operation of the sludge blanket and filter in a serial arrangement without mutual benefit. The volume of reactor required for the gas withdrawal system would be similar to sludge blanket systems, increasing the reactor volume without adding to the biological function.

None of the prior hybrid designs considered the need for the separate means of solids inventory control in the suspended growth and filter zones.

THE HYAN DESIGN

For the treatment of thermal conditioning liquors, a new and different kind of hybrid anaerobic reactor was developed. The HYAN process was developed specifically for the treatment of TCL where solids growth, accumulation and precipitation during anaerobic treatment are of concern. The HYAN process is equally applicable for other high strength wastes that have a potential for excess inert solids accumulation. Such wastes include landfill leachates, pharmaceutical, dairy, food processing, textile, petrochemical, and pulp and paper wastes. Anaerobic filter designs will accumulate high inert solids inventories in the media when treating such wastes, requiring frequent shut-down and the use of difficult cleaning procedures to reduce short-circuiting. The HYAN reactor controls such accumulations by design, keeping its filter media zone free of unwanted solids. The HYAN design includes solids wasting and inventory control capabilities necessary for stable reactor operation.

A conceptual arrangement of the HYAN reactor is shown on Figure 1.

The reactor is in part an anaerobic filter to take advantage of its high performance and resistance to shock loadings. Solids accumulation in the filter and its subsequent effect on performance, however, are controlled by the design and operation of the suspended growth zone below the filter.

The lower suspended growth zone is sized and configured to satisfy multiple functions. Biological growth and treatment takes place producing gas and biomass solids. Inert solids precipitation also occurs. Solids in the feed and the inert solids precipitated are concentrated in the suspended growth zone and are removed through controlled wasting. The biomass inventory is controlled to maintain the rate of COD load per unit mass of biological solids within an acceptable range. Sludge retention time (SRT) can also be calculated and controlled. The ability to remove excess solids from the lower zone allows the control of solids buildup in the reactor and ultimately in the filter bed.

By proper sizing, the gas production rate and organic removal performance in the lower zone can be determined. The height to width ratio of the HYAN and the relative sizing of the suspended growth and filter zones are designed to maintain a continuous flow of gas through the filter zone at a known and desired rate. A minimum gas flow rate is required to agitate the filter zone and to dislodge accumulated solids, allowing their return by gravity to the lower suspended growth zone. The action of the gas rising from the lower zone keeps the filter media clean. A maximum design gas flow rate restricts the loss of solids to the reactor effluent caused by excessive turbulance and the flotation of solids by adsorbed gas bubbles.

The rising gas flow provides vertical mixing between the two reactor zones. Equalized flow distribution and increased mixing of the suspended growth zone are achieved through recirculation of the lower zone contents. Because the recirculation liquid is withdrawn from below the filter zone, the effective liquid velocity into and through the filter zone is not increased. Recirculation flow rate adjustments can be made without causing the rising of poorly settling solids into the filter zone where undesirable accumulation would occur.

By proper design of the gas and liquid upflow rates through the filter zone, and by proper operation of the feed distribution and recirculation systems, the functions of the filter and suspended growth zones are made interdependent as a true hybrid system. The filter zone will achieve high organic removals without accumulating excess solids. By continuously returning solids to the lower zone, high biomass concentrations will be maintained and good organic removal performance will result in the lower zone. The success of these design parameters and process developments has been demonstrated at full scale at the Lakeview Water Pollution Control Plant.

THE LAKEVIEW HYAN DESIGN

The Lakeview facility was constructed by modification of two existing 27 metre diameter digesters. The liquid volume of each reactor is 2910 m³. The cost of $1.6 million included all related construction and control building rehabilitation. The facilities were commissioned in April 1985 and achieved full loading in June 1985. Through much of 1985 and 1986, the reactors have been treating over 6 kg COD/m³/day at an HRT of between 32 and 48 hours, both calculated on the basis of total reactor volume. The facility has generally achieved 72% COD and 80% BOD reductions, while producing between 8,000 and 14,000 m³/day of gas. The fuel gas is used on site to produce steam utilized by other plant processes.

Typical waste characteristics for the Lakeview TCL are shown in Table I. The unfiltered and filtered sample data represent the TCL during the period from 1980 to 1984. Recent analyses have shown that much higher COD and suspended solids concentrations are being loaded to the full scale HYAN reactors. In March and April of 1986, the unfiltered and filtered COD concentrations were 18,280 mg/L and 16,050 mg/L, respectively. Most suspended solids samples have averaged 500 mg/L; however occasional excursions to up to 8,080 mg/L have also occurred.

The reactors were seeded in late March of 1985. Approximately 50% of the reactor volume was filled with anaerobically digested municipal sludge at less than 2% total suspended solids concentration. A feed of 2 litres per second per reactor was commenced on 1 April 1985. Feed rate increases to each reactor were controlled by manual adjustment so that the effluent volatile acid concentration remained below 1,600 mg/L. By the end of June, each reactor was being loaded at 5.8 kg/m³/day of soluble COD, at an HRT of 57 hours. The COD loading exceeded the design at that point.

By early July, the reactors were receiving the full TCL load produced by the sludge thermal conditioning facility. The TCL was warmer than anticipated, and plant effluent water was added to cool the liquid to the desired treatment temperature of 37°C. As the biological sludges grew, performance of the reactors improved. Table II summarizes the trends in performance of the reactors from April to December 1985. The results show an increase in soluble BOD removal from 52% in July to 79% in December. The two reactors were run in parallel at equal loading, and the results for the two are combined. All feed concentrations are for the diluted TCL as fed to the reactors.

In September, October and early November, the feed suspended solids concentrations were abnormally high. Values up to 18,300 mg/L were reported in September. October values ranged from 2,280 mg/L to 19,210 mg/L throughout most of the month. As expected, the reactor effluent suspended solids (SS) concentrations increased during these events. It is significant to note, however, that the reactor effluent SS returned near to its equilibrium within days of each feed solids concentration reduction. The reactors did not experience a loss of soluble COD removal performance during or after the high influent solids events.

Table I. Characteristics of TCL at Lakeview WPCP [4,5]

Parameter (mg/L)	Unfiltered	Filtered
Suspended Solids	150	
Volatile Solids	190	
Total Volatile Solids	4,200	
COD	11,000	10,700
BOD$_5$	5,500	5,300
TKN	900	875
NH4-N		300
SO4		140
TOC		3,770
Metals		
Al	1.7	1.3
Ca	143	138
Cd	.03	.03
Cr	.90	.85
Cu	.40	.40
Fe	28	17
Mn	2.0	1.9
Ni	1.4	1.4
Pb	.15	.11
Zn	2.6	2.4
Co	.07	.07
Volatile Acids		1,430
Acetic		935
Propionic		350
Butyric		135
pH	4.5 to 5.0	
Temperature (°C)	50 to 60	
Alkalinity	1,414	

As an indicator of the magnitude of the influent suspended solids concentrations fed to the reactors, the loadings can be restated. During the month of October alone the suspended solids loading was equivalent to the volume of both reactors at 7.5% suspended solids concentration. Most of these solids are inert and cannot be biologically degraded. Based on tests with filter reactors, such a sustained loading would cause pluggage, short-circuiting and failure of a fixed film media reactor.

Table II. 1985 Operational Results—Lakeview HYAN Reactors

					HYAN Effluent				
	Feed TCL			Loading				Soluble COD	Soluble BOD
Flow L/s	SS mg/L	COD-s mg/L	BOD5-s mg/L	COD-s kg/m³/day	SS mg/L	COD-s mg/L	BOD5-s mg/L	Removal %	Removal %
April	6	320	10400	5740	0.9	490	3250	1090		
May	6	190	12800	7770	1.1	370	4800	2480		
June	15	470	13900	7130	3.1	1080	3410	900		
July	30	320	12990	5250	5.8	560	5930	2510	54	52
Aug.	30	140	10500	3830	4.7	480	4850	1960	54	49
Sept.	24	2800	9250	4890	3.3	1210	4010	2420	57	51
Oct.	35	4780	8170	3000	4.2	5660	3540	880	57	71
Nov.	34	161	8660	2260	4.4	920	2670	530	69	77
Dec.	32	214	11830	4900	5.6	1200	4240	1050	64	79

Note: All COD and BOD$_5$ data are soluble values from filtered samples.

Table III. Operational Results, 7 November to 19 December 1985

| | November | | | | | December | | | | | | | |
	12	14	21	25	27	4	5	10	11	16	19	23	Avg.
Feed													
COD-s					8860	7850	10250	10580	8760	12400	14380	13050	10760
BOD-s	5700	4430	3300	4800	2430	4500						5700	4410
SS	320	390		3490	310	380	360	460		190		300	690
#3 Effluent													
COD-s					2600	1780	4130	2400	2810	3970	4132	5200	3380
BOD-s	640	1050	900	980	470	1070						1050	880
SS	2080	2730		6360	720	1240	1100	1360		1160		1410	2020
#4 Effluent													
COD-s					2270	1530	3640	1740	2690	3800	4210	4580	3060
BOD-s	680	830	900	830	300	1070						1530	870
SS	2550	2910		480	1080	1060	1030	1040		1510		960	1400
Combined Performance - %													
COD-s					73	79	62	80	69	69	71	63	71
BOD-s	89	79	73	81	84	76						78	80

Table IV. Operating Results — March, April, 1986

	20 March	2 April
Feed		
SS	520	480
COD-t	16810	18330
COD-s	16270	15590
BOD-t	5850	6380
BOD-s	4500	5780
#3 Effluent		
Flow-L/s	10	11.75
SS	860	1220
COD-t	5740	6830
COD-s	3980	5000
BOD-t	950	1550
BOD-s	840	1170
VFA	860	1800
#4 Effluent		
Flow-L/s	10	11.75
SS	400	810
COD-t	3070	6100
COD-s	2800	5350
BOD-t	950	1250
BOD-s	520	1030
VFA	430	1680
Combined Performance - %		
COD-t	74	70
COD-s	79	67
BOD-t	84	78
BOD-s	85	81
Organic Load		
kg COD/m³/d	5.0	6.4

The self-cleaning and solids inventory control capabilities of the HYAN design successfully protected its operation during the two-month period of high influent inert suspended solids loads.

The period of high feed solids concentration ended on 7 November 1985, with one brief recurrence on 25 November. Table III shows performance results for a period of detailed analysis from 12 November to 23 December 1985. Whereas Table II includes sampling and analyses two to three times per month by Gore & Storrie, Table III includes all analyses by the Ontario Ministry of the Environment staff at the Lakeview plant. During that period of time, the flow to the reactors averaged 3280 m³/day and varied from 2940 m³/day to 4320 m³/day for extended periods. Flow variations were due primarily to the amount of dilution water added for cooling the TCL to the desired reactor temperature. The resultant hydraulic retention times varied from 32 to 47 hours. Loadings to the reactors averaged 6.1 kg/m³/day as soluble COD. Removals of 71% of soluble COD and 80% of soluble BOD were consistently achieved.

In 1986, the reactors have continued to receive and to treat all the TCL produced at the plant. Organic loadings to the reactors have continued relatively unchanged.

In March 1986 a heat exchanger was commissioned to cool the TCL prior to it being fed to the reactors. This eliminated the need to add dilution water. Although the organic loadings remained constant, the hydraulic retention time (HRT) was increased.

Table IV presents some recent operational results shortly after the cooler was commissioned. As expected, performance of the reactors has improved. The current operating conditions are at 133% of the design organic loading in terms of the mass of COD loaded per cubic metre of reactor volume per day. The flow rate, however, is less than design. The reactors are currently removing 72% of the total COD, 73% of the soluble COD, 81% of the total BOD and 83% of the soluble BOD. These results significantly exceed the intended design performance of 65% COD removal and 75% BOD removal.

Total COD and total BOD performance results are influenced by the suspended solids concentrations in the reactor feed and effluent. For this reason, such data has been excluded from prior results tables. All results are shown in Table IV because of the stable suspended solids level of the feed during this period and to illustrate the effectiveness of the HYAN reactor design. When considering downstream treatment requirements, the reactor effluent total BOD and suspended solids concentrations are of concern. The HYAN design controls these parameters at Lakeview.

By reducing the BOD of the TCL, the Lakeview plant aeration energy requirements have been reduced up to 20%. The feed gas produced by the HYAN process is used on site to create steam necessary for other plant process requirements. The HYAN project was economically viable at the design performance levels. At the current higher performance levels, the project payback has been reduced to about 4 years.

RECENT HYAN DESIGN DEVELOPMENTS

Through continued research and full scale operation, improvements to the HYAN process continue to be made. Gore & Storrie Limited is currently designing other HYAN reactors for TCL wastes. Where the waste strength and characteristics are different from Lakeview, the filter and suspended growth zones are changed in size and proportion accordingly. The changes are based on the HYAN design parameters for sizing, mixing, and solids control developed by the firm.

GS Processes Inc. has been licensed to offer the HYAN process technology and design in the United States. Gore & Storrie provides process design and engineering services to GS Processes Inc. as required.

REFERENCES

1. Young, J. C., and P. L. McCarty, "The Anaerobic Filter for Waste Treatment," *Journal WPCF*, Vol. 41, No. 5 (1969).
2. Gore & Storrie Limited, "Reduction of Heat Treatment Liquor Using the Anaerobic Packed Bed Reactor—A Laboratory Study," Ontario Ministry of the Environment, *Publication E-02* (December 1979).
3. Crawford, G. V., T. Alkema, M. Yue, M. G. Thorne, "Anaerobic Treatment of Thermal Conditioning Liquors," *Journal WPCF*, Vol. 54, No. 11 (1982).
4. Hall, E. R., and M. Jovanovic, "Anaerobic Treatment of Thermal Sludge Conditioning Liquor with Fixed Film and Suspended Growth Processes," *Proceedings of the 37th Purdue Industrial Waste Conference* (1982).

5. Green Bay Metropolitan Sewage District, "Anaerobic Digestion of Heat Treatment Liquors," *Research Report* (1982).
6. Crawford, G. V., and E. R. Hall, "Optimizing the Start up of Anaerobic Packed Bed Reactors," presented at 55th Annual Conference of the WPCF, St. Louis, Missouri (October 1982).
7. Gore & Storrie Limited, "Anaerobic Packed Bed Reactors under Startup and Dynamic Operating Conditions," Unpublished report for Environment Canada and the Regional Municipality of Peel, Ontario (1983).
8. Internal memoranda and research reports, Gore & Storrie Limited (1981 to 1986).
9. Olthof, M., and J. Oleszkiewicz, "Anaerobic Treatment of Industrial Wastewaters," *Chemical Engineering* (November 15, 1982).
10. Lettinga, G. et al., "Design, Operation and Economy of Anaerobic Treatment," *Water Science and Technology*, Vol. 15, Copenhagen (1983).
11. Maxham, J. V., and W. Wakamiya "Innovative Biological Wastewater Treatment Technologies Applied to the Treatment of Biomass Gasification Wastewater." *Proceedings of the 35th Purdue Industrial Waste Conference* (1980).
12. Haug, R. T., et al., "Anaerobic Filter Treats Waste Activated Sludge," *Water and Sew. Works*, 40 (February 1977).
13. Donovan, E. J., et al., "Treatment of High Strength Wastes with an Anaerobic Filter," For presentation at AIChE 86th National Meeting, Houston, Texas, Hydro-Science, Inc., Westwood, N.J. (1979).
14. Vollstedt, T. J., "Treatment of Thermally Conditioned Sludge Supernatant with an Anaerobic Packed Bed Reactor," Zimpro, Inc., *Technical Bulletin 2303-T*, Zimpro Environ. Control System, Rothschild, WI. (1978).

23 TREATMENT OF TANNERY EFFLUENT – A CASE STUDY

T. Damodara Rao, Superintending Engineer
Tamil Nadu Water Supply and Drainage Board
Madras, India

T. Viraraghavan, Professor
Faculty of Engineering
University of Regina
Regina, Saskatchewan
Canada S4S 0A2

INTRODUCTION

India has the largest livestock population (over 300 million) in the world, constituting 13% of the world's livestock population; the country produces approximately 100 million pieces of hides and skins which forms 12% of the total world output [1]. Leather industry is mainly concentrated in states such as Tamil Nadu, Ultar Pradesh, West Bengal, Rajasthan, Maharashtra, Karnataka, Punjab, Andhra Pradesh, and Bihar. Approximately one-third of all leather produced in India is from Tamil Nadu and about 70% of the total exports of leather and leather products are from Tamil Nadu. There are 433 recognized tanneries in Tamil Nadu, out of which 104 units are mechanized. Many of the tanneries are located on the banks of the Palar River in the Vaniyambadi – Ambur region in North Arcot District, and the remaining ones are established at Erode, Tiruchirapalli, Dindigul, and in Madras City and suburbs. Although the tanning industry has been in existence in India for a long time, the problems of environmental pollution from leather industries has been seriously considered only in the last ten to fifteen years. Because of the concentration of most of the tanneries in Tamil Nadu in a sixty kilometre stretch of the Palar River and because of the lack of suitable treatment, tannery effluents pose a serious problem in the region. This paper presents information on tannery effluent characteristics, water pollution aspects of tannery effluents and on a demonstration plant installed at a tannery in Ranipet to treat its effluent.

CHARCTERISTICS OF TANNERY EFFLUENT

Liquid wastes are produced during various operations such as soaking, liming, deliming, pickling, bating, tanning and finishing operations. Generally soak waters, liming wastes and spent vegetable tan liquors are discharged intermittently; spent deliming and bating liquors are discharged once in a day. Washing after different operations contributes to the wastewater flow. The quantity of wastewater discharged from the tanneries is generally 3000 to 3200 litres per 100 kg of skins or hides processed.

Composite Wastewater from a Traditional Tannery

The composite wastewater from a tannery producing vegetable or chrome tanned leather is generally dark brown in colour, with foul smell, alkaline pH, high suspended solids (approximately 3000 mg/L), high BOD (2000 to 3000 mg/L), high salinity with dissolved solids of 5000 to 12000 mg/L, and high COD (400 to 4000 mg/L) [2]. The quantity of wastewater discharged is approximately 30 to 32 litres per kg of hide or skin processed. The usual characteristics of different wastewaters based on a paper by Bhaskaran [3] are presented in Table I. The chrome tan effluent contains considerable amounts of trivalent chromium.

Effluent from Semitanned to Finishing Units

The wastewater discharged from a semitanned to finishing units comprise mainly of wash waters after each unit operation such as stripping, souring, retanning, dyeing and fat liquoring and the periodic discharge of concentrated wastes from each unit operation. The total quantity of wastewater discharged is approximately 45 to 50 litres per kg of raw materials used. Table II shows the typical

Table I. Typical Characteristics of Tannery Effluents [3]

Effluent	pH	Suspended solids, mg/L	BOD, mg/L
Soaking	7.5–8.0	2500–4000	1000–2500
Liming	10.0–12.5	4500–6500	3000–9000
Deliming	3.0–9.0	200–1200	1000–2000
Vegetable tanning	5.0–6.8	5000–20000	6000–12000
Chrome tanning	2.6–3.2	300–1000	800–1200
Composite (including washings)	7.5–10.0	1250–6000	2000–3000

characteristics of the effluents from the various operations and of the composite effluent based on a recent report issued by Environment India [1]. The composite wastewater is brownish in colour and has disagreeable odour; it contains oil and grease of 40 to 80 mg/L and trivalent chromium of 15 to 25 mg/L.

WATER POLLUTION ASPECTS OF TANNERY EFFLUENTS

In Vaniyambadi and Ambur in North Arcot District, the effluents from the tanneries are discharged directly into the Palar River. At Ranipet, also in North Arcot, several tanneries discharge their effluent into drainage courses leading to irrigation tanks. No regular methods of treatment of the wastewater are adopted now except for open ponding due to relatively small size of the tanneries. Most of the tanneries in the region are based on vegetable tanning process.

The Palar River which has surface flow for only a few days a year, has been considerably affected by tannery effluents. Subsurface water quality data (electrical conductivity and chlorides) related to the Palar River in the Vaniyambadi—Ambur region, based on a study conducted by the Tamil Nadu Water Supply and Drainage Board, are presented in Table III [2]. This table shows clearly the deterioration in water quality of the subsurface flows in the Palar River for a distance of approximately 80 km. The subsurface flows in the Palar River are used for providing drinking water for a number of municipalities and villages located on its course. At least 20 km of this river stretch in this region has become unfit for development as a source for drinking water supply. Many municipalities have adopted distant upstream locations in the river, farther from their areas, as water supply sources, to avoid possible pollution from tannery effluents.

Table IV provides additional data on the quality of sub-surface water samples in the Palar River bed near Vaniyambadi based on an investigation conducted by Sastry [4]. Table V shows the quality of groundwater collected from wells on either bank of the Palar River and from wells in the tanneries. These water quality data reflect the same type of influence the effluents from tanneries have on the Palar River subsurface water quality.

TREATMENT OF TANNERY EFFLUENTS

The treatment of tannery effluent would consist of: 1) segregation of salt bearing wastes; 2) equalization; 3) primary treatment (sedimentation); and 4) secondary treatment (usually biological). Both conventional and low-cost methods can be used for the secondary biological treatment of settled tannery wastes. Trickling filter and activated sludge systems are known to be successful in the

Table II. Typical Characteristics of Effluents from Semi-tanned to Finishing Units [1]

Source of Samples	pH[a]	BOD	Boron	Chloride	Oil & grease	Trivalent Chromium
Wastewater from stripping	6.0–6.3	9000–11000	90–100	500–600	Nil	Nil
Wash water from stripping	6.4–6.6	1200–2400	40–50	400–500	Nil	Nil
Wastewater from semi-chroming	3.4–3.3	2000–3000	2.0–3.0	600–750	Nil	600–900
Wastewater from basification	4.2–5.0	1500–1800	2.4–4.0	400–500	Nil	200–400
Wastewater from fat-liquoring and dyeing	3.0–4.0	700–1000	3.0–4.0	80–100	300–400	10–20
Composite wastewater	5.5–6.0	800–1000	4.0–5.0	500–600	30–80	15–25

[a]All values, except pH, are in mg/L.

Table III. Subsurface Water Quality of the Palar River

Distance Along River Course With Respect To the Vaniyambadi Location	Water Quality Characteristics		Remarks
	Electrical Conductivity μmho/cm	Chlorides mg/L	
10 km upstream	700	80	Unpolluted
At Vaniyambadi	4000	1000	Gross pollution
10 km downstream	1400	300	Incomplete recovery
12 km downstream (Ambur)	2200	700	Gross pollution, second dose
15 km downstream	1100	350	Incomplete recovery
17 km downstream	700	100	Incomplete recovery
18 km downstream	1200	250	Gross pollution, third dose
38 km downstream	700	70	Complete recovery
60 km downstream (Ranipet)	850	140	Gross pollution, fourth dose
80 km downstream	750	70	Complete recovery

treatment of tannery effluents [5]. Low-cost treatment methods such as anaerobic lagoons, oxidation ponds, aerated lagoons, and oxidation ditches have been successfully adopted in secondary biological treatment of tannery wastes [6]. Sastry et al. [6] showed that using settled composite waste with soak liquor, the BOD of the waste could be reduced from 1600 to 250 mg/L at a loading rate of 0.16 kg BOD/m³/d with a detention time of 10 days. The BOD of the composite waste without soak liquor was reduced from 1500 mg/L to 190 mg/L by an anaerobic lagoon with a detention period of 10 days, loaded at 0.15 kg BOD/m³/d. Aerated lagoon studies carried out by Sastry et al. [6] showed that tannery waste with or without soak liquor can be successfully treated using an aerated lagoon. A detention time of 3 to 5 days or more is provided depending upon the type of the waste.

Generally the following methods are suggested for secondary biological treatment of settled tannery waste [1]: 1) an anaerobic lagoon followed by an aerated lagoon; 2) an anaerobic lagoon followed by an oxidation ditch; 3) an anaerobic lagoon followed by a waste stabilization pond; and 4) an aerated lagoon followed by an oxidation ditch.

Table IV. Characteristics of Sub-surface Water Samples from Palar River Bed [4]

	Distance from the Stretch of Tannery Waste Discharge[a]			
	Upstream	Within 1.5 km downstream	1.5–3.0 km downstream	3.5–8.0 km downstream
Electrical conductivity, μmho/cm	850–1150	3200–3000	2000–1250	1100–1050
Total dissolved solids (mg/L)	640–740	1900	1200	620–600
Chlorides (mg/L)	88–180	820	450	250–164
pH	7.3–7.7	7.4	7.6	7.6–7.3
Sodium (%)	44–58	73	67	60–28

[a]Descending figures represent the values with increasing distance from the waste discharge points.

Table V. Characteristics of Groundwater Samples on Either Bank of Palar River and from Tannery Wells [4]

		Wells on either bank of Palar River	
	Wells in Tanneries	Above the waste discharge stretch	Below the waste discharge stretch[a]
Electrical conductivity μmho/cm	900–5000	1250–1450	4800–1000
Total dissolved solids, mg/L	630–2620	740–920	3500–640
Chlorides, mg/L	112–1160	180–320	1460–80
pH	7.7	7.7–7.3	7.5–7.8

[a]Descending figures represent values with increasing distance from the waste discharge stretch.

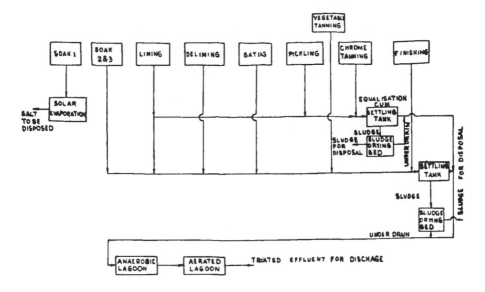

Figure 1.   Schematic of a tannery demonstration plant in one of the tanneries [1].

In any of these treatment schemes, it is necessary to provide for the segregation of soak liquor and solar evaporation in pans. These schemes also include primary sedimentation.

The general design criteria proposed [1] are indicated below.

BOD of the effluent	= 3000 mg/L
SS of the effluent	= 2500 mg/L
Detention time in the settling tank	= 24 h
Detention time in the anaerobic lagoon	= 17.5 d
Detention time in the aerated lagoon	= 5 d
Sludge drying time	= 10 d
BOD reduction in settling tank	= 25%
BOD reduction in aerated lagoon	= 80%
BOD reduction in oxidation ditch	= 95%

Different design criteria are to be adopted in the case of: 1) semi-chrome to finishing units; and 2) raw to finishing units [1].

Demonstration Plant at a Tannery in Rampet

A schematic layout of the various unit operations in a vegetable/chrome tannery along with the treatment scheme adopted is shown in Figure 1 [1]. Soak 1 is segregated and evaporated in a solar evaporation pond 0.5 m deep. Liming liquor and chrome liquor are collected in one equalization/settling tank of 24 hours capacity. The sludge removed from this tank is dried on a separate sludge drying bed. The rest of the wastewater flow is collected in another settling tank; the sludge removed from it is dried separately. The overflow from both the settling tanks is treated in an anaerobic lagoon of 15 days detention followed by an aerated lagoon of 10 days detention. The demonstration unit is performing satisfactorily; it is able to reduce the BOD of the tannery effluent from approximately 2500 mg/L to 50 mg/L.

PROSPECTS FOR THE FUTURE

In order to minimize the adverse impacts of the tannery effluents on the environment especially pollution of surface water and groundwater, it is necessary that tanneries provide sufficient treatment of their effluents. Much work has been carried out in India on the treatment of tannery effluents by low-cost treatment methods; locally developed criteria are available to successfully design and operate such systems. The demonstration plant at Ranipet built on such design criteria is providing excellent performance.

There is a need to study treatment alternatives such as anaerobic filters, rotating biological contactors, etc. to develop suitable treatment under certain circumstances. Techniques to remove as much salt from the salted hides would be advantageous. Solar evaporation of segregated salt bearing wastes is a necessary part of the treatment, and it will be a first step to prevent salinity encroaching on groundwater systems and on agricultural land. There is also a need to develop process operation of "dehairing" without the use of sulphides. The collection and utilization of biogas developed in the anaerobic digestion process should be explored; the possibility of using membrane covers over the anaerobic lagoons to collect biogas should be investigated. Many of these tanneries are small, generating a small quantity of wastewater in a day; the concept of combined treatment of wastewater from a cluster of tanneries needs to be investigated from technical and economical aspects [7].

REFERENCES

1. Sastry, C. A., and Madhavakrishna, W., "Pollution Problems in Leather Industries in India," *Environment India Review Series 2*, Department of the Environment, Gov't of India, New Delhi, India (1984).
2. Govinda Menon, K., Damondara Rao, T., and Swaminathan, R., "Water Pollution Problems in Tamil Nadu-Case Studies," *Proceedings of the Seminar on Environmental Pollution in the context of Present Industrial Developments in India*, Trivandrum, India, The Institution of Engineers (India) (1974).
3. Bhaskaran, T. R., "Treatment and Disposal of Tannery Effluents," *Proceedings, Symposium on Tannery Effluents*, Central Leather Research Institute, Madras, India (1977).
4. Sastry, C. A. *Groundwater Contamination by Tanneries in Vaniyambadi*, Indian Institute of Technology, Madras, India (1985).
5. Thabaraj, G. J., Bose, S. M., and Nayudamma, Y., "Comparative Studies on the Treatment of Tannery Wastes by Trickling Filters, Activated Sludge, and Oxidation Pond," *Bulletin Central Leather Research Institute* (Madras, India), Vol. 8 (1962).
6. Sastry, C. A., Kothandaraman, N., and Murahari Rao, P., "Low-cost Methods for the Treatment of Tannery Waters," *Proceedings, Symposium on Tannery Effluents*, Central Leather Research Institute, Madras, India (1972).
7. Cheda, P. V., Mandlekar, U. V., Handa, B. K., and Khanna, P., "Joint Wastewater Management for a Cluster of Tanneries at Kanpur," *Proceedings of the 39th Industrial Waste Conference*, Purdue University, Indiana (May 1984).

24 APPLICATION OF THE BATCH ACTIVATED SLUDGE PROCESS

Stephen R. Tate, Associate

W. Wesley Eckenfelder, Jr., Technical Director
AWARE Incorporated
Nashville, Tennessee 37228

INTRODUCTION

This paper presents a case study in the application of the batch activated sludge process. This particular application of this technology is somewhat unique in that it was instituted in place of a completely new physical/chemical treatment facility which was found to be incapable of meeting the permit requirements. To complicate the situation, the POTW to which the plant discharges was under consent order and absolutely could not permit excursions beyond specified limits. Plant shutdown was threatened. Although not uncommon to the consulting business, the project carried with it an almost impossible time constraint.

The general scope of the project was to perform wastewater treatability evaluations for development of an upgrade alternative for their existing wastewater treatment system. The major objectives of the study were the following.

- Develop a short-term solution for the plant upgrade
- Develop the short term solution such that a system retrofit, e.g., use of existing tankage, pumps, can be readily accomplished
- Develop sufficient data in the laboratory testing, and subsequent short term system operation, such that a long term solution can be both defined and permanently implemented
- Develop, design, and construct the necessary fix within 60 days

BACKGROUND INFORMATION

The new physical/chemical plant was intended to replace existing aerated lagoons. This is because the aerated lagoons had developed a history of violating the permit conditions for BOD and TSS during the winter months due to a dramatic decrease in biological activity. To initiate the study, historical waste load and treatment records for the aerated lagoons were evaluated. It was evident from the data that the wastestream was biodegradable, high strength, and low volume. Table I presents the average flow and waste load considered appropriate for design. Additionally, the existing treatment system facilities were inspected and evaluated with regard to tank capacities, piping and pumping capability, operational flexibility, and ease of retrofit.

Based on these evaluations, it was determined that the batch activated sludge process was appropriate for treatability evaluations. This process has been found to be applicable in installations where the wastewater is low volume, high strength (in terms biochemical oxygen demand), and readily degrad-

Table I. Design Flow and Waste Load

Parameter	Value
Flow (average)	25,000 gpd
BOD_5	7,000 mg/L
	(1,460 lb/day)
TOC	4,800 mg/L
TSS	150 mg/L
pH	8–10 std units
MBAS	600 mg/L

Table II. Batch Activated Sludge Operating Criteria

Parameter	Lab Scale	Full Scale
Reactor Volume	2 @ 1.63 liters	2 @ 90,000 gal
Feed Volume/Reactor	0.2 liters	12,250 gal
F/M	0.2 day^{-1}	0.2 day^{-1}
MLSS (mg/L)	6,660	6,600
MLVSS (mg/L)	4,600	4,600
BOD (mg/L) avg.	7,000	7,000
Aeration Time (hr)	22	22
Settling Time (hrs)	1	1
Decant Time (hrs)	1	1

able. Additionally, two existing concrete tanks (90,000 gal capacity each) were of appropriate size to potentially serve as batch reactors. Therefore, a treatability testing plan was developed for batch activated sludge.

TREATABILITY

Two batch reactors were setup in the laboratory each consisting of a 4 in. diameter plexiglas column, 2 ft in height, with a side port tap for decant. An air line was placed into the bottom center to two small air stones. The aeration produced from the air stones served also to mix the reactor. The reactors were designed to simulate the two 90,000 gal reactors at the plant.

The batch activated sludge process operates as an extended aeration process with an F/M typically in the 0.1 to 0.2 day^{-1} range to minimize sludge production. A major advantage of the process is its simplicity of operation. The activated sludge reactor serves also as the clarifier such that additional tankage is minimized. Operating criteria for the lab reactors was defined on the basis of the average flow and waste load and obtaining an F/M of 0.2 day^{-1}. Table II presents the operating criteria determined to be appropriate for using both 90,000 gal tanks as parallel batch reactors.

Initially, the reactors were set up at two different loading rates. One column was loaded with 12 in. of sludge and the other with 9 in. of sludge. Following approximately 1 week of operation, the 12 in. sludge depth column was reduced to a 9 in. sludge depth to make the units equal in loading. An activated sludge from a local municipal wastewater treatment plant was used to initially seed the units.

Table III presents a data summary for the treatability study. Figures 1, 2, and 3 present chronological plots of pertinent analytical results including TOC, BOD, TSS, MLSS, MLVSS, and test parameters ZSV (zone settling velocity) and SVI (sludge volume index).

EVALUATION OF RESULTS

For the quick fix improvements to the system, it was desired to achieve effluent concentrations of BOD and TSS less than 800 mg/L and 700 mg/L, respectively. However, the target permit limits (without surcharge) of 300 mg/L BOD and 150 mg/L TSS were desired.

Evaluation of the treatability data collected over four weeks of continuous operation indicated the process to be viable and capable of achieving the desired level of treatment. The first two to three weeks of operation were considered an acclimation period for the sludge. Following this acclimation at approximately Day 15 of operation, excellent treatment results became consistent. From Day 14 through Day 30, the BOD and TSS effluent averages were 254 mg/L and 213 mg/L, respectively. The sludges also exhibited excellent settling, and compaction qualities as evidenced by the ZSV and SVI test results. Oxygen uptake rates, sludge growth, sludge flocculation, settling and compaction all evidenced a healthy, active biomass within a relatively short study period.

The process requirements included nutrient additions (N and P). alkalinity addition (NaOH), and antifoam addition. The estimated dosages for these chemicals were deferred for definition during full-scale operation. Volume requirements for these chemicals were, however, expected to be relatively small and easily added.

PROCESS MODIFICATIONS

It was proposed that the two existing tanks at the facility be fitted to operate as batch activated sludge units. Modifications to the plant included:

Table III. Treatability Data Summary Batch Activated Sludge

Unit	Date	Day of Operation	Influent BOD (mg/L)	Influent TOC (mg/L)	Influent TSS/VSS (mg/L)	Influent MBAS (mg/L)	MLSS (mg/L)	MLVSS (mg/L)	Settled Depth (in.)	ZSV (ft/hr)	SVI mL/g	Feed Vol. (L)	Feed Depth (in.)	BOD_T (mg/L)	Effluent TOC (mg/L)	Effluent TSS (mg/L)	Effluent MBAS (mg/L)	Initial BOD Load (mg/mg VSS)	O_2 Uptake (mg/L-hr)	Buffer Zone (in.)
1	11/7	1	6,530	3,950	—	440	6,620	4,220	6.1	2.2	77	0.22	1.23	230	260	20	—	0.17	31.8	5.9
2	11/7		6,530	3,950	—	440	6,560	4,160	3.4	2.7	72	0.22	1.23	480	410	40	—	0.23	40.2	5.6
1	11/8	2	—	—	—	—	6,590	4,500	6.3	3.3	80	0.39	2.17	795	550	58	7.6	0.28	40.2	3.7
2	11/8		—	—	—	—	5,920	4,120	4.0	3.8	68	0.39	1.57	1,000	620	86	14.8	0.41	35.4	5.0
1	11/9	3	—	—	—	—	6,570	4,450	6.1	0.76	78	0.4	2.13	870	630	80	3.9	0.29	—	5.9
2	11/9		—	—	—	—	6,320	4,340	3.5	3.0	63	0.4	2.09	353	850	86	62.1	0.40	—	5.3
1	11/10	—	—	—	—	—	6,620	4,820	5.7	0.85	71	0.4	1.74	312	1,050	152	25.7	0.27	40.2	6.3
2	11/10		—	—	—	—	6,300	4,560	3.9	3.0	68	0.4	1.98	540	1,070	112	33.0	0.38	34.2	5.1
1	11/11	5	—	—	—	—	6,980	4,740	5.4	2.5	65	0.4	1.77	675	1,180	48	13.1	0.27	60.0	6.6
2	11/11		—	—	—	—	7,220	4,860	3.9	3.5	59	0.4	1.92	652	1,110	60	46.9	0.35	44.4	5.1
1	11/12	6	—	—	—	—	—	—	5.2	—	—	0.4	1.85	—	—	—	—	—	54.6	6.8
2	11/12		—	—	—	—	—	—	4.0	—	—	0.4	1.67	—	—	—	—	—	54.6	4.9
1	11/13	7	—	—	—	—	6,980	4,700	5.5	2.3	65	0.4	1.85	1,590	—	184	—	0.36	—	6.5
2	11/13		—	—	—	—	5,260	3,680	4.3	3.1	91	0.4	1.89	1,365	—	172	—	0.47	—	4.7
1	11/14	8	—	—	—	—	—	—	—	—	—	0.4	—	—	1,700	—	—	—	—	—
2	11/14		—	—	—	—	—	—	—	—	—	0.4	—	—	1,630	—	—	—	—	—
1	11/15	9	7,125	4,300	—	—	6,540	4,280	6.0	0.4	100	0.4	—	1,725	1,560	—	—	0.40	66.0	—
2	11/15		7,125	4,300	—	—	7,060	4,760	4.7	1.4	77	0.4	—	1,530	1,580	—	—	0.36	82.2	3.0
1	11/16	10	—	—	—	—	—	—	—	—	—	0.4	—	—	—	—	—	—	—	—
2	11/16		—	—	—	—	—	—	—	—	—	0.4	—	—	—	—	—	—	—	4.3
1	11/17	11	—	—	—	—	—	—	—	—	—	0.2	—	—	—	—	—	—	—	—
2	11/17		—	—	—	—	—	—	—	—	—	0.2	—	—	—	—	—	—	—	—
1	11/18	12	—	4,200	162/166	—	5,760	4,580	4.0	0.7	76	0.2	—	—	1,040	168	4	0.19	73.8	5.0
2	11/18		—	4,200	162/166	—	5,540	4,100	2.4	5.5	49	0.2	—	—	1,360	196	12	0.21	91.8	6.6
1	11/19	13	—	—	—	—	—	—	—	—	—	0.2	—	—	—	—	—	—	—	—
2	11/19		—	—	—	—	—	—	—	—	—	0.2	—	—	—	—	—	—	—	—
1	11/20	14	—	—	—	—	6,680	4,940	3.6	2.8	60	0.2	—	149	—	304	—	0.17	34.8	5.4
2	11/20		—	—	—	—	6,000	4,560	3.0	3.9	56	0.2	—	307	—	220	22.6	0.19	51.0	6.0
1	11/21	15	—	—	—	—	—	—	—	—	—	0.2	—	67/36a	800a	—	—	—	—	—
2	11/21		—	—	—	—	—	—	—	—	—	0.2	—	142/70a	1,040a	—	—	—	—	—

Table III (continued)

Unit	Date	Day of Operation	Influent BOD (mg/L)	Influent TOC (mg/L)	Influent TSS/VSS (mg/L)	Influent MBAS (mg/L)	MLSS (mg/L)	MLVSS (mg/L)	Settled Depth (in.)	ZSV (ft/hr)	SVI mL/g	Feed Vol. (L)	Feed Depth (in.)	Effluent BOD$_T$ (mg/L)	Effluent TOC (mg/L)	Effluent TSS (mg/L)	Effluent MBAS (mg/L)	Initial BOD Load (mg/mg VSS)	O$_2$ Uptake (mg/L-hr)	Buffer Zone (in.)
1	11/22	16	6,110	4,280	–	–	6,520	4,880	3.8	1.6	66	0.2	–	300	1,080	160	4	0.18	67.8	5.2
2	11/22	16	6,110	4,280	–	–	5,840	4,420	3.1	4.2	56	0.2	–	260	1,660	460	14	0.19	64.2	5.9
1	11/23	17	–	–	–	–	–	–	–	–	–	0.2	–	–	–	–	–	–	–	–
2	11/23	17	–	–	–	–	–	–	–	–	–	0.2	–	–	–	–	–	–	–	–
1	11/24	18	–	–	–	–	–	–	–	–	–	0.2	–	–	–	–	–	–	–	–
2	11/24	18	–	–	–	–	–	–	–	–	–	0.2	–	–	–	–	–	–	–	–
1	11/25	19	9,790	3,500	172	680	7,120	5,360	5.4	0.4	85	0.2	–	250	900	164	19.7	0.16	49.2	3.6
2	11/25	19	9,790	3,500	172	680	6,300	4,700	3.4	3.7	60	0.2	–	330	1,140	124	20.4	0.18	81.0	5.6
1	11/26	20	–	–	–	–	–	–	3.9	–	–	0.2	–	–	–	–	–	–	–	5.1
2	11/26	20	–	–	–	–	–	–	3.5	–	–	0.2	–	–	–	–	–	–	–	5.5
1	11/27	21	8,290	4,800	–	–	6,430	6,070	5.3	–	–	0.2	–	410	480	92	–	–	40.8	3.7
2	11/27	21	8,290	4,800	–	–	6,740	5,360	3.7	–	–	0.2	–	480	495	174	–	–	41.4	5.3
1	11/28	22	–	–	–	–	–	–	–	–	–	0.2	–	–	–	–	–	–	–	–
2	11/28	22	–	–	–	–	–	–	–	–	–	0.2	–	–	–	–	–	–	–	–
1	11/29	23	–	–	–	–	6,860	5,120	–	–	60	0.2	–	165	960	220	–	–	–	5.2
2	11/29	23	–	–	–	–	6,920	5,260	–	–	57	0.2	–	185	1,000	424	–	–	–	5.4
1	11/30	24	–	–	–	–	–	–	–	–	–	0.2	–	–	–	–	–	–	–	–
2	11/30	24	–	–	–	–	–	–	–	–	–	0.2	–	–	–	–	–	–	–	–
1	12/1	25	–	–	–	–	–	–	–	–	–	0.2	–	–	–	–	–	–	–	–
2	12/1	25	–	–	–	–	–	–	–	–	–	0.2	–	–	–	–	–	–	–	–
1	12/2	26	–	–	–	–	8,200	6,120	3.9	2.0	53	0.2	–	–	550	95	–	–	25.8	5.1
2	12/2	26	–	–	–	–	6,560	4,820	2.8	5.3	47	0.2	–	–	645	178	–	–	25.8	6.2
1	12/3	27	–	–	–	–	–	–	4.1	–	–	0.2	–	–	–	–	–	–	–	4.9
2	12/3	27	–	–	–	–	–	–	3.0	–	–	–	–	–	–	–	–	–	–	6.0
1	12/4	28	–	–	–	–	8,600	6,440	3.9	2.9	50	–	–	–	730	184	–	–	33.3	5.0
2	12/4	28	–	–	–	–	6,860	5,240	3.0	5.0	48	–	–	–	650	178	–	–	30.8	6.1
1	12/5	29	–	–	–	–	–	–	4.0	–	–	–	–	–	–	–	–	–	–	–
2	12/5	29	–	–	–	–	–	–	2.9	–	–	–	–	–	–	–	–	–	–	–

aSoluble analyses.

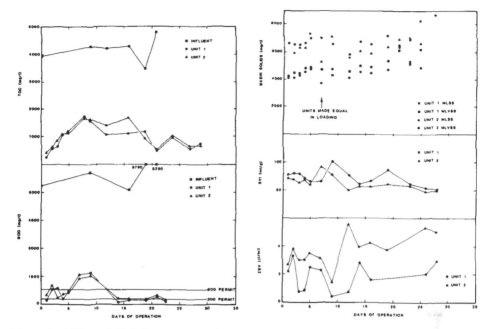

Figure 1. BOD and TOC data for treatability study.

Figure 2. ZSV, SVI, and basin solids data for treatability study.

1. Addition of two, 12,000 gal tanks to serve as wastewater feed holding tanks.
2. Addition of four 15 hp Aire0$_2$ aerators to provide mixing and aeration.
3. Removal of the existing mechanical aerators to limit temperature loss during winter months.
4. Addition of two decant units to remove treated supernatant and discharge to the POTW.
5. Use of two existing 50 gpm pumps for sludge wastage. Another existing tank was modified (sloped bottom and center wall removal) to serve as sludge concentrator. Existing pump was used to pump concentrated waste activated sludge to a tank truck. The waste activated sludge was to be ultimately disposed of at another local POTW.

The proposed modifications resulted in little piping or structural changes to the existing facilities. The proposed modifications were essentially equipment additions required to provide full-scale process operation similar to that obtained during the treatability studies.

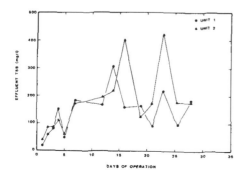

Figure 3. Effluent TSS data for treatability study.

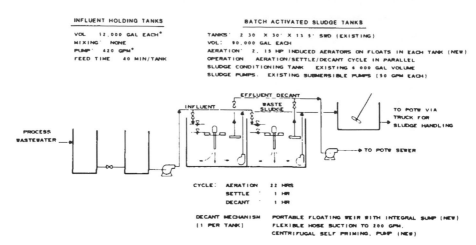

Figure 4.   Schematic flow diagram for retrofit.

Initial sludge seed for the batch reactors was obtained from the POTW. The sludge was to be trucked to the plant and pumped into the tanks. An acclimation period of two to three weeks was to be provided prior to commencement of normal operation as indicated by results obtained from bench-scale treatability tests.

Facilities were also to be provided for addition of anti-foam to the wastewater influent to suppress the anticipated foaming associated with the wastewater.

INTERIM CONCLUSIONS AND RECOMMENDATIONS

It was concluded from the treatability study results that the batch activated sludge process is applicable for treatment of the subject chemical wastewater. It was further concluded that the process could be readily adapted to the existing facilities within a relatively short time frame. It was uncertain at that time as to whether the 300 mg/L BOD and 150 mg/L TSS could be achieved consistently through the winter months. However. the maximum limitations (800 mg/L BOD and 700 mg/L TSS with surcharge) were felt to be consistently achievable.

Based on these results; it was recommended that the plant be given a permit to construct the necessary system modifications for the batch activated sludge process. Figure 4 presents a schematic flow diagram which defined the operational and equipment needs for the retrofit. Figure 5 presents a cross sectional view of the batch reactor.

FULL-SCALE OPERATIONS

The plant modifications were installed during December 1985 and January 1986. State approval of a design report and construction drawings and specifications were required in order to gain a construction permit. The plant was started up in late January 1986.

Given the limited amount of time allowed for treatability, development of design criteria, and installation of the equipment, the application has proven to be quite successful. Ongoing evaluations are continuing to optimize, troubleshoot, and permanentize the installation. Of primary focus are the following:

• Equalization
• Sludge Production
• Evaluation of potentially biologically toxic or inhibitory individual product wastestreams

EFFECT OF FEEDING SCHEDULE

The batch activated sludge process is typically operated at an F/M of 0.1 to 0.2 to minimize excess sludge production and to minimize process stability. Assuming there are no toxic components in the

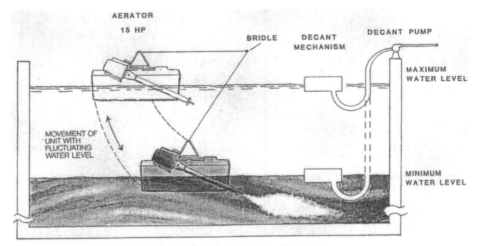

Figure 5.   Batch activated sludge system.

wastewater, the wastewater may be introduced to the basin over a short period of time, i.e., 30 min or over an extended period of time, i.e., 6 to 12 hours. If there is a toxic constituent such as dichlorophenol, the rate of wastewater addition must be less than the rate of degradation to avoid a concentration buildup in the reactor to a toxic level.

The method of batch feeding the reactors has manifested both good and bad effects. Since equalization is currently not provided, the day-to-day waste load has exhibited wide fluctuations in terms of both BOD and COD. Additionally, the production plant typically schedules long campaigns oftentimes with a single product. Although evaluations are just beginning, it is felt that at least one individual product wastestream is shocking the plant. It is uncertain at this time as to whether the feeding schedule should be modified (fed at a slower rate to prevent buildup) or the wastestream is toxic.

The good effects associated with the batch feeding schedule have been excellent sludge settling characteristics. Recently reported literature has indicated that the biosorption process prevails when a high organic driving force exists. This has been found to selectively favor the growth of flocculated bacteria as opposed to filamentous organisms. Biosorption, in simple terms, is the mechanism by which organic substrate is sorbed into the bacteria and then degraded over time. The contact stabilization process is essentially defined by this mechanism. This phenomenon has been evident in the full operations. Figures 6 and 7 present plots of oxygen uptake and residual COD and BOD, respectively, over the aeration cycle of the plant. The plots indicate that the majority of the oxygen consumption occurs over the first few hours of contact, however, the rate of organic substrate decay is somewhat linear over the entire cycle. The flocculated bacteria flourish and are maintained fully aerobic, thereby promoting excellent settling properties in terms of zone settling velocity and sludge volume index.

EFFLUENT QUALITY

Figures 8, 9, and 10 present chronological plots of COD, BOD, TSS, ZSV, and SVI for the plant for the period January 27 through April 30, 1986.

Over the period, the raw influent has exhibited wide fluctuations in organic strength. The average BOD for the period was 4,342 mg/L which has been reduced to an average effluent BOD of 344 mg/L (92% removal). The zone settling velocity and sludge volume index have averaged 6.0 ft/hr and 67.3 mL/gm, respectively.

Deteriorated performance was experienced during two periods. The first problem involved poor resuspension of sludge following the settling cycle. Adjustment of the angle of the aerators corrected the problem. A second period was determined to be caused by a nitrogen deficiency. Upon stepping up the dosage, performance improved almost immediately.

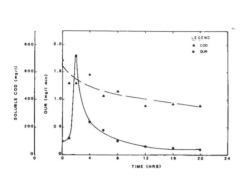

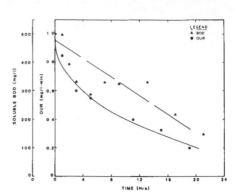

Figure 6. Biosorption during aeration cycle of February 20, 1986.

Figure 7. Biosorption during aeration cycle of February 27, 1986.

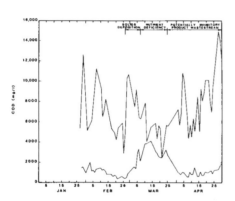

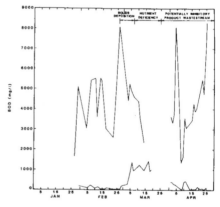

Figure 8. Chronological plot of batch activated sludge COD.

Figure 9. Chronological plot of batch activated sludge BOD.

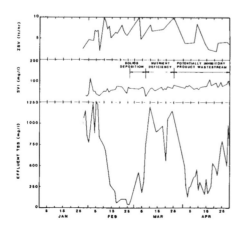

Figure 10. Chronological plot of batch activated sludge effluent TSS, ZSV, and SVI.

The most recent data (April) has given further indication to the need for equalization and study of one individual product wastestream (produced heavily during April). These investigations are continuing at the present time.

SUMMARY

The application of batch activated sludge at this facility has proven to be a rapidly implemented, cost effective solution. Utilization of the existing facilities has been a significant time and cost savings. In addition, the plant has performed within permit compliance through the winter months and offered the company the opportunity to optimize the process and facilities over the coming months. Attached is a design example for the batch activated sludge process.

DESIGN EXAMPLE

The design of a batch activated sludge process is illustrated by the following example.

$$
\begin{array}{ll}
Q & - \; 50{,}000 \text{ gpd} \\
\text{BOD (So)} & - \; 500 \text{ mg/L} \\
\text{TKN} & - \; 2 \text{ mg/L} \\
a & - \; 0.6 \\
a' & - \; 0.55 \\
b & - \; 0.1 \\
\text{F/M} & - \; 0.1
\end{array}
$$

The plant will be operated with 20 hr aeration, 2 hr sedimentation and 2 hr decant. The BOD removal will be:

$$(500 - 10)(8.34)(0.05) = 204 \text{ lb/day}$$

at an F/M of 0.1 the required MLVSS is:

$$204/0.1 = 2{,}040 \text{ lb}$$

and at 85% volatile sludge the MLSS is:

$$2{,}040/0.85 = 2{,}400 \text{ lb}$$

Assuming the sludge has an SVI of 100, the sludge volume will be 1.6 ft³/1b, and the volume required for the settled sludge is:

$$2{,}400 \text{ lb} = 3{,}835 \text{ ft}^3 \text{ or } 28{,}760 \text{ gal}$$

If sludge is to be wasted twice per month, storage must be provided for accumulated sludge.

At an estimated degradable fraction of the VSS of 0.4, the daily accumulation of VSS is:

$$
\begin{aligned}
X_v &= 0.6 Sr - bX_d X_s \\
&= 0.6\,(204) - 0.1 \cdot 0.4 \cdot 2{,}040 \\
&= 40 \text{ lb/day}
\end{aligned}
$$

Storage for 15 days will be:

$$40 \cdot 15 \cdot 1.6 \cdot 7.48 = 7{,}200 \text{ gal}$$

The total volume of basin (excluding freeboard) will be 85,950 gal. This will be a basin 35 ft diameter and 12 ft deep. If 3 ft is provided for freeboard, the operational basin dimensions will be 35 ft diameter by 15 ft depth.

The oxygen requirements can be calculated:

$$
\begin{aligned}
0_2/\text{day} &= a'Sr + 1.4bX_d X_s \\
0_2/\text{day} &= 0.55(204) + 1.4 \cdot 0.1 \cdot 0.4 \cdot 2{,}040 \\
&= 226 \text{ lb/day or } 9.5 \text{ lb/hr}
\end{aligned}
$$

The required HP @ 1.5 0_2/HP-HR is:

$$9.5/1.5 = 6.3; \text{ use } 7.5 \text{ HP}$$

This is equivalent to 150 HP/MG of basin volume which should provide adequate mixing. The nutrient requirement will be:

$$N = 0.123 \cdot X_d/0.8 \, X_v$$
$$= 0.123 \cdot 0.4/0.8 \cdot 40$$
$$= 2.5 \text{ lb/day as N}$$
$$P = 0.026 \cdot 0.4/0.8 \cdot 40$$
$$= 0.5 \text{ lb/day as P}$$

25 DUAL BED ION EXCHANGE REGENERATION – OPTIMIZATION FOR HIGH PURITY WATER SYSTEMS

James A. Mueller, Professor
Environmental Engineering and Science
Manhattan College
Bronx, New York 10471

Warren Riznychok, Senior Associate Chemist

Gordon Bie, Senior Operator
IBM Thomas J. Watson Research Center
Yorktown Heights, New York 10598

INTRODUCTION

The Thomas J. Watson Research Center in Yorktown Heights, New York, is International Business Machines (IBM) basic and applied research division headquarters. Historically, wastewater generated by the research activities has been treated by an Industrial Wastewater Treatment Facility and its effluent recycled for reuse in laboratory sinks, as boiler makeup and feedwater to the sites deionized water systems. In January 1984 a newly constructed High Purity Water System (HPW) was brought on-line and also received recycled industrial waste as a source of makeup water. High purity water, used primarily in semiconductor research at the Center, can have a significant impact on research and development if it does not meet required ionic, organic, and 6 quality specifications.

Due to upgrading of the industrial wastewater treatment plant, the HPW began to use city water directly in January 1985, and by March 1985 the two bed ion exchange process, heart of the HPW system, showed significant reductions in throughput volumes and quality. High pH and conductivity focused attention to the cation unit as the problem. The body of this paper will discuss in detail the steps taken to quickly recover cation bed capacity and to optimize regeneration efficiency.

PROCESS DESCRIPTION

A total water flow scheme is illustrated in Figure 1. Industrial waste characterization in 1982 of the daytime (7:30 A.M.–7:30 P.M.) and nighttime waste demonstrated feasibility of segregation of the waste stream into separate holding tanks, enhancing process operation. Daytime waste undergoes neutralization, biological treatment with rotating biological contactors, settling, disinfection with chlorine, and air stripping. Air stripper effluent quality determines if the pretreated day waste will be discharged to the Westchester County Sewer System or undergo further treatment.

The dilute night waste, which contains 40% deionized water, or pretreated day waste, is further treated by dual media filtration, activated carbon adsorption and two bed ion exchange present in the nighttime waste process flow scheme. Provisions are also in place to pretreat city water through this process area to assure adequate supply for makeup and users. Figure 2 indicates process flow and water treatment possibilities.

High purity water is produced using dual media filtration for gross particulate removal, activated carbon adsorption for removal of organics and chlorine, UV sterilization, two bed ion exchange, mixed bed ion exchange, UV sterilization, 0.6 micron and 0.2 micron membrane filtration (Figure 3). This high quality point of distribution effluent is pumped to the research center where it is repolished by two separate stations. Water that is not required for makeup at the two polishing stations is returned to intermediate storage in the main processing plant.

The polishing station servicing the Advanced Silicon Technology Laboratory and Aisles 1-6 repurifies the water with nonregenerable mixed beds UV sterilization and 0.1 micron filtration. The second

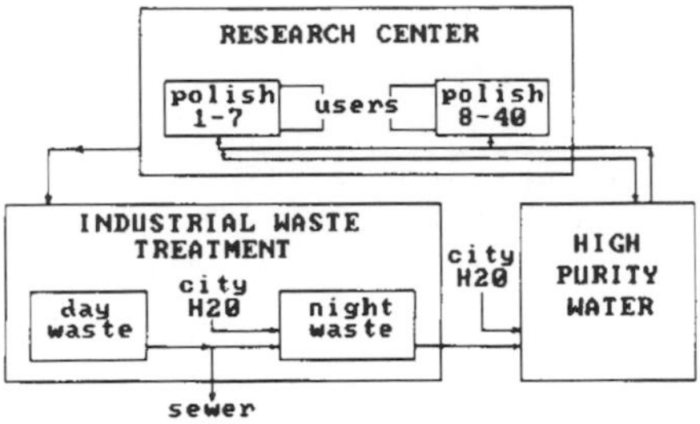

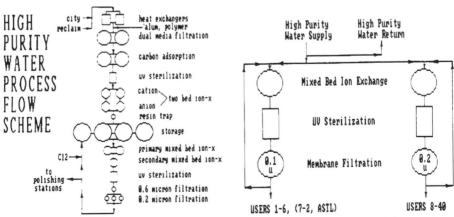

Figure 1. Water and wastewater treatment systems at the IBM T. J. Watson Research Center.

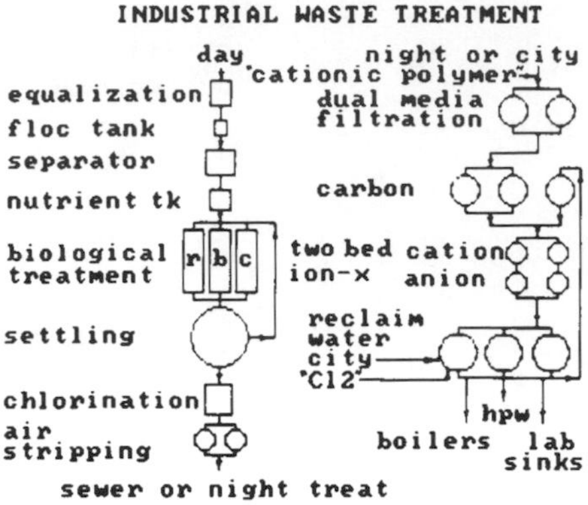

Figure 2. Flow diagram of industrial waste treatment facility.

polishing station has regenerable mixed beds, UV sterilization, and 0.2 micron filtration prior to distribution to users. See Figure 4 for polishing station flow schemes.

PROBLEM SETTING

In January 1985 recycled industrial wastewater was no longer available for makeup to HPW due to an (IWTF) upgrade to accommodate higher volumes and improve effluent quality. A municipal supply from the New Castle Water District became the sole source makeup to HPW. Prior to March, throughput volume was controlled by anion exchange capacity with the beds regenerated when the pH dropped below 5.5 or the conductivity rose above 10 μmhos/cm.

In March of 1985, HPW's two bed ion exchange system was exhibiting losses in throughput capacity while its effluent quality deteriorated with elevated pH and rise in conductivity. Those changes in effluent quality and throughput are seen during days 60 to 100 in Figures 5,6, and 7. Effluent conductivity increased to 180 μm hos/cm and pH showed a significant trend of increasing from 7 to 12 as throughput volume decreased, indicating loss in cation bed capacity.

With sulfuric acid being the chemical used for cation regeneration, it was presumed that calcium sulfate was precipitating in the beds; thus, the regeneration cycle was altered to improve regeneration efficiency. Chemical injection flow rates were raised from 0.8 to 1.0 gpm/ft³. The two-step regenera-

Figure 3. Flow diagram of high purity water plant.

Figure 4. Flow diagram of polishing station.

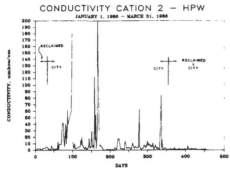

Figure 5. Effluent conductivity for HPW cation #2 ion exchanger from 1/1/85 to 3/31/86.

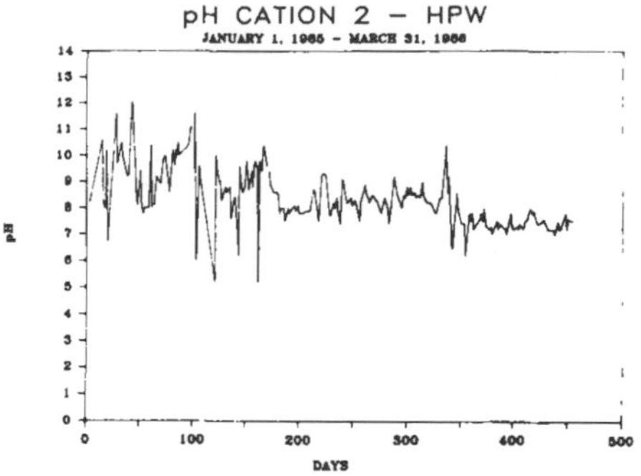

Figure 6. Effluent pH for HPW cation #2 ion exchanger from 1/1/85 to 3/31/86.

tion of 2% and 4% acid at 6.8 lb H_2SO_4/ft^3, as recommended by the manufacturer, was changed to a multi-step regeneration of 0.5%, 1.0%, 2%, and 4% at the same loading rate. Additionally a resin sample sent to the manufacturer for analysis confirmed calcium fouling having an ash content of 85% calcium.

This change in regeneration cycle resulted in no improvement in cation performance. The cation resin was then soaked in 10% HCl for 16 hours. This improved bed capacity but only temporarily. After a subsequent regeneration with H_2SO_4 the cation beds began to fail again. At this time, a decision was made to change the resin and use extended step regeneration and elevated chemical injection flow rates with the new resin.

By June 1985 the fresh resin began to exhibit the same loss of capacity as found in March with high pH and conductivity (Figures 5,6, and 7). Again resin samples showed calcium fouling to be the cause. Samples of sulfuric acid were titrated to determine if the per cent acid meter was reading correctly; they were also checked with a hygrometer. The results presented in Table I show the acid meter was conservative and not the cause of the cation bed failure.

In June 1985 the resin was again changed. Chemical quantities (including HCl soaks and resin changes) used for regeneration during the periods leading up to cation failure and subsequent changes in regeneration chemicals to loss of resin capacity are illustrated in Figure 8.

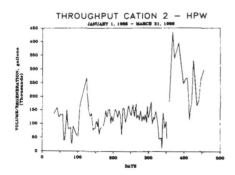

Figure 7. Throughput volume for HPW cation #2 ion exchanger from 1/1/85 to 3/31/86.

Table I. Sulfuric Acid Analysis in June 1985

Nominal	% Acid Meter	% Titration w/v
0.5%	0.53	0.77
1.0%	0.96	1.04
2.0%	1.93	2.12
4.0%	4.15	4.68

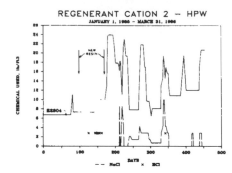

Figure 8.   Chemical usage for HPW cation #2
ion exchanger from 1/1/85 to 3/31/86.

PROBLEM ISOLATION

Having identified the problem as calcium fouling of the cation resin when manufacturer's recommended regeneration specifications were utilized, regeneration sampling was begun to establish exchange capacities and calcium recoveries.

An extended step regeneration using a total of 13 lb/ft³ of sulfuric acid was sampled on June 25, 1985 to provide information on ion recovery during this cycle. The data presented in Figure 9 showed that regeneration had stopped short of removing a large portion of calcium and magnesium although the majority of the sodium appears to have been removed. Also note that earlier regenerations were at levels of approximately 6.8 lb/ft³ and 7.5 lb/ft³ of H_2SO_4 indicating that calcium and magnesium ions were building on the beds during each regeneration. Manufacturers data suggested that 6-9 lb/ft³ of acid would be sufficient to displace all the Ca^{++} and Mg^{++} during regeneration.

A second ion profile of cation #2 was generated on June 28, 1985 using 23.8 lb/ft³ of acid. As illustrated in Figure 10, this time, all of the calcium, magnesium, and sodium loaded on the bed during exhaustion was recovered as well as a large portion of those ions remaining from the prior regeneration. This percentage ion recovery is demonstrated in Table II. Increase in hydrogen ion bed capacity following regeneration using 23.8 lb/ft³ was also achieved (Table III).

The ion concentration profiles have shown that it was not a calcium sulfate precipitate during regeneration that fouled the resin, but in fact, was the inability of sulfuric acid at suggested manufacturers levels to adequately remove ions. This allowed calcium to build on the resin after each regeneration until there was little hydrogen exchange capacity remaining and the bed failed.

SOLUTIONS

Although sulfuric acid is stored in bulk quantity at the site, it was not an economically sound idea to continue to regenerate at the 23.8 lb/ft³ levels. Although HCl worked well during the resin soaks, this was not a feasible alternative because of the incompatability of associated regeneration plumbing, storage tanks, and pumps.

After review of the manufacturer's data on regeneration (Figure 11), it was clear that sodium chloride was an effective regenerant at removing calcium when calcium chloride was the exhausting solution. It was also clear from manufacturers data (Figure 12) and our own ion profiles that H_2SO_4 was very effective at regenerating sodium from the bed. In Figure 13, it appeared that regeneration with NaCl followed by H_2SO_4 would yield regeneration efficiencies at 35-45%, a desired efficiency for optimum operation of these units. Therefore, bench scale tests were performed to determine optimum efficiencies achievable at various regenerant concentrations of NaCl and H_2SO_4.

BENCH SCALE TEST RESULTS

A test column containing 4 inches of Rohm and Haas-Amberlite 120 plus cationic resin was exhausted using the carbon effluent from the HPW plant at hydraulic loadings of 50 gpm/ft². Bench scale column operating conditions were much different than actual (4 ft. depth at 5-8 gpm/ft²) to rapidly obtain an exhausted bed for two step regenerant testing. The column effluent was sampled and tested for hardness (Ca^{++} and Mg^{++}), alkalinity, acidity, pH, and conductivity. Throughput

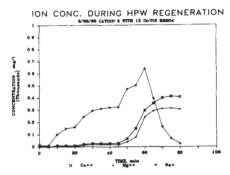

Figure 9. Ion concentration during HPW cation #2 regeneration with 13.0 lb/ft³ of H₂SO₄ on 6/25/85.

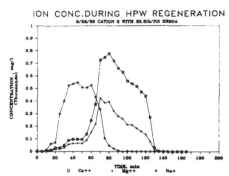

Figure 10. Ion concentration during HPW cation #2 regeneration with 23.8 lb/ft³ of H₂SO₄ on 6/28/85.

Table II. Ion Balance During June 1985 HPW Cation Regenerations

Regeneration			Ion Balance, kg as Ca CO₃			
Date 1985	H₂SO₄ Level lb/ft³	Ion	Removal During Exhaustion	Recovery During Regeneration	Storage on Bed[a]	% Ion Recovery
6/25	13.0					
		Ca^{++}	20.9	6.7	14.2	32
		Mg^{++}	10.5	8.2	2.3	78
		Na^+	11.6	9.2	2.4	79
		Total	43.0	24.1	18.9	56
6/28	23.8					
		Ca^{++}	20.3	21.1	-0.8	104
		Mg^{++}	11.4	16.0	-4.6	140
		Na^+	13.7	11.0	2.7	80
		Total	45.4	48.1	-2.7	106

[a]Bed Volume = 56 ft³.

Table III. Acid Efficiency During June 1985 HPW Cation Regenerations

Regeneration		H₂SO₄ Balance, kg as CaCO₃		%	Increased H⁺
Date 1985	H₂SO₄ Level lb/ft³	H⁺ Used During Regeneration	H⁺ Stored on Bed[a]	Acid Efficiency	Bed Capacity[b] kgrain CaCO₃/ft³
6/25	13.0	337	24.1	7.2	6.6
6/28	23.8	617	48.1	7.8	13.2

[a]Total ion recovery during regeneration.
[b]H⁺ stored on bed (kg as $CaCO_3$) × 15.43 kgrain/kg/56 ft³ bed volume.

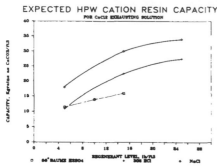

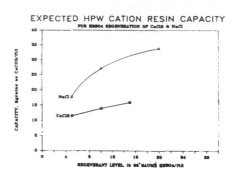

Figure 11. Expected HPW cation resin (amberlite 120 plus) capacity from manufacturers' data for H_2SO_4, HCl, and NaCl regeneration after exhaustion with $CaCl_2$.

Figure 12. Expected HPW cation resin capacity from manufacturers' data for H_2SO_4 regeneration after exhaustion with $CaCl_2$ and NaCl.

volume was monitored both during exhaustion as well as regeneration. The same tests were performed during the regeneration cycle so complete ionic mass balances could be obtained.

The resin was obtained from the full scale bed after regenerating with 23.8 lb/ft³ H_2SO_4 assuring a high hydrogen capacity bed to start with. Two tests were made at each of three regenerant levels, an exhaustion test and a regeneration test.

Figures 14 and 15 represent the exhaustion data versus throughput volume for Test #1. As expected during exhaustion, effluent alkalinity increases or acidity (negative alkalinity) decreases as the bed loses its capacity to donate hydrogen ions. The pH begins to rise until both alkalinity and pH approach influent values.

Calcium and magnesium follow the same pattern rising on exhaustion. due to the pH effect on conductivity, increasing as it is further away from neutral pH, the conductivity decreases during the test as the pH approaches neutral.

Figures 16 and 17 show the results during the regeneration following the above exhaustion. Two distinct peaks in conductivity are seen, the first due to the salt at 5 lb/ft³ and the next due to the acid at 5 lb H_2SO_4/ft³. The salt is highly effective in removing hardness from the resin producing concentrations as high as 24,000 mg/L as $CaCO_3$, about 60% being calcium and 40% magnesium. During the salt injection, any remaining acidity (H^+) is removed from the bed, very slight in these runs since the bed has been almost completely exhausted, causing pH to decrease to about 2. During the 4% acid injection, the effluent acidity never reached 4% acid or approximately 40,000 mg/L. This means that acid injection was stopped too early and, therefore, a significant amount of ion remained on the bed, namely sodium, since during the sodium cycle nearly all Ca^{++} and Mg^{++} were exchanged.

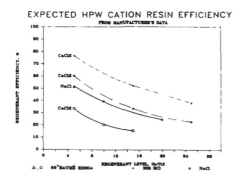

Figure 13. Expected HPW cation resin efficiency from manufacturers' data for various regenerants and exhausting solutions.

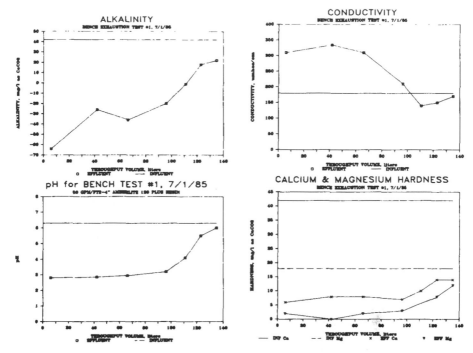

Figure 14. Effluent alkalinity and pH during exhaustion cycle for bench test #1 on 7/1/85 using HPW cation resin.

Figure 15. Effluent conductivity and hardness during exhaustion cycle for bench test #1 using HPW cation resin.

The sodium remaining on the bed would have a pronounced effect on the ion exchange effluent, while in service, due to leaching of the sodium ion. The resulting effluent quality would have high pH and conductivity similar to a resin heavily loaded with calcium ion.

In examining Test 4 and 6 regenerations (Figures 18 & 19) it can be seen that 10 and 12 pounds of acid per cubic foot put the bed in a much more hydrogen ready form for operation. Table IV indicates that column recovery was significant during the bench scale tests with tests 5 and 6 producing over 100% calcium recovery. Some salt remained and was stored on the bed during tests 1 and 2 with complete recovery in the following tests.

Table V summarizes the regeneration efficiences of the bench scale tests. The data indicate that salt regeneration prior to acid will provide acid regeneration efficiencies of 31 to 46%. It should be noted that using salt prior to the acid cycle eliminates the need for step acid regeneration since nearly all of the calcium is displaced in the salt cycle.

FULL SCALE RESULTS

Based on the positive bench scale results, a full scale regeneration was conducted on cation #1 using 10 lb/ft³ of salt and 11.4 lb/ft³ of acid. The results given in Figure 20 indicate that significant hardness recovery was obtained during the salt cycle, the first 4,000 liters of regenerant volume. From the acidity data, it is obvious that the bed was not previously completely exhausted since a significant quantity of acidity was displaced during the salt cycle. The majority of the hardness was removed during the salt cycle with the major portion being calcium. During the acid cycle the acidity rose to above 40,000 mg/L indicating that the major portion of the sodium was removed from the bed. However, during the following exhaustion cycle, the conductivities ranged from 10–20 µmhos/cm compared to a value of less than 5 µmhos/cm when only acid was used for the regeneration, as seen on Figure 5 after day 210. Subsequent regenerations utilized a lower salt loading with variable acid loadings as indicated in Figure 8.

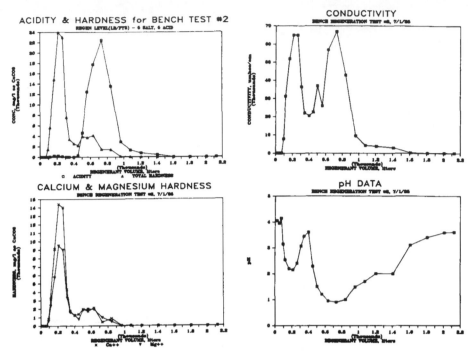

Figure 16. Acidity and hardness during regeneration cycle for bench test #2 on 7/1/85.

Figure 17. Conductivity and pH during regeneration cycle for bench test #2 on 7/1/85.

Figures 21, 22 and 23 show results for the 8/12/85 regeneration using 4.3 lb/ft³ of salt and 21.1 lb/ft³ of acid. As seen in Figure 21, a three stage regeneration was used in this study, the first being salt and the second two being acid. The last acid regeneration increased the acidity to around 40,000 mg/L at an effluent pH of about 0.4 indicating the bed was in equilibrium with the acid regenerant concentration and, thus, substantially completely in the acid cycle. Figure 23 indicates the cumulative hardness for this test, showing that 18 kg of calcium and 9 kg of magnesium, both as calcium carbonate, were recovered. This high acid regeneration produced lower effluent conductivities, as indicated in Figure 5 on day 224.

A continual reduction in the amount of salt and acid utilized during regeneration was initiated, the results for the 11/11/85 (day 315) run shown in Figure 24 at 0.7 lb/ft³ of salt and 8.1 lb/ft³ of acid. Significantly lower hardness recoveries were obtained during the salt cycle, the first 2.8 liters of regenerant volume. Figure 25 shows that the hardness recovery was mainly magnesium at this low salt regenerant level, thus allowing a significant amount of calcium to be stored on the bed.

Figures 26, 27, and 28 summarize the results of calcium balances conducted on the HPW cation units during the summer and fall of 1985. Total hardness recovery varied from 10 to 60 kg as calcium carbonate depending on the regenerant used and the prior history of the bed. On days 210 and 212, when salt was first used on cation beds #1 and #2, the highest hardness recoveries were obtained, mainly in the salt cycle. With decreasing amounts of salt used in subsequent regenerations, most of the calcium recovery occurred in the acid cycle, after day 216, the total quantity being significantly less than what was removed. Figure 27 indicates that during this period, a significant amount of calcium was stored on the bed. As shown in Figure 28, magnesium storage never occurred since the calcium preferentially removed the magnesium from the bed. This calcium buildup allowed lower bed volumes to be processed. Adjusting the salt loading to 10 lb/ft³ on day 337, followed by an overnight HCl soak as shown in Figure 8, was able to regenerate the resin back to full hydrogen exchange capacity.

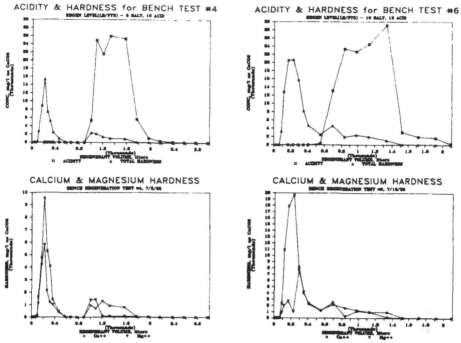

Figure 18. Acidity and hardness during regeneration cycle for bench test #4 on 7/2/85.

Figure 19. Acidity and hardness during regeneration cycle for bench test #6 on 7/15/85.

Table IV. Ion Balance During Bench Scale Cation Tests

Test	Date 1985	Regenerant Level, lb/ft³		Ion	Ion Balance, mg as CaCO₃			% Ion Recovery During Regeneration Cycle		
		NaCl	H₂SO₄		Removal During Exhaustion	Recovery During Regeneration	Storage on Bed	Salt	Acid	Total
0	6/28	0	23.8							
1,2	7/1	5	5	Ca^{++}	3300	2880	420	70	17	87
				Mg^{++}	3200	2100	1100	49	17	66
				Na^+	2630	680	1950	-248	274	26
				Total	9130	5660	3470			62
3,4	7/1	5	10	Ca^{++}	3780	1970	1810	35	17	52
				Mg^{++}	1100	1110	~10	71	30	101
				Na^+	1220	4890	-3670	-171	572	401
				Total	6100	7970	-1870			131
5,6	7/2,7/15	10	12	Ca^{++}	3580	4740	-1160	103	29	132
				Mg^{++}	4170	2040	2130	17	32	49
				Na^+	100	3140	-3040	-4980	8120	3140
				Total	7850	9920	-2070			126

Table V. Regenerant Efficiency and Capacity During Bench Scale Tests

Test	Regenerant Level, lb/ft³		H⁺ Bed Capacity, kgrain CaCO₃/ft³		% Regenerant Efficiency	
	NaCl	H₂SO₄	Exhausted	Regenerated	NaCl	H₂SO₄
0	0	23.8		13.2		7.8
1	–	–	26.6			
2	5	5		16.5	38	46
3	–	–	17.8			
4	5	10		23.1	20	33
5	–	–	22.9			
6	10	12		28.9	24	34

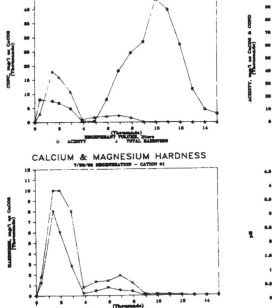

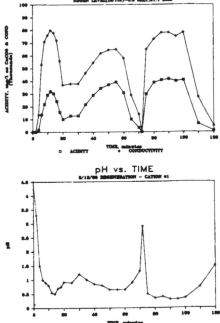

Figure 20. Acidity and hardness during full-scale regneration of HPW cation resin on 7/29/85 using 10.0 lb/ft³ NaCl and 11.4 lb/ft³ H₂SO₄.

Figure 21. Acidity, conductivity, and pH versus time during full-scale regeneration of HPW cation resin on 8/12/85 using 4.3 lb/ft³ NaCl and 21.1 lb/ft³ H₂SO₄.

As shown in Figure 29, during this period the salt regeneration efficiency varied from 30 to 100% while the acid efficiencies varied from 7 to 30%. The acid lowest efficiencies occurred when little to no salt was used prior to acid regeneration.

At the end of December 1985 the source water for the HPW plant was switched from a direct city water feed to a reclaimed water feed from the industrial waste treatment facility with only periodic city water used when the industrial plant was out of operation. Due to the low calcium content of the reclaimed water, salt usage during regeneration in the high purity water plant was discontinued except

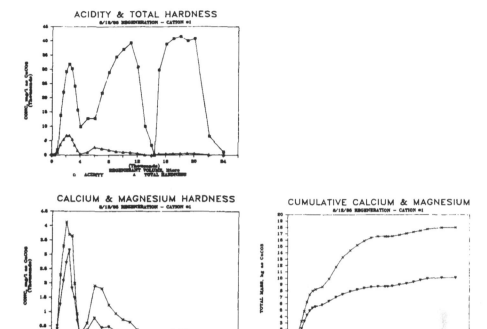

Figure 22. Acidity and hardness during full-scale regeneration of HPW cation resin on 8/12/85 using 4.3 lb/ft³ NaCl and 21.1 lb/ft³ H₂SO₄.

Figure 23. Cumulative hardness recovery during full-scale regeneration of HPW cation resin on 8/12/85 using 4.3 lb/ft³ NaCl and 21.1 lb/ft³ H₂SO₄.

during periods of direct city water feed. Attention was then focused on the IWTF cation exchangers since they were receiving a significant amount of city water along with nighttime wastewater.

Figures 30 and 31 show the conductivity and throughput volumes for the IWTF plants from January 1, 1986 through March 31, 1986. Effluent conductivity and throughput volume are the function of the type of water treated as well as the state of the cation beds.

As indicated in Figure 32, salt injection was begun at the end of January 1986 on day 23. Acid regeneration in the IWTF was relatively constant since it is set by acid storage volume tankage. Figures 33 and 34 show the results of the first salt regeneration at 10.2 lb/ft³ of salt and 6.7 lb/ft³ of acid. Very high quantities of calcium and magnesium were recovered during the salt cycle up to a volume of 4,000 liters due to the significant amount of hardness storage during the prior acid only regenerations.

Figure 35 indicates that the salt cycle provided the major portion of both calcium and magnesium recovery throughout the period with lowest hardness recoveries occurring when no salt was utilized. Regeneration efficiencies for the IWTF (Figure 36) were maintained at around 50% for the salt and at around 30% for the acid, significantly greater than previous results with acid alone. Again lowest efficiencies, approximately 8–10%, occurred when no salt was utilized prior to the acid regeneration.

CONCLUSIONS

At the IBM Yorktown high purity water treatment facilities, a switch in the feed water from a low calcium content to a city water feed where calcium was 55% of the total cations present caused significant regeneration problems for the cation beds. Using the manufacturer's recommended regenerant levels provided incomplete calcium removal from the beds.

Use of a salt regeneration prior to the acid allowed complete calcium recoveries and improved acid regeneration efficiencies.

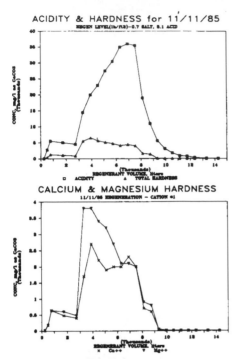

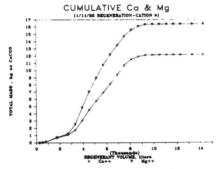

Figure 24. Acidity and hardness during 11/11/85 HPW regeneration using 0.7 lb/ft³ NaCl and 8.1 lb/ft³ H₂SO₄.

Figure 25. Cumulative hardness during 11/11/85 HPW regeneration using 0.7 lb/ft³ NaCl and 8.1 lb/ft³ H₂SO₄.

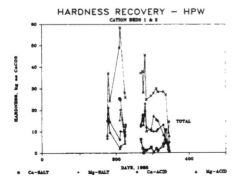

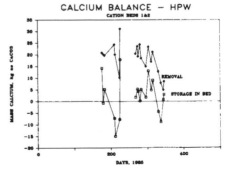

Figure 26. Hardness recovery from HPW cation beds during salt and acid regeneration cycles.

Figure 27. Calcium balance for HPW cation beds.

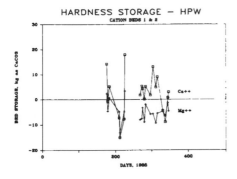

Figure 28. Hardness storage on the HPW cation beds after regeneration.

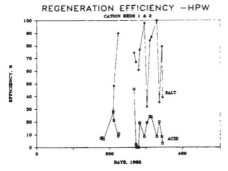

Figure 29. Regeneration efficiency for the HPW cation beds during 1985.

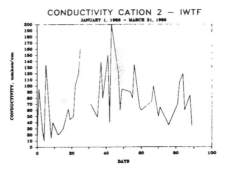

Figure 30. Effluent conductivity for IWTF cation #2 ion exchangers from 1/1/86 to 3/31/86.

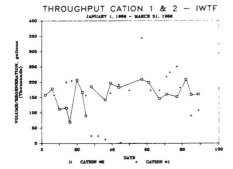

Figure 31. Throughput volume for IWTF cation #1 and #2 ion exchangers from 1/1/86 to 3/31/86.

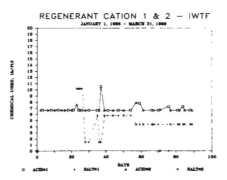

Figure 32. Regenerant chemical usage for IWTF cation #1 and #2 ion exchangers from 1/1/86 to 3/31/86.

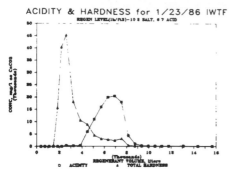

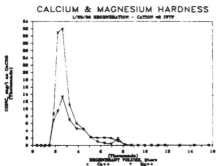

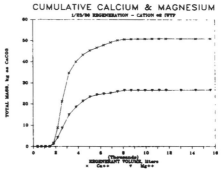

Figure 33. Acidity and hardness during 1/23/86 IWTF regeneration using 10.2 lb/ft³ NaCl and 6.7 lb/ft³ H₂SO₄.

Figure 34. Cumulative hardness recovery during 1/23/86 regeneration using 10.2 lb/ft³ NaCl and 6.7 lb/ft³ H₂SO₄.

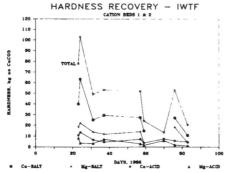

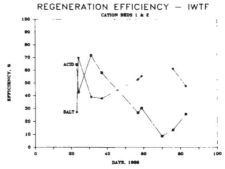

Figure 35. Hardness recovery from IWTF cation beds during salt and acid regeneration cycles.

Figure 36. Regeneration efficiency for the IWTF cation beds during early 1986.

When used in the HPW plant at economic salt levels, some salt leakage occurred, causing an increase in effluent conductivity and an increased load on the mixed bed exchangers. Highly successful operation is presently obtained when used in the industrial waste treatment facility with an acid to salt ratio of about 1.6. Sodium leakage is not a problem due to the back up HPW facility.

The salt-acid regeneration technique is providing long-term stable operation of the high purity water treatment system at the Yorktown facility with substantially higher acid regeneration efficiencies than previously obtainable.

ACKNOWLEDGEMENTS

The authors wish to acknowledge the laboratory assistance provided by the IBM staff, specifically, Nicole Hummel and Judy Lund.

26 EVALUATION OF OXYGEN TRANSFER PARAMETERS FOR INDUSTRIAL WASTEWATERS

Robert C. Backman, Graduate Student

James C. O'Shaughnessy, Associate Professor
Department of Civil Engineering
Northeastern University
Boston, Massachusetts 02115

Fredrick J. Siino, Technical Process Engineer
Mass Transfer Systems, Inc.
Fall River, Massachusetts 02721

INTRODUCTION

The use of aeration devices to transfer oxygen and completely mix activated sludge is an energy intensive process. Due to the never ending increase in electricity costs, the need for cost effective design of aeration equipment is a necessity. Most aeration system performance evaluations have been reported utilizing clean water oxygenation tests. The purpose of this study is to evaluate dirty oxygen transfer parameters for both jet and fine bubble dome aeration systems for a variety of industrial wastewaters.

Specifically, clean and dirty water non-steady state oxygenation tests were conducted using the fore mentioned aeration systems in tanks with similar geometric configurations. Specific objectives of this study were:

1) Determine alpha and beta factors for the following wastewaters:
   a) Dairy
   b) Brewery
   c) Textile
   d) Soft drink
   e) Municipal
   f) Clean water with surfactants.
2) Evaluation of oxygen transfer parameters for the previous listed wastewaters using two different aeration systems under similar operating conditions (same clean water K_La values) and relate impurity levels to alpha.
3) From this study provide insight into the next appropriate step to further define the effectiveness of oxygen transfer for both jet and fine bubble dome aeration systems for industrial wastewaters.

BACKGROUND

The oxygen transfer mechanism in wastewater treatment is usually described by the two-film theory of mass transfer of oxygen from the gas phase to the liquid phase. The basic equation which describes the oxygen transfer process is:

$$\frac{dc}{dt} = K_La(C^* - C) \tag{1}$$

where,

$$(dc)/(dt) = \text{rate of oxygen transfer}$$
$$K_La = \text{overall oxygen transfer coefficient}$$
$$C^* = \text{dissolved oxygen saturation concentration}$$
$$C = \text{dissolved oxygen concentration at time, t}$$

234

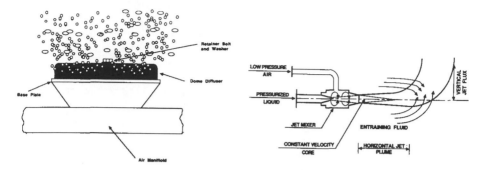

Figure 1. Typical dome diffuser.

Figure 2. Schematic of jet aerator.

K_La is related to the liquid film coefficient, K_L, and the unit interfacial area, a, available for oxygen transfer. The liquid film coefficient is limiting mass transfer factor but is inappropriate to evaluate under most testing conditions. Common practice is to use K_La as a factor describing oxygen transfer.

In computing oxygen transfer requirements for a wastewater, an overall correction factor is used to adjust from field to standard conditions. The equation used is as follows:

$$SOR = \alpha \frac{(\beta C^*walt - CL)}{C^*_{ST}} \theta^{(Tc - 20)} \tag{2}$$

where,

$$
\begin{aligned}
SOR &= \text{Standard oxygen required} \\
\alpha &= \text{Alpha} \\
\beta &= \text{Beta the relative oxygen saturation ratio} \\
C^*walt &= \text{aeration system dissolved oxygen saturation} \\
&\quad \text{concentration at the field conditions} \\
CL &= \text{residual dissolved oxygen level} \\
C^*_{ST} &= \text{aeration system dissolved oxygen saturation} \\
&\quad \text{concentration at standard condition}
\end{aligned}
$$

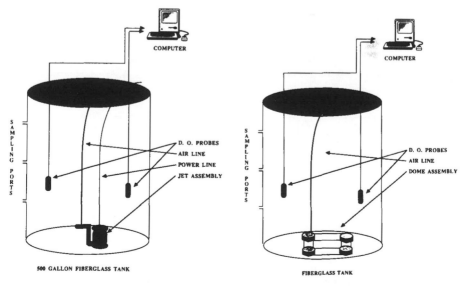

Figure 3. Schematic of fiberglass tank.

Figure 4. Schematic of fiberglass tank.

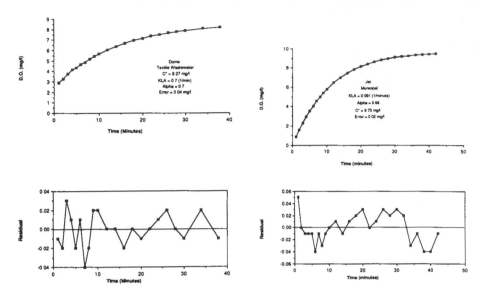

Figure 5.   Oxygenation data for dome system.   **Figure 6.   Oxygenation data for jet system.**

This correction factor accounts for the following specific difference between standard and field conditions

- wastewater constituents
- temperature
- driving force
- barometric pressure

The alpha factor, α, describes the difference between overall transfer rate coefficient in a wastewater to that of clean water:

$$\alpha = \frac{K_L a \text{ (wastewater)}}{K_L a \text{ (clean water)}}$$

Alpha is affected by wastewater constituents such as BOD, COD, suspended solids, surfactants, temperature, aeration device, power level, and basin configuration. The presence of surface active agents (surfactants) in wastewater have been described to have the following effects on alpha [1]: 1) Increased resistance at the gas liquid interface which reduces the overall mass transfer coefficient; and 2) Increased interfacial area per unit volume which contributes to an increase in overall mass transfer coefficient.

The net effect could therefore be an increase or a decrease in overall mass transfer coefficient and alpha. Because of the fore mentioned effects of surfactants, it is plausible that in the presence of surfactants alpha may increase with high surface renewal aeration such as jet aerators and may decrease with low surface renewal devices such as fine bubble aerators [1].

EXPERIMENTAL SETUP

In this study a jet aeration and fine bubble dome system is used to measure oxygen transfer parameters for different industrial wastewaters. The fine bubble dome diffuser is essentially a circular disc with a downward-turned edge. Diffusers are roughly 7 in. in diameter and 1.5 in. high. The dome is mounted on either a PVC or mild steel saddle-type base plate that is solvent welded to an air distribution pipe. A schematic of a typical dome diffuser is shown in Figure 1. Air passing through the fine pore dome results in the release of small bubbles into the bulk liquid.

A jet aerator is a high surface renewal diffused aerator. The jet is a double nozzle device that utilizes pumping of wastewater through the inner nozzle and low pressure air into the outer nozzle (mixing chamber) as shown in Figure 2. This arrangement creates a high velocity gradient within the nozzle

Table 1. Dome Summary

Waste	Test	KLa (1/min)	Alpha	C* (mg/L)	Tabulated C* (mg/L)	Beta	Air Flow (SCFM)	OTR (lb/hr)	AE (%)	OTE (lb/HP-hr)
Clean Water	1	0.098		9.43	9.30	1.01	2.04	0.22	0.11	1.11
	2	0.102		9.01	9.40	0.96	2.08	0.22	0.10	1.10
	3	0.106		9.49	9.50	1.00	2.08	0.24	0.11	1.21
	Avg.	0.102		9.31	9.40	0.99	2.07	0.23	0.11	1.14
	4	0.366		9.20	9.20	1.00	7.12	0.81	0.11	2.70
Brewery	1	0.043	0.42	9.23	9.30	0.99	2.16	0.10	0.04	0.48
	2	0.037	0.36	8.39	9.40	0.89	2.00	0.07	0.04	0.37
	3	0.036	0.35	9.20	9.30	0.99	2.04	0.08	0.04	0.40
	Avg.	0.039	0.37	8.94	9.33	0.96	2.07	0.08	0.04	0.42
Textile	1	0.075	0.72	8.27	9.20	0.90	2.06	0.15	0.07	0.74
	2	0.073	0.70	8.60	9.60	0.90	2.05	0.15	0.07	0.75
	3	0.080	0.77	8.77	9.30	0.94	2.06	0.17	0.08	0.84
	Avg.	0.076	0.73	8.55	9.37	0.91	2.06	0.16	0.07	0.78
Dairy	1	0.040	0.39	6.90	9.20	0.75	2.16	0.07	0.03	0.33
	2	0.034	0.33	6.74	9.40	0.72	2.12	0.06	0.03	0.28
	3	0.037	0.36	7.96	9.60	0.83	2.08	0.07	0.03	0.35
	Avg.	0.037	0.36	7.20	9.40	0.77	2.12	0.06	0.03	0.32
	4	0.090	0.25	8.48	9.40	0.90	7.16	0.18	0.02	0.61
Soft Drink	1	0.035	0.33	8.48	9.20	0.92	2.06	0.07	0.03	0.35
	2	0.027	0.26	8.46	9.20	0.92	2.10	0.05	0.03	0.27
	Avg.	0.031	0.30	8.47	9.20	0.92	2.08	0.06	0.03	0.31
	3	0.098	0.27	9.15	9.40	0.97	2.08	0.22	0.10	0.72
Surfactant 5 ppm	1	0.094	0.90	10.30	10.40	0.99	2.05	0.23	0.11	1.16
	2	0.087	0.84	10.40	10.40	1.00	2.03	0.22	0.10	1.08
	Avg.	0.090	0.87	10.35	10.40	1.00	2.04	0.22	0.11	1.12
Surfactant 15 ppm	1	0.072	0.70	10.10	10.40	0.97	2.03	0.17	0.08	0.87
	2	0.075	0.73	9.90	10.40	0.95	2.03	0.18	0.09	0.90
	Avg.	0.074	0.71	10.00	10.40	0.96	2.03	0.18	0.08	0.88
Primary	1	0.054	0.52	9.75	9.70	1.01	2.03	0.13	0.06	0.63
	2	0.060	0.58	9.16	9.50	0.96	2.02	0.13	0.06	0.66
	3	0.066	0.64	8.61	9.60	0.90	2.05	0.14	0.06	0.68
	Avg.	0.060	0.58	9.17	9.60	0.96	2.03	0.13	0.06	0.66
	4	0.174	0.48	8.10	9.60	0.84	7.16	0.34	0.05	1.13

where wastewater and air are intimately contacted within the nozzle. The high velocity gradient provides fine individual air bubbles with high gas/liquid interface renewal.

The two tanks used in this study, illustrated in Figures 3 and 4, were constructed of fiberglass. The dimensions for the tanks were: 4 ft. diameter, 5.5 ft. high, and 5.1 ft. in water depth. The total volume was 480 gallons. Three sampling ports were located at one third intervals along the tank height.

Dissolved oxygen was measured using membrane probes and values were recorded using a strip chart. Verification of the probe measurement was by the Winkler method. Deoxygenation was obtained by nitrogen gas stripping. Other parameters measured throughout each test run were:

Temperature of wastewater
Air temperature
Ambient pressure
Relative humidity
Air flow rate
Gage pressure

The Jet was run under an aspirating condition using a 0.5 name plate HP submersible pump. Compressed air was used for the fine bubble dome system. Four domes were used in the tank resulting in 3.14 sq. ft. per dome. A gas meter was used to measure flow rate and was verified by a rotameter.

Table II. Jet Summary

Waste	Test	KLa (1/min)	Alpha	C* (mg/L)	Tabulated C* (mg/L)	Beta	Air Flow (SCFM)	OTR (lb/hr)	AE (%)	OTE (lb/HP-hr)
Clean Water	1	0.102		8.70	8.70	1.00	1.28	0.21	0.16	0.59
	2	0.104		9.74	9.50	1.03	1.28	0.24	0.18	0.79
	3	0.104		9.85	9.70	1.02	1.30	0.25	0.18	0.68
	Avg.	0.103		9.43	9.30	1.01	1.29	0.23	0.18	0.65
Brewery	1	0.258	2.49	9.15	9.30	0.98	1.21	0.57	0.45	1.57
	2	0.238	2.30	9.22	9.10	1.01	1.21	0.53	0.42	1.46
	3	0.234	2.27	8.70	9.20	0.95	1.30	0.49	0.36	1.36
	Avg.	0.243	2.35	9.02	9.20	0.98	1.24	0.53	0.41	1.46
Textile	1	0.217	2.10	9.15	9.20	0.99	1.15	0.48	0.40	1.32
	2	0.221	2.13	9.22	9.20	1.00	1.20	0.49	0.39	1.36
	3	0.207	2.00	8.90	9.10	0.98	1.23	0.44	0.35	1.23
	Avg.	0.215	2.08	9.09	9.17	0.99	1.119	0.47	0.38	1.30
Dairy	1	0.181	1.75	8.78	8.80	1.00	1.25	0.38	0.29	1.06
	2	0.186	1.80	8.79	8.80	1.00	1.25	0.39	0.30	1.09
	3	0.165	1.59	8.55	8.90	0.96	1.25	0.34	0.26	0.94
	Avg.	0.177	1.71	8.71	8.83	0.99	1.25	0.37	0.29	1.03
Soft Drink	1	0.052	0.51	9.01	9.20	0.98	1.14	0.11	0.10	0.31
	2	0.056	0.54	10.16	10.20	1.00	1.23	0.14	0.11	0.38
	3	0.048	0.46	10.18	10.20	1.00	1.14	0.12	0.10	0.32
	Avg.	0.052	0.50	0.78	9.87	0.99	1.17	0.12	0.10	0.34
Surfactant 5 ppm	1	0.088	0.85	9.80	10.10	0.97	1.26	0.21	0.16	0.57
	2	0.093	0.90	10.00	10.10	0.99	1.23	0.22	0.17	0.62
	Avg.	0.090	0.87	9.90	10.10	0.98	1.25	0.21	0.17	0.60
Surfactant 15 ppm	1	0.137	1.32	9.95	10.10	0.99	1.25	0.33	0.25	0.91
	2	0.124	1.20	10.05	10.10	1.00	1.28	0.30	0.23	0.83
	3	0.130	1.26	10.10	10.10	1.00	1.23	0.31	0.25	0.87
	Avg.	0.130	1.26	10.03	10.10	0.99	1.25	0.31	0.24	0.87
Primary	1	0.091	0.88	9.73	9.70	1.00	1.20	0.21	0.17	0.59
	2	0.099	0.95	9.69	9.70	1.00	1.20	0.23	0.18	0.64
	3	0.105	1.02	9.50	9.80	0.97	1.19	0.24	0.19	0.66
	Avg.	0.098	0.95	9.64	9.73	0.99	1.20	0.23	0.18	0.63

The initial step in this study was to match clean water K_La values for the two systems. Since the jet was aspirating, its K_La value was used as a set point. Note that the K_La value used is similar to what is found in full scale systems. At least three tests were run for any wastewater. Typical oxygenation data and non-linear regression model fit for both the dome and jet system are shown in Figures 5 and 6, respectively.

RESULTS AND DISCUSSION

Clean water tests for the jet system resulted in a K_La value of 0.103 min^{-1}. This value was also measured for the dome system with a air flow rate of 2.07 SCFM. Dirty water tests for the two systems were run at the similar air and liquid flow rates that produced identical clean water K_La values for the two devices. Tables I and II summarize the results for jet and fine bubble dome system oxygenation tests for the wastewaters shown.

A comparison of K_La values measured for each system for the various wastewaters is plotted in Figure 7. For the Domes, K_La varied between 0.031 and 0.09 min^{-1}; for the Jet it varied between 0.052 and 0.243 min^{-1}. These values along with the clean water results were used to calculate alphas for each wastewater and system. Alpha varied between 2.35 for Brewery wastewater and 0.50 for soft drink waste for the jet. Textile waste had the highest alpha value for the domes and soft drink waste the lowest, 0.73 and 0.30, respectively. As shown in Figure 8, Alpha was significantly higher for all wastewater using the Jet system. Only the surfactant test of 5 mg/L ABS was similar between the two

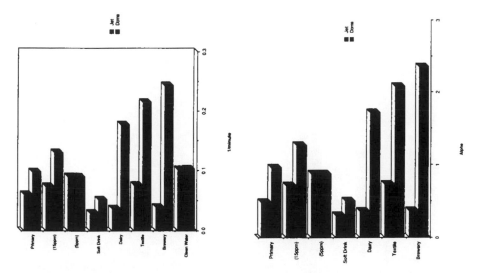

Figure 7.   Comparison of K_La values.   **Figure 8.   Comparison of alpha values.**

systems. Results for municipal wastewater (primary effluent) compared well with reported alpha values in the literature [1].

Comparing the oxygen transfer rates (OTR) in terms of lb per hr, dirty water oxygen transfer showed that the jet system transferred as much as 5 times the amount of oxygen as did the dome system. Figure 9 shows the comparison between the two systems for OTR.

Absorption efficiency for the jet varied between 41% for brewery waste and 10% for soft drink wastewater. The dome absorption efficiency was highest for textile waste, 7%, and lowest for dairy and soft drink waste, 3%. A comparison for absorption efficiency is shown in Figure 10.

Oxygen transfer efficiencies (OTE)(Figure 11) for the domes were in the range of 0.78 to 0.31 lb/HP/hr. The textile waste showed the highest OTE for actual wastewater tested with a value. Jet oxygen transfer efficiencies varied between 1.46 to 0.34 lb/HP/hr. Comparing the results for the two

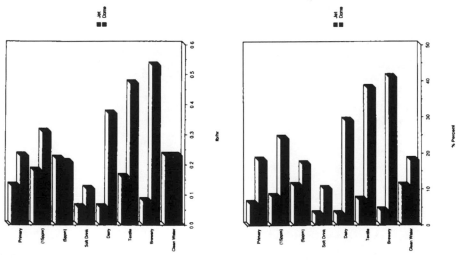

Figure 9.   Comparison of oxygen transfer rates.   **Figure 10.   Comparison of absorption efficiencies.**

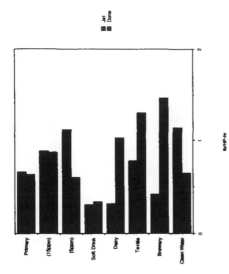

Figure 11. Comparison of oxygen transfer efficiencies.

systems suggests that the cleaner the water (lower contaminant level) the more efficient the dome system in transferring oxygen. On the other hand, the jet was more efficient in transferring oxygen for higher strength wastewater than for clean water. Wastewater characteristics are shown in Table III.

Surfactant tests were run using 5 and 15 mg/L ABS; the results are shown in Tables I and II. The results indicate that possible enhancement of alpha occurs for the jet system with the higher amounts of surface active agents. The domes seem to have depressed alpha values with the addition of surfactants.

Additional runs were made with the dome system using an air flow rate at the high side of the recommended range in an effort to investigate the use of a higher power density (HP/volume) for this type of device. Alpha went down at the higher flow rate, but OTE increased.

SUMMARY

Oxygenation tests were run using a fine bubble dome system and a jet aerator system for five different wastewaters and surfactant waters. Although the system configurations are not indicative to full scale system, several indications are observed:

- Jet system had greater alpha values in higher strength industrial wastewaters.
- Dome system experienced a depressed alpha in the presence of surfactants.
- Alpha enhancement is indicated for the jet aerator in the presence of surfactants.
- The dome system seems to be very efficient in wastewaters that are not high in dissolved constituents.

Although trends are definitely indicated from this study, additional testing is necessary utilizing full scale systems before positive conclusions can be made. Also, further investigation of the effect of surface active agents is necessary before the total effect on oxygen transfer for these systems can be made.

Table III. Wastewater Characteristics

Wastewater	pH	BOD mg/L	COD mg/L	Total Dissolved Solids (mg/L)	Total Suspended Solids (mg/L)	Volatile Suspended Solids (mg/L)
Milk	10.9	970	1267	1100	2850	2060
Brewery	10.2	665	732	1890	1127	668
Primary	8.8	268	504	970	420	401
Textile	5.7	—	—	4370	5078	40
Soft Drink	3.9	720	2861	432	157	152

REFERENCES

1. Bathija, P. R., "Alpha Enhancement in Jet Aeration," *Technology Transfer Seminar on High Efficient Aeration in Wastewater Treatment* (1983).
2. United States Environmental Protection Agency, "Summary Report Fine Pore (Fine Bubble) Aeration Systems," *EPA/625/8-85/010* (1985).
3. Stenstrom, M. K., Hwang, H. J., "The Effect of Surfactants on Industrial Aeration Systems," *Proceedings of the 34th Annual Purdue Industrial Waste Conference*, 902–908 (May 1979).
4. Gibert, G. R., "Measurement of Alpha and Beta Factors," *Proceedings: Workshop Toward an Oxygen Transfer Standard*, EPA-600/9-78-021 (April 1979).
5. Tewari, P. K., Bewtra, J. K., "Alpha and Beta Factors for Domestic Wastewaters," *Journal WPCF*, Vol. 54, No. 9, 1281–1287 (Sept. 1982).
6. Otoski, R. M., et al., "Bench and Full Scale Tests for Alpha and Beta Coefficient Variability Determination," *Proceedings of the 33rd Annual Purdue Industrial Waste Conference* (May 1978).
7. American Society of Civil Engineers, "A Standard for the Measurement of Oxygen Transfer in Clean Water," *Civil Engineering*, (July 1984).

27 WASTEWATER PROFILE OF AN ETHANOL PRODUCTION FACILITY

Kurt W. Anderson, Environmental Supervisor

Dave Szumski, Environmental Technician
New Energy Company of Indiana
South Bend, Indiana 46680

John F. H. Walker, Senior Staff Environmental Engineer
Davy McKee Corporation
Chicago, Illinois 60606

INTRODUCTION

Little published information exists regarding actual operating experience of a major dry-milling-based fuel alcohol plant which would assist the designer or operator of such a plant in developing those modifications that could accomplish desired reductions in overall wastewater loads. It is hoped that this paper will both provide some assistance in these matters and encourage other major fuel alcohol producers to share their experience.

In a time of increasing surplus crop reserves, dwindling fuel resources, widening foreign trade deficits, and record unemployment, the federal government established a new and progressive program, the Alcohol Fuels Program. The program was intended to create a renewable fuel source, reduce foreign trade deficits, stabilize the agricultural market, provide a more environmentally acceptable fuel source, create new jobs, and reduce American dependence on foreign oil. As Barry Commoner put it . . . . "If the government, instead of seeking to reduce agricultural output, were to facilitate the introduction of a crop system capable of producing ethanol as well as food, and suitably integrated with the necessary changes in the oil and automobile industries, farmers would be better off, and the nation would be on the way to a more stable, solar economy."

The New Energy Company of Indiana was formed to play an important role in this new ethanol industry. New Energy Company of Indiana operates a new 60-million-gallons-per-year grassroots production facility in South Bend, Indiana. Design of this facility has incorporated some of the latest technological innovations to ensure efficient and reliable production. The New Energy plant processes more than 22 million bushels of corn per year to produce alcohol, carbon dioxide, and animal feed. The facility has brought over 1,000 new jobs to the state and added needed dollars to the local tax base. The future of this new industry, and the rewards to be realized by the American people, are dependent upon facilities such as the one operated by the New Energy Company.

Process Description and Wastewater Sources

The most common method of producing ethanol-for-fuel begins with conversion of starchy materials to sugars by a process known as saccharification. These sugars are then converted to alcohol via fermentation. The choice of feedstock is dependent upon availability, economics, and byproduct recovery. Corn is the most common feedstock in the United States being widely available, economically feasible, and producing a high quality animal feed byproduct.

Corn-based ethanol plants are of two basic types, wet corn milling and dry corn milling. Most wet corn milling plants were established originally for the production of other types of corn products but have been converted to the manufacture of ethanol-for-fuel as that market has developed. The New Energy Company plant was created expressly for the manufacture of ethanol-for-fuel and is based on a dry corn milling process.

In a modern dry corn milling facility, manufacturing ethanol from corn begins with the grinding of the feedstock in hammermills. Grinding reduces the size of the raw material and breaks up the outer cellulosic wall, making the interior starches more accessible to saccharification. The ground corn is then mixed with water to form a slurry or mash. This mash is then cooked to solubilize the starches.

The solubilized starch slurry is then cooled, and enzymes are added to convert the starches into fermentable sugars. The activities conducted to convert the starches to sugars, and prepare the feedstock for fermentation, are identified under the general heading of mash preparation.

The mash is then placed in a fermenter, and yeast is added to perform the biological conversion of sugars to alcohol and carbon dioxide. At larger facilities, the carbon dioxide generated can be economically recovered as a valuable byproduct. Fermentation processes may be continuous or batch; however, batch operations have become the most common method practiced in the beverage and fuel alcohol industries.

Because of the danger of contamination in the fermentation process, extensive cleaning and sterilization efforts must be practiced to prevent reduced yields. These wash waters are responsible for one of the major wastewater streams produced in the production process.

The fermented mash, termed beer, contains about 11% alcohol. The beer is refined through processes of distillation and dehydration to produce 200 proof fuel grade alcohol. The distillation process begins by separating the alcohol (up to 190 proof) from the water and unfermented materials in a recovery column. In older facilities and converted beverage facilities, two columns, a beer still and rectifier, are used in the distillation process.

To be acceptable as a commercial fuel additive, the 190 proof alcohol stream must then be further concentrated in a dehydration column where a dehydrating agent, or entrainer, is used to remove the last remaining traces of water and produce a 200 proof fuel grade alcohol. The dehydrating agent, usually a solvent material, is then recovered for reuse in an entrainer recovery column. The 200 proof alcohol must then be denatured. The most common denaturing agent for fuel alcohol is gasoline. Once denatured, the alcohol is ready for sale.

The mixture of water and unfermented solubles and solids remaining after distillation of the alcohol is referred to as stillage. Many older or smaller facilities dispose of the stillage by feeding it directly to livestock or by sewering the material. However, modern facilities with larger capacities must practice byproduct recovery both to minimize wastewater loads and to improve the economics of operation.

Byproduct recovery usually entails the collection of the solid material with presses or centrifuges and the drying of the resulting cake into a marketable animal feed, Distillers Dried Grains (DDG). The liquid stream, termed centrate or thin stillage, can then be recovered either as backset to the mash preparation process or, following concentration, as an additive to the DDG, or it can be sewered.

For a plant the size of New Energy's, recovery is the most acceptable alternative. Concentrating the soluble material into a recoverable animal feed supplement is accomplished with evaporators. The concentrated syrup produced in the evaporators is mixed with the cake material from the presses or centrifuges, and the combined material is then dried to produce Distillers Dried Grains and Solubles (DDGS).

The evaporated moisture generated by the evaporators is condensed for heat recovery. The evaporator condensate can then be either sewered or returned to the process. In a large fuel alcohol plant, incorporating byproduct recovery, the evaporator condensate constitutes the largest single wastewater stream. Measures, however innovative, must be taken to reuse this stream and to reduce, if not eliminate, its potentially large impact on the overall plant wastewater load.

Drying operations are extremely varied from plant to plant. Drying of the DDC is normally accomplished with rotary kiln dryers. This type of drying operation produces an exhaust which is heavily laden with particulate matter and which must be controlled. Because of the high moisture content in the exhaust, wet scrubbing is the most common particulate emission control method used. Wet scrubbers reduce the particulate emissions to acceptable levels, but the scrubber blowdown constitutes a considerable wastewater load. Additional control systems such as condensers contribute further to the load. Thus, drying operations are a major source of wastewater.

Other support processes in an ethanol production facility include: steam generation, cooling towers, water treatment, maintenance, and other miscellaneous activities. All of these operations contribute to the plant wastewater load. Unless conservation efforts are practiced in water use and general housekeeping procedures, these sources can significantly impact the plant's overall wastewater load.

A schematic flow diagram for a typical ethanol production facility is presented in Figure 1.

WASTEWATER QUANTITIES AND CHARACTERISTICS

To be able to understand the effects of different operating practices on overall wastewater loadings, the characteristics of the individual contributing wastewater streams within an alcohol production facility must be reviewed. Only plants practicing byproduct recovery will be discussed. Production

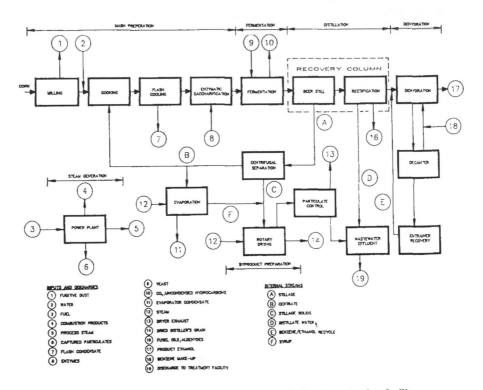

Figure 1.   Schematic flow diagram of a typical ethanol production facility.

economics requires that the design and operation of all major fuel alcohol plants incorporate byproduct recovery.

In a typical, large alcohol production facility that practices byproduct recovery, there are seven principal process wastewater streams. These are: flash condensate; wash waters from the cleaning of process vessels, pumps, and piping; cooling tower blowdown; rectifier bottoms; evaporator condensate; scrubber blowdown; and general housekeeping wastes. Other wastewater streams include: seal water discharge, water treatment blowdowns, boiler blowdown, contaminated stormwater, sanitary wastes, and other miscellaneous minor streams. The principal wastewater streams account for approximately 70% of the flow and 90% of the Biological Oxygen Demand (BOD) discharged from a typical facility. The characteristics of the principal wastewater streams, as identified by three independent sources, are presented in Table I.

Evaporator condensate is the largest single contributor of flow to the overall wastewater load from most alcohol production facilities. Reduction of this flow through recycle or other means is essential to reduction of the total plant effluent. Methods used to reduce evaporator condensate include: increasing the amount of backset to the mash areas, reducing beer gallonage, using reboilers instead of sparger steam, and increasing the efficiency of operation of the presses or centrifuges to reduce solids in the evaporator feed stock.

The simplest recycling practice is to reuse the evaporator condensate in the mash preparation area. However, this practice can cause some problems because of high temperatures and acidity. An alternative is to recycle evaporator condensate back to a process water tank, which can be used to supply a variety of general process needs including wash waters. The process water tank can be used to provide a means for controlling temperature and pH. The recycling of the evaporator condensate for process uses does not reduce organic loading. However, the reuse of an existing wastewater stream for additional needs does reduce the total wastewater volume discharged from a plant.

Other wastewater streams which can be controlled to minimize wastewater discharges include: the flash condensate, the primary flush waters in fermenter cleaning operations, rectifier bottoms, and

Table I. Principal Wastewater Sources (Wastewater Stream Characteristics)

		Flash Condensate	Wash Waters	Cooling Tower Blowdown	Rectifier Bottoms	Evaporator Condensate	DDG Scrub Blowdown	Sub Totals
1.	Radian Corp.[a]							
	Vol. (1/KKg)[d]	Discharged	684	5,524	72	957	586	7,823
	BOD (Kg/KKg)[e]	to Cooling	0.445	0.166	0.090	0.622	0.609	1.932
	SS (Kg/KKg)[e]	Tower	0.27	0.077	0.003	0.011	0.445	0.806
2.	U.S.E.P.A.[b]							
	Vol. (1/KKg)	650	300	Non-contact	920	3,150	Dry	5,020
	BOD (Kg/KKg)	0.585	0.271	Cooling Water	0.618	1.978	Particulate	3.452
	SS (Kg/KKg)	0.012	0.187	Untreated	0	0	Collection Device	0.199
3.	N.E.C.I.[c]							
	Vol. (1/KKg)	Returned	690	670	No	Returned	430	1,790
	BOD (Kg/KKg)	to Process	1.587	0.005	Rectifier	To Process	0.86	2.447
	SS (Kg/KKg)		0.690	0.003	Present		0.22	0.913

[a]Radian Corporation (1982); Source Test and Evaluation Report: Alcohol Facility of Gasohol Production. (Unsupported conceptual estimates)
[b]U.S.E.P.A. (1984); Multimedia Technical Support Document for Ethanol-for-Fuel Industry.
[c]New Energy Company of Indiana; values represent actual data collected during normal operation condition.
[d]Liters per 1,000 kilograms of grain ground.
[e]Kilograms per 1,000 kilograms of grain ground.

scrubber blowdowns. Flash condensate, similar in characteristics to evaporator condensate, can be handled in the same manner as the evaporator condensate. Recycling this stream to the process water tank allows its beneficial reuse in mash preparation and cleaning operations.

Considerable amounts of solids remain behind following the initial draining of a fermenter. The first, or primary, flush of a fermenter will retrieve the majority of these solids. Discharging the primary fermenter flush to the beer wells will enable the recovery of the solid material and greatly reduce the wastewater loads associated with wash waters.

In new "grassroots" fuel alcohol facilities the beer still and rectifier column will probably be combined in a recovery column. This will effectively eliminate the rectifier bottoms as a wastewater stream. In converted beverage facilities, this stream will remain as a low volume, high organic concentration wastewater.

There are two scrubber systems in an alcohol facility: the carbon dioxide (CO_2) scrubber and the DDG dryer scrubber. Because the CO_2 scrubber blowdown contains valuable alcohol, this stream should be discharged to the beer well. Blowdown from the DDG dryer scrubber involves a relatively large volume of organically contaminated wastewater. Methods used to minimize this stream include: the use of cyclones to reduce particulate loadings on the scrubbers, the use of process water as a scrubbing liquor, and the recycling of the scrubbing liquor. The last two control methods may cause some problems with increased odors and maintenance and should be applied with care.

Determining which wastewater reduction practices are applicable to a specific facility requires a complete understanding of individual operations at the facility. Byproduct recovery operations, which contribute the largest portion of wastewater load both in volume and organics, will probably present the best opportunities for reduction. Innovative practices incorporated into the design and/or operation of a facility will help to minimize wastewater loads. A grassroots facility has the best opportunity of adopting and incorporating innovative practices.

WASTEWATER EFFLUENT CHARACTERISTICS

Examination of the characteristics of wastewaters discharged from alcohol production facilities reveals a stream that is high in organic material and solids but a stream that is compatible with treatment processes normally encountered in municipal facilities. In a report published in 1986, the U.S. Environmental Protection Agency estimated that the wastewater from a typical fuel alcohol facility would exhibit a flow of about 16 liters per liter of product produced, a BOD of 24 grams per liter of product produced, and a total suspended solids (TSS) of 8 grams per liter of product produced. This would equate to a daily discharge from a 227×10^6 liters per year (60 million-gallons-per-

Table II. Alcohol Production Facilities (Wastewater Effluent Characteristics)

	Flow (L/L)[h]	BOD (gm/L)[i]	TSS (gm/L)[i]
U.S.E.P.A.[a]	16.0	24.0	8.0
A.N.L.[b]	34.0	24.5	25.5
U.S.E.P.A.[c]	17.5	16.2	11.4
Radian Corp.[d]	21.00	5.17	2.18
S.E.R.I.[e]	7.33	5.86	2.40
Mean[f]	22.5	21.6	15.0
N.E.C.I.[g]	8.2	12.3	4.5

[a]U.S.E.P.A. (1986); Multimedia Technical Support Document for Ethanol-for-Fuel Industry.
[b]Argonne National Laboratory (1980); Environmental Implication of Accelerated Gasohol Production: Preliminary Assessment.
[c]U.S.E.P.A. (1975); Development Document for Effluent Limitation Guidelines and New Source Performance Standards: Miscellaneous Foods and Beverage Point Source Category.
[d]Radian Corporation (1982); Source Test and Evaluation Report: Alcohol Facility of Gasohol Production. (Unsupported conceptual estimates.)
[e]Solar Energy Research Institute (1981); A Guide to Commercial-Scale Ethanol Production and Financing. (Unsupported conceptual estimates.)
[f]Data from Radian Corp. and S.E.R.I. were not included in calculated mean because they are unsupported.
[g]New Energy Company of Indiana: values represent actual data collected during a normal operating condition.
[h]Liters of wastewater per liter of alcohol product.
[i]Gms. of contaminant per liter of alcohol product.

year) alcohol production facility of 11,000 m³/day (2.9 MGD) of flow, 16,500 kg of BOD, and 5,500 kg of suspended solids. This is a considerable load, equivalent to the flow from a community of 29,000 people and the BOD from a community of 182,000 people based on a per capita flow of 100 gallons per day and a BOD of 0.2 pounds per day. Profiles of wastewater effluents reported by various sources are presented in Table II. These data demonstrate the wide range of values that can be encountered.

Differences between the discharge characteristics of individual operating facilities are partially due to variations in design and in operating philosophies. The newer and larger facilities have generally incorporated various water conservation and byproduct recovery techniques. These efforts have resulted in the reduction of wastewater flows to quantities less than 50% of those reported by USEPA (March 1986). A grassroots alcohol production facility should be able to reduce total dry weather wastewater discharges to about 8 liters of wastewater per liter of product, containing approximately 12 grams of BOD per liter of product and 4 grams of TSS per liter of product. Examination of the effluent characteristics data shown in Table III indicate that the New Energy Company is currently achieving such discharge rates. This reflects not only a well-designed facility but a well-operated facility.

WASTEWATER MANAGEMENT PROGRAM

The importance of a good wastewater management program, and its beneficial effect on the total effluent generated from an alcohol production facility, cannot be over-emphasized. The benefits of a well-designed facility can quickly be lost if an effective wastewater management program does not also exist. An effective management program should provide for routine characterization of the final effluent and the various major intra-plant wastewater streams, identification of significant irregularities, data analysis, process review, and spill prevention and response. A successful program will result in effective identification of problems and development of realistic response and correction strategies that can be used in the overall improvement of the plant's operating efficiency.

Characterization of the major wastewater streams, including variability of the streams, trends within the streams, and the effects of different operating practices or conditions on them, is necessary to develop and refine a wastewater management program. Detailed records must be kept about the operating conditions existing during each sampling period. A successful wastewater management program begins with a comprehensive monitoring program.

Table III. New Energy Company (Effluent Characteristics)

Parameter	Mean	April 1986[a]
BOD		
concentration, mg/L	1,500	1,100
kg/day	7,400	4,600
g/L[b]	12.3	6.4
TSS		
concentration, mg/L	500	450
kg/day	2,700	1,900
g/L	4.1	2.6
COD		
concentration, mg/L	2,500	1,750
kg/day	12,300	7,300
g/L	20.5	10.1
Flow		
L/day $\times$ 10^6	4.92	4.16
L/L[c]	8.2	5.8
Production Rate		
(L/day)	600,000	720,000

[a]Major repairs and modification completed during a March campaign have resulted in marked improvement in plant operation and reduction in wastewater loadings.
[b]g/L = grams per liter of product produced
[c]L/L = liters of wastewater per liter of product produced

Any worthwhile program intended to control and minimize wastewater discharges must address spill prevention. Table IV describes the different major process streams of an alcohol production facility in terms of conventional pollutant characteristics. Considering that a spill or leakage of only 8,500 liters (2500 gallons) of product alcohol, or less than 1.5% of the daily production from a 227 million liters (60 million gallons) per year fuel grade alcohol facility, can effectively double the daily BOD discharge rate, spill prevention and response must be a major concern at all alcohol production facilities. The facility must provide systems for spill prevention, spill containment and spill recovery. A Spill Prevention Control and Countermeasure (SPCC) plan should be written for every facility outlining prevention techniques, response activities, and incident investigation procedures. The occurrence of spills can never be completely eliminated. However, through the efforts of a good prevention and response program, problems can be minimized.

No facility nor set of operating practices is beyond improvement. Although most operating procedures do not readily allow for major modifications, minor modifications or improvements can effect marked reductions in wastewater discharges. One operating activity that should be carefully monitored, and adjusted as necessary, is general housekeeping. Water conservation and avoidance of unnecessary discharges to the sewer should be practiced at all times. For the most part, common sense, a general rule that brooms and shovels take precedence over hoses, and a general understanding of the operation and limitations of the treatment facility receiving the wastewaters, provide a solid basis for developing acceptable operating practices.

Table IV. Characteristics of Internal Process Streams

Process Stream	BOD (mg/L)	BOD (lbs/gal)	COD (mg/L)	TSS (mg/L)
Ethanol	800,000	6.85	1,200,000	0
Alcohol (190)	760,000	6.35	1,100,000	0
Mash	150,000	1.25	250,000	160,000
Beer	120,000	1.00	165,000	100,000
Whole Stillage	40,000	0.34	91,000	53,000
Centrate	28,500	0.24	75,000	18,000
Syrup	142,500	1.20	375,000	100,000
Evaporator Condensate	1,000	0.01	1,500	0
Flash Condensate	1,000	0.01	1,500	0

Continued effectiveness of a conservation program requires that the production plant operating personnel actually understand the implications of their acitons, including their errors. Their cooperation in the day-to-day conservation effort is crucial to the programs's success. This means that the ongoing training of operating personnel must include continual updating on the conservation program. It also means that plant management must be visibly committed to the program and encourage its continued implementation.

CONCLUSION

Ethanol is destined to play an important and continued role in America's future. The ethanol industry has an obligation to fulfill this destiny in a productive and responsible manner. To reduce wastewater discharges from these facilities, and thus reduce the associated wastewater treatment burden, innovative practices must be incorporated into the design and operation of new and existing facilities.

Experience to date at New Energy Company's facility in South Bend indicates that significant reductions in overall wastewater loadings can be accomplished by adopting such measures as: the recycling and reuse of presently discharged waste streams, the installation of byproduct recovery operations, reduction in water use (e.g., the use of a reboiler versus sparged steam), the establishment of a comprehensive wastewater monitoring program, and the development of a sound spill prevention and response program.

Such measures will enable a facility to attain wastewater discharge levels that are less than half of those currently seen in the industry. Reduction in discharges to about 8 liters per liter of product, 12 grams of BOD per liter of product, and 4 grams of TSS per liter of product, have been demonstrated at New Energy and are attainable by others.

The New Energy Company's facility in South Bend, Indiana represents a modern, state-of-the-art facility. The innovative designs and operating practices utilized at the New Energy facility provide a model for the industry on which to base further improvements. They demonstrate that fuel grade ethanol facilities can provide the desired boost to America's economy in a productive and responsible manner.

REFERENCES

1. Argonne National Laboratory, *Environmental Implications of Accelerated Gasohol Production*, Distribution Category: Biomass Energy Systems (UC-61a), Argonne Illinois (January 1980).
2. Commoner, Barry, "Ethanol" *The New Yorker* (October 10, 1983).
3. Radian Corporation, *Source Test and Evaluation Report: Alcohol Facility for Gasohol Production*, McLean, Virginia (April 1982).
4. U.S.E.P.A., "Multimedia Technical Support Document for the Ethanol-For-Fuel Industry," *EPA 440/1-86/093* Office of Water Regulations and Standards, Washington, DC (March 1986).
5. U.S.E.P.A., *Draft-Developement Document for Effluent Limitations Guidelines and New Source Performance Standards: Miscellaneous Foods and Beverages Point Source Category*, Office of Water and Hazardous Materials, Washington, DC (March 1975).
6. U.S.E.P.A., "State of the Art: Wastewater Management in the Beverage Industry," *EPA-600/2-77-048*, Office of Research and Development, Cincinnati, Ohio (February 1977).
7. Solar Energy Research Institute, "A Guide to Commercial-Scale Ethanol Production and Financing," *SERI/SP-751-877* (March, 1981).

28 AEROBIC AND ANAEROBIC TREATMENT OF HIGH-STRENGTH INDUSTRIAL WASTE RESIDUES

E. S. Venkataramani, Engineering Associate
Merck & Co., Inc.
Rahway, New Jersey 07065

Robert C. Ahlert, Professor II
Rutgers, The State University of New Jersey
Department of Chemical and Biochemical Engineering
Piscataway, New Jersey 08854

Patricia Corbo, Engineer
DuPont, Savannah River Works
Aiken, South Carolina 29801

INTRODUCTION

The pollution of groundwater by high-strength industrial waste residues, leaching from landfills, is relatively recent and adds a new dimension to existing pollution problems. Many landfills have inadvertently accepted chemical waste for disposal. Over a period of time, waste containers corrode and release solvents, oils, and synthetic chemicals, that constitute potential problems of enormous magnitude and complexity. The list of abandoned hazardous waste sites to be cleaned up under Superfund could grow to between 1500 and 2500 [1]. Possible contamination of groundwater by landfills is most serious in the Northeast. Depending on the wastes deposited at an industrial landfill site, the leachate produced can contain various synthetic and toxic chemicals, including heavy metals and known or suspected carcinogens [2]. Many represent classes of molecules that biologists and biochemists have not investigated previously. Shukrow et al. [3] obtained composition data on leachates, contaminated aquifers, and surface waters in the proximity of twenty-seven sites containing hazardous wastes. Ghassemi et al. [4] developed a data base for thirty different leachates from eleven landfills. Hill et al. [5] investigated the relationships between BOD, COD, and TOC using a measurably larger data file in some industrial waste categories. An extensive discussion of biological treatment of high-strength leachates and industrial residues can be found in a review article by the authors [6]. Recently, Kosson developed an efficient, cost-effective, in-situ bio-reclamation technique for cleanup of hazardous leachates and soil contaminated with waste residues [7].

The present work was designed to establish that aerobic, as well as anaerobic, biological treatment can be used effectively to stabilize organic compounds found in high-strength waste residues.

In the aerobic treatment step, an activated sludge biomass is used to develop microbial populations capable of metabolizing the organic compounds. A research scheme was designed to identify the mode of removal of synthetic organic compounds. The process kinetics were elucidated by following microbial responses to organic species in the high-strength waste.

In the anaerobic treatment step, an anaerobic population was developed from an industrial seed acclimated to two feeds: 1) one containing the industrial residue; and 2) a second, synthetic feed simulating the volatile fatty acid content of the industrial residue. Seed cultures were maintained for nine months. The kinetics of volatile fatty acid uptake and methane production from industrial residue containing other contaminants were compared with those of the synthetic waste to assess the feasibility of application of anaerobic treatment for waste stabilization.

MATERIALS AND METHODS

High-Strength Industrial Waste Residue

The high-strength, complex industrial landfill leachate used in these studies is the aqueous phase of a raw oil/water leachate mixture. This was provided by the USEPA Oil and Hazardous Materials Spills Branch (Edison, NJ). Some of the recognized industrial pollutants found to be present in this waste are illustrated by Table I [3]. Leachate samples designated as EPA-02, 04, and 07 were obtained

Table I. Range of Contaminants Detected in the Wastewater Under Study
(concentrations in μg/L except as noted)

Contaminant	Concentration
Aroclor 1254[a]	70
Aroclor 1016[a]/1242[a]	110 to 1900
Aroclor1016[a]/1242[a]/1254[a]	66 to 1.8 g/L
Benzene[a]	b to 1930
Biphenyl naphthalene	b
Chlorobenzenes[a]	b to 4620
Camphene	b
C$_4$ alkyl cyclopentadiene	b
C$_5$ substituted cyclopentadiene	b
Dichlorobenzene[a]	b to 517
Dichloroethane[a]	180
Dichloroethylene	b
Limonene	b
Methyl chloride[a]	3.1
Methyl Naphthalene	b
Paraffins	b
Petroleum oil	b
Phthalates	b
Phthalate esters	b
Pinene	b
Styrene	b
Tetrachloroethylene[a]	b to 590
Toluene[a]	b to 16,200
Trichloro ethane[a]	b to 490
Trichloro ethylene[a]	b to 7700
Trimethyl benzenes	b
MIBK	2000
Xylene	b to 3300

[a]a priority pollutant.
[b]present.

at different points and times at the same landfill and were used in aerobic biological studies referred to as Phases II, III, and IV, respectively. The anaerobic biological studies were carried out with leachate sample EPA-07A, obtained from a neighboring site. All studies utilized pretreated leachate. Lime flocculation followed by recarbonation and pH adjustment, using sulfuric acid, were used to clarify the wastewater. Typical gross properties of pretreated wastewater are given in Table II; ranges in concentration are cited for some key species. Ultrafiltration analyses revealed that about 80–90% of the organic matter present in pretreated leachate has a molecular weight of 500 or less. It was observed that 40% of leachate DOC is due to volatile fatty acids [8]. It is probable that anaerobic digestion takes place at the landfill site; acid formers are very active, unlike the methanogens that are very sensitive to various environmental parameters. Hence, a high concentration of fatty acids in raw leachate is reasonable.

Acclimated Mixed Culture

An "activated sludge" was developed by aerating secondary sludge, obtained from the Somerset-Raritan (NJ) Sewage Treatment plant that treats a mixture of domestic and light industrial wastewater. The mixed culture was grown on glucose and ammonium sulfate in a mineral nutrient medium, free of added chloride ions. Acclimation of the heterogeneous culture was accomplished by the addition of high-strength wastewater in increasing concentrations with time. The progress of acclimation was monitored daily as DOC removal. A highly stable population was developed in the bioreactor in about three weeks.

For the anaerobic studies, two methanogenic cultures were selected. A leachate digesting culture was selected directly with the leachate as the feed. A volatile fatty acid digesting culture was selected using acetic, propionic, and butyric acids in the ratio found in the leachate. The anaerobic seed was obtained from the Berkeley Heights (N.J.) Sewage Treatment Plant.

Table II. Typical Properties of Pretreated Leachate

Parameter	Average Value
DOC	8–12,000mg/L
COD	23–30,000 mg/L
TKN	1450 mg/L
NH_3-N	1000 mg/L
Total P	14 mg/L
DOC of Fatty Acids	4–5,000 mg/L
TDS	15–17,000 mg/L
Conductivity	13–18,000 μmhos/cm
Sulfate	3400 mg/L
Sulfide	not detected
Nitrate	11 mg/L
Nitrite	2 mg/L
Na,Ca	1,700–17,000 mg/L
Mg,Fe	17–170 mg/L
B,Mn	1.7–17 mg/L
Ni	0.17–1.7 mg/L
Pb,Cr,Si,Al,Cu,Ag	0.017–0.17 mg/L
pH	7.5–9.0
Color	Yellowish Brown
	Not Detected at Levels Listed
Hg	17
As,Te,P,Ti,Cd,Li,Zn,Sr	1.7
Ba,Sb,Ca,In,Bi,Sn,Mo	
V,Nb,Ti,Co,Zr	0.17
Be,Ge	0.017

Analytic Procedure

Most of the analytical tests performed on the leachate are described in Standard Methods for the Examination of Water and Wastewater [9]. Sugar analyses were performed using the DNS method [10] and Somogyi's method [11]. Cell mass analyses were performed gravimetrically as well as by optical density measurements at a wavelength of 540 nm. Organic carbon analyses were performed with an Oceanography International apparatus, with an ampule sealing module, and a Horiba PIR 2000 IR analyses. A 40 ml sample of mixed liquor was centrifuged at 10,000 rpm for fifteen minutes and the clear supernatant used for DOC and sugar analyses. Due to the complex nature of the wastewater, TOC and DOC were chosen as the main performance parameters. Volatile fatty acids and low molecular weight compounds were assayed using a Hewlett-Packard Model 5880A gas chromatograph equipped with a flame ionization detector. Gas analysis was performed by gas chromatography using a thermal conductivity detector.

Experimental Setup

All the aerobic biostabilization experiments of Phases I, II, and III were carried out in wide mouth bottles of 4000 ml capacity equipped with carborundum diffusers [12]. The working volume in all experiments was 2 liters. Air flow was maintained at 1.5 vvm (volume of air/volume of reactor fluid, minute). The reactor contents were mixed with sparged air. Air to the reactor was saturated with water to prevent excessive evaporation. Temperature and pH were maintained at 20 ± 2°C and 6.5-7.5, respectively. All the experiments of Phase IV were carried out in a Micro Ferm Fermenter (New Brunswick Scientific Co., Edison, NJ) with pH, temperature, air flow, and agitation controls. Carbon dioxide evolution was measured with a Horiba Infrared analyzer. Dissolved oxygen concentration was measured with a DO analyzer (New Brunswick Scientific Co., Edison, NJ).

For the anaerobic studies, two liter glass jars, maintained at 37°C by passing water through immersed copper coils, were used as the mother culture reactors. These reactors were mounted on magnetic stirrers. Gas evolved was collected over 1N sulfuric acid.

Toxicity experiments were performed using the culture selected for the volatile fatty acids (VFA culture) and the method developed by Miller for the cultivation of anaerobes [13]. A 50 mL aliquot of

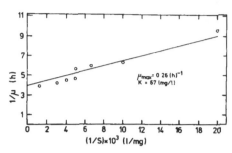

Figure 1. Inverse of specific growth rate as a function of inverse of glucose concentration.

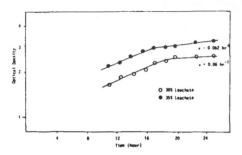

Figure 2. Effect of leachate concentration on the growth rate of an acclimated mixed microbial population with glucose present.

culture was transferred to 100 mL serum bottles filled with gas consisting of 70% nitrogen and 30% CO_2 capped with rubber stoppers and aluminum seals. Five mL of full-strength leachate, with appropriate amounts of yeast extract and phosphorus, were injected into duplicate flasks. This corresponds to approximately one-tenth dilution of leachate, six times higher than the dilution at which the culture was adapted. Gas volume was measured using a water-lubricated, calibrated glass syringe. Samples were taken twice daily and frozen for later analysis. Duplicate control flasks were dosed with synthetic leachate composed of acetic, propionic, and butyric acids and appropriate amounts of nitrogen, phosphorus, and yeast extract.

Anaerobic batch experiments using leachate were performed in duplicate 500 mL serum bottles capped with a rubber stopper containing a glass stopcock through which a hypodermic needle was inserted to remove samples and measure gas. A medium containing (per liter) yeast extract (2 g), 20 mL of 1M phosphate buffer (pH 7.0), resazurin (1 mg), and 1 mL of trace metal solution [14] was boiled to remove oxygen and cooled under nitrogen. After cooling, 65 mL of leachate, 2 g $NaHCO_3$, 0.25 g cystine HCl, and 0.25 g $Na_2S \cdot 9H_2O$ were added and pH adjusted to 7.1 ± 0.1. The 500 mL serum bottles, previously purged with nitrogen, were filled to 350 mL and inoculated with 50 mL of either leachate culture or VFA culture.

RESULTS AND DISCUSSION

Aerobic Biological Studies

Experimental results of Phase I substantiated Monod Growth Kinetics [15]. Figure 1 is a typical Lineweaver-Burk plot; the reciprocal of specific growth rate is plotted against the reciprocal of glucose concentration. The maximum specific growth rate of the mixed microbial population, a composite of the growth rates of different organisms, is 0.26 hr^{-1}; the half-saturation constant is 67 mg/L.

Examples of experimental results from Phase II are shown in Figures 2 and 3. Figure 2 depicts growth rates of acclimated organisms at high concentrations of leachate (30 and 35% by volume) in the presence of glucose (21.5 gm/L). The maximum specific growth rate, 'μ_{max}', of the mixed microbial population is 0.06 hr^{-1}. Figure 3 illustrates the fate of dissolved organic carbon with time during the course of growth of acclimated mixed microbial population on leachate. Table III summarizes the result of growth and biodegradation studies performed; from 72 to 92% removal of dissolved organic carbon was observed. The maximum specific growth rate of an acclimated mixed microbial population lies in the range of 0.08 to 0.1 hr^{-1} for leachate concentrations below 7.5% by volume. For leachate concentrations from 12.5 up to 35% the 'μ_{max}' falls in the range of 0.06 to 0.07 hr^{-1} for synthetic sewage.

Low net growth may indicate co-metabolism and high maintenance energy requirements. Also, natural selection and/or genetic alteration may have taken place during acclimation and is reflected in the altered 'μ_{max}' of the microorganisms. The acclimated population is capable of degrading the organic species present in leachate. The remaining dissolved organic carbon represents less biodegradable organic matter present in the leachate and/or metabolic intermediates and end-products. One possible reason for enhanced degradation of leachate derived organic carbon may be co-metabolism.

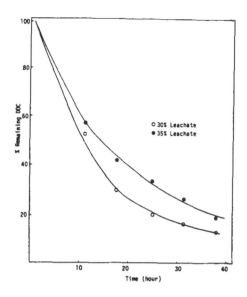

Figure 3. Fate of dissolved organic carbon during the growth of an acclimated mixed microbial population with glucose present.

In Phase III, the acclimated population was subjected to growth and degradation on organic carbon derived solely from leachate. Figures 4 and 5 illustrate results obtained with 15 and 30% leachate, respectively. These plots depict the fate of organic carbon with respect to time. DOC represents the dissolved organic carbon (leachate derived organic carbon) in the medium, and TOC represents the total organic carbon in the system. TOC is the sum of DOC plus the organic carbon in cellular mass and organic carbon adsorbed onto the cell mass (sorption), if any. The data suggest that leachate derived carbon can be used as the sole source of carbon for growth and energy. Co-metabolism is not the sole mode of oxidation of the organic matter present in leachate. The decrease in TOC and DOC with time indicates that removal of organic carbon from the system is due to biological oxidation and not to sorption effects. Also, the difference between TOC and DOC values would represent the organic carbon associated with cellular mass.

Table III. Specific Growth Rates and Efficiency of DOC Removals Obtained in Phase II of Aerobic Biostabilization Studies

Percent Pretreated Leachate	Time of Batch Operation, hr	Maximum Specific Growth Rate 'μ_{max}' hr^{-1}	Percent DOC Removal
2.0	42.5	9.98	88.4
5.0	42.5	0.10[c]	90.7
7.5	42.5	0.082	92.8[a]
10.0	42.5	—	87.2
12.5	42.5	0.07	88.8
15.0	42.5	0.07	84.1
17.5	42.5	0.068	86.8
20.0	35.5	0.07	72.3[b]
22.5	35.5	0.062	79.0
25.0	35.5	0.07	80.6
30.0	38.0	0.06[d]	87.1
35.0	38.0	0.062	80.0

[a]Maximum DOC Removal.
[b]Minimum DOC Removal.
[c]Maximum Specific Growth Rate Observed.
[d]Minimum Specific Growth Rate Observed.

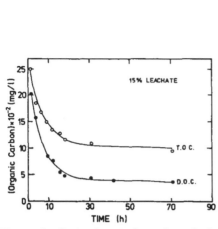

Figure 4. Fate of organic carbon during biodegradation of leachate only.

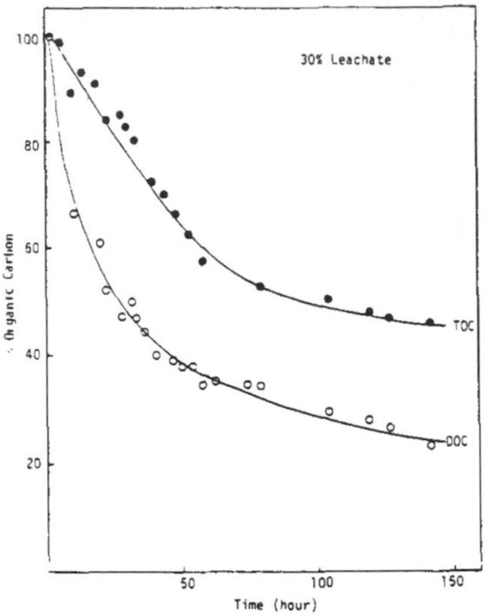

Figure 5. Fate of organic carbon during the course of biodegradation of leachate only.

Stripping experiments demonstrated negligible loss of DOC due to air stripping or evaporation [16]. Table IV summarizes the result of growth rates and efficiency of DOC removal obtained in Phase III of research. From 70 to 88.4% removal of DOC was achieved. The specific growth rates of the mixed culture fall in the range of 0.05 to 0.06 hr^{-1}. The absence of highly fluctuating specific growth rate values indicate stable and well acclimated mixed microbial populations. The presence or absence of glucose does not appear to influence the diversity of the heterogeneous population since similar specific growth rates were obtained in Phases II and III.

In Phase IV, the acclimated population used carbon, nitrogen, and phosphorus derived solely from leachate. Figure 6 represents results obtained with a leachate concentration of 20%. Figure 6 includes five plots that describe variations of pH, cell mass, DOC, TOC, and cumulative carbon in evolved carbon dioxide. The pH increases to a certain point after which it remains steady. It appears that the culture utilizes the fatty acids first, before utilizing other compounds present in the leachate. This

Table IV. Specific Growth Rates and Efficiencies of Organic Carbon Removal Obtained in Phase III of Aerobic Biostabilization Studies

Percent Pretreated Leachate	Time of Batch Operation, hr	Maximum Specific Growth Rate 'μ_{max}', hr^{-1}	Percent DOC Removal
1.0	42.5	—	82.4
2.0	42.5	—	80.7
5.0	42.5	—	86.6
10.0	42.5	—	82.8
15.0	70.5	—	83.0
20.0	70.5	—	88.4[a]
25.0	142.0	0.052	70.2
30.0	142.0	0.05[d]	77.5
35.0	142.0	0.058	72.0
40.0	142.0	0.06[c]	70.0[b]

[a]Maximum DOC Removal.
[b]Minimum DOC Removal.
[c]Maximum Specific Growth Rate Observed.
[d]Minimum Specific Growth Rate Observed.

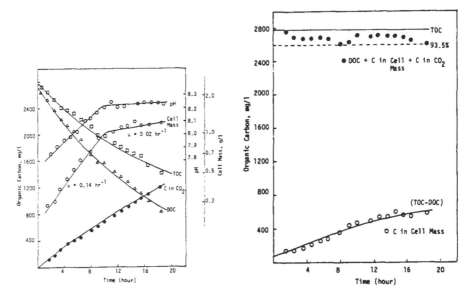

Figure 6. Fate of organic carbon and microbial responses observed during a study with 20% leachate (no pH control).

Figure 7. Carbon balance for the batch study with 20% leachate (no pH control).

result can be inferred from the cell mass plot, also. The latter suggests a diauxic type of growth with two distinct growth phases. The time at which the growth shifts coincides with the time at which the pH value becomes steady. The specific growth rate of the first exponential phase is 0.14 hr^{-1} and that of the second exponential phase is 0.02 hr^{-1}. The dissolved oxygen concentration, during reaction, was maintained above 80% of saturation.

Carbon balance calculations carried out for the system are presented in Figure 7. The solid straight line is the initial total organic carbon value. By conservation of mass, the cumulative carbon content of the system should add up to the total organic carbon with which the system was started. The experimental points shown are the sum of DOC, carbon in cell mass and carbon evolved as carbon dioxide. It has been assumed that 50% of the dry cell weight is due to organic carbon. The solid curved line is the difference between TOC and DOC values during the course of the experiment. The experimental points shown are carbon cell mass. A reasonable agreement between the theoretical and experimental values provide evidence for the biological oxidation of organic species present in the leachate.

The effect of controlled pH on the responses of the acclimated culture was studied, also. pH was controlled by the addition of 1N sulfuric acid and maintained at pH 7.5. The experiment with a leachate concentration of 20% was repeated under these conditions; the results are illustrated in Figure 8. Enhanced specific substrate uptake rate was observed with controlled pH. Hence, all further experiments were conducted at a controlled pH of 7.5.

Table V describes the experimental results obtained in Phase IV of biodegradation studies. It was observed that yield and specific growth rate decreased with increasing leachate concentrations, indicating substrate inhibition. It was evident from various batch experiments that the maximum DOC removal was 90% of the initial DOC; 50% is removed in the first exponential phase, the rest in the second exponential phase. Data obtained with 20, 30, 50, and 100% leachate concentrations were used to obtain the kinetic parameters of the system. Using the values of specific growth rates and corresponding substrate concentrations, the value of the inhibition constant and maximum specific growth rate was determined by nonlinear parameter estimation. The kinetic parameters of the acclimatd heterogeneous population, obtained from these plots, are summarized in Table VI. If it is assumed that the maintenance requirements are negligible, as has been observed for wastewater systems [17-19], it is possible to quantify the role of co-metabolism.

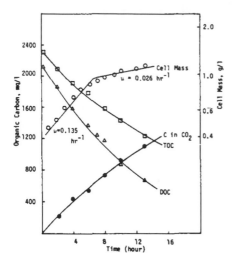

Figure 8.   Fate of organic carbon and microbial responses observed during a study with 20% leachate (with pH control).

Table V. Results Obtained in Phase IV of Aerobic Biostabilization Studies

DOC	Specific Growth Rate, μ, hr^{-1}	Overall Microbial Yield, Y g·cell mass / g·leachate mass	Y_E g·cell mass / g·leachate mass utilized for energy	Specific Substrate Uptake g·leachate carbon / g·cell, hr	Percent DOC Removal
1. First Exponential Phase					
785a	0.12	0.228	0.32	97.5	55.8
1360a	0.14	0.254	0.372	79.0	44.2
1060	0.135	0.252	0.37	113.0	45.4
1710	0.11	0.222	0.31	100.5	40.2
3350	0.085	0.161	0.20	103.5	60.1
4675	0.05	0.111	0.13	95.0	50.2
2. Second Exponential Phase					
628a	0.018	0.26	0.385	77	33.0
1088a	0.020	0.129	0.154	182.5	40.8
848	0.026	0.146	0.179	229	54.0
1368	0.022	0.123	0.145	163	54.3
2680	0.018	0.128	0.152	149	59.8
3740	0.0175	0.120	0.141	142	28.0

aExperiments run with no pH control.

Table VI. Kinetic Parameters of the Heterogeneous Microbial Populations as Estimated from the Results of Phase IV

Parameters	First Exponential Growth Phase	Second Exponential Growth Phase
μ_{max}, hr^{-1}	0.22	0.03
K_i, mg c/L	1770.06	4620.0
M+C	0.37	0.077
	0.38a	0.08a

aFrom L/Y versus L/μ.

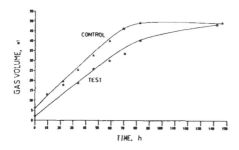

Figure 9.   Gas production for control and test reactors.

Anaerobic Biological Studies

Anaerobic treatment is of particular interest because the major contaminants of the leachate used in this study are precursors to anaerobic methane formation. These contaminants, short chain, volatile fatty acids, and alcohols, are present due to the activity of acid formers in the landfill site. Because these components of the leachate constitute an appreciable fraction of the DOC and are key intermediates in anaerobic digestion, it was of interest to study the performance of methanogens on simulated waste. This was accomplished by selecting a culture for acetate, propionate, and butyrate at concentrations found in the leachate. This culture was maintained in the laboratory for nine months. Gas composition was 71% methane and 29% carbon dioxide and the pH 7.3 ± 0.1.

Toxicity Studies. The objective of this experiment was to compare the rates of utilization of acetate, propionate, and butyrate in leachate containing other organic species to a control using synthetic leachate with no other contaminants present. Figure 9 shows the effect of diluted (one-tenth) prototype leachate on gas production compared to a control receiving synthetic leachate. The control experiment produced a larger amount of gas; it received a higher dose of volatile fatty acids than the test experiment. However, the gas production rates were virtually the same, indicating that inhibition of methanogenesis due to non-volatile fatty acid organics is not a factor. Butyrate and propionate removals show a slight lag, about 20 hrs, followed by rapid uptake, at rates comparable to the controls. The propionate control shows a lag or adaptation period of approximately 40 hrs, possibly because of higher concentration (225 mg/L) as compared to 100 mg/L in the mother culture. The rates of acetate removal are approximately the same in spite of the higher concentration in the control. Overall, no detrimental effects were observed at a leachate dilution of one-tenth using an unadapted culture.

Batch studies. The objective was to determine leachate dissolved organic carbon (DOC) removal, methane production and breakdown, or appearance of specific compounds to assess rate limiting kinetics. The leachate digesting culture and the volatile fatty acid digesting culture exhibit DOC removal for experiments performed in 5% and 10% leachate. Averaged removals are 64.3% for the leachate digesting culture and 69.1% for the volatile fatty acid digesting culture. These values are in excess of 40% removal, expected if only the volatile fatty acids are removed. Thus, an additional 24.3–29.1% of leachate DOC is removed by these cultures. The similarity between DOC reductions indicates the similarity of cultures selected in different manners. It is generally easier to obtain a viable culture by adapting it to specific compounds, in turn to a complex waste, rather than by adapting it directly to the waste.

The rate of methane production for a 5% leachate batch experiment is shown in Figure 10. Methane production was very slow for 200 hrs. This is most likely a result of sulfate-reducing bacteria competing for hydrogen and acetate. The extended lag period followed by rapid methane production may not be attributed to cell growth since methanogens are known for extremely slow growth. The volume of methane in liters was 0.99 L/g and 0.95 L/g for leachate and fatty acid digesting cultures, respectively. This value is slightly higher than typical reported values of 0.91–0.93 g/L (cm^3/Kg). Acetate levels increased followed by uptake, as expected. The formation of acetate is faster than removal. Propionate and butyrate both increased, followed by removal due to uptake. This variability is indicative of the formation of propionate and butyrate from longer chain compounds. The formation of isobutyrate was observed, also, suggesting formation from butyrate metabolism. The fatty acids profiles are illustrated in Figures 11 through 14. Dry cell weight was measured as optical density and correlated with a standard curve. The dry cell weight remained relatively constant over the experimental period. The leachate digesting culture and the volatile fatty acid digesting culture exhibited an average specific DOC utilization rate of 0.154 and 0.211 day^{-1}, respectively. These results are shown in Table VII.

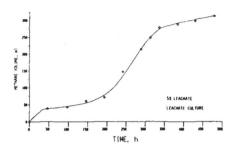

Figure 10.   Methane production.

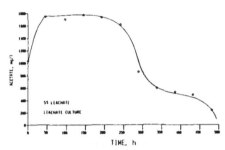

Figure 11.   Acetate concentrations.

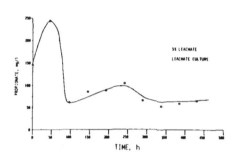

Figure 12.   Propionate concentrations.

These rates are approximately 20% of the specific substrate utilization rates for anaerobic digestion of waste material [20].

Reactor failures at 20% leachate could be a result of one or two effects: 1) methanogenic inhibitors present at appreciable concentrations; and/or 2) the total volatile fatty acid concentration too high in the reactors and toxicity due to the unionized portion. The second phenomena seems more likely because at about 50 hrs the concentration of acetic acid had increased dramatically and pH was about

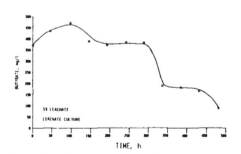

Figure 13.   Butyrate concentrations.

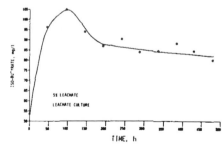

Figure 14.   Iso-butyrate concentrations.

Table VII. Specific DOC Utilization Rates

Reactor	Slope (mg/h)	Correlation Coefficient[a]	Average Dry Weight Conc. (mg/L)	Specific Rate (day^{-1})
5	2.344	0.93	344	0.163
6	1.953	0.98	324	0.144
7	2.290	0.96	300	0.182
8	2.954	0.98	283	0.240

[a]Linear regression of y = mx + b.

7.1, driving the system toward failure [20]. Results indicate that the removal of acetate is the rate limiting step in the anaerobic treatment of leachate. The lags at 5% and 10% dilutions could be explained by removal of acetate as the driving force behind the removal of other compounds [21].

CONCLUSIONS

Aerobic biological studies revealed that a mixed microbial population, acclimated to landfill leachate, degraded 80–90% of the organic species present in the industrial waste liquor, with or without the addition of glucose and other nutrients. Loss of dissolved organic carbon (DOC) is not due to stripping, evaporation, and/or sorption; it is due to biological oxidation. Biostabilization was rapid. Mixed microbial cultures exhibited a diauxic type of growth. As signalled by the increase in pH, during the first exponential growth phase, it is likely that the mixed culture utilizes the fatty acid fraction of the organic solutes in the first exponential phase. Further, it is likely there are at least two groups of organisms and that fatty acid metabolizing organisms have a higher specific growth rate than the others and, hence, show a diauxic response.

Reasonable agreement in the carbon balance provides clear evidence for biodegradation of the organic species present in the leachate. Low sludge yield was observed in this study; this reduces the sludge disposal problem associated with aerobic treatments. The oxygen requirements of the system are quite nominal, also. If it is assumed that microbial maintenance requirements are negligible, as has been observed for wastewater systems, it is possible to quantify the role of cometabolism in the biological oxidation of anthropogenic compounds. The possibility of oxidative assimilation (non-proliferation) is ruled out because quantitative evolution of carbon dioxide, increase in cell mass, and protein content were observed. The ability of the acclimated population to utilize organic carbon and other nutrients solely from leachate further improves process prospects. It was possible to treat highly concentrated waste liquor, i.e., up to 10,000 mg/L of organic carbon. The absence of highly fluctuating DOC values indicates a stable and well-acclimated microbial population.

Anaerobic biological studies demonstrate a DOC reduction of 64.3% for a culture selected with leachate and a reduction of 69.1% for a culture selected for the degradation of acetate, propionate, and butyrate. Specific DOC utilization rates of 0.154 and 0.211 day^{-1} were observed for the leachate and volatile fatty acid digesting cultures, respectively. Cell growth was not observed to any extent during batch experiments. Leachate effects on the cultures were studied through examination of individual volatile fatty acids during batch experiments. The removal of acetate appears to be the rate limiting step. Large concentrations were built up before overall removal was observed. Propionate was more difficult to remove and butyrate was removed without difficulty. Reactor failures were observed for studies with 20% leachate. The failure was likely the result of overloading the system with volatile fatty acids. At 5 and 10% leachate concentrations, no toxicity was observed. Methane was produced at levels of 0.95 to 0.99 L/g DOC (m^3/Kg DOC).

The information obtained in this study clearly demonstrates that aerobic as well as anaerobic biological treatment can be used effectively to stabilize organic contaminants found in high-strength industrial waste residues. The successes in this study and in in-situ biodegradation studies [22], conducted by the authors' research group, further substantiates the significant potential for biological degradation of organic contaminants of industrial origin and the application of microbial treatment as a primary means of residual liquid renovation and disposal.

REFERENCES

1. "Estimates of Superfund Costs Escalate," *Chemical and Engineering News*, 62(51), 19(1984).
2. Staats, E. B., "Waste Disposal Practices – A Threat to Health and the Nation's Water Supply," Report to the Congress of the United States by Comptroller General, *CED-789-120*, U.S. GAO (June 1978).
3. Shuckrow, A. J., Pajak, A. P., and Osheka, J. W., "Concentration Technologies for Hazardous Aqueous Waste Treatment," prepared for Municipal Environmental Research Lab, Cincinnati, Ohio, PB 81-150583, *EPA-600/2-81-019*, (Feb. 1981).
4. Ghassemi, M., Quinliran, S., Haro, M., Metzger, I., Santo, L., and White, H., "Final Report on Compilation of Hazardous Waste Leachate Data," *EPA 68-02-3174*, Work Assignment No. 101, 113 (April 1983).
5. Hill, D. R., and Spiegel, S. J., "Characterization of Industrial Wastes by Evaluation of BOD, COD and TOC," *Journal of Water Pollution Control Federation*, 52, 11, 2704 (1980).
6. Venkataramani, E. S., Ahlert, R. C., and Corbo, P., "Biological Treatment of Landfill Leachates," *CRC Critical Reviews in Environmental Control*, 14, 4, 333 (1984).

7. Kosson, D. S., and Ahlert, R. C., "In-Situ Treatment of Industrial Landfill Leachate," *Environmental Progress* (April 1984).
8. Corbo, P., "Industrial Landfill Leachate Characterization and Treatment Utilizing Anaerobic Digestion with Methane Production," *Ph.D. Dissertation*, Rutgers University, NJ (January 1985).
9. *Standard Methods for the Examination of Water and Wastewater*, 15th Ed., APHA, AWWA, WPCF, Washington, DC (1981).
10. Sumnar, J. B., and Sisler, E. B., "A Simple Method for Blood Sugar," *Arch. Biochem. Biophys.* (1944).
11. Somogyi, M. J., *Journal of Biological Chemistry*, 160, 61 (1945).
12. Venkataramani, E. S., "Aerobic Microbial Treatment of High Strength Industrial Landfill Leachate," AIChE Diamond Jubilee Meeting, Washington, DC (Nov. 1983).
13. Miller, T. L., and Wolin, M. J., "A Serum Bottle Modification of the Hungate Technique for Cultivating Anaerobes," *Applied Microbiology*, 985 (May 1974).
14. Lettinger, G. et al., "Anaerobic Treatment of Wastes Containing Methanol and Higher Alcohols," *Water Research*, 15, 171 (1981).
15. Monod, J., "The Growth of Bacterial Cultures," *Ann. Rev. Microbiol.*, 3, 371 (1949).
16. Venkataramani, E. S., and Ahlert, R. C., "Rapid Aerobic Biostabilization of High Strength Industrial Landfill Leachate," *Journal of Water Pollution Control Federation*, 56, 11, 1178 (1984).
17. Pipijn, P., and Verstraete, W., *Biotechnology and Bioengineering*, 20, 1883 (1978).
18. Esener, A. A., Roels, J. A., and Kossen, N. W. F., *Biotechnology and Bioengineering*, 25, 2803 (1983).
19. Gandy, A., and Gandy, G., *Microbiology for Environmental Scientists and Engineers*, McGraw Hill, Inc. (1980).
20. Anderson, G. K. et al., *Process Biochemistry* (July/August 1982).
21. Mah, R. A., "Methanogenesis and Methanogenic Partnerships," *Phil. Trans. Royal Soc. London*, B297, 599 (1982).
22. Kosson, D. S., "In-Situ and On-Site Treatment of Industrial Landfill Leachates," *Ph.D. Dissertation*, Rutgers University, NJ (April 1986).

Section Three
TOXIC AND HAZARDOUS WASTES

29 APPLICATION OF RISK ASSESSMENT TO SELECTION AMONG SITE REMEDIATION ALTERNATIVES

Eli J. Salmon, Director
Health, Safety and Risk Management
Intellus Corporation
Irvine, California 92715

Richard A. Brown, Technology Manager
Aquifer Remediation Systems
FMC Corporation
Princeton, New Jersey 08540

INTRODUCTION

The Environmental Protection Agency (EPA) mandated [1], that any remedial decisions and strategies related to hazardous substances be scientifically and technologically sound, economically efficient, and socially equitable. This calls for application of risk assessment/management methodologies which the EPA's Administrator recognized [2] as the most important and most difficult role emerging in the 1980's. It becomes necessary to develop well founded and consistent procedures as well as uniform and coordinated approaches that enable deciding if, when, and how remediation of risks arising from hazardous waste sites should be undertaken.

The definition of risk assessment/management by the National Academy of Sciences [3] distinguishes two components, namely:

1. The scientific exercise involved in the assessment of risks.
2. The political, economic, and social aspects of decision making about what action to take.

In the simplest sense, risk assessment is the qualitative or quantitative characterization of potential adverse impacts of particular substances or agents on individuals or populations. It is a function of two measurable factors: hazard and exposure. Risk management, on the other hand, represents the complex judgement and analysis that uses the results of risk assessment to provide a decision about remediation.

In reality, the process of reaching a remediation decision is very complex because of the multitude of considerations that must be optimized. They include:

1. Wide range of chemical and physical agents.
2. Many adverse effects including carcinogenic, mutagenic and teratogenic damages, systemic effects on various organs, and even psychological risks.
3. Various environmental impacts ranging from visibility impairment and crops damage to ecosystem disruption.
4. Different routes of exposure.
5. Multitude of remediation technologies in different states of development.
6. Conflicting remediation objectives.
7. Diversified trade-offs among remediation alternatives.
8. Many uncertainties regarding hazards, environmental processes, and suitability of remediation options to particular projects.

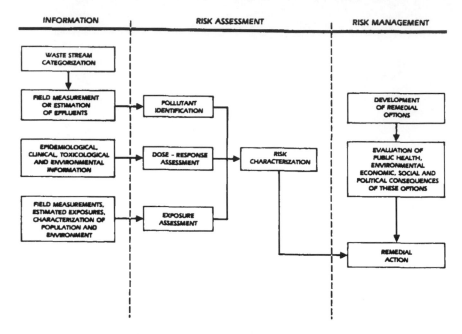

Figure 1. Components of risk assessment.

PROCESS OF RISK ASSESSMENT/MANAGEMENT

Objectives and Components

The objectives of risk assessment/management in the field of hazardous waste sites are:

1. Provide a scientifically and technologically sound, economically efficient, and socially acceptable best remedial response and strategy to overcome the problems associated with hazardous wastes.
2. Provide management with a continuing data base and evaluation on which to review and update all remediation decisions.
3. Ensure documentation and defense of all actions taken, including statement of assumptions and treatment of uncertainties.

The risk assessment/management process consists of three major phases, as depicted in Figure They are:

1. Information and research management. This includes the collection of data by measurements or estimates of waste stream characterization, types and quantities of effluents, epidemiological/toxicological/environmental effects, environmental transport and fate, and exposure characterization of the population at risk. Information is also collected on remediation technologies including technical feasibility, description of applications, constraints, advantages, and disadvantages.
2. Risk assessment. This is the qualitative or quantitative processing and evaluation of information through well-developed procedures and methodologies. The assessment includes hazard identification, characterization of potential health and environmental effects of particular substances on individuals or populations, some quantification of risk, the development of alternative remediation options and examination of trade-offs.
3. Risk management. This is the introduction of value judgement concerning the acceptable risks and trade-offs. This process entails application of economic, engineering, social and political considerations in regards to risk-related information, leading to selection. The alternative preventive and remedial options are analyzed and compared until the most desirable option is selected.

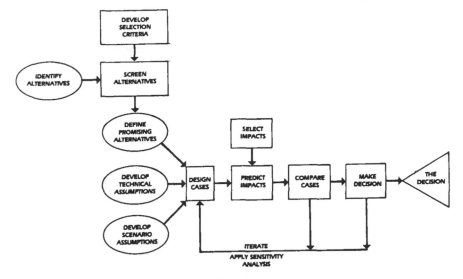

Figure 2.   Stages of analysis.

Scope of the Risk Assessment/Management Process

Salmon [4] applied decision analysis techniques which were developed to address the complexity of technology choice problems to radioactive waste management. The technique evaluates the consequences resulting from selecting any of the various waste management alternatives. Consequences of the selection are then quantified using probabilities and system models. By conducting a broad sensitivity analysis, it is possible to further refine the evaluation.

The same technique is used in this paper to select the most effective remediation action for a hazardous waste site. The evaluations combine several risk assessment/management methodologies, namely:

1. Probabilistic risk assessment. What risks are imposed by the site contamination and how are they influenced by applying alternative remediation actions?
2. Risk/benefit analysis. Are these risks acceptable?
3. Cost/benefit analysis. Can the risks be reduced effectively?
4. Risk/benefit/cost evaluations.How can alternatives for risk reduction be compared?
5. Risk/benefit/effectiveness analysis. At what stage does remediation become ineffective when the marginal cost of continuing remediation exceeds the accruing benefits?

The risk assessment/management technique is outlined in Figure 2.

Identifying Alternatives. Relevant remediations alternatives are identified and described in Table I [5]. Numerous management options are summarized which include biological, chemical, and physical treatments. The strategies range in extremes from "no action" to "abandoning the site and relocating local residents." In between are management options of containing the hazard, partial restoration of the site, complete restoration of the site and full treatment of the contaminated media.

Remediations Criteria. The categories of criteria used for comparing among remediation alternatives are presented in Figure 3. They are arranged according to an hierarchy, which varies from one site to another and depends on the specific situation.

Screening Alternatives. Since the evaluation process is both lengthy and costly, a screening technique is used to reduce to manageable size the number of alternatives under investigation. After determining what is a reasonable number according to availability of resources and existing constraints, screening proceeds in accordance with specific selection criteria based on those in Figure 3. For example, the total cost of the project may be capped at X dollars, or a limitation requires that only demonstrated remediation technologies be used. The specific criteria are divided into "musts" and "wants," and those alternatives that do not meet any of the "must" criteria are rejected. Level of effort and cost criteria are also applied for screening alternatives. Consideration is given to such

Table I. Identifying Remediation Alternatives

General Response Action	Technologies
No Action	Some monitoring and analysis may be performed.
Containment	Capping; groundwater containment barrier walls; bulkheads; gas barriers.
Pumping	Groundwater pumping; liquid removal; dredging.
Collection	Sedimentation basins; French drains; gas vents; gas collection system.
Diversion	Grading; dikes and berms; stream diversion ditches; trenches; terraces and benches; chutes and downpipes; levees; seepage basins.
Complete Removal	Tanks; drums; soils; sediments; liquid wastes; contaminated structures; sewer and water pipes.
Partial Removal	Tanks; drums; soils; sediments; liquid wastes.
On-site Treatment	Incineration; solidification; land treatment; biological, chemical and physical treatment.
Off-site Treatment	Incineration; biological, chemical and physical treatment.
In Situ Treatment	Permeable treatment beds; bioreclamation; soil flushing; neutralization; land farming.
Storage	Temporary storage structures.
On-site Disposal	Landfills; land application.
Off-site Disposal	Landfills; surface impoundments; land application.
Alternative Water Supply	Cisterns; aboveground tanks; deeper or upgradient wells; municipal water system; relocation of intake structure; individual treatment devices.
Relocation	Relocate residents temporarily or permanently.

factors as accessibility of data sources, time available for the evaluation degree of desired accuracy, and elimination of alternatives with considerable costs that do not provide appreciable added benefits.

Design of Cases. For each desired alternative a set of inputs is prepared. These include technical data assumptions as well as scenario assumptions. Technical inputs include:

1. Characterize contaminants
2. Evaluate their mobility
3. Evaluate patterns of soil and groundwater contamination
4. Assess rates of migration and fates of contaminants
5. Evaluate applicability of remedial alternatives

Scenario inputs include:

1. Forecast upcoming federal and local regulations
2. Anticipate developments in remediations technologies and applications
3. Delineate changes in land use practices and prices

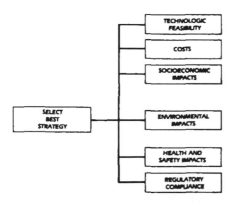

Figure 3.   Remediation criteria.

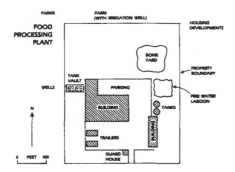

Figure 4.   Map of site and boundaries.

Preparing Scorecards. For each major remediation criterion a scorecard is prepared which compares the alternative remedial actions under consideration. This is carried out by the following process:

1. Identify major potential impacts
2. Determine units of measurement
3. Estimate size of each impact
4. Rank each alternative remediation option in respect to each impact using qualitative or quantitative ranking
5. Reduce and express uncertainties

Selecting the Best Alternative. Prior to decision making, a briefing is held with the decision maker or makers to describe the methodologies and summarize the results of the analysis. At this meeting, the attendees are informed that only major impacts rather than every conceivable impact are included in the analysis. The scorecards are presented and their purpose explained, namely to identify major concerns, comparing them according to some numerical scale. Uncertainties underlying the results are described as well as the manner in which they can alter the values of particular impacts or even their ranking. This information is developed through sensitivity analysis. The decision maker or makers are then left with the task of weighing and comparing the different impacts on the scorecards according to his/their interests. For example, an economist particularly interested in costs may have a different opinion on the relative importance of the various impacts than an environmentalist interested primarily in ecosystems. However, experience indicates that decision makers even in highly controversial cases tend to agree on the relative ranking of the alternatives in respect to any particular impact. In the highly controversial case study of damming the Oosterschelde Estuary in the Netherlands [6], each decision maker reached the same final conclusion, even though for different reasons.

Additional Analyses. If the decision makers cannot reach a decision because of an insufficient number of desirable alternatives, it becomes necessary to design additional cases. The step involves changes in the screening, technical and scenario assumptions, or combination of all three factors. Sensitivity analyses are then applied to the new set of cases. The additional scorecards are presented to the decision makers, and the process of selection is continued until a decision is made.

ILLUSTRATION OF APPLYING RISK ASSESSMENT/MANAGEMENT METHODOLOGY TO A HAZARDOUS WASTE SITE

History and Nature of Problem

The case study represents a 'typical' plant situation which demonstrates the methodology required to define and solve the problem.

A small manufacturing plant in a residential/agricultural neighborhood (Figure 4) was alerted to a potential groundwater contamination problem when fumes were reported in the plant utility building. A survey of the plant found no leaks in readily accessible pipes, valves, flanges, and other process equipment. A review of inventory revealed a probable leak of 5,000 gallons of mixed xylenes from a buried storage tank. This was later confirmed by testing. The affected building was immediately vented to protect personnel, and a site evaluation was initiated.

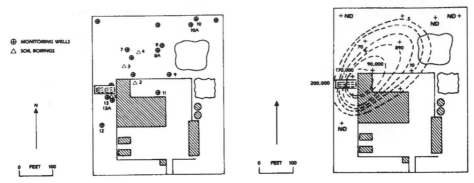

Figure 5. Location of monitoring wells and soil borings.

Figure 6. Extent of xylene in ground water.

Site Description

The plant site and its boundaries are shown in Figure 4. There is much construction over the site as well as several sensitive public health and ecological receptors at risk near the site boundary. These include a housing development, food processing plant, farms and irrigation wells. Geological survey records showed that the subsurface consisted mainly of sand and sandy clay. Depth to groundwater was 15-25 feet. The regional gradient was in a northerly direction with groundwater flow estimated to be a few hundred feet per year. Because of a potentially rapid spread of contaminants, remediation was given an immediate priority.

Site Investigation

Site investigation was carried out in two phases through a combination of monitoring wells and soil borings shown in Figure 5. In the initial phase, wells 1 through 6 were dug. They enabled to determine the extent of soil and groundwater contamination, the free product zone, and the groundwater ingredient. Then a free product recovery system was installed in wells 1 and 2 which consisted of two skimmers. During its thirteen weeks of operation, a total of 3,100 gallons of xylenes were removed from the free product zone. Then additional soil borings and observation wells were installed to define more completely the remaining contamination. They included three cluster wells (#s 8,10,13) to determine the vertical distribution of contamination.

The measurements showed that:

1. 5,100 gallons of mixed xylenes were released.
2. Xylenes were found in three phases, namely:
   - Free product zone – 3,170 gallons
   - Soil (absorbed) – 1,900 gallons
   - Groundwater (dissolved) – 40 gallons
3. Contamination dimensions and volumes were as follows:
   - Free product – 63 cubic yards, i.e., 130 feet downgradient x 100 feet wide x 0.13 feet deep
   - Soil – 1,450 cubic yards, i.e., 130 feet x 100 feet x 3 feet
   - Groundwater – 8,700 cubic yards, i.e., 320 feet x 210 feet x 3.5 feet
4. Contamination was found both below and above water table.
5. Movement of contamination was preferentially downgradient with the groundwater plume shown in Figure 6.
6. Movement of contamination was 300-350 feet/year indicating that out of boundary receptors will become contaminated in less than a year.

Screening Alternatives

There are several management options for dealing with this contamination. These are narrowed down to three, namely:

- Alternative I – Further partial removal of contamination plus containment of remaining portions to prevent the problem from spreading.
- Alternative II – Pump/treat, injecting chemicals to increase solubility of xylenes in water.
- Alternative III – In situ bioreclamation which transforms the contaminant in place to an innocuous form.

In all of these alternatives, partial removal of contaminants took place. As explained 3,100 gallons of free product were removed through pumping.

In Alternative I, further partial removal can be achieved by soil excavation and groundwater pumping. Containment can be achieved either physically through the use of slurry walls or grout curtains or hydrologically with an array of pumping and injection wells. Further description of these methods is provided in the literature [7,8].

The major elements of this alternative are:

1. Reduction of surface infiltration.
2. Upgradient barriers.
3. Down gradient barriers.
4. Integrity of barriers.
5. Odor and gaseous emission controls.
6. Partial leachate and groundwater collection.
7. Storage and handling.
8. Transfer to offsite.
9. Chemical, biological or physical treatment.
10. Monitoring Site.

In Alternative II, the contaminated water is pumped from the hazardous waste site to an on-site treatment plant. Water, containing chemical additives (e.g., detergents), is injected into the soil to flush out absorbed xylenes. The injected water as well as contaminated groundwater are treated. After removal of the xylenes and contaminated chemicals, the water is reinjected into the soil. The operation is repeated several times until an acceptable level of contamination is achieved [7,8].

The major elements of this alternative are:

1. Hydraulic barrier
2. Injecting chemicals to treat leachate
3. Collecting contaminated groundwater
4. Treating water in permeable treatment beds
5. Monitoring site

In Alternative III microorganisms are used to break down the xylenes. Traditionally, biological treatment technology has been used offsite, in surface impoundments, ponds, or treatment facilities. More recently, bioreclamation has been successfully applied on site through the in situ stimulation of indigenous microorganisms by the addition of oxygen and nutrients [9]. As most organic chemicals do not contain all necessary elements for bacterial growth, adequate external sources of oxygen, nitrogen, sulphur, phosphorus, and certain trace minerals must be supplied to the contaminated water or soil in order for the indigenous microorganisms to effectively degrade the organic chemicals present.

The major elements of this alternative are:

1. Site preparation
2. Nutrient balance and transport
3. pH maintenance
4. Soil aeration
5. Monitoring site

Preparing Scorecards

After proceeding with the other phases of risk assessment described in the earlier portion of this paper, six scorecards are prepared (Tables II through VII) plus a "Combined Impacts Scorecard" (Table VIII). The remediation options are compared in respect to six major variables, namely: technological feasibility, costs, socioeconomic – environmental – health impacts and regulatory compliance. The impacts considered in the scorecards are discussed in the following sections.

Technologic Feasibility. This attribute indicates the likelihood that the proposed technology will be suitable to the particular project. Some important components are:

Table II. Technologic Feasibility Scorecard

Item	Containment Plus Partial Treatment	Pump/Treatment	In Situ Biological Treatment
State of Development	Best	Intermed.	Worst
Reliability	Worst	Intermed.	Best
Effectiveness	Worst	Intermed.	Best
Flexibility	Worst	Intermed.	Best
Suitability to Particular Project	Worst	Intermed.	Best

Table III. Costs Scorecard

Item	Containment Plus Partial Treatment	Pump/Treatment	In Situ Biological Treatment
Capital Costs	Worst	Intermed.	Best
Operation and Maintenance	Intermed.	Worst	Intermed.
Pretreatment/Exploration	Worst	Intermed.	Best
Transportation and Disposal	Worst	Intermed.	Best
Monitoring	Intermed.	Worst	Best
Opportunity Costs of Land and Water	Intermed.	Worst	Best
Liability and Compensation	Intermed.	Worst	Intermed.

Table IV. Socioeconomics Impacts Scorecard

Item	Containment Plus Partial Treatment	Pump/Treatment	In Situ Biological Treatment
Deterioration of Property Values During Treatment	Intermed.	Worst	Best
Untreated Contamination	Worst	Intermed.	Best
Disruptions in Land Uses	Worst	Intermed.	Best
Public Concerns	Intermed.	Worst	Best

Table V. Environmental Impacts Scorecard

Item	Containment Plus Partial Treatment	Pump/Treatment	In Situ Biological Treatment
Potential Emissions and Effluents	Best	Worst	Intermed.
Remaining Unbound Contamination	Worst	Intermed.	Best
Odors	Best	Worst	Intermed.
Toxicities of Contaminants	Best	Worst	Intermed.

Table VI. Health and Safety Impacts Scorecard

Item	Containment Plus Partial Treatment	Pump/Treatment	In Situ Biological Treatment
Mortalities	Intermed.	Worst	Best
Morbidities	Intermed.	Worst	Best
Genetic and Teratogenic Risks	Intermed.	Worst	Best
Psychological Impacts	Intermed.	Worst	Best

Table VII. Regulatory Compliance Scorecard

Item	Containment Plus Partial Treatment	Pump/Treatment	In Situ Biological Treatment
Licensing of Operations	Intermed.	Best	Intermed.
Contaminants in Soil	Intermed.	Intermed.	Best
Contaminants in Groundwater	Intermed.	Worst	Worst

1. Developmental state of the technology which ranges from conventionally demonstrated technology at uncontrolled sites to one being only at a conceptual stage of development
2. Demonstrated reliability under widely differing conditions
3. Effectiveness in terms of the technology achieving its objectives both economically and within a short period of time
4. Flexibility to remain effective under changing conditions
5. Suitability to the particular conditions of the project

An index from 1 to 5 which represents the technologic feasibility of each technology is used.

Costs. The various components of costs are evaluated including:

1. Initial site preparation, construction, and capital costs
2. On going operation and maintenance including laboratory and field testing
3. Opportunity costs of the affected land, facilities and waters
4. Contingency funds which are an indicator of uncertainty
5. Potential liability and compensation

Costs are expressed in $ per gallon of xylenes. Money is discounted over time to enable valid comparisons among events that occur at different times.

Socioeconomic Impacts. The major components are:

1. Deterioration of property values during and after treatment due to potential contamination and presence of remediation operations
2. Present and potential disruptions in land uses during and after treatment

These two attributes are measured in annualized millions of dollars. Other components are:

1. Public concerns which reflect the attributes of major interest groups, the media, environmentalists, and other cause-oriented national and local groups. An index which represents the number of members of each interest group multiplied by their attitude levels is used as the unit of measurement [10]. Attitude levels are represented by numerical indexes, ranging from +1 for support, 0 for neutral, -1 for controversial, -2 for action-oriented opposition, and -3 for strong-oriented opposition. An appraisal model based on public surveys and probabilities of public attitudes is described by Beley [10].
2. Untreated contaminants relate to potential hazardous emissions or effluents that may affect the community. The impact is measured by the number of people at risk.

Environmental Impacts. These impacts are measured in terms of the area affected by contamination/potential contamination. The unit of measurement is acres which is an index based on

Table VIII. Combined Impacts Scorecard

Item	Containment Plus Partial Treatment	Pump/Treatment	In Situ Biological Treatment
Technologic Feasibility	Worst	Intermed.	Best
Financial Costs	Worst	Intermed.	Best
Socioeconomic Impacts	Intermed.	Worst	Best
Environmental Impacts	Best	Worst	Intermed.
Health & Safety Impacts	Intermed.	Worst	Best
Regulatory Compliance	Intermed.	Intermed.	Best
Combined Impacts	Intermed.	Worst	Best

value judgement for the specific cases. The index takes into consideration a number of factors such as levels of airborne emissions and liquid effluents, toxicity of the contaminants, and sensitivity of the region at risk.

Health and Safety. Mortality impacts refer to the number of untimely deaths among occupational workers and members of the general public resulting from remediation actions as well as from remaining contamination. Morbidity and genetic impacts refer to the number of years of impairment or life shortening resulting from these factors.

Compliance with Regulations. The attribute refers to easiness of compliance with licensing of remedial actions as well as evaluation of compliance after completion of remediation. Both present and upcoming regulations of federal, state, and local agencies are considered.

An important factor that influences all of the above concerns is the time element of the remedial effort. In general, the emphasis is on quickly eliminating exposure to hazardous substances. Some remedial components achieve almost instantaneous results while others may take a very long time to realize benefits. The time element includes the time it takes to implement a remedy as well as the time it takes to realize beneficial effects.

Making a Selection

In this illustration, "in situ bioreclamation" is clearly the "best" alternative while "pump/treat" is the worst. The in situ treatment has several major advantages over the other alternatives. "Bioreclamation" affects all aspects of contamination, both soil and groundwater. As a result, unlike the other options, it does not require extensive excavation of soil which is severely constrained by depth and by physical construction on this specific site. The bacterial process is also more rapid than the others requiring significantly less time for treatment. Unlike "containment" which requires continuing maintenance and attention, bioreclamation converts the contaminants into innocuous substances. Unlike "pump/treat" which is only partly successful in dissolving and removing soil contaminants while entailing risks of groundwater contamination because of added chemicals, "bioreclamation" is free of these problems.

SUMMARY AND CONCLUSIONS

Remedial investigations and responses to hazardous wastes are often expensive, very complex, and entail many risks. There are numerous alternatives for restoration of sites containing hazardous wastes. They range from such extreme options as abandoning the site and relocating its residents to complete restoration of the site and cleanup of the contamination. There are also on-site and off-site treatment and disposal methods as well as in situ applications. Numerous decisions need to be made during the site remediation process in regards to trade-offs and mitigation options.

This paper describes a risk assessment/management methodology which is effective in selecting among the various alternatives. It illustrates application to an actual case which demonstrates how:

1. The multiple objectives corresponding to concerns involved in comparisons among alternatives are identified.
2. The impacts and interactions involved in each alternative are assessed and compared.
3. The consequences of each alternative are evaluated.
4. A decision is reached concerning which alternative to select.

REFERENCES

1. "Environmental Quality, 1984," 15th Annual Report of the Council on Environmental Quality, Executive Office of the President, Washington, D.C. (1985).
2. "Environmental Quality, 1983," 14th Annual Report of the Council on Environmental Quality, Executive Office of the President, Washington, D.C. (1984).
3. "Risk Assessment in the Federal Government: Managing the Process," National Academy of Sciences, Washington, D.C. (1983).
4. Salmon, E. J., "Decision Framework for Disposal of Radioactive Wastes," Radioactive Waste Management Volume 1, International Atomic Energy Agency, Vienna, Austria, pp. 357-369 (1984).
5. "Guidance on Feasibility Studies Under CERCLA," EPA-540/G-85/003, Washington, D.C. (1985).
6. Goeller, B. F., et al., "Protecting an Estuary from Floods: A Policy Analysis of the Oosterschelde," Rand, Santa Monica, California (1977).

7. "Handbook: Remedial Action at Waste Disposal Sites," Municipal Environmental Research Laboratory, U.S. *Department of Commerce*, NTIS PB82-239054 *Springfield*, Virginia (1982).
8. "Handbook for Evaluating Remedial Action Technology," NTIS PB84-118249, Springfield, Virginia (1983).
9. Brenoel, M., and R. A. Brown, "Remediation of a Leaking Underground Storage Tank with Enhanced Bioreclamation," NWWA Fifth National Symposium and Exposition on Aquifer Restoration and Groundwater Monitoring, Columbus, Ohio (May 1985).
10. Beley, J. R., et al., "Decision Framework for Technology Choice," Volume 1, 2, Electric Power Research Institute, Palo Alto, California (1981).

30 TREATMENT OF HAZARDOUS LANDFILL LEACHATE USING SEQUENCING BATCH REACTORS

Robert G. Smith, Associate Development Engineer
Department of Civil Engineering
University of California at Davis
Davis, California 95616

Peter A. Wilderer, Professor
Technical University of Hamburg-Harburg
Hamburg, West Germany

BACKGROUND

The Georgswerder landfill is located approximately 5 km south of the center of Hamburg, West Germany. It encompasses an area of about 42 ha and contains about $7,000,000 \text{ m}^3$ of compressed refuse, war rubble, and soil. Between 1967 and 1974 about $150,000 \text{ m}^3$ of liquid and semi-liquid industrial and chemical wastes including used oil were deposited in the landfill in open basins and in barrels. The liquid basins were filled with household refuse and covered with soil as were the barrel storage areas.

Following closure in 1979, the landfill area was landscaped and gas extraction wells were constructed. A peripheral ditch was built which intercepts leachate and routes it to a municipal wastewater treatment plant. Leachate containing light-density fluids is passed through a provisional oil separator to separate the water and liquid organic phases for future incineration. The liquid organic phase is collected and stored on-site in barrels. The water phase is discharged to the wastewater treatment plant. Monitoring of the organic phase and ditch sediments revealed the presence of dioxin (2,3,7,8-TCDD) in concentrations up to 42 ppb as well as a wide variety of organic solvents and chlorinated hydrocarbons. Analysis of the water phase revealed no dioxin but significant quanities of organic solvents (benzene, xylene, toluene), phenol, a variety of chlorinated hydrocarbons including trichlorophenol, trichlorethylene, and lindane ($_\gamma$-BHC or HCH) and heavy metals including arsenic.

In 1983, following clear proof of the presence of dioxin in the leachate, the Senate of the City of Hamburg resolved to clean up the Georgswerder landfill. Following extensive monitoring of the existing conditions at the site — geological, hydrogeological, chemical, and biological — development of a cleanup strategy began. At the time of this writing, the detailed cleanup strategy is still under development. However, the strategy will likely include the following elements:

1. Encapsulation of the landfill including vertical retention walls around the landfill and sealing of the surface of the landfill.
2. Extraction of fluids from within the landfill using gravity drains and wells.
3. Treatment of extracted fluids to remove and ultimately destroy hazardous substances using physical, chemical, and biological industrial wastewater treatment methods.
4. Incineration of residues from treatment processes.
5. Ultimate disposal of contaminated solid wastes after solidification.

The concept currently under consideration for the treatment of extracted fluids is shown schematically in Figure 1.

The flotation/sedimentation process will be used to separate oil and associated organics and chemical precipitates from the liquid stream. Many of the toxic organic substances, including dioxin, will be associated with the flotation overflow which will be incinerated directly. Metals, including arsenic and iron will be oxidized and will precipitate as hydroxides and other complexes. This chemical sludge will be thickened, solidified, and incinerated or deposited in an abandoned salt mine.

The water phase effluent from the flotation process will contain significant concentrations of biodegradable organics. Biological treatment of this waste stream will be provided to reduce the organic loading on the active carbon filters and thereby maximize the run-time of the filters. Waste

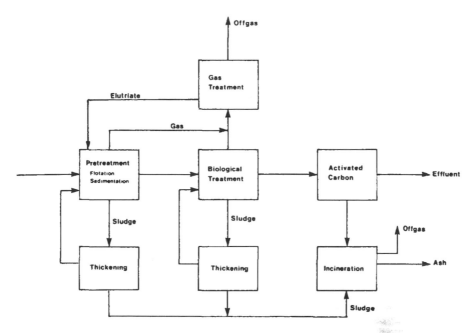

Figure 1.   Schematic of Leachate Treatment Concept.

biological sludge will be stabilized aerobically, and the residue will be incinerated. Offgas from the biological treatment system will be passed through activated carbon filters. Activated carbon will be regenerated by elutriation and the elutriate will be returned for treatment.

Biological treatment of leachate prior to activated carbon filtration has proven successful and cost-effective in several instances [1,2]. The carbon loading to the activated carbon filters is thereby reduced which increases the run-time of the filters and reduces the cost of carbon regeneration and replacement. The Sequencing Batch Reactor (SBR) process was selected for investigation as the method of biological treatment because the process has been successfully applied at other locations for hazardous waste treatment [2,3], and the process in many cases offers the advantages of better stability and greater operational control compared with continuous flow processes [4].

The SBR process is a batch biological process characterized by periodic filling and decanting of reaction basins and separation of the biological solids from the treated effluent in the same reaction basins rather than in separate sedimentation basins. The advantage of the SBR process lies in the ability to periodically change the environmental conditions in the reaction basins in a very controlled manner and thereby select and enrich a microbial population with desired specific metabolic capacities and settling characteristics. Such a specialized microbial population is required for the biodegradation of many leachate constituents.

Two-stage biological treatment is planned to promote the development of two different bacteria populations with specialized metabolic capacities. With a two-stage process it should be possible to maximize the number of organic compounds subject to biodegradation. It is expected that the more readily degradable organics will be metabolized in the first stage reactor and the less-degradable organics will be metabolized in the second stage. Because the substrate concentration will be very low in the second-stage reactor, it is questionable whether biological sludge flocs can be developed and maintained which can be separated by sedimentation. Consequently, a fixed-bed reactor, which provides surfaces for the growth of bacterial films, is suggested for the second stage reactor. Such a reactor operated on the SBR principle would be termed a Sequencing Batch Fixed-Film Reactor.

A silicone-membrane oxygenation system was selected for investigation because it offers the advantage of oxygen supply without the formation of gas bubbles [5]. With this method it is expected that the quantity of volatile organics in the offgas can be minimized because stripping would be eliminated and there would be greater opportunity for biodegradation of volatiles in the reactor. As a result the run-time for the offgas treatment system would be increased.

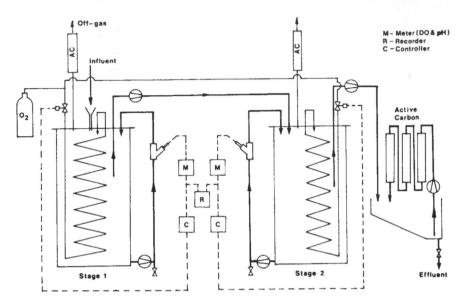

Figure 2.   Schematic of pilot-scale facilities.

OBJECTIVES

The principle objectives of this study were to:

1. Assess the treatability of leachate from the Georgswerder landfill using two-stage SBR biological treatment process in terms of removal of chemical oxygen demand (COD) and total chlorinated hydrocarbons (TOX).
2. Develop suitable process strategies for both first and second stage SBR reactors.
3. Assess the effectiveness and realiability of the silicone membrane oxygenation system.

EXPERIMENTAL FACILITIES

The investigations were carried out with pilot-scale facilities that were housed in a portable container located on the grounds of the Technical University Hamburg-Harburg. A portable container was used to allow the pilot plant to be transported later to the Georgswerder landfill site. The pilot plant included a two-stage SBR system, carbon filtration systems for gas and liquid streams and ancillary equipment. A schematic of the pilot system is shown in Figure 2.

The first-stage SBR was designed to operate as a suspended growth reactor and was equipped with a mechanical mixer, a decant pump, a recirculation pump, computer timers for on-off control, an oxygen and pH control system, and a silicone tubing oxygenation system. The reactor was constructed of glass and had an inside diameter of 150 mm and a height of 900 mm yielding an operating volume of 15 liters. The reactor was covered with a gas-tight stainless-steel lid equipped with gas-tight O-ring seals for the mixing shaft and tubing penetrations. Glass tubing was used for water transport. The oxygen and pH control systems consisted of meters and measuring electrodes mounted in the recirculation stream. Meters were in turn connected to set-point controllers which, in the case of the oxygen system, opened or closed the oxygen supply valve in response to the dissolved oxygen concentration of the reactor contents, and, in the case of the pH system, started or stopped the sodium hydroxide dosing pump in response to the pH of the reactor contents. The silicone membrane oxygenation system consisted of 56.8 m of silicone tubing having an outside diameter of 3.3 mm and a wall thickness of 0.4 mm connected to a pure oxygen cylinder. The tubing was supported within the reactor by means of a notched, stainless-steel column.

The second stage SBR was designed as a fixed-bed reactor and was constructed in the same fashion as the first stage reactor except that there was no mechanical mixer and the silicone tubing support column had a smaller diameter resulting in a tubing length of 46 m. The reactor was filled with

Table I. Summary of Oxygen Transfer Rates, mg/L/min

| System | O_2 Pressure, bar[a] | | | | | |
| | 0.5 | | 1.0 | | 1.5 | |
	Open[b]	Closed[c]	Open	Closed	Open	Closed
1	1.4	1.5	1.6	3.0[d]	2.4	4.3
2			1.6	3.0		
3			1.8	3.2[e]		

[a]Pressure at regulator with all valves closed.
[b]O_2 flowing through silicon tubing.
[c]No O_2 flowing through silicon tubing.
[d]Measured O_2 flux = 4,320 mg/m^2/h versus 4,880 mg/m^2/h theoretical.
[e]Measured O_2 flux = 2,661 mg/m^2/h versus 4,880 mg/m^2/h theoretical.

expanded-clay aggregate (approximately 5 mm in diameter) which surrounded the tubing support column.

The carbon filtration columns were constructed of glass tubing 50 mm in diameter × 900 mm in height. The gas filtration columns were connected to the head space of each reactor with Tygon tubing and were filled with Degussa LV carbon. The effluent gas stream was vented to the outside through ventilating fans. The three water filtration columns were filled with Degussa HWK I carbon and were designed for a filtration rate of 2.5 ml/min/cm^2.

EXPERIMENTAL RESULTS AND DISCUSSION

Pilot-Scale Studies

Pilot-scale studies consisted of a series of tests to determine the oxygen transfer capacity of the silicon membrane oxygenation system and leachate treatability studies conducted in two trials.

Oxygen Transfer Tests

The oxygen transfer capacity of the silicone tubing oxygenation system was measured in both the Stage 1 and Stage 2 reactors. Tests were conducted under several different conditions of oxygen supply pressure and downstream valve positions. Tap water was added to the reactor, and the dissolved oxygen was depleted by adding sodium sulfite and cobalt catalyst to the stirred reactor. The dissolved oxygen concentration was measured by an electrode mounted in the recycle stream, and the value was recorded on a strip chart. The transfer rates were calculated from the slope of the graph of dissolved oxygen vs. time in the range of 3 to 8 mg/L dissolved oxygen. The results are indicated in Table I. The maximum oxygen flux in the Stage 1 reactor approached the maximum value calculated from the permeability of oxygen through silicone membranes. The oxygen flux value observed for the Stage 2 reactor was considerably less than for the Stage 1 reactor because the mixing provided was less complete.

Synthetic Leachate Characteristics

These initial investigations were conducted using a synthetic leachate to minimize effects due to variations in composition known to occur with actual leachate samples and to reduce the risk of exposure of scientific personnel to toxic wastes. The composition of the synthetic leachate used is given in Table II. This composition was developed on the basis of maximum concentrations of principle constituents observed over 3 years of monitoring conducted by the City of Hamburg. Heavy metals were not included because they are expected to be removed during the pretreatment step. The phosphate shown in Table II was not observed in the actual leachate but was added as a supplement to improve the nutrient balance for microbial growth. The synthetic feed was made up daily by adding various stock solutions to a large separatory funnel, diluting with tap water, and shaking vigorously.

Trial 1 StartUp and Operating Strategy

A nitrifying activated sludge from a local wastewater treatment plant was used as the starting sludge for the Stage 1 SBR reactor. The treatment plant receives mostly domestic sewage, and the activated sludge exhibited excellent settling characteristics (SVI = 70). The suspended solids concentration of the

Table II. Synthetic Leachate Components[a]

Compound	Concentration	Unit
Peptone	100	mg/L
Acetate	250	mg/L
Isovaleric Acid	25	mg/L
Humic Acid	500	mg/L
Phenol	15	mg/L
Toluene	10	mg/L
Benzene	10	mg/L
Xylene	10	mg/L
Trichlorethylene	125	μg/L
Trichlorbenzene, 1, 2, 4	350	μg/L
Trichlorbenzene, 1, 2, 3	30	μg/L
Trichlorphenol, 2, 4, 5	1000	μg/L
Ammonium	300	mg/L
Chloride	2052	mg/L
Sulfate	480	mg/L
Phosphate	50	mg/L
Sodium	1049	mg/L
Bicarbonate	800	mg/L
Calcium	200	mg/L
Potassium	41	mg/L
Magnesium	120	mg/L
Manganese	3.5	mg/L
Iron (+ 3)	10	mg/L

[a]COD = 1,170 mg/L.

return activated sludge collected for startup was 7.1 g/L. This sludge was diluted with tap water to 3.0 g/L in the reactor prior to feeding. In addition the Stage 1 reactor was inoculated with approximately 500 ml of tap water filtered through a soil sample obtained from the Georgswerder landfill.

The SBR operating parameters are listed in Table III. The startup feeding strategy was as follows:

day	synthetic leachate	water
1	1 liter	2 liters
2	2 liters	1 liter
3	3 liters	0 liter

Table III. SBR Operating Parameters

	Trial 1		Trial 2			
			Stage 1		Stage 2	
Operating Parameter	Stage 1	Stage 2	Day 1–21	Day 22–72	Day 1–21	Day 22–72
Reactor volume, liter	15	6.2	15	15	6.2	6.2
Cycles/day	1	1	1	1	1	1
Cycle time, h	24	24	24	24	24	24
Feed volume/cycle, liter	3	3	3	3	3	3
Decant volume/cycle, liter	3	3	3	6.2	3	6.2
Retention time, d	5	2.1	5	2.4	2.1	1
Fill time, h	0[a]	0[a]	0[a]	0[a]	0[a]	0[a]
React time, h	20	20	20	20	20	20
Settle time, h	2	2	2	2	2	2
Decant/Idle time, h	2	2	2	2	2	2

[a]Dump fill.

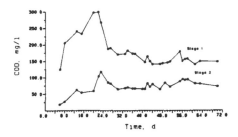
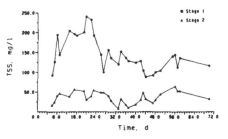

**Figure 3.   Stage 1 and Stage 2 effluent COD.
Concentration versus days of operation.**

**Figure 4.   Stage 1 and Stage 2 effluent TSS.
Concentration versus days of operation.**

Trial 1 Results

Because Trial 1 was basically not successful, only a brief summary and discussion of results are presented. During the first 5 weeks of operation approximately 90% reduction in soluble COD was achieved in the Stage 1 reactor, and over 95% reduction in total COD was achieved through both reactors. The effluent from Stage 2 was nearly colorless and contained less than 20 mg/L suspended solids. However, deterioration in effluent quality in terms of COD, color, and suspended solids began to occur after 5 weeks of operation. The fraction of flocculant organisms in the Stage 1 reactor steadily decreased during the observation period as evidenced by a continual decrease in the observed sludge volume after 30 minutes settling (SV_{30}) and an increase in effluent suspended solids. By the 7th week of operation suspended solids concentrations above 400 mg/L were measured in the Stage 1 effluent. These solids were subsequently fed to the Stage 2 reactor where they were filtered and settled out. Although the Stage 2 reactor continued to achieve substantial COD reduction, the imported solids from the Stage 1 reactor accumulated and eventually started appearing in the Stage 2 effluent.

Clearly the system was not operating as intended. The Stage 2 reactor was designed to operate as a fixed-film reactor, not as a filter. As a result of these observations it was decided to conduct a parallel bench-scale study in an effort to determine the operating conditions necessary to maintain flocculant growth. In addition, the pilot system was restarted with new sludge and the feed composition was changed slightly for a second trial of the treatability studies.

Trial 2 StartUp and Operating Strategy

The same sludge source and preparation procedures described under Trial 1 startup were used for Trial 2 startup. The reactor operating parameters for Trial 2 studies are listed in Table III. After 21 days of operation the initial operating strategy was modified. Instead of decanting 3 liters from Stage 1, 6.2 liters were decanted and stored temporarily in a glass vessel. The usual 3 liters were decanted from Stage 2 into a glass holding vessel. The remaining contents of Stage 2 (3.2 liters) was pumped out and placed in Stage 1. The 6.2 liters of Stage 1 effluent was transferred to Stage 2. Stage 1 then received 3 liters of synthetic leachate as usual. The reason for that change was to decrease the hydraulic retention time in Stage 1 to 2.4 days and in Stage 2 to 1 day. The increased washout rate was expected to selectively favor flocculant growth. In addition, any solids that accumulated in Stage 2 were transferred back to Stage 1. This change in operation had a marked effect on process performance as described below.

Trial 2 Results

The pilot system was operated for a period of 70 days during Trial 2. Effluent values of COD, TSS, and TOX observed over the period for both Stage 1 and Stage 2 reactors are shown graphically in Figure 3, 4, and 5, respectively. Removal efficiencies based on influent and average effluent values over the last 42 days of operation are reported in Table IV for COD, TOC, TOX, ammonia nitrogen, and nitrate nitrogen.

Of particular interest is the pattern of effluent values for COD and TSS which exhibit an increasing trend to the point of 22 days when the decant volume was increased in Stage 1 and the entire contents of Stage 2 was removed each day. Following this change in operation effluent values improved dramatically and remained relatively constant until day 58. At this point it was discovered that phenol had been inadvertently left out of the stock fed solution. Phenol was, therefore, added to the stock

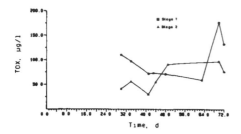

**Figure 5.   Stage 1 and Stage 2 effluent TOX.
Concentration versus days of operation.**

solution. Both reactors exhibited a temporary increase in effluent COD and TSS in response to this new addition. However, after a few days acclimation, the effluent values returned to the previous steady-state values.

To determine the soluble fraction of TOC in Stage 1 and Stage 2 effluents, samples of both effluents were filtered through a membrane filter (0.4 μ) and analyzed. Results are reported in Table V.

Also of particular interest are the characteristics of the sludge in the Stage 1 reactor which are reported in Table VI. As in the Trial 1, the sludge volume (SV$_{30}$) and the mixed liquor volatile suspended solids (MLVSS) decreased steadily after startup. Following the change in decant operation at day 22, the MLSS rapidly disappeared and after a few days there were essentially no mixed liquor solids. Despite this, effluent quality improved markedly as indicated in Figures 3 and 4. Apparently most all bacteria in the first stage reactor were entrapped in the space between the reactor wall and the silicone tubing structure. The bacteria were forced into this space as a result of centrifugal forces generated by stirring. The reactor was, in effect, functioning as a fixed-film reactor rather than a suspended growth reactor as intended. Because treatment efficiency remained high, it was decided to continue the operation of the system in this mode.

The concentration of dissolved oxygen in both Stage 1 and Stage 2 reactors and the pH value in Stage 1 were monitored continuously and recorded. A typical profile of these parameters during an operating period is shown in Figure 6. The dissolved oxygen profile in Stage 1 is characterized by a rapid decrease following feeding. As the concentration drops below the set-point of the control system, oxygen is supplied to the silicone tubing and the DO concentration rises above the set-point.

Table IV. Average Influent and Effluent Concentrations and Process Removal Efficiency – Trial 2

Parameter	Influent	Effluent Stage 1	Effluent Stage 2	Process Removal Efficiency %
COD[a]	1,170	155	75.9	93.5
TOC[a]	380	60.3	31.1	91.8
TOX[b]	1,085	98.8	63.8	94.1
NH$_4$-N[a]	300	<1	<1	99.9
NO$_3$-N[a]	0	175.3	172	—

[a]mg/L
[b]μg/L

Table V. Comparison of Soluble and Total Effluent TOC

Sample	TOC (mg/L)
Stage 1	
Unfiltered	68
Filtered	40
Stage 2	
Unfiltered	36
Filtered	29

Table VI. Stage 1 Reactor Sludge Characteristics — Trial 2

Days of Operation	SV$_{30}$ (mL)	MLSS (g/L)	MLVSS (g/L)	SVI (mL/g)
6	230	–	–	–
8	205	2.97	1.92	106
13	150	2.73	1.37	109
15	100	2.76	1.59	63
20	100	2.36	1.33	75
22	75	2.40	1.08	69
27	10	0.95	0.28	–

Oxygen demand in the reactor eventually begins to reduce the DO concentration once again, and the periodic pattern is repeated until the oxygen demand is satisfied. Following the initial period of very rapid oxygen uptake, which continues for about the first 30 minutes, is a period of moderate oxygen uptake, which lasts about six hours. This oxygen uptake is apparently associated with the nitrification reaction, because the end of this period coincides with the end of the pH control period.

Beginning about day 30, a gradual change in the DO concentration profile in Stage 1 was observed. The height of the DO peaks steadily decreased, indicating the oxygen transfer rate was diminishing with time. Dissolved oxygen profiles illustrating this change are shown in Figure 7. Eventually the oxygen transfer rate was insufficient to meet the initial oxygen demand following feeding, and anaerobic conditions prevailed until the oxygen demand decreased below the transfer rate. To compensate for the apparent loss in transfer efficiency, the oxygen supply pressure was increased to approximately 0.5 bar. The resulting profile with improved oxygen transfer is shown in Figure 7. This observed decrease in the oxygen transfer efficiency of the silicone tubing was apparently due to gas transport limitation at the liquid side of the membrane because of biofilm growth and accumulation of biomass close to the membrane. At the end of Trial 2 the tubing was removed from the reactor for inspection. The silicone tubing was heavily coated with bacteria which could be easily washed off with a soft stream of water. A relatively thin biofilm remained after rinsing. Microscopic examination of that biofilm revealed many individual organisms, very few fiberous or flocculant structures, and few protozoa. Both nitrosomonas and nitrobacter were prominent. Growth of biofilm was not expected because earlier work [5] indicated that biofilms do not develop on silicone tubing when pure oxygen is used and the oxygen concentration at the membrane-liquid interface is high. A grab sample of effluent gas from the silicone tubing taken shortly after feeding was found to contain 92.8% oxygen, 1.9% nitrogen, and 5.3% other gases, indicating little dilution of the oxygen. It must, therefore, be concluded that microorganisms, in particular nitrifiers, are capable of adapting to conditions of high oxygen concentrations.

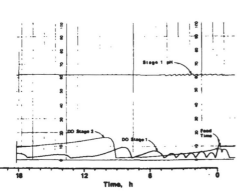

Figure 6. Typical profile of dissolved oxygen concentration and pH in Stage 1 reactor and dissolved oxygen concentration in Stage 2 reactor during reaction period.

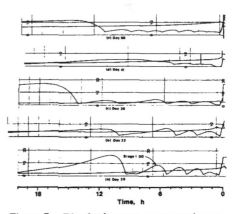

Figure 7. Dissolved oxygen concentration profiles in Stage 1 illustrating decrease in oxygen transfer efficiency during the study period.

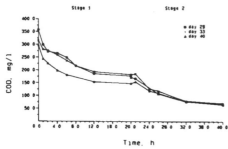

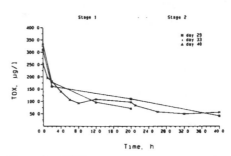

Figure 8.   COD concentration versus reaction time for three different operating days with normal feed.

Figure 9.   TOX concentration versus reaction time for three different operating days with normal feed.

Track studies were conducted to monitor the change in concentration of various leachate constituents during the reaction period. Results for both Stage 1 and Stage 2 for COD, TOX, NH$_4$-N, and NO$_3$-N are shown in Figures 8, 9, 10, and 11, respectively, for three different operating days. Removals of COD and TOX exhibited first-order kinetics characterized by a relatively rapid removal rate at the begining of the reaction period which then decreased during the reaction period. This pattern is repeated in the Stage 2 reactor, indicating that a different population of organisms is responsible for removal in Stage 2. Removal rates in Stage 1 appeared to be improving with days of operation.

Ammonia nitrogen was almost completely removed within 6 hours of operation and the removal appeared to follow zero order kinetics. If it is assumed that a nitrifying biofilm was present on the walls of the silicone tubing, the rate of nitrification per unit surface area can be estimated. Using the average rate of ammonia removal during the first 4 hours of operation, the nitrification rate is estimated to be approximately 205 mg/m$^2 \cdot$ h. This rate is somewhat greater than the range estimated for rotating biological filters used for nitrification of secondary effluents [6]. However, the decrease in ammonia nitrogen concentration is not fully accounted for by the increase in nitrate nitrogen concentration. This imbalance indicates that considerable nitrogen is being removed by mechanisms such as assimilation and denitrification.

In addition to the track studies conducted with normal feed, several studies were conducted with feed spiked with higher concentrations of individual constituents, namely acetate (3× normal), benzene (5× normal), trichlorphenol (4× normal), and ammonia (2× normal). The effects of the increased concentration of each of these constituents on the removal of COD, TOX, NH$_4$-N, and NO$_3$-N were observed by comparing the results of track studies with spiked feed with the results of track studies with normal feed. Based on these comparisons, it was concluded that concentration increases within the ranges studied had virtually no effect on process performance and that the system was operating well within its treatment potential.

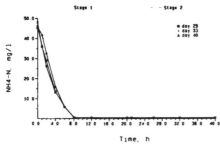

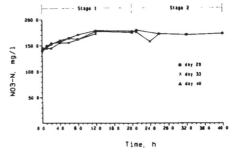

Figure 10.   NH$_4$-N concentration versus reaction time for three different operating days with normal feed.

Figure 11.   Comparison of NO$_3$-N concentration versus reaction time for three different operating days with normal feed.

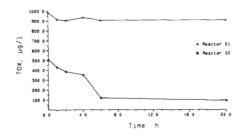

Figure 12. Comparison between TOX concentration versus reaction time for reactor S2 (with bacteria) and reactor S1 (without bacteria). Both received feed spiked with TCP.

A special study was conducted to determine if the removal of TOX observed in the Stage 1 reactor was due to volatilization. Normal feed spiked with trichlorophenol (TCP) was fed to a reactor (Reactor S1) identical to the Stage 1 reactor (Reactor S2) except that Reactor S1 was filled with clean tap water and contained no bacteria. The TOX concentration was measured at several times during the reaction period and the results are compared with those from the TCP track study in Figure 12. Although the spiked concentration in Reactor S1 was much higher than in Reactor S2, there was very little change in TOX concentration observed during the reaction period compared with that occurring in Reactor S2. Based on this comparison, it can be concluded that the removal of TOX observed in the Stage 1 reactor was not attributable to volitalization.

Bench-Scale Study

A bench-scale study was conducted for a period of 25 days to determine the effect of operating parameters on the performance of suspended-growth SBR's and particularly on the sludge characteristics. The operating parameters investigated in the study are summarized in Table VII. Effluent concentrations of COD and TSS in all reactors increased steadily during the period and were much higher than the effluent concentrations in the Stage 1 pilot-scale reactor. As with the pilot reactor, the sludge volume in all bench-scale reactors decreased steadily during the observation period. Although the MLVSS concentration remained relatively constant, an increasing portion of those solids remained in suspension as evidenced by an increasing effluent TSS concentration. It was apparent from these results and pilot study results that none of the operating strategies investigated resulted in

Table VII. Bench-Scale Study Operating Parameters

Operating Parameter	Reactor						
	B1	B2	B3	B4	B5	B6	B7
Reactor volume, l	1.25	1.25	1.25	1.25	1.25	1.25	1.25
Cycles/d	1	1	1	1	1	1	2
Cycle time, h	24	24	24	24	24	24	12
Feed volume/cycle, l	0.25	0.25	0.25	0.25	0.25	0.25	0.25
Retention time, d	5	5	5	5	5	2.5	5
Fill time, h	0[a]	0[a]	0[a]	0[a]	0[a]	0[a]	3
React time, h	20	20	20	20	20	10	20
Anaerobic period, h	0	0	0	3[b]	0	0	0
Settle time, h	3	3	3	3	3	1.75	3
Decant/Idle time, h	1	1	1	1	1	0.25	1
Feed	N[c]-Fe[f]	N[c]+HM[e]	N[c]	N[c]	N[c]+P[d]	N	N[c]

[a]Dump fill.
[b]Stirred, non-aerated period at beginning of react time.
[c]N Feed: acetate   — 275 mg/L
            phenol   — 15 mg/L
            humic acid — 500 mg/L
            inorg. salts — same as Table 2 without Cr and Zn
[d]peptone — 100 mg/L (COD value).
[e]HM-Zinc — 250 μg/L.
    Chrome — 225 μg/L.
[f]-Fe — without Iron.

stable performance of the suspended-growth SBR process, indicating that the suspended-growth activated sludge process is not a suitable method of biological treatment for the leachate in question.

SUMMARY AND CONCLUSIONS

The synthetic leachate used in this study is highly biodegradable. Over 90% reductions in COD, TOC, and TOX and complete nitrification are achievable with the two-stage SBR system.

Stable performance of the Stage 1 SBR as a suspended-growth reactor was not achieved under a variety of different operating and feed strategies.

Operation of the Stage 1 SBR as a "fixed-film" reactor with biomass accumulated in the space between reactor wall and silicone tubing resulted in stable performance for a continuous period of over 40 days.

The silicone tubing oxygenation system is an effective method of oxygenation for the closed, two-stage SBR system. However, the silicone tubing is subject to biofilm growth and accumulation which significantly reduces oxygen transfer capacity.

The system operated well within its treatment potential. Significant increases in the concentration of several leachate constituents resulted in no reduction in effluent quality.

Because of ongoing nitrification, a pH control system is required in the first stage reactor.

A two-stage SBR system is of advantage for biological treatment of the leachate. It appears on the basis of this investigation that both stages should be designed as fixed-film reactors with provisions for removing accumulated solids. The first stage should be equipped with a pH monitoring and control system.

The use of a silicone tubing oxygenation system located external to the reactors should be investigated. Such an arrangement would facilitate the control of biofilm accumulation.

REFERENCES

1. Suckrow, A. J., A. P. Pojak, and C. J. Tonhill, *Hazardous Waste Leachate Management Manual*, Noyes Data Corporation, Park Ridge, NJ, 1082 (1982).
2. Ying, W., R. R. Bonk, V. J. Lloyd, and S. A. Sojka, "Biological Treatment of Landfill Leachate in Sequencing Batch Reactors," Presented at IWPCF Conf. (October 1984).
3. Herzburn, P. A., R. C. Irvine, and M. J. Hanchak, "Treatment of Hazardous Wastes in a Sequencing Batch Reactor," *Proceedings 39th Annual Purdue Industrial Waste Conference* (May 1984).
4. Wilderer, P. A., and E. D. Schroeder, "Anwendung des Sequencing Batch Reactor (SBR) — Verfahrens zur biologischen Abwasserreinigung," *Hamburger Berichte zur Siedlungswasserwirtschaft*, 3 publ. TU Hamburg-Harburg (1986).
5. Wilderer, P. A., J. Brautigam, and I. Sekoulov, "Application of Gas Permeable Membranes for Auxiliary Oxygenation of Sequencing Batch Reactors," *J. Conserv. and Recycling*, 8, 181–192 (1985).
6. Watanabe, Y., K. Nichidome, C. Thanantaseth, and M. Ishegura, *Proc. First Int'l Conf. on Fixed-Film Biological Processes*, University of Pittsburgh, vol. 1, 309–330 (1982).

31 EFFECTS OF ORGANIC FLUIDS ON CLAY PERMEABILITY

Frazier Parker, Jr., Associate Professor

Larry D. Benefield, Alumni Professor
Department of Civil Engineering
Auburn University
Auburn, Alabama 36849

Marshall M. Nelson, Engineer
Soil Conservation Service
U.S. Department of Agriculture
Auburn, Alabama 36830

INTRODUCTION

It has been estimated that there were over 41 million wet metric tons of hazardous wastes produced in the U.S. during 1980. The single largest source is the chemical and allied products industry, which accounts for 62% of all hazardous wastes produced. Approximately 83% of hazardous wastes are treated or disposed in on-site facilities by the industry. The rest is transported to commercial off-site facilities. Secure landfill and treatment are two options used for about 66% of these wastes. The remaining wastes are incinerated, land treated, deep well injected, or recovered [1].

Secure landfills used for disposal of chemical wastes are commonly lined with several feet of compacted clay to provide a supposedly impermeable barrier to assure that chemical wastes placed into the landfills will not contaminate underlying groundwater. Recent research has shown that reliance on the typical low permeability of compacted clay may be justified only if the permeant is water. This research has shown that the impermeability of clay barriers can be permanently destroyed thus destroying their function when exposed to various chemical permeants. Therefore, all chemical materials proposed for disposal into clay-sealed landfills should be identified, and those which adversely affect the impermeability of clay barriers should be disposed of in some other manner. The major obstacle to the development of such a management procedure is the identification of chemicals that adversely affect clay permeability and the identification of clays that may be resistant to degradation. Because permeability tests on clays require large amounts of time (up to a month), considerable operator skill and specialized equipment [2], only a few organic fluids and clay soils have been studied.

The objective of this study was to evaluate the consolidation test as a means of assessing the effects of chemical permeants on clay permeability. Consolidation tests can be performed in approximately one week, the required operator skill is not great and equipment is readily available. The use of the consolidation test is not without distractions, however. Coefficients of permeability will not be precise since the consolidation process depends on the stress-strain state and fluid equilibrium conditions in the soil. Still, for comparative studies permeability values from consolidation tests may be as reliable as other laboratory estimated values. For example, results from a recent study [3] showed actual permeabilities of compacted clay liners as much as 1000 times greater than laboratory predicted permeabilities. Recognizing that actual permeabilities are not likely to be obtained in the laboratory, the consolidation tests become an attractive tool for delineating the effects of organic fluids on clay permeability.

BACKGROUND

A secure landfill is a carefully-engineered depression in the ground for storage of hazardous wastes [4,5]. A bottom liner is normally provided to prevent any hydraulic connection between the wastes and the surrounding environment, particularly the ground water. In many cases the bottom liner is composed of one or more layers of clay. However, synthetic membrane liners are now also required.

Clay liner permeabilities are typically determined as if pure water will be the leachate. In contrast, the actual leachates generated in industrial landfills are most likely to be either highly contaminated

Table I. Soil Sampling

	Soil Series	Location	Soil Association
1.	Chisca	Colbert Co., N. W. Ala.	Limestone Valleys and Uplands
2.	Cecil	Chambers Co., E. Central Ala.	Piedmont
3.	Sumter	Marengo Co., W. Central Ala.	Prairie
4.	Oktibbeha	Marengo Co., W. Central Ala.	Prairie
5.	Conecuh	Bullock Co., S. E. Ala.	Coastal Plain

water or a mixture of organic liquids released by the disposed waste. A typical list of the contents of landfill drums compiled by Anderson [6] contains fifty-four substances and points to the fact that a wide variety of organic liquids have been landfilled in the past.

Green and his associates [7] tested the effects of various solvents on clay columns designed to simulate liners used at chemical waste disposal sites. The solvents tested were benzene, xylene, carbon tetrachloride, trichloroethylene, acetone and glycerol with water as a control. Their investigation showed that certain solvents could cause clay to shrink, crack, and allow rapid transport of contained materials through them.

Brown and Anderson [8] examined the suitability of using water alone to test the permeability of compacted clay liners of hazardous landfills and surface impoundments. They found that traditional permeability tests using water alone qualified four clay soils for lining hazardous waste facilities on the basis of low permeabilities (1×10^{-7} cm/sec). But these same clays underwent large permeability increases when tested with a basic organic fluid (aniline), three neutral polar organic fluids (methanol, acetone and ethylene glycol), and two neutral nonpolar organic fluids (heptane and xylene). They also showed potential for substantial permeability increases when exposed to concentrated organic acid (acetic acid).

To summarize, a knowledge of the permeability (a main criterion used to judge whether a compacted soil liner will prevent movement of leachates below or adjacent to a disposal facility) is needed to determine a liner's suitability for use as a containment system. However, relatively little information exists concerning the impact of organic fluids on clay permeability and no rapid and simple test method is available which will permit evaluation of large numbers of clay soils with a wide range of possible organic fluids.

TEST PROGRAM

Alabama, as is the case for many other parts of the U.S., has a significant hazardous waste disposal problem. Not only must the environment absorb the waste from Alabama industry, but it has become a repository for chemical wastes from other states. For example, it has the largest commercial chemical landfill in the country, a 2,400 acre facility at Emelle in Sumter County that takes in more than 250,000 tons of chemical wastes every year. Considering the scope of the problem, the need for understanding the capabilities of the ground for safely containing hazardous waste becomes obvious.

Soils in Alabama have been grouped by Hajek, Gilbert and Steers [9] into seven associations based on unique characteristics of the soils. Studies of the reactivity of Alabama soils with lime conducted by Moore and Brown [10] showed a wide variability state-wide but with some general correlatibility with the soil associations. The above studies suggest the possibility of the following: 1) variability in the reactivity of clay soils with organic fluids, and 2) potential correlations between soil type and reactivity. The scope of the proposed study was not intended to address possible correlations, but a variety of soil types were selected for testing to illustrate variability and to begin to establish a data base for future correlations.

Samples were obtained from four of the five major soil associations. Samples were identified by their soil series name. These are listed in Table I with their general geographic location and soil association.

Two samples were obtained from the Prairie Association. None were obtained from the Appalachian Plateau Association or the two minor associations (Flood Plains and Coastal Marshes).

The Chisca soil is classified as a thermic Vertic Hapludalf derived from calcareous shale and limestone. Cecil soil is classified as a thermic Typic Hapludult derived from granite and gneiss. Sumter soil is classified as a carbonatic thermic Rendolic Eutrachrept derived from chalk and carbonitic clays. The Oktibbeha soil is classified as a thermic Vertic Hapludalf derived from chalk and Conecuh soil is classified as a thermic Typic Hapludult derived from clayey marine sediments.

Table II. Soil Properties

Soil Series	Primary Clay Mineral (s)	PI	% Sand	% Silt	% Clay	pH	Cation Exchange Cap., me/100 gm
				Soil Property			
Cecil	Kaolinite	22	41	12	47	5.34	1.6
Chisca	Montmorillinite & Kaolinite	54	3	25	72	4.70	38.0
Conecuh	Montmorillinite & Kaolinite	28	24	22	54	4.39	30.4
Oktibbeha	Montmorillinite	79	8	25	67	4.60	18.5
Sumter	Calcareous Montmorillinite	33	11	46	43	7.82	31.8

Table III. Fluid Properties

Fluid	Density, gm/cm^3	Viscosity, Centipoise	Dielectric Constant
		Fluid Property	
Water	1.00	1.00	80.4
Acetic Acid	1.05	1.28	6.2
Xylene	0.87	0.81	2.4
Methanol	0.79	0.54	31.2
Acetone	0.79	0.33	21.4

Properties of the soils are listed in Table II. The soils have relatively large clay contents and high plasticity as indicated by the Plasticity Index (PI). These are normally indications that the soil will have permeabilities acceptable for liner construction or disposal site location.

A 0.01 normal $CaSO_4$ solution was used to simulate natural ground water. In addition, acetic acid, methanol, acetone, and xylene were used. Acetic acid is an organic acid, methanol and acetone are polar compounds and xylene is a nonpolar compound. These are all organic fluids which may be contained in leachates generated at a hazardous waste disposal site. Reagent grade fluids were used for the testing. Certain fluid properties postulated as relevant to permeability are listed in Table III.

Combinations of five soils and five fluids result in the experimental design outlined in Table IV. In addition, tests were run on Oktibbeha soil that had been treated with a particular fluid, dried, and tested with water. These supplementary tests were performed to determine the permanence of the effects of the organic fluids.

Table IV. Experimental Design

Fluids	Cecil	Chisca	Conecuh	Oktibbeha	Sumter
			Soils		
Water (0.01N CaSO$_4$)	1	2	3	4	5
Methanol	6	7	8	9	10
Acetone	11	12	13	14	15
Xylene	16	17	18	19	20
Acetic Acid	21	22	23	24	25

Supplementary Tests: Four supplemental tests were run by mixing Oktibbeha soil with the indicated fluid, curing for 24 hrs, air drying, remixing with water and consolidating.

26. Oktibbeha–Methanol–Water.
27. Oktibbeha–Xylene–Water.
28. Oktibbeha–Acetone–Water.
29. Oktibbeha–Acetic Acid–Water.

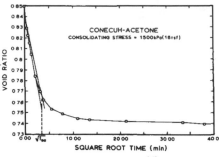

Figure 1.   Plot of typical time$^{1/2}$-void ratio.   **Figure 2.   Plot of typical log$_{10}$ time-void ratio.**

Basic procedures for consolidation testing and analysis, as outlined by Lambe [11] were followed with only minor modifications. Because the samples were of slurry consistency when loading started, fixed ring consolidometers were used. A stainless steel consolidometer and stainless steel porous stones were used with acetic acid to minimize corrosion. Neoprene gaskets and "0" rings in the consolidometers had to be replaced periodically when testing with xylene and acetic acid. Finally, it was necessary to partially seal the consolidometer and to periodically add fluid to the reservoir when testing with the more volatile fluids (methanol and acetone).

Sample Preparation

Samples of the clay soils were air dried and pulverized until all material passed a No. 40 sieve.

Samples of the processed soils were mixed with various fluids and aged for twenty-four hours. The fluid contents produced a slurry of viscous liquid consistency (slightly above liquid limit of soil). The Sumter soil had a high calcium carbonate content and reacted with acetic acid to generate CO_2. This "fluffed" the soil and required additional manipulation to reduce the void ratio prior to testing.

Sample Testing

The slurries were placed in consolidometer rings and large voids removed.

The samples and rings were placed in fixed ring consolidometers, submerged in appropriate fluids and seating pressures of 12 kPa (1/8 tsf) applied with levermatic consolidation machines. Seating pressures were maintained for 24 hours.

The samples were incrementally loaded and rate of consolidation measured.

Data Reduction

Data reduction was computerized using spread sheet and graphic software. Figures 1 and 2 illustrate typical computer generated rate of consolidation plots. Taylor's square root of time curve fitting technique was used to manually estimate the time for 90% consolidation. Coefficient of consolidation, c_v, was computed using the relationship

$$c_v = \frac{TH^2}{t_{90}} \tag{1}$$

where T = dimensionless time factor (0.848 for 90% consolidation), H = drainage distance (one half sample thickness), and t_{90} = time for 90% consolidation. Casagrande's log time curve fitting technique was also used to estimate the coefficient of consolidation for some of the tests. The time for 50% consolidation was obtained from the curves and used in the above equation with T = 0.197. Taylor's technique provided the most reasonable and consistent results, and the reported permeability values were computed using c_v obtained with t_{90} estimates.

For each pressure, Darcy's coefficient of permeability was computed with the equation

$$k = \frac{c_v a_v \gamma_f}{1 + e} \quad \text{(units of length/time)} \tag{2}$$

Table V. Soil Permeability

Fluid	Specific Permeability[a], cm^2				
Soil	Water	Acetic Acid	Xylene	Methanol	Acetone
Cecil	2.0×10^{-13}	4.5×10^{-13}	1.1×10^{-12}	6.0×10^{-13}	1.0×12^{-12}
Chisca	1.6×10^{-14}	1.8×10^{-13}	7.0×10^{-12}	6.0×10^{-13}	3.0×10^{-13}
Conecuh	6.5×10^{-14}	9.8×10^{-13}	5.3×10^{-12}	2.6×10^{-12}	7.0×10^{-13}
Oktibbeha	1.4×10^{-14}	1.2×10^{-12}	9.0×10^{-12}	1.7×10^{-12}	1.3×10^{-12}
Sumter	7.5×10^{-13}	3.0×10^{-15}	1.0×10^{-11}	2.2×10^{-11}	8.0×10^{-12}
Treated Oktibbeha[b]	1.4×10^{-14}	1.2×10^{-13}	1.4×10^{-14}	8.5×10^{-14}	2.0×10^{-13}

[a]Permeability values at void ratio, e = 1.0.
[b]Soil mixed with indicated fluid, dried, and remixed with water for testing.

where k = Darcy's coefficient of permeability, c_v = coefficient of consolidation, a_v = coefficient of compressibility, γ_f = unit weight of fluid, and e = void ratio at the beginning of a load increment. For a given pressure or corresponding void ratio, the coefficient of compressibility was estimated by dividing the change in void ratio by the added increment of pressure.

The absolute or specific permeability was computed with the equation

$$K = \frac{kv}{\gamma_f} \quad \text{(Units of length squared)} \tag{3}$$

where K = absolute or specific permeability, k = Darcy's coefficient of permeability, v = fluid viscosity, and γ_f = fluid unit weight. The influence of the fluid viscosity and unit weight are thus eliminated in the absolute permeability [12].

Results from the 29 consolidation tests (soil and fluid combinations in Table V) were summarized and plotted to illustrate the effects of soil and fluid type. Figure 3 is an example for Conecuh soil. The essentially linear relationships between void ratio and base ten log of absolute permeability is illustrated. The data was also plotted for each fluid as shown by the example for methanol in Figure 4. This figure illustrates the influence of soil type.

DATA ANALYSIS

When organic fluids were mixed with clay soils, dramatic changes in the soil consistency and behavior were observed. The soils, which were highly cohesive when mixed with water, were completely noncohesive when mixed with methanol, acetone, and xylene. Cohesion was greatly reduced when mixed with acetic acid. Wetting of the complete sample was very rapid. Methanol, acetone, and xylene completely destroyed the clay plasticity creating a slurry which acted as a silt having high dilatancy. Acetic acid produced a similar effect but to a lesser degree. Consolidated samples with the organic fluids had little dry clod strength and pulverized easily with hand pressure. This behavior is evidence of the flocculating effect of the fluids and is identical to the aggregated structure observed by Brown and Anderson [8].

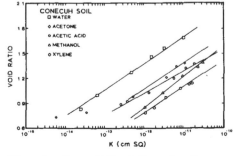

Figure 3.   Permeability of Conecuh soil.

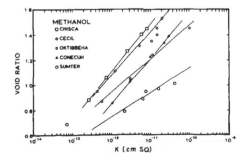

Figure 4.   Permeability of methanol.

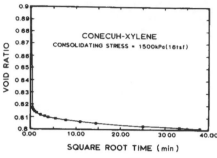

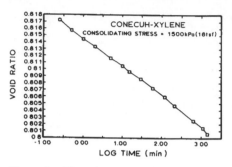

Figure 5.   Plot of time$^{1/2}$-void ratio with organic fluid.

Figure 6.   Plot of log$_{10}$ time-void ratio with organic fluid.

Rate of consolidation curves (void ratio vs. square root of time and log of time) with water and a few other soil-fluid combinations had a typical shape as exhibited by the curves in Figures 1 and 2. However, consolidation occurred very rapidly for most of the samples mixed with organic fluids. In many cases 90% consolidation occurred within 30 seconds compared to 20 minutes or greater for water. Because of the rapid consolidation, many log time versus void ratio curves did not have the shape characteristic of typical consolidation and were difficult to interpret. Square root of time curves were easier to interpret and were used in all computations for c_v. This behavior is illustrated in the curves of Figures 5 and 6. As would be expected, those samples having the more rapid consolidation were also the ones having the higher permeabilities.

Brown and Anderson [8] attributed variations in the permeability of a given soil with different permeants to differences in the microtexture and macrotexture that result from interaction of the soil and permeant. The macrotexture (cracks, channels, or other large voids) that may result when a clay is permeated with a fluid cannot be studied with the consolidation test, but the effect of microtexture can be studied. Dissolution of clay minerals, piping, gas generation, particle interaction and interlayer spacing will affect consolidation characteristics and, thus, permeability. Specific soil-fluid interactions and their effects on permeability will be discussed in conjunction with analysis of test results.

To compare permeability measured with consolidation tests, values at a void ratio of 1.0 were selected. A summary of these values obtained from plots similar to Figures 3 and 4 are contained in Table V. For comparative purposes, corresponding values of Darcy's coefficient of permeability are listed in Table VI.

In order to better visualize differences in permeability, values from Table VI are depicted as a bar chart in Figure 7. The following observations can be made from Figure 7:

a. The permeability for all organic fluids is greater than water for all soils except Sumter where the permeability for acetic acid is smaller. The effects of fluid unit weight and viscosity have been eliminated from these values, thus, indicating that the interaction of the soils with the organic fluids results in a more permeable soil microtexture.
b. In general, the Sumter soil has the largest permeability and the Chisca the smallest.
c. Xylene produced the largest permeability for all soils, except Sumter where methanol produced the largest permeability.

Table VI. Darcy's Coefficient of Permeability for Soils

Fluid Soil	Coefficient of Permeability[a], cm/sec				
	Water	Acetic Acid	Xylene	Methanol	Acetone
Cecil	2.0×10^{-8}	3.7×10^{-8}	1.2×10^{-7}	8.0×10^{-8}	2.3×10^{-7}
Chisca	1.6×10^{-9}	1.5×10^{-8}	7.4×10^{-7}	8.0×10^{-8}	8.0×10^{-8}
Conecuh	6.5×10^{-9}	7.9×10^{-8}	5.6×10^{-7}	3.7×10^{-7}	1.6×10^{-7}
Oktibbeha	1.4×10^{-9}	9.8×10^{-8}	9.0×10^{-7}	2.4×10^{-7}	2.9×10^{-7}
Sumter	7.5×10^{-8}	2.4×10^{-10}	1.1×10^{-6}	3.1×10^{-6}	1.9×10^{-6}
Treated Oktibbeha	1.4×10^{-9}	1.2×10^{-8}	1.4×10^{-9}	8.5×10^{-9}	2.0×10^{-8}

[a]Value for void ratio, e = 1.0. Includes the effects of fluid viscosity and density.

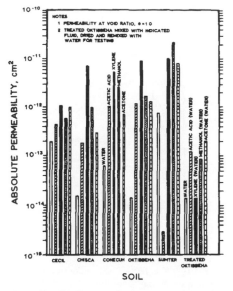

Figure 8. Absolute permeability as a function of clay content.

Figure 7. Bar chart of absolute permeability values.

Figure 8. Absolute permeability as a function of clay content.

d. The effects of the interaction of Oktibbeha soil with acetic acid, methanol, and acetone is partially reversible. This is similar to the response observed by Brown and Anderson [8] and is attributable to irreversible changes in soil fabric. The effects of the interaction with xylene appear to be completely recoverable.

To eliminate the effects of soil grain size and illustrate the relative effects of various fluids, permeability is plotted versus percent clay size particles in Figure 8. The following observations can be made from Figure 8:

a. With water, permeability decreases as clay content increases. This is the expected trend since pore size decreases as grain size decreases.
b. No definite trend was observed for other fluids. Except for Sumter with acetic acid, the pattern for all organic fluids was similar. The soil-fluid interaction may have caused increased flocculation and agglomeration, resulting in an increase in effective particle size. This could explain the apparent diminished influence of particle size.
c. The abnormal response of Sumter with acetic acid is thought to be caused by its high $CaCO_3$ content (from shell fragments). In addition to dissolution and possible clogging by smaller particles as suggested by Brown and Anderson [8], the generation of CO_2 may have resulted in only partial saturation which would have contributed to clogging. In addition, internal pressure created by CO_2 may have resisted the consolidation pressure.
d. With the exception of Sumter with acetic acid, the acetic acid, methanol, and acetone are rather closely grouped.
e. With the exception of Sumter, xylene produced the largest permeability. The effects of the xylene were also completely recoverable upon drying and remixing with water. Xylene was the only nonpolar compound studied. It has the smallest dielectric constant, lowest water solubility, and lowest volatility of the organic fluids tested. Any or all of these may have influenced the soils interactions with xylene.

Permeability has been shown to vary with fluid dielectric constant [12]. As water within the double layer surrounding clay particles is replaced by organic fluids having a lower dielectric constant, the interlayer spacing is reduced allowing attractive forces between particles to more easily overcome the forces of repulsion and a more flocculated, aggregated structure results. A more flocculant structure will have larger pores than a more dispersed structure of the same void ratio and, therefore, will likely have a larger permeability. Permeability and fluid dielectric constant are plotted in Figure 9. Although

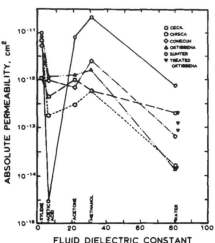

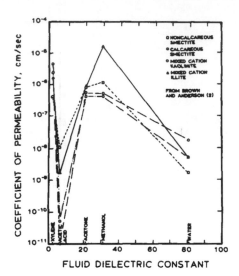

FLUID DIELECTRIC CONSTANT

FLUID DIELECTRIC CONSTANT

**Figure 9. Absolute permeability from consoli-
dation tests as a function of fluid dielectric
constant.**

**Figure 10. Coefficient of permeability from
compacted clay columns as a function of
fluid dielectric constant.**

the available data is not well distributed and there is considerable scatter, there appears to be a trend
of decreasing permeability with increasing fluid dielectric constant. The relatively high values for
Sumter are a result of its high silt (46%) and low clay (43%) contents. Similar results were obtained by
Brown and Anderson [8]. Their results are shown plotted in Figure 10. Permeability for acetic acid is
low indicating that properties other than dielectric constant may be more important.

Permeability was plotted versus soil cation exchange capacity and pH. No clear trends were apparent. From the limited data set, it was concluded that soil cation exchange capacity and pH have no
discernible influence on permeability with organic fluids.

The effects of organic fluids can be demonstrated by comparing their permeabilities with permeabilities for water. Percent differences were computed and are tabulated in Table VII. The percent
differences for each fluid are also plotted versus soil type in Figure 11. The following observations can
be made from Table VII and Figure 11:

a. The percent increase in permeability ranges from 100% for Cecil with acetic acid to
64,000% for Oktibbeha with xylene.

Table VII. Soil Permeability Difference Values

Fluid Soil	% Difference[a] in K With Water			
	Acetic Acid	Xylene	Methanol	Acetone
Cecil	100%	450%	200%	400%
Chisca	1,000%	43,000%	3,600%	1,800%
Conecuh	1,400%	8,000%	3,900%	1,000%
Oktibbeha	8,500%	64,000%	12,000%	9,200%
Sumter	-25,000%[b]	1,200%	2,800%	1,000%
Treated Oktibbeha	750%	0	500%	1,300%

[a]% Difference $= \dfrac{K_f - K_w}{K_w} \times 100$.

K_f = Permeability with Fluid at e = 1.0.
K_w = Permeability with Water at e = 1.0.

[b]% Difference $= \dfrac{K_f - K_w}{K_f} \times 100$.

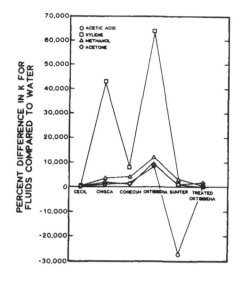

Figure 11. **Difference in permeability as a function of fluid and soil type.**

b. The 25,000% decrease for Sumter with acetic acid is attributable to the reaction of the acid with $CaCO_3$.

c. The dramatic effect of xylene on Chisca and Oktibbeha is thought to be due to their high clay content. The clay mineral particles in these soils, a high proportion of which are montmorillinite, are more chemically reactive than sand and silt particles or other type clay mineral particles.

d. In terms of overall effect on all soils, the fluids would have the following ranking:
   1. Xylene — Largest increase
   2. Methanol
   3. Acetone
   4. Acetic Acid — Smallest increase

For the soils studied by Brown and Anderson [8], the positions of the xylene and methanol were reversed.

CONCLUSIONS

Organic fluids, in general, increase the permeability of clay soils. The magnitude of the increase is fluid and soil specific. Other than the obvious decrease in permeability for water that occurs with decreasing particle size, no definite general trends were observed relating permeability with fluid or soil properties. A weak trend was noted for fluid dielectric constant, but for specific soils and fluids other interactions were more important.

Except for Sumter soil, the relative effects on permeability of the fluid studied were similar to the results obtained by Brown and Anderson [8] from tests on packed clay columns. The magnitude of permeability changes was not as great, however, and is probably due to changes in macrotexture (cracks, channels, etc.) that developed in the permeameter tests.

The consolidation test provides a viable method for evaluating the relative effects of organic fluids on clay soil permeability. It permits efficient evaluation of large numbers of soils and fluids and provides clear delineation of specific interactions. While it does not permit study of the development and effects of macrotexture, it does permit the study of specific soil-fluid interactions on the development of soil microtexture, and thus, the effects on permeability. The use of the consolidation test for providing absolute measures of permeability will require verification and correlation with directly measured values.

ACKNOWLEDGMENTS

This research was supported by the U.S. Department of the Interior through the Water Resources Research Institute of Auburn University and by the Engineering Experiment Station of Auburn University. Contents of this publication do not necessarily reflect the views and policies of the

Department of the Interior or the Engineering Experiment Station, nor does mention of trade names as commercial products constitute their endorsement or recommendation for use by the United States Government or Auburn University.

REFERENCES

 1. Hill, R. D., Shoemaker, N. B., Landreth, R. E., and Wiles, C. C., "Four Options for Hazardous Waste Disposal," *Civil Engineering*, 82 (September 1981).
 2. Daniel, D. E., Trautwein, S. J., Boynton, S. S., and Foreman, D. E., "Permeability Testing with Flexible-Wall Permeameters," *Geotechnical Testing Journal*, ASTM, Vol. 7, No. 3 (September 1984).
 3. Day, Steven R., and Daniel, David E., "Hydraulic Conductivity of Two Prototype Clay Liners," *Journal of Geotechnical Engineering, ASCE*, Vol. 111 No. 8 (August 1985).
 4. Farb, D. G., *Upgrading Hazardous Waste Disposal Sites*, (SW-677), U.S. Government Printing Office, Washington, D.C. (1978).
 5. Fields, T., and Lindsey, A. W., *Landfill Disposal of Hazardous Wastes: A Review of Literature and Known Approaches*, (SW-165), U.S. Government Printing Office, Washington, D.C. (1975).
 6. Anderson, D., "Does Landfill Leachate Make Clay Liners More Permeable?" *Civil Engineering*, 66 (September 1982).
 7. Green, W. J., Lee, G. F., and Jones, R. A., "Clay-Soils Permeability and Hazardous Waste Storage," *Journal of Water Pollution Control Federation*, 53, 1347 (1981).
 8. Brown, K. W., and Anderson, D. C., "Effects of Organic Solvents on the Permeability of Clay Soils," U.S. Environmental Protection Agency Report, *EPA-600/S2-83-016* (1983).
 9. Hajek, B. F., Gilbert, F. L., and Steers, C. A., *Soil Associations of Alabama*, Soil Conservation Service, Auburn, Alabama (November 1975).
10. Moore, Raymond K., and Brown, Glenn C., "Development of Soil Stabilization Guidelines for Alabama Soils," Alabama Highway Department, *HPR Report No. 84* (October 1977).
11. Lambe, T. William, *Soil Testing for Engineers*, John Wiley & Sons, Inc. (1951).
12. Lambe, T. W., and Whitman, R. V., *Soil Mechanics*, John Wiley & Sons, Inc. (1969).

32 AQUEOUS PYROLYSIS OF INDUSTRIAL WASTES

David L. Kincannon, Graduate Student

Don F. Kincannon, Professor
School of Civil Engineering

William L. Hughes, Professor and Director
The Engineering Energy Laboratory
School of Electrical and Computer Engineering
Oklahoma State University
Stillwater, Oklahoma 74078

INTRODUCTION

Industrial wastes present many problems with regard to their disposal or treatment. Improper handling of industrial wastes can lead to pollution problems and public health hazards. Large quantities of industrial wastes are generated continuously which must have adequate treatment facilities or available space for disposal. Solid waste disposal in approved sanitary landfills can require large areas of land. If the industrial waste is a hazardous material, it must be disposed of in a permitted hazardous waste landfill. The land for these landfills must consist of a suitable soil type and an acceptable location that is near the source of waste generation and not in a place that is susceptible to flooding. Public opposition is usually great because nobody wants to have disposal or treatment facilities for wastes located in their vicinity. The opposition is due to the risks of pollution and the possibilities of unpleasant aesthetics and odor problems. This paper presents an investigation into the use of the aqueous pyrolysis method using induction heating that has been developed by Dr. William L. Hughes and co-workers (1) as a method of treating industrial wastes and other solid wastes.

PREVIOUS WORK

Research previously done on pyrolysis has provided helpful information in several areas. Hughes and Ramakumar [1] used aqueous pyrolysis of biomass as a method for generating energy using various biomass waste products as the sources of fuel. They found that the process yielded a gaseous product nearly evenly divided between carbon dioxide and combustible hydrocarbon gases. Their process used induction heating of a batch reactor to pyrolyze cotton, saw dust, newsprint, coal dust, and plant materials [1].

Bohn and Benham [2] in their study of biomass pyrolysis using an entrained flow tubular reactor of a wheat straw feedstock with a steam carrier gas measured the gas yield, gas composition, and process heat of the pyrolysis process. Gas yields of 91% were measured at 950°C, and the process heat was measured in the range of 2300–3000 J/g of pyrolysis gas. The composition of gases was found to be 52% CO, 20% H_2, 11% methane, 8% CO_2, 5% ethylene, and 4% other gases. This study showed that the pyrolysis was strongly influenced by the reactor temperature and not by the steam to biomass ratio [2].

Kemmler and Schlich [3] investigated the use of pyrolysis for the volume reduction of organic wastes. The process was used for pyrolyzing nuclear wastes and spent solvents. The volume reduction of nuclear wastes was 50% and mass reduction was 20%. There was a 7% reduction of volume and nearly 6% reduction of mass of the solvent that was pyrolyzed [3].

PROCESS DESCRIPTION

Aqueous pyrolysis is the conversion of a sample mixed with water into another substance or substances by subjecting the material to high temperature and high pressure [4]. The water is used to replace air in void spaces of the material. Due to thermal fission, the process generally leads to the production of molecules of lower mass. The process results in the decomposition of the material into char, carbon dioxide, methane, and other hydrocarbon gases such as ethane and ethylene. The

293

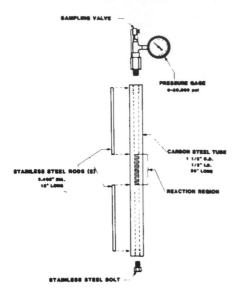

Figure 1.   The reactor vessel.

thermal energy input and the absence of air in the reactor vessel results in temperatures in the range of 300 to 500°C and pressures of 8,000 to 14,000 PSIG being achieved. The thermal energy is provided by electrical induction heating, which consists of a coil of wire wrapped around an electrically-conductive reactor vessel and energized with an alternating current. The resulting magnetic flux causes induced currents to circulate within the walls of the reactor, generating heat, which increases the pressure in the constant volume reactor [5]. Many industrial wastes may be processed into combustible products from the aqueous pyrolysis process, and these products may be reused as a source of fuel for this process. An additional benefit of the aqueous pyrolysis process is the volume reduction of the solid material which is pyrolized [4].

METHODS AND MATERIALS

This investigation of the aqueous pyrolysis method used a technique that consisted of packing a mixture of a waste material and water into the center region of a long, thin cylindrical reactor vessel and then heating the reactor by using a high frequency (approximately 1000 Hz) induction heating process. The vessel, with its contents, was quickly quenched after heating and the contents were recovered. The gaseous products formed in the process were analyzed using gas chromatography.

The Reactor Vessel

The reactor vessel was a 3' long, carbon-steel cylinder tube with a 1 1/2" outer diameter and a 1/2" inner diameter. At one end of the vessel is a pressure gauge and a valve through which passes the gaseous products when they are collected. Two solid stainless-steel spacing rods were placed inside the reactor at each end. These rods are about 15" long and are nearly 1/2" in diameter. The remaining space located in the center of the vessel is the reaction region which has a volume of approximately 18 mL. The material that is to be pyrolyzed was placed in the reaction region. The reaction vessel is shown in Figure 1. A heating coil is arranged so that it encircles the mid-section of the vessel and the reaction region. The heating coil was supplied with electrical energy at about 1000 Hz, from an inverter system, for a short period of time (usually 60 seconds), at a constant rate of electrical input, at a line current of 80 Amps. The thermal input to the constant volume reactor and the presence of water causes a rapid buildup of pressure in the vessel. The process was characterized by a rapid pressure "kick" in which the pressure increased from about 2000 PSIG to pressures above 10,000 PSIG within 10–15 seconds. When the pressure ceased to increase, the reactor vessel was quickly removed from the heating coil and was quenched in a large container of cold water.

Gas Collection System

The device used for the collection of gases consists of a one-gallon capacity stainless-steel cylinder, a vacuum/pressure gauge, a glass trap bottle to collect liquids and suspended solids, and valves, tubing, and fittings (Figure 2). The collection device was prepared for gaseous product collection by evacuation of the cylinder with a vacuum pump and recording of the initial gauge pressure. The reactor vessel, which contained pressurized gases that were to be analyzed, was attached to the collection device. The gases were released into the collection system and any liquids or suspended solids were collected in the bottle trap. Once the transfer of gas was completed, the final gauge pressure was recorded. With the measured change in pressure, an estimate of the volume of gas collected was made. At the gauge end of the gas collection system is a three-way valve and a sampling port from which samples were taken using a hypodermic syringe.

Gas Chromatograph

The analysis of the composition of the gases produced in the aqueous pyrolysis reaction was made using a Perkin-Elmer SIGMA-3 gas chromatograph. This instrument was comprised of a column oven, a Supleco stainless-steel general configuration packed column with Porapak S 100/120 packing material, a hot-wire thermal conductivity detector, a closed loop temperature control by keyboard input, a Perkin-Elmer SIGMA-10 Chromatography Data Station with a printer/plotter for recording the analysis results, and interface and control equipment.

A gas sample was injected into an injection port on the gas chromatograph analyzer. Using helium as a carrier gas, the gases to be analyzed are transported through the heated coil-shaped column where different gaseous materials are separated, which enables the detection of the gases at different times. The results were compared to previously calibrated data for individual control gases that may be found in the samples analyzed.

In the investigation of the destruction of phenols by aqueous pyrolysis, a Perkin-Elmer SIGMA 3B gas chromatograph was used. This instrument consisted of a column oven, a metal packed column (Supleco SP-1240-DA) which separates the priority pollutant phenols, a flame ionization detector (FID) which uses hydrogen (H_2) as the combustion gas, a Perkin-Elmer SIGMA-15 Chromatography Data Station with a plotter/printer, a closed loop temperature control by keyboard, interface, and control equipment, and uses nitrogen (N_2) as the carrier gas.

Processing of Test Materials

The industrial waste test material required some preparation before the aqueous pyrolysis process could be performed. This preparation consisted of making a mixture of the material and water that will produce desirable results and packing the reactor vessel with this mixture. The weight of the solid fraction of the material was measured so that it could be compared to the amount of gas produced. Water was added to most of the materials except for those that already had a high water content such as wastewater sludges. The wastewater sludges tested were dewatered by different amounts to investigate the results obtained due to the varying water contents. The packing of the test material must be done such that air is not trapped inside of the reaction region to ensure that there will be a constant volume reaction. The ends of the reactor vessel were attached and fastened tightly to prevent any unwanted emissions during and after the reaction process. During the induction heating of the vesssel, the pressure change was observed to see if there was a rapid buildup of the pressure, which usually occurs between 45 and 60 seconds into the process. This indicates that a pyrolysis reaction had occurred. After the reaction and collection of the gases produced, the remaining liquid and solid products were recovered for further analysis.

Analysis of the recovered solids consisted of a measure of the percent reduction of the volatile solid content of the material. Volatile solids reduction was determined by taking measured dry amounts of a test material before and after the aqueous pyrolysis reaction and combusting the samples in an oven at 600°C. At this temperature, the volatile organic matter was removed. The difference between the dry weight and the weight of the fixed solids after combustion is the weight of the volatile solids. The values obtained for a material before and after aqueous pyrolysis were compared to determine the extent of the reduction of volatile solids in this process.

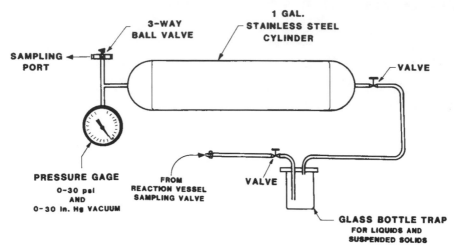

Figure 2.   The gas collection system.

RESULTS

Industrial wastes and other solid wastes were processed using the aqueous pyrolysis method and each were analyzed for the amount of gas produced, the composition of the gases produced, and the reduction of volatile solids. The materials consisted of newsprint, wheat dust, rubber, phenol and dichlorophenol on newsprint, and wastewater sludges of varying amounts of water content. The following symbols were used to identify the waste materials:

NP   – Newsprint
WD   – Wheat dust
R   – Rubber
NP-P – Newsprint with Phenols
WWS – Wastewater sludges

The amount of gas produced was determined by calculating the pressure difference (in Atm) that occurs in the gas collection system when the gases are collected and by comparing this pressure to the volume of the gas collection system at 1 Atm. The volume of the collection system (shown in Figure 2) was estimated as follows:

1 gallon cylinder	– 3785 mL
Glass bottle trap	– 147 mL
Pressure gauge, valves tubes, fittings, etc.	– 14 mL
Total (estimated)	3946 mL

The initial and final gauge pressures were measured in inches of mercury (in. of Hg) and were converted to absolute pressure in units of atmospheres where –29.92 in of Hg is equal to 0 Atm.

The following example shows how the volume of gas produced from an aqueous pyrolysis of a rubber waste was determined.

Initial pressure	= –19.2 in. of Hg
	= 0.3583 Atm. abs.
Final pressure	= –8.4 in. of Hg
	= 0.7193 Atm. abs.
Net pressure change	= 0.361 Atm. abs.
Volume of gas collected (at 1.0 Atm. pressure)	= 3946 mL/Atm. x 0.361 Atm.
	= 1425 mL

Table I. Results of Gas Production and Volatile Solids Reduction

Material	Dry Input Weight, gm	Volume of Gas Produced, mL	Volatile Solids Before	Volatile Solids After	Volatile Solids Reduction
Newsprint (NP)	4.30	371	99%	85%	15%
Wheat Dust (WD)	7.73	1184	87%	42%	52%
Rubber (R)	9.37	1425	44%	26%	40%
Newsprint w/Phenols (NP–P)	4.20	686	99%	65%	34%
Wastewater Sludges					
6% Solids (WWS–1)	1.14	63	72%	62%	14%
12% Solids (WWS–2)	2.14	118	73%	55%	25%
13% Solids (WWS–3)	2.30	710	73%	39%	46%
32% Solids (WWS–4)	5.59	276	73%	65%	11%

Comparison of Input Mass and Volume of Gas Produced

The results of the volume of gas that was produced and the amount of dry input material is given in Table I and Figures 3 and 4. The aqueous pyrolysis process produced 86 mL of gas per gram of newsprint, 153 mL per gram of wheat dust, 152 mL per gram of rubber, and 163 mL of gas per gram of newsprint with phenols. The production of gases using aqueous pyrolysis of wastewater sludges was nearly the same for three of the solid content amounts tested with 49 to 55 mL of gas being produced per gram of dry material. The 13% solids sludge produced 701 mL of gas from 2.3 grams of material which is 309 mL/gm.

Volatile Solids Reduction

The reduction of the organic fraction of the waste materials, the volatile solids, was investigated as a possible benefit of the aqueous pyrolysis process. Results are given in Table I and shown in Figures 5, 6, and 7. The volatile solids of the wheat dust were reduced by 52% and the volatile solids of the rubber waste were reduced by 40%. The newsprint alone had a reduction of volatile solids of 15%, but with phenol and dichlorophenol added to the newsprint the reduction was 34%. Reduction of the

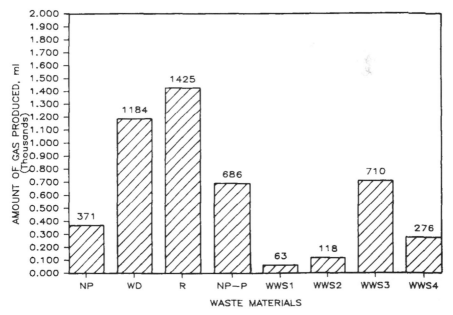

Figure 3.   Amount of gas produced by aqueous pyrolysis.

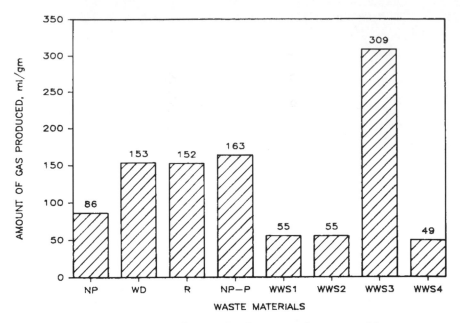

Figure 4.   Amount of gas produced per gram of waste material.

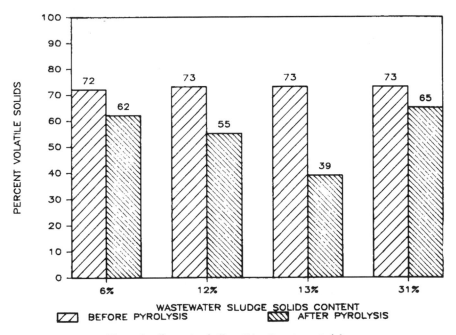

Figure 5.   Percent volatile solids of waste materials.

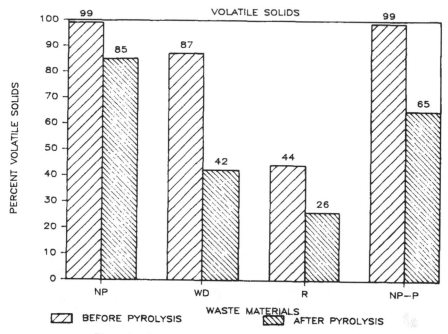

Figure 6.   Percent volatile solids of wastewater sludges.

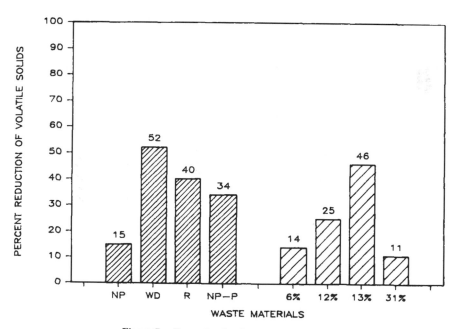

Figure 7.   Percent reduction of volatile solids.

Table II. Composition of Gas Produced

Material	Methane	Carbon Dioxide	Ethane	Ethylene
Newsprint	19%	76%	5%	0%
Wheat Dust	41%	46%	13%	0%
Rubber	69%	5%	24%	1%
Newsprint w/Phenols	0%	100%	0%	0%
Wastewater Sludges				
6% Solids	0%	94%	6%	0%
12% Solids	0%	75%	25%	0%
13% Solids	0%	70%	30%	0%
31% Solids	0%	90%	10%	0%

volatile solids results in a reduction of the overall volume of the waste material so there may be less material which will have to be disposed of. The reduction of volatile solids in the wastewater sludges was dependent upon the water content of the material. The reduction of volatile solids increased as the water content was decreased in the higher water content sludges. However, the volatile solids reduction decreased as the water content decreased 68%.

The sludges having solid contents of 6%, 12%, and 13% had volatile solid reductions of 14%, 25%, and 46%, respectively. The sludge that was dewatered to 32% solids had volatile solids reduced by only 11%. For good reduction of volatile solids, there must be a constant volume reaction in the aqueous pyrolysis process and a rapid pressure buildup in the reactor. Difficulties can be experienced if air is allowed to remain in the reactor because the volume will not be constant and high pressures may not be achieved. Also, if the amount of water used in the aqueous pyrolysis process is too much or too little, there may not be a good reaction.

Composition of Gases Produced

The composition of the gases produced in the process were determined using a gas chromatograph and comparing the results to standard curves developed for several gases. The gases that are most likely to be produced are methane, carbon dioxide, ethane, ethylene, and possibly some small amounts of other higher hydrocarbons. Air was detected in the analysis but was neglected because it existed in the gas collection system prior to the collection of the gases. The air could be removed by evacuating the gas collection system to an absolute vacuum (0 Atm abs.), which would have been difficult, or by replacing the air with helium since helium is the carrier gas used by the gas chromatograph and would not have been detected during the analysis. The results of the gas composition analysis are given in Table II and are shown in Figures 8 and 9. The waste materials produced carbon dioxide and hydrocarbon gases when pyrolyzed. These hydrocarbon gases could possibly be used as a fuel source. The rubber waste had the best results with 95% of the gases produced consisting of hydrocarbons. The composition was 69% methane, 24% ethane, 1% ethylene, and the remaining 5% was carbon dioxide. The wheat dust produced gases that composed of 54% hydrocarbon gases, with 41% methane, 13% ethane, and 46% carbon dioxide. The newsprint produced 76% carbon dioxide, 19% methane, and 5% ethane. The aqueous pyrolysis of the newsprint with phenol and dichlorophenol resulted in the production of a gas consisting only of carbon dioxide. The wastewater sludges produced gases containing carbon dioxide and ethane. The content of ethane of the different solid content sludges correlates with the reduction of volatile solids. The content of ethane of the 6%, 12%, and 13% solids sludge was 6%, 25%, and 30%, respectively, while the ethane produced from the 32% solids sludge was 10% of the gas produced. Only the pyrolysis of the rubber waste produced ethylene gas.

Destruction of Phenol and Dichlorophenol

The aqueous pyrolysis process was investigated as a method for the destruction of toxic pollutants such as phenol and dichlorophenol. The phenols were recovered from the newsprint and water mixture and also from the liquid and solid products after pyrolysis by a microextraction method for analysis by gas chromatography. In this method, sodium chloride is added to the sample, and phosphoric acid is used to lower the pH to 2. Then isopropyl ether is added and the contents are well mixed. The phenol and dichlorophenol are then separated from the mixture by the isopropyl ether

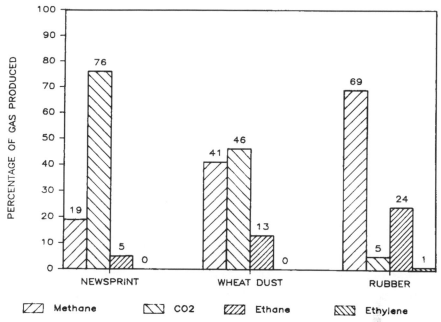

Figure 8.   Composition of gases produced from waste materials.

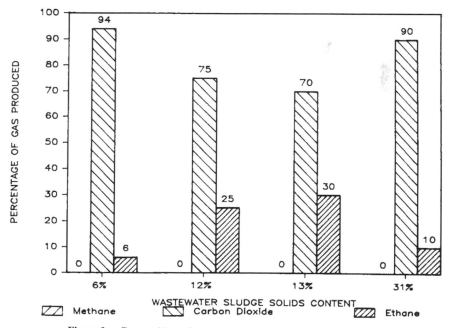

Figure 9.   Composition of gases produced from wastewater sludges.

which floats on top of the mixture. In the chromatography analysis it was found that aqueous pyrolysis did destruct the phenol and dichlorophenol, but there were many other compounds produced as a result of the reaction. These byproducts have not been identified.

DISCUSSION

The aqueous pyrolysis process is a promising method for the treatment and volume reduction of industrial and other solid wastes but needs further investigation. The objective of this study was to investigate the use of the method as a means for processing waste materials so that they can be disposed of in an acceptable and economical manner. This process of pyrolysis has the benefits of volume reduction, volatile solids reduction, and the production of hydrocarbon gases which can be used as a fuel. This process can also convert toxic pollutants into other materials, but this aspect of the study needs further study to determine what are the products that are formed. Aqueous pyrolysis of wastewater sludges produces a safer waste product by breaking down the material into less harmful components. It reduces the health hazard of sludges by eliminating pathogenic organisms. The solid residue byproducts of the aqueous pyrolysis process could possibly be used as another source of fuel, and further analysis of the energy content of the residues are needed. An economic evaluation of this aqueous pyrolysis process has not been made. The feasibility of this process was the purpose in this study. Future work on aqueous pyrolysis will possibly investigate the use of a semi-continuous flow process incorporating heat recovery. At this time, an economic evaluation of the process would be made. The factor which could enhance the production of hydrocarbon gases needs to be studied further. This would involve operational procedures such as the method of packing the reactor, the amount of water used, and the duration of heating and cooling.

Preliminary studies using aqueous pyrolysis on phenols and dichlorophenol along with previous work on other wastes, such as nuclear wastes, show that these types of hazardous wastes may be satisfactorily treated by this process.

In this investigation of aqueous pyrolysis, it was intended to determine the caloric content of the waste material before and after pyrolysis but the non-delivery of a calorimeter made this impossible. This measurement is intended for future study.

REFERENCES

1. Hughes, W. L., and Ramakumar, R. G., "An Experimental Investigation of the Aqueous Pyrolysis of Biomass," *Proceedings of the 1979 Frontiers of Power Technology Conference*, II-1 to II-8, Stillwater, Oklahoma (September 1979).
2. Bohn, M. S., and Benham, C. B., "Biomass Pyrolysis with an Entrained Flow Reactor," *Industrial Engineering Chemistry Process Design and Development*, Vol. 23, 355–363 (April 1984).
3. Kemmler, R., and Schlich, E., "Pilot-Scale Testing of Pyrolysis for the Reduction of Organic Waste," *Nuclear Technology*, Vol. 59, 321–326 (November 1982).
4. Irwin, W. I., "Analytical Pyrolysis, a Comprehensive Guide," *Chromatographic Science Series*, Vol. 22 (1982).
5. "Principles of Transverse Flux Induction Heating," *Light Metal Age*, Vol. 40, 6–9 (December, 1982).

33 BIODEGRADATION OF PENTACHLOROPHENOL IN SOIL ENVIRONMENTS

Chung J. Kim, Graduate Student

Walter J. Maier, Professor
Civil-Environmental Engineering Department
University of Minnesota
Minneapolis, MN 55455

INTRODUCTION

Pentachlorophenol has many beneficial uses as a wood preservative and insecticide. Unfortunately, improper disposal has resulted in contamination of soils and groundwater at numerous locations. Because it is potentially toxic, cleanup has been mandated by regulatory agencies. The costs of cleaning contaminated soils and aquifers by physical-chemical methods are very high and there is interest in finding alternative methods. Engineered systems for biological treatment of soils and groundwater appear to be a promising alternative. It has been estimated that the cost of cleanup using biological treatment would be less than 10% of physical-chemical cleanup costs [1].

Biological treatment of pentachlorophenol containing wastewaters has been reported by several investigators [2–8]. Complete biodegradation using pure cultures and mixed cultures has been demonstrated. Stanlake and Finn [9] isolated and characterized an Arthrobacter sp. capable of utilizing PCP. The specific growth rate of their culture increased with increasing PCP concentration to a maximum of 0.1 hr^{-1} at 130 mg/L but decreased significantly at higher concentration, 0.05 hr^{-1} at 300 mg/L. Crawford et al. [10] isolated a Flavobacterium capable of biodegrading PCP as a sole carbon source or in mixtures with conventional substrates. They report that glutamate grown inoculum capability for PCP biodegradation can be induced by short exposure to PCP. Moos et al. [6] showed that PCP removal in continuous flow activated sludge pilot plants was enhanced by concurrent feeding of conventional organic substrates. Fluctuation in effluent quality was ascribed to substrate inhibition effects. However, Guo et al. [8] reported that activated sludge treatment of PCP containing wastewaters was unreliable. Klecka and Maier [11] isolated mixed cultures growing on PCP as the only source of carbon and energy from industrial wastewater by continuous flow enrichment in chemostats. Kinetic parameters determined in batch experiments showed low cell mass yields and growth rate coefficients, 0.136 g/g and 0.074 hr^{-1}, respectively. The cultures had a high affinity for PCP, as evidenced a low Monod coefficient, $K_s = 0.06$ mg/L. Significant substrate inhibition effects were noted at PCP concentrations above 1 mg/L. Substrate inhibition effects are not uncommon with industrial chemicals and phenolic compounds specifically [12,13].

On the basis of the information presented in the literature, it was concluded that biological treatment for removal of PCP from groundwater and soils should be feasible. However, further testing is needed to define rates and limitations of microbial degradation processes in soils and groundwater. More specifically there is need for engineering oriented studies to describe kinetic parameters that prevail in the presence of soil surface and to establish the effects of transport and availability of nutrients and oxygen. This chapter describes the initial phase of such studies. The research focused on low organic content sandstone material as a model system of subsoil materials.

MATERIALS AND METHODS

PCP Acclimated Enrichment Culture

A mixed culture capable of utilizing PCP as sole carbon and energy source was isolated from activated sludge obtained from the Minneapolis, St. Paul Sewage Treatment Plant. Sludge solids were inoculated into minimal salts medium containing 100 mg/L PCP and incubated in shake flasks. PCP concentration was monitored by UV absorbance at 380 nm. There was an initial lag period of 6–7 days, but essentially all PCP had disappeared in 10 days. The shake flask culture was expanded in chemostats fed with 100 mg/L PCP minimal salts medium at a dilution ratio of 0.25 days^{-1}. PCP was

Table I. Composition of Jordan Aquifer Sand[a]

SiO_2	99.62 wt%	Al_2O_3	0.04 wt%
Fe_2O_3	0.053	CaO	0.014
MgO	0.003	Na_2O	0.01
K_2O	0.01	TiO_2	0.01
MnO	0.001	SrO	<0.01
BaO	<0.01	Carbon	<0.01

[a]Sand was obtained from Frac Sand Co., Jordan, Minnesota

essentially completely removed. Wall growth was observed so that dilution ratio is not a precise measure of specific growth rate coefficient. The chemostat was initially operated for 30 days to obtain a stable and well established culture. Biomass from the chemostat was used as a source of inoculum as described below.

Analytical reagent grade chemicals were used in preparing the medium with the following composition: K_2HPO_4 500 mg/L, $MgSO_4$ 50 mg/L, NH_4NO_3 50 mg/L, plus 50 mL tap water and 950 mL deionized water. The medium was sterilized by filtration through 0.2 μm nucleopore filter. Ninety-nine percent purity PCP was obtained from Sigma Chemical Co.

Soil Materials

Crushed and size fractionated sandstone from the Jordan formation, which underlies Central Minnesota, was used. Jordan sandstone is accessible at an outcrop near Jordan, Minnesota, where it is mined as a commercial source of silica sand. It is composed of essentially pure silica as shown in Table I. Its low organic carbon content makes it more representative of subsoil materials as opposed to surface soils. Jordan sandstone consists of uniform well rounded sand grains that are loosely bonded and easily separated by crushing; over 98% of the material is in the #20–#100 sieve mesh size range, which corresponds to a size range of 0.84–0.14 mm. Smaller mesh size sand was prepared by grinding Jordan sand in a pebble mill, followed by sieve fractionation. As shown in Table II, Fraction A represents material passing #80 sieve and retained on #200 sieve; Fraction B is the material retained on a #500 sieve after washing; Fraction C represents the fines collected by water washing through the #500 screen. Nominal particle size ranges in Table II represent the sieve size openings for 10% and 90% passing, respectively. Fraction C material is in the silt and clay size range; Fraction B consists of silt and fine sand. One significant feature is that the measured quantity of fine material (passing #200 mesh) differs depending on whether the sieving is done by washing or by dry sieving. The Jordan sand showed 0.53 and 0.12 wt% passing #200 sieve (0.075 mm), respectively. This difference is due to sorption of small particles on large sand grains in dry sieving, whereas wet sieving suspends small particles and washes them away.

Adsorption of PCP

PCP is soluble as the phenolate ion but is only slightly soluble below its pKa value of 4.7. The media were buffered above pH 7.2. Adsorption of PCP on sand fractions was tested at three concentration levels and a range of particle concentrations shown in Figure 1. Sand was contacted with PCP solution in shake flasks for several days. PCP concentration in solution was measured before and after

Table II. Size Fractioned Jordan Aquifer Sand

Fraction	Jordan Sand	A[a]	B[a]	C[a]
Sieve Size Range	20–100	80–200	200–500	<500
Sieving		dry	wet	wet
Nominal Size Range, mm	0.14–0.84	0.1–0.2	.03–0.1	.003–.035
Weight % of fines[b]				
Dry wt%	0.12	—	—	—
Wet wt%	0.52	—	24	100

[a]Fractions A, ,B, C prepared by grinding and sieving.
[b]Wt% of material less than 0.075 mm measured by dry sieving or wet sieving through a #200 sieve.

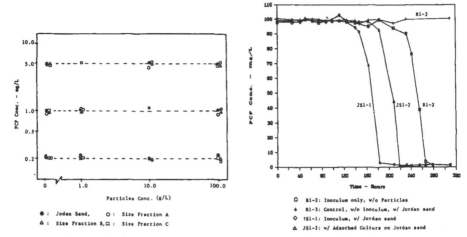

Figure 1. PCP adsorption on particles in a wide range of concentration.

● : Jodan Sand,   ○ : Size Fraction A
△ : Size Fraction B, □ : Size Fraction C

Figure 2. Series 1 batch test. PCP biodegradation in the presence of Jordan sand (20 g/L).

□ R1-2: Inoculum only, w/o Particles
+ R1-3: Control, w/o Inoculum, w/ Jordan sand
◇ JS1-1: Inoculum, w/ Jordan sand
△ JS1-2: w/ Adsorbed Culture on Jordan sand

contacting. As shown in Figure 1, PCP solution concentration was not affected by the addition of up to 100 g/L sand, indicating that PCP was not adsorbed in measurable quantities.

PCP Removal in Shake Flasks

Batch reactor experiments were carried out to measure the effects of sand on PCP removal. PCP medium was contacted with sand fractions and inoculum from the chemostat in flasks that were agitated on a shaker. The shaker action ensures thorough mixing, contacting of the sand and feed solution, and facilitates transport of oxygen into solution. Removal of PCP was monitored by sampling periodically; samples were filtered through 0.2 μm nucleopore membrane filters prior to measuring UV absorbance to determine PCP concentration. 5 mL of chemostat effluent was added as inoculum to 500 mL of medium in each flask directly or by adsorbing the cells on the sand separately. In the latter procedure inoculating solution was added to the dry sand, excess inoculating liquid was drawn off, and the sand was washed with 100 mL of pure water to remove free cell mass. The inoculated sand was then added to the shake flask.

Addition of inoculum resulted in essentially complete removal of PCP in all tests, but an apparent lag phase was observed. As shown in Figure 2, there was a lag of 180 hours in the flask without sand. The presence of 20 g/L of Jordan sand shortened the apparent lag period, more so with direct addition of inoculum to the flask. No PCP removal was observed in the control flask with 20 g/L of sand but no inoculum. The corresponding set of experiments with 20 g/L of Fraction C sand are shown in Figure 3. Fraction C sand shortened the apparent lag time significantly regardless of how the inoculum was added.

Figures 4 and 5 present comparisons of the PCP decay curves for each of the four soil fractions; inoculum was added directly to the flasks in Figure 4; adsorbed inoculum was added in Figure 5. Fraction A sand gave results intermediate between Jordan sand and Fraction C whereas Fraction B showed the smallest effect on lag time. Using the PCP decay curve of the flask without sand addition as a point of reference, it appears that addition of sand and direct inoculation shortened the lag phase in all cases (Figure 4). However, addition of adsorbed inoculum resulted in smaller effects, and there was no beneficial effect with Fraction B sand (Figure 5). Adsorption of inoculating biomass followed by washing probably resulted in some loss of microbial cells. The same volume of inoculum was used in each test, but active cell mass added to each flask depends on the adsorption and retention characteristics of the sand. It appears that sand Fraction B is less effective because it does not adsorb or retain cells as strongly as the other sands. By contrast, Fraction C test data indicate that essentially all the cells are sorbed on this sand fraction.

As a further check on the effect of sand addition, a series of flasks with different amounts of Jordan sand (inoculum added directly to the flask) were carried out. The results show that increasing sand concentration resulted in progressively shorter apparent lag times (Figure 6).

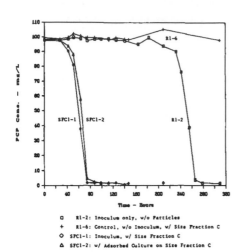

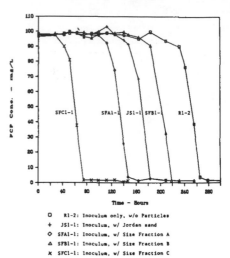

R1-2: Inoculum only, w/o Particles
R1-6: Control, w/o Inoculum, w/ Size Fraction C
SFC1-1: Inoculum, w/ Size Fraction C
SFC1-2: w/ Adsorbed Culture on Size Fraction C

R1-2: Inoculum only, w/o Particles
JS1-1: Inoculum, w/ Jordan sand
SFA1-1: Inoculum, w/ Size Fraction A
SFB1-1: Inoculum, w/ Size Fraction B
SFC1-1: Inoculum, w/ Size Fraction C

Figure 3. Series 1 batch test. PCP biodegradation in the presence of size fraction C (20
g/L).

Figure 4. Series 1 batch test. PCP biodegradation in the presence of particles (20 g/L).

DISCUSSION

The apparent lag period observed at the beginning of the 100 mg/L PCP batch test without sand is characteristic of a substrate inhibited system. Several investigators have reported that PCP is inhibitory at high concentrations [6,9,11]. Inhibition results in slow rates of growth and slow (essentially nondetectable) rates of substrate removal which manifests itself as an apparent lag phase [11–14]. The dramatic reduction in apparent lag time suggests that sand protects the cells from substrate inhibition effects. Adsorption of PCP on the sand is ruled out as an explanation as evidenced by the data in Figure 1 showing that PCP adsorption is negligible. By contrast, Fraction C sand gave essentially

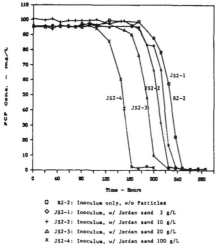

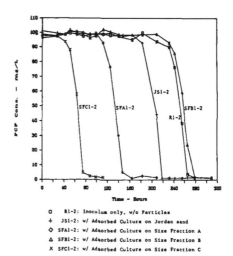

R2-2: Inoculum only, w/o Particles
JS2-1: Inoculum, w/ Jordan sand 2 g/L
JS2-2: Inoculum, w/ Jordan sand 10 g/L
JS2-3: Inoculum, w/ Jordan sand 20 g/L
JS2-4: Inoculum, w/ Jordan sand 100 g/L

R1-2: Inoculum only, w/o Particles
JS1-2: w/ Adsorbed Culture on Jordan sand
SFA1-2: w/ Adsorbed Culture on Size Fraction A
SFB1-2: w/ Adsorbed Culture on Size Fraction B
SFC1-2: w/ Adsorbed Culture on Size Fraction C

Figure 5. Series 1 batch test. PCP biodegradation with absorbed culture on particles.

Figure 6. Series 2 batch test. PCP biodegradation in the presence of different concentrations of Jordan sand.

Table III. Distribution of Small Grain Size Materials in Different Sand Fractions

Size Range, μm	Jordan Sand wt%	Fraction B wt%	Fraction C wt%
0–10	0.14	0	47.5
10–25	0.00	7.2	38.5
25–50	0.05	49.6	14.0

complete sorption of cell mass; it also exhibited the largest reduction in lag phase. It can therefore be inferred that surface interactions between cells and sand grains protect the cell from substrate inhibition effects.

Sand particle size is clearly a significant variable as shown by the fact that Fraction C was more effective than Jordan Sand. However, Fraction B, which is significantly smaller than Jordan sand, was not as effective. It, therefore, appears that the beneficial effects of sand are not correlated with average particle size or total surface.

Hydrometer tests were used to measure the size distribution of small grain size materials. As shown in Table III, Fraction B has essentially no particles below 10 μm because it was prepared by wet sieving. Jordan sand contains significant amounts of less than 10 μm particles in association with the large particles. Fraction C has the highest concentration; it also gave the largest beneficial effect per gram of sand added.

The beneficial effects of particles in the 0–10 μm range was qualitatively confirmed in subsequent experiments using reagent grade 2–10 μm silica particles. By contrast, addition of kaolinite and montmorillonite clay (<2 μm particle size) were not effective and actually retarded PCP removal at 10 g/L levels.

Sand Column Tests

Continuous flow short column tests were carried out to measure rates of transport and removal of PCP in physical environments where the sand grains are stationary, but water, solutes, and microbial cells are free to move. The initial idea was to use shallow depths of sand in order to study performance in a vertically homogenous bed. However, the concept of operating the columns as differential reactors with uniform vertical distribution of active cell mass and minimal concentration gradients did not work out because cell mass sorption results in significant gradients. Short column tests charged with approximately 100 g of sand in a 3.3 cm diameter tube giving a sand depth of approximately 6 cm were, therefore, carried out in order to obtain quantitative descriptions of transport and removal rates based on inlet and outlet measurements. Short column test apparatus are illustrated in Figure 7. Three modes of flow were investigated, saturated upflow, saturated downflow, and unsaturated downflow. The column without sand was operated as a continuous flow reactor without agitation; its liquid level was maintained to approximate the hydraulic detention time in the sand columns.

Solute Transport

Mathematical model descriptions of the movement of solute through a sand column are usually presented with reference to a simplified physical model representing a homogenous aggregation of spherical particle [15–17]. Feed solution is applied at a constant rate, upflow or downflow. The models account for transport of solute by convection, hydrodynamic dispersion, and molecular diffusion. Dispersion and diffusion phenomena depend on concentration gradients; convective transport is actuated by bulk fluid phase flow. It has been shown that transport processes are affected by flow rates, local concentration gradients, soil size characteristics, and soil moisture content. Detailed mechanistic aspects of solute transport will not be discussed in this paper but have been discussed elsewhere [18]. However, the phenomenology of solute transport as it applies to the short column tests will be described briefly in order to set the stage for the subsequent description and discussion of PCP removal by microbial degradation.

Transport of a soluble, and thought to be nonadsorbing, conservative dye (Rhodamine B) was measured in short sand columns. A continuous flow of dye solution was added to a saturated flow column that had been previously saturated with a flow of pure water. A similar experiment was carried out with an unsaturated column that had been equilibrated at low flow conditions. Unsaturation was induced by attaching an appropriate length of capillary suction line at the outlet as shown in Figure 7. Dye concentration in the effluent was monitored by measuring absorbance. Absorbance measurements expressed as the ratio of effluent to feed are shown in Figure 8 as a function of the cumulative volume of solution applied, expressed as total bed pore volume equivalents. Traces of dye appeared in the effluents after passage of 0.5 pore volumes. Dye concentration ratio was approximately 0.5 after passage of 1 pore volume through the saturated column. The saturated column elution curve fits the classical description of movement of a nonadsorbing, conservative solute [18]. The fact that the absorbance ratio of 0.5 occurred at one pore volume displacement indicates that adsorption was not significant [18]. Tailing indicates dispersion. The elution pattern from the unsaturated column is distinctly different both in time and shape. The 0.5 absorbance ratio occurs at 1.4 pore volumes. Furthermore, the tailing effect at the end is much more pronounced. Thus overall residence time of solute appears to be longer in the saturated flow column. The more gradual approach to complete breakthrough suggests that adsorption was a significant factor in the unsaturated flow column. This anomaly is being investigated.

PCP Removal in Short Columns

Short column continuous flow experiments were carried out to measure PCP removal at different flow rates, upflow, downflow, saturated, and unsaturated flow conditions. Approximately 25 mL of chemostat effluent was added as inoculum by layering on top of the column or by mixing with the sand before packing the column; Jordan sand was used. Effluent PCP concentration was monitored.

Typical short column test results, effluent PCP concentration as a function of time, are shown in Figure 9. The unsaturated downflow column gave essentially complete PCP removal after 6 days of continuous flow of 100 mg/L PCP feed at 0.04 mL/min (loading rate of 6.7 cm^3/cm^2/day). By contrast, the saturated flow column did not stabilize nor was PCP removal as efficient. The feed solutions had been presaturated with pure oxygen to increase dissolved oxygen to 30–36 mg/L. The results without any sand are similar to those observed in the saturated soil column. The control sand column without inoculum gave the expected increase in effluent concentration that is indicative of no biological or chemical removal of PCP.

Tests carried out at a lower flow rate of 0.02 mL/min (loading rate of 3.4 m^3/m^2/d) gave similar results. Unsaturated flow conditions gave significantly better removal of PCP (Figure 10).

Column test results using a lower feed concentration, 50 mg/L PCP, but higher flow rate 0.1 mg/min (16.8 m^3/m^2/d) are shown in Figure 11. Operation of the saturated flow column with high DO feed concentrations (presaturated with pure oxygen) improved with time but was not as effective as the unsaturated column operating at the same feed flow rate but lower DO. Comparison of Reactors A and C show that higher DO is beneficial but not as effective as unsaturated flow.

Short Column Test Results

Soil moisture content has long been recognized as an important variable affecting flow of water and transport of solutes through soil columns [17,19,20]. Pressure head, moisture content, and hydraulic conductivity are closely coupled [17]. Hysteretic effects, evidenced by different pressure heads at the same overall moisture content depending on whether the soil is becoming dryer or wetter, show that distribution of water in pore spaces is not a unique function of pressure head. It follows that movement of water and tracers through the column cannot be described by simple convection based models using constant values of permeability and dispersion.

Recent developments in modeling flow and solute transport in unsaturated soils show two trends. One approach is to adapt saturated flow models by incorporating spatial and time variable values of porosity, moisture content, permeability, dispersion-diffusion coefficients, storage-retardation coefficients, and hydraulic conductivity [21] into the basic equations of flow and mass transport. The second approach is to describe flow-transport in terms of physical models that explicitly account for nonhomogeneity. For example, by assuming mobile-immobile water phases [22], or movement of parcels of water [23], or by considering macropore-mesopore flow effects [24], or by using transfer functions to characterize the effects of soil induced variations in transport [25]. Our work has not progressed to the stage where we can identify the best model to correlate our column test results. Nevertheless, reference to literature models provides insight and qualitative understanding of the test results.

Wierenga's [22] model of mobile and immobile water phases assumes that some of the applied water moves through the column without interacting with interstitial soil water. This implies the presence of large, continuous surface, connected pores that serve as conduits for bulk flow. This concept is illustrated in Figure 12. Interconnected large pores serve as the major conduit for bulk flow when the

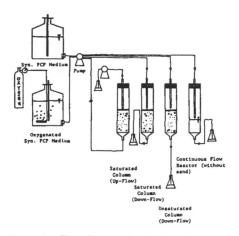

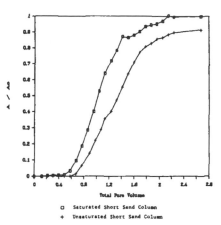

Figure 7.   Flow diagram for short sand column tests.

Figure 8.   Dye transport in a saturated and an unsaturated column. Influent dye absorbance (A_0) = 1.84 at 554 nm.

in the large pores that block flow. As a result, water flows along surfaces and through the fine pore structure.

The more rapid passage of dye through the saturated column observed in Figure 8 is consistent with the concept of flow through large pore conduits. By contrast, dye movement through the unsaturated column is circuitous and retarded. For saturated flow, Freeze [17] has shown that nonreactive, nonadsorptive solutes typically give S shaped, equal tailed, solute breakthrough curves that are centered on the point of arrival of water traveling at the average flow velocity. The shape of the curve. e.g., first appearance and tailing effects, is a measure of dispersion. Displacement of the elution curve in saturated flow systems is usually ascribed to retardation of solute movement due to interaction with surfaces. However, retardation in the unsaturated column (Figure 8) must be the result of differences in flow regime, namely, flow through small pores and surface flow which in turn affect retention in surfaces. Dispersion effects are also more significant and cause the elution curve to spread out. It, therefore, appears that solute residence time in the unsaturated column is longer. Furthermore, the unsaturated flow regime is expected to give more intimate contact between sand surface and the percolating solution and hence greater potential for solute removal by surface adsorbed microorganisms.

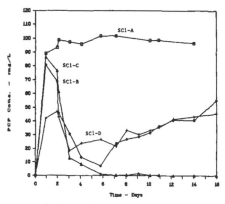

SCl-A: Saturated Column, Control
SCl-B: Saturated Column, Down-flow
SCl-C: Unsaturated Column, Down-flow
SCl-D: Column w/o Sand, Up-flow

Figure 9.   Series 1 short column test. PCP removal in short sand columns (Influent = 100   mg/L   of   oxygenated   PCP medium; flow rate = 0.04 mL/min.).

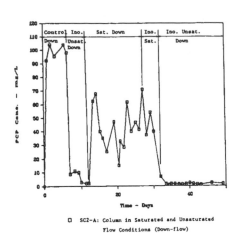

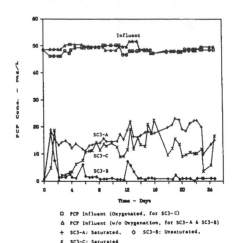

Figure 10. Series 2 short column test. PCP removal by column SC2-A in various flow conditions. Influent = 100 mg/L of oxygenated PCP medium; flow rate = 0.02 mL/min.

Figure 11. Series 3 short column test. PCP removal by saturated and unsaturated short columns. Influent = 50 mg/L of PCP medium with or without oxygenation; flow rate = 0.1 mL/min.

Unsaturated flow columns gave consistently better removal of PCP than saturated columns. The presence of air bubbles is an advantage because it serves as an additional source of molecular oxygen. Biodegradation of PCP requires approximately 0.5 mg/L of oxygen per mg/L of PCP removed. High oxygen demand could, therefore, limit removal at high PCP concentrations. However, data in Figure 11 suggest that there is an intrinsic advantage for PCP removal in the unsaturated column. Column C was fed 50 mg/L PCP solution with 30–36 mg/L DO and had ample oxygen. Measured DO concentrations in the effluent were above 4 mg/L which confirms the presence of excess DO. It, therefore, appears that more effective PCP removal in unsaturated columns is related to flow and transport properties.

Electron microscope examination of sand grains taken from PCP treated columns shows that most bacteria colonize depressions and crevices rather than exposed surfaces on sand grains. The previously mentioned observation that inoculating bacterial cultures were more completely adsorbed on small sand particles than on large particles (Figures 2 and 3) is also indicative of a tendency for bacteria to accumulate in crevices or aggregations of small particles as opposed to the surfaces exposed in large pores. The picture that emerges from these considerations is that unsaturated flow conditions are more favorable for two reasons, namely, better contacting of solute with adsorbed biomass and oxygen availability.

CONCLUSIONS

Biodegradation of pentachlorophenol by acclimated organisms in the presence of soil particles is feasible. Kinetics of biodegradation are enhanced by the presence of small grain sand as evidenced by shorter lag times when treating high concentrations of PCP. Adsorption of PCP on very low organic

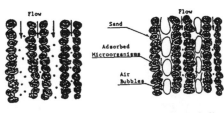

Figure 12. Flow distribution for saturated and unsaturated columns.

content subsoil material is negligible and does not affect transport significantly. However, microorganisms are adsorbed, more so on small size sand grains. It appears that surface interactions between microorganisms and sand are responsible for shorter lag time.

Unsaturated flow conditions give better PCP removal than saturated flow systems. Greater availability of molecular oxygen is one advantage for unsaturated flow. However, flow regime per se is also significant. Unsaturated flow appears to give better contacting of moving solutes with adsorbed microbes and hence more effective PCP removal.

ACKNOWLEDGMENTS

This work was funded by grants from the Unites States Department of Interior through the Water Resources Research Center and the National Science Foundation.

REFERENCES

1. International Conference on New Frontiers for Hazardous Waste Management, Pittsburgh, Penn. (September 15-18, 1985).
2. Dust, J. V., W. S. Thompson, A. Shindala, and N. R. Facinques, "Chemical and Biological Treatment of Wastewater from the Wood Preserving Industry," *Proc. 26th Ind. Waste Conf.*, Purdue Univ. Eng. Ext. Serv. 140: 227-243 (1971).
3. Etzel, J. E., and E. J. Kirsch, "Biological Treatment of Contrived and Industrial Wastewater Containing Pentachlorophenol," *Dev. Ind. Microbiol.*, 16: 287-295 (1975).
4. Jank, B. E., and P. J. A. Fowlie, "Treatment of Wood Preserving Effluent Containing Pentachlorophenol by Activated Sludge and Carbon Adsorption," *Proceedings of the 35th Industrial Waste Conference*, Purdue University, Ann Arbor Science Publishers, Ann Arbor, Mich. (1981).
5. Kirsch, E. J., and J. E. Etzel, "Microbial Decomposition of Pentachlorophenol," *J. Water Pollut. Control Fed.*, 45: 359-364 (1973).
6. Moos, L. P., E. J. Kirsch, R. F. Wukasch, and C. P. L. Grady, Jr., "Pentachlorophenol Biodegradation. I. Aerobic," *Water Res.*, 17: 1575-1584 (1983).
7. White, J. T., "Treating Wood Preserving Plant Wastewater by Chemical and Biological Methods," Environmental Protection Agency Publ., *No. EPA-600/2-76-231*, National Technical Information Service, Springfield, Va. (1976).
8. Guo, P. H. M., Fowlie, P. J. A., Cairns, W. W., and B. E. Jank, "Activated Sludge and Activated Carbon Treatment of a Wood Preserving Effluent Containing PCP," *Technology Development Report EPS 4-WP-80-2*, Water Pollution Control Directorate, Environmental Protection Service, Canada (1980).
9. Stanlake, G. J., and R. K. Finn, "Isolation and Characterization of Pentachlorophenol-Degrading Bacterium," *Appl. Environ. Microbiol.*, 44: 1421-1427 (1982).
10. Pignatello, J. J., M. M. Martinson, J. G. Steiert, R. E. Carlson, and R. L. Crawford, "Biodegradation and Photolysis of Pentachlorophenol in Artificial Freshwater Streams," *Appl. Environ. Microbiol.*, 46: 1024-1031 (1983).
11. Klecka, G. M., and Maier, W. J., "Kinetics of Microbial Growth on Pentachlorophenol," *Appl. Environ. Microbiol.*, 49: 46-53, 1985.
12. Edwards, V. H., "The Influence of High Substrate Concentrations on Microbial Kinetics," *Biotechnol. Bioeng.*, 12: 679-712 (1970).
13. Gaudy, A. F., and Gaudy E. T., *Microbiology for Environmental Scientists and Engineers*, McGraw-Hill Co., New York (1980).
14. Andrews, J. F., "A Mathematical Model for the Continuous Culture of Microorganisms Utilizing Inhibitory Substrates, "*Biotechnol. Bioeng.*, 10: 707-723 (1968).
15. Fried, J. J., *Groundwater Pollution*, Elsevier Publish. Co., New York (1975).
16. Biggar, J. W. and Nielson, D. R., "Mechanisms of Chemical Movement in Soils," *Aerochemicals in Soils*, Edited by I. K. Iskander, John Wiley, New York (1981).
17. Freeze, A. R. and Cherry, J. A., *Groundwater*, Prentice Hall, New Jersey (1979).
18. Kim, C. J., "Biodegradation of Pentachlorophenol in Soils and Groundwater," *Ph.D. Thesis*, University of Minnesota (1986).
19. Smiles, D. E., and B. W. Gardiner, "Hydrodynamic Dispersion During Unsteady, Unsaturated Flow," *Soil Sci.*, Soc. Am. 46 (1982).
20. Bresler, E., "Simultaneous Transport of Solutes and Water Under Transient Unsaturated Flow Conditions," *Water Res.*, 9:4, 975 (1973).
21. Lapalla, E. G., "Recent Developments in Modeling Variably Saturated Flow and Transport," *Symposium on Unsaturated Flow and Transport Modeling*, Edited by Arnold, A. M., Gee,

G. W., and Nelson, R. W. Pacific Northwest Laboratory, U.S. Nuclear Regulatory Commission, Report #NUREG/CP 0030, PNL.-SA-10325, Richland, Wash., 99352 (1982).

22. Wicrenga, P. J., *Solute Transport through Soils*, Edited by Arnold, A. M., Gee, G. W., and Nelson, R. W. Pacific Northwest Laboratory, U.S. Nuclear Regulatory Commission, Report #NUREG/CP 0030, PNL.-SA-10325, Richland, Wash., 99352 p. 211 (1982).

23. Raats, P. A. C., *Convective Transport of Ideal Tracers in Unsaturated Soils*, Edited by Arnold, A. M., Gee, G. W., and Nelson, R. W. Pacific Northwest Laboratory, U.S. Nuclear Regulatory Commission, Report #NUREG/CP 0030, PNL.-SA-10325, Richland, Wash., 99352 p. 249 (1982).

24. Yeh, G. T., and Luxmoore, R. J., *Chemical Transport in Macropore-Mesopore Media Under Partially Saturated Conditions*, Edited by Arnold, A. M., Gee, G. W., and Nelson, R. W. Pacific Northwest Laboratory, U.S. Nuclear Regulatory Commission, Report #NUREG/CP 0030, PNL.-SA-10325, Richland, Wash., 99352 p. 267 (1982).

25. Jury, W. A., and Collins, T. M. *Stochastic Versus Deterministic Models for Solute Movement in the Field*, Edited by Arnold, A. M., Gee, G. W., and Nelson, R. W. Pacific Northwest Laboratory, U.S. Nuclear Regulatory Commission, Report #NUREG/CP 0030, PNL.-SA-10325, Richland, Wash., 99352 p. 319 (1982).

34 GASOLINE RECOVERY IN SOUTHERN MICHIGAN

Richard G. Eaton, Project Hydrogeologist
Michael V. Glaze, Senior Hydrogeologist
O.H. Materials Corp.
Findlay, Ohio 45839

INTRODUCTION

As of the end of 1985, the United States Environmental Protection Agency estimated that there were between 3 to 5 million buried tanks nationwide and that 25% of those were in a leaky condition. Every gas station has at least one and most have several buried tanks. Many major businesses have underground storage tanks for storing liquids ranging in character from gasoline to solvents and spent pickling liquors.

Leaking underground storage tanks constitute one of the largest sources of ground water pollution in the country today.

CASE HISTORY

In southern Michigan, a large factory had an underground gasoline storage tank for the use of intra-plant vehicles. The transmission lines from this tank were fiberglass. The fiberglass transmission lines deteriorated over time and began to leak. No determination of lost gasoline inventory was generated nor any realistic time as to when the loss or losses occurred. Current operating procedures now require that inventories be kept. Periodic tank and line testing of the Petro Tite or Ezy-Chek type is also recommended.

In the illustrated case, however, as in many other instances, gasoline fumes permeated through an 8-inch-thick concrete floor into a work area filled with heavy machinery. Immediate resolution of the vapor phase gasoline was performed by proper ventilation in the effected area. O.H. Materials Corp. (OHM) was retained to locate and to determine the source and abate the effect of the contaminant.

HYDROGEOLOGIC INVESTIGATION

The initial recommendation was to perform a leak test on the tank and ancillary lines to determine the potential leak source or sources. Since the underground tank was of steel construction and upon replacement showed no sign of leak occurrence, the fiberglass transmission line was believed to be suspect and was subsequently replaced. With that acquired information, no leak test was felt to be warranted.

Secondly, installation of monitor wells around the site was accomplished to determine the areal extent of the plume, flow direction and rate of contaminant transport. Because of restraints on well locations due to equipment and in-plant construction, a perimeter placement approach was utilized. A series of 11 monitor wells were installed inside and outside the building. The installed wells were 2-inch-inside-diameter PVC screen and casing. They were installed to a depth of 15 feet. Two-inch-split-spoon samples, taken for lithologic determinations, were screened to determine appropriate screen slot size. Split-spoon samples were also tested with a photoionization detector (PID) to help determine the vertical extent of contamination.

Upon completion of drilling, the wells were surveyed to within 0.01 inch and located on a scaled map. Measurements of static water levels and product thickness were taken for baseline readings. True static water levels were calculated for ground water gradient determination using the following formula modified from Shepherd [1]:

$$S_T = [(1 - D) \times T] + S_W$$

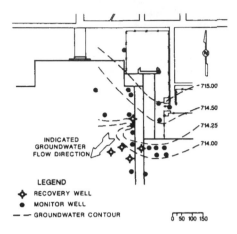

LEGEND
◇ RECOVERY WELL
● MONITOR WELL
— — GROUNDWATER CONTOUR

0 50 100 150

Figure 1. Groundwater elevation map.

where:

S_T = True static water level
S_W = Measured static water level
D = Density of the product
T = Thickness of the product

Water samples were taken for volatile organic analyses to determine baseline on contaminants. Monthly sampling and monthly static water level and product thickness measurements were taken to determine rate and direction of contaminant transport. Figure 1 shows ground water contours and flow direction.

PID readings suggested possible stratification of contaminants. To further check this theory, discrete vertical sampling was performed using a peristaltic pump.

Using a Sandpiper diaphragm pump, a constant rate pumping test was performed on a selected well. "Falling head" permeability testing was performed on two other wells to confirm the pump test findings.

RESULTS OF HYDROGEOLOGIC INVESTIGATION

The soils in this area of Michigan have been effected by and developed almost entirely by the most recent glaciation episode. Glaciation was responsible for a wide variety of soil and surficial deposits and features throughout the area.

This area is characterized by the presence of moraines, till plains, and lake beds. The site area underlying the plant is located in a low-lying till plain.

The soils found at the plant are comprised of stratified glacial drift characterized by interbedded sand and clay till deposits. The upper zone of this soil consists of medium-fine sand with varying amounts of silt of approximately 12 to 18 feet in thickness. Underlying the sand are extensive glacio-lacustrine deposits of sandy clay till. Logs of soil borings in the immediate vicinity of the plant indicate that a fine sand lens of variable thickness occurs within the clay till horizon at a depth of 18 to 20 feet.

Bedrock in the area consists of the Devonian Antrim shale which underlies the glacial drift at an approximate depth of 125 feet. This information is from the Hydrogeologic Atlas of Michigan [2].

Upon reduction of the data acquired in the hydrogeologic investigation, it was determined that the product was confined within the building foundation in the underlying medium to fine grained gray silty sand.

Average depth to water table in this area is 6.5 to 7.5 feet. Gasoline was found to be stratified within the top 4 feet of the water table. This effect is felt to be due to a number of interrelated factors:

1. Seasonal water table fluctuations
2. Vertical and horizontal changes in permeability

3. Ongoing construction dewatering in peripheral areas
4. Non-saturated conditions of solubility
5. Biodegradation with time

The method used to determine permeability from the constant rate test data is derived from Theis [3] as modified by Jacob [4]. The initial calculation used is for transmissivity (T) using:

$$T = \frac{264\ Q}{\Delta S} \tag{1}$$

where:

T = Transmissivity (gpd/ft)
Q = Pumping Rate (gallons/minute)
ΔS = Slope of the distance drawdown plot over one complete log cycle.

A plot of drawdown versus time is made on semi-log paper with Q = 0.5 gal/min and S = 2.4 feet, then T = 55 gal/day/ft. Converting to ft2/day is accomplished by (55 gpd/ft)/(7.48 ft3/gal) = 7.35 ft2/day. Transmissivity is related to permeability by the equation:

$$T = Km \tag{2}$$

where:

K = Permeability (cm/sec)
m = Saturated thickness (cm)

Solving for
K = T/m = $\dfrac{7.35\ ft2/day}{8\ ft}$ = 0.92 ft/day

0.92 ft/day x 30.48 cm/ft x 1 day/86,400 sec = 3.24 x 10^{-4} cm/sec.

Values for permeability were also calculated from the slug tests. This was done as an independent method for determining range and respective validity of those generated parameters. The relationship between the falling head and time is stated by Papadopulos [5]. This equation also calculates transmissivity initially and then utilizes T = Km. The field data is plotted on semi-log paper as (H/H_o) versus time. The value (H) = the imposed gradient at time t_o and H_o = the imposed gradient at the start of the test. The equation used by Papadopulos is:

$$T = \frac{1.0\ (r_c)^2}{t_1} \tag{3}$$

where:

r_c = radius of well casing
t_1 = matching time point (from the H/H_o versus time plot)

T = $\dfrac{1.0\ (3.81\ cm)^2}{200\ sec}$ = 0.0726 cm2/sec

Calculating K from T = Km

K = $\dfrac{7.26\ x\ 10^{-2}\ cm2/sec}{243.84\ cm}$ = 2.98 x 10^{-4} cm/sec

Both of these methods produce similar results which also are consistent with field observations of the on-site soils. The soils are a fine-grained, silty sand of poor drainage and permeability characteristics.

These calculations show permeabilities in the 10^{-4} cm/sec range. This gives us a pumping rate of 0.5 gallons per minute (gpm) with a drawdown of 2.4 feet.

REMEDIAL ACTION

Initially, two possible integrated remedial alternatives were considered. These were biodegradation techniques and physical removal and treatment.

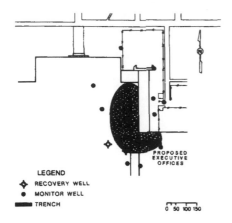

LEGEND
◇ RECOVERY WELL
● MONITOR WELL
▬ TRENCH

0 50 100 150

Figure 2. Phase product plume location.

Biodegradation

Biodegradation was an attractive method since bacterial action can biodegrade major gasoline components in comparatively short times. This area is also mostly contained by the building foundations which aid in controlling migration. The technique consists of adding nutrients and an oxygen source to the substrate to increase bacterial growth. The bacteria use the hydrocarbon as an energy source and degrade the gasoline over time.

Meetings with the Michigan Department of Natural Resources determined that bioreclamation was unacceptable to them for two reasons.

First, the technique requires injection wells which are closely regulated in Michigan.

Second, the State was not convinced that the injected nutrients could be totally contained and no off-site migration would occur.

There remained only physical removal and treatment as an acceptable remedial action.

The time required to remove contaminants from affected soil by pumping and treating the water passing through it is considerably more than that required for bioreclamation. This is due to the propensity of hydrocarbons to adsorb, absorb, and be held by capillary tension in fine grained soils. The majority of this fraction is leached into the ground water very slowly.

Physical Removal and Treatment

The physical removal and treatment of the contaminated ground water and phased product was complicated by the fact that the plume stretched partially under an area proposed for executive office space (See Figure 2).

Under normal conditions, a line-sink of pumping wells with overlapping drawdown cones could be used. In fact, this technique was used in the plant area where it was possible. A series of five pumping wells was installed in the plant; four at the leading edge of the plume in a line-sink configuration, and one in the location of the thickest phased product concentration. Each well was equipped with a double acting diaphragm pump. The contaminated water and phased product were pumped directly into the plant process water waste stream.

In the area slated for executive offices, however, a line sink was not practical. Since construction of the offices was to begin soon, it was decided to arrange for the construction of a trench and sump system to coincide with office construction.

Construction of a trench and sump recovery system inside a building is a complicated matter. Excavation equipment gives off exhaust fumes, and excavated material gives off gasoline fumes in a confined space. Proper ventilation, fire protection, and respiratory protection are essential. Heaving sands below the water table made sheet piling necessary, but an overhead clearance of 25 feet made installation of sheet piling difficult. However, few problems are totally insurmountable.

Before excavation commenced, a plastic curtain was installed enclosing the work area. This would reduce the volume of air needed to be exchanged. Part of one outside wall was removed, and blowers

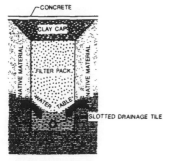

TRENCH CROSS-SECTION

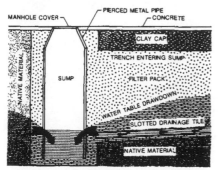

TRENCH AND SUMP LONGITUDINAL CROSS-SECTION

Figure 3. Typical sump and trench hydrocarbon recovery system.

and exhaust fans with explosion proof motors were installed. The air quality was continuously monitored throughout the project with an MSA 260 LEL/0_2 meter and periodically spot checked with an hNU PID to assure personnel safety. If the air quality had reached a dangerous level, all production would have ceased until suitable air quality was restored.

Excavation operations began with cutting the concrete floor into 6 foot by 5 foot sections to enable easier and safer handling. This was done with a walk-behind road saw. These sections of concrete were transported by a front-end loader to a tri-axle dumptruck. After the concrete sections were removed, the soil was excavated around a 12-inch fire main and an 8-inch storm sewer pipe. This enabled a crew to remove a 4-foot section of each pipe. The fire main was not charged, but the storm sewer required plugging to force a reversal of the water flow.

Once the pipes were removed, driving of the sheet piling began. Sheet piling was required because of the heaving sand below the water table. It prevented caving back under the concrete slab floor. Since the trench was to be a maximum of 14 feet deep, sheet piling was driven to 17 feet. The length of time required to drive the sheet pile was extended due to the limiting 25-foot overhead clearance. This limitation required the sheeting to be driven in two vertical sections instead of one. Pile sections were bolted together to minimize fire hazards. As the sheet piling was driven by the vibratory sheet pile driver, visual observations were made to determine if the vibrations were having any effect on the surrounding structures. No adverse effects were noted, but care must always be taken not to crack foundations and walls.

While sheet piling was underway, the trucks and front-end loaders stockpiled the needed pea gravel and clay and constructed a diked staging area, lined with a 4-mil-thick plastic liner, where the excavated material was placed for temporary storage.

After the piling was secured, excavation of a 3-foot-6-inch wide trench was begun, using a rubber tired excavator with a brass-clad bucket to prevent sparking (See Figure 3 for trench details).

The excavation ranged from 12 to 14 feet deep, and the bottom of the excavation was graded to have 2 feet of fall over its 97-foot length to entice water flow to the sump. The excavated material, saturated with a flammable liquid, was directly deposited from the excavator bucket into the waiting buckets of two front-end loaders on a continuous cycle. From the buckets of the front-end loaders, the contaminated soil was deposited into the tri-axle dumptrucks. The trucks delivered the materials to the designated area, where it was stored inside the prepared staging area. Upon completion, the material was tested for hazard classification, covered with 4-mil-thick plastic, and left to await transportation and disposal.

As the trench was completed, a layer of pea gravel about 6 inches thick was placed on the bottom of the trench to serve as a bed for a 6-inch-diameter slotted roll drain pipe. Concurrently, a 9-foot-long section of 24-inch-diameter punch pipe screen and a 5-foot-long section of 24-inch-diameter steel casing was installed at the west end of the trench as a sump. This sump permits the collected liquid to be pumped from the excavation.

When the drain pipe and sump were in place, pea gravel was dumped into the trench to a level approximately 4 feet down from the top of finish grade and leveled. At this time the sheet piling was removed.

After removal of the sheet piles, a front-end loader placed clay into the trench, which was compacted in 1-foot lifts using a walk-behind sheep's-foot compactor. During this stage of the operation the fire main and storm sewer were reconnected.

The final step, before setting the pump, was to install 3/4- inch expansion joint filler around the edges, place 5/8-inch-diameter (Number 5) reinforcement bar on 12-inch centers and pour a 10-inch-thick slab of 3,000 pounds per square inch compressive strength concrete.

After setting a double acting diaphragm pump in the sump and piping the discharge into the plant processing waste water stream, a manhole cover was placed over the sump.

CONCLUSION

So that others may learn from my mistakes, I point out one major fault with this plan. While we know that product is being removed because the hydrocarbon reading went up in the wastewater stream, and while the monitor wells show decreasing concentrations, we have no means to quantify the volume of product removed. Since we never knew how much was lost, this may not be important here, but it might be necessary in another application. In dealing with regulatory agencies, the discussion always *arises* "How clean is clean?" In order to make that determination a number of data points have to be considered:

- Calculation (determination) of lost product volume.
- How much has been recovered?
  A. free phase
  B. solubilized
- For "B.", what concentrations (analytical) have been generated for calculation purposes?

For now, our system works. But anything can be improved, and the next time it will be.

REFERENCES

1. Shepherd, William D., unpublished paper, Environmental affairs, Shell Oil Co., Houston, Texas.
2. Department of Geology, College of Arts and Sciences, Western Michigan University, Kalamazoo, Michigan (1981).
3. Theis, C. V., "The Lowering of the Piezometric Surface and the Rate and Discharge of a Well Using Groundwater Storage," *Transactions, American Geophysical Union*, 16, 519-24 (1935).
4. Jacob, C. E., "Flow of Groundwater," *Engineering Hydraulics* (1963).
5. Papadopulos, I. S., et al., "On the Analysis of "Slug Test" Data," *Water Resources Research*, 9 (1973).

35 INCINERATION OF LIQUID FLAMMABLE WASTES AND BIOLOGICAL SLUDGE IN COAL-FIRED BOILERS

Mark W. Townsend, Utilities Supervisor
Monsanto Company
Nitro, West Virginia 25143

INTRODUCTION

The Monsanto Chemical Company, an operating division of Monsanto Company, has a production facility for the manufacture of rubber chemicals located in Nitro, West Virginia. Residues or co-products generated from these processes are of high heating value and are considered hazardous by RCRA definition. These co-products are reused to generate steam in the stoker-spreader coal-fired boilers, thereby reducing costly outside disposal fees, reducing coal requirements and RCRA regulations.

Additionally, the same boilers are used to incinerate non-hazardous sludge generated from the 2 mgd activated sludge Waste Treatment Plant located on site. The waste sludge is pumped from the holding basin, transferred to a holding tank, and fed to the boilers along with the co-products.

This paper discusses characteristics of the co-products and sludge, design of the systems that feed these streams into the boilers, environmental factors, and cost savings associated with this process.

INCINERATION ALTERNATIVES

In 1976, the Monsanto Nitro plant sought to eliminate the biological treatment of waste residues which caused frequent upsets in the waste treatment system. These residues averaged 15,000 BTU/lb in heat value, therefore, incineration was a logical choice (Table I).

Five alternatives were initially investigated, the first being outside contract incineration (Figure 1). The cost of disposal was $164,000/yr in 1975 dollars. The advantages were no maintenance or operation concerns, no capital outlay, and the method of disposal could be initiated in a very short time. The disadvantages were the high cost and possible liabilities.

A new incinerator with heat recovery was examined. The capital expense was $800,000 in 1975 dollars. Allowing for a ten year depreciation, estimating operating and indirect costs, and subtracting $35,000/yr in steam value, the net cost was $161,000/yr. The advantage of the recovered heat value was outweighed by the disadvantage of high cost, having another system to man and operate, and the potential for refractory, slagging, and salt caking problems.

Another alternative was to modify an existing incinerator at a cost of $75,000 in 1975 dollars. Depreciating this over a ten year period and including operating and indirect costs, the net cost was $29,000/yr. The advantages were reasonable costs; all the residues could be burned in one location, and the project could be operational in six to eight months. The disadvantages were numerous. The incinerator was a high maintenance item and burning the residues had the potential to increase maintenance due to possible refractory, slagging, and salt caking problems. Increased downtime for maintenance would cause production downtime for a unit that required the incinerator to run. Also the total BTU capacity of the incinerator was limiting as was allowable sulfur dioxide emissions.

The fourth alternative was to burn the waste residues in one of the existing gas boilers used to generate process and heating steam for the plant. The capital expense was $90,000 in 1976 dollars. Allowing for a ten year depreciation, estimating operating and indirect costs and savings due to

Table I. Residue Chemical Characteristics

Residue	BTU/Pound	% Ash	% Chlorine	% Sulfur
Building 42	17000	Trace	6.9	1.07
Building 91	16400	0.02	1.0	1.30

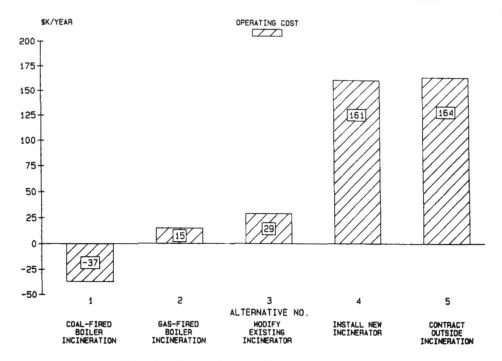

Figure 1.   Comparative cost of residue disposal methods.

reduced natural gas usage and steam generation yielded a net annual cost of $15,000. The advantages were recovering the heat value of the residues with the steam generated and reasonable costs. Also, the capacity for SO_2 emissions was about four times the SO_2 that would be emitted by combustion of the residues thus allowing incineration of increased volumes of residues in the event of product market expansion.

The final alternative investigated was that of residue incineration in the coal-fired boilers also used for generation of process steam.

RESIDUE INCINERATION

Residue incineration in the coal-fired boilers offered many advantages over using the gas boiler. Capital required was $78,000 or $7,800 per year over ten years, $10,000 per year operating costs for a total cost of $17,800 per year in 1976. Additionally, $55,000 worth of steam would be realized for a total savings of $37,200 per year. Gas-fired incineration cost was $15,000 per year, due mainly to additional personnel required to operate the gas-fired boilers.

The residue burning in the coal-fired boilers was a small percent of the total fuel input (2%) whereas the residue would be practically the sole fuel in the gas boiler, thus presenting another disadvantage for gas boiler disposal. The residue would have to be burned at a fast rate to maintain a baseload. Due to the variable supply of residue, this meant the boiler would have to be started up and shut down very frequently which was not desirable from an operational viewpoint.

Whereas the coal-fired incineration of residue required no additional natural gas usage, the gas-fired option required gas consumption for the pilot flame and pre-heatup for each startup.

Better fuel efficiency was achieved on the coal-fired boilers due to the coal fired boilers being equipped with economizers for pre-heating of feedwater.

From a safety standpoint, burning residue in the gas-fired boilers increased the chance of a flame-out which could cause an explosion or a smoke and/or chemical emission. In the coalfired boilers, it would be difficult to have a flame-out.

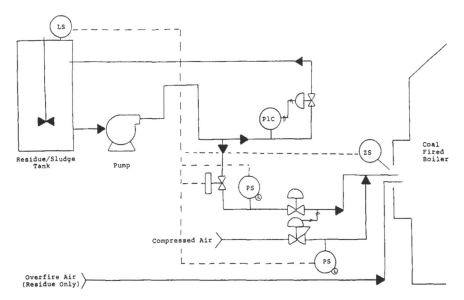

Figure 2.   Typical sludge or residue incineration system.

A project was written to incinerate the residues in the coalfired boilers. An existing 10,000 gallon nickel-clad tank was retrofitted to serve as a storage tank prior to incineration in the boiler. The tank was equipped with a three horsepower 316 stainless steel agitator to blend the different residues and steam heated to keep the residues from setting up as a solid. The temperature was controlled at 85°C. Due to the flammability of the residue, the tank was nitrogen blanketed. The residues are transported from the various production units via tankwagon and pumped into the 10,000 gallon residue tank. A high level probe kicks off the transfer pump to prevent overfilling.

Once the residue reaches the desired temperature, it is pumped by a 316 stainless steel Viking gear pump through steam-traced and insulated Alloy 20 schedule 10 lines into any one of the three coal-fired boilers. Excess residue is recirculated back to the residue storage tank (Figure 2).

Residue going into the boiler flows through a block valve used to isolate the boiler when residue is being burned on one of the other boilers and then through a pressure switch.

After the pressure switch senses the residue pressure, the residue flows through the control valve. Then the residue is combined with compressed air, and this mixture is injected into the boiler above the burning coal bed. The compressed air is controlled by a differential pressure regulator at 10 psi above the residue pressure, thus ensuring enough air for atomization.

Overfire air, used for cinder reinjection and firebox mixing, was later piped over to the area where the burner gun was inserted. This was done to eliminate smoking problems that sometimes occurred when burning residue.

The piping was also designed to be blown out with steam to minimize downtime due to pluggages.

The system also has several safety features. Should the air pressure fail or not stay 10 psi above the residue pressure, the automatic block valve will close and thus stop the feed of residue into the boiler, thus forcing all the residues to recirculate to the storage tank. If the residue tank level falls below 15%, the automatic block valve will close. Also, if the burner gun is not in the proper position, the automatic block valve will not open.

SLUDGE INCINERATION

In 1977, the Monsanto Nitro plant installed a new activated sludge wastewater treatment process. Part of this project included the construction of a digester to thicken the waste activated sludge solids. Excess sludge with a solid content of 1–1½% by weight is periodically pumped into the digester at a rate of about 1000 pounds a day. In the digester, the sludge is concentrated to 8–10% solids by gravity thickening and decanting of water from the surface.

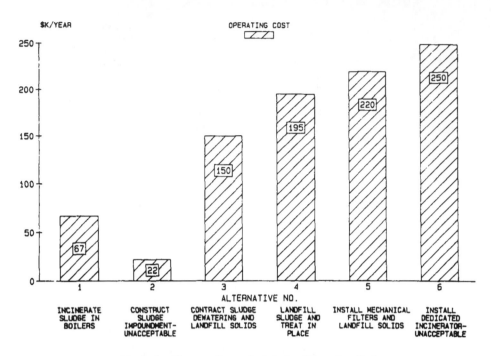

Figure 3.   Comparative cost of sludge disposal methods.

To ensure compliance with NPDES permit limits, in 1980 the sludge age was lowered from greater than 100 days to 30 days. As a result, in 1981 the digester was full. At a cost of $45,000 a sludge lagoon was constructed at the on-site landfill. The sludge was pumped from the digester into tankwagons and transported to the new sludge lagoon. This procedure was repeated in 1982.

Due to space limitations at the landfill, the potential for groundwater contamination, and changes in environmental regulations that could make the sludge hazardous, an alternative to this method of disposal was sought (Figure 3) .

Installation of an incinerator to dry and burn the sludge at the waste treatment plant was considered. This would leave only the ash residue to landfill. An incineration system required approximately one million dollars to purchase and install. Because of the high capital cost and large volume of material gas required, this alternative was not recommended.

One of the several alternatives considered was to install a mechanical filter and associated equipment. The sludge would be dewatered to 30–35% solids and then transported to a hazardous waste landfill. Tests showed at least one pound of filtration aid such as diatomaceous earth and/or flyash was needed for each pound of sludge solids for proper filtration. This system required a capital outlay of $600,000 and cost $220,000 annually to operate.

A similar alternative was to have a contractor once a year to dewater the sludge on-site. The filtrate would be returned to the waste treatment plant and the dewatered sludge hauled to a hazardous waste landfill. No capital would be required, and the annual cost would be approximately $150,000.

Another alternative considered was to continue the present method of disposal. This would involve pumping the sludge from the digester into tank wagons and constructing a one million gallon sludge lagoon at the existing landfill in which to empty the estimated 200 tankwagon loads. Construction of the lagoon and transporting the sludge would cost $45,000 per year. Eventually, the sludge in the lagoons would have to be dewatered or solidified. Dewatering the sludge and sending it to a hazardous waste landfill would add an additional $150,000 to the annual operating cost. Solidifying the sludge in place would cost $100-$150,000 per year. Thus the total annual operating cost would be $145 to $195,000.

Another option was to construct a seven million gallon lagoon at the site of the waste treatment plant. This would provide sludge storage for seven to ten years. Excess water would be pumped back through the waste treatment plant. The cost to have the sludge dewatered and transported to a hazardous landfill would be $1.2 to $1.4 million. Solidification would cost $0.7 to $1.0 million. The capital cost to construct a large sludge lagoon would be roughly $200,000 dollars, with an annual operating cost of $22,000. However, there were disadvantages. The odor from the waste treatment plant would increase due to the large surface area of the lagoon. Additionally, further legislation could declare sludge from chemical plants to be a hazardous waste, thus subjecting the lagoon to hazardous waste storage facility requirements requiring an expensive upgrade. Finally, as a worldwide corporate guideline, Monsanto is moving away from the use of earthen basins.

A final alternative involved the disposal of waste activated sludge by spraying into the fireboxes of the coal fired boilers at a rate of four gallons per minute.

A sludge burning test was conducted and demonstrated that sludge could be sprayed into the fireboxes at rates as high as six gallons per minute without any short-term harm to the boilers.

At sludge feed rates of 6 gpm, some line pluggage occurred. At the design rate of 3 to 4 gpm, minimal pluggage was experienced. The sludge solids ranged from 4-6%, and this range could be met by adjusting the length of mixing time prior to pumping the sludge from the digester and raising or lowering the depth of the pump and mixer.

Consultation with John Zink Company demonstrated the dew point of the stack gas would rise from 261°F to 269°F. well below the stack gas temperature of 293° to 317°F. during the test burn.

At the maximum sludge injection rate of 6 gpm, no wet sludge built up on the grate. Injections at higher rates could create ash handling problems.

When sludge containing 5% solids is injected into the firebox of a boiler, the heat required to vaporize the water is greater than the heat obtained from the combustion of the solids. The net heat loss was calculated to be 1.5 million BTU/hr or less than 5% of steam production. This requirement for additional coal was far outweighed by this method of sludge disposal.

Another concern was that sludge incineration would create an odor problem. The temperature above the grate is approximately 1700°F, thus ensuring destruction of any odor.

Fly ash and bottom ash from the boiler house is hauled back to the coal mine and used for land reclamation. If the incineration of the sludge increased, the concentration of any trace elements to a hazardous level in either the fly ash or the bottom ash, the ash would have to be sent to a hazardous landfill.

Trace element concentrations from the sludge and fly ash were determined. Only barium showed a significant increase in concentration, but it was well below the level that would make the ash hazardous.

Sulfur dioxide production would increase by fourteen pounds per hour when sludge containing 5% solids was incinerated at four gallons per minute in one boiler (Twelve pounds per hour from the sludge and two pounds per hour from the additional coal required). State regulations allow 1.6 pounds per hour of SO_2 for every million BTU's per hour of design heat input for each boiler. Because the SO_2 created by sludge incineration is in addition to the SO_2 produced by the burning of the coal, sludge cannot be incinerated in a boiler when it is operating above a specified limit. Based on 1% sulfur coal and 82% boiler efficiency, one boiler is limited to 91% capacity (or 91,000 lbs of steam/hr) and the other 93% (or 116,000 pounds/hr) when sludge was being incinerated.

A final concern was future regulatory changes may require additional testing.

Annual operating costs for such a system would be $67,000 with a capital outlay of $300,000. Due to the success of the test burn and this being the most environmentally acceptable method of disposal, the decision was made to install permanent facilities.

At the digester, a raft with a 7.5 horsepower submersible centrifugal pump and an 11 horsepower Flygt mixer was installed.

The mixer prevented "bridging" when the sludge was pumped out and would also help to reduce clogging in the piping at the boiler house. Also, the height of the pump and mixer could be raised and lowered for a more consistent sludge solids content. A truck loading dock was installed to facilitate tankwagon loading.

At the boiler house a 10,000 gallon carbon steel vertical tank was installed. This tank was equipped with baffles and a 3 horsepower carbon steel agitator to keep the solids in suspension. Plate coils and insulation were provided and a steam line run to the plate coils to keep the tank contents heated during freezing weather. Level instrumentation was provided and design to shut the sludge feed pump on low level and stop the unloading pump on high level. The alarms were installed in the boiler house control room.

The unloading pump was a 100 gpm centrifugal carbon steel pump. Steam piping was provided to allow the line from the pump to the tank to be blown clear.

A progressive cavity pump was installed to pump from the sludge storage tank into the boilers and sized to provide a flow of twenty gallons per minute through a recirculation line. All sludge piping was electrically traced and insulated.

The inspection port near the back of the two biggest boilers was modified to allow the insertion of the burner gun into the firebox of each boiler, directly across from the residue burning gun. The sludge burner guns were fabricated similarly to the residue burner guns. The burner tips were the same type but larger size than the residue guns. Piping was installed from the recirculation line to each burner gun location and compressed air was supplied to the sludge burner guns. Differential pressure regulators were installed to maintain a constant pressure difference between the air and sludge. These valves prevented excessive air usage and ensured the pressure requirements of the burner tips were satisfied for proper atomization (Figure 2).

An automatic block valve was provided in each sludge feed line that closed if the sludge tank level or air pressure dropped below specified limits. Also, an alarm sounded in the control room. A limit or position switch prevented the automatic valve from opening unless the burner gun was properly inserted in the boiler. Flow indicators were installed near the burner guns and in the control room. Manual throttling valves or "pinch" valves were used to set the injection rate of sludge. Steam piping was installed to allow the sludge feed lines to be blown clean.

CONCLUSION

The incineration of liquid flammable wastes and biological sludge in coal-fired boilers has been successfully demonstrated as a method of disposal at Monsanto's Nitro, West Virginia, plant. In addition to being economical, it is also the most environmentally acceptable method of disposal, eliminating future groundwater contamination liabilities associated with the landfilling or surface impounding of these two wastes. In the case of the liquid flammable residues, another benefit was heat recovery via steam generation. Also, since steam was generated, the residues were exempt from Resources Conservation and Recovery Act regulations falling under the legitimate reuse/recycle classification. Although the burning of sludge required a slight increase in coal usage, the cost avoidance of acceptable alternative disposal methods far outweighed the cost of additional coal.

36 SENSITIZED PHOTOOXIDATION OF BROMACIL: A DYE SURVEY OF POTENTIAL PHOTOSENSITIZING AGENTS

Talbert N. Eisenberg, Graduate Instructor

E. Joe Middlebrooks, Provost & Vice President for Academic Affairs

V. Dean Adams, Director

Center for the Management, Utilization, and Protection of Water Resources

Tennessee Technological University

Cookeville, Tennessee 38505

INTRODUCTION

The fate of herbicides, pesticides, and refractory organics in the environment is of great interest and importance. Their occurrence in natural waters presents a serious problem to public health and safe drinking water. The toxicity and bioaccumulation in the environment of these compounds merit public concern, and their removal prior to reaching natural waters is of the utmost importance.

Bromacil is one of the most important herbicides for non-cropland and citrus control of grasses and weeds [1]. It is a potent and specific inhibitor of photosynthesis [2,3] and is slightly toxic and refractory. The 48 and 72-hr LC^{50} for rainbow trout are 71 and 28 mg/L, respectively. The half life in silt loam soils is approximately 5 to 6 months [4]. Losses from soil due to volatilization and photodecomposition are negligible [5]. Losses from water due to photodecomposition and volatilization are also negligible [6].

Photodecomposition of bromacil in sunlight is negligible because bromacil does not absorb light in the visible spectral region [7]. Furthermore, photolytic cleavage by visible light (e.g., at 600 nm, E = 45 kcal) occurs only on weak covalent bonds [8]. The energy necessary to break chemical bonds between C-C or C-O amounts to 80 and 88 kcal/mole, respectively, and only ultraviolet radiation with energy higher than these values (lambda < 325 nm) can disrupt these bonds in a photolytic reaction. Solar ultraviolet radiation below 270 nm is absorbed by ozone and air and does not reach the earth's surface. For photochemical reactions, it is usually visible light between 400 and 700 nm and ultraviolet radiation between 390 and 350 nm which are active.

Photochemical activity with visible light is associated with colored substances (e.g., dyes). Dye photosensitizers absorb energy in the visible spectral region, and transfer the energy either directly or indirectly to a substrate, which results in the photodecomposition of the substrate. Acher and Saltzman [7], Watts [9], Eisenberg et al. [10] and others have shown that addition of dye photosensitizers to aerated aqueous solutions of bromacil leads to a quantitative and fast sunlight photochemical reaction. The objective of this study was to evaluate 69 different dyes, stains, and indicators at acidic, neutral, and alkaline pH values for the degradation of bromacil.

THEORY

With the absorption of light, a molecule rises from its ground state of lowest energy to an excited state of higher energy in which one of the electrons is at a higher energy level [11]. The activated molecule expends the energy of excitation in one of several ways. The molecule may either emit radiation in the form of fluorescence or phosphorescence, lose its energy as heat by collision with other molecules, dissociate, or take part in chemical reactions. In many chemical reactions, the photosensitizing action of dyes is responsible for key life processes and naturally occurring chemical reactions.

The photosensitizing action of dyes results from the ability of dyes to act either as strong oxidizing or as strong reducing agents, in the presence of reducing or oxidizing substances, with subsequent regeneration [8]. In certain reactions, dyes are predestined sensitizers because the most reactive triplet state 3D is produced in dyes with high efficiency by intersystem crossing from the first excited singlet state 1D. 3D is reactive not only in redox reactions, but 3D can also transfer the energy of the triplet state to other molecules and initiate specific reactions.

In dye-sensitized photooxidations, either an oxygen transfer process or a hydrogen-abstraction reaction is involved. In the oxygen transfer process, the excitation of the dye to the singlet state is followed by intersystem crossing to the triplet state. According to the concepts of Schonberg, Schenk, and others, the reactive species is a metastable dye-oxygen complex produced from the triplet excited dye and oxygen [8].

$$D + h\nu \rightarrow 1D \rightarrow 3D \tag{1}$$

$$3D + O_2 \rightarrow [D \ldots O_2] \tag{2}$$

$$[D \ldots O_2] + A \rightarrow D + AO_2 \tag{3}$$

In contradiction to the above, Kautsky, Egerton, and others have shown that photooxidation can occur in systems with separated dyes and substrates, indicating the production of an oxidizing volatile species [8]. This species may be either hydrogen peroxide (in the presence of water) or a semireduced oxygen molecule $O_2^-\cdot$ which can act as an intermediate in the production of

$$O_2^-\cdot + H^+ \rightarrow HO_2\cdot \tag{4}$$

$$2HO_2\cdot \rightarrow H_2O_2 + O_2 \tag{5}$$

The reaction between oxygen and dye leads to a semioxidized dye radical which is reduced to the ground state by the oxidizable reactant. The semioxidized dye is also produced by an electron dismutation reaction between dye molecules in the excited triplet state and those in the ground state so that the production of semioxidized dye radicals can be appreciable. As in the dismutation process, a semireduced dye is formed which rapidly reacts with oxygen; the reduced $O_2^-\cdot$ also results from this indirect reaction. These reactions lead to the following simple scheme for photosensitized autooxidation reaction.

$$D + h\nu \rightarrow 1D \rightarrow 3D \tag{6}$$

$$3D + O_2 \rightarrow D^+ + O_2^- \tag{7}$$

$$3D + D \rightarrow D^+ + D^- \tag{8}$$

$$D^+ + A \rightarrow D + Aox \tag{9}$$

$$D^- + O_2 \rightarrow D + O_2^-\cdot \tag{10}$$

$$O_2^-\cdot + H^+ \rightarrow HO_2\cdot \tag{11}$$

$$2HO_2\cdot \rightarrow H_2O_2 + O_2 \tag{12}$$

Kautsky, Egerton, and others [8] discuss another possibility for the production of an activated oxygen species by the excitation of the dye. The activated oxygen reacts with organic compounds producing peroxide.

$$D + h\nu \rightarrow 1D \rightarrow 3D \tag{13}$$

$$1D \text{ or } 3D + O_2 \rightarrow D + O_2^* \tag{14}$$

$$O_2^* + A \rightarrow AO_2 \tag{15}$$

The oxygen eventually reacts as the active species, transferring energy from the triplet dye molecule to triplet oxygen (ground state) from which singlet oxygen is produced.

$$D + h\nu \rightarrow 1D \rightarrow 3D \tag{16}$$

$$3D + {}^3O_2 \rightarrow D + {}^1O_2 \tag{17}$$

$${}^1O_2 + A \rightarrow AO_2 \tag{18}$$

Dyes which are sensitizers in such photooxygenation reactions are very reactive toward oxygen (e.g., acridines, thiazines, xanthene-type dyes).

In the hydrogen abstraction reaction, dyes with strong electron- or hydrogen-abstracting groups (derivatives of anthraquinone, flavanthrone, etc.) and compounds containing removable hydrogen are involved. The photoexcited dye is reduced by the oxidizable reactant by a transfer of electrons or hydrogen. The resulting semireduced dye radical is reoxidized to the ground state by oxygen.

$$D + h\nu \rightarrow 1D \rightarrow 3D$$

$$3D + RH \rightarrow DH^* + R\cdot$$

$$R \cdot + O_2 \rightarrow RO_2 \cdot \rightarrow \text{oxidation products}$$
$$DH^* + RO_2 \cdot \rightarrow D + ROOH$$
$$DH^* + O_2 \rightarrow D + HO_2 \cdot$$
$$DH^* + HO_2 \cdot \rightarrow D + H_2O_2$$
$$DH^* + DH^* \rightarrow D + DH_2$$

In the absence of oxygen, the semireduced dye is not regenerated but is transformed to the leuco dye, i.e., photobleached.

METHODOLOGY

Materials

Bromacil [5-bromo-3-sec-butyl-6-methyl-uracil] was obtained as DuPont technical bromacil, a wettable powder containing 95% bromacil. The bromacil concentration was verified against bromacil reference standards obtained from EPA and was found to agree with the manufacturer's claim.

Textile dyes were obtained from Tricon Colors Incorporated. Non-textile dyes and stains were obtained from chemical and biological supply companies. A list of the dyes, stains, and indicators tested is presented in Table I.

Analytical

Quantitative analysis of the bromacil solutions involved liquid-liquid extraction of the compounds from aqueous solution followed by gas chromatographic measurement using a flame ionization detector. Methylene chloride (CH_2Cl_2) was used to extract the bromacil from the water samples. A 10 mL portion of CH_2Cl_2 was added to 100 mL of sample in a 125 mL separatory funnel.. The funnel was inverted, opened to release pressure and closed, agitated, and set aside to allow the CH_2Cl_2 and water phases to separate. The denser CH_2Cl_2-bromacil phase was drawn off into a 40 mL beaker, Na_2SO_4 was added to absorb water, and the CH_2Cl_2-bromacil phase was poured into 1 mL reaction vials and stored at 4°C. Standard bromacil solutions were prepared in the same manner using EPA reference bromacil.

The gas chromatograph was a Hewlett Packard HP 5880A with a Hewlett Packard HP 7673A automatic sampler and tray. A 61 cm x 2 mm (ID) glass column packed with a 1% SP-2250 on 80/100 Supelcoport was used. Operating conditions for the chromatograph were an injector port temperature of 250°C, an initial oven temperature of 160°C for 2 minutes, a program rate of 20°C/min, a detector temperature of 260°C, a nitrogen carrier gas flow of 30 mL/min, a hydrogen flow rate of 30 mL/min, and an air flow rate of 400 mL/min. A 2 μL sample was injected.

Sunlight intensity was measured using an Eppley Radiometer Model 8–46 and an Instrulab 2000 Datalogger. A Fisher Accumet pH Meter Model 800 was used to measure the pH value.

Procedures

Water from a type 1 reagent-grade water system was used. A stock 0.00383 M (1 g/L) bromacil solution in ethanol was used to prepare bromacil standards and samples. A stock 1 g/L dye solution in water or ethanol (if solubility posed a problem) was used to prepare 10 mg/L dye concentrations. The effects of pH and buffer system on ionization of bromacil were determined by liquid-liquid extraction of 15 mg/L solutions of bromacil adjusted to pH values from 3 to 11 with acetate, diphosphate, and carbonate buffers. Quantitation was by comparison of peak area of the sample chromatogram with chromatograms of standards prepared in CH_2Cl_2.

The effects of pH value and different dyes on reaction rates were determined by testing 69 potential photosensitizing agents at pH values of 4, 7, and 10. Samples and standards at pH 4 were buffered with 0.05 M CH_3CO_2H and 0.01 M CH_3CO_2Na, at pH 7 with 0.0194 M NaH_2PO_4 and 0.0194 M Na_2HPO_4, and at pH 10 with 0.0249 M $NaHCO_3$ and 0.0249 M Na_2CO_3. Solutions were aerated prior to sunlight exposure to assure an initial saturation of dissolved oxygen. All experiments were conducted outdoors in sunlight. Solutions of dyes and bromacil were placed outdoors in Pyrex or Kimax vessels. No covering was placed on the sides or the bottoms of the vessels.

Samples were collected at selected time intervals and stored in the dark at 4°C until CH_2Cl_2 extractions were performed. Quantitation was by comparison of peak area of the sample chromatogram with chromatograms of standards. Dyes, which were found to be effective sensitizers, were also

Table I. Dyes, Stains, and Indicators Tested

Common Name	Color Index Name	Source[a]
Acetone		CBS
Acid Red WA		TDM
Acridine Orange		CBS
Acridine Yellow		CBS
Alizarin Cyanine Green GN	Acid Green 25	TDM
Alizarine Red		CBS
Alizarol Cyanine RC		CBS
Alphazurine FGND	Acid Blue 9	TDM
Aniline Blue	Acid Blue 93	CBS
Auromine O Conc. 130%	Basic Yellow 2	TDM
Azorubin Extra Conc. 133%	Acid Red 14	TDM
Bismark Brown 53	Basic Brown 4	TDM
Brilliant Yellow	Direct Yellow 4	CBS
Bromocresol Green		CBS
Bromocresol Purple		CBS
Bromophenol Blue		CBS
Bromothymol Blue		CBS
Carmine (Alum Lake)		CBS
Chrysoidin 3R	Basic Orange 1	TDM
Congo Red		CBS
Cresol Red		CBS
Crystal Violet		CBS
Curcumin		CBS
Diazophenyl Blue		CBS
Dichlorofluorescein		CBS
Eosin Y	Acid Red 87	CBS
Erie Congo 4B (Congo Red)	Direct Red 28	TDM
Erie Fast B Conc.	Direct Brown 31	TDM
Erie Fast Orange A	Direct Orange 26	TDM
Erie Green WT	Direct Green 1	TDM
Eriochrome Black T		CBS
Ethyl Orange		CBS
Ethyl Red		CBS
Evans Blue		CBS
Fast Green FCF		CBS
Fluorescein	Acid Yellow 73	CBS
Fuchsin Acid	Acid Violet 19	CBS
Gentian Violet		CBS
Grand Yellow		CBS
Malachite Green	Basic Green 4	CBS
Marshalls Reagent		CBS
Methyl Red		CBS
Methyl Violet	Basic Violet 1	CBS
Methyl Violet 2B	Basic Violet 1	TDM
Methylene Blue (Fisher)	Basic Blue 9	CBS
Methylene Blue (Tricon)	Basic Blue 9	TDM
Methylene Blue 2B Powder	Basic Blue 9	TDM[a]
Neutral Red	Basic Red 5	CBS
Nigrosin	Acid Black 2	CBS
Phenol Red		CBS
Phloxine B	Acid Red 92	CBS
Propyl Red		CBS
Resorcine Brown R 167%	Acid Orange 24	TDM
Rhodamine B		CBS
Rhodamine WT		CBS
Rhodmaine B Conc. 500%	Basic Violet 10	TDM
Rose Bengal	Acid Red 94	CBS
Safranin O		CBS

Table I. Continued ...

Common Name	Color Index Name	Source[a]
Solantine Turquiose G	Direct Blue 86	TDM
Solantine Yellow 4GL	Direct Yellow 44	TDM
Tartrazine 118%	Acid Yellow 23	TDM
Thionin 1%		CBS
Thymol Blue		CBS
Thyodene		CBS
Victoria Green WB	Basic Green 4	TDM
Victoria Pure Blue BO	Basic Blue 7	TDM
Wool Orange A	Mordant Orange 6	TDM
Wool Violet 4BN	Acid Violet 49	TDM
Wrights Stain		CBS

[a]CBS = chemical and biological supply company
TDM = textile dye manufacturer

tested under reduced conditions by adding 126 mg/L Na_2SO_3 to the samples, omitting aeration, and following the above described procedure.

Data Analysis

The data from the experiments were fit by the method of least squares to the first order equation $C = C_o \exp(kt)$ in which C = bromacil concentration at time t; C_o = initial bromacil concentration at $t = 0$; t = elapsed time; and k is a first order reaction rate constant. The first order equation was linearized to $y = mx + b$ in which $y = \ln(C/C_o)$, $m = k$, $x = t$, and b = intercept. Analysis of variance was used to evaluate the data.

RESULTS

Results from the study are presented in Tables II and III and Figures 1-4. Reaction rates are considerable higher than those reported in the literature [7,9,11] for some of the sensitizers tested. The high values of k may be due to the exposure of the sides of the reactors to sunlight. The sides of a reactor would normally not be exposed to the sun, and the reaction rate constants reported here should therefore be used with caution.

Effect of pH on Ionization of Bromacil

The percent ionization of bromacil was assumed to be the percent of bromacil unextracted in the CH_2Cl_2. The ionization of bromacil is dependent on pH value as shown in Figure 1. Sargent and Sanks [12] reported that at high pH values, substances with ionizable protons will be more completely ionized leaving greater electron density at the site of ionization. The increased electron density enhances reactivity with singlet oxygen, and excess hydroxyl ions are available to take up any hydrogen ions that might be freed if singlet oxygen extracts electrons from the substrate.

Spikes and Straight [13] reported that tyrosine is also photooxidized only when the phenolic group is ionized; however, histidine is photooxidized only when the imidazole group is unionized, and the

Table II. Reaction Rate Constant and Half-Life for Sensitized Photooxidation of 50 mg/L Bromacil. pH = 10, Reactor = 1 L, Sensitizer = 10 mg/L, Sun = 853 W/m², DO = saturated.

Sensitizer	-k (min⁻¹)	A = Co * exp(INTCPT)	t(1/2) (min)	Relative Sensitizing
Dichlorofluorescein	0.0297	48.56	22	0.30
Fluorescein	0.0087	47.15	73	0.09
Riboflavin	0.0659	46.14	9	0.67
Safranin O	0.0050	48.03	131	0.05
Solantine Turquoise	0.0049	49.07	138	0.05
Wrights Stain	0.0988	53.33	8	1.00

Table III. Reaction Rate Constant and Half-Life for Sensitized Photooxidation of
50 mg/L Bromacil. pH = 10, Reactor = 125 mL, Sensitizer = 10 mg/L,
Sun = 849 W/m^2, DO = Saturated

Sensitizer	$-k$ (min^{-1})	A = Co * exp(INTCPT)	t(1/2) (min)	Relative Sensitizing
Safranin O	0.0081	28.34	79	0.01
Tricon MB	0.5154	27.43	1	0.47
Fisher MB	0.5079	27.57	1	0.46
Riboflavin	0.1152	31.15	6	0.10
Rose Bengal	0.7756	31.04	1	0.70
B. Blue 9	0.6866	30.62	1	0.62
D.Blue 86	0.0142	31.26	52	0.01
Dichlorofluorescein	0.0491	29.47	14	0.04
Eosin Y	0.7264	32.26	1	0.66
Fluorescein	0.0304	30.41	23	0.03
Phloxine B	1.1080	29.23	1	1.00
Wrights Stain	0.3306	28.55	2	0.30

rates of photooxidation of methionine, tryptophan, and tyrosine are independent of the percent ionization of the amino and carboxyl groups.

Effect of pH on Reaction Rate

Little or no photooxidation of bromacil was observed at a pH value of 4 for any of the dyes tested. At a pH value of 7, reaction rates began to increase with reaction rate constants ranging from 0.0012 to 0.0032 (min^{-1}) and half-lives ranging from 219 to 573 min for the following dyes: methylene blue, phloxine B, riboflavin, and rose bengal. At a pH value of 10, reaction rates were highest with no bromacil detected for the following dyes at the listed sampling times: 4 min for phloxine B; 8 min for

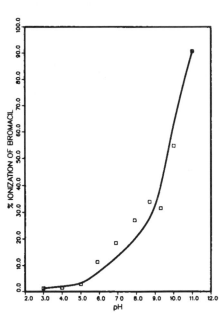

Figure 1.   Effect of pH on ionization of Bromacil.

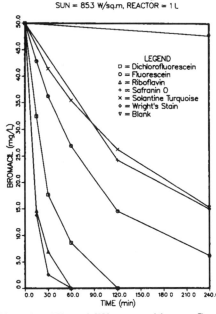

Figure 2.   Effect of different sensitizers on Bromacil removal.

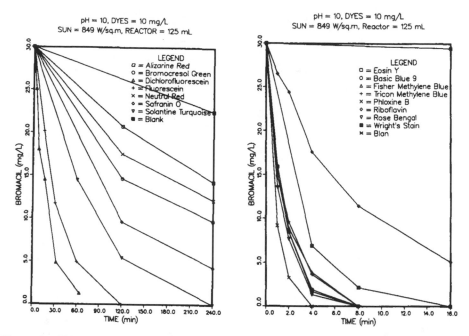

Figure 3. Effect of different sensitizers on Bromacil removal.

Figure 4. Effect of different sensitizers on Bromacil removal.

rose bengal, methylene blue, and eosin Y; 16 min for Wright's stain, 120 min for dichlorofluorescein; and 240 min for solantine turquoise.

Acher and Saltzman [11] observed the same strong pH dependence for a bromacil concentration of 250 mg/L with the sensitizers methylene blue (5 mg/L) and riboflavin (10 mg/L). Little or no degradation occurred at pH values below 5, and the degradation increased at about 6.5% for each pH unit between pH values of 5 and 8 and at about 30% for each pH unit between pH values of 8 and 10. It is interesting to note that their reported degradation rates closely follow the percent ionization rates shown in Figure 1.

Effect of Sensitizer on Reaction Rate

The most effective sensitizers tested were: the basic cationic thiazine dye methylene blue; the basic cationic azine dyes safranine O, neutral red, and riboflavin (vitamin B-2); the acid anionic xanthene fluorane derivatives eosin Y, rose bengal, dichlorofluorescein, fluorescein, and phloxine B; and Wright's Stain, a compound of eosin with oxidized and partly demethylated methylene blue. Textile grade methylene blue was as effective as laboratory grade methylene blue for photosensitizing bromacil. Acridine yellow and acridine orange, which have been reported to be effective sensitizers [13] for certain compounds, were found to be ineffective sensitizers for the photooxidation of bromacil.

Effect of Reducing Agent on Reaction Rate

Sensitized photooxidation of bromacil was observed under anoxic conditions; however, reaction rates were 1 to 2 orders of magnitude less than reaction rates observed with initial saturation of oxygen. It is suspected that oxidation occurred by electron or hydrogen abstraction from bromacil by the dye triplet or semi-oxidized radical with subsequent reduction of the photoexcited dye. Knowles and Gurnani [14] reported that reactive amino acids are similarly partially oxidized in anoxic systems.

Differences in color for the same dyes under aerobic and anoxic conditions were also observed at the close of each experiment. The colors observed in the aerobic and anoxic reactors were: pale orange versus yellow for eosin Y, blue versus clear for methylene blue, pale faded versus strong red for neutral red, pink versus yellow for phloxine B, pink versus yellow for rose bengal, faded versus deep red for safranin 0, and clear versus green tint for Wright's Stain. Bleaching of the sensitizer might be an advantage in the treatment of certain waste streams where color from the sensitizer in the final effluent poses a problem.

CONCLUSIONS

Photooxidation rates for bromacil are dependent on pH value and are maximum at alkaline pH values. The most effective sensitizers were phloxine B, rose bengal, eosin Y, methylene blue, Wright's Stain, and riboflavin. Textile grade methylene blue was as effective as reagent grade methylene blue. Bromacil was partially oxidized in anoxic reactors. Reaction rates were generally 1 to 2 orders of magnitude less than rates in aerobic reactors.

ACKNOWLEDGMENTS

The authors are indebted to the US-Israel Binational Agricultural Research and Development Fund and the Center for the Management, Utilization, and Protection of Water Resources at Tennessee Technological University for financial support.

REFERENCES

1. Martin, H., *Pesticide Manual*, British Crop Protection Council, Worcester, England (1972).
2. Hoffmann, C. E., J. W. McGahen, and P. B. Sweetser, "Effect of Substituted Uracil Herbicides on Photosynthesis," *Nature*, 202:577–578 (1964).
3. Hilton, J. L., T. J. Monaco, D. E. Moreland, and W. A. Gentnter, "Mode of Action of Substituted Uracil Herbicides," *Weeds*, 12:129–131 (1964).
4. Gardiner, J. A., "Synthesis and Studies with 2-C14-Labeled Bromacil and Terbacil," *J. Agr. Food Chem.*, 17:980–986 (1969).
5. Bingeman, C. W., G. D. Hill, R. W. Varner, and T. A. Weidenfeller, *Proc. N. Cent. Weed Contr. Conf.*, 19;42–43 (1962).
6. Moilanen, K. W., and D. G. Crosby, "The Photodecomposition of Bromacil," *Arch. Environ. Contam. and Toxic.*, 2:3–8 (1974).
7. Acher, A. J., and S. Saltzman, "Dye-sensitized Photooxidation of Bromacil in Water," *Jour. Environ. Qual.*, 9:190–194 (1980).
8. Meier, H., "Photochemistry of Dyes," *The Chemistry Of Synthetic Dyes*, Vol. IV, K. Venkataraman, Ed., Academic Press, New York (1971).
9. Watts, J. R., "The Development and Design of Dye Sensitized Photooxidation Waste Stabilization Ponds for Treatment of Industrial Waste," Final Report to BARD; *No.I-171-80*, BARD-Israel (1982).
10. Eisenberg, T. N., E. J. Middlebrooks, and V. D. Adams, "Dye Sensitized Photooxidation of Bromacil in Wastewater," *Proc. 40th Purdue Ind. Waste Conf.*, Butterworth Publishers, Boston (1985).
11. Venkataraman, K., *The Chemistry Of Synthetic Dyes*, Vol. I & II, Academic Press, New York (1952).
12. Sargent, J. W., and R. L. Sanks, "Dye Catalyzed Oxidation of Industrial Wastes," *Jour. Environ. Engin. Div. Am. Soc. Civil Eng.*, 102:879–895 (1976).
13. Spikes, J. D., and R. Straight, "Sensitized Photochemical Processes in Biological Systems," *Ann. Rev. Phys. Chem.*, 18:409–440 (1967).
14. Knowles, A., and S. Gurnani, "A Study of Methylene Blue Sensitized Oxidation of Amino Acids," *Photochemistry and Photobiology*, 16:95–98 (1972).

37 LEAK TIGHTNESS TESTING OF UNDERGROUND TANK SYSTEMS

Richard K. Henry, Environmental Analyst
Lawrence Livermore National Laboratory
Livermore, California 94550

INTRODUCTION

One hundred fifty underground tanks and sumps used to contain petroleum products and potentially contaminated process wastewater have given Lawrence Livermore National Laboratory an unusual opportunity to experiment with a variety of leak test methods and compare applicability of each method. Five types of commercially available Precision Test leak test equipment have been used: Associated Environmental Systems, Hunter Leak Lokator, Horner Ezy-Check, Heath Petro-Tite, and Tank Auditor.

TANK SYSTEMS

Two types of underground tank systems are present at LLNL: tanks and associated piping for the storage of fuel for boilers, emergency generators, and vehicles; and tanks or sumps and associated piping for the retention of potentially contaminated wastewater generated by metal finishing and wet chemistry operations.

The fuel storage tank (Figure 1) is a relatively closed system. The tank is generally constructed of steel, and tank connections typically include the fill pipe, vent line, return line, and a manway. A leak in the fuel storage system is most likely to occur in one of these connections or in the piping itself as a result of improper installation or mechanical failure.

The wastewater retention systems (Figure 2) at LLNL vary considerably in design and in construction materials. Tanks are made of steel, polyethylene, steel-clad concrete, concrete, or fiberglass. The systems involve open-top sumps, drains, complicated and extensive piping, lines for sewer discharge, and pump out lines. Because of greater lengths and complexity of piping, and because the containers are often open at the top, wastewater systems are not only more difficult to test for leaks but are also more leak-prone than the fuel storage systems.

All tanks and sumps were tested if they: 1) do not have secondary containment; 2) are empty or in-use; 3) are intended to contain hazardous or potentially hazardous liquid substances; and 4) are not dedicated exclusively to emergency use.

TANK TESTING METHODS

A program of annual tank testing, using the Precision Test, was initiated in 1985 to determine the tightness of all underground tank systems at LLNL. The definition of a Precision Test, as stated in National Fire Protection Association Code 329, is "any test that takes into consideration the temperature coefficient of expansion of the product being tested as related to any temperature change during the test and is capable of detecting a loss of 0.05 gal (190 mL) per hour" and "should account for all the variables which will affect the determination of the leak rate." These variables include not only temperature changes of the product during the test but also pressure changes, tank end deflection, evaporation, vapor pockets in the tank or piping, and operator error or misuse of the test equipment.

The tests performed on the tanks and sumps at LLNL use volumetric leak detection methods. They identify a leak or determine a leak rate based on the measurement of properties associated with a change in volume. Any factor involved in testing a tank that may mask a change in volume must be strictly accounted for during the actual test. The most important of these is temperature. Temperature effects are determined by the material's coefficient of thermal expansion. Gasoline has a relatively large coefficient of thermal expansion of .0007 per degree Fahrenheit. A reduction in temperature of 0.05 F reduces the volume of gasoline in a 1500 gallon tank by 0.05 gallons. During a tank test this would register as a loss equal to the presently allowed limit of uncertainty per hour. The actual average

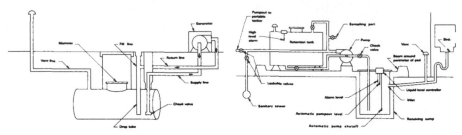

Figure 1.   Example fuel storage system.

Figure 2.   Example wastewater retention system.

temperature of the tank contents is not important nor is the temperature uniformity. It is crucial, however, that the measured temperature changes accurately reflect the average temperature changes of the total liquid volume.

Other variables may cause apparent gain or loss of liquid during a test. Dynamic tank deformation is often a problem when a tank system is subjected to increased pressure, as when it is filled for a test with more liquid than it contains under normal operating conditions. Tank end deflection may appear as a loss of liquid, though it is an actual increase in the volume of the tank. Vapor pockets, if present in the piping or manway, could produce inconclusive test results. Evaporation could contribute to the appearance of a leak where the stored material is highly volatile. Error caused by the complacency or inexperience of the test equipment operator is potentially significant. This potential has been minimized by taking care to evaluate the capabilities of the operators along with the equipment being operated. Since the highest possible accuracy is necessary for these tests, it is important that all factors that may appear as a gain or loss of liquid be recorded.

The aim of each tank test is to determine the leak tightness of the entire system. Fuel systems are tested by filling the tanks and all associated underground piping with liquid to ground level or higher to perform the Precision Test on the system. Wastewater tanks and sumps are tested at a level as high as possible but below the level of the lowest drain discharging into the system. If a leak is discovered in a system test, it is usually followed by a test at a lower level in order to determine if the leak is in the container itself or in the piping. Often, a system may leak but the tank prove to be leak tight. In some wastewater systems it is not possible to test the piping and container together. Piping is complex in these systems, connecting many parts of a building or more than one building. The container and piping must be isolated and tested separately.

FIVE METHODS – DESCRIPTION OF TEST EQUIPMENT AND OPERATING PRINCIPLES

Five types of commercial test equipment have been used by LLNL. All testing was performed by contractors. Test equipment operators were qualified by the respective developers of the test methods or by their representatives. Several tests for two of the methods, A.E.S. and Tank Auditor, were performed or assisted by the developers of the test methods. Tests were performed using the different types of test equipment to permit determination of the best method. Deciding which is "best" includes not only consideration of sensitivity and accuracy but also versatility, tank preparation requirements, time required for the test, scheduling, and cost. The five types of equipment used are known by the following trade names: Petro-Tite, Horner Ezy-Check, Hunter Leak-Lokator, A.E.S., and Tank Auditor.

The Heath Petro-Tite tank test method (Figure 3) uses a 35 gpm pump to circulate the liquid in the tank in order to obtain a uniform temperature change throughout the liquid. The suction line is near the top of the tank. The outlet line is near the bottom, angled both upward and sideways at 45 degrees. The temperature of the liquid is measured by a sensor in the suction line. The actual change in liquid volume is determined by periodically measuring the amount of liquid required to restore it to a fixed reference height. The tank must be full for the test.

The Horner Ezy-Check method (Figure 4) does not circulate the liquid but instead uses a vertically suspended coiled sensor to average the temperatures over the entire tank height. Changes in liquid level are measured and recorded continuously by recording the air pressure required to overcome the static head pressure at the bottom of a submerged air tube.

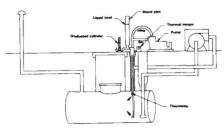

Figure 3.   Petro-Tite.

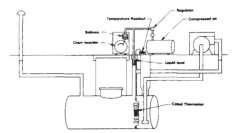

Figure 4.   Horner Ezy-Check.

The Hunter Leak Lokator (Figure 5) uses a single probe, suspended mid-volume in the tank, to measure temperature changes. The output of the probe is recorded on a 10-inch strip-chart recorder with a resolution of 0.01 F.

Level changes of the liquid are measured with a partially submerged sinker, suspended from a balance-type scale. The scale is balanced with sliding weights and a torsion spring. Any subsequent motion of the balance arm is detected with a non-contacting proximity probe and recorded on a second 10-inch strip-chart recorder. Motions of the sinker resulting from level changes create a change in the force applied to the balance, and, subsequently, a change in the balance arm position is recorded. The sensitivity of the recording system, expressed in milliliter/inch, is determined before and after the test by repeatedly lowering a solid with a known volume into the tank.

Associated Environmental Systems (Figure 6), or A.E.S., is a mass balance method employing a pressure transducer placed at the bottom of the tank. A standpipe is used to restrict the zone of liquid level fluctuation to a narrow column of known dimensions. Because pressure is a function of both the height and density of the liquid, as average temperature changes, the changes in liquid height and density in a vertical walled container will offset each other equally. However, in a tank system a reservoir of liquid is contained below the narrow column in which the fluctuations in liquid level are permitted to occur. Liquid height and density will not offset each other equally as temperature changes occur. Therefore, it is also necessary to know the average temperature changes and volume of the liquid in the system to calculate the change in the mass of the liquid and therefore the leak rate. Temperature changes are measured with a single thermistor, also placed at the bottom of the tank. Signals from the pressure transducer and thermistor are processed by a mini-computer and output as a continuous record of changes in average temperature and mass, computed as leak rate.

Tank Auditor (Figure 7), like Leak Lokator, operates on the principle of bouyancy. The Tank Auditor equipment consists of a weight-sensing transducer, an electronic displacement sensor, and a strip chart recorder to monitor level changes in a filled underground tank. A sinker is suspended from the transducer into the product in the tank; a decrease in fluid height will cause a change in bouyancy of the sinker. The resulting mechanical deflection of the weight-sensing transducer is converted to an electrical signal that registers on the chart recorder. Calibration is obtained by the addition of a known volume of the product to the tank followed by the subsequent notation of the number of lines of deflection on the recorder. To compensate for temperature effects, a temperature effects "standard" is used, consisting of a thin column of the tank fluid contained in a copper cylinder. The column of fluid

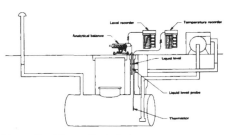

Figure 5.   Hunter Leak Lokator.

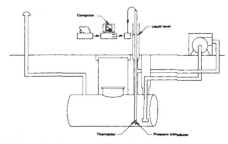

Figure 6.   Associated Environmental Systems (A.E.S.)

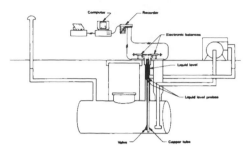

Figure 7.   Tank auditor.

is in the tank and at the tank temperature. The liquid level of the column is monitored by a balance comparable to that of the balance used as the leak detector. Changes in the column are related to changes in the tank.

PROCEDURES FOR FUEL TANK SYSTEMS

To prepare the tank for a test, the tank system is filled with the product to just below ground level the evening before the test is scheduled. Filling the tank to this level permits the entire system, both the tank and all associated underground piping, to be tested. Since the tank is filled several hours before the test, the product is allowed to reach equilibrium with the ambient underground temperature, and tank expansion and end deflection are allowed to occur and stabilize as a result of the increased pressure from the additional volume of liquid. Identifiable vapor pockets are vented. If not vented, a vapor pocket can prevent a successful test by creating erratic, inconclusive data. The problem presented by the presence of vapor pockets in manways was solved at LLNL by drilling a hole through the tops of the manhole covers and installing small valves. Problems caused by the presence of vapor pockets present in other parts of the tank or piping were not observed.

Successful tests also require that tanks be isolated. Several of the fuel storage systems at LLNL have tanks that are interconnected underground by supply lines or vent lines. Tests on these systems yield unsatisfactory results. Monitoring temperature changes simultaneously in two interconnected tanks shows that temperature changes in one tank are independent of temperature changes in the second tank. The result is erratic fluctuations in liquid level. Therefore, valves are used to close connecting piping. Where valves were not present, they were installed.

All tank test methods use equipment, consisting of probes or pipes or both, that must be placed inside an access pipe extending vertically from the ground surface into the top of the tank. Usually the access pipe is the fill pipe. The fill pipe is often connected to a drop tube located inside the tank. Tests performed in systems using the fill pipe/drop tube combination as the tank access pipe have not proven successful. Removal of the drop tube has been necessary. For tanks in which removal of the drop tube was not possible or which had a fill pipe with 90 degree elbows, access to the tank was made by other means. Either access was made through the supply line connection assembly by extracting the supply tube and check valve, or a hole was cut into the manway cover and a standpipe that could extend to ground level was welded onto the manway cover.

PROCEDURES FOR WASTEWATER SYSTEMS

Contrasting with the one-piece fuel tank is the predominant type of wastewater tank. By nature of their construction, most wastewater containers are not sealable at the top. Some of these containers are isolated from other construction. Others are present as sump-like structures in a concrete surface, either at grade, or below grade in a basement or pit.

Coincident with the testing of underground fuel systems in the summer of 1985 was the beginning of a project to test and evaluate methods showing potential applications to testing open-top wastewater containers. This activity was a prerequisite to testing open-top wastewater containers because the demands made on the capabilities of test equipment are greater to test open-top containers than to test the fuel tanks by the procedures defined by the test developers. All the methods considered calculate a gain or loss of liquid into a system by recording fluctuations in the level of liquid in the system during the time of the test. When performed in a relatively narrow fill pipe or stand pipe, the five commercial types of test equipment evaluated are able to perform this task surpassing the level of sensitivity required by the Precision Test. However, the measurement of changes in liquid level in an open-top container means that any apparent gain or loss in the system is spread over a large liquid surface area defined by the vertical walls of the container. The test equipment must be proportionately more sensitive in order to measure level changes in a test of one to two hours duration and still achieve the requirements of the Precision Test.

The test and evaluation project focused on the methods available commercially in the San Francisco Bay Area. These methods are Petro-Tite, Ezy-Check, Leak Lokator, A.E.S., and Tank Auditor.

Petro-Tite and Leak Lokator were not candidates for the task of testing open-top containers. The Petro-Tite test is inappropriate for open-top containers because of its reliance on a stand-pipe filled with liquid. Leak Lokator was not considered because of the relative difficulty scheduling tests and because the test experience itself caused uncertainty regarding the suitability of Leak Lokator for testing open-top containers. In particular, tests on the vehicle fuel tanks at two locations showed that the liquid level sensing and recording system had an extreme tendency to oscillate requiring operator-dependent interpretation for accurate determination of the best straight line drawn through the oscillations to determine the rate of apparent gain or loss.

Two containers with vertical walls were selected for a semi-quantitative evaluation of the ability of A.E.S., Ezy-Check, and Tank Auditor, to measure and record very small changes in liquid level. One of the containers was designated the experimental tank. It is a cylindrical aluminum tank with an inside diameter of 26 inches. It was acquired for the tank testing program and was retrofitted with a needle valve at the bottom to make possible purposely induced measurable leaks. The tank was installed in a concrete pit and wrapped with insulation. The second container is an in-use sump. This sump is a concrete box-shaped structure with inside horizontal dimensions of four feet by four feet.

Ezy-Check was evaluated first. Fuel tank testing experience with Ezy-Check showed the method to be versatile and easy. One hour tests performed in the experimental tank using Ezy-Check equipment demonstrated the sensitivity to measure a leak of .05 gal/hr. However, when the Ezy-Check test was performed on the sump, it was observed that a .05 gallon change spread over the sixteen square foot surface area in the sump caused only two lines of deflection on the recorder chart. Therefore, sixteen square feet was judged to be the largest surface area testable by Ezy-Check in one hour.

The A.E.S. test on the experimental tank determined a leak rate to within 20% of the measured induced leak. A.E.S. was therefore considered capable of testing containers with liquid surface areas up to 3.7 square feet, the area of the 26 inch diameter tank. Because of its use as a pressure transducer, A.E.S. offers the additional advantage of being temperature independent when the test takes place in a vertical-walled container.

Tank Auditor is new in the market and was not available for fuel tank testing at LLNL. However, tests on both the experimental tank and the in-use sump easily accomplished the 0.05 gal/hr criterion for sensitivity. Tank Auditor's use of an electronic balance and the principle of bouyancy showed to be particularly suited to testing open-top containers. The high sensitivity of the test equipment, qualified test technicians, and price made Tank Auditor the method selected for testing most of the wastewater containers.

Several of the Tank Auditor tests on wastewater tanks and sumps have yielded strip charts that cannot be interpreted because the tracings wander on the chart. The character of the irregular tracings appears to be bimodal. One type is characteristic for tracings on underground fuel systems that were shown to have trapped pockets of vapor. Therefore, it is believed that some of the wastewater containers are showing evidence of trapped air in the piping. The cause for the second type is less apparent. Speculation suggests that the cause for the second type is due to changes occurring at the sites where the system is vented to the atmosphere. Outdoors, this may be caused by changes in wind velocity or changes in wind direction with respect to the orientation of the vent orifices. Indoors, this may be caused by the on/off cycling of air conditioning systems. An additional possibility is radio frequency interference with the signal used by the electronic balance.

Testing Considerations

Leak tests in wastewater systems do not have the same factors for possible error as do fuel storage systems. Since many of the wastewater tanks are concrete, tank expansion is not a problem. More piping is involved with wastewater systems, increasing the possible occurrence of vapor pockets. In comparison to gasoline, water has a very low coefficient of thermal expansion, and so temperature changes during the test do not have nearly as dramatic effects on test accuracy. This has been substantiated by the wastewater tests in which temperature was monitored. The predicted and observed effects justify not monitoring temperature during tests when it is difficult or potentially hazardous to the operator to do so.

It should be noted that water used to fill the tanks and sumps for testing is taken out of the ground from pipes and put back into the ground into containers at approximately the same temperature. Several factors specific to each tank system determine the level to which the tank is filled for the test. Since most of the wastewater retention containers are open sumps, it is often not possible to fill them completely. The upper level is frequently the level of the lowest floor drain discharging into the system.

Isolation of the tank from uncontrolled discharges in the building and from problems inherent in the piping is very often difficult to achieve. Two methods are available for isolating a wastewater container for testing: the use of a plumber's plug, and excavation to install a valve in order to isolate the container from the piping.

SUMMARY AND CONCLUSIONS

Fuel Storage Systems

During the summer and fall of 1985, underground diesel fuel and gasoline storage tanks and associated underground piping were tested for leak tightness using performance standards defined by the National Fire Protection Association Code 329 Precision Test. Tests were performed by contracted operators using four commercial test methods: Petro-Tite, Ezy-Check, Hunter Leak Lokator, and A.E.S. Tank systems shown to have leaks in excess of 0.05 gal/hr when filled with liquid to near ground level were retested at a lower level, just above the tops of the tanks. The retests showed no leaks. Therefore, the evidence provided by the tests indicates that the systems which leak when filled for testing do not leak under normal operating conditions.

Wastewater Retention Systems

With the completion of fuel tank testing, a project was undertaken to identify, test, and evaluate methods showing possible application to testing leak tightness in the uncovered open-top containers characteristic of the Laboratory's potentially contaminated wastewater retention systems. Evaluation tests were performed in two containers: a specially prepared aluminum experimental tank capable of measuring induced leaks, and an in-use concrete sump. A.E.S., Ezy-Check, and Tank Auditor were evaluated using these two tanks. Both of the containers have vertical walls and have liquid surface areas of 3.7 square feet and 16 square feet, respectively. The evaluation tests showed A.E.S. to be successful in determining a leak rate of 0.05 gal/hr in the 3.7 square foot experimental tank. Ezy-Check achieved the same success in the 16 square foot sump, as well as in the smaller container. Tank Auditor was shown to be more sensitive than either A.E.S. or Ezy-Check. An upper limit has not been defined for the size of an open-top container testable by Tank Auditor.

During the spring of 1986, wastewater tanks and sump systems were tested for leak tightness. Most tests were performed by Tank Auditor. The prevalent cause of leaks detected by the tests has been shown to be a failed or improper joint connecting piping or other ancillary equipment to the tank or sump.

Inconclusive tests are predominantly caused by either the addition of liquid to the system during the test from an uncontrolled discharge point in the building or from erratic data due to fluctuations in air pressure, sometimes caused by trapped air.

Inconclusive tests are resolved by additional preparation of the tank system prior to testing. For many systems this means isolating the container from the associated piping so that each can be tested separately. Isolation is accomplished by the use of plumber's plugs or by excavation and installation of a valve.

ACKNOWLEDGEMENTS

The author is indebted to his colleagues Richard Sites, Michael Sledge, and Marjorie Gonzalez. Richard Sites coordinated testing activities. Michael Sledge provided editorial support, and Marjorie Gonzalez provided suggestions and assistance to all phases of the project.

REFERENCES

1. Lawrence Livermore Plant Engineering, *Hazardous Substance Storage Structures*, Draft Plant Engineering Standard, Lawrence Livermore National Laboratory, Livermore, CA (1986).
2. Niaki, Shahzad, and John A. Broscious, *Underground Tank Leak Detection Methods*, IT Corporation, Pittsburgh, PA (1985).
3. New York State Department of Environmental Conservation, *Technology for the Storage of Hazardous Liquids*, Albany, NY (1983).

38 ANAEROBIC DEGRADATION OF CHLORINATED SOLVENTS

Cecilia Vargas, Graduate Student

Robert C. Ahlert, Professor

Elizabeth E. Abbott, J.J. Slade Scholar

Marshall G. Gayton, J.J. Slade Scholar

Department of Chemical and Biochemical Engineering

Rutgers, The State University

Piscataway, New Jersey 08854

INTRODUCTION

Many environmentally important man-made compounds are halogenated. The list of halogenated species includes pesticides, plasticizers, plastics, solvents, and trihalomethanes. Of the halogenated compounds, the best known and most studied are the chlorinated compounds. This is because of the highly publicized problems associated with DDT, other pesticides, and numerous industrial solvents [1]. Well contamination by compounds such as chloroform, 1,1,1-trichloroethane, and tetrachloroethylene has been documented [2]. The presence of such compounds in the environment is due to inadequate disposal techniques and accidental spillage.

Many of these chlorinated compounds are quite persistent in the environment. They are not easily degraded aerobically and have been found to break through rapidly in granular activated carbon beds [3]. In contrast to naturally occurring compounds, man-made compounds are less susceptible to biodegradation. Naturally present organisms often cannot produce the enzymes necessary to bring about transformation of the original compound to a point at which the resultant intermediates can enter into common metabolic pathways and be mineralized completely [4].

The metabolic fate of these chemicals under aerobic conditions have been studied widely, and degradation, if any, or persistence patterns established. Anaerobic studies with halogenated compounds have shown several of these compounds are biodegradable. Methane-producing freshwater lake sediment was found to dehalogenate iodo-, chloro-, and bromobenzoates by a reductive reaction [5]. It has been shown that transformations of trihalomethanes can also occur under anaerobic conditions [2]. Lang et al. reported on the biotransformation of some haloaliphatic compounds, suggesting a series of reductive dechlorination reactions [2].

The main interest in biodegradation of chlorinated compounds is in anaerobic digestion. Although there are disadvantages, such as long startup time and incomplete understanding of the microbiology involved, there is more evidence for biodegradation of chlorinated compounds in anaerobic systems than in aerobic systems. Recently, researchers have shown that 1- and 2-carbon halogenated aliphatic organic compounds are biodegradable under methanogenic conditions [2]. There is also field evidence for the long-term transformation of halogenated compounds under anaerobic conditions [2]. No specific research has been done to determine the actual degradation mechanism or the affected group of microorganisms.

In this study, the biodegradability of 1,1,1-trichloroethane (TCA), 1,1-dichloroethane (DCE), and dichloromethane (DCM) under anaerobic conditions was investigated. Also, the acclimation potential of a mixed anaerobic culture to these compounds was assessed.

Although other studies with these compounds have been carried out, only the extent of biodegradation has been studied. In addition to biodegradation studies, acclimation and inhibition at different concentration levels was studied.

Finally, through the variation of conditions to favor certain groups of bacteria, the main population affected by these "toxic" compounds was studied.

MATERIALS AND METHODS

Three different cultures were developed for use in the experiments. They were each adapted to their respective feeds and experimental parameters until stable conditions were obtained. In these studies, acclimation was achieved after about five months. The anaerobic seed was obtained from the Berkeley Heights Sewage Treatment Plant, Berkeley Heights, New Jersey.

Mixed Anaerobic Culture

The mixed culture was developed from the original anaerobic seed. The culture was placed in an atmosphere of nitrogen. Once each week, 210 mL of deoxygenated media containing (per liter), 1 g yeast extract, 2.96 g NH_4Cl, 0.34 g KH_2PO_4, 1 mg resazurin, 0.5 g cysteine hydrochloride, and 1 mL trace metal solution [6] plus 4 mL absolute ethanol was fed to 1.5 liter of culture. This gave a hydraulic retention time of 50 days. This main culture was developed in 2-liter glass aspirator bottles mounted on magnetic stirrers. Water at 35°C is passed through stainless steel coils submerged in the culture broth [7].

Acetogenic Culture

The acetogenic culture was developed from the main anaerobic culture. The pH was adjusted to 5, and the residence time decreased to 10 days in order to wash out the methanogens. The same media used for the mixed culture was diluted accordingly. Once every 2 days, 10 mL of this media plus 0.02 mL of absolute ethanol was fed to 50 mL of culture. Cultures were grown in a nitrogen environment in 100-mL amber serum bottles fitted with hycar septa and aluminum seals. To avoid hydrogen inhibition, the head-space gas was removed and replaced with nitrogen during feeding periods using a calibrated 50-mL glass syringe.

Methanogenic Culture

The methanogenic culture was developed from the main anaerobic culture. Once every 7 days, 7 mL of deoxygenated media containing (per liter) 1 g yeast extract, 2.21 g NH_4Cl, 0.26 g KH_2PO_4, 40.61 g $NaHCO_3$, 1 mL trace metal solution [6], 0.5 g cysteine hydrochloride, and 50 mL $TiC_6H_5O_7$ buffer [8] plus 50 mL H_2 was fed to 50 mL of culture. This results in the same retention time as for the main culture, i.e., 50 days. The culture was grown in a hydrogen environment, in 100-mL amber serum bottles fitted with hycar septa and aluminum seals.

Analytical Methods

Gas production was measured using a water-lubricated calibrated glass syringe. Ethanol and acetic acid were separated and analyzed using a Hewlett-Packard Model 5880A gas chromatograph equipped with a flame ionization detector. TCA was analyzed using an electron capture detector.

Experimental SetUp

All experiments, batch and semi-batch, were performed using 100-mL amber serum bottles. Batch experiments were performed with DCM, DCE and TCA, at concentrations up to 30 mg/L. The toxicant was diluted in absolute ethanol and introduced into the reactors together with the feed at the beginning of an experiment. Two reactors were kept unspiked as controls. The experiment was run until the control reactors no longer produced gas. Gas production was measured daily.

Semi-batch studies were performed with DCE and TCA, using the developed mixed culture. Studies were performed with duplicate control reactors and ten different toxicant concentrations. Toxicants were diluted in ethanol and introduced into the reactors together with the feed at the beginning of the experiment. When the control reactors had ceased gas production, all reactors were refed the same amount of absolute ethanol. Experiments were continued until the control reactors started dying off.

Continuous acclimation studies were performed with TCA, using the separately developed acetogenic and methanogenic cultures. For the acetogenic culture, the toxicant was dissolved in ethanol and introduced into the reactors, at the beginning of the experiment, together with the feed. The reactors were run continuously under the same conditions used in the development stage. During feeding, 10-mL samples were taken and analyzed for substrate and toxicant concentrations. In the methanogenic studies, the toxicant was introduced directly into the reactors at the beginning of the experiment together with the feed. The reactors were run continuously under the same conditions used in the

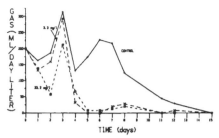

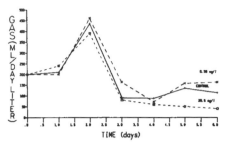

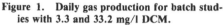

Figure 1. Daily gas production for batch studies with 3.3 and 33.2 mg/l DCM.

Figure 2. Daily gas production for batch sudies with 0.58 and 30.6 mg/L DCE.

development stage. At every feeding, 7-mL samples were taken and analyzed for toxicant concentration. The head space was analyzed for CH_4 and CO_2.

RESULTS AND DISCUSSION

Batch Studies

All batch experiments were run with two controls for each toxicant. The incubation period varied from 1 to 2 weeks, depending on the control reactors. When the control reactors had produced at least the expected amount of gas from a given amount of carbon, the experiments were stopped. The theoretical amount of gas is calculated from the following stoichiometric equation:

$$C_2H_5OH \rightarrow 1/2\ CO_2 + 3/2\ CH_4 \qquad (1)$$

From the theoretical amount of gas calculated, the expected amount of gas, corrected for solubility, is:

$$0.95(CH_4)_{th} + 0.35(CO_2)_{th} \qquad (2)$$

For each set of experiments, the total amount of gas produced by the control reactors did not vary more than 0.5%. The average of the two control reactors was used as a basis of comparison with the test reactors.

Five different concentrations of DCM, ranging from 3.3 to 33.2 mg/L, were studied. The level of toxicity was directly proportional to toxicant concentration. Results for the daily gas production for 3.3 and 33.2 mg/L are plotted in Figure 1.

Of the three compounds tested, DCE was the least toxic. Ten different toxicant concentrations ranging from 0.58 to 35 mg/L were tested. Daily gas production for each of the concentrations followed the control closely. There was no significant inhibition at any of the concentrations. Results for daily gas production for 0.58 and 30.6 mg/L are plotted in Figure 2.

Ten different concentrations of TCA, ranging from 0.4 to 30.1 mg/L, were studied. As with DCM, the degree of toxicity is directly related to toxicant concentration. At 4 mg/L, greater inhibition was observed as a sudden decrease in gas production. At concentrations above 10 mg/L, the microorganisms were severely inhibited. Data for 2 and 30 mg/L are plotted in Figure 3.

Semi-Batch Studies

Semi-batch acclimation experiments were performed with DCE and TCA. Studies were performed with duplicate control reactors and ten different toxicant concentrations. When the control reactors had produced at least the expected amount of gas from a given amount of ethanol, all reactors were refed the same amount of ethanol. For each of the experiments, the daily gas production for the control reactors did not vary more than 0.7%. Thus, the average values of the control reactors were used as the basis of comparison. Experiments continued until the control reactors started dying off. During the course of the experiment, gas production was measured every day. When necessary, 1 N sodium hydroxide was added to adjust the pH to near neutral.

Concentrations of DCE, in the range from 0 to 35 mg/L, were tested. For toxicant concentrations of 25, 30, and 35 mg/L, gas production ceased after 23 days. Feeding of ethanol was discontinued to these three reactors, and they were shelved. However, no recovery occurred. For concentrations under

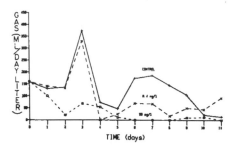

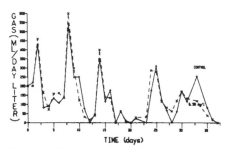

Figure 3.   Daily gas production for batch studies with 0.4 and 30 mg/L TCA.

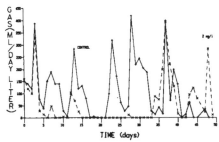

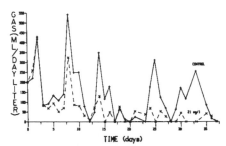

Figure 4.   Daily gas production for semi-batch
study with 0.58 mg/L DCE.

Figure 5.   Daily gas production for semi-batch
study with 21 mg/L DCE.

21 mg/L, growth continued for 37 days with increased inhibition at the higher range. Results for 0.58 and 21 mg/L are plotted in Figures 4 and 5. (The spikes in gas production in the controls are the result of the ethanol fed after the cessation of gas production).

Concentrations of TCA, ranging from 0 to 30 mg/L, were tested. Gas production ceased after 17 days in reactors given doses above 2 mg/L. Ethanol feed to these reactors was discontinued and they were shelved. The reactors given doses of 2 and 4 mg/L resumed gas production after 20 days. No significant inhibition occurred with toxicant concentrations of 0.4 and 0.8 mg/L. Results for the reactors dosed with 2 and 4 mg/L are shown in Figures 6 and 7.

Acetogenic vs. Methanogenic Studies

The toxicant studied in these experiments was TCA, at concentrations ranging from 0 to 27 mg/L. The degree of inhibition was observed as changes in substrate utilization rates, as compared with unspiked controls.

In the acetogenic experiments, the developed acetogenic culture was used with three different toxicant concentrations, i.e., 0.27, 2.7 and 27 mg/L. Substrate (ethanol) utilization is described in Figure 8. It is apparent that the acetogenic population of the anaerobic mixed culture is severely

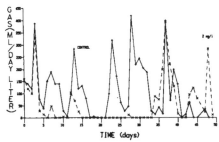

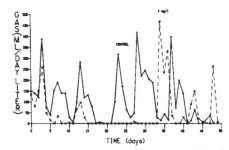

Figure 6.   Daily gas production for semi-batch
study with 2 mg/L TCA.

Figure 7.   Daily gas production for semi-batch
study with 4 mg/L TCA.

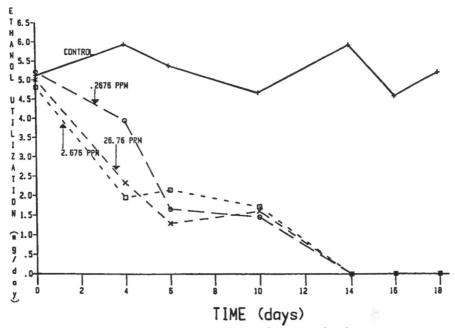

Figure 8.   Substrate utilization rates for acetogenic cultures.

inhibited by TCA at all concentrations. Final analysis of reactor contents showed buildup of ethanol and no detectable changes in toxicant concentrations.

Four different toxicant concentrations, i.e., 2.7, 5.3, 10.7, and 26.8 mg/L, were studied using the developed methanogenic culture. Substrate (hydrogen) utilization rates are described in Figure 9. Hydrogen utilization rates were determined by performing mass balances on head space gas composition. The overall average substrate utilization rate for the reactors remained fairly constant. After 35 days, utilization rates appeared to decline for reactors dosed with 10.7 and 26.8 mg/L TCA. However, the methanogenic culture had much more tolerance to TCA than the acetogenic culture. Final analyses of reactor contents, for initial concentrations of 2.7, 5.3, 10.7, and 26.8 mg/L indicated 100, 96, 80 and 68% removal, respectively, in 35 days.

Modeling Results

To analyze the data obtained with the batch experiments, the activity of anaerobic gasification is defined as:

$$A = v/V \tag{3}$$

where:

A = relative anaerobic gasification activity
v = total volume of gas produced from spiked sludge, mL
V = total volume of gas produced from unspiked sludge, mL

Table I summarizes activity versus concentration data for the three compounds.

Pearson et al. used the following empirical model to analyze combined toxicity effects on anaerobic digestion [9]:

$$A = [1 + (\sum_{i=1}^{m} C_i/K_i)^a]^{-1} \tag{4}$$

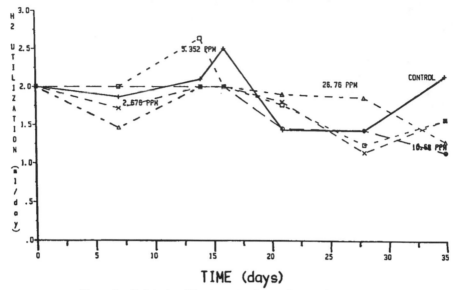

Figure 9.   Substrate utilization rates for methanogenic cultures.

where:

m = number of preservatives
C_i = concentration of ith preservative, mg/L
K_i = half-kill dose, mg/L = value of C_i for which A = 0.5 when m = 1
a = sensitivity exponent

This is simplified, as applied in this study, because m = 1:

$$A = [1 + (C/K)^a]^{-1} \tag{5}$$

Values of K and a are computed by linearizing the equation and performing a least squares linear regression. The results are shown in Figures 10 to 12. The solid line represents the model, while the data are plotted as discrete points.

Calculated half-kill doses are 0.48 mg/L, 26.1 mg/L and 3.8 mg/L for DCM, DCE and TCA, respectively. The calculated values confirm the order of toxicity of the compounds.

Table I. Compound Activities

DCM		DCE		TCA	
C(mg/L)	Activity	C(mg/L)	Activity	C(mg/L)	Activity
3.3	0.42	0.59	1.04	0.40	1.02
9.5	0.36	2.35	1.03	0.80	0.97
19.9	0.30	5.88	1.02	2.00	0.30
26.5	0.33	9.42	0.99	4.00	0.28
33.2	0.32	11.77	1.02	6.00	0.29
		14.12	0.69	10.00	0.23
		21.19	0.63	12.00	0.20
		25.89	0.34	16.00	0.21
		30.60	0.47	20.00	0.18
		35.31	0.38	30.10	0.16

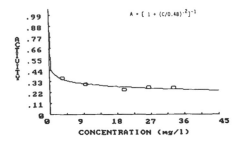

Figure 10. Modeling result for dichlorome-thane.

CONCLUSIONS

A mixed anaerobic population is capable of degrading and/or acclimating to the haloorganic compounds DCM, DCE, and TCA. Half- kill doses, determined from batch experiments, define a relative degree of toxicity for each of the compounds. The microorganisms exhibit a great degree of tolerance for DCE with no apparent inhibition at concentrations up to 35 mg/L.

Acclimation studies with TCA demonstrated that continued periods of zero gas production do not necessarily mean the death of the organisms. Reactors dosed with 2 and 4 mg/L recovered after 20 days of zero gas production. After this lag period, daily gas production was greater than or equal to control reactors. The overall acclimation period was only 33 days, i.e., less than half of the acclimation period of 10 weeks reported by Bouwer et al in studies with methanogenic columns [2]. However, acclimation periods tend to vary greatly for different anaerobic seed cultures. Studies with separate anaerobe populations indicate that the methanogens, known for extreme sensitivity, may be responsible for degradation of these chlorinated compounds. Although no data on similar experiments have been found in the literature, there is some indirect evidence to support these results. An enzyme that dehalogenates chloroacetate exhibits maximum activity at pH 9.5 [10]. At a lower pH, it is denatured. Therefore, the acetogenic environment is not suitable for this enzyme. In a study of reductive dehalogenations of halobenzoates by anaerobic microorganisms, Horowitz et al. reported no dehalogenation in the absence of CH_4 production [5].

The results obtained can explain the recovery of gas production in the mixed populations. If the methanogens were being inhibited, recovery is not likely. The acetogens would continue to utilize the ethanol, and continue to produce hydrogen and acetic acid. During this lag period, both acetic acid and hydrogen would build up. Gas production would not be observed because the methanogens are not able to utilize ethanol as a substrate. The buildup of acetic acid would lower the pH, making conditions unfavorable for the methanogens to recover. Also, increased hydrogen partial pressure would inhibit the acetogens . If the acetogens were being inhibited, as these results suggest, recovery is more likely. During the lag phase, no gas production is observed because the acetogens are inhibited, and ethanol is not being broken down for use by the methanogens. Once the methanogens have degraded the toxicant to below inhibition levels, the acetogens start utilizing ethanol, and gas production resumes.

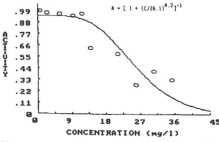

Figure 11. Modeling results for 1,1-dichloroethane.

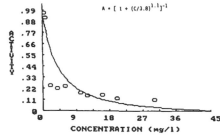

Figure 12. Modeling results for 1,1,1-trichloroethane.

REFERENCES

1. Kobayashi, H., and Rittmann, B. E., "Microbial Removal of Hazardous Organic Compounds," *Environ. Sci. Technol.*, Vol. 16, No. 3, 170–181 (1982).
2. Bouwer, E. J., and McCarty, P. L., "Transformations of 1- and 2-Carbon Halogenated Aliphatic Organic Compounds Under Methanogenic Conditions," *App. Env. Microbiol.*, Vol. 45, No. 4, 1286–1294 (April 1983).
3. Bouwer, E. J., Rittmann, B. E., McCarty, P. L., "Anaerobic Degradation of Halogenated 1- and 2-Carbon Organic Compounds," *Environ. Sci. Technol.*, Vol. 15, 596–599 (1981).
4. Jeris, J. S., and McCarty, P. L., "The Biochemistry of Methane Fermentation Using C^{14} Tracers," *JWPCF*, 178–192 (Feb. 1965).
5. Horowitz, A., Sulfita, J. M., Tiedje, J. M., "Reductive Dehalogenations of Halobenzoates by Anaerobic Lake Sediment Microorganisms," *App. Env. Microbiol.*, Vol. 45, No. 5, 1459–1465 (May 1983).
6. Shelton, D. R., and Tiedje, J. M., "General Method for Determining Anaerobic Biodegradation Potential," *Appl. Env. Microbiol.*, Vol. 47, No. 4, 850–857 (April 1984).
7. Corbo, P., "Industrial Landfill Leachate Characterization and Treatment Utilizing Anaerobic Digestion with Methane Production," *Ph.D. Dissertation*, Rutgers University, N.J. (January 1985).
8. Zehnder, A. B. J., and Wuhrmann, K., "Titanium (III) Citrate as a Nontoxic Oxidation-Reduction Buffering System for the Culture of Obligate Anaerobes," *Science*, Vol. 194, 1165–1166 (1976).
9. Pearson, F., Shiun-Chung, C., Gautier, M., "Toxic Inhibition of Anaerobic Biodegradation," *JWPCF*, Vol. 52, No. 3, 472–482 (March 1980).
10. Motosugi, K., and Soda, K., "Microbial Degradation of Synthetic Organochlorine Compounds," *Experientia*, Vol. 39, 1214–1220 (1983).

39 SOIL AND GROUNDWATER CONTAMINATION AT WOOD PRESERVING PLANTS

John Ball, Professor
Civil Engineering Department
The University of Alabama
University, Alabama 35486

INTRODUCTION

This study is based on four individual sites selected from among several dozen wood preserving plant investigations performed since the effective date of the RCRA regulations [1]. It has been observed that all wood preserving plants investigated can be separated into facilities that have operated with process water surface impoundments and those that have not. Most of the surface impoundments were installed in the early 1970's in response to suggestions from state regulatory agencies who objected to direct discharges or for the purpose of keeping abreast of industry changes. Almost all plants investigated that have had surface impoundments have indicated some groundwater contamination from preservatives. The degree of the problem, however, has ranged from localized conditions at a single monitor well indicating one or more dissolved constituents of a preservative to relatively large areas and to significant depths of free preservative contamination.

THE STUDY

Four plants having significant contamination were selected for discussion based on the availability of information, type of preservative, and other factors such as extent of contamination, soil type, etc. These facilities shall be referred to as Plants A, B, C, and D. Table I presents a summary of several facility characteristics for each of the plants.

Plants A, B, and C began operating prior to 1940 with Plant D starting up in 1973. The older plants began with creosote as the only preservative followed by pentachlorophenol in the 1960's at Plants B and C. The use of the preservative CCA (copper, chromium, arsenic) was started in the 1970's at Plant C. Plant D began operating with creosote and CCA preservatives from initial startup. Although Plants C and D treat with CCA, little contamination has been detected from release of this water soluble preservative. The significant contamination has been from the oil soluble creosote and pentachlorophenol.

Except for Plant B, the surface impoundments began operation between 1972 and 1974. The Plant B pond was constructed in 1974, but an existing drainage canal was used to handle process water from the start of operations. Prior to the early 1970's, process water from Plants A and C was discharged

Table I. Facility Characteristics

	Plant A	Plant B	Plant C	Plant D
Plant Startup	1937	1926	1928	1973
Lagoon Startup	1974	1926	1972	1973
Preservative	Creosote	Creosote Penta[a]	Creosote Penta[a] CCA[b]	Creosote CCA[b]
Number of Cylinders	3	3	4	3
Source of Contamination	1-Ditch 1-Pond	1-Canal 1-Pond	1-Pond	2-Ponds
Total Surface Acres of Source	1.5	1	1	0.5
Total Volume of Source, Cubic Yards	1700	3600	4700	1600

[a]Pentachlorophenol
[b]Copper, Chromium, Arsenic.

347

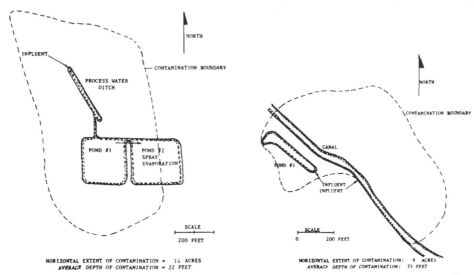

Figure 1.   Extent of contamination at Plant A.   Figure 2.   Extent of contamination at Plant B.

down small drainage ditches and left the properties. The swampy location at Plant B resulted in significant quantities of process water sludge being stored in that portion of the canal on the property. The two ponds at Plant D were constructed prior to beginning operations.

Figures 1, 2, 3, and 4 present a plan view of the four plants indicating the major features of the facilities and the horizontal extent of contamination. Plant A shows the process water ditch which was the initial settling basin for the process water followed by ponds 1 and 2. Prior to closure operations, creosote sludge had filled the ditch and partially filled pond 1. Only clarified process water has reached pond 2. The dashed line indicates the horizontal extent of creosote contamination. The quantities and depths are shown on the figure and in Table II for comparison.

Figure 2 shows the features of Plant B. The process water was discharged into the center of the canal and spread in both directions. The pond was constructed much later and also became contaminated with sludge. The horizontal extent of contamination was less than nine acres. Although the average depth of vertical contamination was 35 feet, contamination has been detected to 55 feet. Additional work will be performed soon to determine the maximum depth of contamination.

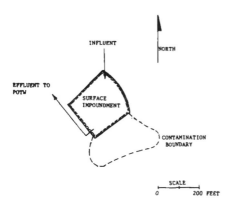

Figure 3.   Extent of contamination at Plant C.

Table II. Contamination Comparison

	Plant A	Plant B	Plant C	Plant D
Horizontal Extent of Contamination in Acres	14	9	13	0.5
Depth of Contamination in Feet				
Average	22	35	60	20
Maximum	22	55	60	20
Estimated Horizontal Ground Water Velocity, Feet/Yr	0.03	0.05	0.2	0.2

Plant C contamination is shown in Figure 3 at one acre to the south of the pond and to a depth of 60 feet. The downgradient contamination is about the same area as the impoundment. The influent into the pond was from the top of the figure with the discharge to a POTW as shown. Prior to POTW discharge spray nozzles were provided to enhance evaporation.

Figure 4 presents the conditions at Plant D. At this site two impoundments were operated until 1980. In 1980, pond 1 was closed and pond 2 cleaned and lined. The contamination at this site has approximately the same areal extent as the surface impoundment which is similarly to Plant C. The depth of contamination at this site is 20 feet deep.

Figure 5 shows the groundwater contours and wells installed at Plant A. The contours indicate a water table that mounds in the area of well W-7 which was thought to be caused by the water levels in the ditch and pond 1. However, the ponds and ditch were cleaned recently and the groundwater contours remained as shown in Figure 5. Additional investigation uncovered what may be a significant leak 20 feet below the surface in the vicinity of well W-7 from a flowing artesian well. This leak has probably accelerated the movement of the contamination downgradient. Steps are being taken to verify this condition and to stop any flow.

Boring logs during construction of the wells indicated a sand strata approximately 20 to 25 feet below the surface and 10 feet below the water table in the area of the contamination outlined in Figure 1. The contamination apparently moved downward from the ditch and pond to the sand strata and then spread horizontally to the extent of the strata. The sand strata does not exist at the outlying wells W1, W3, W6, W10, and W12. The clay appears to prevent significant horizontal movement of the preservative.

Figure 6 shows the groundwater contours for Plant B. The groundwater flow is toward the north and east and is within 10 feet of the surface. Similar to Plant A, this site has a permeable layer of sandy-silt between 20 and 40 feet below the surface. Unlike Plant A, the boring logs indicate this more permeable layer is continuous. In addition there are sandy-silts sloping from the area of the canal and pond downward to the deeper sandy-silt layers. These sloping layers probably provided the conduit for contamination. Some analyses of preservative concentration with depth have indicated increasing

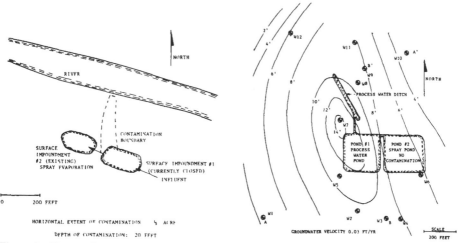

Figure 4.   Extent of contamination at Plant D.   **Figure 5.   Groundwater contours at Plant A.**

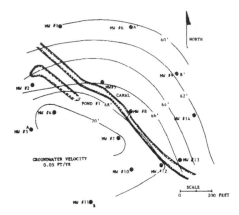

Figure 6.   Groundwater contours at Plant B.

concentrations from the surface to 35 feet and then decreasing concentrations to 55 feet below the surface.

Figure 7 presents the groundwater contours at Plant C. This plant has a water table approximately 25 feet below the surface and two sand layers as shown in Profile A-A' in Figure 8. Although the water samples in wells W1, W2, W3, and W4 have indicated no contamination, all four borings have shown some problem. Borings B-1, B-2, and B-3 indicated contamination in the sand layer that exists about 50 to 60 feet below the surface. The visible contamination was found just above the clay layer. No sign of contamination was found in the lower sand strata in boring B-4. However, considerable concentrations were found just above the water table in the clay as shown in Figure 9. Figure 9 shows the concentration in mg pentachlorophenol and naphthalene per kg dry soil with depth in boring B-4. Most of the contamination was in the clay between the surface and 18 feet deep. Pentachlorophenol and naphthalene are again found in the upper sand layer. No preservative was found in the sandy-clay, lower sand strata or the clay. It is expected that the lower sand strata may become contaminated by horizontal movement with time.

Figure 10 shows the groundwater profile for Plant D sloping toward the river. The water table is approximately 25 feet below the surface. Like the other three plants, the contamination has moved downward from an impoundment through the surface of the water table to a permeable strata and then horizontally. The permeable strata at this site is a sand located below a silty-sand.

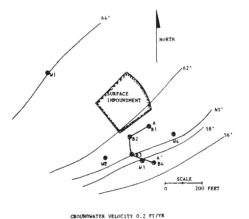

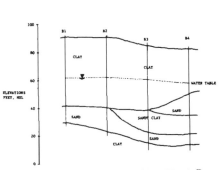

Figure 7.   Groundwater contours at Plant C.   Figure 8.   Geologic profile A-A' at Plant C.

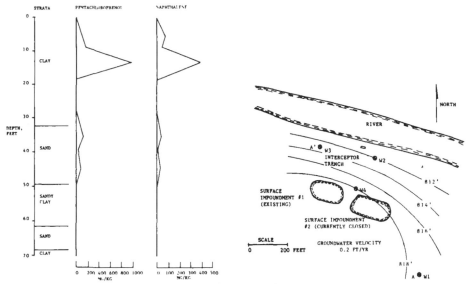

Figure 9. Depth contamination profile for boring B4 at Plant C.

Figure 10. Groundwater contours at Plant D.

SUMMARY AND CONCLUSIONS

All four plants have groundwater contamination that has moved in the downgradient direction of the slope of the water table that covers horizontal areas between one to 13 times the area of the source. In each case the contamination moved vertically downward from the source to the water table and continued to the first most permeable strata. The vertical movement ranged from 20 to 60 feet below the surface. When the most permeable layer was encountered, the contamination would move horizontally in the layer. In two cases, the contamination has moved horizontally up to 500 feet from the source in continuous permeable strata. In another case the movement has been to the extent of the permeable strata, and in the last case a river intercepted the permeable strata approximately 250 feet from the source.

REFERENCES

1. *Federal Register*, Vol. 45, No. 98, 33060-33588 (May 19, 1980).

40 CASE HISTORY–RUBBER TUBING PRODUCTS PLANT WASTEWATER AND RCRA COMPLIANCE

Eric J. Arendt, Corporate Environmental and Safety Manager
Parker Hannifin Corporation
Cleveland, Ohio 44125

Jerome H. Jacobs, Project Manager

Richard Prober, Senior Engineer
Engineering-Science, Ltd.
Cleveland, Ohio 44119

INTRODUCTION

The Parker Hannifin Automotive Connectors Division Plant at Kennett, MO was confronted by several problems concerning their manufacturing processes and the associated wastewater treatment facilities that required a radical revamping of the the water supply and wastewater treatment systems by early 1985. The November 1985 deadline date for Resource Conservation and Recovery Act (RCRA) compliance was rapidly approaching. Insurance coverage required under RCRA could not be arranged. At the same time, new and rather restrictive state NPDES discharge permit limitations had been proposed. In addition, the plant manufacturing processes required increased cooling of contact process water.

The need to resolve these problems led to a complete reevaluation of the existing plant facilities.

BACKGROUND

The Parker Hannifin Corporation Automotive Connectors Division Plant at Kennett, MO produces automotive hydraulic hose using a lead sheath process. It was built in 1965 and acquired by the company in 1984. A staff of about 150 employees operate the plant three shifts daily, six days per week.

Vulcanization of the hose with the lead sheathing is the principal manufacturing process contributing to water contamination and to generation of RCRA listed hazardous waste. Finishing and winding have no wastewater impact, and high-pressure water testing of the finished hose has only a minor impact.

The existing wastewater treatment system is shown in Figure 1.

The plant process waters were collected in a series of below-grade sumps and routed through treatment. The system had been in use since 1975.

Vulcanization spray water (contact cooling) and condensate flowed from the vulcanizer (settling) sump to the filter sump. Wastewater from the lead sheath extruder was pumped to the filter sump. These streams contained lead oxide or hydroxide particulates and had temperatures as high as 140°F due to high temperatures in the processes and to condensation of steam used in the vulcanization.

Effluent from the filter sump flowed by gravity to the boiler room sump and was mixed with acidic spent regenerant from the plant deionized water system. The discharge from this boiler room sump was pumped to the settling pond. Effluent from the settling pond discharged to a local waterway called Buffalo Ditch.

Water from the filter sump was recycled. This combined waste stream was pumped through a 1.0 million BTU/hr shell-and-tube heat exchanger and diatomaceous earth filters. It discharged into the spray sump, the source for vulcanizer spray water. The heat exchanger used non-contact cooling water, which was circulated through a dedicated cooling tower, to reject the process heat to the atmosphere. The filters used diatomaceous earth both as a precoat on paper filter elements and as a body additive during filtration. The spent filter elements were disposed of off-site as a hazardous waste containing lead oxide particulates.

Table I.   Kennett Plant Effluent Quality versus NPDES Limits

	Typical Effluent	NPDES Permit Limits[a]	
		Original	Proposed
pH	7.1–7.7	6.0–9.0	
Suspended Solids (mg/L)	5–37	10	
Oil and Grease (mg/L)	0.9	6	
COD (mg/L)	60	195	
Lead (mg/L)	.02–.20	0.10	.03
Zinc (mg/L	.12–.23	1.00	.10

[a]@ 80,000 gpd flow.

Accumulated sediments were removed from the sumps at scheduled intervals using pressurized water cleanout. Usually, this took place during the annual plant shutdown.

PROBLEMS

The existing wastewater treatment system had been operated subject to the limitations of NPDES Permit Number MO-0001201. The discharges were essentially in compliance with the permit limitations, as can be seen in Table I. In 1985, newer and more stringent permit limitations were proposed, which were beyond the capabilities of the existing wastewater treatment system.

The pond had been operated as a surface water impoundment for hazardous materials (lead in the sediments) on an interim basis, pending final action on the RCRA permit application. In order to achieve compliance with the provisions of RCRA by the deadline date of November 1985, a ground-water quality assessment would be required (Figure 2). Also, privately underwritten liability insurance for the pond was required by the state. Coverage satisfying the state requirements proved to be unavailable.

The pond was sampled and found to have a 1-ft layer of sediments accumulated on the bottom, Analysis of the pond sediments averaged about 16% lead.

Plant operations were known to be affected by limitations inherent in the water and wastewater system. The water cooling capacity was not adequate for the heat load due to vulcanization. Spent

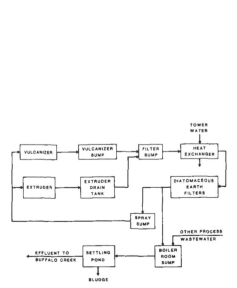

Figure 1.   Schematic of existing wastewater collection and treatment system.

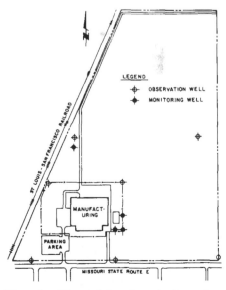

Figure 2.   Observation and monitoring wells at Parker Hannifin, Kennett, Missouri.

Table II.   Determination of Discharge Limits
(Maximum Allowable Conc. in Parker Hannifin Effluent)

	Lead	Zinc
Case 1	4.3	8.6
Case 2	3.6	7.2
Case 3	6.8	3.4
Smallest valve	3.6	3.4
Governing Case #	2	3
Expected Effluent		
Avg.	1.3	2.5
Max.	2.0	5.0

Case 1: Average day 6,000 gpd Parker + 513,000 gpd City, effluent to meet MO water quality interim of 50 μg/L lead, 100 μg/L zinc. No credit for removal in treatment.
Case 2: Maximum day 10,000 gpd Parker + 708,000 gpd City, same effluent criteria.
Case 3: Maximum accumulation in 20 year sludge application 50 lb/acre lead and 25 lb/acre zinc; assuming 100% capture of metals in the sludge.

regenerant from ion exchange demineralization of boiler feed water was disposed of through the plant wastewater treatment system, increasing the hydraulic loading. At times, the discharge was quite acidic as a result of this.

The diatomaceous earth filters required considerable operator attention to manage the pre-coat and body additive operations. Their operational cycles were short, and breakthrough of particles into the spray sump occurred. The spent filter elements containing lead oxide or lead hydroxide particles were awkward to handle and voluminous, hence expensive to dispose of under RCRA regulations.

STRATEGY

Upgrading of the existing wastewater treatment system to meet the new NPDES permit limitation would have required Best Available Treatment technology for removal of lead. Precipitation of lead as the sulfide would have been capable of the desired removal, but that would have involved full-time operators and provisions to eliminate possible emissions of toxic hydrogen sulfide.

The alternative to continuing the existing treatment and direct discharge to Buffalo Ditch was pretreatment and discharge to the municipal sewer system. Such discharges are subject to federal pretreatment standards for the rubber products industrial wastewater category, as well as to applicable local standards. Pretreatment Standards for existing sources have not been promulgated. In the case of this plant, it had to be demonstrated that the discharge of the Kennett City Water and Light Department operated plant would meet NPDES' requirements and that the sludge resulting from treatment would still be acceptable for land application. This can be seen in Table II.

A pretreatment system involving extensive water recycling and reuse was proposed, which would be capable of achieving the city requirements using conventional precipitation of lead and zinc as the oxides or hydroxides. The system is shown in Figures 3 and 4. It replaces the operating labor intensive diatomaceous earth filtration with conventional pressurized sand filtration, and uses a cooling tower instead of noncontact heat exchange. These improvements would result in welcome net savings of operations and maintenance cost.

The new pretreatment system consists of several components which, together, allow the recycling of 80% to 95% of the plant's process water. The plant process water collection sumps were lined with an unplasticized PVC compound and equipped with leak detection monitors. The heated process water reaching the boiler room sump is now pumped to a 5 million BTU/hr prefabricated cooling tower where the temperature is reduced to a maximum of 85°F. The new cooling tower is provided with a 14,000 gallon lined sump and 200 gpm recycle pumps to lower the entering water temperature to a maximum of 120°F, suitable for the PVC tower packing.

The cooled water from this sump is then pumped by dual 200 gpm turbine pumps through a three unit dual media pressure sand filter. The sand filter removes suspended particulates and oils from the recycle water prior to reuse in the production process.

On a periodic basis (every 24 to 48 hours) the pressure sand filters are backwashed, and this reject water is collected in a 6,000 gallon batch treatment tank. This batch treatment tank also receives spent regenerant streams from the plant deionizers. The pH of water in the batch treatment tank is adjusted

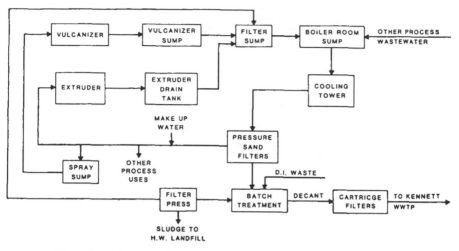

Figure 3.   Schematic of new wastewater collection and treatment system.

to approximately 9, and then about 1 mg/L of cationic polymer is added prior to settling of the floc for about 12 hours (overnight).

The settled supernatant is pumped through cartridge filters prior to discharge to the City of Kennett sewer system. Sludge is allowed to accumulate until a 4-ft^3 cake (containing 30% to 50% dry solids) can be produced by the plate and frame filter press.

Makeup water is automatically added to the process water system at a reduced pressure type backflow preventer, whenever process system pressure falls below makeup (City water) regulated supply pressure. As part of the project, all plant potable water lines were completely separated from any process water uses.

An additional advantage of the pretreatment system is that the pond could be abandoned. Continuing use of the pond in compliance with newly amended standards for RCRA facilities would have required installation of a double lining system to protect against contamination of groundwater. The costs for this, including temporary treatment facilities (similar to the proposed pretreatment system) during construction of the pond lining and removal and disposal of sediments containing lead from the pond, could not have been justified. Further, the uncertainty regarding acceptable liability insurance coverage added to the desirability of closing the pond.

Closure of the pond also involves treatment and disposal of sediments from the pond bottom. Lead oxide and hydroxide particulates are readily removed by processing through the plant wastewater treatment system. Groundwater monitoring would have to be continued in any event, at least until completion of the groundwater quality assessment.

Another problem was that the existing transfer sumps were unlined concrete installed below grade. These sumps would not meet the RCRA requirements proposed in 1985 for containment of hazardous

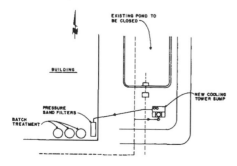

Figure 4.   Layout of new system.

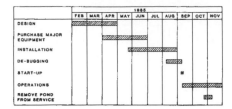

Figure 5.   Design and construction schedule.

wastes. Accordingly, it was prudent to provide linings and leak detection for these sumps as part of the upgrading project.

Design of the improvements began in early February 1985, with startup projected for August of that year (Figure 5). A fast-track schedule was required in order to remove the existing treatment pond from service in time for the November 1985 deadline. Construction was complete by August 1985. Debugging of the equipment commenced on that date and took about three weeks.

STARTUP AND PERFORMANCE

The actual startup of the system was initiated in September 1985. Recycling of treated process water started on that date and has continued uninterrupted since then. The product water quality has been suitable both for discharge into the city sewer system and for reuse, as shown in Table III.

Water usage has been reduced dramatically since the startup of the treatment system. This has resulted in cost savings for the Kennett plant of about $1,000 per month.

The pond was taken out of service in October 1985. Sediments from the pond were processed through the treatment system during the first six months of operation, producing a dry filter cake containing approximately 50% solids. This filter cake, like that produced from normal plant operations, was disposed of as a hazardous waste by landfilling at a site certified for heavy metal sludges.

Table III.   Product Water Quality

	Discharge	Reuse
Total Suspended Solids	7 mg/L	3 mg/L
Chemical Oxygen Demand	70 mg/L	100 mg/L
Oil and Grease	2 mg/L	5 mg/L
Lead (Total)	0.7 mg/L	6 mg/L
Zinc (Total)	0.1 mg/L	0.5 mg/L
pH	8.5 S.U.	8.0 S.U.

41 ILLINOIS HAZARDOUS WASTE RESEARCH AND INFORMATION CENTER: DEVELOPMENT OF AN INDUSTRIAL AND TECHNICAL ASSISTANCE PROGRAM

Frederick L. Doll, Industrial Assistance Coordinator

Daniel D. Kraybill, Industrial Assistance Engineer

David L. Thomas, Director
Hazardous Waste Research and Information Center
State Water Survey Division
Savoy, Illinois 61874

INTRODUCTION

The Hazardous Waste Research and Information Center (HWRIC) was established within the Illinois Department of Energy and Natural Resources (DENR) in 1984 to address hazardous waste management problems within Illinois. It was authorized by the Illinois Assembly through Senate Bill 815 and is funded through general revenues and the Hazardous Waste Research Fund. Based in Champaign, Illinois and administered by the Illinois State Water Survey, it has three basic functions:

RESEARCH — HWRIC funds and performs research to characterize and assess the hazardous waste problem in Illinois and to determine the best solutions to particular hazardous waste problems.
INFORMATION DISSEMINATION — HWRIC runs an information clearinghouse and library specializing in hazardous waste issues. Data on hazardous waste are collected, analyzed, and disseminated in a variety of formats. Information is generally provided on a no-charge basis to the citizens of Illinois.
INDUSTRIAL AND TECHNICAL ASSISTANCE (ITA) — The ITA program provides direct technical assistance to citizens and industries in Illinois with hazardous waste management problems.

The ITA program emphasizes source reduction, recycling, product substitution, and other methods of reducing the amount of hazardous waste generated within a given plant. It also gives guidance with regulatory and permitting processes and makes referrals to qualified consultants and service organizations. This paper describes the establishment of the ITA program and discusses its past accomplishments, current activities, and future plans.

PROGRAM INITIATION

Announcements for openings for an ITA Program Coordinator and Industrial Assistance Engineer were made in the fall of 1984. Interviews began in February of 1985, and the Coordinator was hired in April. The Industrial Assistance Engineer was hired in late May. Both engineers have extensive experience with industrial processes and the environmental disciplines.

Survey of Other State Programs

The first major step was to survey similar activities in other states. A listing of similar programs was developed from material presented at the "Pollution Prevention Pays" Conference held in Raleigh, North Carolina in April of 1985 [1]. Personnel from these programs and other programs were contacted and questioned about their state's programs (Table I).

Two general statements can be made from this survey: 1) Most of the programs dealt with environmental concerns in general, but did not focus on hazardous waste management in particular; and 2) The most effective programs were not associated with any regulatory agency. A location near a large

Table I. Other Technical Assistance Programs

Ontario Research Foundation Mississauga, Ontario	Tennessee Valley Authority Knoxville, TN
Georgia Tech Research Institute Georgia Institute of Technology Atlanta, Ga	Institute for Hazardous and Toxic Waste Management New Jersey Institute of Technology Newark, NJ
Minnesota Technical Assistance Program University of Minnesota Minneapolis, Minnesota	Hazardous Waste Advisement Council Trenton, NJ
State of New Mexico Environmental Health and Energy Dept. Hazardous Waste Program Albuquerque, NM	N. Carolina Dept. of Natural Resources and Community Development Pollution Prevention Pays Program Raleigh, NC
New York Environmental Facilities Corp. Albany, NY	Center for Environmental Management Tufts University Medford, MA
Pennsylvania Technical Assistance Program Pennsylvania State University University Park, PA	

college campus seemed to be the most common location as it gives good access to the resources that go along with a major university.

Initial Tasks

Based in part on conversations with personnel from these programs and to most quickly inform Illinois industry of our program, it was decided that the most effective first action would be to solicit possible hazardous waste generators with a mass mailing. A list of 2537 industrial waste generators was obtained from the Illinois EPA, and a questionnaire (Figure 1) was sent out to them. Basically, the mailing informed the generators of our existence and offered help in several areas.

Of the 2537 generators contacted by mail, 296 (11.7%) responded and 175 (6.9%) requested assistance of some sort (Table II). Primary assistance given included referral to disposal firms (52%), information on new regulations (35%), and direct technical assistance. Follow-up to the returned questionnaires was initially made by telephone. The respondents were asked for more detail about the type of assistance they desired and also were asked four basic questions about their operations: 1) What types of raw materials do you use?; 2) What products do you make?; 3) What processes do you use?; and 4) What wastes do you generate?

Based on the results of this personal contact, assistance was given in the categories listed in Table II. In many cases the appropriate response was as simple as mailing out an information package containing a list of disposers, consultants, laboratories, etc., or a verbal explanation of regulations followed by a backup letter putting that explanation in writing. In fewer instances, a site visit and subsequent plant audit was requested. To date 35 site visits have been made.

Based on the experience gained from these initial contacts, two important points were clear: 1)It is essential to explain to the user immediately upon contact that HWRIC is not part of any regulatory.agency and that the matters discussed will be held in confidence; and 2) Once telephone contact has been made, response to the needs of the user must follow quickly, and the assistance must be of immediate, practical use to the user.

A serious potential problem that can arise in the implementation of an ITA program is the possibility of a breach of confidentiality. Industrial concerns usually have trade secrets, patents and licenses, and production knowledge that they consider confidential, because either the information is necessary to maintain their competitive position and/or is a part of their business assets. In Illinois, the DENR, and all other state agencies are subject to the Illinois Freedom of Information Act (FOIA). The FOIA is a two-edged sword; it enables the government to protect information that it wishes to protect, to a degree, but also gives citizens and outside groups access to that same information by means of formal procedures prescribed under the FOIA regulations. Illinois agencies have taken different approaches to FOIA compliance. The Illinois Environmental Protection Agency and the Illinois Pollution Con-

QUESTIONNAIRE

Type of Industry _____

Name _____

Company _____

Address _____

_____

_____

Phone (_____) _____ No. Employees _____

	YES	NO

1. Does your operation generate hazardous wastes?   ☐   ☐

2. If YES, could your facility utilize technical assistance from HWRIC for hazardous waste management?   ☐   ☐

3. If YES, type of assistance? (check any, some or all of the items below)

* Permitting and regulatory · ☐

Hazardous waste audits and plant survey · · · · · · · · · · · · ☐

Review on-site disposal practices · · · · · · · · · · · · · · · · · ☐

Assist in set-up of systematic hazardous waste
   record-keeping systems · ☐

Research alternatives for off-site disposal · · · · · · · · · · · ☐

Examine process modification to reduce or eliminate
   hazardous waste generation · · · · · · · · · · · · · · · · · · ☐

Research alternatives for on-site hazardous waste treating
   (volume reduction, hazard elimination, etc.) · · · · · · · ☐

Research alternatives for on-site recycling · · · · · · · · · · · ☐

Research alternatives for off-site recycling · · · · · · · · · · · ☐

Assist with information for hazardous waste risk
   and insurance programs · ☐

** Provide technical assistance on laboratory analysis
   of hazardous wastes · ☐

Provide information on or references to specialty consulting engineering firms for specific problems · · · · · · ☐

Provide information on or references to equipment
   vendors for specific applications · · · · · · · · · · · · · · · ☐

Presentation or assistance in organizing in-plant
   educational programs on hazardous wastes · · · · · · · · ☐

Assist with development of in-plant hazardous waste
   handling procedures and develop MSDS (Material
   Safety Data Sheet) collections and routing · · · · · · · · ☐

Provide information on active and passive waste
   exchange activities in the region · · · · · · · · · · · · · · · ☐

Other (please specify) _____

_____

_____

* HWRIC cannot represent firms in the permitting process or enforcement actions, or provide legal interpretations of regulations; it can assist by supplying procedural and informational resources, and coordinate data collection efforts for permitting.

** HWRIC is planning a Hazardous Materials Laboratory for direct technical assistance with both hazardous waste analysis and pilot scale treatment studies within the next few years; present assistance is limited to supply of information and liaison/coordination activities with existing government, university, and private laboratories.

Figure 1.   ITA questionnaire.

Table II. ITA Questionnaire Results

Total Number of Questionnaires Sent = 2537
Total Number of Questionnaires Returned = 296   Response Rate = 11.7%
Numbers of Firms Requesting Assistance = 175   Response Rate = 6.9%

Types of Assistance Given	Number of Consultations
Information on existing regulations	22%
Information on new regulations	35%
Referral to Disposal Firms	52%
Referral to Consultants or Laboratories	24%
Referral to Equipment Vendors	2%
Direct Technical Assistance[a]	32%
On-Site Hazardous Waste Audit	10%

[a]This category consists of suggestions for process changes, use of more efficient types of equipment, recycling options, etc.

trol Board have codified their FOIA regulations in the Illinois Administrative Code. DENR has not but provides a comprehensive internal policy documentation for their line and staff managers to ensure FOIA compliance.

The HWRIC ITA program has not encountered a formal confidentiality problem, although some industrial clients have informally asked us to keep their file information confidential and not circulate it. Agencies setting up ITA programs in other states should examine FOI regulations carefully as part of their initial work plan.

Publicity

Advertising is essential to the success of an ITA program. A great deal of "word of mouth" advertising has taken place as a result of the initial mass mailing. Additionally, ITA personnel have spoken to a large number of trade and professional groups (Table III) to inform them of our existence and to disseminate information on hazardous waste treatment options and changing hazardous waste regulations.

Table III. Trade Associations, Professional Organizations, and Other Groups Addressed by ITA Staff

1985
- Illinois Manufacturers' Association, Environmental Committtee – Chicago
- Chemical Industry Council/Chemical Manufacturers' Association, Environmental Standing Committee – Chicago
- Three Rivers Manufacturing Association, Environmental Committee – Joliet
- Illinois Society of Professional Engineers, Joliet Chapter
- University of Illinois, Cooperative Agricultural Extension, Regional Extension, Offices Directors Meeting – Urbana
- Chemical Coaters Association – Naperville
- Chemical Coaters Association – Peoria
- University of Wisconsin, Madison – 8th Annual Midwest Water Chemistry Conference
- American Business Clubs – Champaign Chapter
- Associated Illinois Soil and Water Conservation Districts, Water Resources Committee – Decatur
- Independent Automotive Service Association – Champaign
- Illinois Society of Professional Engineers – Quincy

1986
- American Electroplaters Society, Educational Chapter – Chicago
- Illinois Farm Bureau Speaker's Program – Decatur
- First International Symposium on Metals Speciation, Separation, and Recovery – IIT – Chicago
- Central Illinois Purchasing Managers' Association – Decatur
- #1 Small Quantity Generator Seminar – Joliet
- Greater Chicago Metropolitan Area Ford Dealers, Parts & Service Managers – Chicago

It is important that ITA personnel be available to speak to trade and professional groups on short notice and that they be able to tailor their presentations to fit the needs of the particular group to which they are speaking. It would not be appropriate, for instance, to give a presentation to a group of environmental managers of major corporations and give the same presentation to a group of auto mechanics.

Several common building blocks can be assembled quickly to form the core of a presentation and a handout package. They are:

1. Items of interest to the particular group being addressed. This is the most important element. It is important to pick a particular topic (e.g., typical hazardous wastes in an auto body shop) and be prepared to talk about it in detail if the group desires.
2. A brochure or pamphlet which describes the capabilities of the ITA program. Different types of brochures are appropriate for different types of groups, but all should include a detachable card which can be sent back to the ITA office with questions at a later date.
3. A questionnaire similar to the one shown in Figure 1.
4. A business card or other material which give the name and phone number of a particular person (preferably the speaker) who can be contacted for assistance.

By giving each group something useful in its first encounter, more followup on their part will be ensured.

The results of both the mass mailing and the speaking engagements have been positive. The existence of the ITA program has been advertised by word of mouth through former users and attendees of various talks. ITA personnel have spoken at 17 locations (Table III) and have assisted 363 individuals, businesses and industries. Of these, 175 learned about ITA through the initial mass mailing, and the remainder have learned about the ITA program through speaking engagements or through word of mouth. After just 10 months, most requests for assistance are now coming from unsolicited call-ins.

Types of firms assisted vary widely in size but tend to be small- to medium-sized manufacturing or service organizations, generally employing less than 500 people. Such firms can seldom afford an environmental staff person, and are often uninformed of the most basic regulatory requirements aside from what they hear from local disposers. Requests for assistance have come primarily from platers, metal working facilities, machine shops, auto body and vehicle maintenance shops, industrial painting operations, and assembly operations.

We have found it important to remain in contact with technical assistance programs in other states. Exchange of information has been helpful in assessing our current activities and in getting new ideas for future activities. We have found that other state programs currently have a very open attitude towards communication with each other and encourage the use of each other's resources.

HWRIC also works with regulatory agencies to the extent that it receives information on current and upcoming regulations from them and assistance in checking out disposal firms for users. ITA staff often poses questions about regulations and permitting to regulatory agencies for users who do not wish to speak with the regulators directly. It is essential that in these cases the user remains anonymous if he so desires. We have found the regulatory agencies to be cooperative in this regard. Trade associations such as the Illinois State Chamber of Commerce are also helpful in determining target groups who may need our assistance.

CURRENT ACTIVITIES

Small Quantity Generators Assistance

HWRIC's ITA program is now attempting to assist Small Quantity Generators (SQGs) in Illinois. It is estimated that thousands of firms in Illinois will be impacted by the new SQG regulations which have been recently finalized by USEPA. Since a comprehensive listing of SQGs is not available, HWRIC is sponsoring a series of compliance seminars aimed at this group of businesses.

The seminars attempt to explain the new regulations in a simplified form that the typical SQG can understand. As with most of the other ITA users, SQGs are generally small firms which lack the staff and/or time to fully review and comply with the myriad of regulations that they are subject to. To fully assist SQGs, HWRIC attempts to bring together speakers from four groups:

1. *Regulatory Agencies*—Personnel from IEPA and USEPA have been most helpful in explaining regulations and providing an "official" representative to field questions about new and impending regulations

2. *HWRIC* — ITA personnel will usually speak on the topic of waste reduction and disposal procedures. They attempt to bridge the gap between the regulators and generators and inform the generators that help is available through HWRIC.
3. *Disposal Firms* — A speaker who has direct on-going dealings in the waste disposal field can provide an excellent perspective for generators.
4. *Trade Organizations* — The Chamber of Commerce and other such organizations can often supply speakers for such an event.
5. *Other State Agencies* — Other state agencies such as the Illinois Department of Labor are helpful in obtaining speakers.

As of this writing, HWRIC has put on one seminar and has two more in the final planning stages. Staff from HWRIC's Information Program have assisted in details of the preparation, publicity, and running of the seminars. The seminars are expected to be conducted on an ongoing basis for the foreseeable future.

Information Collection and Dissemination

ITA personnel are also writing a small quantity generators compliance manual which will parallel many of the things which are said in the presentations. The manual is currently in the draft stage and is expected to be published in the near future. ITA personnel are assembling a database of consultants, laboratories, disposal and hauling firms, and equipment vendors for use in making referrals of users to qualified assistance in the private sector. The ITA program does not have the manpower or the mandate to perform as a consulting firm. The database assists ITA personnel in quickly locating waste management and consulting firms in a user's geographical area and matching the consultants skills with the user's requirements. Initial contact with the subject consultants was made through a mass mailing of a questionnaire and using a mailing list supplied by IEPA. Additional contacts are being made on a continuing basis and the size of the consultants and services database stands at 180 entries.

Assistance to Non-Industrial Groups

Technical assistance has not been limited to assisting business and industry. ITA personnel have also assisted farm communities and other groups with their hazardous waste and community safety problems.

HWRIC is attempting to assist Illinois farmers through an outreach program coordinated with the University of Illinois Extension and the Illinois Farm Bureau. Although this program is still in the early stages, it will assist farmers and agribusiness with waste management problems and particularly with the problem of pesticide disposal.

An example of the type of assistance HWRIC might give would be to organize "milk runs" for pickup of hazardous wastes and expired pesticides at farms or at central receiving stations within a given area. Such a program would be similar to household hazardous waste pickups (which HWRIC is currently assisting a community in conducting) which are currently popular but should be simpler since an agribusiness infrastructure already exists.

In regards to community assistance, HWRIC was approached by a zoning board from a central Illinois community to assess the possible hazards of a "high tech" company which wanted to locate a manufacturing operation in an unused school building located near a residential area. ITA personnel assisted by examining lists of chemicals which were to be used at the site and the types of operations which were to take place. HWRIC felt that the risk to the public was minimal. The firm has since been granted a zoning variance and the operation is in the process of being sited.

In another example, HWRIC was approached by the staff of a suburban Chicago community which had been asked by IEPA to "sign off" on a permit application submitted by an existing industrial waste hauler. The permit was to build a wastewater pretreatment system on its property in that community to commercially treat liquid industrial waste and discharge it to the sanitary sewer. The community was required to certify that its sewers were adequate to receive this type of waste, and their staff lacked the qualified personnel to make such a decision.

The situation was complicated by the fact that the hauler had a poor record of past compliance and also had several alleged violations still pending with IEPA. ITA personnel visited the haulers facility and made recommendations to the community on permit conditions which could help assure safe operation of the facility. The situation is, as yet, unresolved.

FUTURE ACTIVITIES

The Industrial and Technical Assistance program has steadily expanded since its inception. In addition to giving direct technical assistance, ITA personnel, in cooperation with HWRIC's research programs, plan to begin assessing waste reduction, recycling and treatment technologies through a matching grants program. This program will provide up to $10,000 in matching money to firms that wish to assess new technologies or purchase environmental control equipment which may be experimental or which they otherwise may not be able to afford. As part of the matching grant, an evaluation of the technology or equipment will be performed, the results of which will be published as an HWRIC report. Through this program, HWRIC hopes to encourage waste reduction through modification of existing technologies and the development of new technologies. Typical projects being considered are discussed next.

Waste Management Studies. Such a study is not necessarily a test of a new technology but rather a study of a particular industry, the waste it generates, the current management and disposal practices, and possible alternatives to the present practices. An example of this is a study currently being funded by HWRIC and the Energy and Environmental Affairs Division of DENR which examines wastes generated by the electroplating industry in Chicago. In particular, the technology and economics of a centralized recovery facility are being studied.

Pilot Treatability Studies. Such a project would take an existing or new technology and apply it to waste treatment, recycling, or reduction of hazardous wastes. Many technologies already exist which can be easily applied to waste recycling, reuse, or reduction. An example of such a project might be an evaluation of small scale solvent stills for use in auto body shops.

Waste Reduction Studies. These studies would concentrate on modifying production processes to minimize waste production. Although most such projects could be technology intensive, they do not have to be. One suggestion has been a marketing study to be performed by a paint manufacturer which would evaluate the market for a low grade barn paint which could be produced from wash solvents and waste pigments used in existing product lines.

HWRIC is also planning to construct a Hazardous Materials Laboratory (HML). Currently there are no facilities within the state that can offer a safe work place for experimenting with high hazard wastes on a large scale. The HML will include areas for basic, bench scale research and also for applied pilot scale evaluations of treatment, recycling and waste reduction technologies. Currently in the preliminary design phase, the HML is expected to be built and ready for use by late 1988 or 1989.

DISCUSSION

HWRIC's ITA program is currently assisting a wide variety of industries and individuals in Illinois. The program has been successfully established to the point that it could be used as a guide for the establishment of other programs. The elements that are needed to initiate such a program are:

1. A technically competent core staff headed by at least one engineer with extensive experience in all three environmental media (air, water, solid/hazardous waste) and appropriate support staff. Excellent communication skills are essential for all staff.
2. Good communications with regulators, generators, trade groups, transporters and disposers. This is something which is only acquired through time and experience and is really one of the most important elements.
3. The support staff which can assist in publicity, information dissemination, and coordination with other agencies and the legislature.
4. Access to extensive library and technical facilities such as are found at a major university. HWRIC is fortunate in this regard, being close to both the University of Illinois and the State Scientific Surveys.
5. Sufficient funding to gain access to commercial mailing lists, produce descriptive brochures, and put on large numbers of talks and seminars.
6. The ability to operate confidentially.

The ability to conduct pilot studies and applied research are not essential to a successful program but certainly complement it. The ITA program is moving in this direction.

Lastly, but most important, all staff must be highly motivated. They must be ready, willing and able to lend assistance to the potential user. HWRIC staff have aggressively worked to assist Illinois' citizens and industries and stand ready to fund research and demonstration projects and give technical assistance which will solve particular waste management problems.

REFERENCES

1. Workshop on Implementing State Pollution Prevention Programs, Raleigh, NC; Sponsored by Pollution Prevention Pays Program, North Carolina Dept. of Natural Resources and Community Development (April 25, 1985).

42 IN-SITU REMOVAL OF PURGEABLE ORGANIC COMPOUNDS FROM VADOSE ZONE SOILS

Frederick C. Payne, President
Midwest Water Resource, Inc.
Charlotte, Michigan 48813

Charles P. Cubbage, President
Cubbage Environmental Control
Milford, Michigan 48042

Galen L. Kilmer, District Engineer
Groundwater Quality Division
Michigan Department of Natural Resources
Plainwell, Michigan 49080

Laurence H. Fish, Vice President
Custom Products, Inc.
Stevensville, Michigan 49127

INTRODUCTION

Our test site is located in Stevensville, Michigan, 0.9 miles east of the Lake Michigan shoreline, 18 miles north of the Indiana-Michigan border. During mid-1984, a purgeable organic contamination plume was discovered in a useable aquifer, covering approximately 51 acres. Testing showed perchloroethylene as the principal contaminant, reaching levels in excess of 100 ppb in domestic water wells and levels of 800 ppb in nearby industrial production wells. Two light manufacturing facilities were identified by the Michigan Department of Natural Resources as potentially responsible parties (PRP's) under provisions of the Michigan Environmental Response Act.

Custom Products, Inc., one of the two PRP's identified by MDNR, had used PERC as a degreasing solvent prior to discovery of groundwater contamination in the area. The company retained Water Quality Investigators and Midwest Water Resource, Inc. to conduct soil and hydrogeological studies in the vicinity of its sheet metal forming facility.

Soil surface samples were collected in February 1985 for analysis of purgeable aromatic and chlorinated solvents. These tests showed three potential loading points for volatile organic compounds into the soils and groundwater: 1) a location outside the building where solvents had been stored; 2) an interior drywell which drained the paint spray booth; and 3) a location outside an exterior wall where a pipe once exhausted PERC tank sludge which recycling haulers would not accept. The sludge discharge location later proved to be the major loading point and showed 110 mg/ kg PERC in the February 1985 soil surface sampling.

A 9-point triangular grid pattern was established for soil borings outside the building (Figure 1). This grid had a 5-point base leg (A–E) with 25-foot separation between boring points; the center point on the base leg (C) was placed at the sludge discharge point. A second tier of three borings (F-H) was placed parallel to the base leg 25 feet from the wall with 25-foot separation between points. A final boring (I) was placed 50 feet from the wall perpendicular to the known discharge point. Three additional borings (J–L) were placed through the floor of the building, 20 feet from the exterior wall.

The soil borings were conducted with a hollow-stem auger with split-spoon samples taken at 5-foot intervals. Soil samples were placed in Teflon-capped VOA bottles, stored on ice, and carried to Canton Analytical Laboratory, Ypsilanti, Michigan for immediate analysis of purgeable aromatics and purgeable halocarbons. Two borings showed measureable levels of purgeables—the boring located at the sludge discharge point (C), and one of the three borings within 25 feet of the discharge point (B). Table I gives the results of laboratory analysis for these two borings.

All borings showed fine sand throughout the vadose zone, which extended to the 30-foot depth. The volume of contaminated soil was estimated to be between 1,000 and 2,000 cubic yards. More

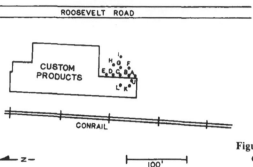

Figure 1. Location of soil borings (A–L) conducted at Custom Products, Inc. May 1 and 2, 1985.

accurate determination of the contaminated volume required additional borings both inside and outside the building. But available information was sufficient to estimate a remedial action budget. Four factors were assessed under the assumption that removal would be accomplished by excavation: 1) A minimum 1,000 cubic yards removal; 2) Destruction of the building, since soil contaminants underlay the foundation; 3) Hauling 180 miles to a hazardous waste landfill; and 4) Lost production in the plant and/or relocation costs. Costs for this plan were estimated at a minimum of $300,000, an unacceptable option for Custom Products, Inc. We were led immediately to consider alternatives to excavation.

Comparison of surface soil sample analysis from February and May showed a decrease from 110 mg/kg to less than 10 mg/kg. This indicated that PERC was evaporating from below, condensing at the cold soil surface in winter and venting to the atmosphere during warm months at a rate which kept PERC levels below laboratory detection limits. During a meeting of the authors in August 1985, it was decided that Custom Products should attempt to take advantage of the venting process and avoid excavation costs.

The system described below was developed as a prototype for enhancement of volatilization of purgeable organics in vadose zone soils. Our goals were first to achieve an acceptable removal of PERC and second, to gather data on the applicability of enhanced volatilization at other sites.

METHODS

The underlying mechanism for this process was the equilibration of vapor phase PERC with liquid PERC on soil particle surfaces and in interstitial spaces. The system forced clean air into the vadose zone at a radius outside the contaminated region and pulled the clean air toward the center of the contaminated area where it was withdrawn under reduced pressure. Removal was accomplished by maintaining PERC vapor pressure in the soil pore spaces at levels less than equilibrium vapor pressure.

Table I. Results of Analysis for Tetrachloroethylene in Soil Samples Collected from Borings B and C, Custom Products, Inc. Sample Collections were Conducted May 1 and 2, 1985. Analysis was Performed by Canton Analytical Laboratory. All values are given as mg/kg dry weight.

Sample Depth Feet	Boring	
	B	C
0–2	NDa	ND
4–6	ND	ND
9–11	ND	700
14–16	ND	5,600
19–21	ND	15
24–26	14	29
29–31	NSb	8.3

aND-Below Detection Limit.
bNS-No Sample Collected.

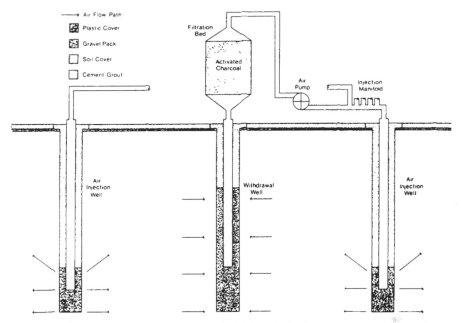

Figure 2.   System configuration for vacuum withdrawal of volatile organic compounds from vadose zone soils.

The removal system construction began with installation of a 5-inch borehole, 25 feet in depth, at the location where sludge had been applied to the soil surface. This boring served as the central vacuum withdrawal well. It was gravel-packed from the 17 to 25-foot depths, and a 2-inch galvanized casing was installed from one foot above grade to the 17-foot depth. The casing was then gravel-packed from the 8 to 17-foot depth and cemented from the ground surface to the 8-foot depth. Given the high pneumatic conductivity of gravel, the pressure drop exerted by the vacuum pump was distributed relatively evenly over the borehole wall.

Five air injection wells were constructed at a 50-foot radius from the withdrawal well on the contaminated side of the building. A sixth injection well was located on the opposite side of the building at a 70-foot radius. Each injection well consisted of a 5-inch diameter, 25-ft deep borehole which was gravel packed from 19 to 25 feet. One and one quarter-inch PVC casing was installed from 1 foot above grade to the 19-foot depth and was then gravel-packed to the 15-foot depth. The injection wells were cemented from the ground surface down to 15 feet to promote clean air loading into the deeper strata.

The surface of the entire site was sealed with 6-mil polyethylene sheeting, which was covered with sand to hold it to the ground surface and protect against puncture. This seal was installed to give us control over the air flow pathway. Given the water table below and the plastic cover above, the pneumatic system was confined so that it insured radial movement of air toward the withdrawal well, rather than vertical movement from the soil surface.

The pumping/filtration station was constructed inside the building. The system configuration is given in Figure 2. Air flowed from the vacuum well through a 2-inch galvanized pipe to a 96 cubic foot filtration bed charged with 1200 lbs granular BPL activated carbon. The air inlet entered the bottom of the filter bed through a liquid trap and exited through the tank top. Exhaust air from the filter bed moved through 2-inch galvanized pipe to a 1-horsepower oilless rotary vane vacuum pump. Vacuum pump discharge air was distributed to the reinjection wells through a manifold with individual valves for each line that allowed balancing of injection pressures.

System performance was monitored upstream and downstream from the filter bed. Air samples were withdrawn into sample bags or collected on activated carbon adsorption tubes and were carried immediately to the laboratory for analysis of PERC content. Samples were collected at 3-hour intervals during the first 72 hours of pumping, at 24-hour intervals from the third through twelfth

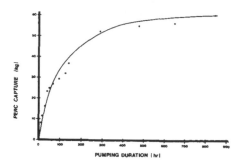

Figure 3. Cumulative PERC capture for Custom Products, Inc. Values are calculated from concentration point estimates and cumulative volume measurements.

day, and at weekly intervals from the twelfth day through present. Air flow was measured by an in-line meter placed on the vacuum side of the system. Readings were made at the time of air sample collections. Vacuum development in the vadose zone was monitored at 5, 10, and 20 feet from the central withdrawal well. These points were set at 15 feet below grade and were linked to water manometers inside the building.

RESULTS

System operation was initiated on December 11, 1985. Vacuum levels on the central withdrawal well rapidly reached 4.5 inches Hg, and the vacuum monitor wells showed pneumatic pressure drop in the vadose zone had stabilized after 17.5 hours. Air flow was stable at 10.42 cfm throughout the experimental period. Backpressure on the injection wells was below detection on standard air pressure gauges which were installed on each of the six return lines.

PERC recovery during the first hours of pumping was extremely high. Liquid PERC moved into the borehole and was captured in the liquid trap in the base of the filter module. Approximately 0.1 gallon was removed in liquid form. Gaseous PERC levels reached 92,000 mg/m³ at 48 hours and declined to 6,000 mg/m³ at 72 hours. Gaseous PERC concentrations in the central withdrawal well remained near 5,000 mg/m³ through the twelfth day. The sample collected on day 19 showed PERC at 1,000 mg/m³, and levels declined to 10 mg/m³ in samples collected on day 35.

Cumulative PERC capture was calculated by integration of concentration data over the time intervals between sample collections, given the constant system air flow of 10 cfm. Results of this calculation are given in Figure 3. The recovery shows an asymtotic decline in accumulation rate to a calculated value of 62 kg after 35 days of pumping.

PERC breakthrough from the carbon filter was detected in samples collected on day 35. Analysis of spent filter carbon showed a 14% capture rate for the carbon filter medium or an estimated PERC yield of 76 kg. Estimation of PERC content in the carbon filter medium was expected to exceed the integration calculation due to underestimation of air flow volume. Given an air volume correction of 15% (the viscosity reduction at 4.5 inches Hg vacuum), the integration yields 72 kg PERC.

Carbon exchange was completed and the system was re-started. PERC levels initially reached 100 mg/m³ in the central withdrawal well and resumed at 10 mg/m³ after 24 hours of pumping. This restart spike is an indication that air exchange rates in the soils exceed the volatilization rate for PERC clinging in liquid form to particle surfaces.

Soil borings were conducted at the PERC sludge loading point after 45 days of pumping. Split spoon samples collected at the 10 and 15-foot levels showed declines of PERC from 700 and 5,600 mg/kg to 0.84 and 0.64 mg/kg, respectively. The sample collected from the 5-foot depth showed an increase from below detection to 0.71 mg/kg. These results are shown in Table II, along with results of pre-pumping sampling at the same location.

DISCUSSION

Substantial removal of vadose zone PERC has been accomplished in the first 45 days of system operation at less than 20% of conservatively projected excavation cost. The system is capable of reaching extremely low levels at this site and will be operated for at least 180 days to test these limits.

Applicability of in-situ vadose zone cleanup to other sites will be determined by four factors:

Table II. Comparison of PERC Levels in Split Spoon Samples Collected at the Sludge Loading Point on Two Dates Prior to System Installation, and after 45 Days of System Operation. All values are mg/kg dry wt.

Depth	Feb 85	May 85	45-Day
0	110	ND	
5		ND	0.71
10		700	0.84
15		5,600	0.64

1. Minimum spill volume — in-situ removal is more cost-effective than excavation for soil volumes in excess of 500 cubic yards. Fixed costs of the system installation and operation exceed excavation and disposal costs for small spills.
2. Contaminant vapor pressure — contaminants subject to this removal process must exhibit vapor pressures sufficiently high to promote rapid evaporation in the soil system.
3. Pneumatic conductivity of soils — soils must be permeable to air movement. Generally, the movement of vapor-phase compounds away from a loading point gives indication of suitable soil permeabilities.
4. Site stratigraphy — sites with strata of varying conductivities present more complicated air flow patterns which must be considered in removal system design.

In-situ removal of soil contaminants through air purging has been established as a potential remedial action tool, but each site must be evaluated individually for feasibility and design.

43 POTENTIALLY RESPONSIBLE PARTIES AND SUPERFUND – AN ANALYSIS OF ISSUES

Michael D. LaGrega, Associate Professor
Department of Civil Engineering
Bucknell University
Lewisburg, Pennsylvania 17837

Allen E. Ertel, Partner
Reed Smith Shaw and McClay
Williamsport, Pennsylvania 17701

James H. Dougherty, Vice President
Roy F. Weston, Inc.
West Chester, Pennsylvania 19380

INTRODUCTION

The cleanup of abandoned hazardous waste sites under the Comprehensive Environmental Response Compensation and Liability Act of 1980 (CERCLA) [1] has been the subject of much controversy over the past several years. Environmentalists and legislators feel that insufficient progress has been made at cleaning up the nations abandoned sites. Superfund was established as a $1.6 billion fund wherein abandoned hazardous waste sites could be cleaned up by the U.S. Environmental Protection Agency using monies from the fund. The original law was intended that EPA would recover costs from potentially responsible parties (PRPs), i.e., industries that generated or handled the waste or owned or operated the site.

The process is initiated by a site being evaluated as severe enough to be placed on a National Priorities List (NPL). The site then undergoes a two stage engineering study to produce a remedial investigation/feasibility study (RI/FS) [2]. Based on this information, the Environmental Protection Agency issues a record of decision (ROD) mandating the exact type of cleanup appropriate to the site. A contract is then issued through the U.S. Army Corp. of Engineers for design of the selected alternative and implementation.

Section 106(a) of the Act enables the EPA to undertake enforcement actions against PRPs which include:

1. Consent orders (a legal document negotiated between EPA and some or all of the PRPs)
2. Administrative orders enforceable through fines of $5,000 per day
3. Seeking an injunction through the district court
4. Litigation including treble damages for the total cost of cleanup

Where PRPs have been identified, EPA attempts to negotiate a consent order wherein a group of PRPs will take over the site and conduct the cleanup.

The purpose of this paper is to overview problems and issues associated with the Superfund program as they apply to potentially responsible parties.

BRIEF HISTORY OF SUPERFUND

CERCLA was enacted to clean up abandoned hazardous waste sites and to pay for damages to natural resources. In enacting this statute, Congress was reacting to public pressure brought on by incidents such as the Love Canal. Additionally Congress wanted to bring order to what was perceived as redundant and inadequate federal laws.

CERCLA was passed in the closing hours of the 1980 Congressional Session. Even though the purpose of the law was clear, the Superfund Bills had been revised and changed extensively during House and Senate hearings and markup sessions. The final version, however, was a compromise put together by various Senators. No conference committee reports were filed about this compromise.

Many of the problems now encountered with the CERCLA law are issues which the Congress never considered or deliberately avoided answering such as the retroactive application of the law, apportionment of costs, and standards for cleanup, (i.e., how clean is clean [3]). Thus, the statute has "acquired a well- deserved notoriety for vaguely-drafted provisions and an indefinite, if not contradictory legislative history" [4].

LIABILITY

One of the first issues to be answered in any CERCLA action is whether or not a company or individual corporate officer has a responsibility to respond. The principle of strict liability eliminates the need to show negligence. Persons dealing with hazardous waste are strictly liable for any damages that occur. There are very limited defenses under CERCLA such as causation by an Act of God, Act of War, or acts or omission of an independent third party.

Section 107(a) of CERCLA provides that response costs may be recovered from four classes of persons: 1) the present owner and operator of a facility; 2) the owner or operator at the time of disposal; 3) generators; and 4) transporters. The courts have taken an expansive view of this section to bring within the scope of liability persons outside the above four categories. For example, in *New York v. Shore Realty Corp.* [5] the state sued Shore Realty Corp. and its officers and stockholders for the cost of cleanup on land it had purchased, even though Shore had not participated in the generation or transportation of the waste. The court held that under CERCLA, Shore was an owner and operator under section 107(a)(1), even though Shore was not involved in operations on the site. The court said "It is quite clear that if a current owner of a site could avoid liability merely by having purchased the site after chemical dumping had ceased, waste sites certainly would be sold, following the cessation of dumping, to new owners who could avoid the liability otherwise required by CERCLA." This reasoning is erroneous because the prior owner would be responsible under 107(a)(2) as a person who was an owner at the time of disposal.

As written, the Shore Realty case interprets section (a)(1) to read an owner *or* operator of a facility is liable. Arguably an innocent purchaser of land on which hazardous wastes have been dumped previously has the responsibility of the clean up. This interpretation does not give any meaning to the difference between the word "and" in Section (a)(1) and the word "or" in (a)(2). In fact, the legislature may not have wanted to impose liability on a present owner of land unless he was also an operator.

If this expansive reading remains the law, how far does the liability run? An owner can be the owner in fee simple, a lessee, an easement owner such as a power company or one owning a right of way, or an owner of a mineral rights. With these extensive possibilities, any entity purchasing an interest in land where there was a possibility of the storage or disposal of hazardous waste must act with extreme caution. Even banks who foreclose on property that turn out to be abandoned hazardous waste sites have been held responsible for cleanup costs [6].

The courts have determined that the seller of material containing hazardous wastes may be liable even though the seller has no direct connection with the buyer's use of the material. For example, a company which sold used transformer oil contaminated with PCBs to the operator of a drag strip, who used the oil for dust control, would be liable for cleanup costs [7].

In another case [8], a district court held a company liable when contaminated soil was excavated and transported to another site. The court held that the liability of the company continued with the contaminated soil to the second site, thus supporting the thesis that the generator is responsible for the hazardous waste until its ultimate destruction. A serious question arises as to the liability of the generator if the waste is taken to a RCRA approved site and a release occurs [9]. Is the generator still liable? Arguably such disposal does not terminate liability for the waste.

Another question concerning continuing liability is the generator's responsibility throughout the entire process of cleanup by another party. Does this depend on whether the responsible parties have contracted with the person performing the cleanup? Arguably, the generators are still secondarily liable under Section 107(a)(3) which provides that a generator is liable if he contracts or arranges for transportation or disposal provided the generator does some act to signify his agreement to the arrangements.

Conceptually and practically, all persons are jointly liable. Joint and several liability means that any one party, even though a minor participant, is liable for the entire costs of cleanup. Thus, the government need only prove one generator is liable (assuming that generator's pocket is deep enough) to recover its entire cost.

Even an individual who works for a corporation and arranges for the disposal of hazardous waste may be liable under CERCLA. In one instance the president and principal stockholder was held personally liable [2]. In one of the more far reaching opinions, a court held liable the following individuals in a chemical company—former president, vice president, shift supervisor and a waste hauler [10]. By this reasoning, individual liability can extend throughout the corporate structure to encompass almost any person who had contact with the hazardous waste.

The principal purpose of CERCLA was to deal with inactive sites, where activities took place many years ago. Consequently, the individual whose activities encompassed the handling or disposal of waste may already have established his liability. This individual liability becomes even more acute for the individual who worked for a business entity which no longer exists. He may be the only responsible person in existence to respond to the damage claims. Coupled with the joint and several liability interpretation of the law, even if other responsible entities are identified, this individual could be required to pay for the entire cleanup.

It should be noted that the evidence of the linkage of a person to a site need not be overwhelming. Many courts have held that the government only need to prove that a generator's waste was disposed of at a site and that the substances that made the defendant's waste hazardous were present at the site [11]. With this sobering view of the responsibilities of cleanup, the natural inclination is to look for someone else to respond to the costs—the insurance company and/or contribution or indemnification by other parties.

INSURANCE

Two important provisions are normally included in a corporation's comprehensive general liability (G.C.L.) policy: the duty to defend and the obligation to pay. Both are important, because in some instances, the transactional costs, i.e., the costs of defense, can be more expensive than the obligation to pay. G.C.L. policy coverage is defined by three different time periods—pre-1966, 1966 to the early 1970s, and thereafter. Each one of these periods applies a different standard, but generally the more recent the policy period, the more restricted the coverage. Coverage has been expanded significantly by court interpretations. Additionally, the duty to defend an insured arises whenever the claim made conceivably could fall within the policy's coverage [12]. It is appropriate to put insurance carriers on notice and ask them to manage the action upon receipt of a notice letter or a request to produce information from EPA.

Pre-1966 policies generally are broadly worded and cover all "accidents." The accident policies were often held to cover claims based on long-term exposure to injurious conditions and substances; thus, if material was deposited prior to 1966, a G.C.L. policy will probably provide coverage. Subsequent to 1966, the G.C.L. policy coverage typically provided "the insurer shall . . . all sums which the insured shall become legally obligated to pay as damages because of bodily injury or property damages caused by an occurrence." "Occurrence" was defined as an accident which resulted in bodily injury or property damage during the policy period. The major problem is when the policy is "triggered," i.e., what causes the policy to apply. There are four current "triggering," theories for personal injuries—exposure, manifestation, injury in fact, and multiple trigger. Exposure theory states injury occurs and coverage is triggered when some cellular damage takes place. The manifestation theory holds that the policy covers from when the injury first manifests itself, even if the cause of injury occurred sometime earlier. The injury in fact theory continues to trigger the policy for each policy period when injury, however small, is found to occur. In one personal injury case the court determined that each company was liable for the damage caused during its policy period; however, since the injuries were not distinguishable, each insurer was jointly and severally liable [13]. The final theory—multiple trigger—says the insurers are liable if any of the preceding theories applies to an insured.

For property damage cases, related theories may apply for triggering the policy:

1. When the complaining party suffers actual damage regardless of when the wrongful act took place [14]. The trigger of actual damage may be: a) the generation of the hazardous substances; b) actual disposal of the hazardous substances; and/or c) discharge or leaking of the hazardous substances.
2. When the effect was felt by the injured party [15].

The case which has provided the broadest coverage was *Township of Jackson v. American Home* [16] which held there were multiple triggers. The court noted "the negligent siting of the landfill, digging beneath the water table, providing inadequate cover, failing to inspect tank trucks, accepting improper and imprudent amounts of liquid," were triggers to policy coverage. Thus, the real interpretation was that any insurer who provided coverage during the life of the landfill essentially covered all claims. This case is currently under appeal.

The general pollution exclusionary clause may be a major stumbling block to coverage from the early 1970s to present. The clause generally states "This insurance does not apply; . . . to bodily injury or property damage arising out of the discharge, dispersal, release or escape of smoke, vapors, soot, fumes, acids, alkalies, toxic irritants, contaminants or pollutants into or upon land, the atmosphere or any water course or body of water; but this exclusion does not apply to such discharge, dispersal, release or escape if sudden and accidental." The major issue under this section is whether the exclusion applies at all, or does the exception remove the exclusion because any event is "sudden and accidental."

The applicability of the sudden and accidental exception depends on which viewpoint is taken. Some courts have held the "sudden and accidental" language means that the damage was not expected nor intended from the standpoint of the insured. For example, in a suit brought against a landfill for contamination by seepage into the groundwater, the court held that the insured intended to deposit the waste in the landfill but never expected it nor intended it to seep into the aquifer, thus it was "sudden and accidental" and the exclusion did not prevent coverage [17].

However, there are limitations imposed on this term, such as when the insured deposited the hazardous material knowing or having reason to expect that it would pollute. In the instance of a landfill, is it "reasonably foreseeable" that it will pollute. If so, then there would be no coverage. A minority have interpreted the clause to be even more restrictive. In a case where the insured dumped waste into the sewer line over a period of years, the court rejected the insured's argument that the injury was "sudden and accidental." Even though the damage was not intended, the regular discharge was not sudden and accidental but intentional, thus the court held the exception did not apply and the insurance company was relieved of any liability [18].

CONTRIBUTION AND INDEMNIFICATION

With the application of joint and several liability to generators of hazardous waste, the natural inclination is to look for other parties to contribute to the cost of the cleanup or provide complete indemnification. The courts that have considered the question have uniformly allowed contribution. No case has yet decided the issue of the apportionment of damages. However, in general liability cases most states require each party to pay a pro rata share of the liability. A minority of states apportion damages based on fault.

The pro rata approach is contrary to the notion of equity because it would treat the small contributor the same as a very large one. Each generator, regardless of how much he deposited at the site, would fund the cleanup equally. On the other hand, it will be difficult for a court to apportion fault since this is a strict liability situation, but costs could be apportioned based on volume of waste deposited or some other criteria.

Corollary to the problem of contribution is the issue of indemnification. Should the party who is at fault indemnify the one who is required to respond because he is strictly liable. The Act is silent on this point, and the only case law is in the area of property owners and their right to indemnification. One court held that the lessor of property should be indemnified by the lessee for costs associated with the cleanup of hazardous wastes which the lessee had deposited [19].

A right to indemnification among owners, operators, generators and transporters is not addressed in the Act. For example, a generator contracts with a transporter to deliver his hazardous waste to an EPA approved site. Instead, it is deposited in a non-authorized disposal site and a cleanup action is brought by EPA. Equity and fairness would say the generator is entitled to indemnification from the transporter, but this issue has yet to be resolved.

MANAGEMENT ISSUES

The Superfund process has not worked well for a number of reasons, not the least of which is the need to develop institutional mechanisms which are understood by all of the parties concerned. Problems include insufficient notice by EPA to the PRPs before decisions are made. This places the PRPs in a position of having a decision made without any input from them and then being asked to pay for the cleanup.

On the other hand, industry has in many cases failed to respond in a timely fashion to notification by EPA that a given industry is a potentially responsible party and may be held liable for the entire cost of cleanup under the principle of joint and several liability. This is sometimes due to a failure to appreciate the severity of the notification letter or to a legitimate feeling on the part of industry that they are in fact not responsible in any way for the site.

In order to have input into the decision-making process, the PRPs should select a steering committee to deal directly with EPA [20]. The agency normally will not negotiate with individual PRPs [21]. The committees generally consist of both attorneys and engineers representing both large and small generators as well as transporters. Additional companies may learn of the steering committee after it has formed, and general practice has been not to exclude any transporter or generator who wishes to join.

The steering committee functions informally on a consensus basis. The basic resources available to the committee are the time and talent of staff donated by the PRPs. As with any volunteer effort, the effectiveness of the committee is a direct function of the capabilities and interest of the individuals involved. Where a company has a significant amount of liability in a site, a senior corporate official should represent the company on the steering committee.

While the identification of PRPs is initiated by the EPA, they lack the resources and incentive to perform an adequate search, and once a PRP steering committee is formed, one of its first tasks is to identify other PRPs who can share the cost of cleanup. The search for additional parties is a time consuming and expensive process involving record searches, interviews with current and former employees of generator and transporter companies and analysis of the waste [22].

Once a significant number of PRP's have been identified, the steering committee attempts to negotiate a settlement with EPA. This process is time consuming. The PRPs want to make sure they understand the situation prior to agreeing to anything. EPA, however, is under pressure to proceed with the site cleanup in as timely a fashion as possible, whether or not the PRPs ever agree to a settlement. Among the issues to be settled:

- Who will actually do the cleanup, EPA or the PRPs? [23]
- What exactly will be done at the site?
- How much will the PRPs pay?
- Can the agency provide protection against future liabilities and litigation?

DEALING WITH THE PUBLIC

A major problem the PRP Committee will have to confront is dealing with the public in the vicinity of the site. At many sites a group has been formed to represent the citizens in the area in dealings with EPA regarding the cleanup of the site.

While the interests of the citizens do vary from site to site, one commonality is the solution that they desire: they want the material removed and taken somewhere else. Removal and transportation of the waste is often the most expensive approach and not necessarily environmentally the most desirable. Additionally, the PRPs may be held responsible for damages that occur from removal actions or for future liabilities at the site that the wastes are removed to.

Ideally the PRP Steering Committee should communicate with the citizens group to better understand their concerns and to attempt to make sure that the citizens groups understand the implications of the solutions that they are demanding. While this appears to be a logical approach, many individual companies making up the PRP Steering Committee may have corporate policies that preclude discussing hazardous waste in public. While this is true to a certain degree for all corporations, those companies serving the consumer market are particularly sensitive to the public relations potential of being involved in a hazardous waste site and may not want information identifying them as a PRP being given to the citizens group. While this may greatly inhibit the ability of the PRP Steering Committee to communicate with the citizens group, it will not protect the identity of potentially responsible parties because the EPA must provide the names of all PRPs to anyone who requests them under the Freedom of Information Act.

An additional factor to be considered is the appearance created when industry fails to be open with the general public. The impression is that the industries involved in the site have "something to hide," which eventually creates a worse atmosphere in which to negotiate. Ideally the decision, of what type of cleanup action is required, is a purely technical decision made within the constraints of the Superfund law. However, the final decision rests with the Environmental Protection Agency, who will consider the political implications created by the citizens group. Thus, in avoiding the citizens group, the PRP's interests may be hurt in the final analysis.

SUMMARY AND CONCLUSIONS

Any individual or corporate entity involved in the generation or transportation of waste or ownership or operation of a site is a potentially responsible party and individually liable for the entire cost of clean up.

Where the courts have acted, it has generally been to increase rather than decrease liability.

The insurance underwriter who provided general comprehensive liability at any time in the process may be responsible for both legal and clean up costs.

The issue of apportionment of costs among PRPs has yet to be decided and in fact one party may be held responsible for all of the costs. However, negotiated settlements have apportioned costs on the basis of waste volume.

If a corporation believes they may be involved in a Superfund site, the following actions are generally advisable:

1. Participate in the PRP committee.
2. Put on notice all present and former insurance carriers.
3. Obtain expert legal counsel and technical consultants.
4. Encourage the PRP committee to initiate contact with citizen group(s) in the area of the site to try to understand their concerns.
5. Search your own records to establish:
   - If your wastes were hazardous.
   - Where wastes were sent.
6. Contact EPA and state agencies to ascertain their understanding of your involvement in the site.

REFERENCES

1. The Comprehensive Environmental Response Compensation and Liability Act of 1980, P.L. 96-510, 42 USC, 9601 ff.
2. Gilardi, E. F., LaGrega, M. D., and Campbell-Loughead, J., "Remedial Action at Hazardous Waste Sites," *Proceedings of the 38th Purdue Industrial Waste Conference*, Purdue University, 243–249 (1984).
3. Hall, John C., J. D., "Designing Risk Analyses to Avoid Pitfalls in Cost Recovery Actions: A Legal/Technical Solution," *Management of Uncontrolled Hazardous Waste Sites*, 313–320 (1984).
4. *U.S. v. Mottolo 23 ERC* 1294, 1296 (D.N.H. 1985).
5. *New York v. Shore Realty Corp.*, 759 F.2d 1032 (2nd Cir. 1985).
6. *U.S. v. Maryland Bank & Trust Co.*, 24 ERC 1193, (D.M.d. 1986).
7. *New York v. General Electric Co.*, 592 F. Supp. 291 (N.D.N.Y. 1984).
8. *Missouri v. Independent Petrochemical Corp.* 22 ERC 1167 (E.D.Mo. 1985).
9. Zeitzew, H., "Liability for Personal Injury Damages: A New Problem for Industrial Waste Generators," *Proceedings of the 39th Purdue Industrial Waste Conference*, Purdue University 353–458 (1985).
10. *U.S. v. Northeastern Pharmaceutical and Chemical Co.*, 579 F. Supp. 823 (W.D. 1984).
11. e.g., *U.S. v. Wade, 20, ERC* 1277 (E.D. Pa. 1983); *U.S. v. South Carolina Recycling and Disposal*, Inc. *20 ERC* 1753, (D.S.C. 1984).
12. *Travelers Indemnity Co. v. Dingwell*, 414 A2d 220 (Me. 1980).
13. *Sandoz, Inc. v. Employer's Liability Assurance Co.* 554 F. Supp. 257 (D.N.J. 1983).
14. *Michigan Chemical Corp. v. American Home Insurance Co.*, 728 F.2d 374 (6th Cir 1984).
15. *Anchor Casualty Co. v. McCaleb* 178 F.2d 322 (5th Cir. 1949).
16. *Township of Jackson v. American Home*, No. 6-29236-8 (N.J. Super. 1984.), appeal pend.
17. *Jackson Township Utilities Authorities v. Hartford A & I Co.*, 186 N.J. Super. 156, 451 A.2d 990 (1982).
18. *Transamerica Ins. Co. v. Sunnes*, 77 Ore. App. 136, 711 P.2d 212 (1985).
19. *Caldwell v. Gurley Refining Co.*, 755 F.2d 645 (8th Cir. 1985).
20. Moorman, J. W., "The Superfund Steering Committee: A Primer," *Environmental Forum* 4, 13–20 (Feb. 10, 1986).
21. EPA Interim CERCLA Settlement Policy, *F.R.*, 50, No. 24 (1985).

22. Rikleen, L. S., Standford, R. L., and Cochroan, S. Robert, Jr., "Site Remediation: The Initial Steps—Structuring a Settlement," *Management of Uncontrolled Hazardous Waste Sites*, 275–280 (1985).
23. Hagger, C. L., Goodwin, Bruce E., and Strickfadden, M. E., "Generator Cleanup vs. Federal Cleanup of Hazardous Waste Sites," *Management of Uncontrolled Hazardous Waste Sites*, 7–10 (1985).

44 SURFACTANT SCRUBBING OF HAZARDOUS CHEMICALS FROM SOIL

Janet Rickabaugh, Research Associate

Sara Clement, Graduate Research Assistant
Department of Civil and Environmental Engineering
University of Cincinnati
Cincinnati, Ohio 45221

Ronald F. Lewis, Microbiologist
U.S. EPA-ORD-HWERL
Cincinnati, Ohio 45268

INTRODUCTION

A lab scale study was conducted to examine the effectiveness of aqueous surfactant solutions for decontaminating soil from a hazardous waste site. The soil used in this study was from the Chem-dyne hazardous waste site in Hamilton, Ohio which was used to store and dispose of pesticides, chlorinated hydrocarbons, solvents, waste oils, and other hazardous wastes from 1975 to 1980. Clean up of the site was initiated in 1980 and most chemical drums were removed by 1983. Soil from the site still remains to be decontaminated. The Chem-dyne soil used in this study was analyzed by gas chromatography for 11 chlorinated hydrocarbons (hexachlorobenzene, dichlorobenzenes, trichlorobenzenes, hexachlorobutadiene, Aldrin, Endrin, Heptachlor, and Dieldrin). The soil contained a total of from 216 to 266 mg/kg of these compounds. Analysis for total chlorinated hydrocarbons (52 peaks) indicated that the total concentration of all chlorinated hydrocarbons in the soil was approximately 2078 mg/kg.

Surfactants were used to promote solubilization of these chlorinated hydrocarbons from Chem-dyne soil. One end of a surfactant molecule is polar (water soluble), while the other end is non-polar (organic soluble) so that they promote aqueous solubilization of sparingly-water soluble compounds. Anionic, nonionic, cationic surfactants and blends of these types were evaluated. Anionic surfactants, the oldest group of commercial surfactants, are usually highly water soluble, but are sensitive to the total ionic strength of the media. The negatively charged polar solubilizing group on anionic surfactants is usually a sulfonate, sulfate, or phosphate functional group. Cationic surfactants are a small group of softening and coating agents with a positively charged solubilizing group which is usually an amino or quaternary nitrogen functional group. Nonionic surfactants do not have a charge on their solubilizing group. They are solubilized by hydrogen bonding at oxygen or hydroxyl groups in the surfactant molecule. Nonionic surfactants usually contain the polyoxyethylene group as their solubilizing group and are now the most commonly used group of commercial surfactants in North America.

LITERATURE SURVEY

Literature on using surfactants for secondary and tertiary oil recovery indicates that a blend of two surfactants, a surfactant and a co-surfactant, is more effective than a single surfactant used alone. Therefore, it was decided to investigate blends of surfactants (nonionic-anionic and nonionic-cationic) as well as anionic, nonionic, and cationic surfactants used alone.

Only a few references to using surfactants for scrubbing soils from hazardous waste sites or spills are available. The concept is still in the lab stage of development. Huibregste et al. [1] developed a mobile soil scrubber to wash contaminated soil. They used 0.1–1% solutions of nonionic surfactants in laboratory shake jar tests to determine the effectiveness of different types of solutions in scrubbing various contaminants (PCB, phenol) out of soils. Initial contaminant concentrations were quite high (61–850 mg/L PCB and 1,000–20,000 mg/L phenol). Removal of phenol by surfactant solutions ranged from 67–97%, while removal of PCB ranged from 20–37%.

377

Botre et al. [2] reported using 2% solutions of anionic, cationic and nonionic surfactants to wash soils contaminated with dioxin (TCDD) from Seveso, Italy. They found that cationic surfactants removed 75% of the TCDD, anionics 60%, and nonionics 45%. They also studied photodecomposition of the TCDD in the surfactant solutions. Photodecomposition of TCDD was complete in 18 hours in a control solution, 8 hours in an anionic surfactant and 4 hours in a cationic surfactant solution.

Ellis, Payne and McNabb [3] used a pair of aqueous nonionic surfactants to clean a test soil spiked with 100 mg/L PCB, 1000 mg/L petroleum hydrocarbons, and 30 mg/L chlorophenols. In bench scale shaker tests and larger scale column soil scrubbing tests, 92% of the PCBs were removed from the soil using a 1.5% total concentration of two nonionic surfactants while a 2% combined solution of these surfactants removed 93% of the petroleum hydrocarbons from the soil. Chlorinated phenols were effectively removed by water alone. Optimum surfactant concentration was found to be 1.5% (total) for the pair. Several methods were evaluated to remove surfactant from the contaminated leachate to allow recycling of the surfactant solution. None of these methods proved entirely successful. Reuse of the surfactant was deemed essential for cost-effective application of this technology in the field.

EXPERIMENTAL METHODS

This research examined the effectiveness of surfactant solutions in removing chlorinated hydrocarbon contaminants from Chem-dyne soils and the use of photolysis to decontaminate the surfactant solutions in lab scale tests.

Thirty-eight commercial surfactants were collected, including anionic, nonionic, and cationic types and blends. Fourteen were selected on the basis of good water solubility, near neutral pH (to minimize soil adsorption), and generally low chloride content (except the cationic). Table I shows the surfactants selected for this study.

These surfactants were tested in lab scale batch and flow-through column soil scrubbing experiments. In batch shake tests 10 gm of contaminated soil was shaken for 2–4 hours in a 250 mL bottle with 200 mL of 1% or 2% surfactant solutions, allowed to sit overnight, filtered and decanted. The contaminated surfactant solution was liquid-liquid extracted (USEPA method 3520) [4]concentrated, and analyzed for chlorinated hydrocarbons by gas chromatography (USEPA method 8120) [4]. Duplicate samples of untreated soil were also soxhlet extracted (USEPA method 3540) [4] and analyzed for chlorinated hydrocarbons by GC. Also in the batch studies, a method to measure TOX (Total Organic Halogen), based on research by McCahill et al. [5], was used. This method involved measuring the increased chloride levels in the sample after photolysis to break the carbon chlorine bonds (organic chlorine). One and 4 gm samples of soil were shaken for 2–4 hours with 30 mL of 0.5%, 1% and 2% surfactant solutions, centrifuged, and decanted. One half of the sample was photoreacted under two 15-watt germacidal lamps for 4 hours. Then the photoreacted sample was analyzed for chloride using a chloride electrode (inorganic + organic chlorine). The other half of the sample (not photoreacted) was analyzed for chloride (inorganic chloride) using a chloride electrode. The chloride value for the

Table I. Surfactants Studied

Type	Name	Company	Chemical Type
Anionic	Biosoft N-300	Stephan	Triethanalomine LAS
Anionic	Surco 60T	Onyx	Triethanol alkyl aryl sulfonate
Anionic	Surco SXS	Onyx	Sodium xylene sulfonate
Anionic	Tryfac 5553	Emery	K salt phosphated alkyl EO
Nonionic	Makon 10	Stephan	Ethyoxylated nonylphenol
Nonionic	Triton DF16	Rohm & Haas	PEO Alcohol
Nonionic	Surfonic N120	Texaco	Nonyl polyethoxyl ethanol
Nonionic	Trydet 22	Emery	PEO ester of fatty acid
Nonionic	Polytergent SL62	Olin	Alkyloxylated alcohol
Nonionic	Merpol SH	DuPont	PEO alcohol
Nonionic	Merpol DA	DuPont	PEO amine
Nonionic	Monolan 2800	Diamond Sham.	Polyoxypropylene block
Blend	Agrimul S300	Diamond Sham.	nonionic/anionic blend
Cationic	Emcol CC9	Witco	diethylmonium chloride

Table II. Batch Studies Average TOX Removals (mg/kg)

	Anionics	Nonionics	Blend	Average
2% Concentration	86	89	113	90
1% Concentration	49	66	82	62
0.5% Concentration	27	33	26	30

un-photoreacted sample was subtracted from the value for the the photoreacted sample to obtain the total organic chlorine (TOX).

The surfactants that performed best in the batch studies were selected for testing in small-scale flow-through columns. The columns consisted of 250 mL burettes containing 50 gm of Chem-dyne soil which was washed with 500 mL of a 2% surfactant solution. The 2% concentration had proved the most effective in the batch studies. Column scrubbing studies were performed with two anionic surfactants (Surco 60T and Biosoft N300), two nonionic surfactants (Makon 10 and Triton DF16), one cationic surfactant (Emcol CC9), and several blends of these surfactants (nonionic-cationic and nonionic-anionic). The columns were evaluated for a period of one week at an average flow of 59 mL/hr, and then rinsed with 250 mL of distilled water. The contaminated surfactant plus the rinse was liquid-liquid extracted and analyzed by GC for chlorinated hydrocarbons. Samples of washed soils and duplicate samples of unwashed soils were soxhlet extracted and also analyzed by GC for chlorinated hydrocarbons.

To investigate photolysis as a means of decontaminating surfactant solutions, samples of contaminated Biosoft (anionic) solution and Emcol (cationic) solution were split into two aliquots after soil scrubbing. One aliquot was photoreacted for 24 hours, liquid-liquid extracted, and analyzed by GC for chlorinated hydrocarbons. The other aliquot was extracted and analyzed without photoreaction for comparison.

Two soil columns were washed with blends of nonionic-cationic surfactants (one with Makon-Emcol, one with Triton-Emcol) for a total period of three weeks. Each column was filled with 100 gm of soil at the beginning of the experiment. The soils were washed with 500 mL of 2% surfactant solutions at an average flow of 61 mL/hr. At the end of each week approximately 25 gm of soil was removed from each column, soxhlet extracted, and analyzed for chlorinated hydrocarbons by GC. A sample of the initial soil used to fill the columns was also soxhlet extracted and analyzed before soil scrubbing. The combined contaminated surfactant solutions and 250 mL distilled water rinses were liquid-liquid extracted and analyzed by GC for chlorinated hydrocarbons weekly. Each week fresh surfactant solutions were used to wash the soil columns.

RESULTS

Surfactant Results

TOX removal studies on small scale batch scrubbing tests of Chem-dyne soils showed little difference between average TOX removals by anionic and nonionic surfactants. The blend showed slightly higher removals of TOX from the soils. Table II shows that the TOX studies indicated that the 2% concentration of all surfactant types removed more TOX from the soils than the 1% and the 0.5% concentrations.

Gas chromatography data from batch studies, listed in Table III, showed the highest removals of a total of 11 chlorinated hydrocarbons using anionic and cationic surfactants.

Tables IV and V show the ranking of the six most effective surfactants by removals of chlorinated hydrocarbons from soils during the batch studies. Table IV shows rank by TOX removal, while Table V shows the rank by removal of a total of 11 target compounds by GC analysis.

Table III. Batch Studies Gas Chromatography Liq-Liq Extraction Data Average Total Removal

	Anionics		Nonionics		Blends		Cationics	
	mg/kg	%	mg/kg	%	mg/kg	%	mg/kg	%
2% Concentration	47.1	20.8	5.7	2.5				
1% Concentration	18.9	8.3	3.7	1.6	1.3	0.6	26	11.5

Table IV. Surfactant rank, TOX Removal Batch Studies

TOX Removed	mg/kg
1. Makon 10 2% Nonionic	150
2. Tryfac 5553 2% Anionic	135
2. Triton DF16 1% Nonionic	135
3. Surco 60T 2% Anionic	112.5
3. Triton DF16 2% Nonionic	112.5
3. Polytergent SL62 2% Non.	112.5
3. Makon 10 1% Nonionic	112.5
4. Tryfac 5553 1% Anionic	97.5
5. Surfonic N120 1% Non.	90.0
5. Polytergent SL62 1% Non.	90.0
6. Trydet 22 2% Nonionic	82.5
6. Agrimul S300 1% Blend	82.5

GC analyses in batch studies showed good removals of hexachlorobenzene, Endrin, Aldrin, and hexachlorobutadiene by most surfactants (except for those showing poor removal of all compounds). The cationic surfactant showed good removals of dichlorobenzenes and trichlorobenzenes as well, while the anionic and nonionic compounds did not remove these compounds as effectively. The cationic surfactants seemed better at removing compounds of lower molecular weight than the nonionic and anionic surfactants.

In the column studies, blends were slightly more effective than any single type of surfactant in removing chlorinated hydrocarbons from the soil. Anionics and nonionics were similar in their removals while cationic surfactants used alone had lower removals than the other types.

In the column studies the contaminated surfactant solutions and the scrubbed soils from the same experiment were extracted by liquid-liquid extraction and soxhlet extraction, analyzed for chlorinated hydrocarbons by gas chromatography, and the results of these two types of extractions were compared. Table VI shows that the soxhlet extractions of soils indicated that more contaminants were removed from the soil than indicated by data from liquid-liquid extractions of the contaminated surfactant solutions.

Emulsion problems were encountered in the liqiud-liquid extraction procedure with 1% and 2% surfactant solutions, particularly with anionic surfactants. Extraction efficiencies of standards in surfactant solutions were erratic, ranging from 0 to 1589%. Table VII shows surfactants ranked by total removal of 11 compounds from the soil. Table VII allows comparison of contaminant removal data from soxhlet and liquid-liquid extractions of samples from the same soil scrubbing experiment. The soxhlet extraction data indicate there is a considerable quantity of contaminants removed from the soils by the surfactants but not measured in the liquid-liquid extract of the wash solutions. This could indicate that the presence of the surfactant caused some of the contaminants to remain in the

Table V. Surfactant Rank, Gas Chromatography Data Batch Studies Removal, Total of 11 Compounds

			mg/kg	%
1. Biosoft N300	2%	Anionic	47.1	22
2. Surco 60T	1%	Anionic	36.9	17
3. Biosoft N300	1%	Anionic	31.5	15
4. Emcol CC9	1%	Catonic	25.7	12
5. Triton DF16	1%	Nonionic	10.5	5
6. Makon 10	1%	Nonionic	2.3	3

Table VI. Column Studies Average Total Removal-11 Compounds (mg/kg) 2% Surfactant Solutions

	Anionic	Nonionic	Cationic	Blend
Soxhlet Ex.	92.4	97.5	69.5	115
Liq-Liq Ex.	8.4	12.6	18.3	18.7

Table VII. Surfactant Rank Column Studies Total Removals 11 Compounds 2% Surfactant Solutions

Removals Soxhlet Extractions	%	mg/kg	Removals Liquid-Liquid Extractions	%	mg/kg
1. Triton-Emcol (Non-Cat)	59	136	1. Triton-Emcol (Non-Cat)	26.5	61
2. Makon 10 (Nonionic)	55	125	2. Emcol (Cationic)	8	18
3. Triton-Biosoft (Non-An)	51	116	3. Makon 10 (Nonionic)	6.5	15
4. Makon-Surco (Non-An)	46	104	4. Makon-Emcol (Non-Cat)	4.7	11
5. Makon-Emcol (Non-Cat)	44	100	5. Triton (Nonionic)	4.5	10
6. Biosoft (Anionic)	41	97	6. Surco (Anionic)	3.8	9

water phase rather than being extracted into the methylene chloride phase during the liquid-liquid extraction process. Because of these difficulties, it was judged that the data from soxhlet extractions of the soils were more reliable than data from liquid-liquid extractions of surfactant solutions.

Much higher contaminant removals were shown in the column studies than in the batch studies. These results may be partially explained by two phenomena. First, while the column data were obtained by soxhlet extractions, the batch data were obtained by liquid-liquid extractions which generally indicated lower contaminant removals. Second, the column data may be higher because the length of time of continuous soil scrubbing in the column studies was one week while it was only one day in the batch studies.

Table VIII shows that removal of the chlorinated hydrocarbons from the soils continued during the entire three week period of the time column studies, but generally at a slower rate during the third week.

Soxhlet data from time column studies also show higher removals of contaminants from soils than the liquid-liquid data indicate took place. In the time studies, the total removal rates and the removal rates of specific compounds generally tended to decrease with time. The cumulative 3 week removal for time studies based on soxhlet data was lower than the corresponding soxhlet removal data for one week column studies. This could be partially due to the higher ratio of soil weight to surfactant volume in the time column studies. Five hundred mL of surfactant solution was used to scrub soil (100 gm, week 1; 73 gm, week 2; 55 gm, week 3) in the time column study while the same volume of solution was used to wash only 50 gm of soil in the initial column study.

Photolysis Results

Preliminary experiments on leachate from one week soil scrubbing columns show that photolysis is promising as a means of decontaminating surfactant solutions. Twenty-four hour photolysis under two 15 watt germacidal lamps (254 nm) destroyed 76% of the 11 chlorinated hydrocarbons in a contaminated anionic surfactant solution (Biosoft) and 100% of these contaminants in a cationic solution (Emcol). Calculations of total removal of all chlorinated hydrocarbons showed that the 24 hour photolysis destroyed approximately 95.4 mg/L of all the chlorinated contaminants in the Biosoft (anionic) solution and 333.7 mg/L in the Emcol (cationic) solution.

CONCLUSIONS.

Surfactant solutions generally increased the efficiency of removal of chlorinated hydrocarbons from Chem-dyne soils over that of water alone. Removals of up to 59% of 11 target compounds were

Table VIII. Average Removals By 2% Nonionic-Cationic Blends

	Week 1		Week 2		Week 3	
	mg/kg	%	mg/kg	%	mg/kg	%
	Weekly Removals of 11 Compounds					
Soxhlet Data	34.5	13	42	16	10	3.5
Liq-Liq Data	20.6	7.8	10.9	4.1	9.9	3.7
	Cumulative Removals of 11 Compounds					
Soxhlet Data	34.5	13	76.5	29	86.5	32.5
Liq-Liq Data	20.6	7.8	31.5	11.9	41.4	15.6

obtained in one week of soil scrubbing with surfactant solutions, while scrubbing with water alone removed less than 1%.

Blends of nonionic-cationic or nonionic-anionic surfactants generally gave better removals of contaminants from Chem-dyne soils than any type of surfactant used alone. Cationic surfactants used alone generally gave lower removals than anionic or nonionic surfactants used alone.

The 2% concentrations of surfactant solutions gave better contaminant removals than the 1% or the 0.5% concentrations.

Standard liquid-liquid extraction procedures did not work well with 1% and 2% surfactant solutions.

Photolysis showed promise as a means of decontaminating surfactant solutions used to wash these soils.

In general, this study showed that the following conditions tend to promote greater removals of contaminants from Chem-dyne soils:

1. High ratios of surfactant volume to soil weight scrubbed.
2. Increased scrubbing times (up to 3 weeks).
3. Higher surfactant concentrations (up to 2%).

DISCLAIMER

Although the research described in this paper was funded wholly or in part, by the United States Environmental Protection Agency, it has not been subjected to Agency review, and therefore does not necessarily reflect the views of the Agency, and no official endorsement should be inferred. This research was funded under EPA Contract No. 68-03-3210-01.

REFERENCES

1. Huibregste, K. R., et al., "Development of a Mobile System for Extracting Spilled Hazardous Materials from Soils", in *Control of Hazardous Materials Spills*, Proc. 1980 Conf., Vanderbilt U. & U.S.E.P.A., (1980).
2. Botre, C. et al., "TCDD Solubilization and Photodecomposition in Aqueous Solutions", *Environ. Sci. & Tech.*, 12, #3, 335–336, (1978).
3. Ellis, W. D., J. R. Payne and G. D. McNabb, "Project Summary, Treatment of Contaminated Soils with Aqueous Surfactants", U.S.E.P.A., H.W.E.R.L., Cincinnati, *EPA/600/S2-85/129, PB 86-122561/AS*, (Dec. 1985).
4. U.S.E.P.A., "Test Methods for Evaluating Solid Waste, Physical/Chemical Methods", *SW-846* (1982).
5. McCahill, M. P., et al., "Determination of Organically Combined Chlorine in High Molecular Weight Organics", *Environ. Sci. & Tech.*, 14, 201–203 (1980).

45 FIRST YEAR'S OPERATING PERFORMANCE OF THE OMEGA HILLS LANDFILL PRETREATMENT ANAEROBIC FILTER

Paul E. Schafer, Project Manager

John L. Carter, Process Engineer
Black & Veatch, Engineers-Architects
Kansas City, Missouri 64114

Gregory C. Woelfel, Regional Engineer
Waste Management, Inc.
Menomonee Falls, Wisconsin 53051

INTRODUCTION

The Omega Hills landfill receives municipal, industrial, and commercial solid wastes and in the past has accepted industrial liquid wastes and sludges. The landfill, which began operation in 1970, is owned and operated by Waste Management of Wisconsin, Inc. Until October 1982, hazardous wastes were co-disposed with the municipal refuse in the RCRA interim-status areas of the landfill. Neither hazardous wastes nor liquid wastes are accepted any longer; however, the EPA classified Omega Hills as a hazardous waste landfill because it did accept hazardous waste in the past. Waste Management has installed an extensive leachate collection system, and the leachate is pumped into a sewer leading to the Milwaukee Metropolitan Sewerage District (MMSD) South Shore Treatment Plant. Although MMSD experienced no difficulties in the operation of the waste treatment plant, the Wisconsin Department of Natural Resources (WDNR) requested that Waste Management provide removal and pretreatment of 200,000 gpd of leachate. In late August 1985 an anaerobic filter was started up to provide pretreatment of the leachate before discharge to the MMSD sewer. This paper will briefly review the pilot plant study and the design of the anaerobic filter and describe the operation of the anaerobic filter since startup.

REVIEW OF PILOT PLANT PROGRAM AND PLANT DESIGN

During the early phases of leachate pretreatment facility design, a pilot plant program was initiated. The pilot plant studies, described previously [1,2], demonstrated that better than 90% BOD removal is feasible at BOD loadings up to 900 lb/d/1,000 cu ft [3]. Based on the results of pilot plant operation, an upflow anaerobic filter was designed. The leachate characteristics determined during the pilot plant program are shown in Table I. Although the WDNR had requested removal of 200,000 gpd, the volume of leachate decreased during the design period, and it became impossible to withdraw 200,000 gpd from the landfill. Therefore, based on pilot plant results, a design basis of 442 lb BOD/d/1,000 cu ft of media was used to provide the capability of treating an average flow of 100,000 gpd at 7.4 days hydraulic retention time (HRT). Process equipment can provide treatment for up to 150,000 gpd, resulting in an HRT of 4.9 days. The pilot plant study supported using an HRT less than 7 days. The design did, however, include provisions for adding a second anaerobic filter. The design basis of the anaerobic filter is shown in Table II.

Leachate is fed just below the media through eight nozzles that are designed to impart a swirling motion to the liquid in the feed zone (Figure 1). Liquid from four recycle withdrawal ports near the top of the digester sidewall is recirculated through the eight feed nozzles to the inlet zone of the reactor to provide additional mixing. A 7 1/2 horsepower recycle pump capable of pumping 700 gpm is used to provide up to a 10:1 recycle ratio (recycle to influent flow). Untreated leachate and a portion of the recycle is pumped through a plate and frame heat exchanger to maintain an operating temperature of 95°F in the reactor. The bottom of the anaerobic reactor is cone shaped with a side slope of 4 to 1 to facilitate removal of sludge. Treated leachate is withdrawn from the reactor via four effluent control boxes which are gravity fed and discharge to the MMSD sewer. Each control box is connected to an odor control system that helps maintain the reactor head gas pressure. A 1/4 horsepower chemical feed pump capable of feeding up to 1.8 gph of 75% phosphoric acid is used for phosphorus addition when

Table I. Pilot Plant Leachate Characteristics

Constituent	Omega Hills Concentration[a]
BOD	38,604
COD	50,930
SS	1,600
Volatile Acids	21,640
Alkalinity	13,790
Hardness	12,230
pH	6.0–6.5
TKN	1,250
Total Phosphorus	4
Sodium	3,200
Potassium	1,200
Calcium	2,730
Magnesium	1,300
Sulfate	1,700
Chloride	3,360
Iron	1,010
Cobalt	0.38
Nickel	1.84
Molybdenum	0.1
Zinc	43.2
Copper	0.06
Lead	0.12
Cadmium	0.08

[a]All units in mg/L, except pH.

it is needed. A 6,000 gallon fiberglass storage tank is used for phosphoric acid storage. Space is also provided for an anhydrous ammonia tank and feed system if considered necessary in the future.

OTHER EQUIPMENT

The leachate pretreatment system was designed with substantial operating flexibility in case leachate characteristics changed or if other industrial wastes are accepted for treatment. A schematic of the unit processes is shown in Figure 2. A 200,000 gallon holding and equalization tank that is 40 feet in diameter and 24 feet sidewater depth provides two days' storage. A 10 horsepower vertical turbine mixer provides mixing of the equalization tank contents. Baffles were installed in the equalization

Table II. Anaerobic Filter Design Basis

Minimum Hydraulic Retention Time, days	4.9
Average Hydraulic Retention Time, days	7.4
Anaerobic Reactor	
Volume, gal	740,200
Diameter, ft	60
Side depth, ft	35
BOD Loading, lb/d/1000 cf[a]	442
Media	
Depth, ft	20
Volume, cu ft	56,500
Surface area, sq ft/cu ft	35
Volumetric void rate	0.95
Maximum Gas Pressure, inches water	12
Design Flow	
Average, gpd	100,000
Peak, gpd	150,000

[a]Loading based on Media Volume, average flow and 30,000 mg/L BOD.

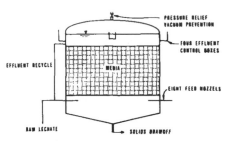

Figure 1.   Schematic of anaerobic filter.

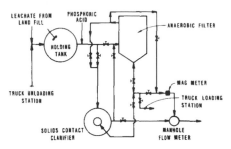

Figure 2.   Schematic of anaerobic pretreatment system.

tank to prevent the formation of vortices. Controls were designed to vary the liquid depth in order to equalize leachate flow as well as its organic strength. Leachate is piped from the landfill directly to the equalization tank; a truck unloading station can be used to pump other wastes to this tank.

A 30 foot diameter and 16 foot sidewater depth solids contact clarifier can be used either for pretreatment before feeding the wastes into the anaerobic filter or to treat the filter effluent before discharge to the sewer. This clarifier can also be used to store sludge from the anaerobic filter. Pretreatment in the clarifier would consist of precipitation of toxic metals that could inhibit the anaerobic process. The clarifier is equipped with a skimming device for removal of grease and oil prior to treatment in the anaerobic system.

LEACHATE CHARACTERISTICS

Figure 3 shows a plot of the volume of leachate removed from the landfill over the past five years. Until 1983 the maximum amount of leachate that could be withdrawn was limited by inadequate collection systems. However, after collection laterals, caissons, hydraulic barriers, and other leachate collection facilities were installed in 1983 and 1984, the volume of leachate withdrawn was the maximum unit volume that can be removed utilizing the upgraded system. Based on the data shown on Figure 3, the maximum amount of leachate available for treatment was slightly more than 150,000 gpd and this volume was available only for about three months. More recently, only about 20,000-30,000 gpd of leachate has been withdrawn.

The leachate characteristics determined during the pilot plant study are shown in Table I. However, more recent evaluations show that the organic strength as well as the leachate volume has decreased significantly. Figure 4 shows the leachate BOD concentration versus time for the past five years. Although there have been periods of lower and then higher BOD concentrations, the overall trend indicates a drop in the BOD. As a result of the decreasing BOD, the organic loading to the anaerobic filter is significantly lower. For example, a leachate BOD of 10,000 mg/L and a flow of 35,000 gpd results in a BOD loading of 50 lb/d/1,000 cu ft of media in the anaerobic filter. This is slightly more than one tenth of the average design loading.

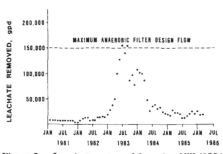

Figure 3.   Leachate removed from landfill (1981 to 1986).

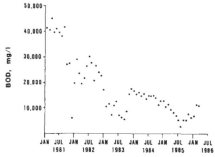

Figure 4.   Leachate BOD concentration (1981 to 1986).

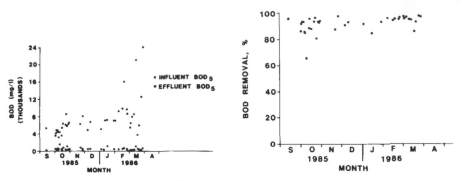

Figure 5. Influent and effluent BOD versus time.

Figure 6. Percent BOD removal versus time.

STARTUP

Construction of the pretreatment facility was completed in August 1985 and startup of the anaerobic plant was initiated. The anaerobic filter was seeded with 50,000 gallons of cattle manure at a cost of $0.05/gal. The manure was cheaper than municipal digester sludge because of a shorter haul distance and did not contain the large quantities of hair found in the municipal sludge. Screening of the municipal sludge for removal of hair and large solids was difficult; hence cattle manure was selected for seed. The manure was hauled in 6,000 gallon tanker trucks and pumped directly to the anaerobic filter. Although the manure was obtained from a farm where liquid manure was disposed of on land, it still contained ground-up straw that caused plugging in the heat exchanger. These problems are discussed later; however, the solution was twofold. First, sludge from the anaerobic reactor was pumped into the clarifier where the straw was removed. The clarifier overflow was returned to the anaerobic filter. The second part of the solution was to install a dual strainer ahead of the heat exchanger to remove straw and other solid particles. Before adding the seed material, the anaerobic reactor was partially filled with water and dilute leachate from another landfill. After adding the cattle manure,the reactor was filled with water. The recycle pumps were turned on to provide mixing and then an acclimation period of several days was provided. Once gas production began, leachate was fed from the holding tank during the day only. Because of the reduced volume of leachate available from the landfill, this procedure was continued for several months. Continuous recycle of the effluent was maintained to keep the reactor contents mixed. Because the manure contained adequate amounts of phosphorus and nitrogen, no supplemental nutrients were added.

ANAEROBIC FILTER PERFORMANCE

Figure 5 shows a plot of the influent leachate BOD and the pretreatment effluent BOD from September 1985 through March 1986. The raw leachate BOD values in February and March ranged from 24,000 mg/L to 3,700 mg/L. However, the wide range of BOD concentrations did not cause a significant rise in the effluent BOD concentrations. The stable effluent quality is the result of the inherent stability of an anaerobic filter as well as the fact that the anaerobic system is underloaded. The percent BOD removals are shown on Figure 6. This figure indicates that the anaerobic filter startup was very rapid, requiring about 45 days as shown by the 90% BOD removal that was achieved during October, the second month after startup. Although there is some variation in the data, there has been a slight trend toward increased BOD removals. During March, the BOD removal averaged about 95%, with the exception of one day.

The BOD loading is shown on Figure 7. The BOD load being received is significantly less than the design loading due to the reduced flow and BOD concentrations discussed previously. In March, the highest loading was about 67 lb BOD/d/1000 cu ft, or about 15% of the design loading. Initially, there was concern that the low flow rates would result in an increased buildup of metals and cause inhibition in the BOD removal. However, this obviously has not occurred as indicated by the 95% BOD removals currently being achieved.

The suspended solids concentrations of the raw leachate and the anaerobic filter effluent are shown on Figure 8. The suspended solids concentrations in the leachate varied widely; however, the effluent suspended solids were less than 1,000 mg/L after October and averaged about 100 mg/L in March.

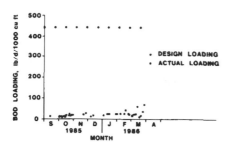

Figure 7.   BOD loading versus time.

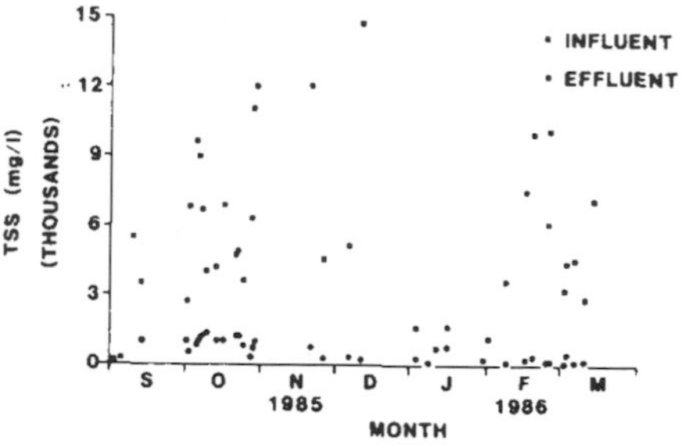

Figure 8.   Influent and effluent TSS versus time.

The very high concentrations of suspended solids in the raw leachate are probably caused by soil or sediment. The unusually low effluent suspended solids concentration in the anaerobic filter effluent is probably due to the long retention time in the anaerobic reactor. As seen on Figure 9, the suspended solids removal during 1986 has been better than 95%, with three exceptions. To date no sludge has been wasted; however, measurements indicate that the bottom cone of the anaerobic filter does contain sludge.

Table III shows the concentrations of cadmium, copper, lead, nickel, and zinc in both untreated leachate and the anaerobic filter effluent. Removals for copper, lead, and zinc were 88%, 84%, and 83%, respectively. Fifty-one percent of the nickel and 20% of the cadmium were removed during this period. Cadmium concentrations are near the detectable limit; therefore, the calculated removal values are unreliable. Nickel appeared not to be efficiently removed at the operating pH of the anaerobic filter, which was attributed to the solubility of nickel salts. The effluent concentrations of these metals are all well below the values required by MMSD's pretreatment standards.

OPERATING PROBLEMS

The chief problem with operation of the anaerobic system is the low volume of leachate available for treatment. Initially, leachate was fed continuously during the day, with no feeding during the night. However, since the temperature of the reactor is maintained by heating the raw leachate before it enters the reactor, this schedule was not satisfactory. To prevent the reactor temperature from dropping overnight, the transfer pumps feeding the anaerobic filter were placed on timer control. This allows 24 hour operation of the heating system as well as providing a more uniform organic load to the anaerobic filter.

The straw in the cattle manure caused plugging in the heat exchanger. The clarifier was very useful in removing large quantities of straw from the system; however, small straw particles continued to plug the heat exchanger. Because of the low influent flow of leachate, it is necessary to recycle some of

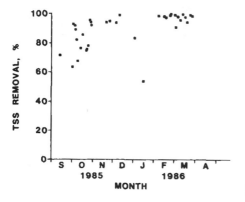

Figure 9.   Percent TSS removal versus time.

Table III. Liquid Waste Inorganic Characteristics (September 9, 1985 through February 8, 1986)

Date	Cadmium Inf. mg/L	Eff. mg/L	Copper Inf. mg/L	Eff. mg/L	Lead Inf. mg/L	Eff. mg/L	Nickel Inf. mg/L	Eff. mg/L	Zinc Inf. mg/L	Eff. mg/L
9/29–10/5	0.1	0.1	1.3	0.4	11.0	1.0	7.5	0.6	10.0	2.1
10/6–12	0.1	0.05	1.8	0.4	9.0	0.5	1.2	0.7	8.7	2.4
10/13–19	0.3	0.3	0.9	0.3	4.0	2.0	1.6	0.9	9.6	1.9
10/20–26	0.3	0.4	0.7	0.4	1.3	1.0	3.0	1.2	7.6	4.2
10/27–11/2	0.2	0.05	2.0	0.3	3.0	0.5	1.6	0.6	15.0	1.10
11/17–23	0.05	0.04	3.45	0.33	5.2	0.2	1.15	0.41	7.5	0.57
11/24–30	0.05	0.03	1.25	0.13	2.1	0.2	0.9	0.58	7.0	0.65
12/1–7	0.01	0.05	0.74	0.14	0.7	0.5	0.89	0.71	2.96	0.86
12/8–14	0.07	0.03	5.03	0.07	3.8	0.3	1.48	0.60	15.09	0.99
12/29–1/5	0.01	4.47	0.10	1.10	1.7	0.1	1.10	0.60	9.33	1.60
1/4–11	0.03	0.01	0.2	0.13	0.3	0.3	0.77	0.63	4.83	1.77
1/12–18	0.07	0.03	0.5	0.10	0.7	0.3	0.83	0.67	5.67	1.80
1/26–2/1	0.07	0.03	0.20	0.07	0.4	0.1	0.97	0.63	6.83	1.28
2/2–8	0.08	0.02	3.23	0.10	1.4	0.2	1.41	0.71	22.4	1.29
Averages	0.10	0.08	1.84	0.21	3.19	0.51	1.39	0.68	9.47	1.61
Percentage Removal	20		88		84		51		83	

the reactor contents to the heat exchanger for maintaining the reactor temperature at 95°F. A double bowl strainer was added to remove the straw and other solids from the recycle stream. The original design considered use of a coil type heat exchanger; however, the plate and frame type heat exchanger was installed to provide greater corrosion protection since it was available with titanium plates and the spiral type was not. The use of a coil type heat exchanger offers the advantages of ease in cleaning and reduced problems with plugging where it can be used. In other installations a coil type heat exchanger will be strongly recommended. The liquid manure provided an excellent seed as shown by the very rapid startup of the Omega Hills system; however, it is recommended that all straw be removed before adding manure to an anaerobic system.

The boiler used to heat the control building and the anaerobic system was to be supplied with gas from the anaerobic filter. Landfill gas was to be used as a supplemental fuel. Excess anaerobic filter gas was intended to be flared; however, the pipe connections were not made to flare the anaerobic gas while the boiler operated on landfill gas. Because of the low BOD load, not enough gas is produced to operate the boiler, and landfill gas is used. To allow the anaerobic filter gas to be flared while operating the boiler on landfill gas, it was necessary to add a small length of gas pipe. Until this pipe was installed, gas was vented at the pressure relief valve on top of the reactor. The only problem encountered during this operation was that during the winter the moisture in the gas caused the relief valve to freeze. This caused an increase of pressure in the anaerobic filter and forced excess liquid out of the reactor.

The startup period for this system was very short with no major difficulties to cause delays in the treatment of the leachate. The basic cause of current operating difficulties is the small volume of leachate available for treatment. WMI is seeking high strength industrial wastes in the Milwaukee area to supplement the feed for the anaerobic filter. Possible sources include wastes from yeast, dairy, rendering, and pharmaceutical industries. Such an arrangement can be advantageous to both WMI and the industry if the cost of hauling and treatment is less than the municipal sewer charges.

CONCLUSIONS

This paper describes the initial pilot plant studies, plant design, and startup of the Omega Hills anaerobic filter. Eight months after startup, operation is stable and treatment results are good. Startup was troublefree and 80 to 90% BOD removal was achieved in less than 45 days. Although the treatment system is underloaded, operation of the anaerobic filter is very stable and 95% removals of both BOD and TSS are achieved under widely varying load conditions. The actual BOD loading on the anaerobic filter is about one tenth of the design loading; however, the removal rates currently being achieved are essentially the same as those achieved with the pilot plant at BOD loadings as much as three times higher than design loadings. Copper, lead, and zinc removals exceed 80%. Because of these results, it is concluded that the anaerobic filter is an excellent system for treating high strength landfill leachate.

REFERENCES

1. John L. Carter, G. Michael Curran, Paul E. Schafer, Russell T. Janeshek, and Gregory C. Woelfel, "A New Type of Anaerobic Design for Energy Recovery and Treatment of Leachate Wastes," *Proceedings of the 39th Industrial Waste Conference*, Purdue University (May 8–10, 1984).
2. John L. Carter, Paul E. Schafer, Russell J. Janeshek, and Gregory C. Woelfel, "Effects of Alkalinity and Hardness on Anaerobic Digestion of Landfill Leachate," *Proceedings of the 40th Industrial Waste Conference*, Purdue University (May 14–16, 1985).
3. John L. Carter, Paul E. Schafer, and Gregory C. Woelfel, "A Case Study of Leachate Collection and Treatment at the Omega Hills Landfill," *Management of Uncontrolled Hazardous Waste Sites*, Washington, D.C. (November 6, 1985).

46 UPTAKE AND RELEASE OF HAZARDOUS CHEMICALS BY SOIL MATERIALS

Ju-Chang Huang, Professor

Brian A. Dempsey, Assistant Professor

Shoou-Yuh Chang, Assistant Professor

Hossein Ganjidoost, Graduate Research Assistant
Department of Civil Engineering
University of Missouri-Rolla
Rolla, Missouri 65401

INTRODUCTION

In many places of the nation, ground water has been contaminated with various concentrations of chloroorganic solvents, such as 1,1,1-trichloroethane (TCA), trichloroethylene (TCE), and tetrachloroethylene or perchloroethylene (PCE). Most of the contaminations have originated from the previous dump of these chemicals into unsealed ponds or landfill sites [1]. Since these organic solvents are highly volatile, a good portion of these chemicals has escaped into the air upon their disposal. However, once covered by soil and refuse, or having seeped into the subsurface environment, little volatilization can occur. As the chloroorganic solvents are seeping into and through the ground, they can be adsorbed by minerals and particulate organic materials. This adsorption process tends to increase the retention time of volatile organic compounds (VOCs) in the soil. They travel more slowly than inert and unreactive salts. Nevertheless, plumes of these solvents have been detected quite far away from the initial site of contamination, indicating only a moderate ability to be adsorbed by soils and minerals [2,3].

In many areas heavily polluted with VOCs, plans have been made or are being considered to pump ground water to the surface and then subject it to air stripping to remove the highly volatile chloroorganic solvents [4]. Although this kind of remedial action appears to be appropriate in removing volatile chloroorganics, it is difficult to estimate the length of pumpage required to achieve adequate cleanup of the subsurface water. This is because one of the major factors governing the future transport and cleanup of these chloroorganic solvents in the subsurface environment is the sorption and desorption reactions of these organics in the saturated and unsaturated soils. For a typical polluted area, a fairly accurate account of the total amount of organic solvents disposed on a given site may be known. However, since a large portion of the disposed VOCs may have been lost into the atmosphere through volatilization, the exact amount of these chemicals entering the ground, their distribution along the soil depth (in the unsaturated zone) as well as in the polluted plume (beneath the ground water table) are normally difficult to define, particularly when the acquired field data show a lack of consistency. This has commonly happened because the volatility of these organic solvents has often introduced a significant degree of experimental errors in the chemical analysis unless proper extraction and analytical protocols are followed. In general, the chlorinated solvents may be present in the pure form (perhaps in droplets held in the soil matrix by capillary forces), adsorbed on the sediment phase, or in the aqueous phase. The chemical distributions between the latter two phases are described by adsorption isotherms. By establishing accurate adsorption and desorption isotherms for a specific adsorbate-adsorbent combination, the following advantages may be derived:

1. The adsorption isotherm will allow an engineer to make an estimate of the total quantity of each organic solvent that has been retained on the saturated sediment phase based on its aqueous concentration found in ground water.
2. The adsorption isotherm will also allow him to check the validity of the field data which have been collected from the polluted plume in any contaminated site. This may be done by comparing the chloroorganic concentrations in both the sediment and the aqueous phases and then determining whether these data coincide with the established isotherm relationships.

Table I. Chemical and Physical Characteristics of the Three Chloroorganics Studied

Compound Type	Name	M.W. (g/mole)	log K_{ow}	Solubility at 20°C (mg/L)	Henry's Constant at 25°C (atm-m3/mole)
PCE	Tetrachloroethylene	165.85	2.88	150	28.70×10^{-3}
TCA	1,1,1-Trichloroethane	133.41	2.17	950	4.92×10^{-3}
TCE	Trichloroethylene	131.39	2.29	1100	11.70×10^{-3}

3. Rate information will permit the engineer to establish the length of time over which pumping must be continued in order to reduce the concentration of contaminants in the ground to safe levels and to determine the extent of future leachings from the aquifer.

Therefore, it is the objective of this study to determine the rate and extent of adsorption of three chloroorganic solvents on selected soils. The specific goals are: 1) to measure the rates of adsorption and desorption of 1,1,1-trichloroethane (TCA), trichloroethylene (TCE), and tetrachloroethylene (PCE) on various soils; 2) to establish adsorption isotherms for these chemicals and then determine the nature and potential usefulness of them; and 3) to evaluate the empirical effects of solids concentrations and possibly slow desorption reactions on the partitioning of the chloroorganics between soil and ground water. The information obtained from this study will be useful for predicting the extent of pollution, confirming the validity of field data, and determining the effectiveness and length of future ground water mining and treatment for the removal of VOCs from ground water.

EXPERIMENTAL PROCEDURES

The three VOCs that were used in this project were reagent grade. The physical and chemical characteristics of these compounds have been identified [5] and listed in Table I. Stock solutions in methanol were prepared as described in *Standard Methods* [6]. Extraction of VOCs from water were made using pentane, and it was found that in an aqueous solution with no head space more than 99.1% of VOCs could be recovered by pentane extraction (see Table II). The GC operating conditions are shown below:

Column Type — Alltech 0.1%AT-1000 on Graphpac-GC, 80/100 mesh
Detector Type — Electron Capture
Carrier Gas — N_2 at 30 mL/min
Temperatures — Injector 150°C
           Detector 220°C
           Column Programming—32°C followed by increase of 12°C/min until
                     reaching 120°C

The chromatograms on two separate columns (6 ft. and 50 cm in length) have shown good VOCs separations and no significant contamination (Figure 1). Standard curves are illustrated in Figure 2.

Table II. Recovery of Chlorinated Organics Using Various Extraction Procedures

Method	Run #	Chromatogram Area (10^{-4})		
		TCA	TCE	PCE
A. Direct extraction	1	44.00	27.19	134.55
from stock solution	2	44.72	27.15	137.07
(in Methanol)	average	44.36	27.17	135.81
B. Extraction from	1	43.66	27.42	132.61
aqueous solution	2	44.29	26.91	138.74
with no head space	average	43.98	27.14	135.68
	% recovery	99.14	100	99.90
C. Extraction from	1	32.31	21.21	92.71
aqueous solution	2	32.71	23.31	93.27
with head space	average	32.51	22.26	92.99
	% recovery	73.29	81.93	68.47

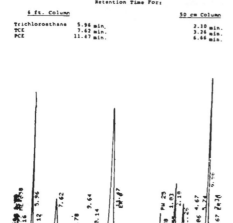

Figure 1.   VOC chromatograms from two separate columns.

As of this time, three soils have been used in these experiments, namely a bentonite and two reference soils from the University of Missouri Soils Testing Laboratory. One of the reference soils contains 5.1% organic material and the other contains 1.9% organic material. This project is continuing, and more soil samples from specific VOC—contaminated sites will be used for the study.

The general experimental design is illustrated in Figure 3. Glassware is cleaned using detergent, chromic acid cleaning solution, thorough rinsing, and oven-dried. From one to ten grams of soil are added to the clean 50 mL centrifuge tubes (step 1 in Figure 3) to achieve final solids concentrations of 20 to 200 g/L. The tubes are filled with distilled water. The soil-water suspension is mixed for 12 hours in order to hydrate the solid surfaces and to permit initial abrasion of the soil particles. After this initial period, a small quantity of VOC in methanol is added through a septum that is located in the cap of the centrifuge tube (step 2). The contents of the tubes are mixed on a rotary mixer for 48 hours (adsorption equilibrium), 10 days (desorption equilibrium), and for various times in the rate experi-

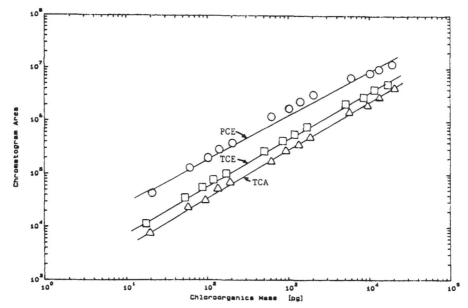

Figure 2.   Standard calibrations for VOC masses versus chromatogram areas.

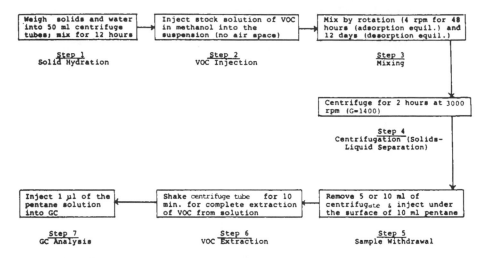

Figure 3.   Sample preparation, mixing, centrifugation, extraction, and analysis.

ments (step 3). The tubes are centrifuged for two hours at 3000 rpm (G = 1400) in a table-top clinical centrifuge (step 4). At the end of this time, 5 or 10 mL of the centrifugate is removed by syringe (step 5). The cap of the tube is loosened so that air can replace the liquid that is being taken for analysis. The removed sample is then injected under the surface of 10 mL of pentane in a 25 mL vial. The vial is capped and agitated for ten minutes using a wrist-action shaker for VOC extraction (step 6). A longer period of extraction has been used when the VOCs are extracted from soil rather than from the clear centrifugate. Finally a 1 μl portion of the pentane solution is injected into the GC (step 7).

Several aspects of this experimental protocol had to be tested for effectiveness. For example, two hours of centrifugation at G of 1400 were used to separate adsorbed VOC from soluble VOC. The rates at which solids were removed from the centrifugate were found to be extremely fast, as shown in Figure 4. Based on total solids, more than 99% of the soil containing 5.1% organic material was separated from the centrifugate after only 10 minutes of centrifugation. This soil has an average particle diameter of 24 μ and an effective size of 6 μ after wetting and abrasion. However, the average diameter of bentonite clay is less than 2 μ and its effective size is less than 1 μ. One hour of centrifugation is required for the separation of 98% of the initial solids from the bentonite. Two hours of centrifugation are sufficient for the separation of greater than 99% of the adsorbent. Additional investigations of experimental techniques that are especially pertinent to the rate and equilibrium studies are included in subsequent sections of this paper.

RATE OF REACTION

The effectiveness of centrifugation in obtaining rate data was first evaluated. This was because we wished to obtain data representing a few minutes of adsorption or desorption reaction while we were using a separation technique that required two hours of centrifugation before the sample was actually extracted. Data shown in Figure 5 indicate that the adsorption or desorption reactions are essentially stopped within the first 5 minutes of centrifugation. In Figure 5 the concentration of soluble PCE after 30 minutes mixing and various times for centrifugation is independent of the centrifugation time. These data are representative of 30 minutes reaction time and are clearly distinguished from 2 hours or 24 hours of mixing time. These results are similar to those obtained by Weber [7] regarding the adsorption of detergents by granular activated carbon. Weber found that the reaction was stopped when mixing was eliminated. Under no-mix conditions, then, the rate determining step is mass transfer (perhaps through bulk solution) and this is slow when the solution and the solids have been physically separated. When mixing is reinitiated, then the incremental removal of detergent again occurred.

The rates of adsorption and desorption of TCE and PCE by the 5.1% soil are shown in Figures 6 and 7. Although no data are shown, the adsorption of TCA, TCE, and PCE on bentonite is also fast and the reactions are completed in a short period of time. The following observations are made from

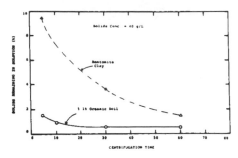

Figure 4. Effect of centrifugation time on solid-liquid separation.

Figure 5. Effect of centrifugation time on percent PCE remaining in solution.

the experimental data: 1) the half-time for adsorption is in the range of hours; 2) equilibration occurs faster for adsorption than desorption; and 3) PCE is adsorbed more rapidly than is TCE on the 5.1% organic soil.

Chemical reaction is considered to be the rate determining step for the adsorption of TCE by the 5.1% organic soil when the suspension is mixed as described above. This conclusion is made because PCE is adsorbed more rapidly than TCE while the rate of molecular diffusion for PCE is slower than for TCE. The conclusion is reasonable but not certain, since solute-adsorbent-solvent interactions also contribute to the rate of dispersion through pores. However, the factors that are typically invoked to explain slow pore diffusion are two and three-dimensional micelle formation, steric inhibitions, and ionic repulsions or attractions. None of these factors can explain the slower rate of adsorption of TCE versus PCE [8]. So, chemical reaction as a rate limiting step is most likely. The rate of adsorption of TCE on the 5.1% organic soil can be described using the data from Figure 6 and the integrated rate equation for a reversible pseudo-first-order reaction: $A \leftrightarrow R$, and $R_o = 0$. Then,

$$\ln (A - A_{eq}) = \ln (A_o - A_{eq}) - k \cdot (1 + 1/K) \cdot t \tag{1}$$

where:

A = concentration of soluble TCE, in $\mu g/L$;
R = concentration of sorbed TCE, in $\mu g/Kg$;
k = forward rate constant in hr^{-1};
K = partition coefficient = 3.39 L/Kg (from Figure 8)

The subscript o indicates initial concentration and the subscript eq designates equilibrium values. Equation 1 is a simplified version of Equation 54 from Levenspiel [9].

For the TCE adsorption during the period between 2 and 12 hours, the overall reversible rate constant for adsorption $(k \cdot (1 + 1/K))$ is 0.156 hr^{-1}, resulting in a half-time for reversible adsorption

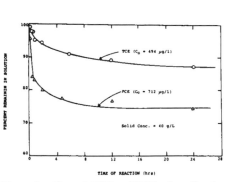

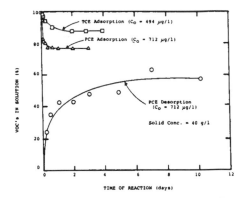

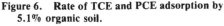

Figure 6. Rate of TCE and PCE adsorption by 5.1% organic soil.

Figure 7. Rates of adsorption/desorption for TCE and PCE by 5.1% organic soil.

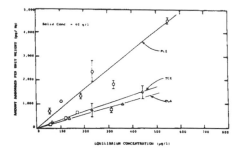

Figure 8. Equilibrium adsorption of PCE, TCE, and TCA by 5.1% organic soil.

of 4.44 hr. Thus, our selected equilibration time of 48 hr. is sufficient, representing more than ten half-times for reversible adsorption of TCE. The individual rate constants are $k = 0.120$ hr^{-1} for the forward reaction, and $k_r = 0.0354$ hr^{-1} for the irreversible desorption of TCE. The calculated half-time for reversible desorption is 15.0 hr. Again, this value is substantially less than the desorption equilibration time of ten days. Actually, the half-times for adsorption are expected to be even faster at higher soil concentrations, since the forward constants are for a pseudo-first-order reaction. The effects of the nature of adsorbent and adsorbate on the rates of adsorption and desorption are presently being studied in further detail.

The adsorption of TCE during the first two hours is faster than predicted by the rate expression that has just been described above and actually accounts for about half of the total adsorption of TCE by this soil. Similarly, the pseudo-first-order rate constant for the adsorption of PCE also appears to decrease with time. The varying rates may be due to the presence of a variety of sorptive sites on the soil. The good fit of data (representing the TCE adsorptions between two and twelve hours) to the integrated Equation 1 is a good indication that a substantial fraction of the adsorption sites are nearly identical.

The rate of adsorption that is reported above is expected to be the slowest case. We have examined TCE, which reacts more slowly than PCE, and we have used a solids concentration that is considerably less than would occur in the ground. Also, the short half-time for PCE desorption indicate that chemical reaction will not limit the rate of partitioning of VOCs between soil and ground water. If mass transfer is rapid, then the VOCs that are dissolved in the ground water at contaminated sites will be in equilibrium with the adsorbed VOCs. As a result, isotherms should be useful for the prediction of the amount of VOCs left in the ground and for the confirmation of analytical results for the total concentrations of VOCs in the ground.

EQUILIBRIUM OF ADSORPTION

The ability to quantitatively recover solute and adsorbate is an experimental requirement for the calculation of adsorption isotherm constants. This ability to recover VOCs from water or soil was investigated. In general, accurate and precise mass balances were obtained for the extraction of VOCs from water or from solids after two days of equilibrium. The mass balances for the desorption experiments were less precise, and the accuracy is questionable. More than 99% of the VOCs in aqueous stock solutions were recovered by extraction with pentane, as shown in Table II. Mass balances were performed after 48 hours of adsorption experiment and also after an additional 10 days of desorption experiment. Results are shown in Tables III, IV, and V. The mass was successfully balanced for the adsorption experiments, with an average of 99.0% of the initial PCE recovered when bentonite was used as adsorbent and 100.9% recovery with the 5.1% organic soil as adsorbent (see Tables III and IV). The precision and accuracy of recovery were not as good for the desorption experiments. An average of 93% of PCE was recovered after desorption from the soil that contained 5.1% organic material (see Table V).

The poorer recoveries of PCE after the desorption experiments are likely due to the longer experimental time (12 days total for adsorption-desorption versus two days total for adsorption only) and the increased sample manipulation. The data in Tables III and IV indicate that there is not any VOC that is adsorbed so strongly after two days that it cannot be extracted with pentane. It is considered unlikely that irreversible adsorption would occur after 10 days either. This suggestion is strengthened by the information that is shown in Table VI. The percentage of the total VOC that is soluble after desorption corresponds with the percent that is soluble after adsorption if it is assumed that only the

Table III. Adsorption Mass Balance for PCE by Bentonite Clay

Sample #	Initial Mass of PCE in The Reactor (μg)	Time of Reaction (hour)	Mass of PCE Left in the Solution (μg)	Mass of PCE Adsorbed by Bentonite (μg)	Total Mass of PCE Recovered (%)
1	14.5	.5	13.81	0.68	99.93
2	14.5	.5	13.20	0.87	97.03
3	14.5	.5	13.2	1.04	98.21
4	14.5	.5	14.28	0.48	101.78
5	14.5	.5	13.63	0.61	98.21
				Ave.	99.03

Reactor Volume = 50 mL
Mass of Soil = 2 g

Table IV. Adsorption Mass Balance for PCE by 5.1% Organic Soil

Sample #	Initial Mass of PCE in the Reactor (μg)	Time of Reaction (hours)	Mass of PCE Left in the Solution (μg)	Mass of Adsorbed PCE by 5.1% Organic Soil (μg)	Total Mass of PCE Recovered (%)
1	16.50	.5	13.51	2.96	99.82
2	16.50	.5	13.50	4.00	106.06
3	16.50	.5	13.48	2.44	96.48
4	16.50	.5	14.20	2.69	102.36
5	35.60	24+	29.90	5.69	99.99
				Ave.	100.94

Reactor Volume = 50 mL
Mass of Soil = 2 g

extractable VOC is still present in the tube (or, that the unextractable VOC has been lost by evaporation or transformation). This can be seen by viewing the very comparable results in columns 2 and 4 of Table VI. Column 3 shows the calculated percentage of total VOC that is soluble at the end of the desorption experiment if it is assumed that no VOC has been lost from the tube during the sample preparation/manipulations and the 12 days of reaction. The values in column 3 are also quite inconsistent among themselves and do not correspond well with the adsorption data (column 2).

Table V. Adsorption/Desorption Mass Balance for PCE by 5.1% Organic Containing Soil

Sample #	Initial Mass of PCE in the Reactor (μg)	Mass of PCE Removed from Reactor (μg)	Mass of PCE Desorbed from 5.1% Organic Soil (μg)	Mass of PCE Remaining 5.1% Organic Soil (μg)	Total Mass of PCE Recovered (%)
1	14.50	10.83	2.73	0.78	98.76
2	14.50	11.29	2.16	0.72	97.72
3	35.6	25.21	5.13	1.67	89.92
4	35.6	26.99	4.45	1.69	90.28
5	7.35	5.46	.94	.31	91.29
6	7.35	5.51	.83	.27	89.93
				Ave.	92.98

Reactor Volume = 50 mL
Mass of Soil = 2 g
Time of Reaction: 12 days for desorption
                  2 days for adsorption

Table VI. Fraction of PCE that is Soluble After Adsorption & Desorption

Sample No.	C_e/C_o After Adsorption	Ce' After Desorption Co' Based on Calculation	Ce' After Desorption Co" Based on Analyzed PCE
1	0.770	0.744	0.778
2	0.803	0.673	0.750
3	0.730	0.494	0.754
4	0.782	0.517	0.725
5	0.766	0.497	0.752
6	0.773	0.451	0.755
$\bar{X} \pm$ Standard Dev.	0.771 ± 0.024	0.563 ± 0.117	0.752 ± 0.017

Note:
    Co = Total PCE mass added to the reactor.
    Ce = Equilibrium soluble PCE after adsorption.
    Ce' = Equilibrium soluble PCE after desorption.
    Co' = Calculated total PCE in the reactor at the start of desorption.
    Co" = Analyzed total PCE in the reactor at the start of desorption.

Karickhoff [10] has reported that the fraction of irreversibly sorbed chemical often increases with incremental time of reaction. However, the inability to extract 100% of the sorbed chemicals has been reported principally for adsorbates of very low solubility, e.g., polynuclear aromatic hydrocarbons [10,11,12] or for reactive organic or inorganic species. The VOC's that were tested in these experiments have moderate solubility in water and are relatively unreactive. Therefore, the incomplete recovery of VOC after 10 days of desorption is interpreted as an experimental error. Conclusions of higher partition coefficients for desorption than for adsorption, as reported in a previous paper [13], are of dubious validity.

Some isotherm data are presented in Figure 8. These and other equilibrium data are consistent with the Henry's form of the adsorption isotherm or $q = KC_{eq}$ where

    q   = mass adsorbed at equilibrium per mass of adsorbent ($\mu g/kg$)
    C_{eq} = equilibrium concentration of solute ($\mu g/L$)

The matrix of partition coefficients (as a function of adsorbent and solute) are shown in Table VII. Average values and standard deviations of the experimental data are indicated for each condition. All of these data represent solid concentrations of 40 g/L. The effect of solids concentrations on the apparent partition coefficients will be described below. The partition coefficients for desorption appear to be identical to those for adsorption experiments if the VOC that cannot be extracted with pentane is considered to be lost from the system.

The trends that are observed in the data in Table VII are consistent with the characteristics of the adsorbates and the adsorbents. First, PCE is substantially less soluble and has a larger K_{ow} than for TCA or TCE (see Table I); the higher K values for PCE are consistent with this fact. Secondly, if all other factors are constant, then the K for non-polar organics should increase with increasing organic content and with increasing surface area of the adsorbent. These tendencies are also observed from the data in Table VII. Several investigators have published equations showing correlations among the

Table VII. Adsorption Partition Coefficients of Chloroorganic Compounds for Bentonite Clay and Two Organic Soils

Chloroorganics Compounds	K in (μg-$kg^{-1}/\mu g$-L^{-1})		
	Bentonite Clay	1.9% Organic Containing Soil	5.1% Organic Containing Soil
TCA	1.8 ± 0.4	1.9 ± 0.6	3.0 ± 0.7
TCE	1.6 ± 0.8	2.0 ± 0.5	3.4 ± 0.8
PCE	5.0 ± 1.0	6.1 ± 1.0	8.2 ± 3.0
Reaction Time	48 hours		
Solid Conc.	40 g/L		
Equilibrium Conc. ranges from	50 to 550 $\mu g/L$		
Temperature	25°C		

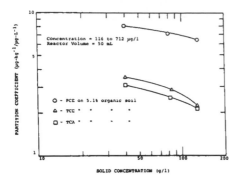

Figure 9.   Adsorption partition coefficient as a function of the solid concentration.

K, K_{ow}, and the fraction of organic carbon in the adsorbent (see e.g., equations 18 through 22 in ref [10]). These predictive equations are most successful when the adsorbent has a substantial portion of organic carbon and when the adsorbate has a limited solubility in water. These are conditions where hydrophobic sorption is likely to be dominant. The predicted values for the adsorption of the VOCs onto the 5.1% organic material (2% organic carbon), using equation 19 from ref [10], are 2.3 L/kg for TCA, 2.8 for TCE, and 7.4 for PCE. These values are only slightly less than the observed values that are listed in the last column of Table VII.

The partition coefficients that are listed in Table VII are substantially less than partition coefficients for the adsorption of VOCs by soils as reported by Urano and Murata [14] who reported K up to 500 L/Kg for TCA and from 200 to 6800 for PCE. Their experimental procedure was different; they measured the fugacity of the VOCs by monitoring the concentration in the head space. Additionally they used oven-dried soils. Nevertheless, their values seem anomalously high when compared to measured partition coefficients for more non-polar adsorbates or when compared to K values for adsorption of VOCs by granular activated carbon (GAC) or other microporous adsorbents. For example, Kong and DiGiano [15] determined K values for VOCs on GAC that are within the range reported by Urano and Murata for adsorption on soils. We know of no other reported values for the adsorption of VOCs by soils.

SOLIDS EFFECTS ON THE ADSORPTION OF VOCs BY SOIL

Several investigators have reported the so-called "solids effect", i.e., decreasing apparent K values with increasing concentration of solids that are used for the adsorption experiments. O'Connor and Connolly [16] have summarized empirical observations of this effect. DiToro and co-workers [12] have explained this effect by invoking irreversible adsorption of a portion of the adsorbate. Others have argued that the effect is due to incomplete removal of adsorbent during solids-liquid separation or due to the presence of soluble complexing agents that can increase the apparent solubility of the adsorbate.

Although none of the above conditions have occurred in the experiments that have been described in this paper, the solids effect has nevertheless been observed. This is indicated in Figure 9. Each point in Figure 9 represents several tests. The concentrations of soil that have been used in these experiments are one to two orders of magnitude greater than the concentrations of adsorbents that have been previously tested [16]. The VOCs that have been used are considerably more soluble than the organic adsorbate for which the solid effect has typically been noted. In this study the effect of solids on the K value for every set of experimental conditions has been consistently observed.

The impact of the solids effect on ground water conditions is extremely important since the concentration of solids in the ground is approximately 2 kg of solids per L of ground water. Thus, the actual partition coefficients for ground water conditions may actually be less than 25% of the values that are shown in Table VII.

EXTRAPOLATION TO GROUND WATER CONDITIONS

Partition coefficients ranging from 1 to 10 L/Kg have been measured for the adsorption of TCA, TCE, and PCE by the three soils examined (see Table VII). These values for K will be lower in the ground water environment due to the solids effect. For K ranging from 0.1 to 5 L/kg and solids concentration of 2 kg/L, then between 17 and 95% of the total VOC in the ground will be adsorbed at equilibrium conditions. The higher the fraction adsorbed, the less dispersion of contaminant that is likely to have occurred in the environment. The lower the fraction adsorbed, the less pumping is required for cleanup of a known contaminated volume. If the K value were substantially higher or lower, then one could predict that nearly all of the contaminant would be adsorbed or soluble, respectively. It is exactly in the range of K between 0.1 and 5 L/Kg, for ground water conditions, that the absolute changes in predicted concentrations are most sensitive to the changes in K value itself. These facts emphasize how critical it is to accurately determine the K value for adsorption of VOCs to soil.

CONCLUSIONS

Isotherms for the adsorption of TCA, TCE, and PCE onto three soils are linear in the soluble concentration range of 0 to 1000 $\mu g/L$ of VOC. That is, Henry's type constants are reasonable mathematical representations for the partitioning of the VOCs between soil and water.

The reaction is reversible. The partition coefficients (K) for desorption are identical to those of adsorption when proper analytical procedure is used. Analytical values for K (at 40 g/L of adsorbent) range from 1 to 10 L/Kg.

Reaction rates are quite fast when proper mixing is provided. The slowest half-times for the reversible adsorption (at solids concentration of 40 g/L) is 4.44 hr while the half-time for reversible desorption is about 15 hrs. Therefore, the partition coefficients may be reasonable representations for adsorption and desorption in the ground water environment.

K values decrease with increasing solids concentration. This effect is significant due to the very high solids concentration in the ground water environment (approximately 2 kg of soil water per L of ground water). Field K values of 0.1 to 5 L/Kg are predicted. The absolute amount of VOC that is adsorbed changes dramatically over this range of K values. Determination of the correct K values is therefore critical in order to predict extent of dispersion as well as time and cost of cleanup.

REFERENCES

1. Byer, H. G., Jr., W. Blankenship, and R. Allen, "Groundwater Contamination by Chlorinated Hydrocarbons: Causes and Prevention," *Civil Engineering*.
2. Roberts, P. V., M. Reinhard, and A. J. Valocchi, "Movement of Organic Contaminants in Groundwater: Implications for Water Supply," *JAWWA*, 74(8):408 (1982).
3. Mackay, K. M., P. V. Roberts, and J. A. Cherry, "Transport of Organic Contaminants in Groundwater," *Environ. Sci. & Technol.*, 19(5):384 (1985).
4. Ball, W. P., M. D. Jones, and M. C. Kavanaugh, "Mass Transfer of Volatile Organic Compounds in Packed Towers," *JWPCF*, 56:127 (1984).
5. U.S. EPA, "Treatability Manual, Vol. I, Treatability Data," *EPA-600-/2-82-001a* (1981).
6. "Standard Methods for the Examination of Water and Wastewater," 16th Edition, APHA (1985).
7. Weber, W. J., Jr., and J. C. Morris, "Kinetics of Adsorption on Carbon from Solution," *J. of the San. Engrg. Div.*, Proc. ASCE, 89(SA2):31 (1963).
8. Weber, W. J., Jr., and J. C. Morris, Closure of "Kinetics of Adsorption on Carbon from Solution," *J. of the San. Engrg. Div.*, Proc. ASCE, 89(SA6) (1963).
9. Levenspiel O., *Chemical Reaction Engineering*, 2nd Ed., Wiley and Sons, NY (1972).
10. Karickhoff, S. W., "Organic Pollutant Sorption in Aquatic Systems," *J. of Hydraulic Engineering*, ASCE, 110(6):707 (1984).
11. Hamaker, J. W., "Adsorption," *Organic Chemicals in the Soil Environment* (C. A. Goring and J. W. Hamaker, eds.), Marcel Dekker, NY (1972).
12. Di Toro, D. M., L. M. Horzempa, M. M. Casey, and W. Richardson, "Reversible and Resistant Components of PCB Adsorption — Desorption Isotherms,"*Environ. Sci. & Technol.*, 16:594 (1982).
13. Huang, J. C., B. A. Dempsey, S. Y. Chang, and H. Ganjidoost, "Sorption and Desorption of Degreasing Chloroorganics With Subsurface Sediments," *Proc. of Petroleum Hydrocarbons and Organic Chemicals in Ground Water*, by Nat'l. Water Well Assoc., Worthington, OH (Nov. 1985).
14. Urano, K., and C. Murata, "Adsorption of Principal Chlorinated Organic Compounds on Soil," *Chemosphere*, 14(3/4):293 (1985).

15. Kong, E. J., and F. A. DiGiano, "Competitve Adsorption Among VOC's on Activated Carbon and Carbonaceous Resin," *JAWWA*, 78(4):181 (1981).
16. O'Connor, D. J., and Connolly, J. P., "The Effect of Concentration of Adsorbing Solids on the Partition Coefficient," *Water Research*, 14:1517 (1980).

47 ORIGINS OF VOLATILE HALOGENATED ORGANIC COMPOUNDS IN CHLORINATED EFFLUENTS FROM WASTEWATER TREATMENT PLANTS CONTAINING DIFFERENT INDUSTRIAL WASTE CONTRIBUTIONS

Joseph V. Hunter, Professor II and Chairman

Gregory Toffoli, Research Assistant
Department of Environmental Science
Rutgers University
New Brunswick, New Jersey 08903

Robert Mueller, Principal Environmental Specialist
New Jersey Department of Environmental Protection
Office of Science and Research
Trenton, New Jersey 08625

INTRODUCTION

Although it has been considered by investigators in the public health area as one of the barriers to the spread of waterborne infectious diseases through the protection of recreational and shellfish uses of receiving waters, wastewater treatment plant effluent disinfection employing chlorination has caused considerable controversy over the past several years.

Excepting such potential considerations as cost, objections to such disinfection through chlorination fall into two categories. The first involves the concerns centered about the release of a "toxic" disinfectant to the receiving waters, that is, "residual" chlorine. The second involves the concerns centered about the oxidative nature of chlorine. As an oxidant, hypochlorite (the reaction product of gaseous chlorine and water) may oxidize an organic compound through the removal of hydrogen or the addition of oxygen or through the replacement of hydrogen by chlorine or even the addition of chlorine to double bonds. The first two reactions are typical of oxidative processes that occur naturally in receiving waters and so are considered relatively unobjectionable. The addition of chlorine to organic molecules, however, may cause the formation of chlorinated organics having objectionable physiological activities — such as odor, toxicity, mutagenicity or carcinogenicity. Thus, it is this second type of reaction that causes concerns over the discharge of such organics to the aquatic environment.

A number of studies [1,2,3] have confirmed the presence of halogenated organics in chlorinated wastewater treatment plant effluents. This was to be expected since the concentration of organic compounds in effluents should act as precursors in a manner analogous to the natural organics in surface waters which can produce substantial quantities of trihalomethanes subsequent to chlorination. As effluents from plants treating a mixture of domestic and industrial wastes would be expected to have a greater diversity of organics than plants treating only domestic wastes, the chlorination of effluents from the treatment of domestic/industrial waste mixtures might be expected to produce a greater diversity and concentration of organics than the chlorination of effluents from domestic wastes only.

A previous paper [4] described an attempt to establish this relationship. In this study, the concentrations of the relatively volatile halogenated compounds were determined in unchlorinated and chlorinated effluents from treatment plants treating wastewaters containing 0%, 25%, and 50% industrial wastewater contributions. Unfortunately, the large number and high concentrations of halogenated organics in the unchlorinated effluents containing industrial waste contributions rendered definitive conclusions impossible, although it could be suspected that if industrial wastes made substantial contributions of precursors for the formation of halogenated organics, this would have been obvious despite the high background levels. To determine whether or not the premise of this study was correct, the study was repeated in a considerably more exacting manner. The results of this study are described in this paper.

PROCEDURES

Six activated sludge treatment plants were selected for this study. Plants number one, two and four treated only domestic wastes from small communities. Plant number three also treated domestic wastes but included was a sizable contribution from office complexes. Both plants five and six treated mixtures of domestic and industrial wastes. Plant number four employed ultraviolet irradiation rather than chlorination for disinfection purposes. This data is summarized in Table I.

The previous study [4] indicated that there could be considerable variation in the background concentrations of volatile halogenated compounds, and other studies [5] indicated that this variation can exist even over short time periods. To account for these variations, each plant was sampled on either five or six consecutive days. The sampling procedure was as follows: 1) Three grab samples separated by intervals of ten minutes were collected from the influent to the disinfection unit; and 2) Starting at one disinfection unit retention time after the initiation of influent sampling, three grab samples were collected from the effluent from the disinfection unit. These samples were also collected at ten minute intervals.

The samples as collected were in completely filled crimp top bottles containing thiosulfate, which were placed in an ice chest and returned to the laboratory for analysis.

To observe the rate of formation (or disappearance) of the volatile halogenated compounds, samples of unchlorinated effluent were obtained from plants one and six, placed in completely filled teflon sealed bottles, placed in an ice chest and returned to the laboratory. These were then raised to 20°C and sufficient chlorine (as sodium hypochlorite) added so that the residual (combined) ranged between one and two mg/L after eight days. The chlorinated effluents were then placed in crimp top bottles and stored at 20°C until analyzed. The contact times employed were 0.5, 1, 2, 4, and 8 days.

Using the extraction procedure of Varna et al. [6], 110 mL of each sample was partitioned into 10 mL of n-pentane. The pentane extract was placed in crimped top vials and run in the automatic mode employing a Varian Model 3700 Gas Chromatograph with a ^{63}Ni electron capture detector. Separation was through a 40 m capillary column coated with 5% phenyl methyl silicone. Only the low boiling volatile halogenated organic compounds were determined. Due to the large blank value using this technique, no attempt was made to determine methylene chlorine in this study. Quantitation was achieved through the use of external standards. If there was no discernible detector response above background noise level, the results were recorded as Not Detected (ND). If a concentration is recorded, the detector response was five times above background levels. Using this criterion, the sensitivities were as follows:

Compound	Sensitivity (μg/L)	Compound	Sensitivity (μg/L)
chloroform	0.8	1,1,2-trichloroethane	1.0
1,1,1-trichloroethylene	2.0	dibromochloromethane	0.1
1,2-dichloroethane	1.6	tetrachloroethylene	0.06
carbon tetrachloride	0.1	1,2-dibromoethane	0.1
trichloroethylene	0.3	bromoform	1.0
dichlorobromomethane	0.1	1,1,2,2-tetrachloroethylene	0.3

Table I. The Industrial Waste Contents and Chlorination Parameters for the Plants Studied

	Parameter		
Plant #	Industrial Waste Contents %	Design Contact Time Min.	Normal Chlorine Residual mg/L
1	0	30	2.0a
2	0	30	1.5a
3	0	20	1.5b
4	0	20	c
5	55	30	1.5a
6	33	30	0.25a

aTotal residual chlorine.
bFree residual chlorine (ammonia removed by nitrification).
cUltraviolet irradiation.

Table II. Effect of Chlorination on the Volatile Halogenated Organic Compounds in the Effluent from Plant Number Three

Volatile Halogenated Organic	Chlorination Tank Influent			Chlorination Tank Effluent		
	9:30 $\mu g/L$	9:40 $\mu g/L$	9:50 $\mu g/L$	9:40 $\mu g/L$	9:50 $\mu g/L$	10:00 $\mu g/L$
Chloroform	ND	ND	ND	7.12	6.82	6.41
1,1,1-Trichloroethane	ND	ND	ND	ND	ND	ND
1,2-Dichloroethane	ND	ND	ND	ND	ND	ND
Carbon Tetrachloride	ND	ND	ND	<0.10	ND	<0.10
1,1,2-Trichloroethylene	ND	ND	ND	ND	ND	ND
Dichlorobromomethane	<0.10	<0.10	<0.10	6.37	7.74	7.25
1,1,2-Trichloroethane	ND	ND	ND	ND	ND	ND
Dibromochloromethane	<0.10	<0.10	<0.10	2.37	2.38	2.68
1,1,2,2-Tetrachloroethylene	ND	ND	ND	0.07	0.08	0.07
1,2-Dibromoethane	ND	ND	ND	ND	ND	ND
Bromoform	ND	ND	ND	ND	ND	ND

RESULTS AND DISCUSSION

Two examples of the data from the plant studies are found in Tables II and III, which give the concentrations of volatile halogenated organics in the three influent (to disinfection) and three effluent samples during one day at plants three and five. As this study resulted in 28 such data sheets, their inclusion here would not be feasible. Instead, these sheets were used to prepare two summary tables. Table IV gives the frequency of occurrence of the volatile halogenated organic compounds in disinfected and undisinfected treated wastewaters, and Table V gives the average concentrations of the volatile halogenated organic compounds in disinfected and undisinfected wastewaters. For the purpose of calculating the average concentrations, not detected was assumed to be zero and values less than the sensitivity were assumed to be half the sensitivity concentrations. Averages thus calculated could easily be under the sensitivity limits of the determinations but are useful for comparative purposes.

As can be observed from Table IV, the only volatile halogenated organic compound not detected in any sample was 1,2-dibromoethane. As was found in the previous study, the frequency of occurrence of the volatile halogenated organic compounds was far greater for the plants treating mixtures of domestic and industrial wastes (plants 5 and 6). Only in plant number three was there a substantial increase in the frequency of occurrence of the volatile halogenated organic compounds (especially trihalomethanes) subsequent to disinfection by chlorination. Plant number four, which employed ultraviolet irradiation instead of chlorination for effluent disinfection, evidenced no substantial differences in frequency of occurrence of volatile halogenated organics before or after effluent disinfection. There was only slight indication of greater frequencies of occurrence of volatile halogenated

Table III. Effect of Chlorination on the Volatile Halogenated Organic Compounds in the Effluent from Plant Number Five

Volatile Halogenated Organic	Chlorination Tank Influent			Chlorination Tank Effluent		
	13:10 $\mu g/L$	13:40 $\mu g/L$	14:10 $\mu g/L$	14:40 $\mu g/L$	15:10 $\mu g/L$	15:40 $\mu g/L$
Chloroform	16.3	15.7	15.9	16.4	16.3	24.3
1,1,1-Trichloroethane	13.2	12.6	12.7	12.8	12.8	14.7
1,2-Dichloroethane	20.0	19.3	21.3	19.1	21.2	23.2
Carbon Tetrachloride	2.92	2.77	2.84	2.73	2.73	2.60
1,1,2-Trichloroethylene	3.97	3.92	2.74	5.01	4.03	3.13
Dichlorobromomethane	1.34	1.16	1.15	6.61	6.49	6.13
1,1,2-Trichloroethane	ND	ND	ND	ND	ND	ND
Dibromochloromethane	0.85	0.95	0.90	4.81	5.03	4.25
1,1,2,2-Tetrachloroethylene	20.0	21.8	20.2	21.7	21.1	9.17
1,2-Dibromoethane	ND	ND	ND	ND	ND	ND
Bromoform	<1.00	<1.00	<1.00	1.79	2.02	4.23

Table IV. The Effect of Disinfection and Industrial Wastes Contents on the Frequency of Occurrence of Volatile Halogenated Organic Compounds

Volatile Halogenated Organic	Plant #1 In (%)	Plant #1 Out (%)	Plant #2 In (%)	Plant #2 Out (%)	Plant #3 In (%)	Plant #3 Out (%)	Plant #4 In (%)	Plant #4 Out (%)	Plant #5 In (%)	Plant #5 Out (%)	Plant #6 In (%)	Plant #6 Out (%)
Chloroform	0	0	0	73	0	100	0	0	100	100	100	100
1,1,1-Trichloroethane	0	0	0	0	0	17	20	27	100	100	100	100
1,2-Dichloroethane	0	0	0	0	0	6	0	0	100	100	100	100
Carbon Tetrachloride	22	78	0	0	0	28	80	73	100	100	100	100
1,1,2-Trichloroethylene	0	0	0	0	0	0	0	0	100	100	100	100
Dichlorobromomethane	100	100	93	100	39	100	33	13	100	100	100	100
1,1,2-Trichloroethane	0	0	0	0	0	0	0	0	0	0	17	17
Dibromochloromethane	83	94	73	100	33	100	0	0	100	100	100	100
1,1,2,2-Tetrachloroethylene	94	89	100	100	0	100	33	27	100	100	100	100
1,2-Dibromoethane	0	0	0	0	0	0	0	0	0	0	0	0
Bromoform	0	0	0	0	0	83	0	0	100	100	0	0

organic compounds subsequent to effluent disinfection in plants one and two. The two plants treating mixtures of domestic and industrial wastes (plants five and six) had such high frequencies of occurrence of the volatile halogenated organic compounds before disinfection that it would not have been possible to observe any increase due to disinfection by chlorination.

As would be expected, the data in Table V indicates that no significant concentrations of volatile halogenated organics are formed during disinfection by ultraviolet irradiation. Plants one and two, which treated only domestic wastewater, likewise formed little or no significant levels of volatile halogenated organic compounds during disinfection, which were the results expected from the first study. On the contrary, plant number three did form significant concentrations of volatile halogenated organic compounds (i.e., trihalomethanes) through effluent chlorination.

Again, as would be expected from the first paper, the plants treating mixtures of domestic and industrial wastewaters had significantly higher background levels of volatile halogenated organic compounds, especially 1,1,2,2-tetrachloroethylene, 1,1,1-trichloroethane, 1,2-dichloroethane, and 1,1,2-trichloroethylene. There were indications of only slight decreases in the average concentrations of nontrihalomethane volatile halogenated organic compounds, which would be expected as these compounds are relatively resistant to oxidation by chlorine [7].

Despite the high background levels observed for the volatile halogenated organic compounds, the more exacting nature of this study indicated that plant number five did produce volatile halogenated

Table V. The Effect of Disinfection and Industrial Wastes Contents on the Concentrations of Volatile Halogenated Organic Compounds

Volatile Halogenated Organics	Plant #1 In µg/L	Plant #1 Out µg/L	Plant #2 In µg/L	Plant #2 Out µg/L	Plant #3 In µg/L	Plant #3 Out µg/L	Plant #4 In µg/L	Plant #4 Out µg/L	Plant #5 In µg/L	Plant #5 Out µg/L	Plant #6 In µg/L	Plant #6 Out µg/L
Chloroform	0	0	0	0.39	0	3.82	0	0	13.43	13.81	0.70	0.83
1,1,1-Trichloroethane	0	0	0	0	0	0.17	0.20	0.27	12.70	12.54	1.55	1.49
1,2-Dichloroethane	0	0	0	0	0	0.19	0	0	41.41	41.28	1.74	1.57
Carbon Tetrachloride	0.01	0.01	0	0	0	0.01	0.04	0.04	4.08	3.81	0.05	0.05
1,1,2-Trichloroethylene	0	0	0	0	0	0	0	0	1.14	1.33	1.06	0.83
Dichlorobromomethane	0.06	0.09	0.10	0.31	0.09	7.27	0.02	0.01	0.67	4.62	0.44	0.32
1,1,2-Trichloroethane	0	0	0	0	0	0	0	0	0	0	0.19	0.24
Dibromochloromethane	0.06	0.10	0.04	0.08	0.06	6.00	0	0	0.35	2.19	0.19	0.21
1,1,2,2-Tetrachloroehtylene	0.04	0.07	0.06	0.09	0	0.10	0.01	0.01	17.62	16.63	8.17	6.16
1,2-Dibromoethane	0	0	0	0	0	0	0	0	0	0	0	0
Bromoform	0	0	0	0	0	0.83	0	0	0.50	0.91	0	0

Table VI. Total Average Concentrations of the Haloforms and Non-Haloforms Volatile Oraganic Compounds in Disinfected and Non-Disinfected Effluents

Plant	Organic Compound Concentrations			
	Other Halogenated Organics		Tri Halo Methanes	
	In μg/L	Out μg/L	In μg/L	Out μg/L
Number 1	0.05	0.17	0.12	0.19
Number 2	0.06	0.09	0.14	0.78
Number 3	0	0.47	0.15	17.92
Number 4	0.25	0.33	0.02	0.01
Number 5	76.95	75.59	14.95	21.53
Number 6	12.76	10.34	1.33	1.36

organic compounds (i.e., trihalomethane) during effluent chlorination. On the other hand, plant number six which also treated a mixture of domestic and industrial wastewaters produced little or no volatile halogenated organics during disinfection.

To better observe the formation of trihalomethanes, the volatile halogenated organics were divided into two groups — the trihalomethanes and the nontrihalomethanes, and the totals of each determined for the disinfection units influents and effluents for each plant. These results are presented in Table VI and again stress the fact that only plants three and five evidenced significant trihalomethane formation due to disinfection employing chlorination. It must be noted here that "significant" is a relative term — these levels would be relatively "insignificant" in the chlorination of a surface water where the formation of several hundred micrograms per liter would be expected.

Chlorination of the upper Passaic River water [8] indicated that trihalomethane formation did not cease until about seven to eleven days. In the short periods of time employed for effluent chlorination, significant trihalomethane formation would not be expected (e.g., in the contact times noted above, the trihalomethane concentrations formed by the chlorination of Passaic River water ranged from 200–600 μg/L). To observe if rate was a factor, the effluents from plants one and six were subjected to laboratory chlorination, and the results of this study are found in Tables VII and VIII.

These results agree well with the "in plant" results. At the plant contact times employed for chlorination, little or no trihalomethanes were formed. However, after eight days, significant levels of trihalomethanes were formed. The formation of trihalomethanes in the laboratory chlorination and "in plant" studies is compared in Table IX. The question arises, however, as to why one plant treating domestic wastes and one plant treating domestic wastes/industrial wastes mixtures produced trihalomethanes during chlorination while the other plants employing chlorination did not.

With the exception of plant number three, all plants that employed chlorination maintained only combined residuals. Plant number three, due to the absence of ammonia, employed free residual

Table VII. Effect of Chlorination and Time on Volatile Halogenated Organic Compound Formation in Effluent from Plant Number One

Halogenated Organic	Concentration in μg/L					
	0 hrs.	12 hrs.	24 hrs.	48 hrs.	96 hrs.	192 hrs.
Chloroform	<0.80	1.21	2.32	4.29	5.94	7.60
1,1,1-Trichloroethane	<2.00	<2.00	<2.00	<2.00	<2.00	<2.00
1,2-Dichloroethane	ND	ND	ND	ND	ND	ND
Carbon Tetrachloride	ND	ND	<0.10	<0.10	<0.10	<0.10
1,1,2-Trichloroethylene	ND	ND	0.55	0.78	0.23	0.43
Dichlorobromomethane	0.18	0.38	0.59	1.38	3.30	5.65
1,1,2-Trichloroethane	ND	ND	ND	ND	ND	ND
Dibromochloromethane	<0.10	0.10	0.12	0.19	0.26	0.49
1,1,2,2-Tetrachloroethylene	0.30	0.28	0.33	0.36	0.32	0.53
1,2-Dibromoethane	ND	ND	ND	ND	ND	ND
Bromoform	ND	ND	ND	ND	ND	ND

Table VIII. Effect of Chloronation and Time on Volatile Halogenated Organic Compound Formation in Effluent from Plant Number Six

	Concentration in $\mu g/L$					
Halogenated Organic	0 hrs.	12 hrs.	24 hrs.	48 hrs.	96 hrs.	192 hrs.
Chloroform	1.15	1.87	3.02	5.25	10.6	7.99
1,1,1-Trichloroethane	<2.00	3.27	3.97	4.86	4.45	5.11
1,2-Dichloroethane	ND	ND	ND	ND	ND	ND
Carbon Tetrachloride	0.22	0.33	0.40	0.45	0.43	0.47
1,1,2-Trichloroethylene	1.05	1.60	1.79	1.92	1.82	1.95
Dichlorobromomethane	0.49	1.35	2.68	4.96	6.83	8.57
1,1,2-Trichloroethane	ND	ND	ND	ND	ND	ND
Dibromochloromethane	0.15	0.29	0.39	0.67	1.15	2.12
1,1,2,2-Tetrachloroethylene	3.91	5.73	7.46	8.91	8.64	9.39
1,2-Dibromoethane	ND	ND	ND	ND	ND	ND
Bromoform	ND	<1.00	<1.00	<1.00	<1.00	1.03

chlorination. Thus, the formation of trihalomethanes in this plant during effluent chlorination was not surprising. Plant number six did not chlorinate to the same residual concentrations as did the other plants, and thus the absence of trihalomethanes is not too surprising. What is surprising was the formation of trihalomethanes during chlorination at plant number five. The bromine containing trihalomethanes here represented 94% of the total trihalomethanes so this could be due to the presence of significant bromide concentrations.

SUMMARY

Disinfection of domestic wastewater treatment plant effluents employing ultraviolet irradiation does not result in the formation of volatile halogenated organic compounds. In addition, disinfection of domestic wastewater treatment plant effluents employing combined chlorine residual likewise results in the formation of little or no volatile halogenated organic compounds. Disinfection of domestic wastewater treatment plant effluents employing free residual chlorine does produce low concentrations of volatile halogenated organic compounds – namely the trihalomethanes.

The results from the disinfection employing chlorination of effluents from plants treating mixtures of domestic and industrial wastes is not so easy to interpret. One plant produced trihalomethanes and the other did not, although both employed combined residual chlorine disinfection. The one that did not used a very low residual, and this could be the reason why trihalomethanes were not formed. Why they were formed in the other plant cannot be determined at present. This could be due to the presence of high bromide concentrations in the wastewater or due to the nature of the organics present. The one obvious observation that can be made is that most of the volatile halogenated organic compounds discharged from such plants were already in the wastewater and not formed during chlorination.

Table IX. Formation of Trihalomethanes During Effluent Chlorination

	Concentrations of Trihalomethanes Formed			
	In Plant Study		Laboratory Chlorination Study[a]	
Trihalomethane	Plant #3 $\mu g/L$	Plant #5 $\mu g/L$	Plant #1 $\mu g/L$	Plant #6 $\mu g/L$
Chloroform	3.82	0.38	7.2	6.84
Dichlorobromomethane	7.19	3.95	5.47	8.08
Dibromochloromethane	5.94	1.84	0.44	1.97
Bromoform	0.83	0.41	0	1.03
Sum	17.78	6.58	13.11	17.92

[a]8-day values.

ACKNOWLEDGMENTS

The work described herein was supported in part by the Office of Science and Research, Department of Environmental Protection of the State of New Jersey, and the New Jersey Agricultural Experiment Station, Rutgers University. Publication No. J-07523-3-86.

REFERENCES

1. Glaze, W. H., and Henderson, J. J., "Formation of Organic Chlorine Compounds from the Chlorination of a Municipal Secondary Effluent," *J. Water Pollution Control Fed.*, 47, 2511 (1975).
2. Glaze, W. H., et al., "Analysis of Organic Materials in Waste Water Effluents After Chlorination," *J. Chrom. Sci.*, 11, 580 (1973).
3. Jolley, R. L., "Chlorine Containing Organic Constituents in Sewage Effluent," *J. Water Pollution Control Fed.*, 47, 601 (1975).
4. Hunter, J. V., Busby, M. M., and Chang, H., "Influence of Industrial Wastes on the Formation of Volatile Halogenated Organics During Effluent Chlorination," *Proc. 40th Industrial Waste Conference*, Purdue Univ., West Lafayette, IN, 631 (1985).
5. Hunter, J. V., et al., "Sources of Pollutants in Surface and Groundwaters", "Water Conference," L. Ciaccio and A. Cristini, Eds., Ramapo College, Mahwah, N.J., p. 1 (1982).
6. Varna, M. M., et al., "Analysis of Trihalomethanes in Aqueous Solutions: A Comparative Study," *J. Amer. Water Works Assn.*, 71, 389 (1979).
7. Lee, Y. S., and Hunter, J. V., "Effect of Ozonation and Chlorination on Environmental Protection Agency Priority Pollutants," *Water Chlorination—Chemistry, Environmental Impact and Health Effects*, Vol. 5, R. L. Jolley et al., Eds., Lewis Pub. Inc., Chelsea, Mich., p. 1515 (1985).
8. Hunter, J. V., and Sabatino, T., "Sources of Halogenated Hydrocarbons in an Urban Water Supply," *Grant Report No. 804394*, U.S. Environmental Protection Agency, Cincinnati, Ohio (1978).

48 CASE HISTORY—USE OF A MOBILE ADVANCED WATER TREATMENT SYSTEM TO TREAT GROUNDWATER CONTAMINATED WITH VOLATILE ORGANIC COMPOUNDS

Jeffery L. Pope, Project Engineer

Richard A. Osantowski, Manager
Corporate Research & Innovation Group
EnviroEnergy Technology Center
Rexnord, Inc.
Milwaukee, Wisconsin 53214

INTRODUCTION

In December 1984, soil borings and test wells drilled at an industrial site located in Wisconsin indicated that soil and the groundwater were contaminated with volatile organic compounds (VOCs).

The soil borings were obtained from the facility's chemical loading and storage area which contained three underground tanks that were to be decommissioned and excavated. Since the inventory records did not show any tank leakages, it was concluded that spillage during loading and unloading over a period of 15 years was the major cause of the contamination.

To determine the extent of organic contamination in the groundwater, nine monitoring wells were installed. Initial groundwater analytical results determined that there were high concentrations of 1,1,1-trichloroethane, carbon tetrachloride, methylene chloride and some other volatile organic compounds. The monitoring system also indicated that the clayey nature of the soil had inhibited any major migration of the contamination.

A four phase remedial action program was initiated in April 1985 to contain the contamination source, treat the polluted groundwater, and excavate and dispose of the contaminated soils and tanks.

BENCH SCALE FEASIBILITY STUDY (PHASE 1)

This phase of the remedial action program included the design, construction and operation of a bench scale air stripper. Figure 1 shows a schematic of the air stripper used. The bench scale air stripping study was initiated:

- To determine the feasibility of reducing the concentration of all the VOCs below the effluent discharge limits.
- To obtain qualitative information for use in the development and implementation of a pilot scale study.

The bench scale air stripper was operated at ambient temperature with two feet of glass Rasching rings (6 mm x 6 mm) as packing. A four gallon sample of contaminated groundwater was obtained from the recovery well. This sample was then continuously recirculated through the air stripper at a rate of 0.6 gpm. The air flow rate was 5 scfm utilizing an in-plant air supply. Effluent samples of treated groundwater were obtained approximately every hour for the duration of the 3 1/2 hour test. After the initial air stripping sequence, it became apparent that suspended matter was building up on the stripper packing. Chemical pretreatment with alum and polymer followed by clarification effectively removed this material.

The analytical results of the bench scale air stripping study are summarized in Table I. The initial total VOC concentration of 301.9 mg/L was reduced to below 3 mg/L after 2 hours of stripping. Since this was less than the effluent discharge limit, Phase 2 of the remedial action program was initiated.

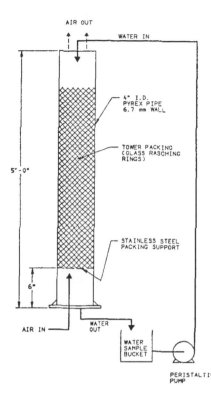

AIR OUT

WATER IN

4" I.D.
PYREX PIPE
6.7 mm WALL

TOWER PACKING
(GLASS RASCHING
RINGS)

5'-0"

STAINLESS STEEL
PACKING SUPPORT

6"

AIR IN   WATER
                OUT

WATER
SAMPLE
BUCKET

PERISTALTIC
PUMP

Figure 1.   Bench-scale air stripper.

PILOT SCALE TREATMENT STUDY (PHASE 2)

It was decided that the Envirex Advanced Water Treatment System Van would be best suited to perform the pilot scale study. This state-of-the-art trailer mounted research vehicle contained all of the necessary physical/chemical treatment equipment required for the study including a laboratory area. Figure 2 presents an isometric drawing of the system. Since the treatment systems housed in the van are prepiped and prewired, minimum manpower was required for setup, operation and shutdown. This resulted in considerable time and cost savings for the project.

In June 1985 the mobile van was transported to the site and a sample of the raw groundwater was analyzed to confirm the characteristics of the contaminants prior to startup and continuous operation of the pilot plant at 10 gpm. Table II presents the results of these analyses. Review of the data indicated that the total VOC concentration had increased more than 3 times that of the initial sample analyzed during the bench testing study (Phase 1). Modifications in the test plan were made to accommodate this increase in total VOC concentration. Unit operations previously selected remained unchanged; however, activated carbon was added as a final polishing step.

The objectives of the pilot study were:

1. To obtain engineering performance data for the design and scale-up of a permanent groundwater treatment system including such information as:
   • groundwater pretreatment requirements prior to air stripping and carbon adsorption
   • VOC removal efficiency of a single pass through the air stripper
   • Activated carbon treatment efficiency, necessary contact time, and pollutant loading.
2. To obtain cost data (operating and capital) and space requirements for the planned full-scale treatment plant

The pilot system was started in mid-June 1985, and operating parameters were optimized during the following weeks. The treatment train is illustrated in Figure 3.

Table I. Bench Scale Air Stripper Results

Parameter	Raw	#1[a]	#2[b]	#3[c]
Acetone	0.17	0.058	–	–
Acrolein	<0.01	–	–	–
Acrylonitrile	<0.01	–	–	–
Benzene	<0.01	–	–	–
Bromomethane	<0.01	–	–	–
Bromodichloromethane	<0.01	–	–	–
Bromoform	<0.01	–	–	–
Carbon Tetrachloride	46.1	3.01	–	–
Chlorobenzene	<0.01	–	–	–
Chloroethane	<0.01	–	–	–
2-Chloroethylvinyl Ether	<0.01	–	–	–
Chloroform	0.11	<0.01	–	–
Chloromethane	<0.01	–	–	–
Dibromochloromethane	<0.01	–	–	–
1,1-Dichloroethane	5.98	<0.01	–	–
1,2-Dichloroethane	<0.01	–	–	–
1,1-Dichloroethene	15.8	0.19	0.17	0.057
trans-1,2-Dichloropropene	<0.01	–	–	–
1,2-Dichloropropane	<0.01	–	–	–
cis-1,3-Dichloropropene	<0.01	–	–	–
trans-1,3-Dichloropropene	<0.01	–	–	–
Ethyl Benzene	<0.01	–	–	–
Methylene Chloride	1.3	0.077	–	–
1,1,2,2-Tetrachloroethane	<0.01	–	–	–
Tetrachloroethene	<0.01	–	–	–
1,1,1-Trichloroethane	226	13.1	1.94	0.97
1,1,2-Trichlorethane	<0.01	–	–	–
Trichlorethene	5.6	0.13	0.12	0.07
Trichlorofluoromethane	<0.01	–	–	–
Toluene	0.34	0.024	0.020	0.011
Vinyl Chloride	<0.01	–	–	–
Total VOC's	301.9	17.0	2.70	1.38

All concentrations are in ppm (mg/L)
"<" preceding a number indicates that parameter is below the 10 ppb detection limit
"–" indicates that this parameter was analyzed for but not detected
[a] 1.5 hours of air stripping
[b] 2.2 hours of air stripping
[c] 3.5 hours of air stripping

Pretreatment

Continuous field pilot testing evaluated selected dosages of alum and polymer addition prior to clarification. Although optimum dosage of 125 mg/L of alum and 1 mg/L of anionic polymer was determined, the sludge production was much less than anticipated from the bench scale treatability tests. It was concluded that any suspended solids present in the groundwater could be effectively removed in a dual media filter without the chemical addition/clarification step. Therefore, the chemical addition/clarification step was eliminated from the design for the full-scale treatment system.

Ferrous iron present in the groundwater tended to partially oxidize to the ferric state after exposure to air in the rapid mix tank used for chemical addition. Post-ferric floc was also observed in the clarified and filtered effluent. To minimize iron buildup on the air stripper packing media, aeration prior to chemical addition was briefly studied. It was concluded from this study that aeration would be a necessary pretreatment adjunct to minimize precipitation of ferric hydroxide on the air stripper media.

The results of the pilot study indicated that aeration for the oxidation of ferrous iron followed by dual media filtration would be the only pretreatment requirements for the full-scale treatment system.

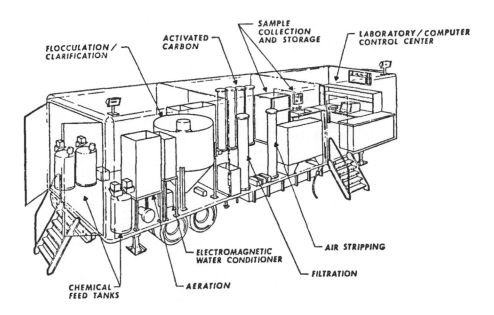

Figure 2.   Envirex mobile advanced water treatment system van.

Air Stripping

Several factors affect the efficiency of air stripping including the air/water ratio, packing type and volume, contaminant levels, etc. For this study, five feet of a one inch diameter packing media were used. Air to water ratios of from 80:1 to 180:1 (scfm/gpm) were evaluated. For the 10 gpm packed tower air stripper, an air to water ratio of 140:1 was found optimum for maximum removal of groundwater VOCs.

Table III presents the results of the air stripper effluent analysis. Although the total VOC removal was approximately 89%, the removal fell short of the 99.5% removal needed to meet the effluent discharge limits. Since the need for additional treatment for organics removal had been anticipated, the activated carbon program was initiated.

Activated Carbon

Addition of the carbon adsorption step was a relatively easy task since the Envirex Advanced Water Treatment System Van was equipped with 4 onboard carbon adsorption columns. A detention time of 30 minutes was studied for removal of residual VOCs. This detention time was based on previous

Table II. Raw Groundwater Analytical Results (Phase 2)

Parameter	Untreated Groundwater (mg/L)
Acetone	13.0
Carbon Tetrachloride	75.0
Chloroform	0.11
1,1-Dechloroethane	10.0
1,1-Dichloroethene	14.0
Methylene Chloride	11.0
1,1,1-Trichloroethane	890.0
Trichloroethene	5.5
Toluene	4.8
Total Volatile Organics	1023.4

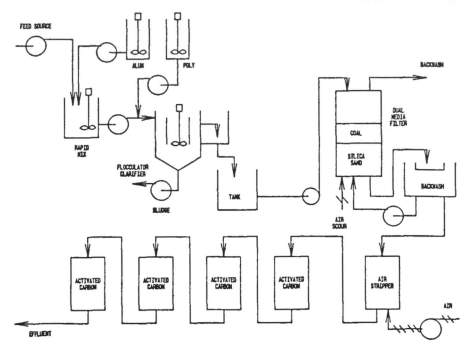

Figure 3. Schematic of the pilot plant treatment system.

experience and carbon isotherms. The treatment system, including the carbon, was operated until 12,000 gallons of raw groundwater was treated. Following treatment, a carbon column effluent sample was obtained from the storage tank holding the treated water and analyzed for all of the regulated discharge parameters. Table IV presents the results of these analyses.

Results of the carbon column effluent sample indicated that all of the metals and organics with the exception of acetone had been reduced to below the effluent discharge limits. The acetone concentration observed was approximately 10 times greater than the concentration observed in the raw groundwater at the beginning of Phase 2.

Table III. One Pass Air Stripper Effluent Results

Parameter	6/12/85 Untreated Groundwater[a]	6/24/85 Air Stripper Effluent[a,b]	% Removal
Acetone	13.0	1.1	91.5
Carbon Tetrachloride	75.0	5.7	92.4
Chloroform	0.11	0.013	88.2
1,1-Dichloroethane	10.0	0.5	95
1,1-Dichloroethene	14.0	2.8	80
Methylene Chloride	11.0	0.85	92.3
1,1,1-Trichloroethane	890.0	103.0	88.4
Trichloroethene	5.5	0.39	93
Toluene	4.8	0.65	86.5
Total Volatile Organics[c]	1023.4	115.0	88.7

[a]All results are in mg/L.
[b]Air stripper operated at ambient temperature.
[c]Sum of the concentrations of the above 9 organic compounds.

Table IV. Carbon Column Effluent Results (Phase 2)

Parameter	7/19/85 Treated Groundwater[a,b]	% Removal	Effluent Permit Limitations[a]
Chromium	0.02	—[c]	—
Copper	0.02	—	6.0
Nickel	0.1	—	4.0
Lead	0.1	—	2.0
Zinc	0.96	—	8.0
Cadmium	0.02	—	0.75
pH, units	7.1	—	5.5
Oil and Grease	2.0	—	100.0
Acetone	150.0	—	—
Carbon Tetrachloride	1.1	98.5	—
Chloroform	0.01	91.0	—
1,1-Dichloroethane	0.034	99.7	—
1,1-Dichloroethene	0.01	99.9	—
Methylene Chloride	0.036	99.7	—
1,1,1-Trichloroethane	0.58	99.9	—
Trichloroethene	0.01	99.8	—
Toluene	0.019	99.6	—
Total Toxic Organics[d]	1.8[e]	—	5.0

[a]All results are in mg/L.
[b]Effluent from system which included a carbon adsorption step.
[c]—Not calculated.
[d]Sum of the concentrations of the above 9 organic compounds.
[e]This result is excluding the final acetone concentration.

The acetone, which had been detected previously in only low concentrations, appeared suddenly in the pilot system feed and effluent. Since the plant does not use acetone as a raw product, it was speculated that the acetone was a byproduct of microbial degradation in the soil of isopropyl alcohol, which is used as a raw material.

Since physical/chemical treatment techniques such as ambient temperature. air stripping and activated carbon adsorption were not effective in removing acetone, an alternative treatment technique was needed to meet the effluent discharge limits. Phase 3 was initiated to evaluate selected treatment alternatives.

ACETONE REMOVAL FEASIBILITY STUDY (PHASE 3)

The feasibility study evaluated acetone removal techniques which could be easily included into the proposed full-scale treatment system design. Six alternative methods were studied on a bench scale to determine their technical feasibility. These methods included ambient temperature air stripping, air stripping at 176°F, biological activated sludge treatment, peroxide addition, chlorine addition, and chlorine addition followed by air stripping. The study concluded that elevated temperature air stripping (176°F) was the most effective.

Shortage of raw groundwater storage capacity and deadlines for full-scale treatment system implementation forced the decision to design a full-scale elevated temperature air stripping system without pilot scale confirmation data.

PERMANENT ELEVATED TEMPERATURE AIR STRIPPING SYSTEM (PHASE 4)

The 10 gpm permanent groundwater treatment system was designed for maximum flexibility and could be operated in either the batch or continuous mode. Thus, if acetone rich pockets of groundwater were recovered, effluent limitations could be met using the multiple pass capability of the air stripper system. The full-scale treatment system schematic is illustrated in Figure 4.

Referring to the Figure 4 schematic, aeration is utilized for oxidation of ferrous iron which is precipitated as ferric hydroxide on the dual media filter. Organics are then stripped at elevated temperature in the packed tower. Notice that, by using the high temperature stripper, it was possible to eliminate the granular activated carbon system from the process train and still meet the required

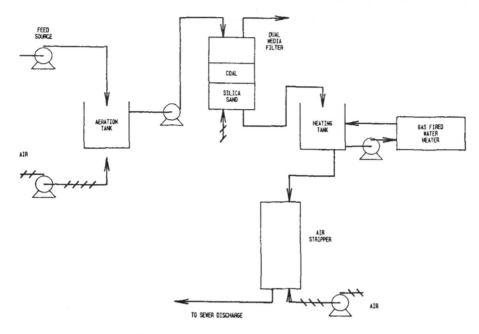

Figure 4.   Schematic of the full-scale groundwater treatment system.

effluent limits. This resulted in a lower capital investment, although operating costs are somewhat higher.

Installation of the permanent treatment system was completed in December 1985. It is housed in a permanent building. The air stripper and all tankage is positively ventilated to the outside. The treatment system will continue to operate until raw groundwater VOC concentrations return to levels consistent with state policy.

SUMMARY

In conclusion, the full-scale system design, based on the bench and pilot studies, was successfully implemented and reduced the total VOCs from as high as 1025 mg/L to less than 5 mg/L. Aeration followed by filtration was found to be necessary pretreatment adjuncts for oxidation and precipitation of ferrous iron. Finally, acetone concentrations of up to 150 mg/L in the raw groundwater were effectively removed using elevated temperature (176°F) air stripping.

49 REMOVAL OF PHENOL FROM A BRINE AQUIFER

David R. Hale, Eastern Regional Manager
Evan K. Nyer, President
DETOX, Inc.
Dayton, Ohio 45459

INTRODUCTION

A portion of the groundwater under a Gulf Coast hazardous waste site has been contaminated over the course of several years. From analysis of the groundwater and history of the site, several sources have contributed to the problem. These sources have been removed, leaving the cleanup of the groundwater the only remaining task to perform. However, this is the most difficult task, even though the plume is confined to the site boundary.

The combined influent water has an average concentration of 15,000 mg/L dissolved solids and 1300 mg/L Total Organic Carbon (TOC) with the main component being 400 mg/L of phenol. For reinjection, this influent must be treated to background quality, which required a final effluent having less than 18 mg/L TOC. The groundwater is naturally brine, and so the dissolved solids will not be removed. It was determined that the optimum pumping rate from the wells would be 23,000 gals/day. At this rate of removal and recharge, the design life of the treatment system was set at 10 years.

The process selection for the treatment system had to consider several important and unique problems connected with the treatment of a brine groundwater. The most critical problem with the design of the treatment system was that the concentration of organics in the groundwater would decrease as the treated water was returned to the ground and forced back to the central wells.

Before the full scale system was put into service, it was decided to run a large scale pilot plant to insure that the assumptions made from the laboratory data were correct. Based on the results from this data, a full scale system was installed. The operation of the full scale biological treatment system over the first two years is reviewed.

PROCESS SELECTION

Several factors had to be considered in the process selection for the treatment system. The most important of these were economics and technical factors. A further consideration had to be given to the fact that the site preferred a relatively simple system that would not require full time monitoring.

Laboratory data showed that the organics in the groundwater could be removed by carbon adsorption or degraded by biological treatment. An economic comparison was then run on the two processes.

The cost of carbon adsorption is directly related to the pounds of organic removed. It takes between 5 and 200 pounds of carbon for each pound of organic removed [1]. For high concentrations of organics, as in this groundwater, the range is usually 5 to 20 pounds of carbon per pound of organic. Using 23,000 gal/day and 1300 mg/L TOC, 249 lbs/day of organics needed to be removed from the groundwater. Using 10 lbs/lb organic for comparison, 2500 lbs of carbon per day were needed. Assuming $0.75/lb of carbon, the operating cost is $1875.00 per day. This number does not include capital costs and other operating costs such as personnel and electricity.

Biological treatment is usually considered as cost per 1000 gallons of water treated. Standard numbers are based on relatively low concentrations of organics and high flow rates. For small flows (less than 100,000 gal/day) and high organic concentrations, a reasonable cost is $0.46 per pound of organic treated. This number is high when compared to text book numbers but is accurate for small scale systems.

Using the same design numbers of 23,000 gal/day and 1300 mg/L TOC the operating cost is $115.00 per day. This figure includes capital and electrical costs, but not personnel. This means that carbon will cost at least 16 times the cost of biological treatment.

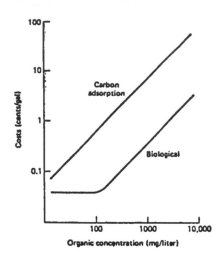

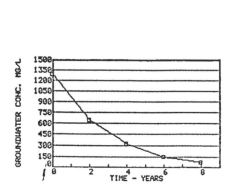

Figure 1.   Change in organic concentration over the life of the project.

Figure 2.   Cost comparison for carbon adsorption and biological treatment.

These cost figures are based on the initial design specifications. The flow through the system will not change during the design life of the project. However, the organic concentration will decrease with time. The groundwater will be removed, treated and returned to the aquifer. The treated recharge water will be used to force the plume back to the central wells. The treated water will mix with the plume as it forces it forward. This mixing will lower the organic concentration of the groundwater that is being removed. Figure 1 shows the expected decrease in organic concentration during the life of the project. These numbers will have to be confirmed after the treatment system is in full operation.

With the change in organic concentration, it was necessary to compare the costs of biological treatment and carbon adsorption over a range of organic concentrations. Figure 2 summarizes this comparison. As can be seen in Figure 2, the relative cost advantage of biological treatment over carbon adsorption does not change until approximately 150 mg/L of organic concentration. The costs do not come close until less than 20 mg/L. From Figure 1, the organic concentration is not expected to reach 150 mg/L until 6 years after the start of the project. From the economic analysis, it was determined that biological treatment of the groundwater was the preferred method.

LIFE-CYCLE DESIGN

There are several designs for biological treatment systems. The easiest to operate are the fixed film designs. However, the activated sludge designs theoretically produce the best effluent. One other consideration was that there were several tanks available on site that could be used as part of the system and that an activated sludge design could use the existing tanks.

Activated sludge was, therefore, preferable on a capital cost basis. However, there were two technical problems with using activated sludge. First, in order to maintain a consistent, low effluent organic level, the activated sludge process requires close operator attention and daily analytical support. These personnel requirements add substantial costs to the operation of the system. It was estimated that an activated sludge design would require at least eight man-hours per day.

The second problem was that an activated sludge system would not be able to adjust to the lower organic concentrations during the design life of the project. Figure 3 summarizes the effect of lower influent organic levels on the sludge age of the treatment system. This analysis assumes: 23,000 gal/day influent flow, 40,000 gal aeration tank, and a 0.25 lbs solids/lb TOC removed yield coefficient.

The activated sludge process relies on the ability of the bacteria to settle. A sludge age between 5 and 20 days is recommended to maintain a good settling sludge. As can be seen in Figure 3, the sludge age quickly goes out of that range as the influent concentration goes down. In the beginning of the project, a large aeration tank was required, and as the influent concentration goes down, the aeration basin must shrink to maintain the proper sludge age.

All of these considerations were combined and a final design developed. Figure 4 is the final design for the groundwater treatment system and shows how the system will look as the influent concentration decreases. The system includes: a first stage activated sludge system, a second stage DETOX H-Series biological reactor, a dual media filter, and a carbon adsorption column. The following are the specifications on each section:

First Stage Biological — The system consists of two 20,000 gal. (15 ft. diameter, 15 ft height) aeration basins, in series, and a hopper bottom clarifier. The first aeration basin has 8 static tube aerators and a 15 hp blower delivering 240 scfm of air. The second aeration basin has 4 static tube aerators and a 5 hp blower delivering 80 scfm of air. The hopper bottom clarifier has 97 sq. ft. of surface area and returns sludge with an air, sludge ejector pump.

Second Stage Biological — The second stage consists of a unique combination of activated sludge and fixed film. The DETOX H-series biological reactor is 10 ft wide, 10 ft high, and 28 ft long. The system maintains the bacteria in the aeration zone by attachment to plastic media. The media is submerged in the water, and the tank is completely mixed. The resulting system is an activated sludge system, but it is no longer limited by the sludge age considerations. The system is also self-regulating and eliminates the need for operator attention.

Dual Media Filter — The filter consists of 10 sq. ft. filtering surface area, air/water backwash with 1 ft. anthracite coal and 1 ft. sand filter media.

Carbon Adsorption — The carbon columns consist of two carbon columns in series with 30 min. residence time in each column.

The resulting system is easy to maintain. There are no moving parts in the entire system with the exception of the blowers and the pumps. The system also requires a minimal amount of operator attention. The first stage biological system is designed to discharge up to 300 mg/L TOC. This flexibility will allow the operators to refine the operation a maximum of three days per week and also minimize the amount of analytical work required.

The DETOX H-Series reactor is self-regulating and requires very little operator attention. The Dual Media Filter and the Carbon Adsorption systems are both fully automated. Manpower requirements are estimated at between 12 and 20 man-hours per week for the entire system and include sampling and analysis as well as operation and maintenance.

The treatment system will also respond to the reduction in influent organic concentration. At 1300 mg/L TOC and above, the entire system will operate. When the influent TOC drops below 900 mg/L, one of the aeration basins in the First Stage Biological Treatment System will be eliminated. When the influent reaches 300 mg/L, the First Stage Biological Treatment System will go out of service. At 100 mg/L, only the carbon will continue treating the groundwater.

FULL SCALE BIOLOGICAL TREATMENT

Based on the results from pilot testing, the full scale treatment system was installed. The first aeration basin of the first stage biological treatment system was set up with 8 static tube aerators. The sludge recycle from the clarifier was switched from the second aeration basin to the first aeration basin, and a six inch line was used to connect the two aeration basins.

A DETOX H-48 submerged fixed-film bioreactor was installed as the second-stage biological treatment. The H-Series reactors come fully assembled to the site. Foundation requirements for placement of the unit at the treatment location simply consisted of two cement footers. After placement of the bioreactors, air and water piping were connected, and the treatment system was ready for operation one day after it arrived on site.

The first and second stage biological reactors have been run for the past two years. The flow during that time has never reached the design flow of 23,000 gal/day. Problems outside of the treatment system have caused this reduced flow. Data from these two years of operation shows the substantial treatment efficiency of the biological treatment system. Figure 5 summarizes the monthly average flow rate to the treatment system. There was no flow during November and December of 1983.

Figure 6 shows the monthly average influent TOC to the treatment system. The influent concentration never reached the design level of 1300 mg/L. The maximum concentration to the treatment system was 1150 mg/L. During the last two years the concentration has slowly dropped to about 800 mg/L. This decrease confirms the life-cycle concentration variation that was expected with the groundwater cleanup.

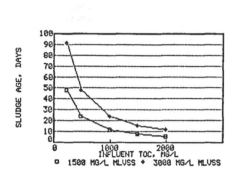

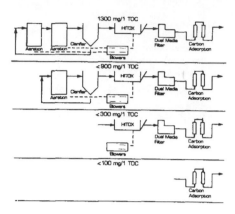

Figure 3. Variation of sludge age with change in influent organic concentration.

Figure 4. Groundwater treatment system.

The biological treatment systems have been able to produce a good quality effluent with the variable influent TOC. Figure 7 summarizes the effluent TOC from the DETOX H-48. The lowest TOC level reached during the past two years was about 150 mg/L. The remaining 150 mg/L TOC is probably non-degradable or refractory. There is a program to develop bacteria that can treat some or all of the remaining organics. However, it is expected that the carbon system will have to remove most of the remaining TOC. During the entire two years of operation, phenol has remained nondetectable.

Figure 8 shows the monthly average percent removal of TOC by the biological systems. The systems have consistently removed between 75% and 85% of the influent TOC. The variation in removal percentages may be due to changes in the specific compound makeup of the plume as it is brought back to the central wells. There may also be slight seasonal variation in the removal. The percent removal is expected to decrease as the influent decreases due to the nondegradable fraction of the organics in the groundwater. It can be concluded that the biological systems are successfully removing most of the TOC from the groundwater. Hopefully, new bacteria cultures can be developed to increase this efficiency further.

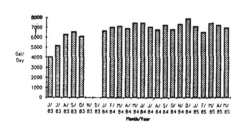

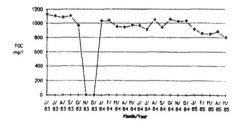

Figure 5. Monthly average flowrate.

Figure 6. Monthly average influent TOC.

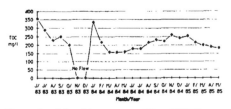

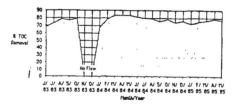

Figure 7. Monthly average effluent TOC.

Figure 8. Monthly average percent TOC removal.

SUMMARY

A portion of the groundwater under a Gulf Coast hazardous waste site has been contaminated with a variety of organic compounds.It was decided to pump the groundwater out of the ground, remove the organic contaminants, and to recharge it back into the ground. Initial influent concentration is 1300 mg/L TOC flowing at 23,000 gal/day. Recharge concentration was set at the background concentration of the groundwater, 18 mg/L TOC. Over the life of the project, the influent concentration will approach the background concentration.

A treatment scheme was developed based on laboratory tests, technical and economic analysis. The final system included the following: a first stage activated sludge system; a second stage DETOX H-series biological reactor; a dual media filter; and a carbon adsorption column. This system was economical and could easily be changed to reflect the changes in the influent concentration.

To insure that the organics in the brine groundwater could be degraded by bacteria, a pilot plant was set-up. The pilot plant was designed so that all of the components would be used on the full scale system. After overcoming salt inhibition and low nutrient concentrations, the pilot plant was able to consistently remove 70% of the TOC.

The full scale plant was installed based on the pilot plant data. The full scale plant has been run for the past two years. During that time the influent TOC concentration has decreased as predicted by the life-cycle design. The full scale plant has consistently removed 75% to 85% of the influent TOC and 99 + % of the phenol with the variable influent conditions. A program is being run to develop new bacteria that can degrade more of the organics left in the water.

REFERENCES

1. Nyer, E. K., *Groundwater Treatment Technology*, Van Nostrand Reinhold Co. Inc., New York, N.Y. (1985).

Section Four
PULP AND PAPER MILL WASTES

50 TREATMENT OF PULP AND PAPER MILL WASTEWATERS FOR POTENTIAL WATER REUSE

William R. Knocke, Associate Professor

Deepak Bhinge, Graduate Assistant

Elizabeth Sullivan, Graduate Assistant

Gregory D. Boardman, Associate Professor
Department of Civil Engineering
Virginia Tech University
Blacksburg, Virginia 24061

INTRODUCTION AND OBJECTIVES

The pulp and paper industry uses large volumes of water in the pulp and bleaching process, with values ranging from 15,000 to 60,000 gallons per ton of bleached pulp [1]. Some of the wastewater pollutants generated by this industry include color, dissolved inorganic solids and organic carbon. Treatment of these wastewaters to reduce the quantity of each pollutant is necessary if the water is to be recycled back to various stages of the process. Also, in certain instances, treatment for the removal of these particular classes of pollutants may be required prior to release in public waterways.

Major sources of color from a pulp mill include the caustic extraction stage in bleaching, the unbleached screenings, and the decker filtrates [2]. Significant amounts of the colored material were felt to originate from lignin and its derivatives. Fuller et al. [3] reported that, in addition to lignin, tannins reacting with iron as well as the natural color bodies of the wood also add to the effluent color level. Schmidt and Joyce [4] showed that the functional groups which contribute the greatest amount of color to pulping and bleaching effluents are aromatic and quinonoid nuclei which may be conjugated with carbonyl and ethylenic groups. Also, the authors indicated that the color bodies present in these effluents vary in molecular weight from less than 400 up to 150,000 mass units.

A variety of wastewater treatment options are available for use in the treatment of pulp and paper effluents. For the control of color and organic carbon concentrations, processes such as chemical coagulation, chemical oxidation, ultrafiltration, biological treatment, and activated carbon addition have been used with varying degrees of success. For the reduction of dissolved inorganic solids concentrations, ion exchange and reverse osmosis have been effectively used in certain cases. Reduction of dissolved species concentrations such as chlorides is important related to corrosion concerns

Table I. Water Quality Requirements for Recycle [8]

Parameter	TAPPI E603 Specifications (mg/L)
Turbidity	25
True Color (Pt-Co units)	5
TDS	250
Total Hardness	100
Free CO_2	10
Fe	0.1
Mn	0.05
Cl	75
Alkalinity	75
Silica as SiO_2	20

Table II. Representative Wastewater Characteristics

Parameter	Biotreated		Bleach Plant		Main Mill[a]	
	Average	Range	Average	Range	Average	Range
TSS (mg/L)	66	38–123	130	12–198	40	–
Temperature (°F)	72	50–94	117	104–126	101	93–116
pH	7.2	6.6–7.6	3.4	2.1–4.9	11.0	9.6–12.3
Color (Pt-Co Units)	2280	2200–2700	2500	1950–2750	1400	390–1960
TOC (mg/L)	175	160–200	370	–	200	140–260
Chlorides (mg/L)	630	500–720	1035	875–1210	32	20–50

[a]After primary clarification

[5]. Also, excess chloride concentrations may reduce the effectiveness of oxidizing agents used in the bleaching process [6,7].

The overall goal of this study was to investigate methods of wastewater treatment which could potentially be used to help produce water for recycle in pulping and bleaching systems. The specific objectives involved assessing the effectiveness of chemical coagulation, activated carbon (both powdered and granular), and ion exchange for the removal of dissolved color, total organic carbon (TOC) and chlorides from pulp and paper wastewaters. The impact of process variables such as pH and chemical dose on treatment efficiency were also considered. Success of treatment was based upon comparison of treated water quality against specifications published by the Technical Association of the Pulp and Paper Industry (TAPPI); representative values are listed in Table I.

METHODS AND MATERIALS

Sample Collection

The wastewaters used in this study were generated at the Union Camp Corporation Mill in Franklin, Virginia. At the time of this study, the Franklin facility produced 1800 tons per day of bleached and unbleached Kraft paper and board and produced approximately 25 to 30 million gallons per day of wastewater. Three different wastewater sources were sampled: a) the pulp mill waste stream ("main mill"); b) the bleach line waste stream ("bleach plant"); and c) the combined main mill and bleach plant effluents which had been biologically treated in an aerated stabilization lagoon (termed "biotreated"). Each waste in this study contained high concentrations of color, total organic carbon, and dissolved inorganic solids; representative values are included in Table II.

Nalgene carboys were used for sample collection, transport and storage. Samples were collected and transported by Union Camp employees and graduate students from Virginia Tech. Samples were stored in a 20°C storage room and/or an ambient temperature laboratory (18–22°C). Collection of samples was done approximately once every two to three weeks.

Chemical Coagulation Studies

All coagulation/flocculation studies were performed using a standard six-place Phipps and Bird laboratory stirrer. Usually, one-liter waste samples were used, although sample size varied in certain experiments between 500 mL and 1500 mL. Standard stock solutions of aluminum sulfate, ferric chloride, ferric sulfate, and lime were prepared prior to performing each experimental test. Chemical dose ranges utilized during the study were from 100 to 1500 mg/L of each coagulant. The coagulant was added to the square sample vessels with the stirrer operating at a speed of 150 to 175 revolutions per minute (rpm). Also, the desired test pH was adjusted at this time using the addition of either sulfuric acid or sodium hydroxide. After two minutes of rapid mixing, the sample was flocculated using a speed of 20 rpm for thirty minutes. After flocculation, a thirty or sixty minute sedimentation period was provided followed by supernatant sample collection. The treated samples were then analyzed for a variety of parameters including color, TOC, and pH.

Activated Carbon Studies

Both powdered and granular activated carbons were used to generate adsorption isotherms for color and organics removal. Carbon utilized were Nuchar SA and Nuchar WV-L (8x30 mesh), both provided by the Westvaco Corporation of Covington, Virginia. Carbon isotherms were developed on both raw wastewater samples and samples which had received prior treatment with either aluminum sulfate, lime, or ferric salts. Carbon dosages from 20 mg/L to 10 g/L were investigated. Samples were agitated in sealed 250 mL flasks for a twenty-four hour period. After this period of mixing, the samples were removed from the agitator, filtered, and then analyzed for residual color and TOC.

A granular activated carbon column testing method was also developed; carbon quantities for the column were estimated using isotherm capacity data and a desired effluent color level of 20 Pt-Co units. Empty bed contact times of ten to twenty minutes were used. Wastewater applied to the column had received prior chemical coagulant treatment resulting in an applied color level less than 150 Pt-Co color units. The effluent quality produced by the column was monitored as a function of time. Color, TOC, and pH were measured until the column clogged due to excessive head loss buildup.

Ion Exchange Studies

The synthetic ion exchange resin used during this study was Amberlite IRA-68, which is manufactured by the Rohm and Haas Company of Pennsylvania, IRA-68 is a weakly basic, gelular acrylic resin with functional groups of tertiary amines; a more detailed description is provided elsewhere [9]. The resin was supplied in the hydroxyl form and required conversion to the bicarbonate form in order to accomplish chloride ion removal. The resin was selected for study based upon its reported value for selective chloride ion removal [10].

Wastewater samples applied to the ion exchange column had received prior chemical treatment for color and organics removal. Granular activated carbon treatment was also provided to certain ion exchange influent samples. The flowrate utilized was 0.15 to 0.20 bed volumes per minute (based upon manufacturer's recommendations). Effluent samples were collected as a function of operating time and measured for residual color, pH, sulfate, and chloride concentrations. Sulfate values were evaluated since it was felt to be the major competing anion present in the waste samples. The exchange column study was terminated when exhaustion was achieved or when excessive head loss was generated.

Physical and Chemical Analyses

The method of measuring color was that recommended by the National Council of Air and Stream Improvement (NCASI) which involved measuring sample absorbance at a wavelength of 465 nm [11]. Sample pretreatment included pH adjustment to 7.6 and filtration through a 0.8 um filter. Absorbance measurements were made using a Beckman DU-6 ultraviolet spectrophotometer; color levels were ascertained by comparison to color standards prepared using a #500 APHA platinum-cobalt color standard.

Adjustments of sample pH for color standardization were made using either sulfuric acid or sodium hydroxide. All measurements of sample pH were made using an Accumet Model 610 pH meter and probe.

Total organic carbon (TOC) concentrations were determined using a Dohrmann Model DC-52 organic carbon analyzer equipped with an ultra low organics module. The oxidizing agent used in the system was potassium persulfate. When necessary, samples were diluted for analysis using water produced in an ultra-pure (Mill-Q) water treatment system. An ion chromatograph was utilized to analyze raw wastewater and treated waste samples for chloride and sulfate concentrations.

A limited number of waste samples were analyzed for the molecular weight distribution of organics present. Ultrafiltration cells purchased from the AMICON Corporation were used to separate molecular weight ranges from 500 to 30,000 mass units. Samples were diluted to a TOC of less than 10 mg/L using ultra-pure (Mill-Q) water to minimize ultrafilter membrane fouling. The applied and filtrate TOC was measured across each AMICON ultrafilter and the resulting information used to generate a molecular weight distribution of the organics present.

RESULTS AND DISCUSSION

The following subsections highlight the results obtained from a variety of treatment studies completed using each of the three waste streams generated at the Union Camp facility. Space limitations dictate that the paper present typical results for one waste stream only. Discussion with respect to how all treatment processes functioned when treating all three waste streams has not been included. Most of the data presented relate to the waste stream which currently receives aerobic biological stabilization at the plant site. This wastewater is considered the prime candidate for treatment for recycle based on temperature and/or thermal heat balance considerations.

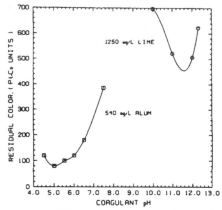

Figure 1. Effect of coagulation pH on the removal of color from biotreated pulp and paper wastewaters.

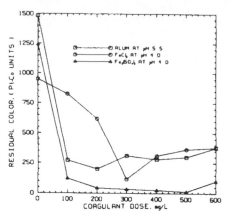

Figure 2. Use of aluminum and ferric salts under optimum pH conditions for the removal of color from main mill wastewaters.

Chemical Coagulation Studies

Numerous chemical coagulation studies were completed using each of the three waste streams of interest. In general, the results indicated that efficient (greater than 90% removal) of color could be obtained when using aluminum sulfate or ferric salts; lime addition for color removal was much less efficient and required much higher chemical dosages. Each of the coagulants was studied under a variety of pH conditions; typical results are presented in Figure 1. Alum coagulation was optimum in the pH range between 5.0 and 5.5; ferric coagulation was optimized at pH 4.0. These values correspond nicely to the reported pH ranges needed to optimize color and organics removal in water treatment facilities. In comparison, lime coagulation was optimized in the pH range between 11.5 and 12.5; however, optimum lime treatment often yielded an effluent color in excess of 600 to 700 Pt-Co color units.

Figure 2 presents typical data obtained with alum and iron salts when used under optimum pH conditions. Required dosages were found to be a direct function of influent waste color concentration; typical optimum dosages ranged from 200 to 600 mg/L for the main mill and biotreated waste streams. The bleach plant waste stream often required dosages in excess of 1,000 mg/L to optimize treatment. This difference might be due in part to the production of lower molecular weight organics following the bleaching operations. It is characteristic of chemical coagulation systems that removal efficiency is a direct function of the molecular weight of organics present.

A summary of optimum treatment systems is provided in Table III for the metal ion coagulants considered in this study. The results indicate that mass removal of color causing organics requires large dosages of chemical coagulants which, in turn, results in the production of large amounts of chemical sludge that require dewatering and ultimate disposal facilities. Experiments conducted during this study indicate that these sludges will gravity thicken to less than 3% dry solids; mechanical dewatering will result in a final cake solids concentration less than 15 or 20% dry solids.

The removal of TOC by chemical coagulation was found to be effective, although removal efficiencies were always less than that obtained with respect to color removal. Figure 3 presents typical data from these studies. Maximum TOC removal efficiency was typically in the 75 to 80% range when color removal efficiency was correspondingly in the 95 to 99% range. In certain studies, greater than 90% color removal resulted in less than 40% TOC removal. This indicates the presence of a significant pool of organic compounds which 1) do not contribute to the color level of the waste stream; and 2) are not amenable to removal by coagulation.

Figure 4 shows typical data obtained when comparing the molecular weight of organic compounds present before and after coagulant addition. Compounds with molecular weight of 5000 mass units and greater are efficiently removed; compounds less than 5000 mass units in weight are only margin-

Table III. Summary of Chemical Coagulation Results

Coagulant	Optimum	Optimum Dose (mg/L)	Final Color Pt-Co Units	Optimum Removal
1. BIOTREATED EFFLUENT				
Alum	5.5	400	80	96
FeCl$_3$	4.0	300	26	99
Lime	12.0	1800	750	73
2. MAIN MILL SOURCE				
Alum	5.0	200	200	92
FeCl$_3$	4.0	300	115	95
Fe$_2$(SO$_4$)$_3$	4.0	500	20	100
Lime	12.0	3000	225	91
3. BLEACH PLANT EFFLUENT				
Alum	—	1500	400	83
Lime	—	2000	1000	24

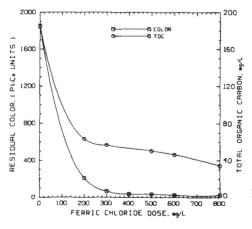

Figure 3.   Relationship between color and TOC removal when using iron salts to coagulate biotreated pulp and paper wastewaters.

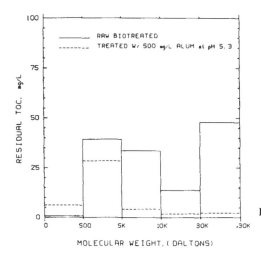

Figure 4.   Molecular weight distribution of organics present in biotreated pulp and paper wastewaters before and after alum coagulation.

Table IV. Biotreated Effluent Treated with Powdered Activated Carbon (PAC)[a]

PAC (mg/L)	Color After Treatment (Pt-Co units)			Percent Color Removal		
	pH 2	pH 4	pH 7	pH 2	pH 4	pH 7
20	250	1380	2200	89	37	0
50	560	1450	2100	74	34	4
100	490	1340	1950	78	39	11
200	540	1200	1865	75	45	15
500	125	900	1475	94	59	33
1000	25	655	1200	99	70	45
2000	20	390	460	99	82	79

[a]Raw color 2200 Pt-Co units.

ally removed by chemical coagulation. A complete recycle system must include treatment methods other than coagulation which will more efficiently remove these low molecular weight organic compounds.

Activated Carbon Treatment Studies

Typical results showing the response of biotreated pulp and paper wastewater to the addition of powdered activated carbon are given in Table IV. The pH utilized during carbon treatment had a significant impact on color removal efficiency noted; lower pH conditions resulted in much more efficient carbon utilization. This result was found to be true with all three of the wastewaters considered. Also, Table IV indicates that the use of powdered activated carbon as a primary color removal method would be very expensive due to the extremely high dosages of PAC required. In comparison, Figure 5 shows that PAC used in conjunction with a chemical coagulant can provide extremely low residual color levels at dosages much less than the values listed in Table IV. Data presented in Figure 5 also show that PAC might be more efficiently used in a two-state coagulation process versus being added initially with the primary chemical coagulant.

Representative results from granular activated carbon (GAC) column treatment studies are shown in Figure 6. In general, GAC treatment resulted in approximately 85 to 90% removal of the applied color; TOC removal was less efficient and averaged in the range of 50 to 60%. Column studies indicated that it was possible to predict GAC uptake of color and TOC from isotherm results. In each

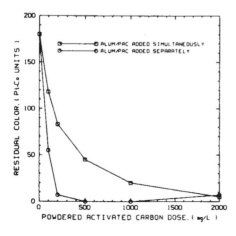

Figure 5. Beneficial impacts of powdered activated carbon addition for the removal of color from biotreated pulp and paper wastewaters (initial alum dose = 400 mg/L).

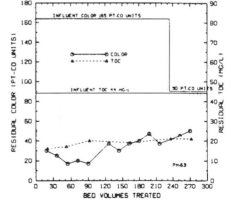

Figure 6. Typical results obtained when treating pulp and paper wastewater in a granular activated carbon column (waste pretreated by alum coagulation).

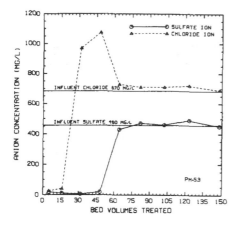

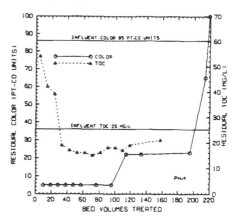

Figure 7.   Use of ion exchange for the removal of chlorides and sulfates from biotreated wastewaters (waste pretreated by alum coagulation and granular activated carbon).

Figure 8.   Removal of color and total organic carbon by ion exchange media (column received alum pretreated wastewater).

study carbon capacity was equal to or exceeded the capacity predicted from Freundlich isotherm plots of batch adsorption data. One problem noted with GAC column studies was the buildup of significant head loss during column operation. Although all wastewater samples applied to the column were prefiltered, significant head loss developed during each study. Pilot-scale systems will be needed to better investigate the cause of such head loss and how it may be minimized in full-scale implementation of the process.

Chloride Removal Using Ion Exchange

Experiments aimed at the selective removal of chloride ions from biotreated wastewater samples did not prove successful in spite of resin manufacturer's claims for resin specificity. Typical results are shown in Figure 7. During exhaustion of the resin, the total equivalents of anions exchanged was within 5% of the resin exchange capacity reported by the manufacturer. The major problem observed was that the resin did not provide specificity for chloride ions in the presence of significant concentrations of divalent anions such as sulfate. The data show chloride removal occurring in the initial stages of treatment followed by release of this exchanged chloride during later stages of treatment resulting in column effluent concentrations greater than present in the influent. A mass balance on chloride ions during the course of an experiment indicated less than 10% of the available exchange sites were occupied by chloride ions when sulfate ions were present in the column influent.

To further investigate this phenomenon, a portion of the biotreated wastewater was treated with barium, resulting in the precipitation of barium sulfate and the corresponding reduction of soluble sulfate concentrations to less than 5 mg/L. When sulfate concentrations were low, the column was very efficient for chloride removal with observed exchange capacities equaling manufacturer's rated capacity. The practical result of these findings is that chloride removal in the presence of multi-valent anions cannot be efficiently accomplished in waters that contain significant concentrations of sulfate.

Figure 8 presents data showing that ion exchange systems may produce a side benefit of color removal during treatment; TOC removal was very marginal indicating a selective removal preference for those organic compounds which cause color in the waste stream.

SUMMARY AND CONCLUSIONS

The overall objective of this study was to determine if pulp and paper wastewaters could be treated to a level of quality acceptable for recycle and reuse.

Metal ion coagulants such as aluminum and iron salts are efficient for the removal of color from pulp and paper wastewaters, especially when pH conditions of coagulation are optimized. However, treatment requires large dosages of coagulant, resulting in the production of significant quantities of waste sludge which must be disposed of.

The removal of total organic carbon (TOC) was less efficient when considering chemical coagulation with typical removal percentages in the 40 to 60% range and a maximum of 80% observed. The molecular weight distribution of organics present had a significant role in determining TOC removal efficiency.

Powdered activated carbon can be utilized in conjunction with metal ion coagulants to lower residual color levels to less than 10 Pt-Co color units. Dosages required are high. Granular activated carbon could also be utilized as a polishing step for color and TOC removal. Carbon column capacities were accurately predicted based upon batch adsorption isotherm data.

Selective ion exchange for chloride removal was not found to be nearly as efficient as suggested by resin manufacturers. Instead, in the presence of competing anions such as sulfate, chloride removal was insignificant. Further research into segregation of waste streams and/or alternate methods for chloride removal is necessary to better address this problem.

ACKNOWLEDGMENTS

The authors wish to express their sincere appreciation to Norman Shroyer, David Breed, and Phil Pagoria of the Union Camp Corporation for their help and financial support of this project. The help of Dr. Edward Bryan and the support of the National Science Foundation through the Presidential Young Investigator Award Program are also appreciated.

REFERENCES

1. Sanks, Robert L., "Ion Exchange Color and Mineral Removal From Kraft Bleach Wastes," *Environmental Protection Technology Series No. EPA-R2-73-255*, USEPA, Washington, DC. (1973).
2. Dugal, H. S., et al., "Color Removal in Ferric Chloride-Lime *TAPPI*, 59, 9, 71 (1976).
3. Fuller, R. R., et al., "Operating Experience with an Advanced Color Removal System," *TAPPI, 59, 9, 66 (1976)*.
4. Schmidt, R. L., and Joyce, T. W., "An Enzymatic Pretreatment to Enhance the Lime Precipitability of Pulp Mill Effluents," *TAPPI*, 63, 12, 63 (1980).
5. Matthews, John E., "Industrial Reuse and Recycle of Wastewaters — Literature Review," *Environmental Protection Technology Series No. EPA/6002-80-183*, USEPA, Washington, DC. (1980).
6. Reeve, D. W., Pryke, D. C., and Tran, H. N., "Corrosion in the Closed Cycle Mill," *Proceedings Fourth International Symposium on Corrosion in the Pulp and Paper Industry*, Swedish Corrosion Institute, Stockholm, Sweden, May 30–June 2, 1983, pp. 85–90 (1983).
7. Ahlers, P. E., "Closing of Bleaching Systems," *Pulp and Paper Industry Corrosion Problems*, Vol. 2, National Association of Corrosion Engineers, Houston, Texas, pp. 23–26 (1977).
8. Berger, H. F. "Evaluating Water Reclamation Against Rising Cost of Water and Effluent Treatment," *TAPPI*, 49(8):79A (1966).
9. Amberlite IRA-68 Technical Notes, IE-255, Rohm and Haas Company (May 1979).
10. Downing, D. G., Kunin, R., and Pollio, F. X., "Desal Process — Economic Ion Exchange Systems for Treating Brackish and Acid Mine Drainage Waters and Sewage Waste Effluents," *Chemical Engineering Symposium Series*, Vol. 64, No. 90 (1968).
11. "An Investigation of Improved Procedure for Measurement of Mill Effluent and Receiving Water Color," *NCASI Technical Bulletin No. 253* (December 1971).

51 CONDITIONING OF PULP AND PAPER SLUDGE USING DIRECT SLURRY FREEZING

M. Z. Ali Khan, Assistant Professor
Department of Civil Engineering
King Abdul Aziz University
Jeddah
Saudi Arabia - 21413

INTRODUCTION

Surveys [1] of the pulp and paper mill industry indicate that the wastes generated from different processes involved in producing pulp and paper are generally quite high in sulfates, sulfites, solids, etc. The BOD tends to be high especially with wastes containing sulfites. These side streams are usually treated physically, chemically or biologically [2-6] to produce effluents which meet water quality standards for discharge into the receiving water body. This treatment results in a residue called "sludge," which contains a lot of chemicals, especially sulfates and sulfites, in colloidal suspension and dissolved forms, making this sludge less amenable to settling, thickening, conditioning, dewatering, etc. before final discharge. For final discharge of the sludge, it is very important to reduce its volume by removing the associated water by physical or mechanical means. Lower volumes mean lower disposal costs.

Most of the sludge treatment processes [6-22,47-55,57-60,62-64] presently used have their difficulties and their drawbacks, particularly with regard to certain types of sludges. Many investigators [23-46,56,61] had previously considered natural and indirect freeze-thaw processes on laboratory or large size pilot plant scale experiments as an alternative in conditioning and dewatering of domestic sludges. Most of the studies, as mentioned above, indicated that both the natural and indirect processes were quite effective in improving the sludge drainage characteristics but were not considered cost-effective due to higher costs, poor quality of filtrate, and other operational problems. To overcome or reduce these problems a direct slurry freezing process [67,68,69] was developed for conditioning and disposal of the waste activated sludge from sewage treatment works serving residential areas. The process was evaluated as quite economical [69]; the filtrate was much better than the other freezing processes and either similar or better than the conventional processes. The direct slurry freezing process was not tested for the industrial sludges, especially the pulp and paper waste sludge. History and development of the sludge freezing processes is presented in Table I.

It is the purpose of this research to attempt to evaluate the feasibility of the direct slurry freezing process for conditioning, dewatering and disposal of sludge obtained from the treatment of pulp and paper waste streams.

SAMPLING AND TESTING

The sludge samples were collected from the outlet of the return sludge line from the secondary clarifier of the activated sludge process utilized for treating the side-streams of the Wesvaco Pulp and Paper Mill at Covington, Virginia. The samples were transported directly to the laboratory, located at Blacksburg, Virginia, and stored by refrigeration for a few hours before testing and analysis. The analysis, before and after direct slurry freezing, was conducted according to *Standard Methods* [66]. The Beckman 215 analyzer was used for total carbon measurements and the pH was measured using a Leeds and Northrup probe.

FREEZING PROCESS DESCRIPTION

The experimental sludge at a certain solids concentration is allowed to flow from a sludge tank "ST" (surrounded by ice) through a flow regulator "FR" into the freezer "F" where liquid butane, flowing through the heat exchanger "H" and flow regulator "R," comes in contact with the sludge (intimate contact is achieved by mechanical mixing). After heat exchange between butane and sludge, the slurry

Table 1. History and Development of Sludge Freezing Process

Process	Year	Reference
1. Natural Freezing and Drainage	1929	[23]
	1973	[27]
	1979	[28]
2. Indirect Freezing (Mechanical) and Dewatering (Expensive, no Heat Recovery)	1950	[25]
	1970	[26,35,36]
	1974	[45,49,50]
	1976	[39,42,43,44]
	1965	[31]
	1967	[34,61]
	1969	[32]
	1971	[24,33]
	1980	[57]
	1983	[58,59,60]
3. Direct Freezing (Solid + Slurry) and Dewatering	1974	[67]
	1976	[68]
	1978	[69]

frozen sludge is pumped through the condenser Coil "C_1" to the heat exchanger "H_1" from where it flows to heat exchanger "H_2" and finally to the butane stripper "BS" for recovery of dissolved butane from the supernatant for reuse if necessary. The hot butane vapors from the freezer "F" are compressed by vacuum-pressure pump "VF" condensed on the Coil "C_1" and recovered as liquid for reuse in the freezer. The slurry frozen sludge in the heat exchanger "H_1", condenser Coil "C_1", and heat exchanger "H_2" liquifies the butane vapors. The schematic representation is given in Figure 1.

PROCESS THEORY

Liquid butane has a boiling point lower than the freezing point of water. Butane extracts the heat of vaporization from the sludge thus inducing freezing and forming ice-slurry. If the freezer operates under a vacuum of 3 cm of mercury, the butane boiling point is lowered to about -1.5°C to insure better freezing. Most of the energy is recovered and reused.

EXPERIMENTAL DESIGN

The process described in Figure 1 is for a continuous flow process and was developed by Khan [67,68]. For the purpose of this study, only batch studies were conducted by putting about 2 liters of pulp and paper sludge having total solids of about 1.33% in the freezer. The liquid butane flow rate was about 10 mL/min., and the butane contact time or residence time was 5-6 hours. A mechanical

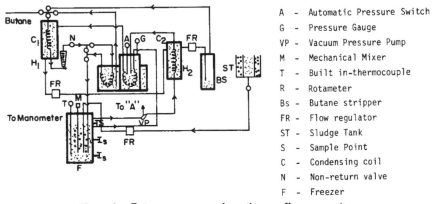

A	-	Automatic Pressure Switch
G	-	Pressure Gauge
VP	-	Vacuum Pressure Pump
M	-	Mechanical Mixer
T	-	Built in-thermocouple
R	-	Rotameter
Bs	-	Butane stripper
FR	-	Flow regulator
ST	-	Sludge Tank
S	-	Sample Point
C	-	Condensing coil
N	-	Non-return valve
F	-	Freezer

Figure 1.   Butane recovery and continuous flow apparatus.

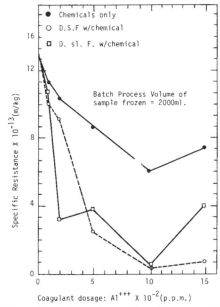

Figure 2. Effect of coagulant dosage on the specific resistance of pulp and paper waste sludge.

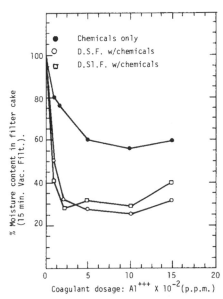

Figure 3. Effect of coagulant dosage on filter cake quality of the pulp and paper waste sludge.

mixer was used to prevent the solids from freezing. At the end of 6 hours the slurry frozen sludge was taken out of the freezer and allowed to thaw at room temperature; different parameters were then tested to check the improvement in the sludge characteristics.

The following set of experiments were conducted:

1. direct slurry freezing without chemicals (D. sl. F w/o chem.).
2. direct slurry freezing using Al^{3+} coagulant dosages of 100 to 1500 mg/L.(D.sl.F w/ chem).
3. direct solid freezing without chemical (D.S.F w/o chem.).
4. direct solid freezing using Al^{3+} dosage of 100 to 1500 mg/L (D.S.F. w/chem.).
5. original sludge with chemicals but without any freezing (O.S. w/chem., w/o F.).

The effectiveness of direct freezing (D.F.) was determined by the specific resistance test as described by Coakley and Jones [65], the sludge settling characteristics by calculating the interfacial settling velocity and sludge volume index (S.V.I) as presented by Mohlman [62], gravity drainage on sand beds [64], and the quality of the filtrate and supernatant by T.O.C. analyzer.

EXPERIMENTAL RESULTS

Results indicate that the addition of chemicals to the sludge, without freezing, shows a reduction in specific resistance from 13×10^{-13} m/kg to about 7×10^{-13} at a chemical dosage of 1000 mg/L of Al^{3+}, but solid freezing and slurry freezing reduced this specific resistance to $< 1 \times 10^{-13}$ m/kg (Figure 2).

The cake moisture (15 min of vacuum filtration) after slurry and solid freezing was about 30%, but the unfrozen sludge still had more than 60% moisture (Figure 3). The rate of solids production (lbs/ ft2/hr) in the case of solid freezing was twice as much as in the case of slurry freezing (Figure 4). The optimum conditioning effect seemed to be at about 1000 mg/L dosage of alum. There was practically little improvement by slurry freezing without the addition of any chemicals.

The direct slurry freezing resulted in about 50% volume reduction after 30 min settling thus producing an improvement in the thickening and settling process. The original sludge without any chemical addition or freezing did not settle at all, whereas solid freezing gave results very close to slurry freezing (Figure 5).

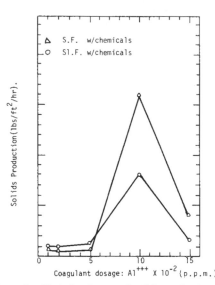

Figure 4.   Variation in rate of solids production with coagulant dosages with vacuum system.

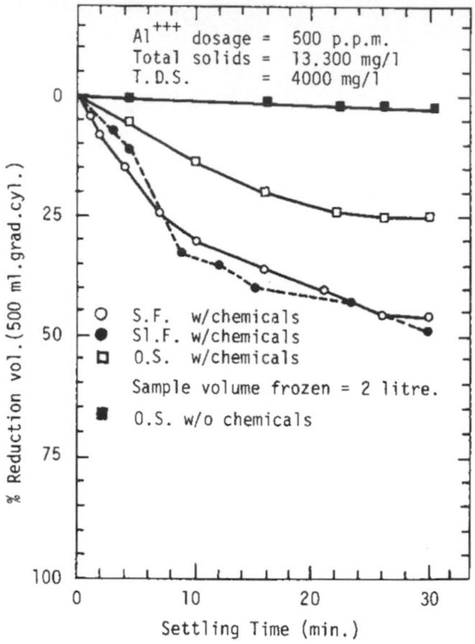

Figure 5.   Settleability of pulp and waste sludge after direct slurry freezing.

The gravity drainage of directly frozen sludge, after 24 hours, indicated that the moisture content left was about 92.5% while for the unfrozen sludge the same value was about 97.5% (Figure 6). After the gravity drainage of 1 day, any further moisture content reduction occurring was considered due to airdrying (Figure 7).

Airdrying for 7 days resulted in further reduction of moisture content to about 77% and 86% for the directly frozen slurry and the unfrozen sludges.

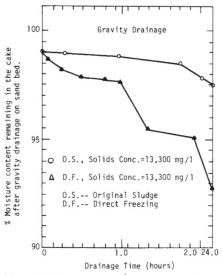

Figure 6.   Bench-scale sand-bed studies. Dewatering due to gravity drainage of pulp and paper waste sludge.

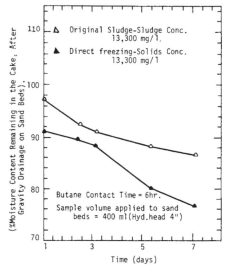

Figure 7.   Bench-scale sand-bed studies. Dewatering due to air-drying of pulp and paper waste sludge.

Table II. Cost Comparison of Sewage Sludge Handling and Conditioning Processes

Process	Reduction in sludge COD (percent)	Sludge Solubili- zation	Supernatant and Filtrate Quality		Cost per metric ton of dry solids	Ref. No.
			pH	Quality		
1. Direct Slurry Freezing	35	Low	7 – 8	Good	6 – 20	[67,68,69]
2. Direct Solid Freezing	50 – 70	High	7 – 7.5	Poor	5 – 35	[67,68,69]
3. Anaerobic Digestion	60 – 70	High	6 – 7	Poor	15 – 20	[30]
4. Aerobic Digestion	30 – 70	Low	4 – 7	Good	15 – 30	[30]
5. Heat Treatment	–	High	5 – 7	Poor	35 – 283[a]	[55]
6. Chemical Addition	20 – 40	Low	6 – 6.5	Moderate	10 – 35	[30]
7. Land Spreading	–	–	–	–	35	[51]
8. Chlorine Conditioning	–	Moderate	5 – 6.5	Poor	32 – 35	[70]
9. Dewatering Heat Drying and Bagging	–	–	–	–	135	[30]
10. Heat Drying and Barging	–	–	–	–	170	[30]
11. Composting	–	–	–	–	41	[30]
12. Land Application to Strip Mini.	–	–	–	–	167	[48]
13. Aerobic Dig. & Press. Filt.	–	–	–	–	30 – 60	[71,72]
14. Gamma Radiation Treatment.	–	–	–	–	25	[30]

[a]$35/metric ton of dry solids for 91 metric ton/d installations and $283/metric ton of dry solids for 0.9 metric ton/d installation.

MECHANISM OF FREEZE AND THAW CONDITIONING

Improved dewatering characteristics as measured by sand bed and vacuum filter dewatering rates [24,25,26] due to freezing (direct, indirect, or natural) have been attributed to several factors, some of which are:

1. Freezing causes cellular dehydration and enhances flocculation [67,68,69].
2. Freezing of sludges results in the segregation of the solute both dissolved and suspended in a region between the ice crystal domains [29].
3. Freezing destroys the sliminess of biological sludges and results in irreversible damage to the cellular structure.
4. Slower freezing resulted in larger ice-crystal formation and better sludge conditioning [29].
5. Quick thawing has been found to be lethal to bacterial die-off and sludge-water separation.

ECONOMICS OF THE PROCESS

Khan [67,68], while utilizing this process for the waste activated sludge from sewage treatment works, indicated that the cost of conditioning and disposal of sludge, without the cost of dewatering equipment, would be about $20-30 per ton of dry solids. This cost is quite close to the other processes, such as chemical conditioning ($8-25), direct solid freezing ($10-30), wet air oxidation ($30-35), anaerobic digestion ($15-20) and heat treatment ($8-25). A comparison of sludge handling and conditioning costs for various processes is presented in Table II.

DISCUSSION OF RESULTS

The improved settling rates (after freezing and thawing of sludge), reduction in specific resistance, and the better solids production rate can be attributed to the fact that during the process of slow freezing, the freezing ice exerted a pressure of 12 atm. [25]. The dehydration, during slow freezing, allows the internal cell water to move out thus bringing the internal proteins so close as to form S-S bonds thus resulting in an irreversible reaction. This would result in cellular precipitation and floccu- lation upon thawing of the frozen sludge.

In the case of drainage, the initial faster rates were due to the filtration of free water, after which there was a lag time during which the sludge particles compacted together and then released the water held in between them, resulting in a secondary increase in drainage but at rates slower than the initial rate. In the case of air drying, initial moisture reduction was higher because the moisture could move to the surface for evaporation, but after seven days there was no further appreciable improvement. The higher dissolved content in the filtrate and supernatant, after direct slurry freezing and thawing, was probably due to the release of internal cellular salts during dehydration.

SUMMARY

It appears that the direct slurry freezing process is feasible for the conditioning of biological waste sludges generated from the treatment of the waste streams from pulp and paper mills. The conditioning and disposal cost (not including the dewatering equipment cost), using chemicals such as Al^{3+}, is about \$20–30 per ton of dry solids in the sludge. This cost is comparable to or less than the cost of the other conditioning and disposal methods.

One of the major advantages of this process, which the other freezing processes do not have, is that the sludge can be conditioned on a continuous flow basis, and it can be used out in cold as well as hot climates [67,68,69].

REFERENCES

1. Warrick, L. F., "Pulp and Paper Industry Wastes," *Wastes. Ind. Eng. Chem.*, 39, 670 (1947).
2. Ng., K. S., J. C. Mueller, and C. C. Walden, "Ozone Treatment of Kraft Mill Waste," *Water Pollution Control Federation*, 1742 (1978).
3. Barton, C. A., and J. F. Byrd "Joint Treatment of Pulping and Municipal Waste," *Water Pollution Control Federation*, 988 (May 1973).
4. Peterson, R. R., "Design for Criteria for High Purity Oxygen Treatment for Kraft Mill Effluents," *Water Pollution Control Federation*, 2317 (Sept. 1975).
5. Voelkel K. G., and R. W. Deering, "Joint Treatment of Municipal and Pulping Effluents," *Water Pollution Control Federation*, 634 (April 1974).
6. Zimmerman, F. J., "New Waste Disposal Process," *Chemical Engineering*, 65, 117 (1958).
7. Vogler, J. F., and W. Rudolf, "White Water Sludge Drainage Factors, "*Sewage Industrial Wastes*, 23, 699 (1951).
8. Brecht, W., et al., "Studies on the Dewatering of Water—Water Sludges from Paper Mills," *Papier 18*, 741 (1965).
9. Coogan, P. W., "Incineration of Sludges from Kraft Pulp Mill Effluents," *TAPPI*, 44A (1965).
10. Follet, R., and H. W. Gehm, "Manual of Practice for Sludge Handling in the Pulp and Paper Industry," *Technical Bulletin, No. 190*, National Council for Stream Improvement, New York (1966).
11. Andrews, G., et al., "Effluent Sludge Dewatering Practiced by Two Pulp and Paper Mills of Mead Corpn., *TAPPI*, 50, 99A (1967).
12. Swets, D. H., L. Pratt, and E. E. Metcalf, "Thermal Conditioning in Kalamazoo, Michigan," *Water Pollution Control Federation*, 575 (March 1974).
13. Valee, R. P., et al., "Field Consolidation of High Ash Papermill Sludge," *Journal Geotech. Eng. Div., Proc. ASCE*, 100, G.T3, 309 (1974).
14. Scherlev, A., "Dewatering of Biological Chemical Paper Mill Wastewater Sludge with the 'System Hiller' KHD Centrifuge," *Water Resources Abs.*, 7, 12, W74-06392 (1974).
15. Rusenfeld, A. S., et al., "Pulp and Paper Mill Sludge Disposal by Combustion," *TAPPI*, 56, 10, 97 (1973).
16. Oledale, J., "Use of the Multiroll Press for Dewatering Clarifier Sludge," *Paper Trade Jour.*, 158, 1, 26 (1974).
17. Strittmatter, G., "Removal of Residual Water Sludges," *Water Resources Abs.*, 7, 10, W74-05263 (1974).
18. Hurkin, M., et al., "Bacterial Protein from Pulp Paper Mill Sludge," *TAPPI*, 57, 131 (1974).
19. Kozich, R. F., "Clarification of White Water by Vacuum Filtration," *Water Resources Abs.*, 7, 23, W74-12415 (1974).
20. Strittmatter, G., "The Treatment and Removal of Wastewater Residual Sludges in the Paper Industry," *Water Resources Abs.*, 7, 23, W74-12417 (1974).
21. Mathian, M., "Residual Sludge Treatment Generalities and Applications in the Paper Industry," *Chem. Abs.*, 81, 24, 158280Y (1974).
22. Hukuo, K., et al., "Artificial Light Weight Aggregate from the Waste Mud of Silica Sand Factories and Waste Fluid of Pulp Factories," *Chem. Abs.*, 80, 10, 52102F (1974).
23. Babbitt, H. E., and H. E. Schlenz, "The Effect of Freeze Drying on Sludges," Illinois Engineering Experiment Station, *Bulletin No. 198*, p. 48 (1929).

24. Sewerage Commission City of Milwaukee, "Evaluation of Conditioning and Dewatering Sewage Sludge by Freezing," *Water Pollution Control Research Series*, 11010, EVE 01/M1.
25. Clements, G. S., R. J. Stephenson, C. J. Regan, "Sludge Dewatering by Freezing with Added Chemcials," *Journal and Proceedings Institute of Sewage Purification Journal, Part 4*, p. 318 (1950).
26. Cheng, C., D. M. Updegroff, and L. W. Ross, "Sludge Dewatering by High Rate Freezing at Small Temperature Differences," *Environmental Science and Technology*, Vol. 4, p. 1145 (1970).
27. Penman, A., and D. W. Vanes, "Winnipeg Freezes Sludge, Slashes Disposal Cost," *Sanitary Engineering, ASCE.*, Vol. 43, p. 65 (1973).
28. Rush, R. J., and A. R. Stickney, "Natural Freeze-Thaw Sewage Sludge Conditioning and Dewatering," Canada Environmental Protection Service Report, *EPS 4-WP-79-1* (January 1979).
29. Ezek, G., Ho-Ming Tong, and Carl C. Gryte "On the Mechanism of Dewatering Colloidal Aqueous Solutions by Freeze-Thaw Processes," *Water Research*, Vol. 14, pp. 1079 (1980).
30. "Sludge Treatment and Disposal—Process Manual Design," *EPA-625/1-79-011*, U.S.E.P.A., Cincinnati, Ohio (1979).
31. Benn *et al.*, "Sludge Concentration by Freezing" *Water Sewage Works*, 112, 401–406 (1965).
32. Benn *et al.*, "Disposal of Sludge by Freeze-Thaw Processes," *Filt. Separation*, 6, 383–389 (1969).
33. Baskerville, R. C., "Freeze and Thaw as a Technique for Improving the Dewateribility of Aqueous Suspensions," *Filt. Separation*, 141–144 (1971).
34. Katz, W. J., et al., "Freeze Methods for Conditioning of Activated Sludge," Presented at the 16th Southern Water Resources and Pollution Control Conference, Duke University, pp. 119–128 (April 1967).
35. Gruber, F., "Water Removal from Presedimentation and Filtration Sludges," *German Patent*, 1, 809, 772 (1970).
36. Wood, C. W., et al., "Dewatering of Sludge," *British Patent*, 1, 182, 019 (1970).
37. Shimada, T., "Freezing Dehydration Process of Organic Sludges," *Chem. Abs.*, 87, 24, 18888/m (1977).
38. Rush, R. J., "Natural Freez-Thaw Sewage Sludge Conditioning and Dewatering," *Canada Envron. Prot. Ser. Tech. Dev. Rep. EPS 4-WP-79-1* (1979).
39. Wilhelm et al., "Freeze Treatment of Alum Sludge," *Journal American Water Works Assn.*, 66, 312 (1976).
40. Oyama, M., "Freezing Dehydration of Sludge," *Chem. Abs.*, 85, 18, 13 0091W (1976).
41. Murakamai, M., et al., "Effect of Addition of Some Flocculants on the Freezing of Treatment Sludges from Sewage Treatment Plants," *Chem. Abs.*, 85, 6, 36839z (1976).
42. Yamamoto, S., "Freeze-Dry Treatment Sludge," *Chem. Abs.*, 84, 4, 21810p (1976).
43. Nagel, O., et al., "Conditioning and if Necessary, Dewatering Sewage Sludge and Similar Dewaterable Sludges," *Chem. Abs.* 85, 18, 130114f (1976).
44. Seya, M., et al., "Dewatering of Industrial Sludges," *Chem. Abs.*, 85, 2, 10066v (1976).
45. Matsumara, T., et al., "Dewaterability of Municipal Sewage Sludge by Freeze-Thawing Process," *Chem., Abs.*, 80, 16, 871895 (1974).
46. Abson, J. W., "Dewatering of Biological Sludges," *Chem. Abs.*, 81, 14, 82105x (1974).
47. Foster, D. L., et al., "Wastewater Sludges—A Fertilizer Substitute," *Pub. Works*, 107, 8, 79 (1976).
48. Alter, J. H., "Chicago's Program for Using Sludge to Reclaim Land," *Compost Sci.*, 17, 4, 22 (1976).
49. Hubbs, S. A., et al., "Optimization of Sludge Dewaterability in Sludge Disposal Lagoons," *Journal Amer. Water Works Assn.*, 66, 658 (1974).
50. Westerhoff, G. P., et al., "Water Treatment Plant Wastes Disposal," *Journal Amer. Water Works Assn.*, 66, 319, 378, 441 (1974).
51. Troemper, A. P., "The Economics of Sludge Irrigation, *Proc. Natl. Con. on Municipal Sludge Management, Information Transfer, Inc.*, Washington, DC 115 (1974).
52. Miner et al., "Biological Sludge Dewatering Practices in the Pulp and Paper Industry," *Proc. 30th Ind. Wastes Conf.* Purdue Univ., Ann Arbor Sci., Publ. Inc., Ann Arbor, MI 601 (1977).
53. Bauer, J., "Modern Separation Processes for the Elimination of the Wastewater and Sludge Problem in Pulp and Paper Manufacture," *Chem. Abs.*, 86, 14, 95417d (1977).
54. Koziorowski, B., et al., "Properties of Sludges from Aerated Wastewater Ponds," *Chem. Abs.*, 87, 12, 80220v (1977).
55. Ewing L. J., et al., "Total Cost of Heat Treatment of Wastewater Sludges," *Proc. 3rd Natl. Conf. on Sludges Management and Utilization, Information Transfer, Inc.*, Rockville, Md., 179(1977).
56. Tera, T., et al., "Designing Procedures for Sludge Treatment Apparatus Using Freeze Separation System," *Chem. Abs.*, 87, 14, 106455p (1977).

57. Miner, R., "Experience with Belt Filter Presses in the Pulp and Paper Industry," *Journal Water Pollution Control Federation*, 52, 2389 (1980).
58. Möller, U.K., "Reaction: Chemical Conditioning" in "Sludge characteristics and Behavior," J. B. Carberry, and A. J. Englande, Jr. (Eds.), *NATO ASI Ser. E, No. 66*, Martinus Nijhoff Publications, The Hague, Neth. (1983).
59. Eden, G. E., "Modern Trends in Sludge Management: Sludge Conditioning," *Water Science Technology*, 15, 37 (1983).
60. Cornier, J. C., et al., "Recent Developments in Sludge Conditioning and Dewatering," *Water Science Technology*, 15, 49 (1983).
61. Katz, W. J., et al., "Freezing Method for Conditioning Activated Sludge, *16th Southern Water Resources and Pollution Control Conference*, Duke Univ. (April, 1967).
62. Mohlman, F. W., "The Sludge Index," *Journal Sewage Works*, 6, 119 (1934).
63. Nebiker, J. H., "Drying of Wastewater Sludge in the Open Air," *Journal Water Pollution Control Federation*, 39, 608 (1967).
64. Nebiker, J. H., et al., "An Investigation of Sludge Dewatering Rates," *Journal Water Pollution Control Federation*, 41, R255 (1969).
65. Coakley, P., and B. R. S. Jones, "Vacuum Sludge Filtration (I). Interpretation of Results by the Concept of Specific Resistance," *Sewage and Industrial Wastes*, 28, 963 (1972).
66. APHA, AWWA, and WPCF, *Standard Methods for the Examination of Water and Wastewater*, 13th edition. American Public Health Association, Washington, DC (1971).
67. Khan, M. Z. Ali, "Principles and Techniques for Conditioning of Waste Activated Sludge by Direct Slurry Freezing," *Ph.D. Dissertation*, Virginia Polytechnic and State Univ., Blacksburgh, Virginia, U.S.A. (1974).
68. Khan, M. Z. Ali, C. W. Randall, and N. T. Stephens, "Direct Slurry Freezing of Waste Activated Sludge," *Virginia Water Resources Center Bulletin, No. 94* (1976).
69. Randall, C. W., "Butane is Nearly 'Ideal' For Direct Slurry Freezing," *Water and Wastes Engineering*, p. 43 (March 1978).
70. Williams, T. C., "Phosphorus is Removed at Low Cost," *Water and Wastes Engineering*, 13, 11, 52(1976).
71. Nelson, O. F., "Operational Experience with Filter Pressing," *Water Pollution Control Federation—Deeds and Data* (March 1978).
72. Bizjack, G. J., *et al.*, "Wausau Solves Dual Problem Using Filter Press," *Water and Wastes Engineering*, p. 28 (Feb. 1978).

52 EFFECTS OF PAPERMILL WOOD ASH ON CHEMICAL PROPERTIES OF SOIL

Lewis M. Naylor, Senior Research Associate
Department of Agricultural Engineering
Cornell University
Ithaca, New York 14853

Eric J. Schmidt, Senior Environmental Engineer
Georgia-Pacific Corporation
Atlanta, Georgia 30348

INTRODUCTION

The pulp and paper industry has used wood waste as fuel on a wide scale since the 1960's and has recognized that wood can be an economical and readily available source of energy. However, the ash residual from such commercial wood burning facilities can also be a valuable source of crop nutrients and lime for agricultural uses. This report discusses results and provides recommendations for the use of wood ash in agriculture.

Pulp and paper industries utilize enormous quantities of timber resources annually for production of paper products. Since a portion of the tree is unusable in production operations, these residuals such as the bark from production operations must be managed through other methods. Burning such residues known as hog fuel in wood fired boilers for production of steam and electricity is economical, environmentally sound, and energy conserving. However, ash is generated from burning the hog fuel. Agricultural use of this ash along with the fly ash from air pollution control can represent the final step in this resource recovery program.

Wood ash has been known to be an important source of potash and lime for many years [1,2], although a recent comprehensive reference makes little reference to it [3]. In the 1938 Yearbook of Agriculture [2], wood ash was suggested to "rate as a potash material with a comparatively high lime content, some phosphoric acid and magnesium, and small amounts of other elements." Unleached hardwood ashes were suggested to contain upward of 6% potash in the form of carbonate, 2% phosphoric acid and 30% lime. Thus, with the increasing use of wood as a fuel and the subsequent need to dispose of the ash, it is important to reconsider the use of wood ash for agricultural purposes and to put such use on a quantitative basis.

The overall objective of this research was to quantify through laboratory and field studies the agronomic value of paper mill wood-derived ash as a fertilizer and an alternative liming material. Specific objectives of this research were to: 1) examine the characteristics of the wood ash in terms of macronutrient content (N, P, K, Ca, Mg) and trace mineral content; 2) assess the availability of these crop nutrients in the soil where the wood ash is incorporated; and 3) quantify the liming value of the wood ash when incorporated into soil.

MATERIALS AND METHODS

The experimental lime equivalences of the wood ash and the effects on the extractable nutrient content of soil were investigated initially in soil incubation studies.

Wood ash samples used in the research were supplied by the Lyons Falls Paper Company (formerly a Georgia-Pacific Corp. facility), Lyons Falls, NY. The limestones used as controls were Limecrest Pulverized Limestone (Limestone Products Corp., Sparta, NJ) and Modern Rotary Kiln Hydrated Lime (Millard Lime and Stone, Annville, PA), and potash (Muriate of Potash, Agway). The soils used in the incubation studies were Mardin silt loam (coarse, loamy, mixed mesic Typic Fragiocrept) and Burdett silt loam (fine, loamy, mixed, mesic Aeric Ochraqualfs).

The experimental design was two soils by three liming/fertilizer treatments by six application rates with three replications for a total of 108 pots. The entire quantity of soil to be used in each experiment

437

was screened (1 mm stainless steel) and homogenized for 30 minutes in a large double shell mixer. Wood ash and the commercial limestone materials were mixed with 3.0 kg of each soil in amounts of 0, 3.0, 6.0, 12, 24 and 48 g, dry basis. These treatment rates were approximately equivalent to 0, 2.24, 4.5, 9.0, 17.9 and 35.9 metric tons/hectare (0, 1, 2, 4, 8 and 16 tons/acre). The commercial potash fertilizer material was added to 3.0 kg of each soil in amounts of 0, 0.10, 0.20, 0.40, 0.80 and 1.60 g. These treatment rates are approximately equivalent to 0, 34, 68, 135, 270 and 540 kg/ha of potassium. Each treatment was mixed in bulk for 30 minutes in a small double shell mixer, divided into three 1 kg replicate samples, and each sample placed into a 20 cm plastic pot with drain holes. Pots were incubated at 25°C for 60 days with periodic watering to simulate wet/dry cycles and to permit physical/chemical reactions with the agricultural soils. At the end of the incubation period, soils were sampled and 10 g samples were analyzed individually for water pH and extractable nutrients using Morgan's solution [4,5].

The field study was conducted in Lewis County, New York on a Hartland very fine sandy loam (coarse silty, mixed mesic Dystric Entrochrept) of 2 to 6 percent slopes. The field had lain fallow for two years following corn. In preparation for the study, the soil was disced and weeds were killed using Round Up. Plots 6 m by 244 m received 0, 9.4, 11.7. 17.3, 22.6, 31.2, and 49.8 tonnes/ha (dry basis). The ash (72.4 ± 4.0% solids) was applied on June 24, 1985, by manure spreader and disced in. Applications rates were determined from 5 trials per plot in which the ash was collected in pans placed in the spreading lane and weighed. The weight of the ash per pan was averaged and converted to an application rate in tonnes/ha.

Soil samples were collected from the plots prior to ash application and again on October 10, 1985, about 4 months following treatment. These samples, each a composite of 6 to 8 cores (0–15 cm), were collected at 60 m intervals within each plot. Analysis of variance indicated no significant difference in extractable nutrients between plots prior to ash application.

Ash, limestone, and potash samples were analyzed for 1:1 water pH, and total (HNO_3/HCl digestion) Cd, Cr, Cu, Ni, Pb, Zn, Fe, Al, Mn, Ca, Mg, Na, and K by atomic absorption and P by colorimetric determination. Nitrogen was determined by micro-kjeldahl [6]. Soil samples were analyzed for Morgan's solution exchangable Ca, Mg, K, Fe, Mn, and Al [4].

Data were examined statistically for outliers using the method of Dixon [7], and means were compared using analysis of variance and regression procedures. Statistical differences between treatment means were evaluated using Duncan's New Multiple Range Test [8].

RESULTS AND DISCUSSION

Composition of Wood Ash, Agricultural Limestones and Potash Fertilizer

The wood ash samples used in the laboratory and field experiments contained, 27.0 and 12.8% Ca, 3.08 and 1.66% K, 0.79 and 0.33% P, and 1.55 and 0.81% Mg, respectively (Table I). Analyses of two other ash samples in 1984 (data not shown) indicated that the ash used in the field study was probably more typical of ash produced at this mill. This difference is probably due to variation in the types of wood burned in the boilers. It is well known, for example, that hardwoods generally contain a greater amount of Ca, Mg and K than softwood [9]. Both sources of ash contained less Ca and Mg than the limestone, but at least 10 times as much K. The ash would be rated as about an 0-1-2 ($N-P_2O_5-K_2O$) fertilizer plus lime.

Ash samples also showed differences in mineral content, although both contained acceptable concentrations of non-essential metals. The Cd contents were 7.9 and 4.2 mg/kg, compared to 0.7 mg/kg for agricultural limestone. Cu and Zn, both plant nutrients, were higher in the ash than in the limestone, whereas the Ni and Pb contents of the ash were similar to those in limestone. The potash fertilizer (muriate of potash, i.e., KCl) had a potassium concentration of 45.1%. This is equivalent to 54.1% as K_2O. The pH of the sample, 9.2, would suggest that a portion of the potassium may be in the form of an oxide.

Fineness of the ash (duplicate, 250 g samples) from the wood fired boilers was determined by sieve analysis (Table II). These samples were collected over a period of 5 months and analyzed individually. Results showed that 65 ± 10% of the ash would pass a 100 mesh sieve. This fraction would be effective as a neutralizing material due to its high pH and would react rapidly with soil acidity due to its fineness. The material collected on the 20 and 60 mesh sieves was virtually all unburned carbon (charcoal). This fraction consisted of about 25% of the total sample by weight, but about 80% by volume. It is also apparent from Table II that the principal source of fineness variability is in the fraction coarser than 20 mesh, i.e., the unburned carbon.

Table I. Chemical Composition of Wood Ash, Commercial Agricultural Limes and Commercial Potash Fertilizers

	Composition[a]				
Parameter	Wood Ash[b] 1 $5-12-84$	Wood Ash[b] 2 $6-24-85$	Agricultural Limestone $CaCO_3$	Hydrated Lime CaO	Potash Fertilizer KCl
	%				
Total-N	0.05	0.12	0.01	0.02	0.02
Organic-N	0.05	0.12	0.01	0.02	0.02
NH_4-N	0	0.00	0	0	0.01
P	0.79	0.33	0.06	0.05	0.05
K	3.08	1.66	0.13	0.14	45.08
Ca	27.00	12.80	31.40	55.38	0.04
Mg	1.55	0.81	5.09	0.80	0.13
Na	0.27	0.20	0.07	0.05	1.41
Al	1.59	1.69	0.21	0.26	0.00
Fe	1.11	1.32	0.29	0.11	0.01
	mg/kg				
Cd	7.9	4.2	0.7	0.2	0.2
Cr	21.1	9.1	6.0	10.0	2.0
Cu	90.3	40.0	10	12.0	2.0
Mn	12,700	6,600	453	41.1	5.0
Ni	49.1	11.6	20.0	34.0	10.0
Pb	72.2	38.0	55	45.1	21.0
Zn	381	200	113	4.0	2.0
pH	12.7	11.9	9.9	12.5	9.2
Total Solids, %	99.71	72.4	100	99.87	99.83

[a]Means of duplicate analyses.
[b]Wood ash samples were provided by the Lyons Falls Paper Co., Lyons Falls, NY. Sample 1 was used in the laboratory study and Sample 2 in the field study.

Chemical analysis of the ash by sieve size (Table II) suggested that screening out the charcoal would enhance the analysis of the material as a fertilizer and as an alternative liming material. The mean calcium content of the material finer than 60 mesh was 17.7% relative to 1 to 3% for the coarser material.

Laboratory Study

Soil pH. The wood ash contained 270 kg Ca (670 kg expressed as $CaCO_3$) per tonne, and had a pH of about 12. These characteristics suggested good potential of the material as an alternative liming material.

The results of the incubation study indicated that wood ash incorporated into the soils improved soil pH at all application rates (Table III). The pH of the unlimed Burdett soil (4.8) was increased to 6.1 with an ash addition equivalent to 17.1 tonne/ha and to 7.0 with an ash addition of 35.9 tonne/ha.

Table II. Physical and Chemical Analysis of Wood Ash by USS Sieve Size

USS Sieve	Ash Analysis, %				Sieve Analysis
	P	Ca	Mg	K	
< 20[a]	0.06	1.1	0.4	1.1	13 ± 10
20 − 60	0.12	3.2	0.3	1.8	12 ± 1
60 − 100	−	−	−	−	10 ± 2
> 60	0.64	17.7	1.3	2.9	75 ± 11
> 100	−	−	−	−	65 ± 10

[a]Material retained on screen was charcoal.

Table III. Effect of Wood Ash and Limestone on pH of Two Soils (Laboratory Study)

		Application Rate, Tons/Hectare, Dry Basis					
		0	2.24	4.5	9.0	17.9	35.9
Sample	Soil			Soil pH			
Wood Ash	Mardin	5.1	5.3	5.6	6.0	6.6	7.2
Ground Limestone	Mardin	5.1	5.6	5.9	6.5	6.9	7.2
Wood Ash	Burdett	4.8	5.0	5.2	5.5	6.1	7.0
Ground Limestone	Burdett	4.8	5.1	5.5	6.1	6.7	7.3

For the Mardin soil (untreated pH = 5.1), addition of 17.9 and 35.9 tonne/ha increased soil pH to 6.6 and 7.2, respectively. A soil pH of 6.2 to 7.0 is desirable for most crops [10,11]. For routine on-farm use, however, it is important to standardize the neutralization potential of the ash relative to other common liming materials.

Liming materials may be compared based on their effective neutralizing value (ENV). The ENV is the percent effectiveness of a particular limestone relative to a standard limestone (pure $CaCO_3$) with an ENV of 100. The ENV of the wood ash was estimated from the results of the soil incubation study in which a known limestone was used as a comparison. The soil pH achieved was a linear function of the logarithm of the application rate for both materials. This relationship follows from the definition of pH: the negative logarithm of the hydrogen ion concentration. The regression equations developed from the results were:

Mardin Soil

$$pH = 1.59 \text{ Log } R_{ash} + 4.62 \qquad r^2 = 0.98 \text{ ash}$$
$$pH = 1.40 \text{ Log } R_{lime} + 5.09 \qquad r^2 = 0.99 \text{ limestone}$$

Burdett Soil

$$pH = 1.63 \text{ Log } R_{ash} + 4.21 \qquad r^2 = 0.92 \text{ ash}$$
$$pH = 1.86 \text{ Log } R_{lime} + 4.37 \qquad r^2 = 0.95 \text{ limestone}$$

where: R_{ash} = ash application rate, tonne/ha
R_{lime} = limestone application rate, tonne/ha

From the regression equations, it can be shown that to achieve a pH of 6.2 for the Burdett soil would require 9.7 tonne/ha of limestone (ENV 83) and 17 tonne/ha of the wood ash. Thus, an ash application of 17 tonne/ha would provide an equivalent soil pH neutralization as 9.7 tonne/ha of the ground limestone (ENV 83). From this comparison, the ENV of the ash was estimated:

$$ENV \text{ (ash)} = ENV \text{ (control lime)} \times \frac{\text{control lime application to achieve pH 6.2}}{\text{ash application to achieve pH 6.2}}$$

For the limestone (ENV = 83) used as a control in this comparison, the wood ash was estimated to have an ENV of:

$$ENV \text{ (wood ash, sample 2)} = 83 \times \frac{9.7}{17} = 47$$

To achieve a pH of 6.5 on the same soil would require 14 tonne/ha of the limestone or about 26 tonne/ha of the wood ash resulting in an experimental ENV of about 45. Similar ENV results were obtained using the Mardin soil in the experiment. The experimental ENV was estimated at 52 to 55 for achieving soil pH of 6.2 and 6.5, respectively. Thus for this sample of wood ash, the ENV was estimated to be in the range of 45 to 55.

Crop nutrients. The concentration of soluble nutrients in the Morgan's solution extract was considered to be that available for plant uptake in northeastern soils (6,7). Results are shown in Table IV.

Extractable potassium was a linear function of the amount added to the soil. The regression slopes suggest that 18 to 35% of the added potassium from wood ash would be available, contrasting with 63 to 76% availability of the potassium in the commercial potash fertilizer (Table V). This appears to indicate that K derived from the ash is less available to plants than that from the commercial potash fertilizer. However, this may be related to the lower extractability of K by Morgan's solution as pH increases [12].

Table IV. Effects of Wood Ash on Soil Chemical Properties, Laboratory Study

	Application Rate, Tonnes/Hectare, Dry Basis					
	0	2.24	4.5	9.0	17.9	35.9
	Extractable Nutrients, kg/ha[a]					
Burdett Soils						
P	0.6±0 d	1.6±0 d	1.7±0.97 c	2.2±0 b,c	3.0±0.65 b	4.5±0 a
K	100±1.9 f	128±2.4 e	145±1.1 d	165±3.4 c	216±6.0 b	307±5.7 a
Mg	133±5 e	164±16 d	166±6.0 d	191±3.9 c	245±18 a	220±3.4 b
Ca	1520±45 e	1980±120 d	2190±17 d	2720±79 c	4060±247 b	5760±88 a
Fe	50±5.8 a	45±2.6 b	34±64 c	24±0 d	12±.65 e	10±.65 e
Al	356±27 f	327±7.2 e	273±6.2 d	217±2.4 c	139±3.9 b	94±2.4 a
Mn	16±1.4 d	15±1.1 c	12±.65 b	10±0 a	15±1.4 c	28±1.7 e
Mardin Soils						
P	6.7±0.6 d	7.9±0 c	7.9±.6 c	9.0±0 b	10±.6 b	11±0 a
K	166±20 e	179±5.2 e	213±3.6 d	268±5.6 c	363±10 b	547±15 a
Mg	162±13 d	173±16 c,d	202±4.7 c	235±25 b	256±7.8 a,b	275±22 a
Ca	1560±149 e	1830±153 d,e	2250±84 d	3010±92 c	4320±174 b	6020±620 a
Fe	102±11 a	83±6.7 b	74±3.9 b	52±1 c	31±2 d	22±2 d
Al	226±17 a	199±10 b	179±5.5 c	140±1.1 d	103±3.4 e	89±11 e
Mn	35±5.7 b,c	27±4.5 c	28±3.9 b,c	26±.65 c	38±3.9 b,c	53±11 a

[a]Mean ± std. dev. of 3 replicates. Means in rows followed by the same letter are not statistically different, p = 0.05

Table V. Regression Slopes of Extractable Nutrients and Soil pH in Ash Amended Soils.

		Soil		
		Laboratory		Field
Material	Parameter	Mardin	Burdett	Hartland
		Regression Slope, R_p[a]		
Ash	K	0.35**[b]	0.18**	0.39**
Potash	K	0.76**	0.63**	—
Ash	Ca	0.44**	0.47**	0.62**
Limestone	Ca	0.49**	0.58**	—
Ash	Mn	0.052*	0.029 N.S.	0.068**
		Regression Slope, R_{pH}[c]		
Ash	Al	–66*	–121**	–121*
Ash	Fe	–38**	–19*	–5.6**
		Regression Slope, R[d]		
Ash		1.59**[2]	1.63**	1.24**
Limestone		1.40**	1.86**	—

[a]R_p = Addition rate of parameter in kg/ha.
[b]Significance levels: **p = 0.01, *p = 0.05, N.S. p>0.05
[c]R_{pH} = Soil pH
[d]Logarithm of ash addition, dry tonne/ha

Table VI. Effects of Wood Ash on Soil Chemical Properties, Field Study

Hartland Soil	Application Rate, Tonnes/Hectare, Dry Basis						
	0	9.4	11.7	17.3	22.6	31.2	49.8
				Soil pH			
pH	6.1 ± 0.3	6.4 ± 0.04	6.6 ± 0	6.7 ± 0.2	6.8 ± 0.2	7.1 ± 0.1	7.4 ± 0.06
				Extractable Nutrients, kg/ha[a]			
P	3.2 ± 1.2	4.7 ± 1.1	2.6 ± 0.7	5.0 ± 1.6	3.4 ± 0	5.2 ± 0.6	5.6 ± 1.1
K	105 ± 21	192 ± 21	164 ± 25	274 ± 65	226 ± 43	339 ± 141	417 ± 51
Mg	65 ± 19	109 ± 16	98 ± 21	137 ± 29	133 ± 32	154 ± 45	189 ± 7.8
Ca	1750 ± 626	2750 ± 500	2660 ± 563	3040 ± 775	3580 ± 971	4320 ± 1210	5740 ± 667
Fe	27 ± 6.6	24 ± 3.3	22 ± 4.6	24 ± 7.4	25 ± 2.2	21 ± 1.7	22 ± 7.6
Al	546 ± 182	325 ± 69	376 ± 243	360 ± 238	297 ± 120	322 ± 55	231 ± 35
Mn	7.4 ± 3.5	15 ± 4.0	8.6 ± 0.6	18 ± 4.8	17 ± 4.7	23 ± 8.5	29 ± 7.4

[a]Mean $\pm$ std. dev. of 5 replicates.

When the wood ash is used as a liming material at 20 tonne/ha to improve soil pH, extractable potassium (25% extractable) would be increased by about 150 kg/ha. Based on regression results the 150 kg/ha increase would be equivalent to that supplied by 215 kg/ha of potash fertilizer K (70% extractable) or to about 355 kg/ha of commercial potash fertilizer containing 60% K_2O. Thus, the wood ash used as an alternative liming material would provide important amounts of potassium. Extractable Ca and Mg was a linear function of the addition rate of the mineral. The regression slopes for extractable Ca were similar for the limestone and the ash, both rich sources of Ca (Table V).

Aluminum, Fe, and Mn solubility are moderated by soil pH. This moderating effect is important because these minerals can be phytotoxic, although Mn and Fe are also essential micronutrients. The wood ash contained about 1.6% Al, 1.2% Fe, and 1.0% Mn. Hence ash application increased the total concentration of these minerals in the soil. As soil pH increases, the solubility of Al, Fe, and Mn decreases [13]. Since addition of the ash to soil increases soil pH, separation of the effects of pH (decreasing solubility) and the addition of the mineral (increasing the amount in the soil) is difficult.

An inverse relationship between soil pH and extractable Al and Fe was observed and is evident statistically from the regression slopes (Table V), demonstrating the powerful effect of soil pH on extractable Al and Fe even though up to 570 kg/ha of Al and 398 kg/ha of Fe was added. Thus, while ash application added Al and Fe to the soil, the resulting higher soil pH was the dominant factor affecting extractability of these minerals.

The complex interaction of soil pH and amounts of Mn added to soil was evident. Soil pH controlled Mn extractability until 114 kg/ha of the mineral had been added (about pH 5.5 to 6.0). and thereafter extractable Mn increased and appeared to be controlled by Mn additions.

Field Study Results

Soil pH. The mean pH of the untreated soils (6.1) in the field experiment was increased by incorporation of the ash (Table VI). The mean pH of the treated plots varied from 6.4 at the lowest treatment rate (9.4 tonnes/ha) to 7.4 at the highest treatment rate (49.8 tonne/ha). Regression analysis indicated that soil pH was a linear function of the logarithm of the ash application rate:

$$\text{Soil pH} = 1.24 \text{ Log R} + 5.21 \qquad r^2 = 0.85 \qquad (1)$$

The regression slope is smaller than those obtained in the laboratory study (Table V) because of the lower Ca concentration in the ash used in the field study.

The regression slope suggests that an ash application of 10 tonne/ha will improve the pH of this soil to about pH 6.5. However, an application of 100 tonne/ha would increase the pH only an additional 1.24 units to about 7.7. Thus, applications moderately in excess of that recommended from a preapplication soil test would not be expected to have a deleterious effect on soil pH and consequently on availability of micronutrients.

Crop nutrients. Incorporation of the ash substantially improved extractable potassium in the treated soil. Extractable potassium in the untreated soils averaged 105 kg/ha. Incorporation of the ash improved the potassium to as high as 417 kg/ha at the highest application rate. While substantial variation in extractable potassium was noted within treatments, regression analysis indicated a linear relationship between extractable potassium and applied potassium from the ash. The regression equation suggested that 39% of the applied potassium was extractable using Morgan's solution:

$$K_{extr}, kg/ha = 0.39 K_{appl}, kg/ha + 118 \qquad r^2 = 0.75 \tag{2}$$

An ash application of about 28 tonne/ha would adjust this soil to pH 7, appropriate for an alfalfa seeding. Such an application would provide about 460 kg/ha of potassium, or nearly 180 kg/ha of extractable potassium for crop use.

Changes in extractable Al and Fe in the field study were similar to those observed in the laboratory incubation experiment (pH was the dominant factor affecting extractable Al and Fe). Although up to 841 kg/ha Al and 657 kg/ha Fe were applied with the ash in the field experiment, extractable amounts were decreased with increasing soil pH.

$$Al_{extr}, kg/ha = -121 \text{ soil pH} + 1130 \qquad r^2 = 0.12 \tag{3}$$
$$Fe_{extr}, kg/ha = -5.6 \text{ soil pH} + 61 \qquad r^2 = 0.19 \tag{4}$$

In contrast with the results of the laboratory study, a strong relationship between extractable Mn and applied Mn was evident from the field data. Ash application increased extractable Mn from 6.6 kg/ha(mean) in the untreated soil to 26 kg/ha (mean) in the highest treatment (328 kg/ha Mn applied). The regression relationship suggests that about 7% of this applied Mn was extractable.

$$Mn_{extr}, kg/ha = 0.068 Mn_{appl}, kg/ha + 8.2 \qquad r^2 = 0.65 \tag{5}$$

Thus, for this soil the applied Mn was the predominant factor affecting Mn extractability throughout the range of ash application.

CONCLUSIONS

The results of the field study supported those obtained in the soil incubation experiments in the laboratory. Wood ash incorporated into the soil improved soil pH. Soil pH was a function of the logarithm of the ash application rate. The regression slopes from the field and laboratory studies (Table V) differed slightly (about 1.2 versus 1.6, respectively) and this difference is related to the higher concentration of Ca in the ash used in the laboratory. An experimental ENV of the ash used in the laboratory study was estimated to be about 50.

Extractable K and Ca were linear functions of the application rate of the mineral (Table V). Extractable K was about half of that for commercial potash fertilizer. It was felt that the lower values may have been associated with the higher pH, produced with ash addition. Extractable Ca was similar for the ash and the limestone. Results in the field were similar to those obtained in the laboratory. Extractable Al and Fe were strongly influenced by soil pH as indicated by the regression slopes in both the field and the laboratory (Table V). This relationship existed even though ash applications supplied large amounts of Al and Fe to the soil. Mn relationships were less clear. In the laboratory ash addition reduced Mn extractability initially, but at the higher application rates Mn extractability increased. In the field, extractable Mn was increased throughout the range of ash applications. Therefore, results of this research suggest that the paper mill wood ash can supply agronomically important amounts of plant nutrients and can also serve as an alternative liming material.

ACKNOWLEDGMENTS

The authors thank Al Hoffman, Rich Weber, and Dan McGough of the Lyons Falls Paper Co., Lyons Falls, NY for their invaluable assistance in providing time and equipment during the course of this research. Jim Johnson is thanked for his planning and supervision of the field work. A special thanks also goes to Ken Mayhew on whose farm the field study was conducted. Financial support of this research was provided by George-Pacific Corp., Atlanta, GA.

REFERENCES

1. Agee, A., *Right Use of Lime in Soil Improvement*, Orange Judd Co., New York (1919).
2. United States Department of Agriculture, "Soils and Men," USDA Yearbook of Agriculture, pp. 518–519 (1938).
3. Barber, S. A., "Liming Materials and Practice," R. W. Pearson and F. Adams (ed.), *Soil Acidity and Liming*, Am. Soc. of Agron., Madison, WI, pp. 125–160 (1967).
4. Greweling, T. and M. Peech, "Chemical Soil Tests," *Department of Agronomy Bulletin 960*, Cornell University, Ithaca, NY (1965).
5. Lathwell, D. J. and M. Peech, "Interpretation of Chemical Soil Tests," *Department of Agronomy Bulletin 995*, Cornell University, Ithaca, NY (1964).
6. Doty, W. R., M. C. Amacher and D. E. Baker, "Manual of Methods," Soil and Environmental Chemistry Laboratory, Department of Agronomy, *Information Report 121*, Institute for Research on Land and Water Resources, The Pennsylvania State University, University Park, PA (1982).
7. Snedecor, G. W. and W. G. Cochran, *Statistical Methods*, The Iowa State University Press, Ames, IA (1980).
8. Steel, R. G. D. and J. H. Torrie, *Principles and Procedures of Statistics with Special Reference to the Biological Sciences*, McGraw-Hill Book Co., Inc., New York (1960).
9. Ellis, E. L., "Inorganic Elements in Wood," W. A. Côté, Jr. (ed.), *Cellular Ultrastructure of Woody Plants*, Syracuse University Press, Syracuse, NY, pp. 181–189 (1965).
10. *Cornell Field Crops Handbook*, New York State College of Agriculture and Life Sciences, Cornell University, Ithaca, NY (1978).
11. Lathwell, D. J. and T. W. Scott, "A Study of Soil Fertility on New York Farms," *Cornell Extension Bulletin 1063*, New York State College of Agriculture and Life Sciences, Cornell University, Ithaca, NY (1966).
12. Naylor, L. M. and J. C. Dagneau, "Cement Kiln Dust—A Resource Too Valuable to Waste," I. J. Kugelman (ed.), *Toxic and Hazardous Wastes, Proc. 17th Mid-Atlantic Ind. Waste Conf.*, Technomic Publ. Lancaster, PA, pp. 353–366 (1985).
13. Arkin, G. F. and H. M. Taylor, "Modifying the Root Environment to Reduce Crop Stress," *ASAE Monograph No. 4*, American Society of Agricultural Engineers, St. Joseph, MI (1981).

Section Five
DAIRY WASTES

53 RBC KINETICS IN TREATING DOMESTIC AND INDUSTRIAL DAIRY WASTEWATER UNDER LOW AND HIGH ORGANIC LOADING CONDITIONS

Rao Y. Surampalli, Senior Environmental Engineer
Water Management Division
U.S. Environmental Protection Agency
Kansas City, Kansas 66101

E. Robert Baumann, Professor of Civil Engineering
Department of Civil Engineering
Iowa State University
Ames, Iowa 50011

INTRODUCTION

Fixed-film biological processes found early application in wastewater treatment, but their use declined with the advent and wide scale use of the activated sludge process. However, with the development of plastic media in the early 1960's, interest in fixed-film processes has grown once again. During the late 1960's this new interest in plastic media led to the development and commercialization of rotating biological contactors (RBC's) which provided many of the advantages of old rock-media trickling filters without some of their disadvantages. Because of the new media developments and the smaller energy requirements of RBC treatment process, compared with those of activated sludge processes, engineers employed RBC process extensively for wastewater treatment during the mid and late 1970's.

However, the use of the RBC process for wastewater treatment declined in the early 80's due to various operational problems and because of structural problems relating to shaft and media failure. Numerous reports in the literature and RBC facility surveys [1,2,3,4,5] have reported difficulties with the initial stages of RBC systems resulting in heavy biomass growth, the presence of nuisance organisms such as Beggiatoa, and a reduction in organic removal rates. These problems have been attributed to excessive organic loadings that result in low dissolved oxygen conditions which subsequently lead to development of Beggiatoa growth and deteriorating process efficiency. Beggiatoa organisms compete with heterotophic organisms for oxygen and for space on RBC media surfaces. Their predominance can result in an increase in the concentration of biomass on the media while at the same time causing substantial reduction in organic removal per unit area.

A nationwide RBC teleconference [5] publicized the results of an EPA study. A significant finding of the study was that heavy biological growth due to high organic loadings and the resulting low dissolved oxygen levels leading to the development of heavy sulfide oxidizing organism growths were in part responsible for the failure of shafts and media observed at a number of plants. The EPA study suggested that RBC plants should be designed with adequate flexibility including wastewater recirculation, positively-controlled alternate flow systems such as step feed or use of an enlarged first stage, supplemental aeration, and other such means of operating flexibility to optimize performance. Consequently, the authors designed a study with an objective to better understand the kinetics and performance of RBC's operated with supplemental aeration and an enlarged first stage under high and low organic loading conditions.

To achieve the study objective, it was necessary to have an RBC treatment plant that had at least two parallel trains so that one train could be used as a control and the other train could be used to use and evaluate the effectiveness of supplemental aeration and an enlarged first stage. A full-scale RBC plant having two parallel trains used to treat industrial and domestic wastewater in a town with a population of 6000 was used. The new RBC plant had four stages in each train and had a problem

445

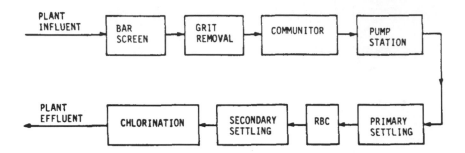

Figure 1.   Wastewater treatment plant flow diagram.

meeting its standard secondary effluent limits because of the overloaded conditions in the first two stages resulting in part from low dissolved oxygen levels in the influent wastewater. This overloaded RBC plant was considered to be ideal for the investigation of RBC kinetics and its overall performance under low and high organic loading conditions with the use of supplemental aeration and an enlarged first stage.

MATERIALS AND METHODS

RBC Facility

The full-scale RBC wastewater treatment plant consisted of a manual bar screen, comminutors, a grit chamber, a raw sewage pump station, two circular primary clarifiers, rotating biological contactors (two trains with four stages in each train), three rectangular final clarifiers, two chlorine contact chambers and two anaerobic sludge digesters. A schematic plant flow diagram is shown in Figure 1. This RBC facility was designed to meet standard secondary limits of 30 mg/L BOD_5 and 30 mg/L suspended solids.

The research at this RBC facility was conducted in two separate phases. The Phase I study was designed and conducted to investigate the effectiveness of adding supplemental aeration alone on RBC kinetics and overall plant performance. The schematic layout of the RBC units is shown in Figure 2. All of the north RBC units were provided with supplemental air by installing fine-bubble reef diffusers manufactured by Environmental Dynamics, Inc. Four reef diffusers (2 ft. x 1.5 ft.) were provided in each RBC stage, and a total of 16 diffusers were installed in the four stages. The air was supplied from an existing 320 cfm blower using PVC and flexible piping. These fine-bubble diffusers had high oxygen transfer capacities with air flow rate capacities of 0 to 30 cfm/ft^2.

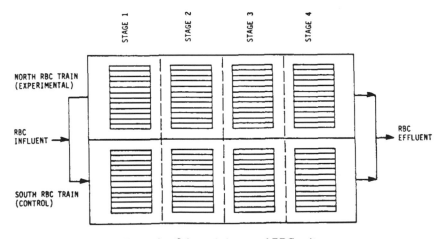

Figure 2.   Schematic layout of RBC units.

In the Phase II study, the combined effect of supplemental aeration and use of an enlarged first stage on RBC kinetics and performance was investigated by removing the wooden baffles between the first and second stages of the north RBC train. The supplemental aeration provided in the north RBC train during the Phase I study was continued, and the south train, without supplemental air but with an enlarged first stage, was used as control. An enlarged first stage can be considered as partial step feed to equalize the organic load in the first two stages.

Sampling and Analyses

The study at this plant was conducted for approximately six months. Eleven 24-hour composite samplers were used to collect influent and effluent wastewater samples from each RBC stage in both trains. The composite samples were flow proportioned prior to analysis. The collected composite samples were analyzed for soluble COD, ammonia nitrogen, suspended solids, and volatile suspended solids on a daily basis. Samples were also analyzed for soluble BOD_5 (inhibited, to suppress nitrification) once a week. Stage oxygen uptake rates were measured periodically. Wastewater temperature, pH and dissolved oxygen levels were measured in each stage at about 2 a.m. and in the afternoon at the time of peak loading. The biomass thickness in the RBC stages was measured periodically, and the growth conditions were observed and noted on a daily basis. The biomass thickness was measured by scraping biomass from a known area of the media. Knowing the biofilm weight and the area measurements, the biofilm thickness was calculated. All other tests were conducted according to *Standard Methods* [6].

The relationship between 5-day soluble BOD ($SBOD_5$) and SCOD was as follows:

$$SBOD_5 = 0.6 \text{ SCOD} - 11.06 \text{ Influent to RBCs} \qquad (1)$$

$$SBOD_5 = 0.21 \text{ SCOD} - 1.43 \text{ Effluent from RBCs} \qquad (2)$$

RESULTS AND DISCUSSION

During the Phase I and Phase II studies, the RBC performance was investigated separately under low and high organic loading conditions with the intent to find an optimum organic loading range above which oxygen limiting conditions and zero order kinetics would prevail. In the beginning of each phase and during the transition from low to high organic loadings, an acclimation period of 10 to 15 days was allowed to attain a steady-state operating condition. During the higher loadings, the plant loading was increased in an incremental fashion over a period of time by adjusting the degree of pretreatment provided for an industrial waste originating at a dairy plant.

Startup

After installation of aeration equipment in the north RBC train, air was introduced through the sixteen fine-bubble reef diffusers (four in each stage) from a 320 cfm existing blower. During the first few days of air use, the biomass from the north RBC units sloughed off heavily, creating a solids overloading problem in the final clarifiers. It was noticed that biomass slough off was significantly less in the second week of operation. Since the biomass was thick and heavy in all stages prior to air use, it was decided to continue the aeration for three weeks prior to initiation of sampling. Intermittent 24-hour composite sampling of each stage was carried out to observe when steady-state conditions prevailed. The sampling results indicated steady-state conditions existed at the beginning of the third week. During the entire period of this study, steady-state conditions were based primarily on the effluent COD of each stage.

After a few weeks of operation in both phases, the biomass associated with the north RBC train turned gray in color in the first two stages and a brownish red tan color appeared in the last two stages, particularly during the Phase I study. Beggiatoa growth disappeared from all stages in the north RBC train with supplemental air and use of an enlarged first stage. However, the south RBC train which was used as a control had heavy, thick biomass growth and Beggiatoa was prevalent in all stages, particularly in Phase I study.

Wastewater Characteristics

Influent wastewater characteristics for the Phase I study are shown in Tables I and II for the low and high organic loading rate studies, respectively. The wastewater coming to this RBC plant is a mixture of domestic and industrial dairy waste. During lower loading studies, the soluble COD varied between 86 and 210 mg/L with a mean of 164.4 mg/L. During the higher loading studies, the mean

Table I. RBC Influent Wastewater Characteristics at Lower Loadings During the Phase I Study

Parameter	Mean	Range	Standard Deviation
Flow, mgd	0.707	0.618–0.912	0.063
Soluble COD, mg/L	164.4	86–210	35.6
Ammonia-N, mg/L	24.8	16–42	7.50
SS-mg/L	107.0	68–240	40.6
VSS-mg/L	84.8	62.2–192.0	29.6
DO-mg/L	1.23	0.80–1.50	0.18
pH	7.33	7.20–7.50	0.09
Temperature, °F	72.3	70–75	1.40

Table II. RBC Influent Wastewater Characteristics at Higher Loadings During the Phase I Study

Parameter	Mean	Range	Standard Deviation
Flow, mgd	0.852	0.636–1.423	0.242
Soluble COD, mg/L	430.0	240–690	152.9
Ammonia-N, mg/L	22.3	18–25	2.71
SS,mg/L	149.2	74–234	40.7
VSS,mg/L	123.9	72.5–184.9	29.6
DO,mg/L	1.07	0.80–1.54	0.22
pH	7.26	6.8–7.4	0.19
Temperature, °F	71.3	70–72	0.90

value of soluble COD was 430 mg/L and varied from 240 to 690 mg/L. As can be seen from Tables I and II, flows to the plant remained within the design limit of 1.1 mgd except for several days during the higher loadings when the flows exceeded the design limit with a maximum flow of 1.4 mgd. The mean flows during low and high organic loadings were 0.707 and 0.852 mgd, respectively. Overall, the flows remained high during the higher organic loadings. Mean influent ammonia nitrogen values remained approximately the same during both low and high organic loading periods. However, during low loadings the ammonia nitrogen varied between 16 and 42 mg/L, whereas it varied only between 18 and 25 mg/L during higher loadings. Sufficient phosphate was measured in the wastewater to fulfill the nutritional requirements of the RBC process. The RBC influent total suspended solids varied from 68 to 240 mg/L with mean values of 107 and 149 mg/L at the lower and higher organic loadings, respectively. The volatile suspended solids in the influent were approximately 80% of the total and increased a little during the higher organic loading studies. Wastewater temperature during the Phase I studies varied between 70 and 75°F.

The influent wastewater characteristics during the Phase II study did not vary much from those observed during Phase I except for the soluble COD concentration. Tables III and IV list the influent wastewater characteristics observed during the Phase II study. At lower organic loadings, the RBC mean influent soluble COD concentrations in the Phase II study were lower than those observed in the Phase I study. This was because of lower waste production at the dairy plant. The influent ammonia nitrogen concentrations were approximately the same in both phases; the mean values in Phase II were

Table III. RBC Influent Wastewater Characteristics at Lower Loadings During the Phase II Study

Parameter	Mean	Range	Standard Deviation
Flow, mgd	0.786	0.573–1.20	0.161
Soluble COD, mg/L	239.2	130–300	50.4
Ammonia-N, mg/L	22.1	20.5–26	2.1
SS,mg/L	182.0	100–332	55.2
VSS,mg/L	133.6	85–249	41.2
DO,mg/L	1.00	0.60–1.30	0.22
pH	7.17	7.00–7.35	0.09
Temperature, °F	68.5	66–70	1.50

Table IV. RBC Influent Wastewater Characteristics at Higher Loadings During the Phase II Study

Parameter	Mean	Range	Standard Deviation
Flow, mgd	0.706	0.407–0.935	0.138
Soluble COD, mg/L	362.5	270–540	76.4
Ammonia-N, mg/L	20.4	15.5–25.0	3.5
SS,mg/L	207.6	146–348	52.9
VSS,mg/L	159.5	109.5–233.2	33.6
DO,mg/L	1.02	0.90–1.30	0.12
pH	7.06	6.8–7.2	0.10
Temperature, °F	66.9	63–70	2.10

in the range of 20 mg/L. Phase II wastewater temperatures were lower than in Phase I but not sufficiently lower to affect the process efficiency. The wastewater temperatures varied between 66 and 70°F at lower loadings; at higher organic loadings, they varied from 63 to 70°F. During Phase I, the wastewater temperature was always above 70°F.

Kinetics

One of the objectives of this study was to investigate the kinetics involved in wastewater treatment using the RBC units. Using the mean sampling results from the Phase I and II studies in which the soluble COD concentrations in each stage were measured as a function of time, an attempt was made to fit these mean data into several kinetic models as described below:

Zero order	$-r_s = k$	Plot of S versus time
First order	$-r_s = kS$	Plot of Log S versus time
Second order	$-r_s = kS^2$	Plot of 1/S versus time

where the k's are not the same and dimensions vary with the order of the reaction.

A zero order kinetic model would be applicable when the appearance of substrate from stage to stage or at equal time intervals is constant. A zero order rate model would yield a straight line when the substrate concentration, S, is plotted against time or stage. One way of showing that the process is first order with respect to substrate removal is to plot the substrate remaining, S, as a function of retention time or stage on semilog paper. A resulting straight line indicates first order kinetics and the slope of this line is the reaction rate constant. A second order rate model would yield a straight line when the reciprocal of substrate concentration, or 1/S, is plotted against time or stage.

Figure 3 shows that the experimental data provided a reasonable fit to the first order kinetic model. Figures 4 and 5 show plots of the data in Phase I to determine their fit to zero and second order kinetics. It can be seen from these plots that the data fit first order kinetic models at higher organic loadings with supplemental air (Figure 3) and a zero order model without supplemental air (Figure 4). The zero order behavior in the RBC units without supplemental air was due to oxygen-limiting conditions at higher organic loadings that prevailed in the south RBC units because of low dissolved oxygen concentrations in the stage wastewaters. Also, the attached biomass on the south RBC units was thick and heavy. This could have caused mass diffusion limitations both in terms of substrate and oxygen.

At lower organic loadings, Figure 3 suggests that first order substrate removal occurs with supplemental air and without the supplemental air the data approximately fit a second order kinetic model as can be seen in Figure 5. However, there is a distinct slope change in Figure 3 following the first stage of the system, particularly at lower loadings. These slope changes correspond to a change in the rate of substrate utilization in the subsequent stages. Most of the substrate is removed in the first stage with the first stage removal being much higher than the removal in the remaining stages. Such phenomena indicating higher substrate removal in the first stage was observed by several other investigators [7,8,9,10]. However, these substrate removal rates decrease and approach a constant limiting value at the higher organic loadings indicating saturation with substrate, oxygen limitation, or both. At these high loading conditions, the data show a linear relationship of substrate removal with stage which indicates zero order kinetics as was observed in Figure 4 without the supplemental air.

The second order kinetics, as was observed in Figure 5 at lower organic loadings without the supplemental air, was also observed by Opatken [11]. Opatken showed that second order kinetics

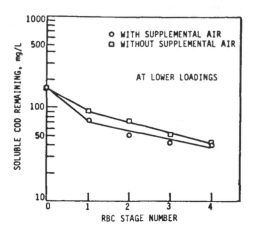

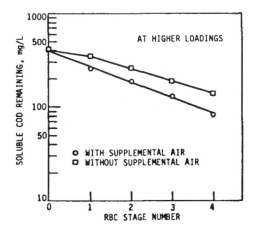

Figure 3.   Soluble COD remaining versus stage for first order kinetics during the Phase I study.

could be applied to describe RBC substrate removal, particularly at lower organic loadings. He indicated that a primary advantage of second order rate reactions was that one kinetic rate constant could be used to describe substrate removal throughout the RBC system. However, this second order kinetic concept cannot predict the substrate removal at higher organic loadings where zero order kinetics occur because of substrate saturation and/or oxygen limitation. In a subsequent section, the importance of total organic loading concepts as a design parameter rather than the use of first, second and zero order models for design purposes will be discussed.

Kinetic models for zero order, first order, and second order substrate removal were described earlier. In the Phase II study, after monitoring the stage soluble COD concentrations, an attempt was made to fit the obtained mean data into various kinetic models. Figures 6, 7 and 8 show the plots for first-order, zero-order, and second-order kinetics, respectively.

It can be seen from these plots that the data from both the south and the north RBC units fit a first-order model at lower organic loadings. In the Phase I study, first order kinetics were observed only in the units where supplemental air was provided and second order kinetics were observed without the supplemental air. At lower loadings the influent soluble COD concentrations in Phase II were higher than those observed in Phase I. However, the observed first-order kinetics without the air were due to the enlarged first stage that divided the incoming load more evenly among the first two stages rather than on the first stage only. Also, the mixed-liquor dissolved-oxygen levels increased to a certain extent because of the reduced load on the enlarged first-stage. However, the slopes of these linear relationships showing first-order kinetics are different as seen in Figure 6; the steeper slope with supplemental air indicates higher substrate removal.

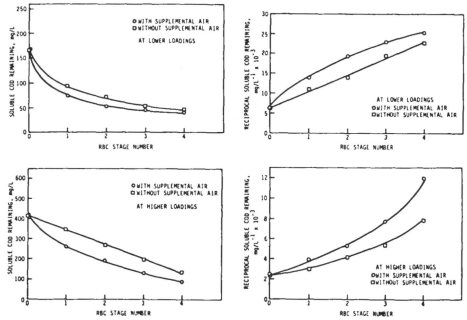

Figure 4.   Soluble COD remaining versus stage for zero order kinetics during the Phase I study.

Figure 5.   Reciprocal of soluble COD remaining versus stage for second order kinetics during the Phase I study.

Figure 7 shows that under higher organic loadings, zero-order kinetics were observed with the enlarged first stage without supplemental aeration. However, the north RBC units with supplemental air and an enlarged first stage indicate first-order substrate removal as observed in Figure 6. Similar kinetics were observed in the Phase I study at higher organic loadings with and without the supplemental air. In the Phase II study, the zero-order kinetics observed in the absence of air at higher organic loadings suggest that baffle removal to create an enlarged first stage to lower the organic load was not sufficient to overcome the oxygen limitation problem. In the north RBC units, this oxygen limitation at higher loadings was eliminated with the use of the supplemental air.

The results in this study suggest that supplemental air in the initial RBC stages, where the organic loadings are high, is essential to overcome potential oxygen limitations at both low and high organic loading rates. The use of an enlarged first stage is helpful to some extent at lower organic loading rates, but, at higher loading rates, the use of an enlarged first stage only will not be enough to overcome oxygen limitation problems.

RBC Performance

In both phases of this research study, the carbonaceous content of the wastewater was measured in terms of soluble COD. Both the flow rate and organic concentration have definite relationships with organic removal rates and process efficiency. It has been observed that removal rates and process efficiency were indeed dependent on the total organics applied to the RBC rather than its concentration or flow rate. One advantage of using the total organic loading concept in design is the ability to predict organic removal rates and treatment efficiency at any loading condition, irrespective of whether the RBC units are functioning under zero, first, or second-order kinetics. Organic removal relationships are established in terms of the total organics applied and the loading points or conditions at which zero order kinetics occur can be observed. These points correspond to loading conditions where the system changes from a biochemical reaction limiting process to an oxygen limiting process. Several authors [5,7,12,13] have suggested the use of total organic loading concepts to determine the required surface area for RBC design and to predict process performance. In this study, the total organic loading approach has been used to evaluate RBC performance.

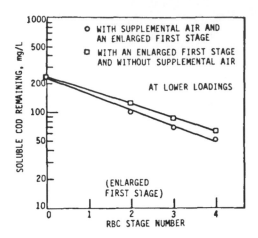

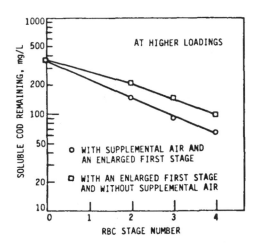

Figure 6.   Soluble COD remaining versus stage for first order kinetics during the Phase I study.

Figure 9 shows the total RBC plant performance observed during the Phase I study at overall COD loading rates with and without supplemental air at lower loadings. Both curves in Figure 9 have high correlation coefficients and approximately the same (0.988 and 0.986) slope. In similar relationships, Dupont and McKinney [7] have obtained a slope of 0.893 based upon BOD_5 data collected from a fullscale RBC plant treating municipal wastewater. Pano and Middlebrooks [8] reported slopes similar to those observed in this study using COD data from pilot plant analysis.

At lower organic loadings most of the soluble COD is removed in the first stage, leaving the subsequent stages in a starving mode that depend on the mixed-liquor particulate COD for survival. In other words, after the first stage, the limiting nutrient is the soluble substrate, and this could be the reason why the RBC performance is the same at overall COD loading rates with and without supplemental air. However, it should be recognized that RBC systems with supplemental aeration will have higher COD removal rate capabilities when substrate is not limiting as was observed in the first stage (see Table V), and this will be shown further in the study results at higher organic loadings in the subsequent discussion.

The significance of the supplemental aeration which provides higher stage dissolved oxygen levels shows up well in the results at higher organic loadings as indicated in Figure 10. Figure 10 shows the relationship between the overall SCOD load applied and removed at higher organic loadings with and without the supplemental air. The linear relationship observed with supplemental air suggest that there was no oxygen limitation. This is further substantiated by the fact that the SCOD removal data

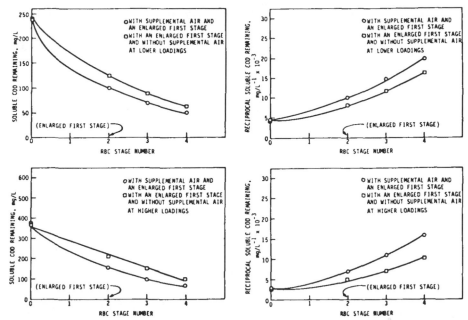

Figure 7. Soluble COD remaining versus stage for zero order kinetics during the Phase II study.

Figure 8. Reciprocal of soluble COD remaining versus stage for second order kinetics during the Phase II study.

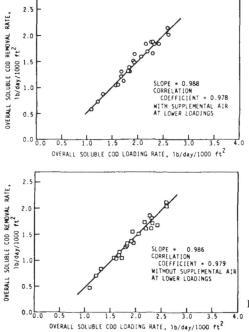

Figure 9. Overall soluble COD removal versus loading at lower loadings during the Phase I study.

Table V. Summary of Soluble COD Removal Efficiencies (%) During the Phase I Study

Stage	With Supplemental Air		Without Supplemental Air	
	At Lower Loadings	At Higher Loadings	At Lower Loadings	At Higher Loadings
Stage 1	55.7	39.9	43.9	19.0
Stage 2	29.3	24.2	21.7	24.8
Stage 3	15.3	33.2	28.3	26.3
Stage 4	8.7	36.3	14.3	27.5
Overall	75.8	80.7	72.7	67.4

conform to first order kinetics removal as observed earlier in Figure 3 and the maintenance of higher dissolved oxygen concentrations as shown in Figure 11. The RBC stage dissolved oxygen concentrations varied between 1.5 and 5.23 mg/L depending upon loading conditions with supplemental aeration. In the absence of supplemental air, the dissolved oxygen remained less than 1 mg/L in all the stages, particularly at higher loadings. The correlation coefficient and slope representing the linear relationship in Figure 10 when supplemental aeration was used were 0.988 and 0.788, respectively. This slope was significantly less than that found in Figure 9 for operation under lower organic loading conditions. However, without the supplemental aeration, the removal rate becomes constant around a loading rate of 8 lbs/day/1000 ft² suggesting oxygen limitation at higher loadings. Earlier, Figure 4 also indicated zero order kinetics at higher loadings without the supplemental aeration.

Table V summarizes the stage soluble COD removal efficiencies obtained during the Phase I study. First-stage efficiencies always remained higher with supplemental air. The significance of supplemental aeration at higher loadings is evident in both the higher overall and higher first-stage removal efficiencies. First-stage soluble COD removal efficiency was 39.9%, double the efficiency observed without the supplemental air. The overall efficiency was 80.7% which was also considerably higher than the efficiency of 67.4% obtained without the air. Also, significant ammonia nitrification took

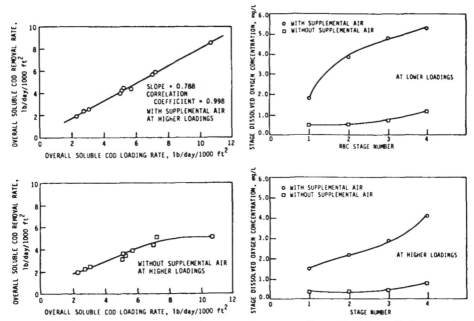

Figure 10.   Overall soluble COD removal versus loading at higher loadings during the Phase I study.

Figure 11.   Stage mixed-liquor D.O. concentrations during the Phase I study.

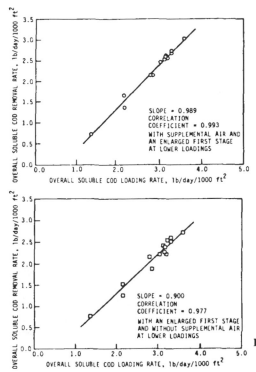

Figure 12. **Overall soluble COD removal versus loading at lower loadings during the Phase II study.**

place in the RBC units that were provided with supplemental air, whereas the ammonia nitrification was insignificant in the absence of supplemental air.

During Phase II, the overall organic loading removal relationships were once again linear at lower loadings as shown in Figure 12. The slope was 0.989 with supplemental air and 0.900 without supplemental air. The overall soluble COD removal rates shown in Figure 12 suggest that there was some difference in removal rates with and without the air. Also, first order kinetics were observed, as shown in Figure 6, during the lower loadings with and without supplemental aeration. Similar observations were made at lower organic loadings rates during the Phase I study except that second order kinetics were observed without the air. This is due to existence of substrate limiting conditions after the first-stage as most of the soluble COD removal takes place in the enlarged first stage. However, it is important to recognize here that the first-stage removals were significantly higher with supplemental air (Table VI) as were observed in Phase I. Figure 13 shows the stage dissolved oxygen concentrations observed during the Phase II study. Once again RBC units with supplemental air had signifi-

Table VI. **Summary of Soluble COD Removal Efficiencies (%) During the Phase II Study**

Stage	With an Enlarged First Stage and With Supplemental Air		With an Enlarged First Stage and Without Supplemental Air	
	At Lower Loadings	At Higher Loadings	At Lower Loadings	At Higher Loadings
Enlarged[a]				
Stage 1	58.3	59.0	47.9	43.8
Stage 3	31.9	38.4	30.4	28.3
Stage 4	25.0	30.5	28.6	33.1
Overall	78.7	82.4	74.1	73.0

[a]Stage 1 and 2 combined.

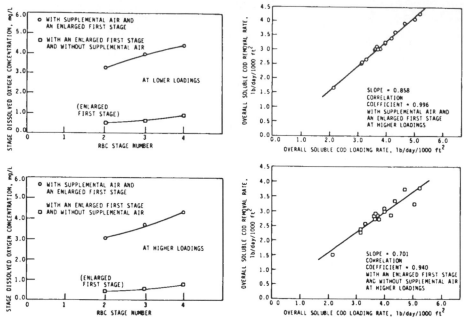

Figure 13. Stage mixed-liquor D.O. concentrations during the Phase II study.

Figure 14. Overall soluble COD removal versus loading at higher loadings during the Phase II study.

cantly higher dissolved oxygen concentrations compared to units that were not provided with supplemental air.

Figure 14 shows the observed overall soluble COD removal rates with and without the supplemental air at higher organic loadings. As can be seen, the relationships were linear and the correlation coefficients were high. The slopes with and without the air were 0.858 and 0.701, respectively. The steeper slope with supplemental air demonstrates that higher removal rates were possible with supplemental air. Also, Figure 7 indicated zero order kinetics at higher organic loadings in the absence of supplemental air. With supplemental air, the soluble COD removal rates increased and percent removals increased as the loading increased. However, as the loading increased, the soluble COD removal rate and percent removal decreased without the supplemental air. A comparison of overall soluble COD removal rates observed during the Phase I and Phase II studies suggest that, at higher organic loading rates, soluble COD removal rates were higher with an enlarged first stage (stage 1 and 2 combined) than with four separate stages. This comparison further substantiates that step feeding in RBC systems would be helpful when operated at higher organic loading rates. Table VI shows a summary of soluble COD removal efficiencies observed during the Phase II study. As shown, the enlarged first stage produced overall soluble COD removal efficiencies significantly higher with supplemental aeration, particularly at higher organic loadings. Ammonia nitrification in Phase II decreased significantly due to the creation of enlarged first stage even in the presence of supplemental air. This further suggests that the plug-flow mode is preferable for ammonia nitrification. In the absence of supplemental air, ammonia nitrification did not change from what was observed in Phase I.

SUMMARY AND CONCLUSION

The results of this study indicate that first-order kinetics can be maintained in the presence of supplemental air due to the maintenance of high dissolved oxygen concentrations in the stage mixed-liquor. The maintenance of thinner biofilms on the media further enhances the diffusion of oxygen and substrate into inner layers of the biofilms thereby increasing the overall process efficiency. The creation of an enlarged first stage by combining the first two stages by removal of baffle between them is helpful in increasing the dissolved oxygen concentrations in the beginning stages at lower organic loadings; however, at higher organic loadings, the creation of an enlarged first stage alone is not helpful and provision of supplemental aeration is necessary to alleviate oxygen limitation problems and to maintain first order kinetics. The results also indicate that for ammonia nitrification, the plug flow mode is preferable over the enlarged first stage mode.

ACKNOWLEDGMENTS

This study was conducted as part of the work of the Iowa State University Engineering Research Institute. The early assistance of a former faculty associate, Rudy J. Tekippe, in planning this study is appreciated. Dr. Tekippe has since returned to James M. Montgomery and Associates. Special appreciation is due to the elected officials of the city for making their plant available for this study, and to the City Manager, Patrick Callahan, and plant superintendent, Charles Pietscher, and the plant operators for their cooperation, without which the study would have been impossible.

Views expressed in this article are those of the authors and should not be construed as an official position of the Environmental Protection Agency.

REFERENCES

1. Chesner, W. H. and Iannone, J. J., "Review of Current RBC Performance and Design Procedures," Report prepared for U.S. EPA under Contract No. 68-02-2775.
2. Hitdlebaugh, J. A. and Miller, R. D., "Full-Scale Rotating Biological Contractor for Secondary Treatment and Nitrification," *Proceedings of the First National Symposium on RBC Technology*, Champion, Pennsylvania (February 1980).
3. Surampalli, R. Y., Tekippe, R. G., and Baumann, E. R., "The Value of Supplemental Air in Improving RBC Performance," *Proceedings Second International Conference on Fixed-Film Biological Processes*, Arlington, Virginia (July 1984).
4. U.S. EPA, Region VII, *In-House Memo on Inventory and Analysis of RBC Facilities*, Kansas City, Missouri (April 1983).
5. U.S. EPA, "A Nationwide RBC Teleconference," Cincinnati, Ohio (September 1983).
6. *Standard Methods for the Examination of Water and Wastewater*, 16th Edition, American Public Health Association, Washington, D.C. (1985).
7. Dupont, R. R. and McKinney, R. E., "Data Evaluation of a Municipal RBC Installation," *Proceedings of the First National Symposium on Rotating Biological Contactor Technology*, Champion, Pennsylvania (February 1980).
8. Pano, A. and Middlebrooks, E. J., "Kinetics of Carbon and Ammonia Nitrogen Removal in RBC's," *Journal Water Pollution Control Federation*, 55, 956 (1983).
9. Stover, E. L. and Kincannon, D. F., "Rotating Disc Process Treats Slaughterhouse Waste," *Industrial Wastes*, 22, 33 (1976).
10. Surampalli, R. Y. and Baumann, E. R., "Role of Air in Improving First-Stage RBC Performance," *Prodings of the National Conference on Environmental Engineering*, ASCE, Boston, Massachusetts (July 1985).
11. Opatken, E. J., "Rotating Biological Contactors—Second Order Kinetics," *Proceedings of the First International Conference on Fixed-Film Biological Processes*, Kings Island, Ohio (April 1982).
12. Poon, C. P. C., Chin, Y. L. and Mikucki, W. J., "Factors Controlling Rotating Biological Contactor Performance," *Journal Water Pollution Control Federation*, 51 601 (1979).
13. Wilson, R. W., Murphy, K. L., and Stephenson, J. P., "Scale-Up in Rotating Biological Contactor Design," *Journal Water Pollution Control Federation*, 52, 610 (1980).

54   CELROBIC® ANAEROBIC TREATMENT OF DAIRY PROCESSING WASTEWATER

Anthony M. Sobkowicz, Environmental Section Manager
Badger Engineers, Inc.
Cambridge, Massachusetts 02142

INTRODUCTION

This paper reports on an onsite pilot plant test of the Celrobic® anaerobic biological process, treating the segregated wastewater from production of soft Italian cheeses. Most prior studies of the anaerobic treatment of wastes from the production of cheese have focused on the treatment of whole whey, the byproduct of all cheese production [1,2,3]. A large scale study has also been performed on a dilute cheese dairy waste [4].

In this dairy the whey produced in cheese manufacturing is concentrated to produce a marketable byproduct. The waste available for treatment contains only small quantities of whey along with whole milk and milk fats and substantial quantities of chemicals used in cleaning solutions and differs dramatically from whole whey alone.

The objectives of the onsite pilot program were: 1) to identify and segregate waste streams of higher strength suitable for anaerobic treatment; 2) to demonstrate that the waste could be continuously treated in an anaerobic system to produce an effluent suitable for discharge to the overloaded POTW, and 3) to demonstrate that the anaerobically treated effluent could be further treated by a trickling filter.

WASTE DESCRIPTION

Milk is transported to the cheese dairy from dairy farms via bulk transport tankwagons. In the receiving area the milk is pumped through chilling coils to storage tanks called "silos". The stored milk is pumped through pasteurizers where it is subjected to high-temperature quick pasteurization and then on to tanks where it is mixed with rennet and bacterial cultures to start cheese formation. The curd formed is cut to separate most of the solids formed from the whey liquor. The curd is then pumped to the DMC process where it is drained, matted and cured to become mozzarella cheese. The raw cheese is heated to drive out any remaining water and, while still in a plastic state, is forced into molds where it is cooled to promote solidification. The molded cheese is ejected into tanks containing strong brine where it is allowed to ripen for market. After ripening the cheese is packaged and refrigerated for shipment.

The whey byproduct is pumped to a multiple effect evaporator where it is concentrated to a slurry containing about 75–80% solids. The whey concentrate is used primarily as an additive and extender in a variety of food-grade products. Cream that is separated from the milk is also sold as a byproduct.

Since all of the equipment that comes into contact with milk must be cleaned every day, cleaning is the principal source of contaminated wastewater in the dairy. Figure 1 illustrates the uses for water in the cheese dairy and shows how they become contaminated. Clean-In-Place (CIP) systems which recover and recirculate cleaning solutions are used to minimize the water and chemical use associated with equipment cleaning. This dairy uses four recirculating systems to service the following areas:

- milk receiving
- milk silos
- milk pipelines
- autovats
- dmc machine

The cleaning cycle includes a prerinse, washing, sanitization, and postrinse. The prerinse reuses the water collected from the prior postrinse which is then sewered. The alkaline cleaning solution that is used for the principal washing cycle is recaptured, and generally speaking, only the "dragout," the

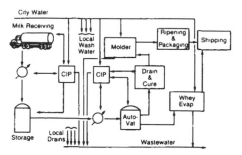

Figure 1.   Cheese plant water system.

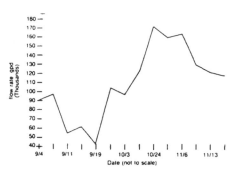

Figure 2.   Dairy wastewater effluent flow.

residue that wets the wall, is lost and represents the purge from the system. This purge is made up by passing forward water from the postrinse. The postrinse must use unrecirculated water of potable quality to ensure that sanitary conditions are maintained.

Once-through water is used to cool the molder head where the cheese is formed into blocks. This water contacts the cheese and so must be of high quality. Since it becomes contaminated through this contact, it requires treatment as well.

In the whey concentrating evaporators, large quantities of overhead condensate are collected. This water is relatively pure so that some of it is used for boiler feed water and for pump seal water. The pump seal water is sewered after once-through use.

Washdown of the floors and incidental cleaning represents another important source of waste-waters. While normally low in contamination, floor washings can be an enormous source of organics when spills occur. Figure 2 shows that the total wastewater from the plant was highly variable in flow with an average of 115,000 gpd and a range of 0.043 to 0.170 mgd. The strength of the waste, shown in Figure 3, is such that the organic contaminant level was somewhat low for success with an anaerobic system. Figure 4 shows the quantities of wastewater that were associated with the various manufacturing activities. The most heavily contaminated streams are the floor drains and the whey room rinse water.

SEWER SEGREGATION

In considering the possible application of the Celrobic high rate anaerobic system to pretreat the cheese dairy wastes, it became obvious that the segregation of the more contaminated wastewaters would mean a more economical overall process [5]. Toward that end the various source wastewaters

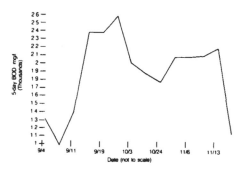

Figure 3.   Dairy wastewater effluent BOD.

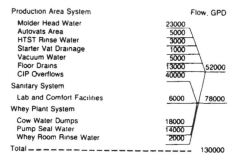

Production Area System	Flow, GPD	
Molder Head Water	23000	
Autovats Area	5000	
HTST Rinse Water	3000	
Starter Vat Drainage	1000	
Vacuum Water	5000	
Floor Drains	13000	52000
CIP Overflows	40000	
Sanitary System		
Lab and Comfort Facilities	6000	78000
Whey Plant System		
Cow Water Dumps	18000	
Pump Seal Water	14000	
Whey Room Rinse Water	2000	
Total		130000

Figure 4.   Wastewater sources.

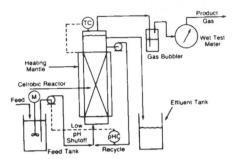

Figure 5.   Schematic of Celrobic pilot plant.

were evaluated and a segregation scheme was devised. With this scheme, about 85% of the organic matter available was concentrated into about 35% of the water.

CELROBIC PILOT PLANT DESCRIPTION

The Celrobic 20-liter pilot unit which was used to treat the strong waste was patterned after the units that were used to provide design information for the existing full-scale systems [6,7,8]. Similar pilot plants have been used to evaluate the treatability of a host of wastewaters considered as candidates for Celrobic treatment. A schematic diagram of the Celrobic pilot plant is shown in Figure 5. The system includes:

- waste collection/reactor feed tank
- reactor
- reactor feed effluent and recycle
- offgas monitoring system

WASTE COLLECTION/REACTOR FEED TANK SYSTEM

After the strong waste stream was segregated, it was sampled with a compositor to obtain a daily sample representative of the average wastewater expected as feed to a full scale facility. The feed tank was large enough to store sufficient feed for more than 24 hours of pilot plant operation at the highest anticipated loading. It was agitated to insure a homogenous feed and cooled to prevent excessive premature degradation. When required, alkalinity adjustments were performed in the feed vessel.

Reactor

The pilot reactor is a 20-liter, glass reactor packed with one- inch polypropylene pall rings. The reactor dimensions are six- inch diameter by four feet high. The working volume of the reactor was approximately 21 liters.

The temperature of the reactor was controlled at 98°F by a blanket heating mantle connected to an automatic temperature controller. Feed is introduced, along with recycled liquor, below the packing and is uniformly distributed using a splash plate.

The reactor effluent and offgas exit the reactor together and are separated in the disengaging section on top of the packed reactor.

Influent, Effluent, and Recycle Systems

The feed is combined with the recycle and introduced to the bottom of the reactor. Treated wastewater is discharged from the top of the reactor directly to the effluent storage tank. Recycle is drawn from the top of the vessel at a lower level. A peristaltic pump controls the recycle rate.

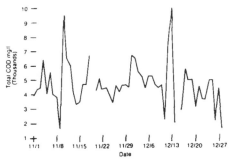

Figure 6.   Segregated strong waste total COD.

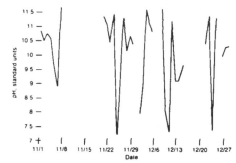

Figure 7.   Segregated strong waste pH.

Offgas Monitoring System

The offgas from the gas/liquid separator vessel flows through a water sealed trap to a wet test meter to measure the amount of gas produced and is then discharged to the atmosphere. A caustic bubbler system (not shown) was used to monitor the acid gas content of the produced gas.

PILOT PLANT OPERATION

The pilot plant program was designed to last 12 weeks.

Because of the need for a rapid study, the Celrobic pilot plant was filled with packing that was coated with biological growth from laboratory studies on a synfuels waste. That waste contained high levels of low molecular weight organic acids, alcohols, and ketones and was not dissimilar from a milk waste after acidogenesis had occurred.

The early portion of the onsite pilot study was hampered by the lack of onsite analytical support and by mechanical problems associated with product sampling and feed pumping. Once daily analyses of COD became available, the feed could be regulated sufficiently to prevent excessive shocking of the biomass.

As shown in Figures 6 and 7, the segregated strong waste daily composite had a COD that varied from 1600 to 10,000 mg/L with a pH that varied from 7 to 11.5. Figures 8,9,and 10 illustrate the performance of the pilot plant systems for the period from November 1 to December 28.

Later in the study, highly concentrated "whey sludge," a material containing 30,000 to 100,000 mg/L of Chemical Oxygen Demand, was added to the feed to maintain a uniform organic loading to the reactor. The figure shows that the unit was able to handle this additional waste load and produce a treated wastewater suitable for discharge.

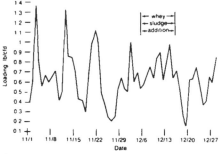

Figure 8.   Celrobic reactor loading during pilot plant study.

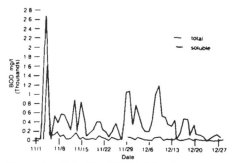

Figure 9.   Celrobic effluent BOD during pilot plant study.

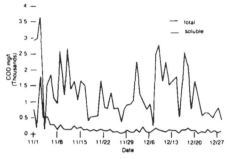

Figure 10. Celrobic effluent COD during pilot plant study.

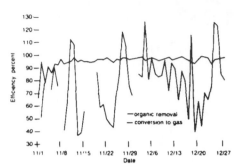

Figure 11. Celrobic reactor efficiency during pilot plant study.

PERFORMANCE

As the loading to the Celrobic reactor became relatively stable, the performance of the unit improved. The effluent contained coarse and fine solids. The coarse solids settled rapidly while the fine solids would not settle out. Overall solids levels in the effluent, and the COD associated with them, fluctuated in an unpredictable way so that soluble COD and BOD were monitored to more closely assess the success of the treatment method. Effluent soluble COD averaged 195 mg/L in the latter days of the study while effluent soluble BOD approached 80 mg/L.

Based on effluent soluble COD, the organic removal efficiency of the pilot plant was measured at 98%. As you can see in Figure 11, the conversion to gas fluctuated approximately between 50 and more than 100% due in part to the accumulation of biodegradable materials within the reactor and in part to aberrations caused by the sampling time lag between sampling the feed and the effluent.

SETTLING TESTS

As testing proceeded, it became clear that removal of solids from the Celrobic effluent would be necessary in order to achieve the objectives of the program. In order to assess the impact of clarification, a series of settling tests were conducted. The results of the tests in terms of the impact of settling on the Celrobic effluent COD and BOD are shown in Figures 12 and 13. The coarse solids settled out rapidly but the turbidity did not. The resultant clarified COD averaged 630 mg/L, and the clarified BOD averaged 165 mg/L.

TRICKLING FILTER TESTS

The operators of the Publicly Owned Treatment Works (POTW) into which the cheese dairy waste was discharged expressed concern that the treated effluent from the Celrobic unit might not degrade properly in their trickling filter system; that is, they questioned whether easily degradable components would be removed while more difficult to remove organics would remain. To allay these concerns, a

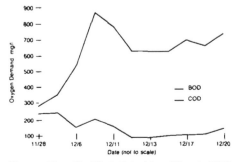

Figure 12. Clarified Celrobic effluent BOD and COD during pilot plant study.

Figure 13. Celrobic pilot plant data summary.

	Untreated Waste	Celrobic Effluent	Settled Effluent
Total COD	4400	1750	630
Soluble COD	960	195	195
Total BOD	2075	700	165
Soluble BOD	450	80	80
Suspended Solids	690	1000	230
pH (Range)	7-11.5	6.5-7.6	6.5-7.6

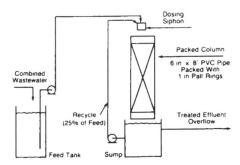

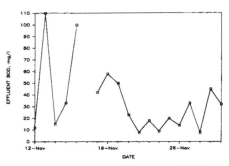

Figure 14. Schematic of trickling filter pilot plant.

Figure 15. Trickling filter effluent BOD on Celrobic effluent.

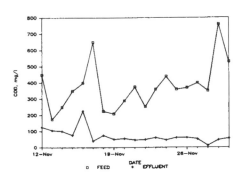

Figure 16. Trickling filter effluent COD on Celrobic effluent.

pilot trickling filter test was set up and conducted on a waste consisting of POTW primary settler effluent and Celrobic treated dairy waste. It was not intended that the test model the POTW but simply that it demonstrate that the discharge requirement of the POTW could be met by aerobic trickling filter on the Celrobic treated waste. The trickling filter (Figure 14) was set up in a six-inch diameter by seven foot section of PVC piping. Hydraulic loading to the unit was set based on the average hydraulic loading to the full scale unit and, like the full scale unit, a 25% recycle stream was maintained. The one-inch polypropylene pall ring packing was acclimated on the municipal trickling filters and quickly came up producing an effluent with an average total BOD of about 30 mg/L and a COD of about 60 mg/L (Figures 15 and 16).

CONCLUSIONS

This test program demonstrated that a Celrobic Unit can continuously treat a strong dairy waste at a loading of 0.75 lb/cfd (12 kg/m³d) to produce an effluent suitable for discharge to a conventional aerobic process. It also demonstrated that the addition of chemical buffers was not essential with waste and that solids produced in the reactor were the principal source of COD and BOD in the reactor effluent.

REFERENCES

1. Switzenbaum, M. S., S. C. Danskin, "Anaerobic Expanded Bed Treatment of Whey", *Proc. of the 36th Industrial Waste Treatment Conference*, Purdue University (May 1981).
2. Parker, C. D., "Methane Fermentation of Whey", *Proc. of the Second National Symposium on Food Processing Wastes* (March 1971).
3. Boening, P. H., and Larsen, V. F., "Anaerobic Fluidized Bed Treatment of Whey", *Proceedings of the 38th Industrial Waste Treatment Conference*, Purdue University (May 1983).
4. Samson, R., et al, "Dairy Waste Treatment using Industrial-Scale Fixed-film and Upflow Sludge Bed Anaerobic Digesters: Design and Start-up Experience", *Proc. of the 39th Industrial Waste Treatment Conference*, Purdue University (May 1984).

5. Moores, C. W. and A. M. Sobkowicz, "Anaerobic Treatment Cuts Treatment Costs", *Pollution Engineering* (May 1986).
6. Ragan, J. L., "Celanese Chemical Company Experience with Anaerobic Filters", Dept of Energy/ EPA Seminar, Howey-in-the-Hills, Florida (January 1980).
7. Harvey, L. M., and J. C. Rubiano, "The Anaerobic Approach: A Better Answer", *Industrial Wastes* (March 1983).
8. Okkes, A. G., "Afvalwater anaerob behandelen", *PT/Procestechniek*, Netherlands (April 1984).

Section Six
PLATING WASTES

55 IRON PROCESS FOR TREATMENT OF Cr(VI) WASTEWATERS

Alan R. Bowers, Assistant Professor

Carlos A. Ortiz, Graduate Student

Roberto J. Cardozo, Graduate Student
Department of Civil and Environmental Engineering
Vanderbilt University
Nashville, Tennessee 37235

INTRODUCTION

The technology for treatment of Cr(VI)-bearing wastewaters has been well established. Conventional methods are based on the reduction of Cr(VI) to Cr(III) followed by precipitation as $Cr(OH)_3$ at high pH, ca. 8.0 to 10.0 [1-11]. Although other techniques such as ion exchange and activated carbon adsorption have been successfully demonstrated in the laboratory, they have not been employed to any significant degree in full-scale operations [12-15].

A variety of reducing agents have been suggested for Cr(VI) reduction, including sulfides, bisulfites, metabisulfites, ferrous sulfate, and sulfur dioxide [1,7,8,11]. Although quite effective as reducing agents, the cost of these chemicals may not be entirely justified due to the availability of equally effective waste materials, such as scrap iron, for the reduction of Cr(VI). Furthermore, the cost of reducing Cr(VI) with conventional chemicals is greatly increased by side reactions with dissolved oxygen, which is ubiquitous in aqueous systems that are in contact with the atmosphere [16, 17].

Typically, the chemical dosage needed for Cr(VI) reduction is two times the stoichiometric requirement [11,16]. Based on this ratio, a comparison of chemical costs for Cr(VI) reduction is shown in Table I.

The costs shown in Table I indicate that scrap iron has the potential to be significantly more economical than conventional means of Cr(VI) reduction, e.g., > 10 times cheaper than the least expensive alternative. In addition, the iron solubilized by oxidation may act as a conditioning agent for the resulting $Cr(OH)_3$ sludge and, thus, produce a greater overall benefit in cost that would not be

Table I. Cost Comparison of Chromium (VI) Reducing Agents

Reducing Agent	Chemical Costs[a] ($/Kg Cr)
SO_2	0.85[b]
$NaHSO_3$	1.15[c]
FeS	3.33[d]
$FeSO_4$	1.25[e]
Scrap Iron	0.05[f](0.08)

[a]Costs calculated assuming required dosage equal to 2 times the stoichiometric requirement[11].
[b]$210/ton for SO_2 [20].
[c]$0.174/lb for $NaHSO_3$ [20].
[d]$0.45/lb for FeS [20].
[e]$130/ton for $FeSO_4$ [20].
[f]$1.00/100 lb for scrap iron, current market price in Nashville, TN., based on telephone survey of local dealers, August, 1985. Value in parenthesis represents 3.5 times the stoichiometric requirement as found necessary in this work.

realized by the sulfur-based reducing agents. Iron(II)-based reducing agents would produce Fe(III) as a conditioning agent as well. However, since the reduction of Cr(VI) to Cr(III) requires 3 electrons, these agents will produce much greater quantities of sludge than the scrap iron process.

Previously, Gould [18] examined the reduction of Cr(VI) using relatively pure iron wire (J.T. Baker reagent grade iron wire, $\geq$ 99.9% Fe). This work showed that iron surfaces were effective in reducing Cr(VI) to Cr(III) under typical reducing conditions (pH 2.0 to 3.0). However, the excess soluble Fe, as Fe(II) or Fe(III), produced due to reactions with dissolved oxygen was not investigated, nor was the influence of excess Fe on the final sludge characteristics determined.

This work examines the use of scrap iron filings to treat Cr(VI)-containing metal plating wastewaters. The kinetics of iron particle reaction with Cr(VI) and O_2 as well as the settling characteristics and specific resistances of the sludge produced by this process have been evaluated for comparison to conventional treatment schemes.

EXPERIMENTAL MATERIALS AND METHODS

Iron Filings

Scrap iron filings were obtained at random from a metal working shop on the Vanderbilt University campus. The filings were a typical 4140 steel, i.e., 0.38 to 0.43% C, 0.75 to 1.0% Cr, 0.8 to 1.1% Mo, 0.035% P, 0.025% S, 0.2 to 0.35% Si [19]. The filings were initially sieved so that no particles larger than 0.84 mm (US sieve no. 20) were used in the experimental work. The remaining particles were categorized according to their size distribution as shown in Figure 1. Total surface area was estimated by size and shape analysis and was found to be about 75 cm^2/g ($\pm$ 15).

Materials and Reagents

All chromium solutions were made up with stock solutions prepared from CrO_3 (99.9% Aldrich Chemical Co., Inc.) or Cr_2O_3 (Aldrich, Gold-Label, 99.999%). Synthetic wastewater containing 1 $\times$ 10^{-3}M Chromium (52 mg/L) was used for all experiments. Background ionic strength was maintained at 3.0 $\times$ 10^{-1}M Na_2SO_4 as a swamping electrolyte.

The pH was adjusted by using 1M and 0.1M solutions of NaOH and H_2SO_4 prepared from concentrated reagent grade stocks (Fisher Scientific, Inc.). Experimental systems were diluted by no more than 1% due to addition of acid or base solutions. Hydrated lime (Fisher Scientific, Inc.) was also used to adjust the pH in some designated precipitation experiments.

Chemical Analyses

The pH values during all experiments were recorded with an analog pH meter (Fisher Accumet Model 610). The probe was standardized daily with NBS buffers (pH - 7.0, and either 4.0 or 10.0, Fisher Scientific, Inc.).

The chromium analyses were carried out in accordance with *Standard Methods* [21]. Chromium (VI) was measured directly by the absorbance of the purple Cr(VI)-diphenylcarbizide complex at 540 nm using a Spectronic 20 UV-VIS spectrophotometer (Bausch and Lomb, Inc.). Total chromium was

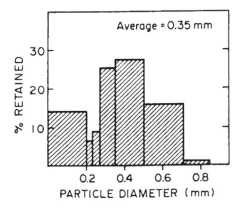

Figure 1.   Sieve analysis for iron particles.

determined by oxidizing the samples with $KMnO_4$ and then measuring the Cr(VI) by the above procedure. The Cr(III) was taken as the difference between the two measurements.

Iron was analyzed by the phenanthroline method outlined in *Standard Methods* [21].

Specific resistances were determined by the Buchner Funnel test as described by Vesilind [22]. A vacuum pressure of 200 mm of mercury was applied to the sludge samples. The small particle size of some of the samples made it necessary to use a filter aid (Hyflo Super-Cel, Johns-Manville, Inc., Denver, CO).

Jar tests were run to compare settleabilities. One-liter samples were stirred at 20 rpm for 10 minutes and then allowed to settle quiescently for 30 minutes. Grab samples of the supernatant were then taken at a fixed depth (5 cm) and analyzed for total Cr(III). i.e., soluble and particulate by acidification prior to chemical analysis.

Batch Kinetics

The rates of Cr(VI) reduction and O_2 reaction were studied in completely-mixed batch reactors. Two grams of iron filings were placed in 1 liter of synthetic wastewater, previously adjusted to the desired initial pH value, and grab samples were taken at various times for chromium analysis. The procedure was the same for studying the reaction of O_2 except that only the background electrolyte solution was used, i.e., no Cr(VI) was available. The effect of solution mixing was determined using baffled one liter reaction vessels with 1 1/2" x 5/8" stirring bars. Stirring rpms were measured using a strobe light (Strobotac type 1531, General Radio Co., Concord, MA). It was found that the rate of Cr(VI) reduction was independent of mixing beyond 1,000 rpm in this system. Therefore, all experiments were run beyond this limit, 1500 $\pm$ 50 rpm so that mixing did not play a role in the results or analysis.

Continuous Flow

A continuous flow reactor was operated in the completely-mixed mode. Samples were collected continuously using an automatic fraction collector (Buchler LC100).

RESULTS AND DISCUSSION

Iron exists in aqueous solution in one of two oxidation states, either $+II$ or $+III$, with the most stable state (in equilibrium with O_2 or Cr(VI)) being Fe(III) [23, 24]. The reaction from the ground state to the $+III$ oxidation state will proceed in two steps, or for Cr(VI):

$$3Fe^o + 2HCrO_4^- + 14H^+ \rightarrow 3Fe^{+2} + 2Cr^{+3} + 8H_2O \tag{1}$$

and,

$$3Fe^{+2} + HCrO_4^- + 7H^+ \rightarrow 3Fe^{+3} + Cr^{+3} + 4H_2O \tag{2}$$

and, for O_2:

$$2Fe^o + O_2 + 4H^+ \rightarrow 2Fe^{+2} + 2H_2O \tag{3}$$

and,

$$4Fe^{+2} + O_2 + 4H^+ \rightarrow 4Fe^{+3} + 2H_2O \tag{4}$$

Equations 1 through 4 are written for the chemical species which are most likely to predominate under the conditions of pH and Cr(VI) concentration in typical plating wastewaters [13, 24].

Rate Kinetics

The rates of Cr(VI) reduction and Fe oxidation by oxygen were examined in completely-mixed batch reactors. The results of these rate studies are shown in Figure 2. The reaction appears to be zero order for both Cr(VI) and O_2, and the rates increase to a maximum value as the pH decreases. Previously, Espenson and King [25] reported that reactions between Fe(II) and $HCrO_4^-$ were extremely complex being second order with respect to Fe(II) and varying between first and second order with respect to $HCrO_4^-$. Also, Singer and Stumm [26] showed that oxidation of Fe(II) with O_2 exhibited a first order dependence on Fe(II) and that the rate increased with increasing pH. The zero order rate of reaction obtained for the particulate iron in these experiments indicates that the surface oxidation of molecular iron to Fe(II) is the limiting reaction step for both Cr(VI) and O_2.

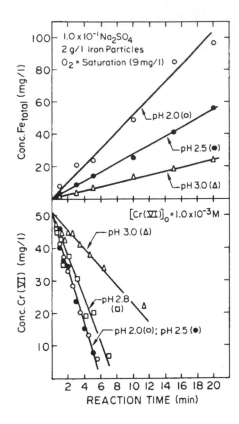

Figure 2. Batch kinetic data for reaction of D.O. (upper) and Cr(VI) (lower) with iron particles at various pH values.

The rate equations of interest are those for Cr(VI) reduction and total iron production. These may be expressed from the rates Cr(VI) and O_2 reactions, or:

$$\frac{dCr(VI)}{dt} = -k_{Cr} \tag{5}$$

k_{Cr} is the observed zero order rate constant for the reaction between the iron particles and Cr(VI) at a particular pH value and iron particle concentration (mg/L/min).

and,

$$\frac{dFe}{dt} = k_{Cr} + k_{DO} \tag{6}$$

where k_{DO} is the observed zero order rate constant between dissolved oxygen (at saturation concentration) and the iron surface at a particular pH value (mg/L/min). The observed rate constants must depend on the solution pH and the iron surface available for reaction (note that only 10% of the iron was used up during the course of the batch reactions, i.e., surface area constant). The observed zero-order rate constants may then be expressed by the following relationship:

$$k_{obs} = \frac{k_1[Fe]\{H^+\}}{k_2 + \{H^+\}} \tag{7}$$

in which k_{obs} is the observed rate constant, either k_{Cr} or k_{DO} (mg/L/min), k_1 is the maximum zero-order reaction constant (mg/g/L/min), k_2 is the half-velocity constant (M), $[H^+]$ is the activity of the hydrogen ion in solution (M), and [Fe] is the concentration of iron particles in solution (g/L). Note that k_1 and [Fe] may be expressed in terms of iron surface area instead of grams for application to other particle distributions. Then, Equation 7 may be inverted and linearized as follows:

Table II. Rate Constants for Aqueous Reactions Between Iron Particles and Cr(VI) or O₂

Surface Reactant	k_1 (mg/g/min)	k_2 (M)
Cr(VI)	10.5 (13.4 × 10⁻²)ᵃ	1.0×10^{-2}
O₂ᵇ	3.8 (5.0 × 10⁻²)ᵃ	1.1×10^{-2}

ᵃvalues in parentheses represent k_1 based on the iron surface area rather than weight (ca. 75 cm²/g); units are in mg/cm²/min.
ᵇO₂ is at the solubility limit (about 9.0 mg/L).

$$\frac{[Fe]}{k_{obs}} = \frac{1}{k_1} + \frac{k_2}{k_1}\frac{1}{\{H^+\}} \tag{8}$$

Values of k_{obs} were obtained from the slopes of the zero order lines in Figure 2 and a plot of $[Fe]/k_{obs}$ versus $1/\{H^+\}$ is shown in Figure 3. The data fit reasonably well with Equation 8 and a summary of the rate constants is presented in Table II.

The kinetic results presented here conflict somewhat with the earlier work of Gould, who reported that the rate of Cr(VI) reduction was half-order with respect to Cr(VI) and H^+ concentrations [19]. However, Gould worked with relatively pure iron wire which appeared to have a rust preventative coating. Other differences may be accounted for by the iron particle size, which was smaller in this work, i.e., Gould's surface area was 20.78 cm²/g versus 75 cm²/g in this work.

Continuous Operation

Since the reaction for Cr(VI) reduction is zero order, then no advantage will be gained by designing a process in the plug flow versus the completely-mixed mode. A bench-scale system was then designed as a continuous flow completely-mixed process in order to evaluate the effectiveness of the iron particles during continuous operation. The results of the continuously operated system are shown in Figure 4. Predictions based on Equation 5 and 6 (dashed lines in Figure 4) show good agreement with the observed data, although somewhat more iron was oxidized into solution after the initial hour than was predicted.

One major objective of this process was to reduce the amount of total iron introduced as a result of the treatment scheme and significantly reduce the sludge production compared to ferrous sulfate

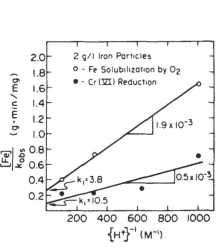

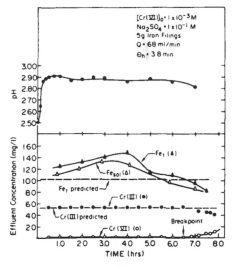

Figure 3.   Analysis of pH-dependent rate constants for the reaction of O₂ and Cr(VI) with iron particles.

Figure 4.   Completely-mixed continuous flow reactor conditions as a function of time.

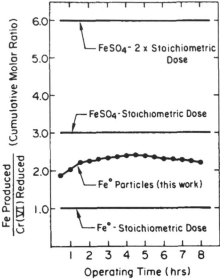

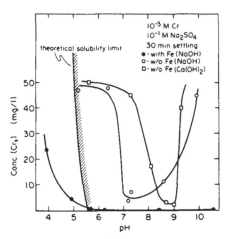

Figure 5. A comparison of cumulative molar ratios of Fe produced to Cr(VI) reduced in continuous flow reaction systems.

Figure 6. A comparison of precipitation, coagulation, and settling data for Cr(III) precipitated with NaOH and Ca(OH)₂ versus Cr(III) produced by reduction of Cr(III) with iron filings and co-precipitation of Cr(OH)₃ and Fe(OH)₃. Note: theoretical solubility for Cr(OH)₃ calculated from constants reported in Sillen and Martell [32].

reduction. The cumulative ratio of total iron produced by Cr(VI) reduced was evaluated over the 8-hour operating period for the continuous system. This ratio is compared to the stoichiometric requirements for Fe° and Fe^{+2} as well as the typical operational dosage of two times the stoichiometric amount for Fe^{+2} and presented in Figure 5. Even including the additional iron produced by the reaction with O_2, the iron filings produced substantially less aqueous iron than the stoichiometric requirement for $FeSO_4$, i.e., 2.2 vs 3.0 for the cumulative ratio after 8 hours of continuous operation.

Effect of Solution Fe on Precipitant Characteristics

Iron, usually applied as ferric chloride or ferrous sulfate, provides an excellent substrate for co-precipitation of a variety of metals and is frequently used as a conditioning agent for organic or inorganic sludges [27,28,29]. In addition, the resulting iron oxide or hydroxide particles are excellent adsorbents for more soluble "free" and complexed metals [30,31]. The use of ferrous sulfate or iron filings as a reducing agent will result in one or more of these "free" added benefits to the treatment process. However, the final total dry weight of the precipitated solids will be increased significantly due to the amount of highly insoluble ferric ions that are produced during the reduction of Cr(VI).

To evaluate and compare the effectiveness of Cr(VI) precipitation processes, a sample of synthetic wastewater (containing Cr(VI) initially) was first reduced with iron filings, and another synthetic wastewater was prepared with Cr_2O_3(Cr(III)). A comparison of these two systems after precipitation and settling is shown in Figure 6 based on total chromium as soluble and residual particulate matter. The wastewaters containing Cr(III) without the iron produced by oxidation show a typically narrow pH-operating range, ca. pH 7.2 to 8.0 for NaOH neutralization and 8.5 to 9.1 for Ca(OH)₂ neutralization. On the other hand, the iron reduction process produces an effluent which exhibits a much broader operating range, ca. pH 5.5 to 10.5. In addition, the iron system shows much better chromium removals over the entire pH range and removal even exceeds the theoretical solubility of Cr(OH)₃ at pH < 5.5. [32]. This implies that the iron hydroxide precipitate not only improves the coagulation and settling of the Cr(OH)₃ but is significantly adsorbing Cr + 3 ions from solution as

Table III. Comparison of Specific Resistances of Sludge Prepared from the Iron Process and Without Iron

Sludge Conditions	Specific Resistance (sec^2/g)
w/o iron (NaOH precipitated)	5.4×10^9
with iron (NaOH precipitated)	5.3×10^8
w/o iron (polymers added)	
0.70[a]	1.5×10^9
0.95	2.2×10^8
1.35[b]	0.8×10^8

[a] values represent the best performing polymer from a group of three which were screened, units are in lbs/ton of dry solids.
[b] Approximate optimal dosage.

Table IV. Pertinent Characteristics of Various Cr(VI) Reduction-Precipitation Schemes

Reducing Agent[a]	Excess Sludge Production	Conditioning Benefits	Handling Hazards[c]
FeS	yes	yes	1[d]
FeSO$_4$	yes	yes	1[e]
NaHSO$_3$	no	no	2[f]
SO$_2$	no	no	3[g]
Iron	yes[b]	yes	0[h]

a: ranked according to highest costs first (from Table I)
b: produces less excess sludge than FeSO$_4$
c: compounds ranked from 0 to 3 according to the Acute Local Toxic Hazard Rating, with 0 being no hazard and 3 being highest hazard [34]
d: possible H$_2$S generation (Hazard Rating = 3) [36]
e: slight to moderate irritant [36]
f: slight to moderate irritant [36]
g: may cause permanent injury after short exposure to small quantities [36]
h: hazardous only upon chronic exposure to dust [36]

well. Wilson et al. [33,34,35] have shown that Fe(III)-hydroxide is an effective adsorbing colloid for chromium and a variety of other metals.

Comparison of Dewatering Characteristics

The dewaterability of the two sludge types was judged based on the specific resistant values obtained from Buchner Funnel Tests [22]. Both sludges were precipitated using caustic (NaOH) to maintain uniformity. Three polymers (ones recommended by manufacturers for inorganic sludges) were compared as conditioning agents for the sludge produced without iron (III) and the best one was examined at several dosages. A comparison of the specific resistances is presented in Table III. The iron(III) sludge appears to be much more dewaterable than the sludge without iron, i.e., specific resistance without iron is 10x greater than the sludge with iron. However, with the addition of polymers, the specific resistance of the sludge without iron was decreased greatly and the iron(III) sludge was surpassed, i.e., specific resistance without iron but conditioned with polymers was 6.6x less than sludge with iron. This would imply that the sludge without the iron(III) may be conditioned and dewatered without the excess sludge produced by the iron process, but only with the added cost of the polymers. The excess sludge may be warranted on a cost basis in many situations where sludge disposal is not a problem due to the offsetting chemical costs for reduction of Cr(VI) and the costs of added polymers.

SUMMARY AND CONCLUSIONS

Scrap iron filings have been successfully demonstrated as a reducing agent for Cr(VI) in acidic solutions. The chemical costs for Cr(VI) reduction using this process are substantially lower than conventional reducing agents (see Table I).

In addition, the solubilized iron provides a well-conditioned sludge product. Both settleability and specific resistance of the resultant Cr(OH)$_3$ sludge were improved dramatically by co-precipitation with the Fe(OH)$_3$ byproduct. Also, these and other studies indicate that the Fe(OH)$_3$ precipitate serves as a co-precipitate and adsorbent for "free" and complexed metals in aqueous solution.

A comparison of the characteristics of the iron process with conventional reduction/precipitation schemes is shown in Table IV. This comparison also indicates that large iron particles (not dust) create no special handling or health hazard in the work environment as well.

Considering the cost, conditioning, and handling advantages of the iron-Cr(VI)-reduction process, it may have a variety of useful applications within the metal-finishing industry. This is particularly true in situations where the excess sludge production would not contribute significantly to the overall treatment and sludge disposal costs.

REFERENCES

1. Chamberlain, N. S., and Day, R. V., "Technology of Chrome Reduction with Sulphur Dioxide," *Proceedings of the 11th Industrial Waste Conference*, Purdue University, 129 (1956).
2. Rouse, J. V., "Removal of Heavy Metals from Industrial Effluents," *J. Environ. Engrg. Div., ASCE*, 102, 5 (1976).
3. Dean, J. C., Bosqui, F. L., and Lanouette, K. H., "Removing Heavy Metals from Wastewater," *J. Environ. Sci. Technol.*, 6, 518 (1972).
4. Thomas, J. J., and Theis, T. L., "Effects of Selected Ions on the Removal of Chrome (III) Hydroxide," *J. Water Pollut. Control Fed.*, 48, 2032 (1976).
5. Bowers, A. R., Huang, C. P., and Chin, G. L., "Predicting the Performance of a Lime-Neutralization/Precipitation Process for the Treatment of Some Heavy Metal-Laden Industrial Wastewaters," *Proceedings of the 13th Mid-Atlantic Industrial Waste Conference*, Ann Arbor Science, Ann Arbor (1981).
6. Hannah, S. A., Jelus, M., and Cohen, J. M., "Removal of Uncommon Trace Metals by Physical and Chemical Treatment Processes," *J. Water Pollut. Control Fed.*, 49 (1977).
7. Higgins, T. E., and Termaath, S. G., "Treatment of Plating Wastewaters by Ferrous Reduction, Sulfide Precipitation, Coagulation and Upflow Filtration," *Proceedings of the 36th Industrial Waste Conference*, Purdue University, 462 (1981).
8. Higgins, T. E., and Sater, V. E., "Results of Pilot-Plant Tests on Treating Electroplating Wastes by Alkaline Reduction, Precipitation, and Upflow Filtration," *Environ. Prog.*, 3, 12 (1984).
9. Barnes, G. E., "Complex Metal Finishing Wastes Licked by Effective Chemical Treatment," *Wastes Engineering*, 8, 124 (1957).
10. Serota, L., "Science for Electroplaters: Treatment of Chromate Wastes," *Metal Finishing*, 55, 65 (1957).
11. "Control and Treatment Technology for the Metal Finishing Industry: Sulfide Precipitation," U.S. EPA (1980).
12. Bowers, A. R., and Huang, C. P., "Activated Carbon Processes for the Treatment of Chromium(VI)-Containing Industrial Wastewaters," *Prog. Water Technol.*, 12, 629 (1980).
13. Huang, C. P., and Bowers, A. R., "Activated Carbon Process for Treatment of Wastewaters Containing Hexavalent Chromium," *EPA-600/2-79-130*, U.S. Environmental Protection Agency, Cincinnati, Ohio (1979).
14. Paulson, C. F., "Chromate Recovery by Ion Exchange," *Proceedings of the 7th Industrial Waste Conference*, Purdue University, 36, 6, 209 (1952).
15. Tabakin, R. B., and Ciancia, J., "An Ion Exchange Process for Recovery of Chromate from Pigment Manufacturing," *EPA-670/2-74-044*, U.S. Environmental Protection Agency, Cincinnati, Ohio (1974).
16. Kunz, R. G., Hess, T. C., and Yen, A. F., "Kinetic Model for Chromate Reduction in Cooling Tower Blowdown," *J. Water Pollut. Control Fed.*, 52, 2327 (1980).
17. Taylor, C. R., and Qasim, S. R., "More Economic Treatment for Chromium-Bearing Wastes," *Proceedings of the 37th Industrial Waste Conference*, Purdue University, 189 (1982).
18. Gould, J. P., "The Kinetics of Hexavalent Chromium Reduction by Metallic Iron," *Water Res.*, 16, 871 (1982).
19. Joseph T. Ryerson and Sons, Inc., *Stocks and Services*, Minneapolis, MN (1976).

20. *Chemical Marketing Reporter*, 228 (July 27, 1985).
21. *Standard Methods for the Examination of Water and Wastewater*, 15th Edition, American Public Health Association, New York, NY (1980).
22. Vesilind, P. A., *Treatment and Disposal of Wastewater Sludges*, Ann Arbor Science, Inc., Ann Arbor, MI (1975).
23. Stumm, W., and Morgan, J. J., *Aquatic Chemistry*, 2nd Edition, John Wiley and Sons, Inc., New York, NY (1981).
24. Pourbaix, M., *Atlas of Electrochemical Equilibria in Aqueous Solutions*, Pergamon Press, Inc., New York, NY (1966).
25. Espenson, J. H., and King, E. L., "Kinetics and Mechanisms of Reactions of Chromium(VI) and Iron(II) Species in Acidic Solutions," *J. Amer. Chem. Soc.*, 85, 3328 (1963).
26. Singer, P. C., and Stumm, W., "Acid Mine Drainage: The Rate Determining Step," *Science*, 167, 3921 (1970).
27. Pauling, L., *The Chemical Bond*, 3rd Edition, Cornell University Press, Ithaca, NY (1960).
28. Ives, K. J., *The Scientific Basis of Flocculation*, Sijthoff and Noordhoff, The Netherlands (1978).
29. Dick, R. I., "Sludge Treatment," in *Physicochemical Processes*, W. J. Weber, editor, John Wiley and Sons, Inc., New York, NY (1972).
30. Davis, J. A., "Adsorption of Trace Metals and Complexing Ligands at the Oxide/Water Interface," *Ph.D. Dissertation*, Stanford University, Stanford, Ca (1977).
31. Davis, J. A., and Leckie, J. O., "Effects of Adsorbed Complexing Ligands on Trace Metal Uptake by Hydrous Oxides," *Environ. Sci. Technol.*, 12, 1309 (1978).
32. Sillen, L. G., and Martell, A. E., *Critical Stability Constants*, Special Publication No. 17, The Chemical Society, London (1964).
33. Curran, B. L., Kennedy, R. M., Clarke, A. N., and Wilson, D. J., "Electrical Aspects of Adsorbing Colloid Flotation. X. Pretreatments, Multiple Removals, Interferences, and Specific Adsorption," *Sep. Sci. Technol.*, 14, 669 (1979).
34. Huang, S. D., and Wilson, D. J., "Hexavalent Chromium Removal in a Foam Flotation Pilot Plant," *Sep. Sci. Technol.*, 19, 603 (1984).
35. Thackston, E. L., Wilson, D. J., Hanson, J. S., and Miller, D. L., Jr., "Lead Removal with Adsorbing Colloid Flotation," *J. Water Pollut. Control Fed.*, 52, 317 (1980).
36. Sax, N. I., *Dangerous Properties of Industrial Materials*, 4th Edition, Van Nostrand Reinhold, Co., New York, N.Y. (1975).

APPENDIX

Notation

Cr(VI)	=	hexavalent chromium concentration (mg/L)
Fe	=	soluble iron concentration (mg/L)
[Fe]	=	iron particle concentration in solution (mg/L)
$\{H^+\}$	=	activity of the proton (M)
k_{Cr}	=	observed zero-order rate constant for the reaction between hexavalent chromium and iron particles (mg/L/min)
k_{DO}	=	observed zero-order rate constant for the reaction between dissolved oxygen and the iron particles (mg/L/min)
k_{obs}	=	observed zero-order rate constant as a function of $\{H^+\}$ (mg/L/min)
k_1	=	maximum specific zero-order rate constant mg/min/g)
k_2	=	half zero-order reaction rate constant (M)
t	=	time (minutes)

56 A CASE HISTORY OF REMOVING CADMIUM FROM ELECTROPLATING WASTEWATER USING CALCIUM SULFIDE

James D. Edwards, Associate

John W. Cammarn, Senior Chemical Engineer
Burgess & Niple, Limited
Columbus, Ohio 43220

INTRODUCTION

United States Environmental Protection Agency (U.S.EPA) Effluent Guideline Limitations for the Electroplating and Metal Finishing Point Source Categories limit the allowable discharge of cadmium to navigable waters and to Publicly Owned Treatment Works (POTW). Communities have often placed additional restrictions on cadmium discharges in an effort to protect sewage treatment system performance and to extend the usable life of sludge land application sites.

The following is a case history of the design, construction, and operation of a 330-gallon per minute (gpm) electroplating wastewater treatment facility for a large midwestern manufacturing facility. The facility uses both rack and barrel plating lines to electroplate chrome and cadmium. The treatment plant replaced an existing facility and was constructed to conform with U.S.EPA 40 CFR Part 433 regulations for pretreatment prior to discharge to a 2.7 mgd municipal wastewater treatment system. The local municipality has adopted a pretreatment ordinance which complies with U.S.EPA standards.

The wastewater treatment plant is one of the largest in the United States utilizing the calcium sulfide process to remove cadmium. The treatment plant utilizes two stage alkaline chlorination for cyanide destruction, sulphur dioxide for hexavalent chrome reduction, sodium hydroxide precipitation, calcium sulfide, precipitation and effluent filtration. The resulting sludge is dewatered in a plate and frame filter press followed by disposal off site as a hazardous waste.

The treatment plant was installed by using the design/construction method. Burgess & Niple Service Corp., a wholly owned subsidiary of Burgess & Niple, Limited, served as the general contractor. Burgess & Niple, Limited, served as the design engineer. The project was awarded in April 1984 and was on-line in March 1985. Construction cost was approximately $1,800,000 including design engineering, construction management, interconnecting sewers, equipment, buildings, and startup assistance.

WASTEWATER TREATMENT PROCESS DESCRIPTION

The manufacturing facility has separate gravity sewer systems for cyanide, hexavalent chromium, general rinse water and concentrated acid/alkaline dumps. These sewers are connected to the wastewater treatment plant headworks which include a two stage alkaline chlorination system for cyanide destruction, a two stage system for chrome reduction, an influent equalization tank, and a batch holding tank for concentrated acid/alkaline dumps.

The headworks tanks are below grade and are isolated from the remainder of the treatment plant. The tanks have a separate ventilation system which is not connected with the main treatment plant. Access to the tanks is through Bilco doors located in the top slab of the tanks. The slab also serves as the inflow pump room and an area for the bulk liquid caustic storage tank, dilute caustic day tank, and feed pumps.

An emergency holding tank was provided as a part of the headworks. This tank can receive treated cyanide and chrome waste and general rinse water in the event of a power failure or influent pump shutdown.

A schematic of the wastewater treatment system is shown in Figure 1. Major design criteria for the headworks system is as follows:

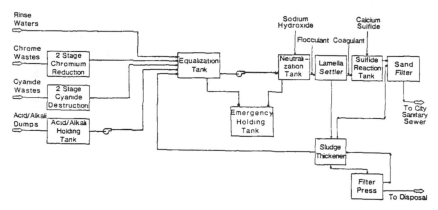

Figure 1.   Schematic of treatment system.

CYANIDE DESTRUCTION SYSTEM

1st Stage Tank Dimensions	9' x 9' x 5.75' SWD
2nd Stage Tank Dimensions	9' x 9' x 5.75' SWD
Working Volume Each Tank	3,500 gallons
Average Detention Time, Each Tank	60 Minutes
Mixer Horsepower, Each Tank	1 horsepower
1st Stage Chlorinator Capacity	1,500 lb/day
2nd Stage Chlorinator Capacity	1,000 lb/day

HEXAVALENT CHROMIUM DESTRUCTION SYSTEM

1st Stage Tank Dimensions	9' x 9' x 5.75' SWD
2nd Stage Tank Dimensions	9' x 9' x 5.75' SWD
Working Volume Each Tank	3,500 gallons
Average Detention Time, Each Tank	60 minutes
Mixer Horsepower, Each Tank	1 horsepower
Sulfonator Capacity	100 lb/day

CONCENTRATED ACID/ALKALI DUMP TANK

Tank Dimensions	10' x 19' x 5.75' SWD
Working Volume	8,200 gallons
Mixer Horsepower	1 horsepower
Transfer Pump Capacity	200 gal/hr

EQUALIZATION TANK

Tank Dimensions	13' x 19' x 5.42' SWD
Working Volume	10,000 gallons
Mixer Horsepower	1 horsepower
Transfer Pumps	2 each
Transfer Pump Capacity	350 gpm avg.

Effluent from the sulfide reaction tank flows by gravity to a Parkson DynaSand® filter which removes the cadmium sulfide and other suspended solids which were not removed in the Lamella clarifier. Sludge from the DynaSand Filter flows by gravity to the sludge thickener. Treated effluent from the filter flows through an effluent metering and sampling station and is discharged to the municipal sanitary sewer system.

The hydroxide and sulfide sludges are combined and thickened in a gravity sludge thickener. The tank is located below grade outside the treatment building. Supernatant from the tank is returned by gravity to the wastewater influent equalization tank. Thickened sludge is pumped from the tank by two air operated diaphragm pumps to a Perrin plate and frame filter press. Filtrate from the press is returned to the thickener and ultimately to the influent equalization tank. The dewatered sludge cake is E.P. toxic and is disposed in a secure landfill.

Major design criteria for the treatment process is as follows:

Table I. Design Conditions

Parameter	Concentration Range
Total Cyanide	30 – 300 mg/L[a]
Total Chrome	50 – 100 mg/L[b]
Hexavalent Chrome	30 – 90 mg/L[b]
Total Cadmium	12 – 180 mg/L
Total Copper	0.1 – 2 mg/L
Total Zinc	1 – 8 mg/L

[a]To cyanide destruction system.
[b]To chrome reduction system.

NEUTRALIZATION TANK

Neutralization Tank Dimensions	10 dia. x 17.5' h
Working Volume	8,800 gallons
Mixer Horsepower	1 hp

FIRST STAGE CLARIFIER

Parkson Lamella	Model No. 1135/55
Typical Effluent Total Suspended Solids Concentration	20 mg/L
Typical Sludge Concentration	0.1 wt. percent.

SULFIDE REACTION TANK

Sulfide Reaction Tank	10' dia. x 14.5' h
Working Volume	7,000 gallons
Mixer Horsepower	1 horsepower

EFFLUENT FILTRATION

Parkson DynaSand	Model No. DSF-64
Typical Effluent Total Suspended Solids Concentration	5 mg/L
Typical Sludge Concentration	0.1 wt. percent

GRAVITY THICKENER

Gravity Thickener	30" dia x 8' SWD
Solids Loading Rate	0.8 lb/day/ft²
Hydraulic Loading Rate	100 gpd/ft²
Sludge Concentration	3-5 wt. percent

EMERGENCY HOLDING TANK

Tank Dimensions	43.5'x14.75'x10.5'SWD
Maximum Volume	50,000 gallons
Transfer Pump	1 each
Transfer Pump Capacity	105 gpm avg.

CAUSTIC STORAGE TANK

Storage Tank Dimensions	8' Dia. x 14'
Volume	5,000 gallons
Day Tank Volume	500 gallons
Circulating Pumps	2 each
Circulating Pump Capacity	5 gpm

TYPICAL TREATMENT PLANT INFLUENT COMPOSITION

The design influent conditions for design of the facility are given in Table I.

The equalization tank lift pumps transfer waste to the caustic neutralization tank, which serves to form cadmium and chromium hydroxide along with hydroxide compounds of the remaining heavy metals. The major component being removed is chrome hydroxide. The neutralization tank is located on the upper level of the treatment building in order to allow treated wastewater to flow by gravity through the remainder of the treatment system. Effluent from the neutralization tank flows to a Parkson Lamella® clarifier, which is equipped with flash mix and flocculator tanks to allow polymer addition and conditioning before settling. Sludge from the Lamella is discharged by gravity to a 30 foot diameter gravity sludge thickener which is located outside the treatment building.

Effluent from the Lamella flows to the sulfide reaction tank where the calcium sulfide slurry is added. The cadmium reaction with the sulfide is rapid. A coagulant is added to assist in flocculating the extremely fine cadmium sulfide particles. The sulfide reaction tank is covered and has a separate vent to the atmosphere in order to eliminate any buildup of sulfide odor in the treatment building.

SLUDGE DEWATERING

Perrin Plate and Frame Filter Press	1,200 mm
Number of Chambers	40
Dewatered Sludge Solids	30 wt. percent
Press Cycle Time	4.5 hours

CALCIUM SULFIDE TREATMENT SYSTEM

In view of the U.S. EPA 40 CFR Part 433 cadmium limitation of 0.26 milligrams per liter (mg/L) for discharge to POTWs, several alternatives to hydroxide precipitation were evaluated. Because of the large expected chromium concentrations, a two-stage hydroxide/sulfide process was determined to be most effective. Chrome and cadmium would be removed as hydroxide with the remaining cadmium being removed as a sulfide. Laboratory scale bench tests confirmed the suitability of the two stage process design.

The use of soluble sulfide compounds to produce sulfide ions for heavy metal precipitation has been practiced for some time. These systems use sodium hydrosulfide, sodium sulfide, or hydrogen sulfide. Each of these materials is relatively difficult to handle, the first two being hygroscopic solids and the third being a corrosive gas. Each releases a measure of hydrogen sulfide which has the characteristic of "rotten egg" odor in low concentrations and which is lethal in high concentrations.

Permutit Company has developed the proprietary Sulfex® process which uses iron sulfide to produce the sulfide ions for reaction with the cadmium. Due to the low solubility of iron sulfide, significant excess, in the order of 2 to 4 times the stoichiometric amount, must be added to achieve the required cadmium removal. The excess iron sulfide carries through the settling and dewatering systems as an essentially inert solid, thereby increasing the sludge generated when compared to either soluble sulfide or hydroxide precipitation.

General Electric Company has published a series of articles on their development of a system for producing calcium sulfide which can be used to remove heavy metals by sulfide precipitation. Development of the process was a result of their need to achieve extremely low copper concentrations in wastewater from a wire drawing operation. Advantages of the calcium system included little or no hydrogen sulfide odor, a sludge that was easier to coagulate and settle and reduced sludge volumes when compared to the iron sulfide system. Since the initial publication of their research, General Electric has obtained a Patent for the process and licenses its use.

The General Electric system uses hydrated lime and either sodium hydrosulfide or sodium sulfide to produce the soluble sulfide. Because General Electric had no facilities installed or in operation on a large scale, this facility was the first and required design of the calcium sulfide mixing and feed system. General Electric's research indicated that when sufficient calcium sulfide had been added, the wastewater pH rose sharply thereby indicating that the metallic reactions were complete. Laboratory testing done during the design phase failed to indicate a consistent indicator point, and therefore, this process control method was abandoned.

A schematic flow diagram of the sulfide system is shown on Figure 2. The calcium sulfide slurry is made batchwise approximately once per operating shift. Bagged hydrated lime is fed by a volumetric feeder and is slurried to approximately 0.55 wt. percent in a 600 gallon storage tank. As the lime is being fed, either sodium sulfide or sodium hydrosulfide is manually weighed and added to the slurry tank. The stoichiometric ratio of calcium to sulfide is maintained at 1.1–1.2 by regulating the time the volumetric lime feeder runs and the amount of sulfide added. The lime slurry from the volumetric feeder flows by gravity directly into the calcium sulfide slurry tanks. While the hydrated lime and sulfide are being charged, dilution water is added to make up the proper volume to be fed during the shift. The storage tank is agitated continuously to thoroughly mix the lime and sulfide during preparation and storage.

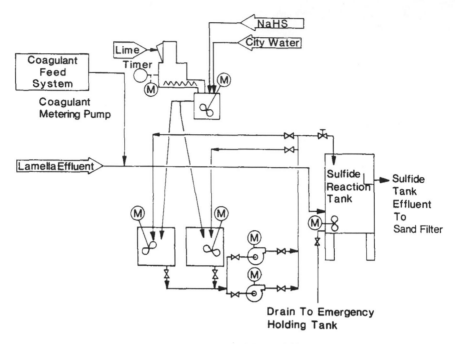

Figure 2. Schematic of calcium sulfide system.

Two calcium sulfide batch tanks were provided, one being on line and one being prepared for the next shift. Two centrifugal pumps were provided, one in use and one standby, to pump the calcium sulfide slurry up to the sulfide reaction tank. In order to reduce plugging, the calcium sulfide slurry is pumped at a high rate through a circulating line and the excess is returned to the calcium sulfide batch tank. A control valve regulates the amount of calcium sulfide fed to the sulfide reaction tank. Effluent sulfide concentration is tested and the sulfide dose regulated to maintain an excess.

Lime and sodium sulfide are stored in a chemical feed room with a separate ventilation system. The calcium sulfide batch storage tanks and feed pumps are also located in a separate room which is ventilated. Typical lime doses are 46–50 lb/1,000 gallons with sodium sulfide/hydrosulfide doses in the range of 34 to 38 lb/1,000 gallons in the batch tanks. The molar ratio of calcium to sulfide in the batch tanks is approximately 1.0 to 1.5. Typical calcium sulfide concentration in the sulfide reaction tank is 27 mg/L. The molar ratio of sulfide to cadmium in the sulfide reaction tank is 9 to 1.

Once in the reaction tank the calcium sulfide reacts according to the following mechanisms:

$$Ca(HS)_2 \rightarrow Ca^{++} + 2HS \tag{1}$$

$$Cd^{++} + HS \rightarrow MS + H^+ \tag{2}$$

$$HS + OH \rightarrow H_2 + S^= \tag{3}$$

$$Cd^{++} + S^= \rightarrow CdS \tag{4}$$

$$OH + H^+ \rightarrow H_2O \tag{5}$$

The calcium hydroxide particles provide nucleation sites for the precipitation of the cadmium sulfide resulting in a somewhat larger particle. However, the particles which are produced are still insufficient to provide good settling. A coagulant is added to the sulfide reaction tank to assist in producing larger particles which can be removed in the effluent filter.

SYSTEM PERFORMANCE

The following results for the calcium sulfide system were obtained during acceptance tests which were performed at the wastewater treatment plant:

Parameter	Sulfide System Feed	Sulfide System Effluent
Total Cadmium	3.3 mg/L	0.17 mg/L
Total Chrome	0.26 mg/L	0.09 mg/L
Hexavalent Chrome	0.02 mg/L	0.005 mg/L
Total Copper	0.17 mg/L	0.05 mg/L
Total Nickel	0.05 mg/L	0.04 mg/L
Total Lead	0.07 mg/L	0.04 mg/L

These removals reflect both the precipitation and removal of the metallic sulfides through the effluent sand filter as well as the removal of the remaining chrome hydroxide which remained after the first stage precipitation and clarification.

Removal rates were calculated for both the first stage hydroxide and second stage sulfide systems on the operating plant and are as follows:

Parameter	Hydroxide Removal Percent	Sulfide Removal Percent	Overall Removal Percent
Total Cadmium	86	95	99.3
Total Chromium	93.5	64	97.8
Total Copper	—	71	71[a]
Total Zinc	—	87	87[a]

[a]Influent concentrations less than 0.5 mg/L to first stage.

CONCLUSIONS

The calcium sulfide system is a safe, easy-to-operate means of treating electroplating wastewaters containing cadmium. Effluent dissolved cadmium concentrations are well below the limitations established by U.S. EPA Part 413 and 433 regulations. Effluent quality is dependent upon the success of coagulating and filtering the resultant cadmium sulfide particles which are produced in the treatment step.

REFERENCES

1. Kim, B. M., "Wastewater Treatment Process," *United States Patent 4,329,224* (Issued May 11, 1982).
2. Kim, B. M., and P. A. Amodeo, "Calcium Sulfide Process for Treatment of Metal-Containing Wastes," *Environmental Progress*, Vol. 2, No. 3 (August 1983).
3. Kim, B. M., "Treatment of Metal Containing Wastewater with Calcium Sulfide," *The American Institute of Chemical Engineers*, No. 0065-8812-81-4842-0209 (1981).
4. Higgins, Thomas E., and Stephen G. TerMaath, "Treatment of Plating Wastewaters by Ferrous Reduction, Sulfide Precipitation, Coagulation and Upflow Filtration," *Proceedings of the 36th Industrial Waste Conference, Purdue University* (May 1981).

Section Seven
OILFIELD AND GAS PIPELINE WASTES

57 DELIQUIFICATION OF OILFIELD DRILLING SLURRIES AND PRODUCTION SLUDGES

Stephen D. Field, Assistant Professor
Department of Civil Engineering

Andrzej Wojtanowicz, Assistant Professor
Department of Petroleum Engineering

Kevin Cange, Graduate Research Assistant
Department of Civil Engineering
Louisiana State University
Baton Rouge, LA 70803

INTRODUCTION

Recent concern over the past improper disposal of liquid and solid wastes has led to the creation of several regulatory systems to assure proper waste management. Although waste drilling fluids associated with the exploration and production of oil and natural gas are not classified as hazardous wastes, the potential for mismanagement of the waste drilling fluids exists. Waste drilling slurries (muds) and production sludges from exploration and production activities have typically been discharged into earthen surface impoundments, or pits. The waste drilling fluids consist of drilling mud(s), borehole cuttings, and various other wastes generated at the drill site [1]. Consequently,the pits may contain a variety of materials, including heavy metals, crude oils, diesel oils, high salt concentrations, and bentonite (mud) suspensions.

Regulations have been developed by the State of Louisiana requiring proper storage, treatment, and disposal of waste drilling fluids. Because of the large number of pits in existence (over 10,000 in Louisiana), volume reduction is of critical concern. This paper presents results of chemical conditioning, vacuum filtration, and pressure filtration techniques for deliquifying water-base and oil-base drilling muds and production sludges.

Reserve Pits

The majority of on-shore oil and gas exploration and production wells in Louisiana are drilled using water-base muds. Water-base muds are colloidal suspensions of low quality clays, bentonite, and water, and commonly are used for shallowwell drilling. These muds may contain a variety of additives to improve the required characteristics for drilling purposes. Barite, or barium sulfate, is used as a weighting agent to increase the density of the mud. Increasing the density of the mud serves to keep borehole cuttings from settling in the mud. Lignite and chrome-lignosulfonates are used as thinners. A variety of corrosion inhibitors may also be used in the muds including sodium hydroxide and zinc carbonate. In addition to these and other additives, materials such as lead from pipe dope and crude and diesel lubricating oils may be unintentional constituents of the water-base muds.

During surface drilling, water is added to control the mud viscosity, which dilutes the drilling mud, resulting in excess volumes. The mud, along with cuttings removed from the mud via shaker screens, hydrocyclones and centrifuges for desanding and desilting, and mud cleaning , traditionally have been disposed in reserve pits. This mud and associated cuttings are non-toxic and could be returned to the environment in many cases with minimum management practices such as neutralization (if required) and land spreading (dispersing).

oil, contain a variety of proprietary additives and typically require much more sophisticated treatment prior to disposal on-site.

In most cases, deep wells are drilled first with water-base muds, and then the oil-base muds are introduced at greater depths. Past practices have been to use a reserve pit for storing waste drilling muds from both deep and shallow operations, thereby contaminating the water-base mud.

In Louisiana, on-site closure options for reserve pits using land treatment are regulated based on oil and grease content, pH, metals content (As, Ba, Cd, Cr, Pb, Hg, Se, Ag, Zn), soluble salts and cationic distribution. Moisture content requirements are added for on-site burial or trenching. Pit contents not meeting the regulatory criteria must be transported to permitted commercial oilfield treatment and storage facilities. Liquid discharge to surface waters is regulated by the National Pollution Discharge Elimination System (NPDES).

Production Pits

When a well has been put into production, the oil from the well is usually associated with varying quantities of salt water, sand and silt. Older production wells typically contain more salt water produced per barrel of oil than newer wells. The oil is separated from the water at the well using sedimentation and flotation processes. Sand and silt are removed by sedimentation processes at the well site. The sand , silt, and salt water, along with crude oil not completely separated, is discharged to a pit for storage and additional separation. Salt water and rainwater collected from the pit is typically deep-well injected after being treated by chemical treatment and filtration to prevent injection well formation clogging. Oil that collects on the surface of the pit is collected and reclaimed when quantities are great enough to skim. Solids are left to accumulate in the pit until closure.

Production pits differ from reserve pits in that they usually contain no chemical constituents other than crude oils and salt water, which limit their disposal opportunities. Both reserve pit and production pit treatment would benefit from effective solids-liquid separation treatment for volume reduction prior to final disposal.

METHODS

Samples of reserve pit muds and production pits were collected from various pits around the State of Louisiana. These samples were characterized according to source and analyzed for pH, alkalinity, chlorides, percent solids (by weight and by volume), and percent oil. Percent oil and percent solids by volume were determined by retort analyses. All analyses followed documented procedures [2,3] for water and solid portions.

Reserve pit and production pit samples were initially screened for dewaterability using various combinations of pH adjustment and polymer addition. pH adjustment using sulfuric acid, hydrochloric acid, ferric chloride, ferric sulfate and alum was examined. Polymer addition evaluation addressed 36 different polymers from five different manufacturers. The polymers included cationic, anionic, and nonionic with varying charge densities and molecular weights. Manufacturer specifications typically characterize polymer products with molecular weight in the range of 1–10 million as flocculants, and those with molecular weights of 50–100 thousand as coagulants. Oil soluble anionic and nonionic polymers were labeled emulsion breakers. According to these definitions, polymers available included 14 cationic, 9 anionic, 3 nonionic, and the remaining unidentified with respect to charge. Of these, 3 would be classified as coagulants, 5 as emulsion breakers, 25 as flocculants; the remaining were unclassified. Preliminary screening of polymers was conducted qualitatively for sample water release, and character of floc as soft or hard with reference to withstanding filtration through filter fabric. Fifty milliliter aliquots were mixed by hand with a spatula to disperse the chemicals; care was taken to avoid over-mixing and destruction of formed floc.

Based on the preliminary screening of pH adjustment and polymer addition, a typical treatment performance was used to evaluate 9 different filter cloths. Cloths evaluated included a range of thread counts from 24 by 21 to 110 by 45. Performance was judged by character of the cake formed in terms of dryness and release from the filter cloth, penetration of solids into the filtercloth (blinding), and quality of filtrate in terms of solids carrythrough.

Bench scale evaluations on a pressure filtration apparatus were initiated for those samples and pH-polymer treatment combinations that showed promise. The pressure filtration unit was a mechanical screw press equipped with top and bottom surface filter cloth and drainage collection devices and a pressure gauge to measure pressure applied. Measurements for performance were free drainage, pressure drainage and cake solids concentrations. Maximum pressures were limited to 30 psi. Time of

Table I. Characteristics of Representative Drilling Muds and Production Pits

Pit Description	pH	Alkalinity (mg/L as CaCO$_3$)	Chlorides (mg/L)	% Solids (by wt.)	% Oil (by vol.)
Non-dispersed water-base mud	8.0	1,700	120	10	0
Dispersed water-base mud A	11.4	7,800	500	21	0
Dispersed water-base mud B	9.2	13,800	2,500	43	1
Offshore barge (mud & production sludge)	7.5	450	7,200	48	8
Production sludge	6.7	a	27,000	53	8

a Unobtainable.

drainage for the test was similar to that typically experienced in belt filter operations, 30 seconds free drainage. Pressure was applied until no more water release occurred. Bench scale testing for detailed analyses of pH-polymer dose response curves was performed on selected samples with the previously determined most successful treatment combination for each sample.

Based on bench scale performance tests, full scale testing was performed on a belt filter press (2 meter Winkel press) and a vacuum extractor (0.3 meter). Sample selection was limited to a production sludge and a water-base mud due to lack of sample volumes easily obtainable and time available for equipment testing. The results are compared to centrifuge performance on the same samples.

RESULTS

The results of analyses of selected samples are presented in Table I. The non-dispersed water-base mud contains no thinners or other known additives, representing the least chemically complex type of mud to be treated. The two dispersed water base muds are chrome lignosulfonate muds from different sources, A & B. These samples demonstrate the variability in character of similar muds, specifically oil content and solids content in water-base mud systems. The offshore barge sample contains a mixture of mud and production sludges of unknown origin. The production sludge represents a sample collected from the bottom of an active pit used to separate salt water, solids, and oil.

Filter cloth evaluation demonstrated that the highest thread count (110 by 45) tested performed the best. This filter cloth produced the best results in terms of filtrate quality, cake form, and reduced floc penetration into the filter cloth, and was used throughout the remainder of the bench and full scale testing.

The bench scale testing of pH adjustment and polymer addition provided preliminary optimum performance data for different types of polymers for each sample. These results, presented in Table II, indicate that more than one type of polymer may give satisfactory results. Typically, the anionic and nonionic polymers work equally well; the cationic polymers are less successful on all the muds tested. The nonionic polymers are favored over the anionic polymers primarily because of their tendency to produce a clearer filtrate and less blinding of the filter cloth. pH adjustment is required on the reserve mud pit samples for satisfactory performance, but it is less critical for the production sludges. Emulsion breakers were used for those samples containing oil. The emulsion breakers, as can be seen by the results of dispersed water-base mud B in Table II, do provide flocculation of the sample to some degree. For the production sludges with larger fractions of oil, the emulsion breakers serve to separate oils from the filtrate; the flocculants provide the primary means of chemical treatment.

Table III presents the results of full scale testing of the belt filter press, vacuum extractor and centrifuge on two samples, the dispersed water-base mud, and the production sludge. The performance of both the belt filter press and the centrifuge were superior to the vacuum extractor in terms of cake solids and hydraulic loading. The feed rate for the vacuum extractor, when trebled to account for difference in belt width from the filter press, still falls below the performance of the other equipment. As noted in Table III, for the production sludge the belt press was only tested under the one condition listed and should not be considered optimized.

Table II. Performance of Flocculants on Various Types of Pit Samples

Sample	Polymer Type	Polymer Dosage (lb/ton)	pH Adjustment	Volume Reduction (%)	Final Cake Solids (%wt)	Remarks
Non-dispersed	+	N/A	–	–	–	No success
water-base mud	0	11.6	5.5, Fe$_2$(SO$_4$)$_3$	71	46	–
(20%) solids)	–	10.5a	5.5, Fe$_2$(SO$_4$)$_3$	42	34	–
Dispersed water-	+	19.9a	7, HCL	62	45	–
base mud A	0	6.6	7, HCL	56	47	Some blinding
(21%)	–	5.0a	7, HCL	54	50	Some blinding
Dispersed water-	+ +EB	1.4a	6, HCL	60	61	EB: emulsion
base mud B						breaker
(43%)	0 +EB	1.4	6, HCL	54	61	–
	– +EB	3.4a	6, HCL	58	61	Blinding, cloudy flt.
	Emulsion					
	breaker	0.03a	6, HCL	50	54	Blinding, cloudy flt.
	only					
Offshore barge	+	N/A	–	–	–	–
(mud & produc-	0 +EB	4.7	None	39	64	–
tion sludge)	– +EB	3.9a	None	44	72	–
(39%)						
Production	+ N/A	–	–	–	–	No success
sludge	0 +EB	1.8	6, FeCl$_3$	37	80	–
(53%)	- +EB	5.1a	6, HCL	4	55	–

a Densities assumed, no product information available.

DISCUSSION

The variety of drilling mud characteristics is not fully depicted in Table I. Of the 21 samples collected during this study from different pits around Louisiana, the pH varied from 6.7 to 11.9. In a study performed by the Shell Oil Company [1], 309 pits were sampled demonstrating a pH range from less than 3 to greater than 10. The majority (74%) of the 309 pits also showed oil and grease levels less than 1%, and it was estimated that approximately 50% of all pits have a need for treatment due to excessive salt contamination. Additional conclusions from the Shell study indicate that only 16% of the pits sampled would require special management for barium; less than 3% of the pits would require special management for any other metals. From this study it is apparent that salt contamination presents the major factor necessitating special management techniques for most pits. Solids-liquid separation is the most viable solution to removing salts from the muds for volume reduction prior to liquid and solids disposal.

The performance of polymers during bench scale studies indicated that the better polymers were the nonionic and moderately charged anionic ones. Dosages for the two types were similar. The cationic polymers were moderately charged and worked only for the dispersed mud systems. High and super

Table III. Optimum Full Scale Equipment Performance

Pit Type	Unit	Feed Solids (%, w/w)	Feed Rate (GPM)	Polymer Dosage (lb/ton)	Cake Solids (%, w/w)	TSS Effluent (mg/L)
Non-dispersed water-	Belt press	9	50	18.1	45	160
base mud	Centrifuge	13	50	8.5	45	110
	Vacuum Extractor	10	5	13.6	21	150
Production sludge	Belt pressa	33	30	8.0	54	3,970
	Centrifuge	25	20	4.8	63	100

a Only condition tested for production sludge, no coagulant pretreatment.

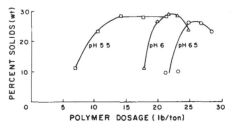

Figure 1. Influence of pH on nonionic, high molecular weight polymer performance for the treatment of non-dispersed water-base mud. Initial solids content = 10.5%.

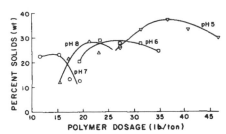

Figure 2. Influence of pH on nonionic, high molecular weight polymer performance for the treatment of dispersed water-base mud A. Initial solids content = 9.7%.

charged polymers did not work as well with and without pH adjustment. The molecular weight of the better performing polymers were in the range of 5 million for the nonionic polymers and 15 million for the anionic polymers. The cationic polymers were also of high molecular weight.

The better performance of the higher molecular weight polymers indicate that the bridging mechanism is important in slurry dewatering [4]. This advantage comes from longer polymer length but may be negated with increasing molecular weights where less chains per unit weight and viscosity increases reduce polymer effectiveness. The molecular weight range found effective in this study is similar to the upper limit of effectiveness for coal slurries containing appreciable amounts of bentonite [4].

The influence of pH on polymer performance is demonstrated in Figures 1 and 2. In Figure 1, the non-dispersed water-base mud demonstrates reduced polymer dosage per dry ton of solids at a pH of 5.5. This effect was demonstrated regardless of pH adjustment chemical used; however, ferric sulfate and ferric chloride seemed to be slightly more effective, possibly due to the weight of the iron.

Figure 2 demonstrates an increased effectiveness as determined by percent solids in the cake but with increasing polymer dosage. This result may indicate that perhaps a moderately charged polymer would perform better. These indications have yet to be demonstrated through bench scale testing in this study.

The pH influences on polymer performance are suspected to be primarily the result of altering the surface chemistry of the solids particles. This may be explained through the lack of increased effectiveness of increasingly higher charge densities of anionic polymer with increasing pH of the samples tested. It would be expected that anionic polymer effectiveness would change if polymer chemistry and rheology at different pH levels were influencing performance.

Figure 3 presents the effects of pH adjustment only on an oil-mud obtained from offshore drilling activities not included in data presented thus far. This sample, for which limited supply prevented complete description, demonstrates that pH adjustment alone accounts for some deliquification and solids concentration. This sample, having an initial pH of 11.5 prior to treatment, was unaffected by polymer addition in dewatering performance. The effect of pH adjustment on treating oil-base muds is well known, and efforts to improve treatment with polymers have not been successful.

The effectiveness of pH adjustment on treatment of production sludges is negligible. As evidenced in Table II, the offshore barge mud and production sludge were treated without pH adjustment, whereas the production sludge required minor adjustment from pH 6.7 to pH 6.

The effect of slurry solids concentration on polymer dosage is presented in Figure 4. This figure contains the results of the different mud and slurry types presented earlier as well as full scale tests. Looking first at only the non-dispersed waterbase mud results, a noticeable relationship of polymer dosage with initial slurry concentration appears. Both bench scale and full scale data are in agreement. With other types of muds and sludges added, the same general trend is supported. This type of relationship, wherein decreasing polymer dosage is required with increasing solids concentration, has been reported [5] for municipal wastewater sludges. The opposite effect has also been reported [4] with the explanation being the difficulty in dispersing the flocculant in higher solids suspensions. The initial solids content of the samples tested did not show any consistent trend in influencing the percent solids in the cake for each sample.

The results of full scale tests on the non-dispersed waterbase mud and the production sludge demonstrate that similar performance between the belt filter press and centrifuge can be expected. As

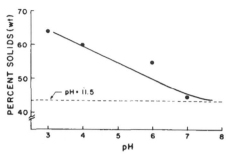

Figure 3. Effect of pH adjustment on an oil-base mud from offshore drilling activities. Initial solids content = 43.5%.

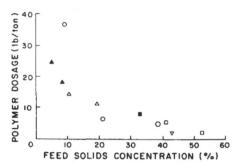

Figure 4. Effect of solids concentration on nonionic polymer dosage. Open symbols represent bench scale treatment; closed symbols represent full scale tests. △ = non-dispersed water-base mud; ○, ▽ = dispersed water-base muds A and B, respectively; ◯ = offshore barge mud and production sludge; □ = production sludge.

indicated in Table III, the vacuum belt extractor did not perform well by comparison. The inferior performance is believed to be the result of a surface limitation at or near the belt caused in part by cake formation, which limits the effective pressure differential required to remove the liquid from the slurry. The belt press employs two surfaces and supplies shear forces when passing over rollers which prevent a dominant surface phenomenon from occurring as long as the belt does not blind.

The performance of the belt filter press was optimized for the non-dispersed water-base mud as shown in Figure 5. Hydraulic loading to the press was varied and the belt speed and tension varied at each flow rate. Polymer dosage was found to be unaffected by loading rate; belt speed was found to be limited only by the distribution of the material on the belt. For the non-dispersed water-base mud a strong floc was produced enabling belt tensions to be used up to those limited by the strength of the belt, which was approximately 60 bars. In addition, a "nip roller" was used to provide a final "squeeze" prior to cake discharge. As can be seen in Figure 5, cake solids concentrations are optimized at approximately 50 gallons per minute while filtrate suspended solids increase with increasing loading. These results are typical of belt filter press performance trends [5].

As noted in Table III, belt press performance on the production sludge was limited to one test run. Time limitations for access to the press (three days for testing) prevented testing different operational conditions and optimizing performance. Limitation of the production sludge treatment effectiveness occurred by the formation of a soft cake which extruded from the belts under belt tensions greater than 20 bars. An optimum chemical treatment scheme for this sample was not determined. The

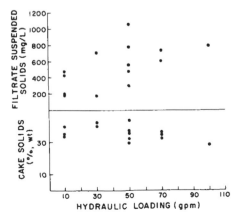

Figure 5. Two meter belt filter press performance using nonionic polymer to treat non-dispersed water-based mud.

extrusion problems may be associated with the high oil content of the production sludge making strong cake formation impossible due to its oily nature.

The centrifuge has been commonly used in oilfield practices for cleaning mud. The results of this study indicate that the use of a belt filter press can provide similar results. The polymer dosage for belt presses is above that for centrifuges; however, power requirements for the press are less. Based on similar performance of the two systems, a cost-effective analysis should be performed to select the more suitable equipment. Recycle of washwater for the belt press should be considered in this analysis. Although not tested at full scale, recycle of the wash appeared possible provided clarification and skimming of the wash is provided prior to recycle.

CONCLUSIONS

Solids-liquid separation is a viable technique to be used to clean up reserve pits resulting from oil and gas well drilling and production sludges from well operation. Since approximately 50% of the reserve pits and most of the production pits contain salt contamination which prevents onsite disposal techniques, volume reduction and removal of salt water from solids will enhance solids disposal options.

Suitable chemical treatment was obtained in bench scale experiments for most samples tested. The bench scale experiments demonstrated that nonionic and anionic polymers have greater applicability to the samples tested, although oil-base muds still show little promise of being treated with polymers.

The belt filter press can perform as well as a centrifuge. It does require a higher polymer dosage than the centrifuge but requires less power. Recycle of the washwater for the belt press should be investigated when performing cost analyses.

ACKNOWLEDGMENTS

This research was sponsored by grants from the Louisiana Office of Conservation, Department of Natural Resources, and the LSU Center for Energy Studies under Grant No. 85-02-12.

REFERENCES

1. Freeman, B. D., and L. E. Deuel, *Guidelines for Closing Drilling Waste Fluid Pits in Wetland and Upland Areas*, Shell Oil Company, New Orleans, LA (1983).
2. APHA, *Standard Methods for the Examination of Water and Wastewaters*, 16th Edition, American Public Health Association, Washington, DC (1985).
3. Baroid, *Apparatus and Procedure for the Field Testing of Drilling Muds, Oil and Gas Fluids*, Baroid Division of NL Industries, Inc., Houston, TX (1975).
4. Connelly, L. J., and P. F. Richardson, "Coagulation and Flocculation in the Mining Industry," *Proceedings of the Solid/Liquid Separation and Mixing in Industrial Practice Symposium*, AIChE, Pittsburgh, PA (1984).
5. USEPA, "Dewatering Municipal Sludges, Design Manual, "Office of Research and Development, MERL, Cincinnati, OH, *EPA-625/1-82-014* (1982).

58 CHARACTERIZATION AND TREATMENT OF SPENT HYDROSTATIC TEST WATER IN THE NATURAL GAS INDUSTRY

Forrest R. Sprester, County Public Health Engineer
City of El Paso
El Paso, Texas 79936

Fernando Cadena, Associate Professor
Civil Engineering Department
New Mexico State University
Las Cruces, New Mexico 88003

Donald J. Fingleton, Project Manager
Energy and Environmental Systems Division
Argonne National Laboratory
Argonne, Illinois 60439

INTRODUCTION

Meyer [1] estimates that over one million miles of collection and distribution pipelines are presently in use by the natural gas industry in the United States. These lines are used to transport natural gas from the production fields to the consumer. The Institute of Gas Technology [2] reports that gas pipelines are periodically tested for physical integrity using any one of the following procedures: acoustic, hydrostatic, isotopic, radiographic, and ultrasonic techniques.

Hydrostatic testing of natural gas pipelines is one of the most popular means for leak detection since it offers the additional advantage of removing obstructions in the pipeline. The main intent of such testing is to identify weaknesses in the pipeline by hydrostatically pressurizing a segment of pipeline. Minimum test pressures have been established by industry associations and the U. S. Department of Transportation for various operating pressures. Water used in hydrostatic tests is often disposed without treatment into temporary infiltration (percolation) pits or into any nearby ditch or body of water. Land disposal of these wastes is usually followed by site recovery using a soil layer to cover the remaining sludge. Natural gas pipelines have been recently recognized as sources of contamination to groundwater [3]. Some preliminary guidelines to minimize the environmental impact of wastes generated by the natural gas industry during hydrostatic testing of transmission pipelines are presented in this document.

HYDROSTATIC TEST DESCRIPTION

Description of hydrostatic test procedures for natural gas pipelines have been reported in the literature beginning in the late 1950's and early 1960's by Brooks [4,5], Heineman [6], and Gray [7]. A more recent review of operational procedures is presented by the Institute of Gas Technology [2].

During a hydrostatic test, a pipeline is divided into a test section, or sections, ranging in lengths of up to 15 miles. The length of these sections depends upon the position of existing valves, availability of water, and differences in elevation along the pipeline. In order to minimize the amount of air in the pipeline, water is injected into the line behind a foam or rubber "pig" in the shape of a ball or bullet, as shown in Figure 1. Following the filling of the line, the pressure in the line is increased to the pre-established test pressure using a positive displacement pump. Once test pressure is successfully reached, the pressure is maintained for a minimum of eight hours before discharging the water from the pipeline. In order to squeegee out as much water as possible, the Institute of Gas Technology [2] recommends discharge velocities greater than 3 to 5 miles per hour. Alcohol is oftentimes run between two pigs after test completion [2]. This additive removes any remaining water from the line and prevents valve "freeze-ups" and hydrate formation.

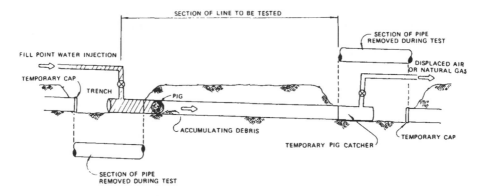

Figure 1. Schematic of hydrostatic test.

It is obvious that the source of fill water used in the hydrostatic test affects the quality of the discharge. The final discharge point, or points, may occur at any location in the transmission line including unplanned points as a result of pipe failure.

It is common practice to run several pigs down a line prior to hydrostatic testing to scrape or squeegee out hydrocarbons accumulated during normal pipeline operation; however, such practice is not totally effective in removing all the fluid coatings.

The efficiency of a pig to wipe wastes deteriorates with distance travelled. The quality of the water immediately behind the pig is inferior to the water near the fill point due to contaminants not removed by the pig. The variation in water quality throughout the pipeline is subjectively illustrated in Figure 2 as indicated by an increase in turbidity in a completely filled section of pipeline. The variability of water characteristics poses a problem in sample collection in that samples must be taken at different times during discharge to adequately describe the character of the discharge. Two major classes of hydrostatic tests for natural gas pipelines are: newly constructed pipelines and existing pipelines. Wastewater characteristics and recommended waste management practices for wastes generated in hydrostatic testing of new and existing natural gas pipelines are presented.

ANALYTICAL METHODS

An extensive experimental study on properties of spent hydrostatic test waters from natural gas pipelines was conducted by Barkley [8] who reported analytical results from 66 hydrostatic tests from several major gas pipelines. The list of significant parameters in the present publication was selected based on Barkley's results. All samples in the present study were collected at or near temporary pig catchers and preserved and analyzed following protocol by the U. S. Environmental Protection Agency [9] and in accordance with *Standard Methods for the Examination of Water and Wastewater* [10].

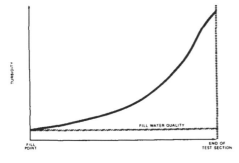

Figure 2. Variation in turbidity throughout test section.

Samples in this study were collected at the start of the discharge and at 10- to 15-minute intervals thereafter in all monitored discharges. The number of samples collected in each test varied from six to fifty-three depending on the total volume of the discharge. The following analyses were conducted in the field immediately after sample collection: dissolved oxygen, immediate oxygen demand, specific conductance, pH, and turbidity.

Laboratory analyses of the samples included: chemical oxygen demand, oil and grease, total iron, nitrate, and total dissolved solids. Several of the samples collected were analyzed for organic compounds using gas chromatography/mass spectrometry (GC/MS). GC/MS results on the same samples have been presented and interpreted in a prior paper by Eiceman et al. [11].

CHARACTERISTICS OF SPENT HYDROSTATIC TEST WATERS

Newly Constructed Pipelines

New lines accumulate soil, welding solids, lumber, mill scale, paint overspray, and other wastes which are left in the line during installation. During cold weather testing, chemicals, such as alcohols or glycol compounds, are introduced into the pipeline to prevent freezing during the test. The quality of the spent water is therefore affected by the presence of such additives. A field investigation of three hydrostatic tests of new pipeline (Tests A, B, and C) was conducted in detail. The discharge from each line was observed by one of the writers from beginning to conclusion. Test A involved 24 miles of newly painted, 34-inch diameter pipe. The test section was subdivided into five equal sections due to the scarcity of water in the area. After completion of a successful test in one section, about 1.2 million gallons of water were transferred to the next section, and so on. Thus, the fill water was relatively well mixed prior to discharge.

A 12-mile section of 24-inch diameter painted pipe was hydrostatically tested during Test B. The test was conducted in two sections. First, a two-mile section was tested and the fill water was transferred to the remaining ten-mile section. This test also required 1.2 million gallons of fill water.

A partially painted pipeline was tested during Test C. The total length of this 16-inch pipeline was 5.5 miles. The entire section was tested at one time. About 1.5 miles of the 5.5-mile line was not internally painted. In fact, this new section had experienced considerable corrosion after exposure to the weather for several years before its use in the actual construction of the pipeline.

A summary of quantitative chemical and physical analyses for Tests A, B, and C are provided in Tables I, II, and III, respectively [12]. O&G concentration was determined for Tests B and C only, on a total of 15 samples. This parameter exceeded 30 mg/L in six of the nine samples collected for Test B. COD was determined on a total of 31 samples. The absence of dissolved oxygen in all samples was probably due to chemical oxidation within the enclosed pipeline. As expected, total iron concentrations were highest in Test C, which contained 1.5 miles of corroded pipe.

Existing Pipelines

The major source of contamination in spent hydrostatic waters from existing pipelines is the accumulated product carried in the line over past years of operation. Although the gas industry has devised numerous processes to separate the methane from the liquid ethane, propane, butane, pentane, and various other large molecular weight hydrocarbons, such processes are not 100% efficient. According to Bergman et al. [13], large molecular weight petrochemicals in the gas pipeline tend to deposit on the internal pipeline walls due to retrograde gas condensation.

Total concentrations of C11 to C19 hydrocarbons near 100 mg/m³ have been detected in consumer distribution lines [14]. Natural gas retrograde condensate also contains a variety of inorganic compounds such as hydrogen sulfide, nitrogen oxides, salts, acids, and other corrosives [15]. Substantial concentrations of nascent ferrous sulfide are commonly found in pipelines transporting gas containing large quantities of hydrogen sulfide (sour gas). Davani [14] reports that the total concentration of measured polycyclic aromatic hydrocarbons (PAH) in retrograde condensate varies from 0.1 to approximately 6.0 mg/m² of internal pipeline surface. Other hydrocarbons typically found in gas pipelines include benzene, toluene, xylenes, compressor engine oil, and organic and inorganic corrosion inhibitors.

Field investigation of two hydrostatic tests of used pipeline carrying unprocessed natural gas was conducted by one of the writers. Both pipelines tested were 8.4 miles long, and 14 and 16 inches in diameter (Tests D and E, respectively). The source of fill water for both tests was potable ground water. Both sections of line had been used extensively to transport natural gas. Considerable accumulation of liquids and solids existed on the walls of each pipeline as observed on the sections of pipe cut

Table I. Analytical Results for Test A

Parameter	Units	Fill Water Quality	Discharge Water Quality
COD	mg/L	<10	28–162
Conductance	umhos/cm	1,310	1,365–1,420
Dissolved Oxygen	mg/L	7.64	1.3–4.8
Nitrate	mg/L	0.1	0.02–0.64
Oil and Grease	mg/L	NA	NA
pH		8.1	6.88–8.52
Settleable Solids	mg/L	ND	ND–403
TDS	mg/L	NA	NA
TSS	mg/L	ND	1.7–230
Total Iron	mg/L	0.44	0.2–35.5
Turbidity	FTU	<10	220

NA means no analysis conducted. ND means non-detected.

Table II. Analytical Results for Test B

Parameter	Units	Fill Water Quality	Discharge Water Quality
COD	mg/L	<10	22–43
Conductance	umhos/cm	2,400	2,300–2,800
Dissolved Oxygen	mg/L	8.2	1.4–8.2
Nitrate	mg/L	0.05	0.04–0.28
Oil and Grease	mg/L	ND	ND–115
pH		8.0	6.38–8.00
Settleable Solids	mg/L	NA	NA
TDS	mg/L	3,616	3,800–5,050
TSS	mg/L	0.02	10–370
Total Iron	mg/L	0.07	1.9–153
Turbidity	FTU	<10	20–>500

NA means no analysis conducted. ND means non-detected.

out to connect the pig catchers. An average thickness of one inch of semi-solid condensate was measured on the invert of the pipelines. The condensate on the invert alone corresponds to approximately 3,500 cubic feet of concentrated petrochemicals. Condensate and spent hydrostatic test water are discharged without treatment at the pig catcher end of the line.

Approximately 175,000 gallons of fill water were required for Test D, and about 370,000 gallons for Test E. The analytical results for Tests D and E are reported in Tables IV and V, respectively. The COD results for both tests are plotted as a function of percent of the discharge in Figure 3. These

Table III. Analytical Results for Test C

Parameter	Units	Fill Water Quality	Discharge Water Quality
COD	mg/L	<10	26–450
Conductance	umhos/cm	1,450	1,350–1,630
Dissolved Oxygen	mg/L	8.2	2.15–9.6
Nitrate	mg/L	NA	NA
Oil and Grease	mg/L	<0.5	0.5–16.9
pH		7.01	6.85–7.09
Settleable Solids	mg/L	ND	5–10
TDS	mg/L	700	700–790
TSS	mg/L	0.35	135–1,677
Total Iron	mg/L	2.5	17–300
Turbidity	FTU	ND	50–>500

NA means no analysis conducted. ND means non-detected.

Table IV. Analytical Results for Test D

Parameter	Units	Fill Water Quality	Discharge Water Quality
Benzene	mg/L	NA	25–38
Chloride	mg/L	220	241–512
COD	mg/L	12	2,800–19,100
Conductance	umhos/cm	1,360	1,300–2,100
Dissolved Oxygen	mg/L	NA	ND
IOD	mg/L	ND	2.6–4.4
Nitrate	mg/L	ND	13–52
Oil and Grease	mg/L	ND	NA
Settleable Solids	mg/L	ND	0.2–10.0
Sulfate	mg/L	95	155–550
Sulfide	mg/L	ND	9.6–30.8
TDS	mg/L	759	6,316–13,382
TSS	mg/L	ND	872–5,432
Total Iron	mg/L	0.8	196–532
Turbidity	FTU	15	>500

NA means no analysis conducted. ND means non-detected.

COD results are very high but agreeable with the values reported by Barkley [8] for similar tests. The COD of the condensate was not analyzed by conventional water chemistry protocol. However, chemical oxygen demand of these petrochemical wastes is likely to be greater than the highest reported values in Tables IV and V.

Field measurements indicate that the pH values of the wastes at the beginning of the test were below 3.0. Nevertheless, no other reliable measurements were obtained at other times due to failure of the pH electrodes. It is not clear at the present time if interferences present in the wastes may have caused instrument failure.

Several hundred organic compounds are known to exist in wastes generated during hydrostatic testing of existing gas pipelines. Eiceman and co-workers [11] have studied the composition of spent waters from tests D and E. Their work encompassed a comprehensive search of aromatic compounds present in the discharges. According to these researchers, the discharge water contained "estimated minimum concentrations of benzene of 25 to 38 mg/L." These researchers also found elevated concentrations of alkylated benzenes and alkylated disulfides in the wastes. Their results are by no means comprehensive, suggesting the existence of many other not-identified compounds. For instance, investigations by the Federal Energy Regulatory Commission show that polychlorinated

Table V. Analytical Results for Test E

Parameter	Units	Fill Water Quality	Discharge Water Quality
Benzene	mg/L	NA	NA
Chloride	mg/L	220	234–421
COD	mg/L	12	2,100–>100,000
Conductance	umhos/cm	1,360	1,350–1,400
Dissolved Oxygen	mg/L	NA	ND
IOD	mg/L	NA	NA
Nitrate	mg/L	NA	2.5–12.0
Oil and Grease	mg/L	ND	NA
pH		6.98	NA
Settleable Solids	mg/L	ND	12–18
Sulfate	mg/L	95	112–232
Sulfide	mg/L	ND	ND–5.6
TDS	mg/L	759	954–3,046
TSS	mg/L	ND	2,892–9,352
Total Iron	mg/L	0.8	396–900
Turbidity	FTU	15	>500

Na means no analysis conducted. ND means non-detected.

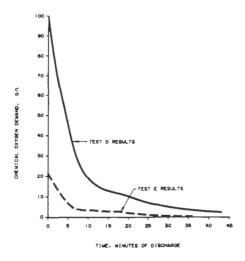

Figure 3.   COD as a function of minutes of discharge.

biphenyls, PCB, have been identified in natural gas distribution systems [16]. Processing of natural gas removes a large fraction of the heavier molecular weight components found in untreated gas. Even though retrograde condensation may be reduced after processing, such condensation is not completely eliminated.

The New Mexico Environmental Improvement Division has conducted a chemical characterization study of spent test waters from a processed natural gas pipeline [17]. Results from the NMEID study are summarized in Table VI. Columns numbered 1, 2, 3, and 4 correspond to samples collected after approximately 12, 37, 63, and 87% of total discharge duration, respectively. It is interesting to note that the concentrations of most parameters in this study tend to decrease with time as shown in Figure 1. However, the concentration of phenols increases drastically towards the end of the test. This anomaly may be a result of the decrease in solubility of phenols in acidic conditions (acidic wastes are typically found in higher concentrations at the beginning of the test and tend to decrease as the line is washed by cleaner water towards the end of the test).

RECOMMENDED WASTE MANAGEMENT PRACTICES

Application of standard petrochemical wastewater treatment technology to the control of wastes generated during hydrostatic testing is limited to instances when wastewater facilities are relatively close to the point of discharge. Due to remoteness, large volumes, and infrequency of the wastes, on-site treatment is recommended for these types of wastes.

Table VI. Analytical Results of Hydrostatic Test Water of Processed Natural Gas Pipeline

Parameter	Units	Fill Water Quality	Discharge Water Quality 1	2	3	4
Benzene	mg/L	NA	ND	ND	10	13
Ethyl Benzene	mg/L	NA	2.6	2.7	5.9	1.7
Hydrocarbon Fuel	mg/L	NA	500	136	290	50
Iron	mg/L	0.35	43	12	30	8
Lead	ug/L	<5	160	170	23	<5
Oil and Grease	mg/L	NA	8.2	53	136	11.7
Phenols	mg/L	NA	0.5	0.2	1.6	27.9
Toluene	mg/L	NA	24	24	33	18
TDS	mg/L	787	792	782	770	1071
Total Org. Carb.	mg/L	NA	37	37	116	311
TSS	mg/L	<1.0	23	25	64	120

NA means no analysis conducted. ND means non-detected.

Discharges from used pipelines are very complex mixtures of a vast number of chemical compounds at varying concentrations. Therefore, a single, simple waste treatment scheme, adequate for all discharges, may not be easily developed without obtaining additional information. Volume minimization is typically the simplest and most economical waste management practice for any type of industrial waste. This principle may be easily applied to hydrostatic testing of pipelines. For example, prior to filling a new pipeline, a relatively small volume of water (up to 3000 gallons) could be placed between two pigs and forced with natural gas or nitrogen through the length of the pipeline. This precleaning technique would collect the majority of the debris or condensate, deposited in the line during construction or operation, respectively. Detergents may be added in order to more effectively remove condensate. The resulting slug of wastewater would then be sampled and analyzed on site for turbidity. If the wastewater has a turbidity greater than 500 FTU, the process should be repeated. The slugs of wastewater should be confined in a tank truck or lined pit. Another alternative for reducing the volume of highly concentrated waste consists of testing shorter sub-sections of pipeline and reusing the same water in each consecutive test. A lined pond to receive the pre-cleaning wastewater from newly constructed pipelines may be adequate. The wastewater could be treated and disposed using natural aeration, sedimentation, and evaporation prior to discharge. The resulting solids could be buried in place with the liner.

A simple filtration operation may remove a significant fraction of the suspended solids present in spent test waters from new pipelines. Filter equipment commonly used during the filling of the line may be used for discharge treatment as well. Such equipment is used during the filling of the line to reduce the potential of injecting unwanted materials into the new line. This filtration procedure was used during the discharge of Test B. The pressure differential across the 100 mesh filter was observed to increase from a few psi at the beginning to 400 psi at the end of the discharge. This pressure differential is a clear indication that the filter was capable of removing a significant fraction of the suspended solids present.

Direct discharge of spent hydrostatic test water from a new pipeline to surface water is not recommended. A small pit should be constructed to hold 10 to 20% of the spent test water and an overflow provided to allow land spreading over as wide an area as possible. This practice would reduce the potential for erosion and would allow for some removal of the iron, solids, and oil and grease. Direct discharge of wastewaters from used pipelines into receiving waters and uncontrolled land disposal of the wastes should not be practiced.

A more practical solution for discharges in remote areas may be onsite treatment. For instance, a temporary, lined treatment basin may be built to retain the spent hydrostatic test water. A single liner is all that is believed necessary for the temporary basin.Field treatment in such basin may include adsorption, coagulation, flocculation, sedimentation, and volatilization.

Certain materials, such as straw and clays, cast into the wastewater treatment basin, are known to adsorb the oils and other organics and settle to the bottom of the basin along with other sediments. Barkley [8] conducted several laboratory bench tests which suggested that straw was an economical sorbent for these types of wastes. The clay fraction in soils is also an excellent adsorbent phase for large molecular weight organic compounds.

Hazardous purgeable organics, for instance, single- or double-ring aromatic hydrocarbons, are removed by volatilization in agitated conditions. Cadena et al. [17] report that the rate of volatilization of toxic compounds from aqueous solutions increases with the reaeration rate. This rate is increased with turbulence and decreased with liquid depth. Therefore, in order to take maximum advantage of natural volatilization, a temporary basin should be built as shallow as possible and in a location where maximum wind effects are experienced.

Wastewater confined in temporary basins may be released, preferably by land spreading, after the hazardous organic compounds have settled, decomposed, or volatilized. Total evaporation of the volatile wastes and water in the basin is a likely alternative for arid locations. Presettled wastewater from hydrostatic testing of existing pipelines may be easily applied to the land using one of several spray irrigation techniques. Spraying the wastewater into the air also allows for oxidation, volatilization of light organics, evaporation, and spreading over a large area. Light suspended solids, such as iron filings, would be trapped at or near the soil surface, and the wastewater would infiltrate the soil surface only within the upper one to two feet of soil.

Frequently, gas pipelines are located in or near oil/gas production fields. Many gathering lines, typically of short length and small diameter (two to eight inches) are located in a single producing field. Wastewater discharges from hydrostatic tests of gathering lines may be transported to secondary recovery wells for deep well injection.

A general treatment scheme may be developed by a pipeline company that routinely conducts such tests in one climatic region. Such a scheme may then be reviewed for a general permit application.

CONCLUSIONS

Large volumes of wastewater are generated by the Natural Gas Industry during the hydrostatic testing of new and existing transmission pipelines. In most instances these waters are discharged without treatment on land or into a receiving body of water.

Spent hydrostatic test waters from newly constructed pipelines contain construction debris, over-spray paint, paint solvents, oil and grease, suspended and dissolved iron, and at times antifreeze additives. Substances identified in spent test waters from existing pipelines include high concentrations of: benzene, ethyl benzene, ferrous sulfide, hydrocarbon fuel, lead, nitrate, oil and grease, polynuclear aromatic hydrocarbons, sulfide, iron, phenols, and toluene. Polychlorinated biphenyls, PCB, may be found in some spent test waters when this compound has been used as a compressor lubricant. Many other not-identified hydrocarbons are also present in the discharges. Sour gas transmission lines generate acidic wastes with pH values as low as 3.0.

Although the waste waters may be of considerable strength, particularly for used pipelines, several procedures may be used to reduce environmental impact. These include pre-cleaning the pipeline before filling with water, selecting the least toxic chemical additives for the fill water, and using the smallest amount of water possible.These procedures may not be sufficient to reduce to acceptable levels the concentration of toxic compounds, such as benzene and toluene, which will require further treatment. Due to the remote location of the majority of hydrostatic tests, the following waste management alternatives were suggested: 1) wastewater volume reduction; 2) construct a total containment basin and field treat the wastes; 3) land application; 4) use of the wastewater for secondary petroleum recovery; 5) treatment at an existing facility; 6) use of erosion control devices; and 7) recovery of concentrated petrochemicals in refineries.

REFERENCES

1. Meyer, R. A., ed., *Handbook of Energy Technology and Economics*, John Wiley and Sons, Inc., New York (1983).
2. Institute of Gas Technology, "Symposium Papers and Related Information on Nondestructive Testing for Pipe Systems," IIT Center, Chicago, Illinois (June 7-10, 1976).
3. U.S. Environmental Protection Agency, "Ground Water Protection Strategy," *EP 1.2:G91/7*, Office of Ground-Water Protection, Washington, DC (1984).
4. Brooks, L. E., "Hydrostatic Testing of Pipelines," *Journal of the Pipeline Division, American Society of Civil Engineers*, 83:PL3 (1957).
5. Brooks, L. E., "Autographic Control for High-Pressure Pipeline Testing," presented at the Petroleum Engineering Conference, Los Angeles, CA, *Paper No. 64-PET-40* (1964).
6. Heineman, W. P., "Testing of Pipe and Pipelines," Presented at ASCE Transportation Engineering Conference, Minneapolis, MN, *Preprint No. 211* (1965).
7. Gray, J. S., "Retesting Pipeline Justifies Higher Allowable Discharge Pressures," *Oil and Gas Journal*, pp. 122-135 (1965).
8. Barkley, W. A., "Characterization and Treatment Evaluation of Hydrostatic Test Water Effluent," prepared for American Gas Association, *AGA Project PR 85-62*, Washington, D.C. (1973).
9. U.S. Environmental Protection Agency, "Handbook for Sampling and Sample Preservation of Water and Wastewater," *EPA-600/4-82-029j*, Environmental Monitoring and Support Laboratory, Cincinnati, OH (1982).
10. *Standard Methods for the Examination of Water and Wastewater*, 15th Edition, American Public Health Association, New York, NY (1981).
11. Eiceman, G. A., C. S. Leasure, and B. D. Baker, "Characterization of Discharge Water from Hydrostatic Testing of Natural Gas Pipeline Using High Resolution Gas Chromatography/Mass Spectrometry," *International Journal of Environmental Analytical Chemistry*, 16:149 [1983].
12. Texas Department of Water Resources, "Summary of Hydrostatic Test Water Analyses from Permit Information," Verbal Communication with Mr. Bill Taylor, Austin, TX (1984).
13. Bergman, D. F., Tek, M. R., and Katz, D. L., "Retrograde Condensation in Natural Gas Pipelines," *Final Report*, Pipeline Research Committee, American Gas Association, Project No. RR26-69 (1975).
14. Davani, B., "Presence, Movement, and Fate of Polycyclic Aromatic Hydrocarbons in Wastes from the Production of Natural Gas," *Doctoral Dissertation*, New Mexico State University (1985).

15. Webb, B. C., "The Art of Pigging," The American Society of Mechanical Engineers, Energy Technology Conference, Houston, TX (1978).
16. Hoffman, R. H., Comments Presented by the Federal Energy Regulatory Commission, Environmental Evaluation Branch to the Regional Administrator, USEPA, Region VI, on the Draft General NPDES Permit for Hydrostatic Testing of Natural Gas Transmission Pipeline in the State of Arkansas, Louisiana, Oklahoma, and Texas (October 28, 1983).
17. Sisneros, K. M., "Comments on National Pollutant Discharge Elimination System (NPDES) Draft on General Permit for Natural Gas Transmission Pipeline Hydrostatic Test (PHT) Discharges," Presented by the New Mexico Water Pollution Control Bureau to Mr. Joe Ramey, Director, New Mexico Oil Conservation Division (May 12, 1983).
18. Cadena, F., Eiceman, G. A., and Vandiver, V. J., "Removal of Volatile Organic Pollutants from Rapid Streams," *Journal of the Water Pollution Control Federation*, 56:460 (1984).

Section Eight
FOOD WASTES

59 DESIGN AND OPERATION OF A CRANBERRY WASTEWATER TREATMENT SYSTEM

Frederic C. Blanc, Professor

James C. O'Shaughnessy, Associate Professor
Department of Civil Engineering
Northeastern University
Boston, Massachusetts 02115

Ken Carlson, Project Manager

Paul Weisman, Vice President
LEA Group Engineers/Architects/Planners
Boston, Massachusetts 02111

INTRODUCTION

Cranberries are a major agricultural crop in Massachusetts and a few other states. During the six to eight week harvest season, the cranberries are delivered to cranberry receiving facilities which wash and sort this fruit to prepare it for shipping or processing. This paper presents information on how such a seasonal cranberry receiving wastewater is being treated at a new cranberry receiving facility. It includes the results of treatability studies and pilot studies performed for the aerated lagoon—land application treatment system which was selected as well as design information and first season operational data for the aerated lagoon operation.

PROCESS OPERATIONS

Cranberries which are marketed as fresh fruit and are picked dry represent only a small portion of the total harvest. The bulk of the fruit is hydraulically harvested from flooded bogs. This water-harvested fruit is brought into the receiving station by truck, weighed on a truck scale, and dumped into a bulk holding tank or directly into a conveyor type system which leads to the cleaning operation. Most of the water from the hydraulic harvesting operation has drained from the truck by the time it arrives at the receiving station; however, the fruit is still wet. In addition to water, the berry surface is also coated with a slime layer of fine soil particles, algae, and other matter. The bulk of water-harvested fruit also contains pieces of vines, tiny leaves, and other grit-like debris which must be removed before storing or processing the fruit into juices or sauces.

Processing operations at a cranberry receiving location consist of physical and hydraulic operations which remove dirt and other matter from the fruit. Flumes, vibrating and rotating screens, brushes, and spray type rinses are all used to clean the berries. A good amount of the process water used in the receiving operation is recirculated during the day.

Such processing operations produce the bulk of the wastewater from the receiving operation. Additionally, there are drippings from solid waste storage areas, the truck receiving area, and discarded damaged berries. Such wastes are generally low in volume but high in strength.

WASTEWATER CHARACTERIZATION

To obtain the wastewater characteristics and treatability information for the wastewater treatment design of a receiving facility to be constructed, treatability studies and a waste sampling program were conducted at an existing receiving facility during the 1984 harvesting season. Total water usage and individual wastewater flows associated with fruit receiving operations were measured and sampled. Based on this sampling program, a profile of cranberry receiving wastewater was constructed. The

497

Table I. Typical Wastewater Characteristics for the Major Wastewater Sampling Locations in Cranberry Receiving Operations

Locations	pH	BOD₅ mg/L	COD mg/L	Susp. Solids mg/L	VSS mg/L
Central Floor Sump	3.2	925	2,030	470	440
Recirculation Lines 1 and 2	3.1	1,430	4,010	560	512
Recirculation Lines 3 and 4	3.2	880	2,655	425	395
Berry and Trash Bin Drippings	2.9 – 3.5	2,000 – 9,000	2,000 – 32,000	100 – 1,000	100 – 850

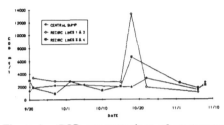

Figure 1.   COD concentrations of wastewater samples.

Figure 2.   Relationship between soluble BOD and soluble COD for influent wastewater samples.

total water usage for cranberry receiving at the facility tested averaged at approximately 100,000 gallons per day during the active part of the harvest season. The two main types of wastewater flow are floor drainage to sumps and receiving process wastewater which is recirculated and discharged to wastewater treatment on a daily basis.

Table I lists typical wastewater characteristics for the major wastewater sampling locations. The values listed generally represent average values. Figure 1 depicts the variation in COD concentrations from the major wastewater sampling locations with time. The COD concentrations ranged from 1,000 mg/L to 13,000 mg/L. The average COD concentration of all samples was 2,745 mg/L. Soluble COD percentage ranged from 46% to 87% with an average of 68% soluble COD. The COD and BOD concentrations for a number of samples are plotted in Figure 2. The average COD/BOD ratio was approximately 2.5 to 1 for all samples analyzed. The data in Figure 2 indicates that a good relationship exists for soluble COD and BOD. However, the relationship for total BOD and COD is quite poor. Due to the nature of the wastewater and the variation in suspended solids, only soluble COD and soluble BOD were expected to exhibit a good ratio. The suspended solids variations ranged from 200 to 1400 mg/L with an average value of 475 mg/L. Figure 3 presents the suspended solids variations in the three major waste streams.

Figure 4 indicates the variation in BOD₅ concentration for the major wastewater samples during the study. The average BOD₅ value for the locations shown in Figure 4 was 1030 mg/L. Approximately 80% of this BOD₅ is due to the soluble fraction. The ratio between five day BOD and ultimate BOD varied from sampling location to sampling location. For the wastewater composites used in the pilot

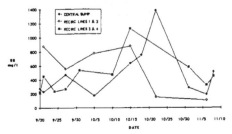
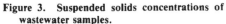
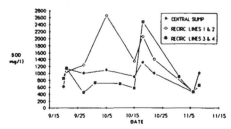

Figure 3.   Suspended solids concentrations of wastewater samples.

Figure 4.   BOD concentrations of wastewater samples.

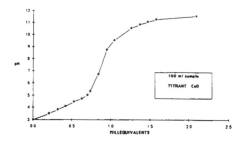

Figure 5.   Titration curve for recirculation line 4. Sample date = 10/1/84.

studies, the five day BOD represented 60 to 70% of the ultimate BOD. The biochemical oxygen demand of the smaller waste streams such as trash bin drippings was much higher than that of the recirculation water. A value of 5000 mg/L was typical of the BOD_5 from such locations.

All wastewater streams sampled during this study had low pH values. The values ranged from a low of 2.8 to a high of 5.5 pH units. The pH values for the samples collected from the central sump and recirculation lines were in the same range. The pH range of 3.0 to 3.5 represents the value of most of the wastewater samples analyzed. The wastewater from the receiving facility requires neutralization before biological treatment.

Data for a titration curve of the recirculation line 4 wastewater are plotted in Figure 5. This sample was collected on October 1, 1984 from one of the recirculation lines. The data indicates that 7.5 millequivalents per litre are required in order to neutralize this sample to a pH of 5.5. Approximately 10 millequivalents per litre are required to reach a neutral pH range. Using CaO, and based on a water usage of 100,000 gal/day and 25,000 barrels of cranberries processed, the calculated lime requirement is between 0.007 to 0.01 lb per barrel of cranberries processed.

In addition to neutralization, nutrient addition is required for cranberry wastewaters prior to treatment. Table II depicts the available nitrogen and phosphorus in the major wastewater streams.

Waste production during the harvest season was evaluated on a production unit basis to obtain waste generation factors. In the cranberry industry, the production unit is the barrel. One barrel of cranberries weighs 100 pounds. The volumetric wastewater generation unit can vary greatly from 0.5 to 4 gallons per barrel of cranberries received per day. The BOD_5 generation factor in this case was 0.1 lb BOD_5 per barrel processed while the suspended solids production was 0.022 lb. SS per barrel.

TREATABILITY STUDIES

As part of the overall wastewater treatment evaluation, field pilot treatability studies were collected using rotating biological contactors, and laboratory pilot studies were conducted to simultate the performance of an aerated holding lagoon. This paper will discuss the aerated holding lagoon pilot studies which were utilized in the final design.

The wastewater utilized for the pilot testing program was taken from a 24 foot diameter 4 foot deep above ground pool which was erected as a waste holding tank. Wastes from the central sump and the two recirculation lines were pumped to this pool which had a working capacity of 13,000 gallons and could supply an influent feed with constant wastewater characteristics for periods of up to one week. The wastewater in the pool was neutralized to an approximate pH of 6 and spiked with sufficient nutrients. Whenever the wastewater storage pool was refilled, a portion was withdrawn, transported to the laboratory, refrigerated, and then used as influent feed for the laboratory pilot aerated holding lagoon studies.

Table II. Nitrogen and Phosphorus Concentrations in Wastewater Samples[a]

Sample	NO_3 (mgN/L)	NH_4 (mgN/L)	TKN (mgN/L)	Ortho-P (mgP/L)	Total-P (mgP/L)
Central Sump	0.7	0.5	1.2	0.66	1.3
Recirc. Tank 2	0.9	1.0	7.0	0.63	0.68
Recirc. Tank 3	1.2	0.3	12.0	0.59	0.65
Plant makeup water	1.2	<0.02	<0.05	0.06	0.22

[a]Based on average of samples from 10/5/84 and 10/29/84.

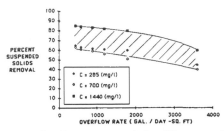

Figure 6. Percent suspended solids removal versus overflow rate for raw wastewater.

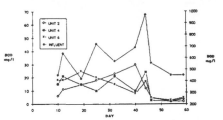

Figure 7. Influent and effluent BOD concentrations for units 2, 4, and 6.

A number of settling column tests were run on site. These batch column tests use a 6-inch diameter, 6-foot deep settling column. The samples tested had suspended solids concentrations ranging from 285 mg/L to 1,440 mg/L. The range of suspended solids removal was from 40 to 80 percent with final suspended solids concentrations ranging from 90 mg/L to 250 mg/L. Results of the settling tests are presented in Figure 6.

PILOT AERATED LAGOON UNITS

Six pilot laboratory reactors were operated to obtain design information for the aerated holding lagoons. The units were fed intermittently on a daily basis to simulate the expected daily discharges of recirculation water in the full-scale facility. Two units, units 1 and 2, were set up as 20-liter completely mixed units with a 20-day hydraulic retention time. Both units were seeded with activated sludge from a municipal treatment plant which had been receiving cranberry processing wastewater. Unit 1 was started with an initial mixed liquor suspended solids concentration of 300 mg/L while Unit 2 was started with an initial mixed liquor suspended solids concentration of 200 mg/L. Units 3 and 4 were set up at similar mixed liquor suspended solids concentrations to Units 1 and 2; however, Units 3 and 4 were operated as variable volume aerated storage lagoons which accumulated wastewater resulting from a feeding of 1 liter per day. The only effluent removed from Units 3 and 4 was the 300-milliliter volume required for the weekly operating performance analysis. Units 5 and 6 were operated on a variable volume basis similar to that for Units 3 and 4; however, the initial volume in Units 5 and 6 was 10 liters. Unit 5 was seeded with activated sludge to an initial mixed liquor suspended solids concentration of 200 mg/L while Unit 6 received no initial mixed liquor seed. The reason for operating Units 5 and 6 in this manner was to simulate a lagoon which accumulated a considerable amount of wastewater before neutralization, nutrient addition, and aeration would be initiated. All laboratory lagoon pilot units were operated for a sixty day period which simulated the length of a normal harvest season.

Operation of the laboratory pilot units in this fashion generated aerated lagoon data for hydraulic retention times from 10 to 60 days. Each of the six reactors received an influent which was pH adjusted to 5.5 with sodium bicarbonate and received supplemental nitrogen and phosphorus addition.

Figure 7 illustrates the influent and effluent BOD_5 concentrations during the operation of units 2, 4, and 6. The data for units 1, 3, and 5 produced essentially the same type of plot. Table III presents some key operational parameters for Unit 6 illustrating how the treatment process might develop in the initial year. One of the consequences of a seasonal cumulative aerated lagoon operation is a continually decreasing food to microorganism ratio, usually occurring in most startup operations. Throughout the study, the food to microorganism ratio of the reactors ranged from 0.02 to 1.10 lb BOD/[lb MLVSS-day]. A typical value for any of the units was 0.1 lb BOD/[lb MLVSS-day] with the F:M ratio being calculated as:

$$\frac{\text{lb } BOD_5 \text{ applied per day}}{\text{lb MLVSS under aeration}} \tag{1}$$

At the end of the study the mixed liquor volatile suspended solids concentration was usually in the 300 to 400 mg/L range with 75% of the suspended solids being volatile.

Effluent BOD_5 concentrations in the aerated pilot lagoons were generally in the 10 to 20 mg/L range after the first 10 days of operation. After 45 days of operation, all the BOD_5 concentrations remained in the under 10 mg/L range.

Table III. Key Operational Parameters for Unit 6

Day	Influent BOD (mg/L)	Effluent BOD (mg/L)	Percent BOD Removal	MLSS (mg/L)	MLVSS (mg/L)	F:M
10	450.0	18.0	96.0	85.0	60.0	0.46
12	640.0	18.0	97.2	70.0	35.0	1.10
19	420.0	25.0	94.2	50.0	35.0	0.52
25	720.0	20.0	97.2	90.0	70.0	0.35
32	570.0	15.0	97.4	100.0	75.0	0.22
40	690.0	8.0	98.7	300.0	225.0	0.07
44	970.0	17.0	96.9	295.0	220.0	0.09
46	555.0	5.0	99.1	315.0	225.0	0.05
54	450.0	3.0	99.3	350.0	240.0	0.03
59	450.0	5.0	98.9	390.0	285.0	0.03
Avg.	591.5	13.4	97.5	204.5	147.0	0.29

Suspended solids production was found to vary between 0.15 and 0.5 pounds of suspended solids produced per pound of BOD_5 removed. The estimated solids production chosen to represent the process based on the data was 0.28 pounds of suspended solids produced per pound of BOD_5 removed.

DESIGN

Collection System

Wastewater, consisting of recirculated washwater and fresh water rinse generated during the fruit cleaning process, flows through a rotating drum screen for solids removal, then is discharged to a holding tank prior to recirculation. Overflow from the holding tank flows by gravity to a duplex submersible pumping station which discharges through a 4 inch force main to the infuent manhole in front of the Operations Building. The holding tank is also completely emptied to the pumping station on a regular basis. Wastewater is directed to one of the two lagoons at the influent manhole. The major design consideration was to provide the ability to utilize either or both of the lagoons for wastewater storage and treatment. To accomplish this, two shear gates were provided in the influent manhole. From here, wastewater flows directly into the selected lagoon(s) through one of two 12-inch influent lines. Figure 8 shows the approximate facility layout.

Aerated Lagoons

Two lagoons are presently provided for wastewater storage and treatment. Several construction materials were considered for the lagoons, including concrete, and hypalon, PVC, and polyethylene liners. After cost, installation and performance were evaluated, polyethylene was selected as the most acceptable liner material. The total required lagoon volume of 1 million gallons was determined from an evaluation of receiving station fruit processing capabilities for initial design and estimated future conditions.

Two lagoons, each with a nominal capacity of 500,000 gallons, were designed to provide maximum flexibility with space allowed on the site layout for two additional future lagoons. Each lagoon is approximately 180 ft. × 105 ft. at the top with an effective depth of 12 ft. Slopes are 3:1. Bottom dimensions are 90 ft. × 15 ft. with a slope of approximately 0.02 feet/foot. Two 30 HP floating mechanical aerators are provided in each lagoon. The major lagoon design consideration was to provide enough capacity in the lagoons to hold and treat the 8-week operating season flows while discharging the initial permit limit of only 25,000 gpd.

Two simplex submersible pumping stations were provided: one for recirculating the lagoon contents through the reaction tank for chemical addition and the second for transferring the contents to the final clarifier. The pumps are interchangeable and the pump chambers are preceded by valve chambers for isolating pumps in both the pumping stations.

Nutrient and pH adjustment chemicals are added to the lagoons by a manually operated recirculation system using a submersible pump, reaction tank, and distribution box. Readily available bagged chemicals are mixed in three individual chemical solution tanks, one each for lime (pH control),

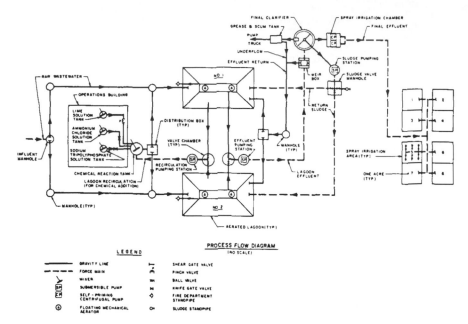

Figure 8.   Treatment plant process flow diagram.

ammonium chloride (nitrogen) and sodium tripolyphosphate (phosphorus). The nitrogen and phosphorus solutions are added to the reaction tank by manually adjusted ball valves. The lime solution is added by a manually adjusted pinch valve. The volume of chemical to be added is determined by the operator based on both influent wastewater and lagoon conditions.

Land Application System

Following clarification, the effluent flows to the spray irrigation pump chamber where duplex self-priming centrifugal pumps discharge to the land application system. An application rate of 1 inch/acre/week was established in the groundwater discharge permit, corresponding to a flow rate of approximately 25,000 gal/acre/week. For maximum design and operational flexibility, the spray system is set up in a grid pattern of 8 one-acre plots. Each acre contains 27 spray nozzles on surface piping which can be broken down and stacked in the off-season. Flow to each plot is controlled by a manually operated irrigation valve. Automatic air evacuation valves are located at the high points in the buried distribution piping, and manual drain valves are located at the low points. Each acre is moderately wooded with both pine and hardwoods over groundcover and brush. Soils are predominantly sand with groundwater at a depth in excess of 10 feet. Effluent is applied to a different acre each day, such that approximately 25,000 gal/acre is applied each week (1 inch/acre/week).

Initial Operation

Beginning on September 23, 1985, the Receiving Station commenced normal operation and began receiving cranberries for processing. Simultaneously, process water from the receiving station was received at the wastewater treatment facility. Two weeks later, on October 7, 1985, sufficient volume was present in Lagoon No. 1 to start the aerators and add nutrients. For data description purposes, October 1st was chosen as the first day of operation despite the fact that insufficient aeration and nutrients were not added. Analysis of the influent confirmed the previous pilot study conditions: the influent had a low pH (approximately 3.4) and was deficient in the nutrient nitrogen and phosphorus. The typical influent ammonia−nitrogen and total Kjeldahl nitrogen concentrations were <0.5 mg/l and 4.0 mg/l, respectively. Lime, ammonium chloride, and sodium tripolyphosphate were utilized for adjusting the pH and supplying nitrogen and phosphorus, respectively. A total of 950 lbs. of lime, 1050 lbs. of ammonium chloride, and 100 lbs. of sodium tripolyphosphate were added to lagoon 1 during the operating period.

Table IV. Typical Wastewater Characteristics for the Influent and Effluent to the Aerated Lagoon During the Initial Operating Season

Parameter	Influent	Effluent
COD settled, mg/L	2,640	60
COD filtered, mg/L	1,920	48
BOD settled, mg/L	535	35
BOD filtered mg/L	530	28
Suspended solids, mg/L	190	80
Volatile suspended solids, mg/L	137	75
pH	3.3	7.2

Because the volume of cranberries received did not approach the receiving facility capacity during this initial harvest season, only one lagoon was used for waste treatment. Lagoon No. 2 was filled with potable water in accordance with local fire department requirements prior to commencement of facility operation and remained unused during the operation period.

The typical waste strength parameters for the influent to the lagoon are presented in Table IV. After the addition of nutrients and aeration on the 7th day of operation, the biological suspended solids began to develop in the lagoon. Figure 9 depicts the increase in total suspended solids in the lagoon with time. Increases in suspended solids after the 29th day of operation are due to a combination of biological suspended solids reduction and sludge return from the clarified lagoon effluent which was discharged to the land disposal area. The typical influent suspended solids value presented in Table IV does not reflect the contribution of the periodic recirculation tank discharges which occurred on an almost daily basis. Recirculation tank dumps delivered a wastewater with a suspended solids concentration of approximately 400 mg/L.

The soluble organic substrate concentration in the lagoon is represented by Figure 10 which depicts the five day BOD values for filtered samples at various days during the operating period. In general, the influent soluble BOD_5 of 530 mg/L was reduced to less than 30 mg/L after the biological floc became established in the lagoon. Table IV lists some of the other effluent parameters in addition to BOD.

Figure 11 depicts the ammonia nitrogen concentrations in the lagoon from shortly after the nutrients were added until the aerators were turned off for the winter shutdown period. By October 29,

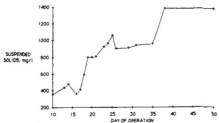

Figure 9.   Lagoon suspended solids concentration versus time.

Figure 10.   Lagoon filtered BOD_5 versus time.

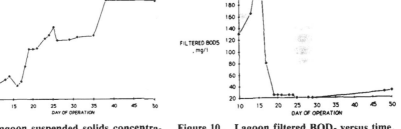

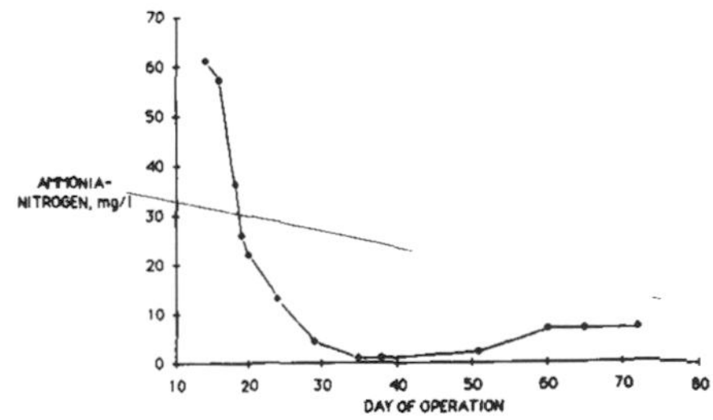

Figure 11.   Ammonia nitrogen in lagoon.

1985, the sufficient quantities of nitrogen had been utilized by the biological process to meet the discharge permit requirements and allow the spraying of the effluent. The data indicates that added nitrogen was incorporated into the biomass and later released due to endogenous activity in the biomass. This caused the ammonia-nitrogen concentration levels to increase after the harvest ended. Further increases in the soluble nitrogen content of the lagoon might pose an operational problem in that the total nitrogen concentration might be higher than the maximum allowable 10 mg/L nitrogen limit in the land application discharge permit.

During the month of November, a total of 249,940 gallons of effluent were sprayed over a 22-day period, an average of 11,360 gallons per day. In December the spraying period was curtailed due to the cold weather, and only 12 days of spraying were realized. Spraying of the effluent was discontinued on December 12th due to the persistent subfreezing weather. During this period approximately 168,980 gallons of effluent were disposed of. Overall, a total of 418,920 gallons were treated and disposed of on the spray fields during the operating season. Approximately 391,000 gallons of wastewater, including an estimated 50,000 gallons of sludge, remained in the lagoon over the winter season.

60 FULL-SCALE OPERATION OF AN ACTIVATED SLUDGE PLANT TREATING CORN WET MILLING WASTEWATER

Jerry D. Allen, Senior Environmental Engineer
A. E. Staley Manufacturing Company
Decatur, Illinois 62525

David W. Sass, Process Engineer
A. E. Staley Manufacturing Company
Loudon, Tennessee 37774

Enos L. Stover, Professor
School of Civil Engineering
Oklahoma State University
Stillwater, Oklahoma 74078

INTRODUCTION

The A. E. Staley Manufacturing Company, whose corporate headquarters are in Decatur, Illinois, is one of the nation's largest processors of corn. Major corn wet milling facilities are located in Decatur, Illinois; Loudon, Tennessee; Morrisville, Pennsylvania; and two in Lafayette, Indiana. Products include corn sweeteners, corn syrup, corn starches, corn oil, power alcohol, and animal feeds.

The corn wet milling process, which is shown pictorially in Figure 1, produces significant quantities of wastewater. Flows from the major Staley facilities are in the range of 1.0–2.5 mgd containing 10,000–25,000 lb BOD/day. Discharge, in all cases, is to a POTW which requires pretreatment.

All of the Staley wastewater pretreatment plants include, at least in part, activated sludge. The pretreatment plants normally remove the organic constituents readily. However, poor sludge settle-

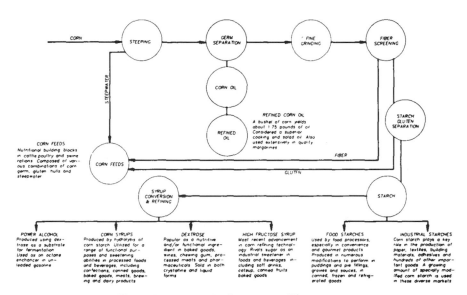

Figure 1.   Schematic of corn wet milling process.

505

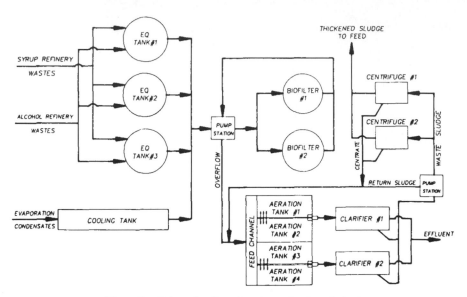

Figure 2.   Schematic of wastewater pretreatment plant.

ability, not always associated with filamentous organisms, is common to the Staley pretreatment plants as well as throughout industries with high-carbohydrate-type wastewater [1]. Filamentous bulking has occurred and been attributed to low aeration basin dissolved oxygen concentrations, both high and low food-to-microorganism ratios, and inadequate nutrients [2,3].

A study was conducted at several of the Staley pretreatment plants in an effort to determine how best to operate to avoid settleability problems. This paper describes the results of that investigation at one such plant and the steps that are being taken in response to it.

DESCRIPTION OF WASTE PRETREATMENT PLANT

The plant selected for this chapter is that associated with the Staley corn wet milling facility in Loudon, Tennessee. The plant treats an average of over 2.2 mgd containing about 23,300 lb BOD/day. A flow diagram of the pretreatment plant is presented in Figure 2.

Approximately 45% of the total volume, which contains only 30% of the total organic load, flows to three 320,000 gallon equalization tanks. The remaining volume and load, which are extremely variable, flow through a 128,000 gallon cooling tank. Flows from the equalization tanks and the cooling tank are combined and pumped over the first of the two-stage pretreatment process, two 62-foot diameter by 16 foot (packing) deep biofilters. Wastewater is recirculated over these at a rate of 1 gpm/ft² in order to control filter flies. This results in a recycle ratio of 4:1 and an organic removal efficiency of 58%.

Partially treated wastewater from the biofilters flows on to the activated sludge process. Return activated sludge, nutrients, and influent are mixed in an aerated channel which distributes flow to four 320,000 gallon aeration basins. (Note: The fourth basin was put on line in March, 1985.) Aeration is provided by a fine bubble dome diffuser system. Figure 3 shows the daily flow and load to the aeration basins following equalization and treatment in the biofilter. Average organic load to the activated sludge process is 9,900 lb BOD/day. Average removal efficiency is 87%.

Mixed liquor from the activated sludge process overflows to two 100-foot long, 30-foot wide, and 10-foot deep (water) travelling bridge clarifiers. Final effluent is combined with the facility's domestic wastewater and pumped to the City of Loudon POTW for treatment prior to discharge to the Tennessee River. Sludge is either recycled to the aeration basins or wasted to two horizontal, solid bowl centrifuges for thickening. It is then slurried with steepwater, evaporated, and dried as part of the animal feed stream. Figure 4 shows daily average effluent quality. Overall organic removal efficiency is 95%.

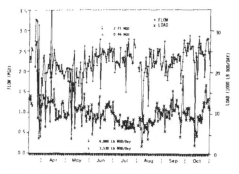

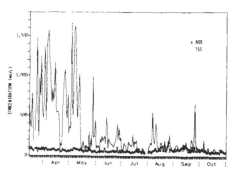

Figure 3. Flow and BOD load to aeration sys-   Figure 4. Effluent quality.
tem.

Periodic problems with sludge settleability prompted a study to determine how to better control the process to minimize these problems. The remainder of the paper will describe this study.

ANALYSIS OF PLANT OPERATION

The first step was to collect plant operating data and determine significant correlations between the various operating parameters. Specifically, it was hoped that significant correlations could be identified between sludge settleability as indicated by the sludge volume index (SVI) or the final effluent solids concentration and other plant operating parameters. These parameters could then be more carefully monitored and controlled.

Initial data was collected for the period March 14, 1985 (the date when the fourth aeration basin was placed in service) through May 1985. This was later updated to include data through October 1985. Results presented are those for the entire data collection period. These are not significantly different from those for the initial period.

Data was entered on an IBM PC using Lotus 1-2-3, stored on floppy disks, and transferred to an IBM 4381 system for multiple regression analysis. This analysis was performed using the Statistical Analysis System (SAS) from SAS Institute, Inc. of Cary, North Carolina.

The portion of the statistical system used regresses a dependent variable against a list of independent variables both individually and in combination. Both the SVI and effluent suspended solids concentration were used as dependent variables. Table I lists the independent variables. To represent load fluctuations, variables were established which represent the percent change in load for a particular day compared to averages for the previous 1 through 6 days.

The most significant single correlation ($R^2 = 0.66$) was between the sludge volume index (SVI) and the food-to-microorganism (F/M) ratio as indicated on Figure 5. The greatest correlation ($R^2 = 0.36$) using effluent solids as the dependent variable was with the sludge age. It was, therefore, concluded that the SVI was a sensitive indication of plant operation and that control of the F/M ratio could provide operational reliability and stability.

The second step was to estimate an optimum F/M ratio. This was done by plotting SVI versus 1/(F/M), as shown on Figure 6. Two distinct sets of data are apparent from the Figure. The lower region (higher F/M and lower SVI) is from a period of time following a severe nitrification/denitrification problem (shown on Figure 5 as the period of high F/M's starting the first part of August). Following this event, the plant was reseeded with municipal sludge, ammonia addition was limited, and polymer was used in the clarifiers. Wasted sludge had to be landfilled instead of used in the animal feed stream.

Table I. Independent Variables for Statistical Analysis

BOD (mg/L and lb/day) and Flow to Aeration
Flow and BOD of the various influent streams
Aeration Basin pH, temperature, DO, and MLSS
F/M Ratio
Sludge Retention Time

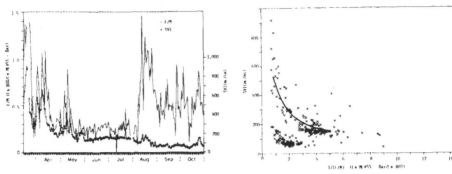

Figure 5.   Correlation between SVI and F/M.   Figure 6.   SVI versus 1/(F/M).

Because of these factors, this data is not considered representative. Performing a nonlinear regression on the remainder of the data indicates a best-fit ($R^2 = 0.75$) curve of the form:

$$SVI = 516 * (F/M)^{0.8} \tag{1}$$

This predicts a F/M ratio below 0.3 lb BOD/lb MLVSS/day is necessary to maintain an SVI below 200 mL/gm. Although there were not enough data points at lower F/M ratios to adequately define the curve, a lower limit of 0.15 lb BOD/lb MLVSS/day is necessary, based on experience, for good settleability.

With these analyses completed, several operational changes were made.

A. Flow from the equalization tanks is automatically controlled to compensate for changes in flow through the cooling tank. (Previously, flow from an equalization tank was set based on organic content and remained constant until it was empty. This resulted in considerable variations in organic loading.) The next step will be to install an on-line TOC monitor so that the flow from the equalization tanks can be controlled to compensate for changes in total organic load through the cooling tank. This will provide even better hour-to-hour control over the total organic load to the plant.

B. A target aeration basin solids concentration (MLSS) is established on a daily basis in order to maintain an optimum F/M ratio.

C. The return sludge (RAS) flow rate is manually adjusted in order to maintain the target solids concentration according to the mass balance:

$$Q_{RAS} = \frac{Q/1440 \times MLSS}{SS_{RAS} - MLSS} \tag{2}$$

where: Q_{RAS} = flow rate of RAS (gpm)
      Q = influent flow rate (Gal/Day)
      MLSS = concentration of mixed liquor suspended solids (mg/L)
      SS_{RAS} = concentration of solids in RAS (mg/L)

D. The wasting rate is manually adjusted on a daily basis in order to maintain a sludge bed depth in the clarifiers and/or a target recycle concentration.

RESULTS

Figures 3 through 5 show the operational results since these changes were made about June 1, 1985.

By controlling the variability of the influent loading as much as possible and then varying the aeration basin solids concentration to match the remaining influent variability, and thus maintain an optimum F/M ratio, sludge settleability (SVI) was considerably improved and more consistent.

In practical terms, there is a limit to how much and how fast the MLSS concentration can be changed. During normal operation, the MLSS concentration can only be changed about 500 mg/L over a 24-hour period by increasing either the recycle or the wasting. At the desired range of F/M ratios, this corresponds to a maximum increase or decrease in influent load of about 3000 lb BOD/day. Thus, the primary operating parameter for adjustment when using F/M control is the influent BOD loading.

PROCESS MODELING AND KINETICS

The final step in the overall evaluation process was to find a kinetic model which reasonably predicted the operation of the activated sludge plant. This could then be used in conjunction with a previously selected model of the biofilters to predict and optimize plant performance under various conditions.

Under steady state conditions, the substrate utilization rate in a reactor is described by the equation:

$$\frac{dS}{dt} = \frac{Q(S_i - S_e)}{V} \tag{3}$$

Where: $\dfrac{dS}{dt}$ = substrate utilization rate (mg/L/day)

   Q = influent flow rate (MGD)
   S_i = influent substrate concentration (mg/L)
   S_e = effluent substrate concentration (mg/L)
   V = reactor volume (MG)

Of the many mathematical representations of the substrate utilization rate, two, the Eckenfelder [4] and the Stover/Kincannon [5] models, were selected for further evaluation. This is primarily due to the fact that these, in one form or another, incorporate the F/M ratio.

The substrate utilization rate used in Eckenfelder's multiple zero order model takes the form:

$$\frac{dS}{dt} = \frac{K'_e X S_e}{S_i} \tag{4}$$

where: K'_e = Eckenfelder's modified model substrate removal rate constant (mg/mg/day)
   X = MLVSS concentration (mg/L)

Substituting into Equation 3 and rearranging into a linearized form gives:

$$\frac{Q S_i (S_i - S_e)}{X V} = K'_e S_e \tag{5}$$

The Stover/Kincannon monomolecular model defines the substrate utilization rate as:

$$\frac{dS}{dt} = \frac{U_{max} \dfrac{Q S_i}{X V} X}{K_B + \dfrac{Q S_i}{X V}} \tag{6}$$

where: U_{max} = Stover/Kincannon maximum specific utilization rate (lb BOD/lb MLVSS/day)
   K_B = Specific substrate loading at which the rate of substrate utilization is one-half the maximum (lb BOD/lb MLVSS/day)

Substituting into Equation 3 the linearized model takes the form:

$$\frac{X V}{Q(S_i - S_e)} = \frac{K_B}{U_{max}} \frac{X V}{Q S_i} + \frac{1}{U_{max}} \tag{7}$$

The evaluation consisted of plotting the plant operating data using the respective linearized forms and performing a linear regression to determine the appropriate factors and correlations. These are shown on Figures 7 and 8 for the two models.

It became very obvious from the plots and correlations that the Stover/Kincannon model (Figure 8) very adequately ($R^2 = 0.99$) depicts actual plant operation. This model is being used in conjunction with the model for the biofilters to estimate plant capacity, to determine when and if to put aeration basins on and off-line, and to predict plant performance under various conditions.

CONCLUSIONS

For easily degraded wastes, the SVI is a sensitive indication of operational changes.

For plants with variable loadings, F/M control provides consistent operation.

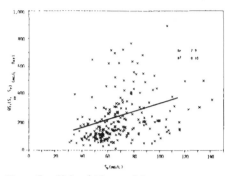

Figure 7. Eckenfelder model.

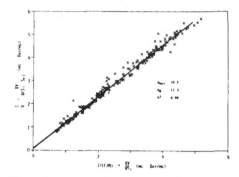

Figure 8. Stover/Kincannon model.

For a given set of operating parameters, there is an optimum range of the F/M ratio which produces good sludge settleability.

The Stover/Kincannon model very adequately describes the operation of an activated sludge process.

In addition, the following hypothesis is presented for discussion and, hopefully, further investigation. For a given aeration volume and kinetic characteristics, there is a maximum amount of variability (over a given time span) which can be tolerated by an activated sludge system without significantly affecting operation. For a given influent variability, this then determines the minimum equalization capacity required.

REFERENCES

1. Stover, E. L., "Start-up Problems at a Plant Treating Food Processing Wastewater," *Journal of the Water Pollution Control Federation*, 52, 2, 249 (1980).
2. Eikelboom, E. H., "Filamentous Organisms Observed in Activated Sludge," *Water Research*, 9, 365 (1975).
3. Jenkins, D., Richard, M. G., and Neethling, J. B., "Causes and Control of Activated Sludge Bulking," presented at the Annual Conference, South African Branch, Institute of Water Pollution Control, Port Elizabeth, Republic of South Africa.
4. Adams, C. E., W. W. Eckenfelder, and J. C. Hovious, "A Kinetic Model for Design of Completely Mixed Activated Sludge Treating Variable Strength Industrial Wastewater," *Water Research*, 9, 37 (1975).
5. Stover, E. L., "Process Control Strategies for Troubleshooting Activated Sludge Plants," presented at the Virginia Water Pollution Control Association's Seminar on Controlling Activated Sludge: Approaches and Case Histories, Richmond, (November 25, 1984).

61 LOW-RATE ANAEROBIC TREATMENT OF A POTATO PROCESSING PLANT EFFLUENT

Robert C. Landine, Head Environmental Engineering Division

Stephen R. Pyke, Environmental Engineer

Graham J. Brown, Environmental Engineer

Albert A. Cocci, Senior Environmental Engineer
ADI International, Inc.
Fredericton, New Brunswick, Canada E3B 4Y2

INTRODUCTION

In 1985, ADI was commissioned to design and conduct a seven month anaerobic pilot plant treatability study. The on-site pilot was to simulate a proposed full-scale, anaerobic process, and was to be operated at loadings and hydraulic retention times (HRT) anticipated in full-scale operation. The full-scale and pilot anaerobic technology chosen was the ADI-BVF. This is a low-rate digestion process which has proven advantages in the treatment of potato processing wastewaters. Advantages which made this technology the choice process included: ability to treat high concentration of SS, effective and reliable wastewater treatment, ability to absorb shock loads, economical construction and operation, digestion of waste solids from aerobic processes, energy production from biogas, and simplicity of operation.

The objectives of this study were to evaluate the treatability of the potato processing wastewater (including peel waste), as well as evaluate the performance of the technology for the purpose of verifying full-scale design criteria. Wastewater treatment plant personnel carried out the day-to-day operation of the plant. Supervision and sample analysis was provided by ADI.

PILOT STUDY

Apparatus

The pilot plant, illustrated schematically in Figure 1, consisted of: 1) an 840 litre (220 gallons) liquid volume BVF pilot reactor, 0.61 metres (24 inch) diameter by 3.3 metres (10.8 feet) long complete with heaters, thermostats, thermometers, insulation, manual mixers, influent and effluent headers, and sludge recycle piping, 2) two 200 litre raw wastewater and feed storage drums, 3) variable-speed, peristaltic pumps (raw wastewater, feed, and recycle) with timer controls, 4) variable-speed, feed mixer, and 5) wet-tip biogas meter.

Operation

The reactor was seeded with 116 litres (31 gallons) of anaerobic sludge obtained from a municipal treatment plant primary digester, 45 litres (12 gallons) of raw wastewater and 679 litres (179 gallons) of tap water, and it was heated to 28°C (82°F). Following seeding, the reactor was nursed through a startup phase (day 1 to 63) during which raw wastewater, peel waste, nutrients, and alkalinity were added in either 10 or 20% increments until full feed strength was attained. To promote reactor maturity during startup, therapeutic anaerobic seed sludge dosages were added directly to the reactor on day 30 (20 litres or 5 gallons), day 36 (20 litres or 5 gallons), and day 44 (30 litres or 8 gallons). Operating temperature was increased to 35°C (95°F) on day 42.

Following startup, the study proceeded through a steady-state operation phase at 35°C (95°F) from day 64 to 154, and at 28°C (82°F) from day 155 to 189. Operating temperature was reduced from 35 to 28°C (95 to 82°F) over a seven-day period (day 155 to 162) at a rate of approximately 1°C per day (2°F per day). Following steady-state, two stress loading periods (day 190 to 194 and day 200 to 203) were conducted at 28°C (82°F). The pilot then remained idle until completion of the study and dismantling on day 213. Further reactor operation details are provided in Table I.

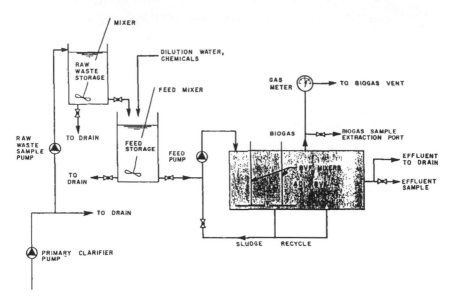

Figure 1.   Schematic of pilot plant.

The pilot plant was operated on a day-to-day basis for the duration of the study. Daily, a 24-hour raw wastewater composite was collected, by peristaltic pump, in the raw waste storage tank. On the following day, this composite was used in preparing a new feed in the feed storage tank. The feed consisted of raw wastewater, dilution wastewater (during startup only), blended peel waste, nutrients (NH_4Cl, KH_2PO_4), and alkalinity ($NaHCO_3$). A 150 litre (40 gallon) feed volume was prepared during startup and steady-state operation, and a 200 litre (53 gallon) volume was prepared during stress loading.

Nutrient additions continued into steady-state operation until they were eliminated completely on day 85. Alkalinity additions of varying dosages were maintained throughout the study.

A double-head, peristaltic feed pump was adjusted to deliver the required feed volume from storage to the reactor, thus establishing the desired HRT. The feed mixer was operated simultaneously during the feeding cycle to ensure a homogenous feed mixture. The second pumphead recycled bottom sludge from the reactor to the influent header, where it was blended with incoming feed. The recycle-to-feed ratio was maintained at 1:1 throughout the study. Reactor contents at the influent end were manually mixed on a daily basis.

Biogas was collected from the top of the reactor and metered using a wet-tip gas meter.

Raw wastewater generally consisted of that waste produced from potato processing. However, from day 125 to 153, raw wastewater consisted of waste produced solely from pea processing and had a distinguishing green colour and pea odour.

Analyses

The test schedule for routine analyses conducted during the study is presented in Table II. Biochemical oxygen demand (BOD), suspended solids (SS), volatile suspended solids (VSS), ammonia nitrogen (NH_3-N), total kjeldahl nitrogen (TKN), total phosphorus (total-P), pH, and alkalinity analyses were conducted in accordance with *Standard Methods*. Chemical oxygen demand (COD) was performed using Hach digestion equipment, volatile acids (VA) by a method described by Ripley, Boyle and Converse [1], biogas (CO_2, O_2) using an Orsat gas apparatus, and hydrogen sulfide (H_2S) using a Drager gas sampling pump and Drager type H_2S tubes.

Table I. Pilot Plant Operation Summary

Mode of Operation	Day	Comments
Startup	1	Start feed at 10 % strength (90% dilution water); 28°C; nominal feed rate = 105 L/d (HRT = 8 d); chemical addition = 5 mg/L N, 1 mg/L P, 1.5 g/L NaHCO$_3$; peel waste addition = 1 mL/L
	3	1 kg NaHCO$_3$ added directly to BVF contents
	8	20 % feed strength; 1 kg NaHCO$_3$ added directly to BVF contents; chemical addition = 10 mg/L N, 2 mg/L P; peel waste addition = 2 mL/L
	9	0.5 kg NaHCO$_3$ added directly to BVF contents
	10	1 kg NaHCO$_3$ added directly to BVF contents
	15	30 % feed strength; chemical addition = 15 mg/L N, 3 mg/L P; peel waste addition = 13 mL/L
	17	1 kg NaHCO$_3$ added directly to BVF contents
	22	40 % feed strength; 1 kg NaHCO$_3$ added directly to BVF contents; chemical addition = 20 mg/L N, 4 mg/L P; peel waste addition = 4 mL/L
	25	Chemical addition = 3 g/L NaHCO$_3$
	27	Additional 0.5 kg NaHCO$_3$ added to feed
	28	Additional 0.5 kg NaHCO$_3$ added to feed
	29	50 % feed strength; chemical addition = 25 mg/L N, 5 mg/L P; peel waste addition = 5 mL/L
	36	60 % feed strength; chemical addition = 30 mg/L N, 6 mg/L P; peel waste addition = 6 mL/L
	42	Temperature increased to 35° C
	43	70 % feed strength; chemical addition =35 mg/L N, 7 mg/L P; peel waste addition = 6.5 mL/L
	50	80 % feed strength; chemical addition = 40 mg/L N, 8 mg/L P, 1.5 g/L NaHCO$_3$; peel waste addition = 7 mL/L
	57	100 % feed strength; chemical addition = 50 mg/L N, 10 mg/ L P; peel waste addition = 9 mL/L
	60	1 kg NaHCO$_3$ added directly to BVF contents
Steady-state	64	Start of steady—state 7 d/wk operation, 35°C, nominal feed rate = 105L/d (HRT = 8 d)
	67	Peel waste addition = 11 mL/L
	71	Chemical addition = 33 mg/L N, 7 mg/L P
	73	Peel waste addition = 13 mL/L
	74	1 kg NaHCO$_3$ added directly to BVF contents
	75	Peel waste addition = 15 mL/L
	78	Peel waste addition = 17 mL/L chemical addition = 16 mg/L N, 3 mg/L P
	80	Peel waste addition = 19 mL/L
	85	Chemical addition = 0 mg/L N, 0 mg/L P, 0 g/L NaHCO$_3$
	89	Peel waste addition = 21 mL/L
	92	Peel waste addition = 23 mL/L, 0.5 kg NaHCO$_3$ added directly to BVF contents
	93	0.5 kg NaHCO$_3$ added directly to BVF contents
	94	1 kg NaHCO$_3$ added directly to BVF contents
	95	Peel waste addition = 25 mL/L
	99–112	Feed stopped due to factory shutdown
	113	Full feed strength resumed
	125	Pea production at factory commenced
	129	1 kg NaHCO$_3$ added directly to BVF contents, chemical addition = 3.3 g/L NaHCO$_3$
	134	Peel waste addition = 29 mL/L
	137	Peel waste addition = 32 mL/L
	153	Pea production at factory stopped

Table I Continued

Mode of Operation	Day	Comments
Steady-state continued	155	Start of steady-state 5 d/wk operation (weekend shutdown), temperature decreased to 28°C, nominal feed rate = 105 L/d (HRT = 8 d)
	171	Chemical addition = 1.7 g/L NaHCO₃
Stress Loading	190–194	28°C, nominal feed rate = 123 L/d (HRT = 6.8 d)
	195–199	Feed stopped, pilot idle
	200–202	28°C, nominal feed rate = 185 L/d (HRT = 4.5 d)
	203	Feed stopped, pilot idle
	213	Study completed, reactor dismantled

Table II. Routine Analysis

Parameter	Frequency[a]	
	Influent	Effluent
COD	W (C)	W (C)
BOD	W (C)	W (C)
SS	W (C)	W (C)
VSS	W (C)	W (C)
pH	D (G)	D (G)
Alkalinity	TW (G)	TW (G)
VA	TW (G)	TW (G)
Temperature	D (G)	D (G)
NH₃-N	M (C)	M (C)
TKN	M (C)	M (C)
Total-P	M (C)	M (C)

[a]Biogas composition analyzed weekly on a grab sample.
  D = daily, W = weekly, TW = twice weekly, M = monthly,
  C = weekly composite of daily grabs,
  G = instantaneous grab

Results

The temporal variations of influent and effluent COD, BOD, and SS are presented in Figures 2, 3, and 4, respectively. The temporal variations of effluent pH, VA, and alkalinity are presented in Figures 5, 6, and 7, respectively. The COD, BOD, and SS values reflect the results of tests on weekly composites; pH, VA, and alkalinity values represent results of grab samples.

Average results for influent and effluent COD, BOD, SS, VSS, pH, VA, alkalinity and temperature during steady-state operation at 35°C (95°F) and at 28°C (82°F) are presented in Table III. The average feed rate during steady-state operation was 106 L/d, resulting in an HRT = 7.9 d. Average

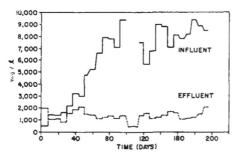

Figure 2.   COD of influent and effluent.

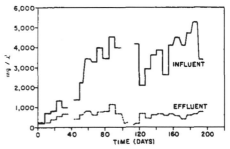

Figure 3.   BOD of influent and effluent.

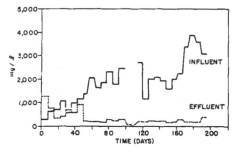

Figure 4.   SS of influent and effluent.

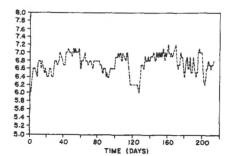

Figure 5.   Effluent pH.

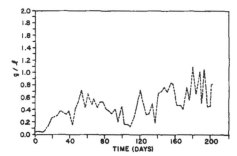

Figure 6.   Effluent volatile acids.

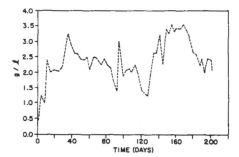

Figure 7.   Effluent alkalinity.

Table III. Average Pilot Plant Results

Mode of Operation	Parameter[a]	Influent	Effluent	Percent Removal
Steady-state	COD	7,605 (1,128)[b]	1,304 (196)	82.4
(day 64 to 154)	BOD	3,492 (727)	636 (237)	81.1
Temp = 35°C (95°F)	SS	1,951 (430)	216 (37)	88.6
HRT = 7.9 d	VSS	1,866 (427)	185 (29)	89.7
	pH	3.6 to 7.1	6.0 to 7.0	
	VA	1,554 (609)	440 (197)	
	Alkalinity	1,293 (967)	2,277 (551)	
Steady-state	COD	8,480 (584)	1,281 (213)	84.8
(day 155 to 189)	BOD	4,528 (439)	569 (95)	87.4
Temp = 28°C (82°F)	SS	3,007 (756)	208 (38)	92.4
HRT = 7.9 d	VSS	2,922 (753)	200 (33)	92.5
	pH	5.8 to 7.4	6.4 to 7.2	
	VA	1,441 (671)	706 (215)	
	Alkalinity	1,621 (700)	3,150 (424)	
Stress Loading #1	COD	8,500	2,100	75.3
(day 190 to 194)	BOD	3,393	757	77.7
Temp = 28°C (82°F)	SS	3,060	239	87.3
HRT = 6.8 d	VSS	3,010	342	88.6
Stress Loading #2	COD	8,600	1,500	82.6
(day 200 to 202)	BOD	4,280	733	82.9
Temp = 28°C (82°F)	SS	3,690	130	93.7
HRT = 4.5 d	VSS	3,610	226	93.7

[a]Temperature expressed in °C, pH in standard units, all others in mg/L.
[b]Values in parentheses represent sample standard deviation.

Table IV. Raw Wastewater, Peel Waste (Characterization)

Parameter[a]	Raw Wastewater	Peel Waste
COD	3,524 (1,287)[b]	140,350 (27,998)
BOD	1,464 (613)	62,158 (9,156)
SS	1,203 (590)	72,148 (16,969)
VSS	1,184 (592)	71,971 (16,793)
TS	2,546 (350)	135,138 (14,932)
TVS	1,864 (300)	121,621 (14,318)
pH	3.4 to 7.0	
Alkalinity	90 (86)	
VA	975	
TKN	128 (18)	
NH$_3$-N	33 (6)	
Total-P	34 (6)	

	Day 1 to 7	Day 120
Total Sulfur	16.5	—
Sulfide	0	—
Cl	61.2	64.5
Ca	43.6	43.7
Cd	0.02	<0.01
Co	<0.02	<0.02
Cr	<0.02	<0.02
Cu	0.21	0.10
Fe	2.85	3.43
Hg (ppb)	<1	<1
K	149	142
Mg	12.7	17.2
Na	51.5	113
Ni	<0.02	<0.02
Pb	0.23	<0.05
Se	<0.01	<0.1
Zn	2.00	0.3

[a]pH in standard units, all others in mg/L.
[b]Values in percentages represent sample standard deviation.

COD, BOD, and SS loadings were 1.0, 0.48, and 0.29 kg/m³/d (62.3, 29.9, 18.1 lb/1000 ft³/d), respectively.

Specific methane generation rate during the same period was determined at 0.35 m³ CH$_4$ (STP)/kg COD removed (5.61 ft³ CH$_4$/lb COD removed). Average biogas CH$_4$ concentration was 61.0% by volume and average H$_2$S concentration was 977 ppm (std. dev. = 357, max. value = 1812 ppm).

Results of influent and effluent COD, BOD, SS, and VSS during stress loading are presented in Table III. The average feed rate during stress loading #1 was 123 L/d (HRT = 6.8 d), and 185 L/d (HRT = 4.5 d) during stress loading #2. COD, BOD, and SS loadings during stress loading #1 were 1.24, 0.50, and 0.45 kg/m³/d (77.3, 31.2, and 28.2 lb/1000 ft³/d), respectively, and during stress loading #2 were 1.89, 0.94, and 0.81 kg/m³/d (118, 58.6, and 50.5 lb/1000 ft³/d), respectively.

Specific methane generation rates for stress loadings #1 and #2 were 0.34 and 0.36 m³ CH$_4$ (STP)/kg COD removed (5.45 and 5.77 ft³ CH$_4$/kg COD removed), respectively.

Nitrogen and phosphorus data collected during steady-state operation produced average influent values for TKN, NH$_3$-N, and total-P of 215, 59, and 40 mg/L, respectively, and average effluent values of 219, 163, and 35 mg/L, respectively.

Characterizations of raw wastewater, peel waste, and anaerobic sludge (obtained from reactor upon completion of the study) were conducted and are presented in Tables IV and V.

During the course of the entire study, approximately 31 kg of TVS were fed to the reactor. Upon completion of the study, the reactor was dismantled, and a net TVS accumulation (solids remaining less solids added as seed) of 5.0 kg was determined. Therefore, approximately 16% of volatile solids fed over the 213-day period were not anaerobically digested. It should be noted that a total of 4.8 kg

Table V. Reactor Anaerobic Sludge Characterization

Parameter[a]	
COD	76 000
BOD	9 000
SS	55 200
VSS	30 600
TS	66 189
TVS	34 625
pH	7.0
VA	2 710
Alkalinity	6 667
TKN	2 590
NH_3-N	536
Total-P	266
Total Sulfur	264
Sulfate	40.0
Cl	122
As	0.72
Ca	1 260
Cd	0.40
Co	0.30
Cr	3.72
Cu	36.0
Fe	976
Hg (ppb)	<0.01
K	525
Mg	251
Mo	0.5
Na	1 340
Ni	2.35
Pb	21.2
Se	<0.1
Zn	73.2

[a]pH in standard units, all others in mg/L unless otherwise indicated.

of VSS left the reactor via the effluent over the 213-day period. In other words, 9.8 kg of solids represents the combined sum of undigested solids plus biological solids produced and solids lost in the effluent over the 213-day study period.

DISCUSSION

Steady-state results presented in Table I indicate that the potato processing wastewater, including peel waste , is well suited to anaerobic treatment using this low-rate process. Removals of COD, BOD, and SS remained consistently high throughout the study. Serious physical or biological problems were not encountered during the study.

During steady-state operation at 28°C (82°F), COD, BOD, and SS effluent concentrations averaged 1281, 569, and 208 mg/L, respectively. At an average HRT of 7.9 days, average removals for COD, BOD, and SS were 82.4, 81.1, and 88.6%, respectively. Steady-state operation at 35°C (95°F) produced similar, although slightly lower, removals. Better performance at 28 (82) versus 35°C (95°F) was attributed to the following: 1) during the initial stages of 35°C (95°F) operation, the reactor had not yet achieved full maturity, 2) a two-week factory shutdown during 35°C (95°F) operation interrupted feed and thus delayed reactor development, and 3) factory pea processing (days 125 to 153) produced a change in wastewater characteristics during 35°C (95°F) operation, and a period of reactor adjustment was observed.

Throughout the study, NaHCO$_3$ was added to the feed in dosages ranging from 1.5 to 3.3 g/L. During startup, NaHCO$_3$ provided alkalinity buffering against potentially-damaging VA concentrations. Supplements of NaHCO$_3$ continued following startup to correct a low, feed pH, the result of raw wastewater acidification during 24-hour storage prior to use. In full-scale, alkalinity supplements would be anticipated during startup.

However, at full maturity, an effluent alkalinity of at least twice the effluent VA concentration, without supplement, can be anticipated to provide effective buffering.

Nitrogen and phosphorus supplements were supplied to the feed to ensure adequate nutrient availability, but by day 85 were eliminated entirely. Average influent TKN and total-P concentrations of 215 and 40 mg/L, respectively, provided an influent COD/N/P ratio of approximately 500/13/2.5. This ratio exceeds our commonly accepted minimum value for anaerobic digestion of 500/5/1. There was also excess nitrogen and phosphorus in the reactor effluent. Nutrient supplement during startup or regular operation would not be required in full-scale.

The steady-state average specific methane generation rate corresponded to the common theoretical value 0.35 m³ CH_4 (STP)/ kg COD removed (5.61 ft³ CH_4/lb COD removed).In practice, 90% of the theoretical value would be anticipated, the result of a portion of COD being converted to biomass rather than biogas, the reduction of raw waste sulfur to sulfide, and dissolution of methane in the liquid effluent. In this case, the higher values are believed to be a reflection of the influent COD concentrations actually being greater than found by test results; it is not uncommon for COD results to be in error on the low side due to the difficulty of catching all the SS sampled from a barrel which contains liquid of high SS concentrations. If this reasoning is correct, the actual loadings and removals were, in fact, somewhat higher than calculated and reported herein.

Stress loading operation indicated the ability of the process to accept short-term peak loading. Hydraulic loads 1.75 times average steady-state operation resulted in only slight reductions in COD and BOD removal efficiencies.The average specific methane generation rate remained at 0.35 m³ CH_4 (STP)/kg COD removed (5.61 ft³ CH_4/lb COD removed). A reduction in biogas CH_4 content was noted during the stress loading periods (43.9% CH_4 on day 193 and 45% on day 203); however, 70% CH_4 content was observed shortly following a return to normal loading conditions.

Solids investigation indicated 68% of influent volatile solids were anaerobically digested. It is believed that the brevity of the study limited solids from greater percent destruction. A value of 90% or more would not be unusual for this process over a longer term, e.g., two years, such as would be provided in full-scale design.

At the end of the study the consistency of the reactor sludge was 6.6% total solids. The sludge contained 1.6% protein on a wet weight basis or 25% on a dry weight basis.

FULL-SCALE PLANT

A 45,000 m³ (12 USMG) capacity ADI-BVF is part of a full-scale potato processing wastewater treatment facility scheduled for completion in May 1986. The BVF will provide anaerobic pretreatment for estimated average daily raw waste COD and SS loads of 39,000 kg (85,000 lb) and 19,300 kg (42,400 lb), respectively. Subsequent treatment is provided by an activated sludge system consisting of an aerated basin and secondary clarification.

On the basis of an average daily waste flow of 5680 m³/d (1.5 mgd US), the BVF will provide an HRT of 7.9 days and estimated average effluent COD, BOD, and SS concentrations of 1000, 500, and 500 mg/L, respectively. The anticipated full-scale BVF operating temperature is 27 ± 2°C (81 ± 4°F). A flexible, floating, high-density polyethylene cover is incorporated for biogas collection. The cover is fitted with 0.5-inch ethafoam insulation to reduce heat loss through the cover.

Biogas generated within the BVF is collected beneath the cover and conveyed under vacuum to a separate utilization system. Biogas production is estimated at an average of 15,000 m³/d (0.53 × 10⁶ ft³/d) and will fuel two 425 kW engine-generators. Electrical energy produced is to be used at the waste treatment plant and exported to the potato processing plant. Engine-generator waste heat recovery is to be utilized to maintain BVF temperature.

Following seeding, heating and alkalinity addition, the fullscale BVF will proceed through a startup phase until full raw waste addition is achieved. Full load operation is expected in the fall of 1986. A future paper will be prepared which addresses the full-scale BVF startup and operating performance.

CONCLUSIONS

Results of the pilot plant study illustrate the capabilities of the low-rate process in achieving high removals in treating the potato processing wastewater investigated. Stress loadings indicated an ability to easily handle short-term hydraulic and organic loadings nearly twice steady-state values.

Removals and loadings for steady-state and stress loading #2 operations are as follows:

	Hydraulic Feed Rate (L/d)	HRT (d)	COD Loading (kg/m³/d)	COD (%)	Removals BOD (%)	SS (%)
Steady-state	106	7.9	1.0	84.8	87.4	92.4
Stress	185	4.5	1.9	82.6	82.9	93.7

For every kilogram of COD removed in the process, approximately 0.35 m³ CH_4 (STP) was collected (5.61 ft³ CH_4/lb COD removed).

A 68% destruction of influent volatile solids occurred over the 213-day study.

The results of the pilot study were used in designing a fullscale ADI-BVF system in which the biogas is recovered and used in an engine-generator to produce electricity, and heat is recovered to increase the temperature of the fermenter.

REFERENCE

1. Ripley, L. E., Boyle, W. C., and Converse, J. C., "Improved Alkalimetric Monitoring for Anaerobic Digestion of Poultry Manure," *Proceedings of the 40th Purdue Industrial Waste Conferencew* (May 1985).

62 APPLICATION OF THE TWO-PHASE ANAEROBIC FLUIDIZED BED PROCESS TO THE TREATMENT OF CORN PROCESSING WASTEWATER

Dilip J. Kothari, Development Engineer

Alan Y. Li, Technology Manager
Dorr-Oliver, Inc.
Stanford, Connecticut 06904

Thomas J. Kontol, Environmental Mgmt. Supervisor
American Maize-Products Company
Hammond, Indiana 46320

INTRODUCTION

The corn wet milling industry processes corn to produce starch, oil, corn syrup, and animal feed. Normally, clean corn is steeped in a dilute solution of sulfuric acid to loosen the hull, soften the gluten, and dissolve minerals or organic matter in the kernel. Next, the corn is coarsely ground and cycloned to separate the germ from which oil is extracted. The kernel residue is finely ground and screened to separate the soluble starch and gluten from the fiber and hull, which are used as feed additives. Starch is centrifugally separated from the gluten and marketed either as starch, modified starch, or hydrolyzed into corn syrup or corn sugar. Gluten is also used as a feed additive. Both the feed-producing and starch-manufacturing process evolve a process wastewater containing approximately 3% of the corn in soluble form. Reuse programs, which are known as "bottling up," have reduced plant losses to less than 0.5% of the dry corn raw material.

Major wastes from corn processing plants include the volatile organics entrained in the evaporator condensate resulting from the evaporation of steep water, sweet water resulting from refining of corn syrup and dextrose, starch filtrates from the preparation of modified starches, and wastes from "bottling up" processes [1]. Other in-plant liquid waste sources will vary depending on the products made and processes used in a particular plant [2].

The Hammond, Indiana corn processing plant of American Maize-Products Company manufactures modified and unmodified starches, corn syrups, corn syrup solids, dextrins, and feed co-products. Corn processing waste streams can include steep water condensates, sweet water, starch filtrates, carbon regeneration water, and gluten. These waste streams are currently disposed of in two ways:

1. A major portion of the corn processing wastewater is treated in an on-site system consisting of activated sludge and polishing lagoons, chemical treatment, and final clarification. After clarification, the effluent is discharged to Lake Michigan. Waste activated sludge (WAS) from the aerobic system is thickened and removed by centrifugation.
2. At times, waste streams, representing a significant organic load, are sent directly to the Hammond, Indiana publicly-owned treatment works (POTW).

The Hammond POTW sewer charges, assessed on flow, 5-day biochemical oxygen demand (BOD_5) and total suspended solids (TSS), are very significant. American Maize has investigated various alternatives to reduce sewer charges and to reduce the load on its existing aerobic system. Among these alternatives are product recovery, hauling away raw wastewater for disposal elsewhere, expanded aerobic treatment capacity, and anaerobic pretreatment. Anaerobic pretreatment has been chosen as the most cost effective and practical option.

Anaerobic pretreatment has proven to be a very attractive means of removing large quantities of organic material from highly concentrated waste streams prior to on-site aerobic treatment [3] or discharge to a local POTW. Significant reductions in aeration power costs or sewer surcharges can be achieved with anaerobic pretreatment as well as production of large quantities of methane-rich

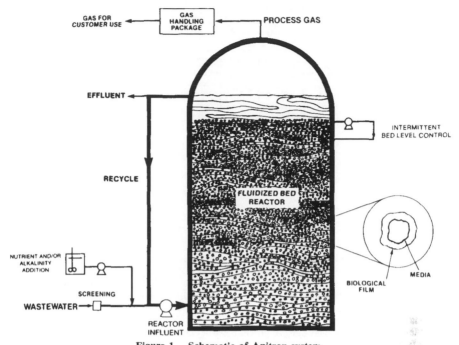

Figure 1.   Schematic of Anitron system.

process gas. Accordingly, various commercial anaerobic processes are currently being applied to pretreat a broad spectrum of industrial wastewaters.

The Anitron™ System

Dorr-Oliver's Anitron™ system (Figure 1) is the commercial embodiment of the fixed-film fluidized bed process providing a means of optimizing the volumetric efficiency of the anaerobic process while ensuring the maximum tolerance to shock load conditions. The patented fluidized bed process involves passing wastewater upward through a bed of media, such as sand, at a velocity sufficient to expand the bed. The fluidized media thus provide a vast surface area for biological film growth resulting in a biomass concentration normally an order of magnitude greater than that in a suspended growth system. This high biomass concentration allows utilization of a reactor much smaller than those used for suspended growth systems. Besides being compact, other inherent properties of the fluidized bed process include:

- maximum contact between liquid and media
- minimal substrate diffusional resistance
- no channeling or plugging in the reactor
- ability to control and optimize biological film thickness.

Because of its high efficiency and compactness, the Anitron system is ideal for integration with aerobic processes for complete effluent treatment. The anaerobic process can reduce the bulk of the organics going to the aerobic process, which provides effluent polishing. Since the anaerobic microorganism yield is much lower than the yield for aerobic microorganisms, pretreatment can reduce sludge production significantly. In the case of pretreatment prior to a low sludge-producing aerobic process, such as extended aeration, the Anitron effluent will enter the aerobic process directly and further reduction of waste activated sludge through anaerobic treatment may not be necessary (Figure 2). However, when a higher rate aerobic system, such as the one at American Maize, follows the Anitron system, further reduction in the quantity of sludge through WAS recycling back to the Anitron system may become desirable. A clarifier is introduced in this treatment scheme to capture nondegradable solids as well as sheared biomass and to concentrate them for final disposal as digested solids (Figure

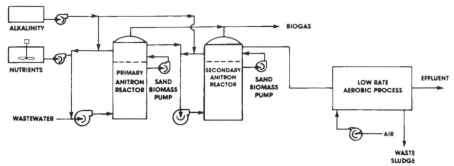

Figure 2.   Two-phase Anitron system pretreatment with low rate aerobic polishing.

3). This scheme also provides a single wasting point for the entire wastewater treatment plant simplifying the sludge wasting practice.

Anaerobic methane fermentation is essentially a two-phase process in which complex organic compounds are sequentially converted through intermediates to methane. A recent development in optimization of the Anitron system is two-phase operation. This two-stage configuration allows maintenance of optimal environmental conditions for acid formation in a primary stage and methane production in a secondary stage. Results from numerous pilot plant studies confirm the superior performance and flexibility in operation of the two-phase anaerobic concept over other single-phase reactor technologies [4,5,6]. The Anitron system is particularly well suited to the staged operation as no unit process for bacterial separation is required.

A two-phase pilot plant study was undertaken by Dorr-Oliver and American Maize in November of 1984. The principal objectives of the pilot plant program were to derive process information for anaerobic treatment of corn processing wastewater and the design of a full-scale plant. Specifically, the objectives were to determine the:

1. Optimal volumetric loading in terms of kg COD/m^3-day.
2. Ability of the two-phase Anitron system to digest recycled waste activated sludge from American Maize's aerobic system.
3. Effect of organic shocks of the type actually experienced in operation of the corn wet milling plant on the performance of the two-phase Anitron system.
4. Size of the clarifier required to concentrate waste digested sludge in the Anitron effluent.

This paper will present the results obtained from the pilot plant studies conducted over a nine-month period and discuss integration of the two-phase Anitron system with the existing treatment scheme.

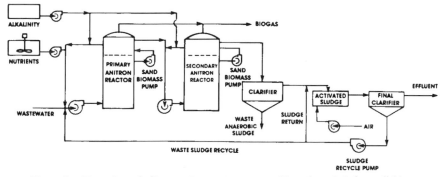

Figure 3.   Two-phase Anitron system pretreatment with activated sludge polishing.

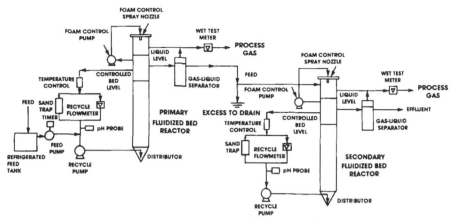

Figure 4.   Schematic of two-phase Anitron pilot plant.

PILOT PLANT FACILITIES AND OPERATION

The Anitron system treatability unit is Dorr-Oliver's pilot-scale anaerobic fluidized bed reactor, designed and developed for a treatability study of wastewater and for derivation of information for the design of full-scale systems. It is a scaled-down version of the Anitron system put together in a convenient package for easy shipment, installation, and operation.

The two-phase Anitron system treatability unit (Figure 4) consists of two clear PVC reactors, each 10.2 cm in diameter and approximately 3 m high to the effluent outlet port. Additional pilot plant components include a refrigerated feed tank, feed and recycle pumps, a means for foam control, gas-liquid separator, wet-test meters for gas flow measurement, and various other instrumentation. Temperature control of each reactor is achieved by adjusting the temperature of the recycle stream using an electrically controlled heat exchanger. Automatic reactor pH control is achieved with addition of sodium hydroxide. The treatability unit also provides for gas, liquid, and solid media sampling.

The two-phase Anitron pilot plant was started up using seeded sand media from a full-scale Anitron plant. Reactors were further seeded with waste digester sludge from the Hammond POTW.

Feed (Table I) to the pilot plant reactors was prepared on-site by mixing steep water condensates, starch filtrates, carbon regeneration water, gluten, and sweet water in appropriate proportions to simulate the expected wastewater. To ensure proper nutrients, nitrogen (in the form of urea), phosphorus (phosphoric acid) and a micronutrient mixture were added to the feed. During the combined treatment of process wastewater and waste activated sludge, the WAS was mixed with the feed. The

Table I. Characteristics of Feed Wastewater to the Anitron Pilot Plant

Parameter[a]	Corn Processing Wastewater, mg/L	Corn Processing Wastewater With WAS, mg/L
Total COD	9,028	9,488
Soluble COD	6,428	5,861
Total BOD$_5$	5,900	5,849
Soluble BOD$_5$	4,691	3,922
TSS	2,181	2,495
VSS	1,986	2,251
Total Alkalinity as CaCO$_3$	0	0
TKN	41	49
TP	48	49

[a]All numbers except TKN are averages of fifteen sets of data. TKN numbers are averages of six sets. TKN and TP numbers are reported after subtracting the nutrient addition.

Table II. Summary of Test Program

Test Program	Program Description	System Volumetric Loading (VL) Rate[a], kg COD/m³/day	System Hydraulic Retention Time[a] (HRT), hours	Average Feed Wastewater COD, mg/L
Phase 1	Steady-state evaluation for corn processing wastewater			
	Run 1	16.0	11.9	8,840
	Run 2	18.7	11.5	9,310
Phase 2	Steady-state evaluation for mixture of corn processing wastewater and WAS			
	Run 3	20.2	12.1	10,193
	Run 4	25.4	7.9	8,430
Phase 3	Shock Load Tests			
	Lower Shock Loading	120	9	44,900
	Higher Shock Loading	195	9	71,800
Phase 4	Clarification Tests	—	—	—

[a]VL and HRT are based on total fluidized bed volume in the two reactors.

environmental conditions in each reactor remained constant throughout the experimental program. The temperature in each reactor was approximately 35 C. The pH was approximately 6 in the first-stage reactor and 7 in the second-stage reactor.

To ensure proper operation and performance of the pilot plant and to derive useful information for process design, the pilot plant was monitored for operating conditions such as pH, temperature, bed height, feed rates, etc. at least once per day. Samples of the feed, effluents, and gas were taken routinely from the pilot plant for laboratory analyses. The analyses and their frequency varied with the various phases of the study. All liquid samples were analyzed in accordance with standard procedures [7]. Gas samples were analyzed using GowMac Model 550 gas chromatograph with thermal conductivity detector.

Four experimental phases were completed during the operation of the pilot plant as summarized in Table II. During the first two phases, data was derived for each experimental run once steady-state conditions were achieved. Results were derived for each run over a period of two to four weeks by analyses of samples taken around the pilot plant reactors at a frequency of twice per week. The individual sets of results were averaged to represent the results for each run.

RESULTS AND DISCUSSION

Phase 1 — Steady-State Evaluation With Corn Processing Wastewater

The pilot plant was operated at two volumetric loading rates and the corresponding performance results for the treatment of corn processing wastewater are presented in Table III. As the volumetric loading rate was increased from 16.0 to 18.7 kg COD/m³-day, effluent soluble COD increased from 772 to 1072 mg/L. Corresponding effluent BOD$_5$ values were 397 and 504 mg/L. Volatile acids also increased from 358 to 530 mg/L.

Phase 2 — Steady-State Evaluation With A Mixture of Corn Processing Wastewater and Waste Activated Sludge Digestion

As a part of the integration of the Anitron system into the existing aerobic wastewater treatment system, WAS from the aerobic system was mixed with corn processing wastewater in the pilot plant to determine the degree to which the WAS could be stabilized and its impact on system performance. The WAS represented 8% of the total wastewater flow. Table IV presents performance results for Phase 2. As the volumetric loading rate was increased from 20.2 to 25.4 kg COD/m³-day, effluent soluble COD increased from 1077 to 1163 mg/L. Effluent soluble BOD$_5$ also increased from 673 to 903 mg/L. With the increase in the volumetric loading rate, volatile acids also increased from 457 to 709 mg/L.

Table III. Two-Phase Anitron System Performance Results for the Treatment of Corn Processing Wastewater

Parameter	Run No. 1	Run No. 2
System Volumetric Loading Rate, kg COD/m³-day	16.0	18.7
Feed Values, mg/L		
COD	8,840	9,310
BOD$_5$	5,770	5,196
Effluent Values, mg/L		
Soluble COD	772	1,072
Soluble BOD$_5$	397	504
Volatile Acids (as Acetic)	358	530
Removal, %		
COD	91.5	88.5
BOD$_5$	93.1	91.7

Note: All values stated are means during the experimental period.

A major portion of the effluent COD and BOD$_5$ consisted of volatile acids in all four runs. These volatile acids are readily biodegradable and, based on past experience, expected to be degraded in long-term operation as the methanogenic population in the reactors increases.

As shown in Table V, methane yield was substantially lower at low volumetric loadings than the theoretical value of 0.35 m³ CH$_4$/kg COD removed. It is believed that this is a result of the inaccuracy of the wet test meters at low gas flows produced at lower loadings and loss of pressure in the reactors during frequent sampling and maintenance work on the reactors. This is supported by the observation

Table IV. Two-Phase Anitron System Performance Results with a Mixture of Corn Processing Wastewater and Waste Activated Sludge

Parameter	Run No. 3	Run No. 4
System Volumetric Loading Rate, kg COD/m³-day	20.2	25.4
Feed Values, mg/L		
COD	10,193	8,430
BOD$_5$	6,211	5,306
Effluent Values, mg/L		
Soluble COD	1,077	1,163
Soluble BOD$_5$	673	903
Volatile Acids (as Acetic)	457	709
Removal, %		
COD	89.4	86.2
BOD$_5$	89.2	83.0

Note: All values stated are means during the experimental period.

Table V. Methane Yield, Biomass Yield and Effluent Volatile Acids During Steady-State Evaluations

Test Program	Methane Yield, m³ CH$_4$/kg COD Removed	Biomass Yield kg VSS/kg COD Removed	Effluent Volatile Acids, mg/L
Phase 1			
Run 1	0.29	0.07	358
Run 2	0.21	0.08	530
Phase 2			
Run 3	0.26	0.07	457
Run 4	0.31	0.09	709

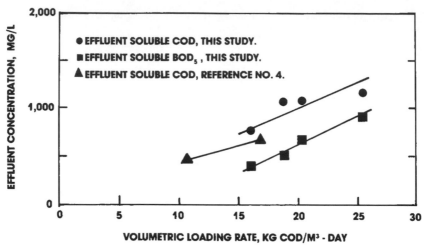

Figure 5.   Effect of reactor volumetric loading rate on two-phase Anitron system performance.

that as the volumetric loading rate was increased the measured methane yield also increased. At a loading of 25.4 kg COD/m³-day, the methane yield was 0.31 m³ CH₄/kg COD removed, which is close to the theoretical value. Biogas quality ranged from 63 to 73% methane. Net biomass yield during all runs ranged from 0.07 to 0.09 kg VSS/kg COD removed.

Increasing volumetric loading from 18.7 (Run 2) to 20.2 kg COD/m³-day (Run 3) as a result of WAS addition to the feed did not increase the effluent soluble COD concentration (1072 mg/L as compared to 1077 mg/L). The average volatile acid concentrations were also similar in both runs, indicating that the WAS addition had no effect on the process performance.

The performance of the two-phase Anitron pilot plant in this study is compared to a previous study [4] where cheese whey permeate represented the feed (Figure 5).

Phase 3—Shock Load Results

In order to determine the ability of the two-phase Anitron system to withstand potential organic shock loads resulting from short-term increases in wastewater concentration, pilot shock load tests were conducted. This condition was simulated by feeding the pilot plant a higher concentration feed prepared by using steep water, which is the major source of organics in corn processing wastewater. Two levels of shock loads were tested (Table II). Waste activated sludge addition to the feed was continued during this phase of the study.

Before and after each shock test, the pilot plant was operated at a volumetric loading rate of 24 kg COD/m³-day and a system HRT of nine hours. During the shock load, feed concentrations were increased for four hours and normal loading was resumed at the end of the shock loads. Effluent samples were taken frequently during and after the shock load.

As shown in Figure 6, during the lower shock load test, the primary reactor effluent COD peaked at 7,960 mg/L four hours after the start of the shock. The secondary reactor effluent COD peaked at 1,730 mg/L six hours after the start of the shock. After approximately 24 hours, the reactors had recovered substantially and no effects of the shock could be detected from the analysis of the pilot plant effluent after 48 hours.

A higher shock load applied to the system one week following the first shock load (Figure 7) produced a similar response. Effluent COD in the primary reactor peaked at 10,550 mg/L four hours after the start of the shock while the secondary reactor effluent COD peaked at 6,460 mg/L ten hours after the initiation of the shock. Both reactors recovered substantially 24 hours after the initiation of the shock load and recovered completely after 48 hours.

These shock load test results demonstrate the ability of the two-phase Anitron system to absorb sudden organic shock loads of up to eight times the operating load (24 kg COD/m³-day). The test results also correlate well with previous shock loading experience. A recovery period of 11 hours was reported [8] for a shock load of 150% influent COD increase for four hours duration in a single-phase

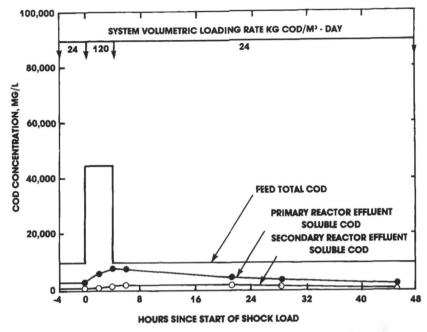

Figure 6.   Response of two-phase Anitron system to lower shock load.

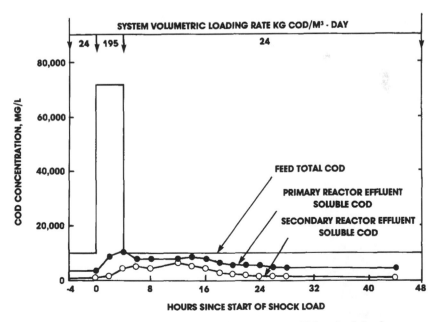

Figure 7.   Response of two-phase Anitron system to higher shock load.

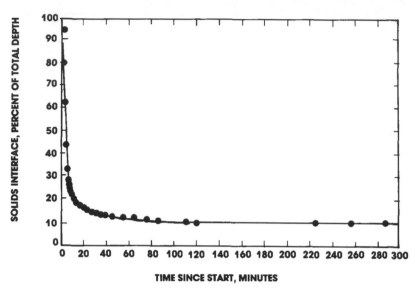

Figure 8.   A typical Anitron effluent solids settling curve.

anaerobic fluidized bed pilot plant operating at a volumetric loading rate of 5 kg COD/m³-day. The tolerance of the anaerobic fluidized bed system to short-term temperature reductions and hydraulic overloadings is also reported [8].

Phase 4 – Clarification Test Results

During the course of the pilot plant study with waste activated sludge addition in the feed, a number of clarification and thickening tests were performed on the final effluent. The purpose was to investigate the effectiveness of a conventional clarifier in removing suspended solids in the effluent, particularly the undigested WAS, and to obtain design information for sizing the clarifier. Clarifier underflow would provide a single wasting point for anaerobically digested sludge for a coupled anaerobic/ aerobic system by recycling the waste activated sludge back to the Anitron system for stabilization and volume reduction as discussed previously.

Standard Dorr-Oliver sedimentation test procedures were employed. Pilot plant effluent was collected, and suspended solids were concentrated to approximately three times the original concentration for testing the zone and compression settling. Samples were stirred to suspend all solids and poured into one-liter graduated cylinders. Settling rates were measured by frequent readings of solid liquid interface with time. After 30 minutes of settling, samples were withdrawn from each cylinder one inch below the liquid surface using a pipet so as not to disturb the interface. Figure 8 illustrates a typical settling curve. The overflow suspended solids concentration obtained was averaging 192 mg/L and underflow concentration ranged from 1.3% after a 30-minute settling time to 2% after a two-hour settling time. Sludge volume index (SVI) was calculated to be 70 mL/gram, which is comparable to those reported in the literature [9] at similar initial solids concentrations. An SVI less than 100 mL/ gram denotes good settling characteristics for activated sludge [10].

DESIGN IMPLICATIONS

Based on the pilot plant results discussed previously and Dorr-Oliver's experience with a full-scale Anitron system, a two-phase Anitron system was designed for the anaerobic pretreatment of American Maize's corn processing wastewater at their Hammond, Indiana plant. This Anitron system is designed to be followed by the existing aerobic system as the polishing step prior to discharge of the wastewater to Lake Michigan.

In addition, the Anitron system is designed to handle waste activated sludge from the aerobic plant for digestion and volume reduction. Clarification of Anitron effluent will provide a single wasting point for anaerobically digested sludge for the entire wastewater treatment system as illustrated previously in Figure 3.

SUMMARY AND CONCLUSIONS

The anaerobic biological fluidized bed system is applicable to the treatment of corn processing wastewaters. At a volumetric loading as high as 25.4 kg COD/m³-day, reductions of more than 86% COD and 83 % BOD₅ can be achieved.

The Anitron system can treat additional WAS at 8% of the total wastewater flow with no effect on process performance.

Methane yield varied from 0.21 to 0.31 m³/kg COD removed with methane content ranging from 63 to 73% in the biogas. Methane yield is considered low due to the inaccuracy of wet test meters at low flows and loss of pressure in the reactors during frequent samplings and maintenance work on the reactors. It has been demonstrated in this study that as the gas production rate increases, the measured methane yield also increases due to more accurate gas measurements.

Biomass net yield varied from 0.07 to 0.09 kg VSS/kg COD removed on the basis of effluent VSS.

Shock loads as high as eight times the operating load (up to 195 Kg COD/m³-day) were absorbed by the two-phase Anitron system without adversely affecting the system. The effluent quality returned to normal in approximately one day.

Suspended solids resulting from waste biomass in the effluent can be efficiently concentrated with the use of a conventional clarifier. The effluent suspended solids concentration following clarification was less than 200 mg/L.

ACKNOWLEDGMENTS

The authors gratefully acknowledge the contributions made to the pilot plant program by the Environmental Laboratory personnel of the American Maize-Products Company and by the Dorr-Oliver Development Center personnel.

REFERENCES

1. Nemerow, N. L., *Industrial Water Pollution*, Addison-Wesley Publishing Company (1978).
2. Sutton, P. M., Langley, D. F., Warner, K. E., and Park, K., "Oxitron System™ Fluidized Bed Wastewater Treatment Process: Application to High Strength Industrial Wastewater," *Proceedings, 34th Annual Purdue Industrial Waste Conference*, West Lafayette, Indiana (1979).
3. Carpenter, W. L., and Berger, H. F., "A Laboratory Investigation of the Applicability of Anaerobic Treatment to Selected Pulp Mill Effluents," *TAPPI Proceedings*, Environmental Conference, p. 173 (1984).
4. Sutton, P. M., and Li, A, "Single-Phase and Two-Phase Anaerobic Stabilization in Fluidized Bed Reactors," *Water Science and Technology* (G. B.), 15, 133 (1983).
5. Cohen, A., Breure, A. M., Van Andel, J. G., and Van Deusen, A., "Influence of Phase Separation on the Anaerobic Digestion of Glucose-I, Maximum COD Turnover Rate During Continuous Operation," *Water Research*, 14, 1439 (1980).
6. Cohen, A., Breure, A. M., Van Andel, J. G., and Van Deusen, A., "Influence of Phase Separation on the Anaerobic Digestion of Glucose-II, Stability and Kinetic Response to Shock Loadings," *Water Research*, 16, 449 (1982).
7. *Standard Methods for the Examination of Water and Wastewater*, 15th Edition, APHA-AWWA-WPCF, Washington, DC (1980).
8. Bull, M. A., Sterritt, R. M., and Lester, J. N., "Response of the Anaerobic Fluidized Bed Reactor to Transient Changes in Process Parameters," *Water Research*, 11, 1563 (1983).
9. Schroepfer, G. J., and Ziemke, N. R., "Development of the Anaerobic Contact Process-II, Ancillary Investigations and Special Experiments," *Sewage and Industrial Wastes*, 31, 697 (1959).
10. Campbell, H. W., Rush, R. J., and Tew, R., "Sludge Dewatering Manual," *Research Report No. 72*, Ministry of the Environment, Ontario, Canada (1978).

63 CHARACTERIZATION AND TREATMENT OF WASTEWATER GENERATED FROM SALINE AQUACULTURE OF CHANNEL CATFISH

Ricardo B. Jacquez, Associate Professor

Paul R. Turner, Assistant Professor

Hamdy El-Reyes, Graduate Research Assistant

Chih-Ming Lou, Graduate Research Assistant
Department of Civil Engineering
New Mexico State University
Las Cruces, New Mexico 88003

INTRODUCTION

Background

The Southwest desert of the United States is characterized by arid climate and limited availability of freshwater. Because of this freshwater limitation, aquaculture, which is very water intensive, has not developed into a prominent industry in New Mexico. However, production of aquatic animals could be accomplished economically using intensive recirculating systems and brackish groundwaters. Total New Mexico groundwater reserves (fresh and brackish) have been estimated at 2.5×10^{13} m^3 of which 75% is brackish (1000 mg/L total dissolved solids (TDS)) [1]. Brackish groundwaters in New Mexico contain up to 200,000 mg/L TDS but most of the supply is much less than 100,000 mg/L [2]. Such is the case in Roswell, New Mexico where the groundwater has a TDS concentration of 14,000 mg/L. The water is not usable for direct human consumption or conventional agriculture without extensive dilution or desalination but has the potential to be used for culturing salt tolerant fish species. Some species of fish grow well in brackish water while other species such as channel catfish could be cultured only in freshwater or in slightly brackish water (10,000 mg/L TDS) [3,4].

Research Objectives

The specific objectives outlined for the study were as follows:

1. Determine the pollution load generated by an intensive catfish culture system operating at a salinity range between 1000 to 9000 mg/L TDS.
2. Evaluate the effect of high salinity on biological nitrification in submerged filters.

Scope of Study

The first phase of the study was conducted at the Roswell Test Facility (RTF) located in Roswell, New Mexico. Fingerling channel catfish *(Ictalurus punctatus)* were grown in circular raceways containing brackish groundwater ranging in salinity between 1000 to 9000 mg/L TDS. The catfish growth studies were conducted over a six month period. Water samples to characterize the pollutional load generated by the catfish were collected and analyzed during the last three months of the study.

The second phase of the study was conducted in the laboratory of the Civil Engineering Department at New Mexico State University (NMSU). The effect of high salinity on biological nitrification was evaluated by operating five, laboratory scale, submerged, nitrification filters using brackish groundwater from the RTF. The five salinity levels which were evaluated included TDS concentrations ranging from 1000 to 9000 mg/L. In addition to the filter studies, respirometer studies were performed to determine the influence of high salinity on the exertion of nitrogenous biochemical oxygen demand.

Table I. Concentrations of Ions Contained in Roswell City Water and the Brackish Well at the RTF

Constituent	Concentration, mg/L	
	City Water	Brackish Water
Sodium, Na	67	4,449
Potassium, K	1	23
Calcium, Ca	184	525
Magnesium, Mg	51	156
Chloride, Cl	97	6,948
Sulfate, SO_4	451	1,488
Bicarbonate, HCO_3	238	190
Dissolved Solids	1,055	14,240
Total Hardness, $CaCO_3$	670	1,950
pH	7.65	7.63

MATERIALS AND METHODS

Wastewater Characterization

Fingerling channel catfish were cultured in circular fiberglass raceways 1.39 m in diameter and which held a working volume of 760 liters of water. Water flowed into each tank at a rate of 5.7 L/min resulting in an exchange rate of 2.3 hours. Salinity levels selected to conduct the catfish growth studies included TDS concentrations of 1000, 3000, 5000, 7000, and 9000 mg/L. Each salinity level was studied in triplicate yielding a total of 15 culture tanks. The salinity levels were created by blending Roswell drinking water with brackish water at the RTF. The characteristics of the two separate water sources are shown in Table I. The water fed into each tank was heated to a temperature of 26–28 C and was sprayed at an angle to create a clockwise circular flow pattern and surface aeration. Supplemental aeration to each raceway was supplied by three, 12 cm airstones connected to an air compressor. The initial fish stocking level was 100 catfish having a mean length of 157 mm and mean weight of 35.6 gm.

Water samples were collected as 24 hour composites made by randomly sampling each tank from the five salinity groups. Since each salinity group consisted of triplicate tanks, an individual tank was sampled at a frequency of every three hours or eight times in a 24 hour period. During collection, all samples were preserved by refrigeration at 4 C. Water quality was characterized by monitoring COD, PO_4-P, Organic-P,SS, VSS, BOD_5, NH_3-N, Organic-N, and NO_3-N. All water quality analyses were performed in accordance with the procedures outlined in *Standard Methods* [5] with the exception of COD. COD was measured by a modification of the Hach mini vial procedure. Modification of the Hach procedure was necessary because of the high chloride content of the saline water. Under the final experimental conditions, the highest chloride content expected was 4600 mg/L. Chloride interference was removed by adding mercuric sulfate ($HgSO_4$) to precipitate mercuric chloride ($HgCl_2$). Since mercuric sulfate and mercuric chloride are essentially insoluble in the COD reagent-water mixture, the digested samples required setting overnight to allow the precipitate to completely settle. All COD analyses were made by adding approximately 92 mg of $HgSO_4$ per two mL sample.

Nitrification Study

The nitrification study was conducted using five laboratory scale submerged upflow filters. Each filter was constructed of 4.4 cm diameter (inside) plexiglass, 17.1 cm in length, and packed with 6 mm diameter ceramic beads to a depth of 15.9 cm. The clean filter void volume was measured as 130 mL. The filters and the synthetic wastewater were maintained at a temperature of 30 C to coincide with that used to culture the catfish at the RTF. The wastewater used throughout the experimental program consisted of a synthetic feed made by combining saline water with tap water (400 mg/L TDS) in varying ratios to obtain TDS concentrations of 1000, 3000, 5000, 7000, and 9000 mg/L. Reagent grade $(NH_4)_2SO_4$ and KH_2PO_4 were added to yield NH_3-N and PO_4-P concentrations of 15 and 3.75 mg/L, respectively. Based on toxicity of NH_3 to catfish [6], all nitrification filter experiment were conducted using an NH_3-N concentration of 15 mg/L. Each filter was seeded with a nitrifying culture developed for municipal wastewater secondary effluent and soil. Effective culture development and seeding required approximately 14 weeks. Oxygen was provided by aerating the feed solution with

Table II. Summary of Conditions Created by Varying the Hydraulic Loading Rates During Experiment No. 2

Filter Hydraulic Loading Rate, m/day	Initial Nitrogen Loading Rate kg NH$_3$-N/m^3/day	Void Volume Detention Time min	Hydraulic Passes Per Day
4.7	0.42	26	55
9.5	0.83	13	111
18.9	1.66	6.5	221
28.4	2.49	4.3	333
37.9	3.32	3.3	443

compressed air. The pH of all filters was monitored frequently and was periodically adjusted to 7.0 to 7.2 using 1 N Na$_2$CO$_3$. NO$_3$-N concentrations were measured by the Hach Chemical Company cadmium reduction method. Total and volatile solids were measured using the procedures outlined in *Standard Methods* [5].

Experiment No. 1. The objective of this experiment was to determine the influence of salinity on the nitrification process under a fixed loading condition. During this experiment the initial synthetic feed was continuously recycled for a period of eight days at a hydraulic loading rate of 4.7 m/day. Samples were collected once per day and were analyzed for their NO$_3$-N concentration. Distilled water was added every day to make up losses due to evaporation.

Experiment No 2. The objective of this experiment was to determine the combined effect of salinity and hydraulic loading rate on the nitrification process. During this experiment the initial synthetic wastewater was continuously recycled through the filters for a 24 hour period. At the end of 24 hours, samples were collected and analyzed for their NO$_3$-N concentrations. A fresh wastewater was made and was then recycled for an additional 24 hours. Five hydraulic loading rates were evaluated. A summary of the loading conditions is presented in Table II.

Experiment No. 3. The objective of this experiment was to evaluate the influence of salinity on the nitrification process under a fixed hydraulic loading condition. The synthetic wastewater was continuously recycled for a period of 48 hours at a hydraulic loading rate of 18.9 m/day. Salinity levels of 3000, 5000, and 7000 mg/L were tested. Samples were collected every six hours and were analyzed for their NO$_3$-N concentrations. Distilled water was added periodically to make up for losses due to evaporation.

Experiment No. 4. The objective of this experiment was to simulate a continuously operating nitrification system operating in a flow through/recycle mode. Only two salinity levels were evaluated, 3000 and 5000 mg/L TDS. Both filters were operated at a hydraulic loading rate of 18.9 m/day. Recycle ratios (Q$_r$/Q) that were evaluated included 0.33, 1.0, and 3.0. Samples were collected from the final effluent line once per day and were analyzed for their NO$_3$-N concentration. At the end of this experiment, the packing material from all five filters was removed to determine the total and volatile solids concentrations of the filter contents.

Respirometer Study

A respirometer study was conducted using a Pneumatic Computerized BOD (PC/BOD) Respirometer developed by Drohbyczer [7]. Continuous exertion of nitrogenous BOD was measured over a period of five days. Salinities of 1000, 3000, 5000, 7000, and 9000 mg/L TDS were tested. The synthetic wastewater contained 15 mg/L NH$_3$-N and 3.75 mg/L PO$_4$-P. Seed for each sample was provided by adding 50 mL of the feed waters being recycled through the nitrification filters.

RESULTS AND DISCUSSION

Wastewater Characterization Study

On three different dates 24-hour composite samples were collected from the fish culture tanks at the RTF. The samples were analyzed to determine nutrient production (excreted metabolites) by the catfish. Combining the results of water quality characterization with fish weights at the times of sampling generates unit nutrient production data. As shown in Figures 1 and 2, nutrient production

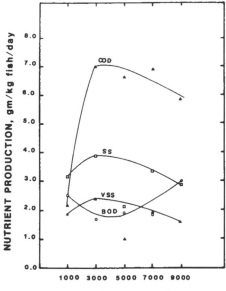

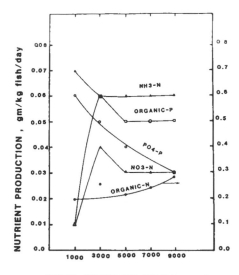

Figure 1. Nutrient production by catfish cultured under a salinity range of 1,000-9,000 mg/l TDS.

Figure 2. Nutrient production by catfish cultured under a salinity range of 1,000-9,000 mg/l TDS.

was variable but was influenced by salinity. In general, between salinities 1000–3000 mg/L TDS nutrient production increased to a maximum and thereafter decreased or remained constant. The increase in nutrient production between the salinity range 1000–3000 mg/L TDS indicated that catfish growth was normal or slightly stimulated while the decrease in nutrient production indicated that the growth of the catfish in the salinity range above 3000 mg/L TDS was adversely effected. The results of the catfish growth study confirmed this finding. At salinity levels of 7000 mg/L TDS and higher, the growth of the catfish was found to be significantly less than at salinities ranging between 1000–5000 mg/L TDS. As the growth rate decreased, the fish generated less metabolic wastes resulting in a decrease in nutrient production.

As seen from the data in Table III, the average concentration of each water quality parameter was consistently low due to the high dilution rate. With respect to ammonia levels, it is clear that the goal of minimizing toxicity was accomplished. Under the experimental conditions the average unionized

Table III. Summary of Water Quality Characteristics and Nutrient Production for Saline Water Catfish Culture

Parameter	Average Concentration mg/L	Nutrient Production, gm/kg fish/day	
		Average for Present Study	Literature Based
COD	38	5.73	1.94
BOD_5	14	2.21	2.84
PO_4-P	0.4	0.05	0.04
Organic-P	0.3	0.04	0.16
NH_3-N	0.4	0.06	0.36
NO_3-N	0.2	0.03	0.26
Organic-N	1.4	0.20	0.05
SS	21	3.08	3.41
VSS	12	1.76	5.23

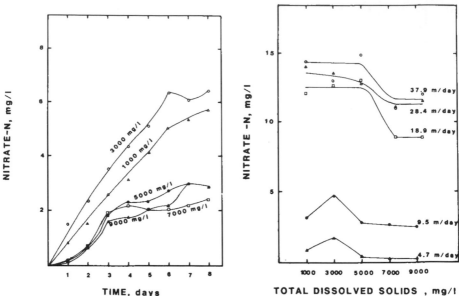

Figure 3. Influence of salinity on nitrate production in the submerged filters at a hydraulic loading rate of 4.7 m/day.

Figure 4. Influence of hydraulic loading rate and salinity on nitrate production in the submerged filters.

ammonia concentration was less than 0.004 mg/L and never exceeded 0.01 mg/L. These concentrations are far below the levels reported to significantly decrease the growth rate of catfish [6]. It was also observed that the majority of the average nutrient production rates obtained during this study were considerably lower than rates reported in the literature for fresh water systems where the TDS concentrations were consistently less than 500 mg/L. Of particular interest is the result that significantly less NH_3-N was produced under the saline conditions. The difference between the fresh and saline water rates was not uniformly consistent, but the findings of the current study averaged 57% less than those obtained from the literature. This finding is significant because with lower nutrient production (specifically ammonia) a reduced water rate could be used to culture the catfish. In a state like New Mexico where saline water is three times more abundant than fresh water, the culture of an aquatic species like catfish could be beneficial from the standpoints of effective water use as well as water conservation. Culturing catfish in a 3000–5000 mg/L TDS saline water would produce a marketable food product from an otherwise unusable resource while at the same time conserving what is probably the state's most precious resource, water.

Nitrification Study

Seeding of the filters was found to be a slow process. In total, 14 weeks were required to effectively seed the five filters. Initially it appeared that seeding was occurring rapidly because a white material was accumulating quickly in the filters yet the rate of nitrification was very slow. Nitrate production was steady but slow indicating limited biological activity. On several occasions nitrite-nitrogen (NO_2-N) was qualitatively checked to determine if the culture may have been deficient in nitrite oxidizing bacteria, *Nitrobacter*. NO_2-N was not detectable. After noticing that a similar white material was accumulating in the feed containers used to recycle the synthetic waste, it was determined that the white material was soluble in dilute hydrochloric acid. The solubility in the acid suggested that this material was an inorganic chemical precipitate rather than biological growth. This suggestion was to be confirmed later in the study when the contents of the filter were analyzed for their volatile solids content. It was also observed that the amount of the precipitate accumulating in the filters increased proportionately to the salinity of the water.

Experiment No. 1. The results of this experiment are presented in Figure 3. The filters operating at salinity levels 1000 and 3000 mg/L TDS consistently produced twice as much NO_3-N as the filters operating at salinity levels of 5000–9000 mg/L TDS. The data suggests three possibilities: 1) the three filters operating under the higher salinities were accumulating a chemical precipitate and were therefore limited in their biological activity as compared to the two lower salinity levels; 2) an increase in salinity from 1000 to 3000 mg/L TDS stimulated nitrification; and 3) the salinity levels 5000–9000 mg/L were inhibiting nitrification.

Experiment No. 2. The results of this experiment are shown in Figure 4. Two observations are made from the data: 1) under all hydraulic loading conditions the filters operating at salinity levels of 1000 and 3000 mg/L TDS produced an average of 80% more NO_3-N than the filters operating at salinity levels 5000–9000 mg/L TDS; and 2) the filters operating at hydraulic loading rates of 18.9 to 37.9 m/day produced an average of six times more NO_3-N than the filters operating at hydraulic loading rates of 4.7 and 9.5 m/day. The results of this experiment tend to confirm the suggestions from Experiment No. 1 that salinity levels 7000–9000 mg/L TDS were inhibitory to nitrification and that a salinity near 3000 mg/L TDS was optimum for nitrification. The results also begin to suggest that biological activity at the higher salinities was reduced due to accumulation of chemical precipitate in the filters. As the amount of precipitate increased within the filter, void space available for bacterial growth was decreased and the growth of the nitrifiers was directly inhibited by the minerals contained in the precipitate.

As seen in Figure 4, progressively increasing the hydraulic loading rate increased the degree to which nitrification was achieved. The maximum achievable NO_3-N concentration was 15 mg/L. At hydraulic loading rates of 37.9, 28.4, and 18.9 m/day the maximum NO_3-N concentrations at 1000–5000 mg/L TDS averaged around 14.2, 13.3, and 12.9 mg/L, respectively. These concentrations represent 86–95% completion of nitrification. At hydraulic loading rates of 4.7 and 9.5 m/day the maximum average NO_3-N concentrations were 1.7 to 4.3 mg/L, respectively. These concentrations represent 11–29% completion of nitrification. The filters operating at the higher loading rates were more effective in completing nitrification because the mass transfer of substrate (NH_3-N) and dissolved oxygen was higher. Under the higher loading conditions the nitrifying bacteria could oxidize more NH_3-N because more oxygen was available. Under the lower loading conditions oxygen became the limiting factor.

Experiment No. 3. After evaluating the results of Experiment No. 2, filters receiving TDS concentrations of 1000 and 9000 mg/L were eliminated from the testing program. A loading rate of 18.9 m/day was selected based on the results obtained during Experiment No. 2 and was considered to be capable of achieving a reasonable level of efficiency in an actual nitrification system. The results from Experiment No. 3 are presented in Figure 5. A comparison of the three nitrate production curves shows that the filter receiving synthetic wastewater containing 5000 mg/L TDS produced nitrate at a much faster rate than the filters receiving 3000 and 7000 mg/L. Nitrification rates (day^{-1}, base 10) for each filter system were 0.54, 1.76, and 0.42 for TDS levels 3000, 5000, and 7000 mg/L, respectively. The filter receiving 5000 mg/L TDS reached its maximum nitrate concentration 12 hours before the filter receiving 3000 mg/L and approximately 24 hours before the filter receiving 7000 mg/L. Based on the results of this and the previous two experiments, the decision was made to eliminate the filter treating a salinity of 7000 mg/L TDS in the remaining experiment.

Experiment No. 4. The results of this experiment are shown in Figure 6. As seen from the data, at a hydraulic loading rate of 18.9 m/day a recycle ratio of 1.0 was found to optimize nitrification in the filters. Increasing the recycle ratio to 3.0 did not improve the performance of the filters. Although the filter receiving a TDS concentration of 3000 mg/L consistently produced an average of 15% more NO_3-N than the filter receiving 5000 mg/L, the difference was not considered to be significant.

A review of all previous filter data indicates that the difference in the degree of nitrification between salinity levels 1000, 3000, and 5000 mg/L TDS was not significant. In general, the results indicate that under the salinity range 1000–5000 mg/L TDS, achievement of 75–100% of maximum nitrification was possible. This finding is compatible with the results obtained by Jones and Hood [8]. In their study it was determined that for a freshwater isolate of a *Nitrosomonas* species, optimum nitrification occurred at a salinity range of 3000–5000 mg/L TDS. At a salinity range of 10,000–30,000 mg/L TDS nitrification progressively decreased from 55–5% of maximum level. In the current filtration study the original inoculum was enriched from secondary sewage effluent and soil and, therefore, was considered to be a mixture of freshwater species.

Speculation as to which specific ions inhibited nitrification at a TDS level above 5000 mg/L was difficult because of the complex makeup of the saline water (see Table I). A summary of the concentrations of the major ions contained in the synthetic wastewaters used in this study is presented in

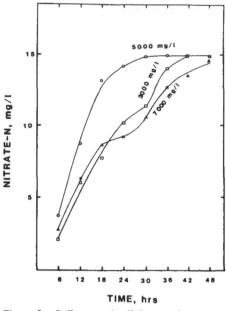

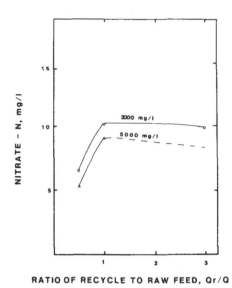

Figure 5. Influence of salinity on nitrate production in the submerged filters at a hydraulic loading rate of 18.9 m/day.

Figure 6. Influence of recycle ratio and salinity
on nitrate production in the submerged filter
at a hydraulic loading rate of 18.9 m/day.

Table IV. Based on information available in the literature [9], it could not be concluded that specific cations shown in Table IV contributed to the inhibition of nitrification. Two elements which showed potential to act in a toxic manner were the heavy metals cadmium (Cd) and selenium (Se), but information specific to their toxicity to nitrifier species was not available in the literature. It is very likely that the inhibition resulted from increased osmotic pressure. Since the seed culture was obtained from a freshwater source, the nitrifiers were not able to function properly at the higher osmotic pressures created by the higher salinity levels.

At the end of the filtration study, the contents from each filter was removed and analyzed for percent volatile solids. Volatile solids were measured as an indication of biological mass. The percent volatile solids of the filter contents was extremely low ranging between 6–9%. Volatile solids in most biological filters range between 75–85%. The low percentages further substantiate the previous hypothesis that accumulation of inorganic solids reduced the level of biological activity within the filters and was a contributing factor to the inhibition of nitrification at the higher salinity levels. Since

Table IV. Chemical Composition of Synthetic Wastewater Used During the Submerged Filter Nitrification Study

	Total Dissolved Solids, mg/L				
Constituent	1,000	3,000	5,000	7,000	9,000
Sodium, Na	312	937	1,562	2,187	2,812
Potassium, K	2	5	8	11	15
Calcium, Ca	37	111	184	258	332
Magnesium, Mg	11	33	55	77	99
Chloride, Cl	488	1,464	2,440	3,415	4,391
Sulfate, SO_4	102	305	508	712	915
Zinc, Zn	0.03	0.10	0.17	0.24	0.30
Silver, Ag	0.001	0.003	0.01	0.01	0.01
Cadmium, Cd	0.002	0.01	0.01	0.01	0.02
Selenium, Se	0.18	0.53	0.88	1.23	1.58

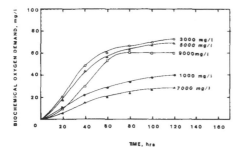

Figure 7.   Exertion of nitrogenous BOD at a salinity range of 1,000-9,000 mg/L TDS.

the overall difference in percent volatile solids was not significant, it is difficult to reach a definite conclusion from this data regarding salinity and its influence on the growth of nitrifying bacteria.

Respirometer Study

The results of the respirometer study are presented in Figure 7. The results of the BOD test were basically similar to the results of the filtration experiments. The salinity level 3000 to 5000 mg/L TDS exerted the maximum BOD while a salinity of 7000 mg/L exerted the minimum BOD. On a stoichiometric basis the biochemical oxygen requirement for complete nitrification is 4.6 mg O_2 per mg NH_3-N [10]. Since each sample contained 15 mg/L NH_3-N, the maximum theoretical BOD which could be exerted was 69 mg/L. A summary of the final BOD_5 and the corresponding NO_3-N concentrations are shown in Table V. This analysis shows that only the salinity level of 7000 mg/L TDS deviated significantly from stoichiometric requirements. Excluding a salinity level of 7000 mg/L TDS, the average oxygen demand was 4.7 mg O_2 per mg NH_3-N. Overall the results of the respirometer study tend to verify the results of the filtration experiments. Salinity levels of 3000 and 5000 mg/L TDS optimized nitrification while salinity levels greater than 5000 mg/L tended to inhibit nitrification.

Final Summary

The results of this study have demonstrated that a salinity range of 1000 to 5000 mg/L TDS was not inhibitory and may have even improved nitrification while a salinity level greater than 5000 mg/L was clearly inhibitory to the nitrification process. This finding was compatible with the results of the catfish growth study which showed that catfish could not be feasibly cultured at a salinity level greater than 5000 mg/L TDS. Therefore, the upper limit on salinity for overall success of culturing channel catfish (from a standpoint of culture efficiency as well as wastewater treatment efficiency) using saline groundwater from Roswell, NM is 5000 mg/L TDS. In future applications to treatment of wastes generated by saline water catfish culture, the submerged filter has good potential for being used to optimize water usage and minimize ammonia toxicity to the fish. A potential treatment system might include sedimentation to remove suspended solids and a submerged filter to achieve nitrification. Suspended solids removal would be essential as a pretreatment step to prevent filter clogging. Filter clogging is still a potential problem because the saline water used in this study produced a chemical precipitate which could eventually fill the filter void space and cause flow obstruction. Past experi-

Table V. Analysis of Data from the Respirometer Study Conducted at a Salinity Range 1,000-9,000 mg/L TDS

Total Dissolved Solids, mg/L	BOD_5 mg/L	Final NO_3-N Concentration, mg/L	Theoretical BOD_t, mg/L[a]	Final Requirements mg O_2/mg NH_3-N[b]
1,000	42	9.2	42	4.6
3,000	74	14.7	68	5.0
5,000	69	14.8	68	5.0
7,000	29	13.7	63	2.1
9,000	60	13.4	62	4.5

[a]Determined by multiplying final NO_3-N by 4.6.
[b]Determined by dividing BOD_5 by final NO_3-N.

ences with submerged upflow nitrification filters indicate that clogging by bacterial growth is also a potential problem [11]. Once in operation, the filters could be hydraulically loaded at a rate of 18.9 m/day and a recycle ratio of 1.0.

Under this loading condition a filter could produce a treatment efficiency of 65–75% oxidation of ammonia. This treatment efficiency should consistently maintain the unionized ammonia (NH_3) concentration below the inhibitory level of 0.15 mg/L [6] and allow the water to be recycled through the catfish culture system at a fairly extensive rate.

CONCLUSIONS

Nutrient production rates measured for channel catfish cultured in saline water were approximately 57% less than nutrient production rates reported for freshwater systems.

A salinity level of 3000–5000 mg/L TDS was found to be the optimum salinity range for nitrification.

Salinity levels greater than 5000 mg/L TDS were found to be inhibitory to the nitrification.

A specific ion(s) in the saline water from the RTF was not identified as the source(s) of inhibition to nitrification. Inhibition exerted by the saline water was caused by high osmotic pressure.

A hydraulic loading rate of 18.9 m/day was found to be acceptable for a submerged upflow filter nitrifying a high salinity wastewater.

At a hydraulic loading rate of 18.9 m/day, a recycle ratio of 1.0 was found to optimize nitrification in the submerged upflow filter.

ACKNOWLEDGMENT

Support for the research was provided by the departments of Civil Engineering and Fishery and Wildlife Sciences at New Mexico State University, the Roswell Test Facility and the New Mexico Water Resources Research Institute.

REFERENCES

1. Bahr, T. G., and Herman, R. P., "Water Problems and Research Needs for New Mexico," New Mexico Water Resources Research Institute, *Report No. 135*, Las Cruces, New Mexico (1981).
2. Hood, J. W., and Kister, L. R., "Saline Water Resources of New Mexico," *USGS Water Supply Paper 1601*, Santa Fe, New Mexico (1962).
3. Allen, K. O., and Avault, J. W., Jr., "Effects of Salinity on Growth and Survival of Channel Catfish *Ictalurns Punctatus*," *Proc. Ann, Conf. Southeast Assoc. Game and Fish Commission*, 23: 319–323 (1969).
4. Lewis, S. D., "Effect of Selected Concentrations of Sodium Chloride on the Growth of Channel Catfish," *Proc. Ann. Conf. Southeast Association Game and Fish Commission*, 25: 459–466 (1971).
5. *Standard Methods for the Examination of Water and Wastewater*, 14th Ed., Amer. Pub. Health Assn., Washington, D.C. (1976).
6. Piper, R. G., *Fish Hatchery Management*, U.S. Dept. of Interior, Washington, D.C. (1982).
7. Drohbyczer, Aleksander, "Use of Oxygen Uptake Data for Parameter Estimation and Control of Bio-reactors," *Ph.D. Dissertation*, New Mexico State University (1985).
8. Jones, R. D., and Hood, M. A., "Effects of Temperature, pH, Salinity, and Inorganic Nitrogen on the Rate of Ammonium Oxidation by Nitrifiers Isolated from Wetland Environment," *Microbial Ecology*, 6 (4): 339–347 (1980).
9. Meiklejohn, J., "Some Aspects of the Physiology of the Nitrifying Bacteria," *Symposium on autotrophic microorganisms*, pp. 68–83, Cambridge Univ. Press, London (1954).
10. Gaudy, A. F., and Gaudy, E. T., *Microbiology for Environmental Scientists and Engineers*, McGraw-Hill Book Co., New York, N.Y. (1980).
11. Haug, R., and McCarty, P., "Nitrification in Submerged Filters," *Journal of Water Pollution Control Federation*, 44, 11 (1972).

Section Nine
COAL, COKE, AND POWER PLANT WASTES

64 DEVELOPMENT OF A MIXED AEROBIC CULTURE FOR DEGRADING COAL-TAR

Jeffrey D. Enzminger, Graduate Student

John V. Lepore, Graduate Student

Carolyn Gleason, Graduate Student

Catherine Dreyer, Graduate Student

Robert C. Ahlert, Professor
Department of Chemical and Biochemical Engineering
Rutgers, The State University of New Jersey
Piscataway, New Jersey 08854

INTRODUCTION

Coal-tar has been generated as a byproduct of coking and coal-gas works since the beginning of the 19th century. Where it is improperly landfilled, coal-tar may pose a threat to groundwater resources. At St. Louis Park, Minnesota, a suburb of Minneapolis, residues from a tar distillation and wood preserving plant, operated between 1918 and 1972, contaminated the principal local aquifer, forcing the closure of a number of wells up to two miles from the source. A hydrocarbon phase with an associated total organic carbon content of 6000 ppm was discovered moving downwards through the aquifer [1].

Coal tar is a heterogeneous substance containing up to 10,000 species of which 462 have been characterized [2]. Selected coal tar constituents are shown in Table I. The largest distillate fraction is the pitch, which is the residue after distillation up to 647 K. Tar pitch contains polynuclear aromatic hydrocarbons (PAHs) of from four to 30 aromatic rings which may be substituted and homo- or heterocyclic. Higher molecular weight compounds may not be fully condensed. Table II lists some of the four to six ring compounds contained in the pitch fraction, several of which are carcinogenic.

The toxicity of coal-tar has been well documented with respect to skin contact and fume inhalation [3]. There is sufficient evidence that topical application causes skin cancer and that inhalation of tar fumes causes lung cancer. Coal tar exposure has also been connected to bladder cancer and several PAHs found in coal tar are known carcinogens. Little is known concerning the health effects of chronic exposure to dilute aqueous solutions of tar constituents.

Of the remediation options at coal-tar-contaminated sites, on-site treatment is the most desirable since the expense and hazards of transportation off-site are avoided. One on-site treatment option under consideration is biological decomposition. However, the low aqueous solubility and high average molecular weight of the tar compounds present a major obstacle to biodegradation. In this paper, preliminary results on the biological degradation of a gasification-tar are presented, focusing on the ability of a surfactant to improve the biological removal of PAHs.

MATERIALS AND METHODS

Maintenance Culture Development

The mixed culture used in this work was derived from sewage sludge and was developed in two 1-liter fill-and-draw reactors with 750 mL working volumes. Half of the reactor volume was replaced with fresh medium daily; at the same time 1 g of gas-tar was added to the reactors. Reactors were stirred with magnetic stirring bars and aerated using coarse fritted gas dispersion tubes.

Table I. Coal-Tar Constituents [2]

	Coke Oven (%)	Lurgi Gasifier (%)
Acenaphthene	1.05	0.57
Anthracene	0.75	0.32
Benzene	0.12	0.02
Carbazole	0.60	0.22
o-Cresol	0.25	1.14
m-Cresol	0.45	1.83
p-Cresol	0.27	1.51
Diphenylene Oxide	—	0.57
Ethylbenzene	0.02	0.04
Fluorene	0.64	0.62
High-Boiling Tar Acids	0.83	11.95
Medium-Soft Pitch	63.5	33.1
a-Methylnaphthalene	0.65	0.63
b-Methylnaphthalene	1.23	1.05
Naphtha	0.97	3.02
Naphthalene	8.80	2.01
Phenanthrene	2.66	0.28
Phenol	0.61	0.97
Styrene	0.02	0.01
Tar Bases	2.08	2.50
Toluene	0.25	0.05
o-Xylene	0.04	0.05
m-Xylene	0.07	0.07
p-Xylene	0.03	0.03
Xylenols	0.36	5.55

The maintenance medium contained (per liter of H_2O): 2 g $(NH_4)_2SO_4$, 0.3 g $MgSO_4$, 5 g KH_2PO_4, 1.5 mg $FeCl_3.6H_2O$, 30 mg $MnSO_4.H_2O$, 25 mg $CaCl_2$, 30 mg yeast extract (Difco), 0.5 g glucose, 1.5 g sodium succinate, 1 g salicylic acid. The pH was adjusted to 7 with 1 N NaOH.

Measurement of Carbon Dioxide Evolution

Carbon dioxide generation experiments were carried out in 1 liter stirred reactors as described above. The mineral medium contained (per liter H_2O): 0.5 g $(NH_4)_2SO_4$, 0.1 g $MgSO_4$, 0.5 mg $FeCl_3.6H_2O$, 10 mg $MnSO_4.H_2O$, 50 mM phosphate buffer adjusted to pH 7. The carbon source was added at 0.1%. The innoculum was 3% of the working volume. Effluent gas was run through a Horiba PIR 2000 Infrared CO_2 analyzer connected to a strip chart recorder.

Batch carbon dioxide generation experiments were also carried out on evaporated coal tar. Evaporated tar was prepared by allowing a portion of tar to volatilize to a constant weight over three days during which time the tar mass was reduced by 15%. The tar residue was dissolved in toluene and coated on the bottom and sides of a reaction vessel, and the toluene was allowed to evaporate.

Table II. Pitch Components [2]

4 Ring Compounds	5 Ring Compounds	6 Ring Compounds
Chrysene	Picene	Dibenzpyrenes
Fluoranthene	Benzopyrene	Dibenzfluoranthenes
Pyrene	Benzofluoranthenes	Benzoperylenes
Triphenylene	Benzoperylenes	
Naphthacene		
Benzanthracene		

Compounds with as many as 20–30 condensed rings are postulated.

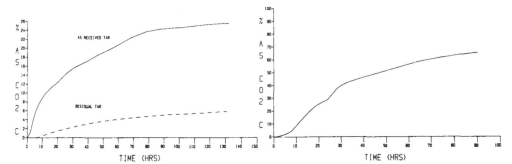

Figure 1.   Mineralization of tar to CO₂.   **Figure 2.   Naphthalene mineralization to CO₂.**

Extraction Experiments

Solutions of 0.1% gas-tar in water or in 0.1% Triton X-100 were sonicated for ten minutes in 125 mL Erlenmeyer flasks. The aqueous and tar phases were separated by centrifugation. The aqueous phase was extracted 5 times and the tar phase twice with 30 mL methylene chloride. Extracts were pooled, filtered through a 0.5 micron filter and concentrated to 5 mL on a rotary vacuum evaporator. The concentrated extracts were separated by reverse-phase liquid chromatography and detected by UV absorbance (Series 4 liquid chromatograph and LC-95 uv detector, Perkin-Elmer, Norwalk, CT). An LC-PAH column 15 cm x 4.6 cm with 5 μm packing was used (Supelco, Bellefonte, PA). The mobile phase was acetonitrile/water: 55/45 for 1 minute followed by a 7 minute linear gradient to 100% acetonitrile held for 7 minutes. Flow rate was 2 mL/min. with injection size 5 μL.

Gas-tar PAH Biodegradation in Shake-flasks

Shake flask studies were run in 250 mL flasks with 50 mL liquid volume. The medium used was the same as for the CO_2 generation experiments. Coal tar was added at 0.1%. Sixteen flasks were prepared: eight received 0.1% glucose and eight received 0.1% Triton X-100. All flasks were sonicated for 10 minutes. Replicate flasks from each group were immediately acidified to pH 2 and stored at 5 C. The remaining flasks were shaken at 200 rpm on a G33 rotary shaker (New Brunswick Scientific, Edison, NJ) at ambient temperature. Two flasks each were acidified at 0, 2, and 5 days. After five days, all flasks were neutralized and extracted five times with 30 mL of methylene chloride. The extracts were processed as described above.

RESULTS AND DISCUSSION

Results of carbon dioxide evolution experiments are shown in Figure 1. Twenty six percent of the as-received tar was converted to CO_2 carbon over 130 hours. A black residue coating the teflon stirring bar and the circumference of the reaction vessel at the liquid surface indicated that the tar was not completely biodegraded. The carbon dioxide evolution from the residual gas-tar following volatilization was considerably less than from the as-received tar. This suggests that the most easily degradable substrates in the tar are the lower molecular weight compounds such as benzene, toluene, xylene, and naphthalene. These compounds accounted for approximately 14% of the as-received tar.

The ability of the mixed culture to grow on naphthalene is shown in Figure 2. Naphthalene mineralization was much more extensive than that of coal tar. This reinforces the hypothesis that the lower molecular weight hydrocarbons are preferentially degraded.

The main limitations to biodegradability of the gas-tar may be the high molecular weight or low aqueous solubility of most of its components (Table II). Several studies have been conducted on the biodegradability of various PAHs [5]. While PAHs of 2–3 rings can serve as sole sources of carbon for biodegradative microorganisms, larger PAHs appear to be biodegraded only through co-metabolic action (Table III). Work by Wodzinski [6,7,8] suggests that PAHs are not degraded from the solid phase. Thus, lack of degradation of the higher molecular weight PAHs may be due to their low solubilities (Table IV).

The high molecular weight tar components found in the pitch fraction are likely to be highly resistant to aerobic biodegradation. While work on degradation of high molecular weight hydrocar-

Table III. Aqueous Solubility of PAHs [4]

Compound	MW	Solubility (ppb)	# of Rings
Naphthalene	128	31,700	2
Acenaphthene	154	3,190	3
Fluorene	166	1,980	3
Phenanthrene	178	1,290	3
Anthrecene	178	73	3
Fluornathene	202	260	4
Pyrene	202	135	4
Benz(a)anthracene	228	45	4
Chrysene	228	2	4
Benzo(a)pyrene	252	3.8	5

Table IV. Biodegradation of PAHs by Pure Cultures [5]

As Sole Carbon source	Through Cometabolism
Benzene	Benzo(a)anthracene
Toluene	Benzo(a)pyrene
Xylenes	
Naphthalene	(Cometabolic Substrates were Biphenyl and Succinate).
Phenanthrene	
Anthracene	

bons is scant, results from landfarming studies of petroleum sludges suggests that materials such as asphaltics accumulate in the soil. Such materials may persist in landfarms for years [9].

Biodegradation of coal tar may only be applicable to the readily leachable components. Thus, it was of interest to examine the extent of degradation of various gas-tar PAHs. Shake-flask gas-tar degradation was performed with glucose as a supplemental carbon source. Figure 3 shows results of liquid chromatography of flask extracts after 0, 2, and 5 days. Peaks are labeled based on retention times determined from standard PAH solutions. The reduction in naphthalene peak height in the control flask is probably due to volatilization (compare with Figures 5 and 6). There was a significant reduction in peak height for the lower weight PAHs suggesting extensive biodegradation of those PAHs. While degradation of PAHs eluting later than benz(a)anthracene appeared to occur, proof must await better resolution of the these compounds. These results suggest that it may be possible to stabilize the gas-tar by degrading the lower molecular weight and, therefore, more mobile components. However, the stability and hazards of the residue must be analyzed to determine if partial biodegradation will be an acceptable remediation alternative for coal-tar wastes.

Chemical dispersants have been tested as a means of increasing the rate of degradation of oil spills by increasing the surface area available for microbial attack. Dispersants have been shown to increase

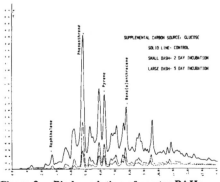

Figure 3.   Biodegradation of gas-tar PAHs.

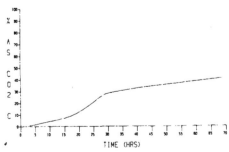

Figure 4.   Mineralization of Triton X-100 to CO$_2$.

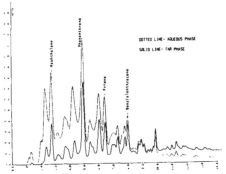

Figure 5.   Triton X-100 extraction of gas-tar.   **Figure 6.   Aqueous extraction of gas-tar.**

the microbial transformation of PCBs [9] and ellipticines [10]. Klevens [11] showed that aqueous solubilities of PAHs could be increased substantially by surfactant addition. In this work, the effect of a surfactant, Triton X-100, on the extent of gas-tar PAH degradation was tested.

First, the biodegradability of the surfactant was determined. Carbon dioxide evolution from Triton X-100 as sole carbon source is shown in Figure 4. Based on dry cell mass and CO_2 evolution, approximately 60% of the Triton X-100 added was converted to either CO_2 or cellular carbon. Lack of complete conversion may be due either to complexation of the surfactant with silicon anti-foam (Antifoam A, Sigma Chemical, St. Louis, MO) or to the recalcitrance of the branched aliphatic moiety in the surfactant molecule.

Two experiments were performed with Triton X-100. First, the ability of Triton X-100 to affect the partitioning of PAHs between aqueous and tar phases was examined. The results of the partitioning experiments are shown in Figures 5 and 6. Addition of Triton X-100 results in a substantial partitioning of tar PAHs into the aqueous phase. Lower molecular weight compounds were affected to a greater extent than the higher molecular weight PAHs.

A second experiment was performed to determine the effect of Triton X-100 on the extent of gas-tar PAH degradation. Results of a shake flask experiment in which 0.1% Triton X-100 was incorporated as a supplemental carbon source is shown in Figure 7. The three HPLC traces represent samples acidified at 0, 2, and 5 days. As before, a substantial decrease in the peak heights corresponding to the lower molecular weight PAHs was observed, particularly for phenanthrene. However, less degradation of the higher molecular weight PAHs appeared to occur relative to the glucose-supplemented samples.

CONCLUSION

Twenty-six percent of a gasification-tar residue was converted to carbon dioxide by a sewage derived mixed microbial culture without addition of a supplemental carbon source. The lower molecular weight and more volatile components, such as benzene, toluene, and naphthalene, were degraded preferentially. Naphthalene was utilized by the mixed culture as sole carbon source.

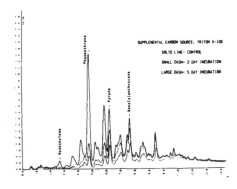

Figure 7.   Biodegradation of gas-tar PAHs.

Glucose supplemented biodegradation of gas-tar in shake-flasks resulted in extensive conversion of lower molecular weight PAHs. The use of a dispersant, Triton X-100, did not appear to improve the extent of PAH degradation. The selective degradation of the more mobile compounds in coal-tars may be an acceptable means of treatment, if it can be shown that the tar residues are stable.

REFERENCES

1. Hult, M. F., and M. E. Schoenberg, "Preliminary Evaluation of Ground-water Contamination by Coal-tar Derivatives, St. Louis Park Area, Minnesota," *U.S. Geological Survey Water-Supply Paper 2211* (1984).
2. McNeil, D., "Tar and Pitch," In: Mark, H. F., Othmer, D. F., Overberger, C. G., and G. T. Seagorg, eds, Kirk-Othmer, *Encyclopedia of Chemical Technology*, 3rd ed., J. Wiley & Sons, NY, Vol. 22, pp. 564–600 (1983).
3. "Industrial Exposures in Aluminum Production, Coal Gasification, Coke Production, and Iron and Steel Founding," Volume 34 of: *Evaluation of the Carcinogenic Risk of Chemicals to Humans*, International Agency for Research on Cancer, pp. 65–99, 101–131 (1984).
4. Fu, J-K, Luthy, R. G., and D. A. Dzombak, "Adsorption of Polycyclic Aromatic Hydrocarbon Compounds onto Soil and Transport of Naphthalene in Unsaturated Porous Media," *DOE/PC/30246-1557*, pp. 5–6 (1983).
5. Cerniglia, C. E., "Microbial Transformation of Aromatic Hydrocarbons," In: Atlas, R. M., ed, *Petroleum Microbiology*, Macmillan Inc., NY, pp. 99–128 (1984).
6. Wodzinski, R. S., and D. Bertolini, "Physical State in Which Naphthalene and Bibenzyl are Utilized by Bacteria," *App. Micro.*, 23, p. 1077–1081 (1972).
7. Wodzinski, R. S., and J. E. Coyle, "Physical State of Phenanthrene for Utilization by Bacteria," *App. Micro.*, 27, p. 1081–1084 (1974).
8. Dibble, J. T., and R. Bartha, "Effect of Environmental Parameters on the Biodegradation of Oil Sludge," *Appl. Env. Micro.*, 37, 729–739 (1979).
9. Liu, D., "Enhancement of PCBs Biodegradation by Sodium Ligninsulfonate," *Wat. Res.*, 14, 1467–1475 (1980).
10. Chien, M. M., and J. P. Rosaza, "Microbial Transformations of Natural Antitumor Agents: Use of Solubilizing Agents to Improve Yields of Hydroxylated Ellipticines," *App. Env. Micro.*, 40, 741–745 (1980).
11. Klevens, H. B., "Solubilization of Polycyclic Hydrocarbons," *J. Phys. Coll. Chem.*, 54, 283–297 (1950).

65 SLUDGE REDUCTION IN COAL-FIRED POWER PLANT FLUE-GAS DESULFURIZATION WASTEWATER TREATMENT

Yoshihiro Etoh, Senior Researcher

Tadashi Takadoi, Manager, Technical Section

Ikuo Itoh, Vice-Manager, Technical Section
Kurita Water Ind., Ltd.
Atsugi 243-01, JAPAN

INTRODUCTION

The most important problem facing coal-fired power plants is environmental pollution. The flue gas from coal-fired plants contains larger quantities of SO_x, NO_x, and dust than that from oil-fired plants. These pollutants are one cause of air pollution and acid rain. Flue-gas desulfurization in coal-fired power plants is mostly carried out by the wet lime and limestone process. The process removes the pollutants to satisfactorily low levels, but the water circulating in the system must be drawn off mainly to prevent accumulation of chloride ions coming from the coal and make-up water, which cause corrosion. The blow-down water is the so-called flue-gas desulfurization wastewater. The wastewater contains fluorides, heavy metals, organic and inorganic COD components, and other pollutants. Since 1974 we have been studying removal of those pollutants and have developed a treatment process consisting of coagulation (mainly for fluorides and heavy metals removal) and adsorption (mainly for COD removal). The process has already been adopted at many power plants. However, this process (hereafter designated as the conventional process) has the problem that large amounts of chemicals have to be added, and, therefore, a large amount of sludge forms in the coagulation stage. Since 1980 we have been trying to improve the process and have succeeded in reducing the amount of chemicals and sludge by half through the effective use of the aluminum and magnesium already contained in the wastewater.

This paper discusses the treatment of wastewater from the flue-gas desulfurization process in coal-fired power plants, focusing on the improved method.

WASTEWATER CHARACTERISTICS

The wastewater quality varies depending on the type of desulfurization equipment and the kind of coal. Table I shows how the quality of wastewater from desulfurization plants varies with the kind of coal. As can be seen in Table I, the boron concentrations vary by a factor of 10 with South African coal having the lowest and Australian coal the highest. This is important because fluoride removal depends on the concentration of boron in the wastewater. The amount of fluorides in the form of fluoroborate ions (BF_4^-) increases as the boron concentration increases, which makes fluoride removal difficult. However, the presence of a certain amount of aluminum ions in the wastewater precludes interference of boron. Therefore, the addition of aluminum may be required when there are extremely high concentrations of boron or low concentrations of aluminum.

Table II shows the typical quality of the wastewater from the desulfurization plant of the wet lime and limestone process including dust separation.

CONVENTIONAL PROCESS

A basic flow diagram for the conventional process is illustrated in Figure 1 though there may be some variations for different plants. A settling basin often precedes the raw water tank, and fly ash in the wastewater is settled by a polymer coagulant yielding raw water with a SS content of less than 500 mg/L.

Table I. Wastewater Quality and Kind of Coal
(Matsushima Power Station, Electric Power Dev. Co. Ltd.)

No. Coal Parameter[a]	A China	B South Africa	C Australia	D South Africa 70% China 30%	E South Africa 76% China 24%	F South Africa	G North America 20% South Africa 80%
Appearance	Gray	Gray	Gray	Gray	Gray	Gray	Gray
pH (−)	2.1	2.3	1.4	1.5	1.8	2.5	2.9
COD Mn	39.0	70.0	38.3	50.0	57.1	57.9	83.8
SS	1,290	9,160	356	2,680	3,970	−	−
Conductivity	1.02×10^4	1.04×10^4	2.47×10^4	1.43×10^4	1.29×10^4	−	−
SO_4^{2-}	3,280	4,480	3,520	4,670	4,960	−	−
Cl^-	520	700	2,250	410	490	−	−
Ca	500	840	448	510	583	636	654
Mg	430	660	400	505	621	579	421
Al	212	482	250	265	375	360	332
F	490	1,130	314	560	575	800	540
B	53.5	12.9	103	14.9	15.4	10.0	12.9

[a]Unit: Conductivity in μS/cm, others in mg/L.

Removal of Fluorides and Heavy Metals

Fluorides adsorb onto aluminum hydroxide flocs formed in the 1st reactor and are removed by settling. Figure 2 shows the adsorption of fluoride by aluminum hydroxide flocs. The adsorption is affected by the quality of the wastewater such as salt concentration, pH, temperature, etc. Aluminum

Table II. Typical Quality of Wastewater from Desulfurization Equipment

pH (−)		$1 \sim 3$
COD Mn (mg/L)		$20 \sim 120$
	Recalcitrant	$(15 \sim 50)$
SS (mg/L)		$300 \sim 10,000$
	After thickener	(less than 500)
F (mg/L)		$100 \sim 1,300$
SO_4^{2-} (mg/L)		$3,000 \sim 10,000$
Cl (mg/L)		$400 \sim 5,000$
Ca (mg/L)		$200 \sim 1,000$
Mg (mg/L)		$400 \sim 1,500$
Al (mg/L)		$100 \sim 800$
B (mg/L)		$10 \sim 130$

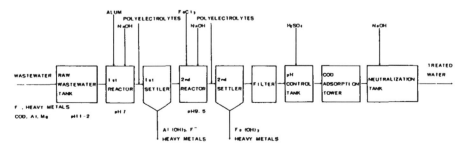

Figure 1.   Schematic of the conventional process.

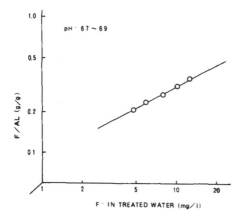

F⁻ IN TREATED WATER (mg/l)

Figure 2. Fluoride adsorption by Al(OH)₃ of wastewater in Table IV.

sulfate (alum) is added to supplement the aluminum already in the wastewater to reduce fluoride concentration down to 10 mg/L or less. For instance, the aluminum required for wastewater containing 500 mg/L of fluorides will be about 1600 mg/L as calculated from Figure 2. Generally, an aluminum dosage of 800–1500 mg/L will be necessary since raw wastewater usually contains 100–800 mg/L of aluminum. Most heavy metals are removed in the 1st settler together with the fluorides. Any remaining heavy metals are removed in the 2nd settler by coagulation with ferric chloride at a high pH.

Removal of COD

Desulfurization wastewater contains COD components not removable with the coagulation treatment. After elaborate analytical investigations they were found to be sulfur compounds such as dithionate ($S_2O_6^{2-}$) and hydroxylamine disulfite ($N(OH)(SO_3)_2^{2-}$). Various treatments in Table III were tried to remove them. Activated carbon, ozone, chlorination, and UV irradiation were quite ineffective. The most efficient was a synthetic adsorbent. Figure 3 is a flow diagram of the COD removal stage employing the synthetic adsorbent.

After removal of the fluorides and heavy metals, wastewater is filtered and fed to the COD adsorption towers, which are packed with the synthetic adsorbent. The adsorbent can selectively remove the COD components from wastewater containing high concentrations of salts. After satura-

Table III. Treatability in Soluble COD$_{Mn}$ in Wastewater[a]

Treatment	COD Mn in Raw Water	COD Mn in Treated Water	Removal Ratio	Remarks
Activated Carbon	23 ~ 67 mg/L	22 ~ 62 mg/L	1.5 ~ 15% (7%)	Not affected by kinds of activated carbon, pH and SO_4^{2-}
Ozone	12 ~ 90 mg/L	11 ~ 80 mg/L	1 ~ 11% (5.6%)	
Ozone + Activated Carbon	21 ~ 57 mg/L	19 ~ 53 mg/L	7 ~ 12% (10%)	
Chlorine Oxidation	12 ~ 90 mg/L	10 ~ 83 mg/L	10 ~ 20% (14%)	
Hydrogen Peroxide + Ultraviolet Ray	25 ~ 61 mg/L	23 ~ 59 mg/L	5.7 ~ 6.5% (6.2%)	
Electrolysis	58 mg/L	58 mg/L	0	
Activated Sludge	58 mg/L	58 mg/L	0	
Simple Heating	58 mg/L	58 mg/L	0	
Cation Resin	59 mg/L	59 mg/L	0	Dowex – 50W
Anion Resin	59 mg/L	6 mg/L	90%	Dowex – 1, Unregenerable
Synthetic Adsorbent	16 ~ 61 mg/L	2 ~ 4 mg/L	90 ~ 94%	Regenerable

[a]Wastewater had been treated by coagulation and filtration.

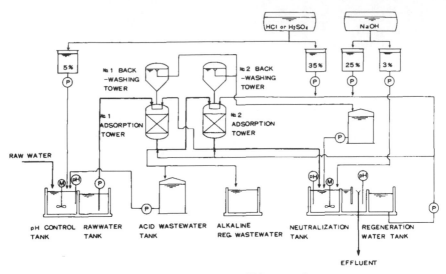

Figure 3.   Flow diagram of COD removal stage.

tion the adsorbent is regenerated with solutions of caustic soda and sulfuric acid (or hydrochloric acid). The regeneration gives a sharp elution curve, and the strong fraction (1–1.5 Bed Volumes, COD 1000–3000 mg/L) is extracted and disposed of after neutralization, which entails spraying on coal for combustion. Otherwise, it is decomposed by acidification and heating. The weak fractions of the eluate, i.e., the fore- and post-fractions of the strong fraction, are returned to the raw water tank for adsorption. This treatment process always yields an effluent with a COD concentration of less than 10 mg/L.

IMPROVED PROCESS

Purpose and Idea

The conventional process above explained can certainly remove fluorides, heavy metals and COD. However, much coagulants are needed in the coagulation stage, and the added coagulants such as alum and ferric chloride produce a great deal of sludge. Hence, we got the idea of removing fluorides and heavy metals without adding sludge-forming chemicals by making effective use of the aluminum and magnesium already contained in the wastewater.

The effect of pH on the removal of fluorides, aluminum, and magnesium was studied with the wastewater shown in Table IV by adding sodium hydroxide. As shown in Figure 4, fluorides were reduced to about 20 mg/L at a pH of around 7 by adsorption onto aluminum hydroxide. The aluminum was almost completely removed, but most of the magnesium remained in solution.

Thus, the best process seems to be one in which most of the fluorides are removed in the 1st stage at a neutral pH with aluminum in the wastewater, and the remaining fluorides are further precipitated in the 2nd stage with magnesium hydroxide at a higher pH.

Table IV. Wastewater Quality (Shimonoseki Power Plant)

Parameter[a]		Parameter[a]	
pH	1.7	Mg	184
SS	218	Al	682
F	223	SO_4^{2-}	5,800
Ca	720		

[a]Unit = mg/L.

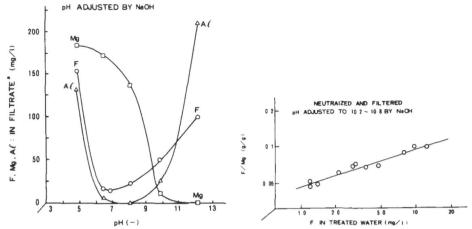

Figure 4. Effect of pH on removal of F, Mg, and Al. (a = No. 5A filter paper filtrate).

Figure 5. Fluoride adsoption by Mg(OH)$_2$.

Fluoride Removal by Magnesium

The capacity of magnesium to remove fluorides was examined with the above mentioned waste-water after neutralization and filtration. Magnesium chloride (MgCl$_2$.2H$_2$O) and sodium hydroxide were added to the water and fluoride and magnesium concentrations in the filtrate (No. 5A filter paper) were measured. Figure 5 shows the relationship between the fluoride concentration in the filtrate and the ratio of the amount of fluorides removed to the amount of magnesium removed (F/Mg). As can be seen in Figure 5, fluoride concentration can be reduced to as low as 1 mg/L by magnesium, but the adsorption capacity was about 1/4 that of aluminum. Next, methods of control-ling the amount of magnesium removed, or in other words, the quantitative removal of fluorides, were investigated. Keeping the pH constant does not result in a constant amount of magnesium being removed as the magnesium concentration in the influent wastewater fluctuates. Therefore, we decided to try controlling the dosage of sodium hydroxide in order to fix the amount of magnesium removed. As can be seen in Figure 6, there was a linear relationship between the dosage of NaOH and the amount of Mg removed. This relationship is based on the reaction shown in Eq. 1.

$$MgSO_4 + 2NaOH \rightarrow Mg(OH)_2 + Na_2SO_4 \tag{1}$$

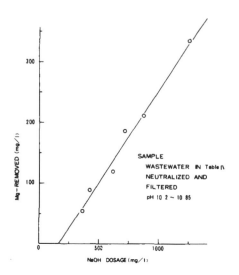

Figure 6. Relationship between NaOH dosage and Mg removal.

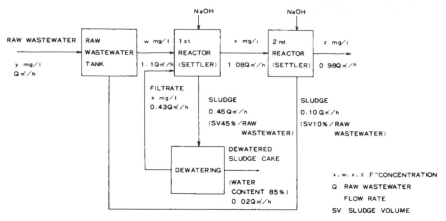

Figure 7.   Conceptual diagram for designing Mg circulation method.

Thus, the amount of Mg, i.e. the amount of F to be removed can be controlled to a prescribed value by controlling the dosage of NaOH. The precipitated magnesium hydroxide can be returned to the raw wastewater tank and used as a neutralizer as shown in Eq. 2.

$$H_2SO_4 + Mg(OH)_2 \rightarrow MgSO_4 + 2H_2O \tag{2}$$

Consequently, an increase in magnesium precipitation does not result in an increase in the amount of NaOH required in the total system. The magnesium returned to the raw wastewater tank is reused for fluoride removal. Even when the wastewater has a low magnesium concentration, it can be enriched by the circulation of the magnesium precipitated.

Magnesium Circulation Method

The fluoride concentrations and dosage of NaOH in the 1st and the 2nd stages were calculated for a system involving the recycling of sludge containing magnesium hydroxide. Figure 7 is a conceptual flow diagram for the calculation where values in units of mg/L are fluoride concentrations. The material balance in the raw wastewater tank can be expressed as Eq. 3.

$$1.1wQ = yQ + 1.08xQ - 0.98zQ \tag{3}$$

where $1.1wQ$ is the amount of fluorides going out of the raw wastewater tank, yQ is that in the raw wastewater, and $1.08xQ-0.98zQ$ is that in the return sludge. The unknown values which must be obtained are w (the influent F concentration of the 1st reactor) and x (the effluent F concentration of the 1st reactor). Assuming an Al concentration of 680 mg/L in the raw wastewater, the relationship between w and x, which is an equilibrium line, can be obtained from Figure 2, and an operating line can be obtained by letting the F concentration of the raw water $(y) = 223$ mg/L and that of the treated water $(z) = 10$ mg/L in Eq. 3. Figure 8 is a diagram to obtain fluoride concentrations in the influent and effluent of the 1st reactor. The point of intersection of the equilibrium line and the operating line in Figure 8 gives the values of w and x, that is, 205 mg/L and 11 mg/L, respectively. The amount of NaOH required in the 2nd reactor in order to obtain $z = 10$ mg/L will be 200 mg/L from Figure 5 and 6. These values can vary widely depending on the concentrations of Al and F in the raw wastewater. Accordingly, calculations should be made assuming the worst case, a high F and a low Al content in the raw wastewater. Calculations can be made for heavy metals in the same way.

Figure 9 is a flow diagram for wastewater treatment with magnesium sludge recycling. In this system the minimum amount of magnesium required to remove the F is circulated, and the excess, which is dissolved in the treated water, is discharged. The magnesium discharged as sludge comes only from the 1st reactor precipitated at a neutral pH, and the amount is extremely small. The only chemicals necessary are NaOH and a polymer flocculant. The number and size of chemical tanks and pumps can be reduced from those required for the conventional process. This process is called the Mg circulation method and is presently in operation at 7 plants.

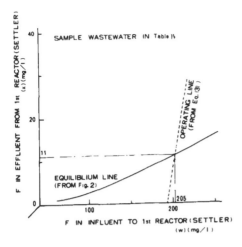

Figure 8.   Diagram to obtain fluoride behavior in Mg circulation method.

TESTS WITH AN ACTUAL PLANT

Operation of Plant

Two treatment lines using the conventional process had been installed and were operating at the Shin-Ube power plant of the Chu-goku Electric Co. To evaluate the Mg circulation method, piping for sludge return and a constant feed apparatus for NaOH were installed on one of the lines. The other line was operated as usual in the conventional method as a control.

In the Mg circulation method the only chemical added was NaOH to the lst and the 2nd reactors. The lst reactor was adjusted to a pH of 7 (measured values of 6.6–7.45) while the 2nd reactor was fed with 1000 mg/L of NaOH which was previously determined to be the appropriate amount by beaker tests. All the sludge in the 2nd settler, which amounted to 60 m3/day, was returned to the raw wastewater tank. The raw wastewater flow was about 300 m3/day for each line.

In the line using the conventional process, the lst reator was fed with 5000 mg/L of alum and the pH was adjusted at 7 with NaOH (measured values of 6.4–7.6) while 150 mg/L of $FeCl_3$ was added to the 2nd reactor and the pH was adjusted to 9.5 (measured values of 9.5–10.0).

Test Results

Figure 10 shows fluoride concentrations in the raw and treated water. The fluoride concentrations in the effluent from the No.2 settler of the Mg circulation method and in the effluent from the No. 1 settler of the conventional process (about the same as the effluent from the No.2 settler) were almost always less than 10 mg/L. Since the fluoride concentrations were about the same for both methods, it was easy to compare sludge production and the cost of chemicals on the same basis. The magnesium circulation method removed 270 mg/L of magnesium on average, and the fluoride to magnesium-

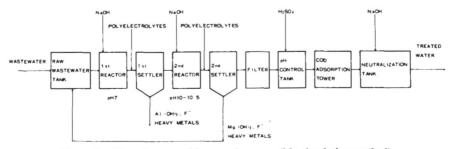

Figure 9.   Flow diagram of improved process (Mg circulation method).

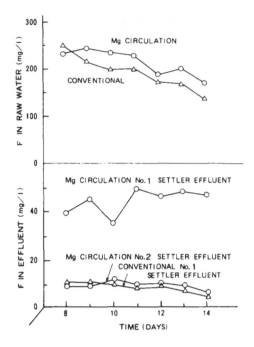

Figure 10.   Behavior of fluoride.

removal ratio (F/Mg) was 0.12–0.14, a little higher values than those shown in Figure 5. A further reduction in the fluoride concentration, if desired, is possible by increasing the amount of NaOH added to the 2nd reactor in the Mg circulation method or by increasing the amount of alum to the lst reactor in the conventional process.

The amount of sludge produced and dosage of NaOH over a two week period are compared in Figures 11 and 12, respectively. The sludge production in the Mg circulation method was calculated from the concentraion of suspended solids in the lst reactor. The sludge production in the conventional process was derived from the sum of the suspended solids in the lst and the 2nd reactors. The Mg circulation method definitely produced less sludge than the conventional process. The Mg circulation method requires a higher NaOH dosage in the 2nd reactor than the conventional process, but the sludge can be reused for neutralization of the raw wastewater, and therefore, the overall NaOH consumption is less than that of the conventional process by the amount required for neutralization of the alum in the conventional process. Table V compares the two methods in terms of the cost of chemicals and sludge production. The costs for the Mg circulation method were 68% less and sludge production was reduced by 56%. The amount of the reduction for both these items depends on the quality of the wastewater, but the Mg circulation method can certainly reduce them by half as

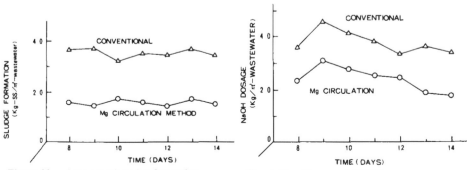

Figure 11.   Amount of sludge formation.   **Figure 12.   Dosage of NaOH.**

Table V. Comparison between Conventional Process and Mg Circulation Method

Item	Process	Conventional	Mg Circulation
Chemical Costs[a]	NaOH	51	32
	Alum	45	0
	FeCl$_3$·6H$_2$O	4	0
	Total	100	32
Sludge Formation (kg SS/m³)		3.55	1.57

[a]Chemical costs are presented as % of the total for the conventional process.

compared to the conventional process. Heavy metals were reduced below the effluent standards with both methods.

CONCLUSION

A coal-fired power plant desulfurization wastewater treatment process has been developed. It consists of coagulation for the removal of fluorides and heavy metals and adsorption for the removal of COD. The process has been further improved to reduce sludge production and the cost of chemecals. Seven plants using the improved process which is called the Mg circulation method are now operating satisfactorily. This method is applicable to other F-containing wastewaters, and is used for the treatment of wastewater from the manufacture of semiconductors.

We are presently considering further improvements and the following two points are under study: one is the automatic control of the chemical dosage by instrumentation of the F measurement to reduce chemical costs, and the other is the development of a boron treatment process. These studies will further improve the desulfurization wastewater treatment process.

66 AUTOTROPHIC BIOOXIDATION PROCESS FOR TREATMENT OF COAL GASIFICATION WASTEWATERS

Frank J. Castaldi, Senior Engineer
Radian Corporation
Austin, Texas 78766

INTRODUCTION

Biological treatability studies were conducted on wastewaters from a non-tar producing coal gasification process as part of a characterization program to develop design and environmental data for synthetic fuels plants based on this technology [1,2]. The results of this experimentation produced a series of wastewater characterization and treatability documents which established basis of design data for the following treatment unit operations/processes: 1) coagulation/clarification for removal of small diameter (<2 microns) carbon particles (fines) present in the wastewaters after bulk settling of dense solids suspensions found in the quench liquors/gas cooling condensates [3]; 2) chemical conversion and fixation of cyanide to less toxic forms [3]; 3) steam stripping for removal of dissolved acid gases and ammonia [4]; and 4) activated sludge for ammonia, thiocyanate, and organic carbon removal [5,6].

This paper presents the results of a study which developed basis of design data for an autotrophic activated sludge process to treat process wastewaters from an ash agglomerating fluidized-bed gasification process. The technology also is generally applicable to the treatment of process wastewaters from entrained-bed gasifiers and should be viewed as a technology innovation for treatment of low tar, oil, and phenol coal gasification quench liquors and gas cooling condensates. Previous research has indicated that a variety of aerobic bacteria are capable of degrading thiocyanates during the biooxidation of coke oven wastewaters and fixed-bed coal gasification raw product gas quench condensate. One of the purposes of this study was to examine the feasibility of developing an obligate autotrophic population of bacteria that could perform similar metabolic functions in order to achieve thiocyanate degradation and attain simultaneous nitrification in a single bioreactor.

CHEMOAUTOTROPHIC BIOOXIDATION

Chemoautotrophic organisms are dependent on chemical energy sources and employ carbon dioxide as the principal carbon source. The use of CO_2 as a carbon source by chemotrophs is always associated with the ability to use reduced inorganic compounds as energy sources. This ability is confined to the bacteria and occurs in a number of specialized groups that can use reduced nitrogen compounds (NH_3, NO_2^-), ferrous iron, reduced sulfur compounds (H_2S, $S^=$, $S_2O_3^=$), or H_2 as oxidizable energy sources.

By definition, a chemoautotroph must possess two special biochemical capacities: the ability to derive Adenosine Triphosphate (ATP) and reducing power from the oxidation of a reduced inorganic compound and the ability to use CO_2 as its principal or sole source of carbon, an attribute that implies possession of specialized enzymatic machinery. The two groups of chemoautotrophic bacteria examined during the study were the nitrifying bacteria and sulfur oxidizing bacteria.

The nitrifying organisms fall into two separate physiological groups: bacteria that oxidize ammonia to nitrite and bacteria that oxidize nitrite to nitrate. The most common ammonia oxidizer in soil is *Nitrosomonas* and among the nitrite oxidizers the most common soil form is *Nitrobacter*. The chemoautotrophic sulfur-oxidizing bacteria are placed in a single genus, Thiobacillus. These organisms can grow at the expense of elemental sulfur, and many can use thiosulfate as well.

The process of converting ammonia to nitrates involves the destruction of alkalinity, which, in turn, means a drop of pH, and it follows, an inhibition of the nitrification reaction. From the equations presented below, it can be seen that two equivalents of alkalinity are consumed per equivalent of ammonia nitrogen oxidized. Theoretically, 7.14 mg of total alkalinity are destroyed per mg of ammonia nitrogen oxidized to nitrate.

$$NH_4^+ + 2O_2 \rightarrow NO_3^- + 2H^+ + H_2O \tag{1}$$

$$2H^+ + 2HCO_3^- \rightarrow 2H_2O + 2CO_2 \tag{2}$$

The theoretical oxygen requirement, based on the biochemical equations of nitrification, is defined as 4.57 mg of oxygen required per mg of ammonia nitrogen nitrified. It is assumed that this oxygen demand must be satisfied by atmospheric molecular oxygen, furnished and dissolved by conventional aeration equipment.

Nitrifying bacteria have more specific environmental requirements than do the heterotrophic bacteria responsible for carbon removal. These environmental requirements include specific factors which influence the metabolism and biosynthesis of the microorganisms. In particular, attention is given to pH, mixed liquor dissolved oxygen concentration, temperature, and the possible presence of toxic materials.

The nitrifiers are sensitive microorganisms that are subject to severe upsets should environmental factors be unfavorable. Compounds which are most often associated with toxicity to nitrifiers include heavy metals, cyanides, halogenated compounds, phenols, mercaptans, and amines. The nitrifiers also are subject to substrate inhibition. Both *Nitrosomonas* and *Nitrobacter* have been inhibited by the presence of free ammonia and free nitrous acid at varying concentration levels.

Members of the genus Thiobacillus grow aerobically with inorganic sulfur compounds as the energy source. All members of the genus with the exception of one are strict autotrophs. Most species of Thiobacillus grow best with thiosulfate as the energy source, but sulfide, sulfur, and thiocyanate also can be used by certain species for growth. These bacteria grow best between the pH range of 6.0 and 7.6 and the temperature range of 25°C to 35°C. The optimum pH range for thiocyanate degradation by bacteria is between pH 6.7 and pH 7.2. An alkaline pH tends to inhibit the biodegradation of thiocyanate, and this is the result of a physiological inhibition (the upper tolerance limit of the bacteria) and not a toxic effect. Temperatures above 37°C inhibit the growth of Thiobacillus.

Thiocyanates serve as both a source of energy and as a source of carbon and nitrogen for certain groups of thiocyanate degrading bacteria. It is hypothesized that thiocyanate is first biochemically hydrolyzed to cyanate and sulfide as follows:

$$SCN^- + H_2O \rightarrow HS^- + HCNO \tag{3}$$

and cyanate is then hydrolyzed further to carbon dioxide and ammonia:

$$HCNO + H_2O \rightarrow NH_3 + CO_2 \tag{4}$$

and sulfide is oxidized to sulfate:

$$HS^- + 2O_2 \rightarrow HSO_4^- \tag{5}$$

Based on laboratory observations, the overall reaction is:

$$SCN^- + 2H_2O + 2O_2 \rightarrow CO_2 + SO_4^= + NH_4^+. \tag{6}$$

WASTEWATER CHARACTERIZATION

The wastewaters for the biological treatability study were collected during a test burn of a non-tar producing coal gasification process development unit. The wastewaters were steam-stripped without pH adjustment. Characterization data for the pretreated wastewaters are presented in Table I. These data characterize a wastewater that is low in total organic carbon (TOC) and has moderate to high levels of ammonia (NH_3-N) and thiocyanates (SCN). The observed magnitude of the chemical oxygen demand (COD) includes contributions from both the organic carbon and SCN concentrations of the wastewater.

Data for feedwater 20-day biochemical oxygen demand (BOD_{20}) are also presented in Table I. These BOD_{20} values were representative of a seeded influent sample, and the observed experimental concentrations were determined to be approximately 75% of the theoretical (ultimate) oxygen demand of the wastewater as estimated from the average ammonia, thiocyanate, and organic carbon concentrations.

Thiocyanate contributes the largest portion of the chemical oxygen demand (COD) of the wastewater. The oxygen demand produced by the ammonia in the wastewater is not accounted for in the

Table I. Bioreactor Feedwater Quality

Parameter	Mean (mg/L)	Range (mg/L)
TOC	38	29–44
COD	178	145–242
BOD_{20}	1,163	917–1,539
NH_3-N	253	238–270
NO_3-N	<0.5	<0.5–3.4
NO_2-N	<0.1	<0.1–0.21
SCN	130	116–150
Total CN	1.19	0.8–1.56
Total P	4.95	2.5–6.0
TDS	294	220–490
Alkalinity ($CaCO_3$)	57	13–191
pH	6.91	5.4–8.15

COD test because dichromate is not reduced by ammonia or any ammonia liberated from the breakdown of proteinaceous matter.

The wastewater contains moderate levels of fluoride (17 mg/L), chloride (499 mg/L), sulfate (244 mg/L), and reduced sulfur compounds in addition to the other pollutants discussed. The feedwater BOD_5 concentration was approximately 250 mg/L. The majority of the trace metals were present at low concentrations (<0.5 mg/L). Elements with measurable concentration levels (>5 mg/L) at low concentrations in this non-tar producing gasifier wastewater; however, it was added to the bioreactor feedwater as a supplemental nutrient for microbial cell growth.

BENCH-SCALE EQUIPMENT

Continuous-flow reactor systems were used in the laboratory to determine the applicability and kinetics of the activated sludge process. The continuous-flow laboratory system was designed to provide a steady supply of raw wastewater through a biological reactor and to permit a continuous withdrawal of the treated effluent.

The system shown in Figure 1 combines both aeration chamber and upflow clarifier into a single unit and was found to be useful in the laboratory for bench-scale experimentation because it eliminates the need for a separate clarifier and an additional pump to control the return sludge. The test reactor was made of plastic and had an aeration chamber volume of ten liters.

Air was applied to the bench-scale aeration chambers through a diffuser stone. This air supply was provided by an air compressor and was regulated at flow rates in the range of 3,000 to 7,000 cm^3/min. These high air flow rates were necessary to adequately mix the aeration basin contents and maintain high dissolved oxygen levels in solution in the absence of mixers. The air supply was scrubbed with water to saturate it prior to introducing it into the reactor in order to minimize evaporation. The water scrubbing was preceded by filters that serve to collect oils and other impurities which were present in the air supply.

STEADY-STATE EVALUATION

The principal functions of the treatability study were to ascertain the degree of biodegradation possible by conventional activated sludge processes and to develop the basic criteria which would establish proper operation of autotrophic biological treatment facilities.

The necessary coefficients to mathematical models which establish basic operation criteria were determined by operating bench-scale biological reactors at various substrate loadings and evaluating each system for substrate removal, sludge production, and oxygen requirements. Substrate removal, sludge production, and oxygen utilization in continuously-fed, completely mixed bioreactors were evaluated during this study at water temperatures which varied between 22°C and 23.5°C.

Sludge settleability and thickening tests were performed throughout the treatability study in order to determine surface loading rates on secondary clarifiers which would provide both effective clarification and thickening during normal activated sludge process operation.

The day-to-day performance and/or operating parameters of TOC, COD, and NH_3-N removal and specific oxygen utilization were used as indicators of steady-state operation. Figure 2 depicts the daily

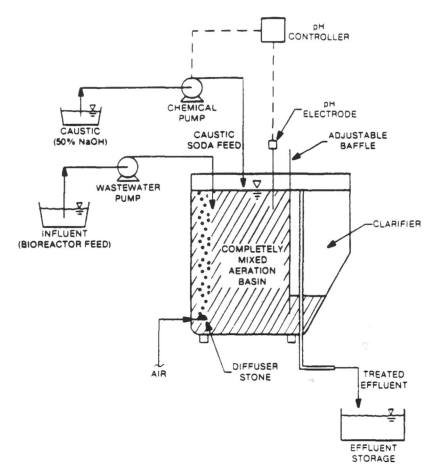

Figure 1.   Bench-scale activated sludge unit.

record of these parameters for a test bioreactor. Coincident, consistent levels of these parameters provided the criterion for establishing which portions of the data record for each reactor should be included in the evaluation. The operating period selected for steady-state analysis was a six week period toward the end of the test run. In the subsequent design parameter evaluation, other analytical data generated during this period for the test bioreactors were presumed to be useful, except for obvious outliers occasionally encountered.

The mean values and standard deviations of the bioreactor mixed liquor operating parameters examined during the treatability study are presented in Table II. Substrate utilization expressed as the magnitude of the calculated food-to-microorganism ratio (F/M) with respect to ammonia, thiocyanate, TOC, and BOD_{20} is presented in Table III. The BOD_{20} utilization values were estimated from the experimental data by recognizing that more than 75% of the influent ultimate biochemical oxygen demand for the wastewater is generated by ammonia-nitrogen oxidation. Therefore, the observed NH_3-N concentration of the wastewater should correlate well with the influent BOD_{20} value.

Oxygen utilization was expressed as immediate demand, initial demand, equilibrium demand, and endogenous demand. An explanation of these terms is presented in Figure 3 which depicts typical bioreactor oxygen uptake curves for steady-state operation. In general, the apparent oxygen demand as measured in the first 30 seconds of the oxygen uptake analysis was considerably different than that monitored in the later portion of the test. Normally, the observed magnitude of the oxygen uptake rate is monitored in the first two minutes of the analysis. These values which are defined as the "initial

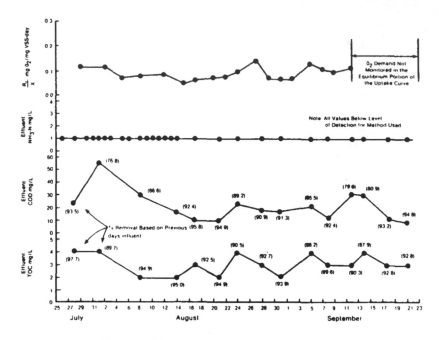

Figure 2.   Bioreactor steady-state performance data.

demand" were measured between 15 seconds and two minutes of uptake time. The apparent magnitude of the specific oxygen utilization rate in this portion of the curve was usually one-eighth that of the observed immediate demand. When the specific oxygen utilization was determined in the portion of the uptake curve below the two minute mark, a third magnitude of this parameter was obtained. This portion of the curve was defined as the "equilibrium demand" and probably represents a substrate limiting environment. When the oxygen uptake analysis was performed after 10 minutes of bioreactor aeration without the introduction of feed, the observed specific oxygen utilization value was presumed to be characteristic of the endogenous demand.

The magnitudes of the immediate, initial, equilibrium, and endogenous demands varied from 1.8 to 2.9, 0.24 to 0.40, 0.08 to 0.14, and 0.06 to 0.09 mg O_2 per mg VSS per day, respectively. The initial demand form of the oxygen utilization parameter was more variable than the equilibrium demand parameter that was used to assess steady-state performance. These data suggest that the observed

Table II. Mixed Liquor Characteristics

Parameter	Bioreactor 1	Bioreactor 2	Bioreactor3
Sludge Age, days	58.5 ± 6.0	104.3 ± 14.5	150.3 ± 18.7
MLSS, mg/L	$2,706 \pm 180$	$3,641 \pm 169$	$4,733 \pm 113$
MLVSS, mg/L	$2,128 \pm 152$	$2,937 \pm 182$	$3,772 \pm 88$
pH	7.17 ± 0.16	7.1 ± 0.24	7.23 ± 0.05
Temperature, °C	22.2 ± 0.9	22.5 ± 0.9	22.0 ± 0.7

Table III. Substrate Utilization

Parameter	Bioreactor 1	Bioreactor 2	Bioreactor 3
mg NH_3-N/mg VSS-day	0.149 ± 0.009	0.107 ± 0.005	0.084 ± 0.005
mg SCN/mg VSS-day	0.067 ± 0.006	0.051 ± 0.005	0.039 ± 0.004
mg TOC/mg VSS-day	0.017 ± 0.003	0.013 ± 0.002	0.01 ± 0.002
mg BOD_{20}/mg VSS-day	0.619 ± 0.088	0.45 ± 0.06	0.35 ± 0.037

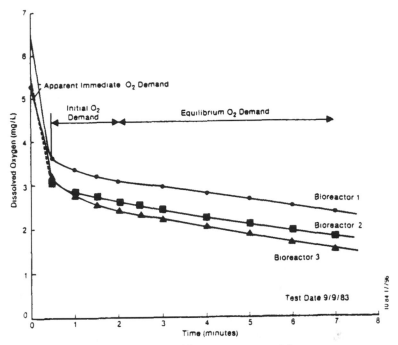

Figure 3.   Steady-state bioreactor oxygen uptake curve.

specific oxygen utilization rates for the wastewater, as measured in a substrate-rich environment, can vary considerably.

The experimental data also suggests that the design parameters that present the most practicable assessment of bioreactor oxygen utilization are the immediate and initial oxygen demand values because neither the equilibrium nor the endogenous demand occur in a substrate-rich environment. However, the apparent variability associated with both the immediate and initial oxygen demand values is substantial, and this variability seems not to be predicted by the observed pollutant levels in the bioreactor since effluent quality was usually quite consistent throughout the study. It is surmised that most of the variability in the oxygen demand may be due to the sensitivity of the nitrifier population to changing wastewater characteristics.

The effluent qualities of the three test bioreactors are indicated by the data presented in Table IV. The steady-state period was generally characterized by excellent pollutant removal with comparable

Table IV. Steady-State Effluent Quality

Parameter	Bioreactor 1	Bioreactor 2	Bioreactor 3
BOD_{20}, mg/L	5.6	5.1	5.0
TOC, mg/L	3.0	2.8	2.7
COD, mg/L	17	17	18
NH_3-N, mg/L	<1	<1	<1
NO_3-N, mg/L	286	286	286
NO_2-N, mg/L	0.13	0.12	<0.1
SCN, mg/L	<1	<1	<1
Total CN, mg/L	0.89	0.90	0.86
TDS, mg/L	2,929	3,002	2,976
Alkalinity ($CaCO_3$) mg/L	32	29	36
pH	7.17	7.1	7.23
Total P, mg/L	2.8	2.8	2.8

effluent values observed for all three bioreactors with respect to ammonia, thiocyanate, TOC, COD, and BOD_{20}. In general, all of the microbial substrate-specific pollutants were at reduced levels in the effluent after treatment.

Trace elements data for the bioreactors indicated little apparent difference between the influent and effluent quality associated with biooxidation treatment because the majority of the trace elements were present at low concentrations. Sodium was present in the bioreactor effluent at elevated concentrations (>850 mg/L) because of the addition of sodium hydroxide (NaOH) to the test reactors for pH control during biooxidation. However, accumulation of some pollutants in the biosludges was also apparent from the trace element data.

Elements that were at elevated concentrations in the sludge (relative to the water phase) are aluminum, barium, calcium, iron, potassium, magnesium, phosphorus, silicon, and zinc. Elements that were also bioaccumulated but to lower levels are cadmium, chromium, copper, manganese, nickel, lead, and titanium. The presence of high concentrations of iron (120 μg/g), copper (10 μg/g), and nickel (3.4 μg/g) in the biosludge may indicate a potential for the accumulation of complexed cyanides in the sludge. This may present a possible constraint on the handling of wastewater treatment sludge from a commercial-scale treatment facility.

The presence of elevated concentrations of trace elements in the biosludges results from the long operational sludge ages of the autotrophic activated sludge process. Long bioreactor sludge ages promote the accumulation of certain pollutants in the waste activated sludges. Therefore, an apparent trade-off exists between the system stability needed to guarantee year-round nitrification that results from long sludge age operation and the potential effects of bioaccumulation which could require that the waste sludges be handled as a hazardous material. However, sludge produced from a long cell residence time autotrophic system will be a small quantity on any given day. Therefore, these sludges are most likely to be wasted from the process seasonally, depending on mixed liquor requirements during either winter or summer operation.

DESIGN PARAMETERS

The experimentally determined coefficients for sludge production, oxygen utilization, and substrate removal are presented in Table V. The graphical form of the sludge production formulation is shown in Figure 4. The growth-yield coefficient was derived by applying linear regression analysis to data relating to mass of volatile solids wasted during the study to the substrate utilization characteristics of the three test bioreactors.

It is apparent that the sludge yield for the autotrophic activated sludge process is quite small. Typical values of the sludge yield coefficient for carbonaceous substrate removal vary between 0.4 and 0.9 mg VSS per mg substrate (BOD) utilized. Test data for the wastewater produced a BOD-based sludge yield value that is an order of magnitude smaller than a heterotrophic process.

The low sludge yield of a chemoautotrophic process is a major factor in process design because it specifies an operational system that does not require a large capital investment for sludge handling facilities. This type of biological treatment process should permit a seasonal sludge wastage schedule that would depend on mixed liquor population dynamics. Periods of cold weather that specify a slower microbial metabolism will require that sludge not be wasted in order to increase the active mixed liquor solids concentration to meet normal pollutant loadings. Sludge wastage rates during warm weather operation are dependent on pollutant loading and normal operation kinetic factors. Under certain conditions the loss of biological solids with decanted liquors from the process clarifier might balance microbial yield, thus requiring no sludge wastage during normal operation.

A graphical representation of oxygen utilization as a function of the BOD_{20} utilization is presented in Figure 5. This function was developed from initial oxygen demand data. The formulation more closely meets the theoretical oxygen requirements if the 99% (upper) confidence limit of the data set is used to generate the oxygen utilization coefficients.

Substrate utilization kinetics for both ammonia and thiocyanate removal were evaluated using the Lawrence-McCarty Model. In general, the data indicate that the observed kinetic function is zero order (i.e., $K_s \ll S$). Therefore, it was then possible to calculate a maximum specific substrate removal coefficient, k_o, for both ammonia and thiocyanate from the combined substrate utilization data sets of the three test bioreactors. These values were presented in Table V. The observed BOD_{20} first-order rate constant was determined to be 0.09 L per mg VSS per day.

The operational sludge ages of the test bioreactors varied between 58 and 150 days, and the apparent activated sludge settling quality as measured by the zone settling velocity showed a deteriora-

Table V. Summary of Design Parameters

Parameter	Y mg VSS/mg Substrate	b Day^{-1}	r
Sludge Production			
Biochemical oxygen demand[a]	0.038	0.007	0.997

Parameter	a′ mg O$_2$/mg Substrate	b′ mg O$_2$/mg VSS/day	r
Oxygen Utilization			
Biochemical oxygen demand[a]			
Initial Demand (mean value)	0.528	0.055	0.992
Initial Demand (maximum value)[b]	0.739	0.131	0.999

Parameter	Eckenfelder Model k_1[c] L/mg VSS/day	Lawrence-McCarty Model K_o[d] mg Substrate/mg VSS/day
Substrate Utilization Kinetics		
Biochemical oxygen demand[a]	0.09	—
Ammonia-nitrogen (mean value)	—	0.118[e]
Ammonia-nitrogen (maximum value)	—	0.19
Thiocyanate (mean value)	—	0.052[f]
Thiocyanate (maximum value)	—	0.083

[a]20 day BOD.
[b]99% confidence limit of the true mean value.
[c]First-order removal rate coefficient.
[d]Maximum specific substrate removal coefficient (zero order).
[e]SD = ± 0.028 mg NH$_3$-N/mg VSS/day.
[f]SD = ± 0.012 mg SCN/mg VSS/day.

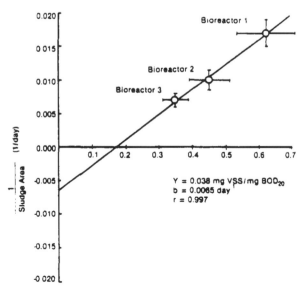

Substrate Utilization
(mg BOD$_{20}$ Removed/mg VSS/day)

Figure 4.   Sludge production.

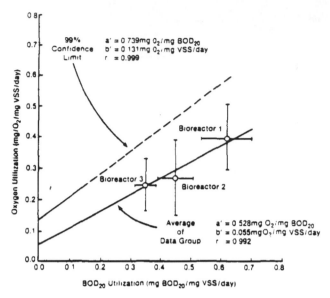

Figure 5.   Oxygen utilization.

tion at higher sludge ages. Typical bioreactor sludge settling curves during steady-state operation depicted particle-liquid interface, hindered settling at sludge ages greater than 100 days.

The experimental relationship between the zone settling velocity of the mixed liquor solids and the initial concentration of the mixed liquor is presented in Figure 6. This relationship demonstrates the degree to which mixed liquor concentration impacts upon sludge settleability. The apparent deteriora-

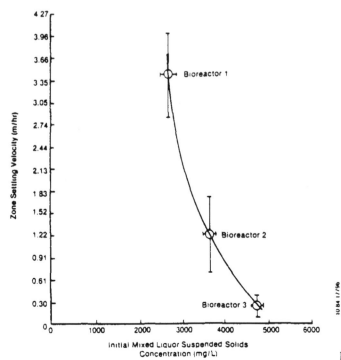

Figure 6.   Sludge settleability.

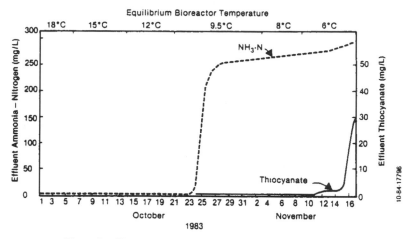

Figure 7.   **Temperature impact on autotrophic biooxidation.**

tion in sludge settleability is quite pronounced for a relatively narrow range of mixed liquor concentrations. This is not usually true of activated sludges developed from the biodegradation of carbonaceous BOD substrates. However, poor settleability is not uncommon of activated sludges that contain a large population of nitrifiers. This effect may also be a function of sludge age as was previously noted.

TEMPERATURE EFFECTS

A series of tests were performed with a continuously fed, completely mixed bioreactor to identify the effects of operation at lower ambient temperatures on bioreactor substrate removal performance. These tests were conducted at equilibrium bioreactor temperatures of 22°C to 5.6°C. Bioreactor operating conditions during the temperature study were similar to the mixed liquor characteristics (Table II) and the substrate utilization levels (Table III) established for bioreactor 3 during steady-state testing.

This bioreactor was then subjected to a series of temperature reductions, and performance data were recorded after the mixed liquor had equilibrated to the test temperature. At least three days were allowed for temperature adjustment before performance data were obtained from the test bioreactor. Data collection continued for at least one week after the bioreactor had equilibrated to the test temperature. However, performance data may not represent a steady-state condition because the actual testing period at each experimental temperature was too short to assess long-term impact.

The test data indicated an apparently severe impact upon ammonia removal efficiency at a mixed liquor temperature of approximately 10°C. This effect is depicted in Figure 7. At this temperature bioreactor performance deteriorated rapidly, reducing both the level of ammonia utilization and the observed alkalinity usage. The volatile cell mass also was reduced dropping from 79% to 63% of the total suspended solids under aeration. The observed average ammonia utilization was reduced from 0.086 mg NH_3-N per mg VSS per day to 0.024 mg NH_3-N per mg VSS per day. These effects intensified as the operational temperature was lowered further during the study. At an operational temperature of 6°C, ammonia utilization ceased. These effects occurred despite bioreactor operation at sludge ages in excess of 100 days.

The microbial washout rate increased during this period, and this may have been due to cell lyse. The severity of the temperature impact would imply a toxic shock effect rather than an effect due solely to temperature. This also would imply that there exists a coincidental and possibly synergistic effect between temperature and substrate (free ammonia) toxicity with nitrification of wastewaters that contain high concentrations of ammonia (i.e., > 100 mg/L).

The test results show that nitrate formation was strongly inhibited at temperatures of 10°C or less. However, at no time during the study did nitrite buildup in the mixed liquor of the test bioreactor.

Table VI. Nonsubstrate Ion Stress Test Operation

Parameter	Control Bioreactor	Stress Bioreactor
Substrate Utilization		
(mg NH$_3$-N)/(mg VSS-day)	0.108	0.103
(mg SCN)/(mg VSS-day)	0.048	0.046
Sludge age, days	155	137
MLSS, mg/L	3,672	8,341
MLVSS, mg/L	2,726	2,913
Sludge volume index, mL/gm	62	24
Total dissolved solids, mg/L	2,872	7,286
Chloride, mg/L	499	3,087
Fluoride, mg/L	17	275
Boron, mg/L	9.7	146
pH	7.0	6.95
Temperature, °C	20.5	20.8

The observed thiocyanate removal efficiency deteriorated below a 6°C bioreactor temperature (see Figure 7). There was no apparent reduction in thiocyanate biodegradation as the temperature was reduced until this temperature was reached, despite elevated levels of ammonia in the bioreactor.

NONSUBSTRATE ION STRESS

Pollutants which are common to most non-tar producing coal gasification process wastewaters are chloride, fluoride, and boron. These nonsubstrate ions may exist at elevated concentrations in gasifier quench liquors and gas cooling condensates and are considered a potential source of problems for biological treatment. Their actual concentration in a given wastewater is coal dependent with different equilibrium concentration ranges for each coal type (e.g., bituminous, sub-bituminous, or lignite).

Since the quench liquors associated with a non-tar producing coal gasification process are likely to be cycled during the quenching operation, certain ions (i.e., Cl$^-$, F$^-$, BO$_2^-$) may be concentrated in the process wastewater. Control of these ions in the wastewater would largely be through predetermined quench liquor blowdowns which would be designed to restrict the concentration of a given pollutant (e.g., chloride) to within the prescribed limits. Given the nature of this cycling operation, certain pollutants in quench liquors from a commercial-scale facility would inevitably reach concentrations above those measured in wastewaters from a process development unit.

A nonsubstrate ion toxicity study was performed with bioreactor activated sludges to document the impacts of increased concentrations of chloride, fluoride, and boron in the wastewater on biooxidation treatment efficiency. The study was conducted with two continuously-fed, completely mixed bioreactors. One reactor served as a control receiving feed with the ambient levels of the subject pollutants. The stress bioreactor was fed a wastewater with elevated concentrations of chloride, fluoride, and boron.

The bioreactor operating conditions during the nonsubstrate ion toxicity study are presented in Table VI. The associated bioreactor performance data are summarized in Table VII. Little difference between the effluent ammonia and thiocyanate qualities of the control and stress bioreactors is apparent. Total dissolved solids levels in the stress bioreactor increased by 250%, apparently without impacting biological treatment efficiency.

Experimental data indicated the presence of a chemical precipitate in the stress bioreactor mixed liquor which increased the density of the activated sludges. This precipitate (probably CaF$_2$) required that fluoride salts be added to the stress bioreactor feed in excess of the target concentration.

Elevated concentrations of chloride, fluoride, and boron appeared to have little effect on ammonia and thiocyanate removal efficiency during biooxidation of the wastewater. Therefore, relative to the subject pollutants, the impact of coal-type on autotrophic biological treatment performance should be negligible.

CONCLUSIONS

Gasifier wastewaters pretreated by steam stripping for ammonia reduction and acid gas removal can be successfully treated biologically with the activated sludge process. An ammonia nitrogen

Table VII. Nonsubstrate Ion Stress Test Performance

Parameter	Influent Quality		Effluent Quality	
	Control Bioreactor	Stress Bioreactor	Control Bioreactor	Stress Bioreactor
TOC, mg/L	35	35	2	2
COD, mg/L	172	184	15	40
NH_3-N, mg/L	255	255	<1	<1
NO_3-N, mg/L	<0.5	<0.5	276	278
NO_2-N, mg/L	<0.1	<0.1	0.13	0.11
SCN, mg/L	128	127	<1	<1
Total CN, mg/L	1.19	1.17	1.05	1.0
Alkalinity ($CaCO_3$), mg/L	106	97	28	30
pH	6.95	6.61	7	6.95

concentration of 285 mg/L and a thiocyanate concentration of 150 mg/L can be biooxidized at a neutral pH with a mixed culture of nitrifying and sulfur-oxidizing bacteria.

Concentrations of 3000 mg/L chloride, 285 mg/L fluoride, and 150 mg/L boron in the wastewater did not inhibit either the nitrifying or sulfur-oxidizing bacteria in the process. The presence of low levels of free and complexed cyanide in the wastewater did not have an adverse impact on ammonia or thiocyanate removal during biooxidation. Consequently, an autotrophic activated sludge process can be used to treat similar wastewaters derived from the gasification of different coals.

A study of temperature effects on the biodegradation of ammonia and thiocyanate indicated that the nitrifying bacteria were severely impacted at a bioreactor operational temperature below 10°C. This effect was thought to be largely the result of ammonia inhibition. The observed thiocyanate removal efficiency deteriorated at a bioreactor operational temperature below 6°C. However, there was no apparent reduction in thiocyanate biodegradation as the temperature was reduced until this temperature was reached, despite elevated levels of ammonia in the test bioreactor.

Although the settleabilities of the activated sludges were generally good, the apparent activated sludge settling quality as measured by the zone settling velocity showed a deterioration at higher sludge ages. Operational sludge ages for the experimental process varied between 60 and 150 days. Generally, these sludge ages were considered necessary to maintain system stability for complete nitrification and to minimize sludge production.

Heavy metals were concentrated in the activated sludges during experimental testing. However, the degree of bioaccumulation associated with the presence of metal cyanide complexes in the wastewater could not be determined from the experimental data.

ACKNOWLEDGEMENTS

This work was supported by the Environmental Protection Agency, EPA Contract No. 68-02-3137. Mr. William J. Rhodes, with the Air and Energy Engineering Research Laboratory at Research Triangle Park, was the Project Officer.

REFERENCES

1. Castaldi, F. J., and S. L. Winton, "Treatment System Design for Process Wastewaters From Non-Tar Producing Coal Gasification Technology," *GRI 85/0124* (June 1985).
2. Mann, R. M., G. E. Harris, W. R. Menzies, A. V. Simonson, and W. A. Williams, "Environmental, Health and Safety Data Base for the KRW Coal Gasification Process Development Unit," *GRI 85/0123* (June 1985).
3. Trofe, T. W., "Basis of Design Report for Suspended Solids Removal and Cyanide Pretreatment for Treating Wastewater from the KRW Energy Systems Gasifier Process Development Unit," Report prepared by the Department of Energy's Morgantown Energy Technology Center, *DOE/MC/21046-1803, GRI 85/0124.2* (April 1985).
4. Skinner, F. D., and B. J. Hayes, "Steam Stripping of KRW Energy Systems Gasifier Process Development Unit Wastewater," Report prepared for the Department of Energy's Morgantown Energy Technology Center. *DOE/Mc/21046-1802, GRI 85/0124.1* (April 1985).
5. Castaldi, F. J., "Acclimation of a Mixed Culture of Microorganisms to Treat Wastewater from

the KRW Energy Systems Gasifier Process Development Unit," *DOE/MC/21046-1801, GRI 85/0124.3* (April 1985).
6. Castaldi, F. J., "Biological Treatability of KRW Energy Systems Gasifier PDU Wastewaters," *EPA/AEERL, EPA-600/7-85-019* (May 1985).

67 PDU SCALE NITRIFICATION/DENITRIFICATION OF PRETREATED COAL GASIFICATION WASTEWATER

John R. Gallagher, Research Associate

Richard E. Shockey, Research Engineer

Charles D. Turner, Professor of Civil Engineering

Gale G. Mayer, Research Supervisor
Energy Research Center
University of North Dakota
Grand Forks, North Dakota 58202

INTRODUCTION

The University of North Dakota Energy Research Center (UNDERC) is involved in a wastewater research program using gasification condensates. The principal goals of this research are to develop treatability and environmental data on effluents from coal gasification. Previous and on-going work has focused on the reuse of gasification wastewaters. Treatment options examined include solvent extraction, steam stripping, activated sludge treatment for organics removal, and nitrification/denitrification.

Research at UNDERC has examined the effect of stripped gas liquor (SGL) and biologically treated SGL on the operation of pilot scale cooling towers. Use of SGL as cooling water, without further treatment, led to high levels of biological fouling, decreased heat transfer, and the stripping of large amounts of volatile organic compounds and ammonia to the atmosphere [1]. Biological treatment, without nitrogen removal, of the SGL resulted in reduced foaming and virtually eliminated heat transfer problems under deposit corrosion and stripping of volatile organics in a cooling tower test. However, the stripping of ammonia was approximately the same as without the biological treatment. Ammonia stripping was 90.6% with biologically treated SGL at influent ammonia concentrations ranging from 780 to 1250 mg/L.

Cooling tower tests with the SGL demonstrated that biological treatment to reduce the organic loading was sufficient for satisfactory operation of the tower at 10 cycles of concentration. However, ammonia stripping remained at high levels, which could result in industrial hygiene and nuisance odor concerns, environmental damage, and, after longer periods of acclimation, lead to biofouling in the cooling tower from nitrification. These concerns, coupled with that of discharge to receiving waters, prompted research into biological nitrogen removal.

MATERIALS AND METHODS

Wastewater

The Great Plains Gasification Plant (GPGP) uses Lurgi dry ash technology to gasify lignite. A major byproduct of this operation is a gas condensate stream contaminated with organics (phenolics, alcohols, fatty acids), inorganics (ammonia, thiocyanate, sulfide), and oils and tar. At GPGP the aqueous stream is solvent extracted using the Lurgi Phenolsolvan process and steam stripped with ammonia recovery using the Phosam W process [2]. UNDERC obtained fresh SGL immediately after steam stripping. The fresh SGL was trucked the 440 kilometers and stored in mixed, nitrogen-purged polyethylene tanks.

Biological Treatment

In order to separate the effects of organic removal from those of nitrogen removal, the SGL was pretreated at UNDERC using a complete-mix activated sludge system described in detail by Gallagher *et al.* [3]. The operational parameters of this treatment were: solids retention time of 3 to 10 days, 3

Table I. Composition of Gasification Condensate from the Great Plains Gasification Plant (mg/L)

Parameter	Raw Condensate[a]	Stripped Gas Liquor[b]
Organics		
COD	21,330	2,770
TOC	6,930	870
BOD$_5$	—[c]	1,150
Phenol	2,490	10
o-Cresol	450	7
m-Cresol	770	4
p-Cresol	580	4
Methanol	600	320
Ethanol	ND[d]	ND
Acetone	520	ND
2-Propanol	ND	ND
Acetonitrile	130	5
Propanol	ND	ND
2-Butanone	100	ND
Propionitrile	<10	ND
Catechol	180	200
5,5-Dimethylhydantoin	ND	ND
5-Ethyl,5-methylhydantoin	ND	ND
Acetate	490	435
Formate	160	110
Butyrate	ND	30
Propionate	110	20
Inorganics		
TKN (NH$_3$-N)	4,200	—
Ammonia	6,060	760
Thiocyanate	10	9
Cyanide	<1	5
Sulfide	46	ND
Thiosulfate	110	8
Sulfate	54	30
Chloride	<5	6
Phosphate	ND	207
Alkalinity (CaCO$_3$)	17,200	2,120
Conductivity (umhos/cm)	25,600	—
pH	8.6	9.3

[a]Represents a single analysis.
[b]Represents the mean of fourteen 6,400 liter loads.
[c]— = not analyzed.
[d]ND = not detected.

day hydraulic retention time, and dissolved oxygen concentration controlled between 2 and 3 mg/L. Influent and effluent wastewater composition is presented in Tables I and II.

Nitrification/Denitrification Test System

The nitrification/denitrification test system used in this research is shown schematically in Figure 1. The reactors were constructed from vinyl-coated 208 liter (55 gallon) steel drums. Three of the four steel drums were later changed to polyethylene drums. An overflow line was installed on the tank reactors to provide a working volume of 170 liters (45 gallons). The reactors to be used for nitrification were fitted with flexible fine-bubble air diffusers. The flow of air was controlled with rotameters. The dissolved oxygen concentrations in the nitrification mixed liquor were maintained between 4 and 6 mg/L. All four reactors were equipped with mechanical stirring. The denitrification reactor was equipped with a plexiglass cover to reduce oxygen transfer at the surface.

Clarifiers were constructed of polyethylene funnels with a total volume of 4 liters and an operational volume of 3.4 liters. Return sludge was recycled hourly with a peristaltic pump controlled by a

Table II. Characterization of Influent/Effluent Wastewaters[a]

Parameter[b]	Nitrification Influent	N1	N2	N3	DN[c]
SRT	—	37	21	12	10
pH	7.5	7.6	7.6	7.6	7.9
COD	839 ± 40	—	—	—	—
TOC	154 ± 4	—	—	—	—
BOD_5	24 ± 2	—	—	—	—
NH_3-N	523 ± 21	51 ± 16	56 ± 15	151 ± 23	37 ± 19
NO_2-N	184 ± 33	10 ± 5	2 ± 0.6	25 ± 15	6 ± 5
NO_3-N	97 ± 29	738 ± 9	738 ± 42	623 ± 37	61 ± 33

[a]Mean concentrations plus or minus the standard error of the mean during the operating period.
[b]All values in mg/L except pH and SRT (days).
[c]The mean values for DN exclude a period of upset during accidental usage of petroleum ether as a supplemental carbon source. N1 effluent is the influent to DN.

timer. Peristaltic pumps were also used to add feed and caustic to the nitrification reactors and to add methanol to the denitrification reactor.

Stored SGL was fed to the organic removal reactor at a flow rate of approximately 220 mL/min. The clarified effluent from this operation flowed into a sump basin. Feed for the nitrification reactors was pumped from this sump basin at a rate of 39 mL/min to each reactor to give a hydraulic retention time of 3 days. Feed to the denitrification reactor was clarified effluent from the nitrification reactor operated at the longest solids retention time (SRT).

The addition of an exogenous carbon source to the denitrification reactor was required since biological pretreatment resulted in a very low carbonaceous BOD. Methanol was used because it is an efficient energy source [4] and results in a lower amount of biomass production [5]. Additionally, methanol can be readily produced at gasification facilities. Methanol addition was based on data from the EPA nitrogen control manual and other literature reports [4, 6]. The stoichiometric quantity of methanol for complete removal of nitrate-N is estimated as 2.47 (methanol:nitrate-N, mass/mass). An aqueous solution of methanol was continuously pumped into the denitrification reactor at a rate designed to give the 2.47 ratio.

Monitoring/Control of the Test System

The nitrification reactors were designated N1, N2, and N3, and the denitrification reactor was designated as DN. Volumetric solids wasting was used to establish nominal solids retention times (SRT). Nominal SRTs were 80, 50, 30, and 10 days for N1, N2, N3, and DN, respectively. The actual SRTs were calculated from a mass balance of the solids through the system, since the volumetric method ignores solids lost in the effluent. Volumetric wasting was performed daily throughout the operation.

Antifoam was needed rarely since the organic content of the feed wastewater to the nitrification and denitrification reactors was very low. The antifoam agent used was a dilute solution of Union Carbide

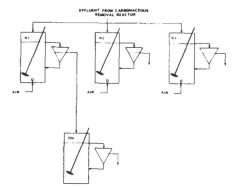

Figure 1. The UNDERC PDU Nitrification-Denitrification System.

SAG5441, a silicone-based antifoam consisting of an emulsion of 30% active ingredient in water. The manufacturer reports that this antifoam has no biochemical oxygen demand. The temperature of the mixed liquors was controlled by maintaining the room temperature at a constant 26°C. Phosphorus addition was not necessary since the GPGP wastewater had a mean o-phosphate concentration of about 10 mg/L.

The pH of the feed to the nitrification reactors was not controlled, but was in the range of 7.0–8.0. The pH of the nitrification mixed liquor was maintained at 7.6 $\mp$ 0.2 by addition of alkali. Initially, potassium hydroxide was used for pH control but was later switched to sodium bicarbonate. Saturated solutions of these compounds were pumped into the mixed liquor by a peristaltic pump connected to a pH control system. The pH of the denitrification reactor was not continuously controlled; however, a 50% solution of phosphoric acid was added as needed to bring the pH to about 7.0.

Analyses were conducted in the UNDERC waste analysis laboratory using *Standard Methods* [7]. Ammonia analysis was by distillation/titration while analyses of nitrite, nitrate, and phosphate were by ion chromatography. Samples for all analyses, except solids, were filtered through 0.45 μm pore size filters and, thus, represent soluble portions. Solids were determined gravimetrically.

RESULTS

Wastewater Characterization

The composition of GPGP raw gasification condensate and the SGL are shown in Table I. The analysis of the raw wastewater is based on one sample while the SGL represents the mean of fourteen 6400 liter loads of fresh SGL.

Inspection of the analytical data on the SGL reveals that the pretreatment at GPGP removes about 99.6% of the phenol and 87.5% of the ammonia. Compounds in addition to phenols and ammonia are generally reduced in proportion to their volatility. One anomaly in these data is the large increase in phosphate concentrations after pretreatment. The phosphate source is the ammonia recovery (Phosam W) process which uses phosphoric acid.

The COD and TOC closures on the SGL, as estimated from the theoretical COD and TOC of identified compounds, was 44.7% and 41.7%, respectively. The unidentified fractions are composed of a large number of compounds present in amounts at or near the detection limit. Additionally, the dark color taken on by this wastewater after exposure to oxygen results from oxidation of catechol, with other constituents, to form a colored polymer. This polymer may account for as much as 32.8% of the TOC resulting in a TOC closure of 74.6% [8].

The chemical characterization of influent and effluent wastewaters to the nitrification and denitrification systems are shown in Table II. The nitrification influent COD concentration was 839 mg/L, and total organic carbon (TOC) was 154 mg/L. The BOD, however, was only 25 mg/L. This indicates that most of the COD and TOC are refractory carbon, not readily amenable to biodegradation. The influent also shows a high mean concentration of nitrite and nitrate. This is the result of nitrification in the organic removal activated sludge reactor. Overall nitrification in this reactor was minimal. However, during a period of very high SGL ammonia concentrations, the rate of nitrification in this reactor increased dramatically even though the solids retention time (SRT) was only 4 days.

The ammonia and nitrite concentrations in the nitrification effluents, shown in Table II, reflect upsets due to spikes in influent ammonia and are not representative of effluent ammonia concentrations achieved under typical influent ammonia concentrations. The mean effluent nitrite and nitrate concentrations in the DN reactor also reflect system upsets.

NITRIFICATION

Process Performance

As would be expected, the best performance was noted in the reactor with the longest SRT, and performance decreased with decreasing SRT. The nominal SRTs established and maintained by volumetric mixed liquor wasting were 80, 50, and 30 days. Actual SRTs based on a mass balance of solids, as calculated for steady-state operation, were 37, 21, and 12 days, respectively.

Figures 2 through 4 show the concentration of ammonia-N in the influent and effluent of the nitrification reactors. The curve for influent ammonia is the same in these figures since the influent is the same to all reactors. The influent ammonia reached concentrations of greater than 990 mg/L as NH_3-N whereas ammonia-N concentrations of about 400 mg/L were typical. The high observed

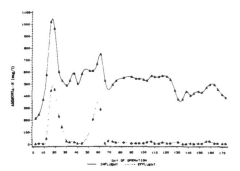

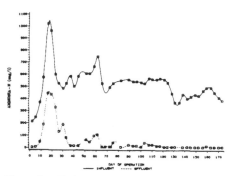

Figure 2. Variation in the ammonia-N concentrations in the influent and effluent of reactor N1 (solids retention time of 37 days).

Figure 3. Variation in the ammonia-N concentrations in the influent and effluent of reactor N2 (solids retention time of 21 days).

ammonia-N concentrations are the result of process upsets at GPGP. For example, in Figure 2, reactor N1 (SRT = 37 days) shows ammonia concentrations during two episodes well above otherwise low levels near a mean of 51 mg/L. These increases are due to failures in the pH adjustment system resulting in the addition of excessive amounts of potassium hydroxide and an increase in a mixed liquor pH of greater than 12.

In Figure 3, reactor N2 (SRT = 21 days) also shows a large increase in ammonia concentrations, beginning on day 6, over normal effluent concentrations. This increase was also due to a failure in the pH adjustment system. The mean effluent ammonia-N concentration for this reactor was 56 mg/L (Table II). Reactor N3 (SRT = 12 days) ammonia concentrations did not experience any pH adjustment problems, as shown in Figure 4. However, the effluent ammonia concentrations in this reactor showed increases that correlated with high concentrations of ammonia in the feed since this reactor did not stabilize with regard to effluent ammonia concentrations until about day 110 of operation. The mean effluent ammonia-N concentration in this reactor was 151 mg/L. After stabilization, the effluent ammonia-N concentration dropped to about 30 mg/L. The overall mean ammonia removal for these reactors was 90.4, 89.3, and 71.2%, respectively, for N1 through N3.

The mean nitrite-N concentration in the influent was 184 mg/L, reflecting the occurrence of nitrification in the organic removal reactor despite the 4 day SRT. Nevertheless, effluent nitrite-N concentrations were low throughout the operating period for all three nitrification reactors. The mean nitrite-N concentrations were 10, 2, and 25 mg/L for reactors N1, N2, and N3, respectively. Effluent nitrite-N was detectable only when reactor upset occurred such as during the pH adjustment problems.

Figures 5 through 7 show the variation in influent and effluent nitrate-N in reactors N1 through N3, respectively. Nitrate-N concentrations in the influent had a number of increases over normal concen-

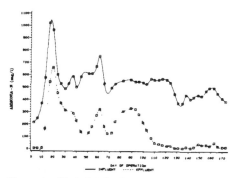

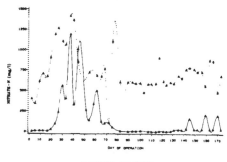

Figure 4. Variation in the ammonia-N concentrations in the influent and effluent of reactor N3 (solids retention time of 12 days).

Figure 5. Variation in the nitrate-N concentrations in the influent and effluent of reactor N1 (solids retention time of 37 days).

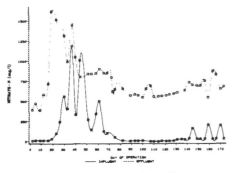

Figure 6. Variation in the nitrate-N concentrations in the influent and effluent of reactor N2 (solids retention time of 21 days).

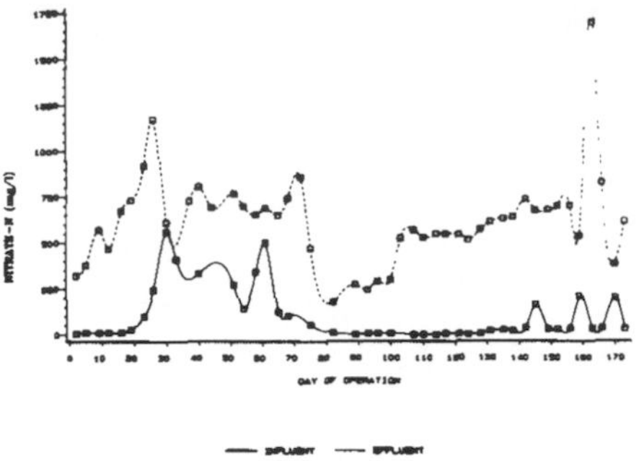

Figure 7. Variation in the nitrate-N concentrations in the influent and effluent of reactor N3 (solids retention time of 12 days).

trations. These increases are related to high concentrations of ammonia in the influent and are a result of nitrification occurring in the organic removal reactor. Under normal conditions nitrate-N concentrations in the effluent from reactors N1 through N3 were steady. The observed variability reflects sharp changes in influent ammonia and nitrate.

The mixed liquor suspended solids concentrations were used as indicators of biomass. Figure 8 shows the concentrations of volatile suspended solids from the mixed liquor of the three nitrification reactors. The solids concentrations in all three reactors show a dramatic decrease during periods of high influent ammonia concentrations, apparently resulting from inhibition. Reactors N1 and N2 recovered quickly after the upset but required considerable time to reach steady-state. The N1 and N3 reactors required approximately 60 days to reach steady-state while the N2 reactor reached steady-state after only 30 days.

The ratio of mixed liquor volatile suspended solids (MLVSS) to mixed liquor suspended solids (MLSS) varied according to SRT. The mean ratio of MLVSS to MLSS, expressed as percent, was 82.7, 88.0, and 91.5%, respectively, for N1 through N3. These differences are apparently due to the higher percentage of nonviable biomass associated with longer SRTs. The mean solids concentrations for the operating period are given in Table III.

Design Parameters

The kinetics of nitrification were calculated from data obtained during steady-state operation of the three reactors. Steady-state is defined, for these purposes, as a time when the mixed liquor solids and effluent nitrate concentrations were relatively stable and effluent nitrite was near zero. The ammonia removals for these reactors, for the period for which kinetic data were calculated, were 97.7%, 96.9%, and 87.4%, respectively, for N1 through N3. The determination of kinetics was complicated somewhat by the occurrence of nitrification in the organic removal activated sludge reactor which provided feed to the nitrification reactors. This was accounted for by converting incoming nitrite into ammonia-N equivalents based on their relative oxidation states.

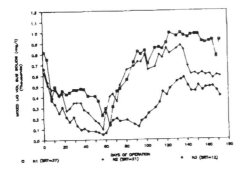

Figure 8. Variation in the mixed liquor volatile suspended solids (MLVSS) concentrations in reactors N1 through N3 (SRTs of 37, 21 and 12 days, respectively).

**Table III. Nitrification/Denitrification Reactor Performance Indicators —
Mixed Liquor Determination[a]**

Parameter	Mixed Liquor			
	N1	N2	N3	DN[b]
MLSS (mg/L)	821 ± 42	633 ± 34	353 ± 29	377 ± 27
MLVSS (mg/L)	679 ± 38	557 ± 32	323 ± 28	338 ± 26
Sludge Volume Index (ml/g MLSS)	59.8 ± 3.2	61.6 ± 2.3	79.4 ± 5.0	19.6 ± 5.2
Settleable solids (ml/L)	44.8 ± 1.7	37.1 ± 2.0	27.0 ± 1.8	9.0 ± 2.4
Oxygen Consumption (mg O_2/g MLVSS/hr)	19.6 ± 2.1	40.4 ± 5.1	66.2 ± 6.7	—

[a]Mean value plus or minus the standard error of the mean during the operating period.
[b]The mean values for DN exclude a period of upset during accidental usage of petroleum ether as a supplemental carbon source. N1 effluent is the influent to DN.

Equation 1 was used to calculate the reaction rate constant, K_1. This equation assumes a first order reaction and produces a line with a slope of K_1 when $(So-Se)/(\bar{x}\,\theta)$ is plotted against Se.

$$\frac{(So-Se)}{\bar{x}\,\theta} = K_1\,Se \qquad (1)$$

where, So = influent NH_3-N, mg/L
  Se = effluent NH_3-N, mg/L
  $\bar{x}$ = MLVSS, mg/L
  θ = hydraulic retention time, days
  K_1 = reaction rate constant, 1/mg-day.

A plot of steady state nitrification data according to Equation 1 yielded a line with a least squares correlation coefficient of 0.6984 providing a K_1 of 0.0033 1/mg-day.

Equation 2, the Hanes modification of the Michaelis-Menten equation was used to calculate the maximum specific substrate utilization rate, U_m, and the Michaelis-Menten constant, K_m, the point at which the reaction rate is half of the maximum value [9].

$$Se \Big/ \frac{(So-Se)}{\bar{x}\,\theta} = \frac{Se}{U_m} + \frac{K_m}{U_m} \qquad (2)$$

where, U_m = maximum specific substrate utilization rate, day^{-1}
  K_m = Michaelis-Menten constant, mg/L.

A plot of Equation 2, using the nitrification data yielded a least squares correlation coefficient of 0.8778 and a U_m of 0.3164 day^{-1} and K_m of 8.709 mg/l.

Equation 3 is used to obtain the biomass yield coefficient and the endogenous decay coefficient:

$$\frac{1}{\theta_c} = -K_d + Y_g \frac{(So-Se)}{\bar{x}\,\theta} \qquad (3)$$

where, θ_c = solids retention time, days
  Y_g = biomass yield coefficient, mg MLVSS/mg NH_3-N
  K_d = endogenous decay coefficient, day^{-1}

A plot of this equation using the nitrification data yielded a least squares correlation coefficient of 0.91 and provided a Y_g of 0.3434 mg MLVSS/mg NH_3-N and a K_d of 0.018 day^{-1}. Equation 4 is used to calculate the endogenous respiration coefficient and the oxygen coefficient:

$$O_R = K_e' + Y' \frac{(So-Se)}{\bar{x}\,\theta} \qquad (4)$$

where, O_R = specific oxygen uptake rate, mg O_2/mg MLVSS-day
  K_e' = endogenous respiration coefficient, day^{-1}
  Y' = oxygen coefficient, mg O_2/mg NH_3-N.

Table IV. Nitrification Kinetic Coefficients

Coefficient	Units	This Study	Others (Reference)
K_1	L/mg-day	0.0033	0.00052 [11]
U_m	day^{-1}	0.3164	0.3–3.0, 1.2, [12, 13]
K_m	mg/L	8.709	3.6, 2.86, 1.0 [13, 14, 15]
K_d	day^{-1}	0.018	0.12, 0.0038, [13, 16]
Y_g	mg MLVSS/mg NH$_3$-N	0.3434	0.11–0.32, 0.127 [12, 13]
K_e'	mg O$_2$/mg MLVSS/day	0.11[a]	—
Y'	mg O$_2$/mg NH$_3$-N	4.0658	4.33–4.57, 4.19[b] [6, 9]

[a]Determined experimentally.
[b]Theoretical.

The plot of Equation 4 using the date from the nitrification reactors yielded a least squares correlation coefficient of 0.719 and provided a Y' of 4.0658 mg O$_2$/mg NH$_3$-N and K_e' of –0.1276 mg O$_2$/mg MLVSS-day. The negative value from graphical determination probably resulted from differences in viability at different growth rates. As the SRT increased, the viable fraction of the MLVSS decreased, shifting the line and resulting in a negative K_e'.

The kinetic values obtained, shown in Table IV, generally agree with values found by other investigators. The only exception to this is the K_m where the calculated value of 8.709 is 2.4 times greater than the highest K_m found by Stratten and McCarty [10]. This may be because of the high influent ammonia concentrations and/or the fact that these reactors had a long period of acclimation and operation (total acclimation and operation of 12.5 months). Thus, a well acclimated biomass was obtained that had enzyme systems optimized for uptake at high substrate concentrations.

DENITRIFICATION

The sole purpose of the denitrification work was to demonstrate process feasibility with this wastewater. Therefore, the system was set up at a single SRT and hydraulic retention time (HRT) with the only variable being the influent. The influent for the denitrification reactor was clarified effluent from the N1 nitrification reactor. This reactor, which was operated at a 37 day SRT, produced the most stable effluent of the three nitrification systems. BOD and TOC analyses were not performed on the influent to and effluent from the denitrification reactor. However, as Table II indicates, the influent to the nitrification reactors had a mean BOD$_5$ of 24 mg/L. The BOD was probably reduced to even lower levels after passing through the nitrification reactor. Therefore, the only carbon and energy source for the denitrifying bacteria would be the supplementary organic materials.

Ammonia concentrations did not change across the denitrification reactor. Influent nitrite-N concentrations were low (mean of 10 mg/L) with one large excursion above the normally low concentrations. Variation in effluent nitrite-N concentrations were generally minimal with a mean effluent nitrite-N concentration of 6 mg/L. Two increases occurred in effluent nitrite-N concentrations where values were observed above the normally low levels. The first increase resulted from an upset in N1 [feed to the denitrification reactor (DN)] and the second from incomplete denitrification resulting from accidental usage of petroleum ether instead of methanol for a supplemental carbon source.

Figure 9 shows the nitrate-N concentrations over the operating period. There is a gap in effluent nitrate data corresponding to the period when petroleum ether was accidentally used as feed to the reactor. When the petroleum ether addition was discovered, methanol was reinstituted as feed (day 95). Effluent nitrate-N concentrations fell to near zero from day 100 through day 156. At about day 156, methanol addition was stopped to determine if sufficient organic matter was present to allow for denitrification without supplementation. Nitrate concentrations immediately increased to influent values. The overall mean effluent nitrate-N concentration, excluding the period of petroleum ether usage, was 61 mg/L. Under normal steady-state conditions the nitrate-N concentrations were generally much lower, below 1 mg/L with methanol supplementation.

The variation in the mixed liquor solids is shown in Figure 10. The periods when petroleum ether was added and when methanol supplementation was stopped are apparent in the changes in solids concentrations. It is interesting to note that the biomass survived the petroleum ether addition and recovered fairly quickly after returning to supplemental methanol feed. The volatile suspended solids (MLVSS) were 90% of the total suspended solids (MLSS) in the DN reactor.

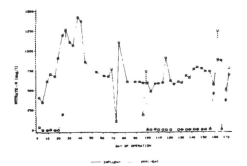

Figure 9.   Variation in the nitrate-N concentrations in the influent and effluent of the denitrification reactor.

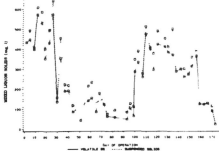

Figure 10.   Variation in the mixed liquor total suspended and volatile suspended solids concentrations in the denitrification reactor.

CONCLUSIONS

Great Plains Gasification Plant's (GPGP) stripped gas liquor, pretreated by biological oxidation, can be nitrified. Complete nitrification can be obtained at influent concentrations of greater than 500 mg/L as NH_3-N.

The highest removal rates and best system stability were found at the highest SRT. The low SRT reactor (12 days) did not attain stable operation until after approximately 110 days of operation. The ammonia removals during steady-state operation were 97.7%, 96.9%, and 96.9%, respectively, for reactors N1 through N3 (SRTs of 37, 21, and 12 days).

Influent ammonia-N spikes up to 1020 mg/L were observed. At these high levels, inhibition was observed at all three SRTs. However, when concentrations fell to typical values (500 mg/l NH_3-N), the systems recovered rapidly.

Denitrification of biologically treated and nitrified GPGP SGL was complete in an activated sludge reactor using a 4 day HRT, 10 day SRT, and a 2.47:1 molar ratio of methanol to NO_3-N as the supplemental carbon source.

Nitrate-N concentrations in the influent greater than 750 mg/L were completely denitrified demonstrating a maximum observed specific removal rate of 92.8 mg NO_3-N/mg MLVSS/day.

ACKNOWLEDGMENT

This research was funded by U.S. Department of Energy Cooperative Agreement DE-FC21-83FE60181.

DISCLAIMER

This report was prepared as an account of work sponsored by the United States Government. Neither the United States nor any agency thereof, nor any of their employees, makes any warranty, express or implied, or assumes any legal liability or responsibility for the accuracy, completeness, or usefulness of any information, apparatus, product, or process disclosed, or represents that its use would not infringe privately owned rights. Reference herein to any specific commercial product, process or service by trade name, mark, manufacturer, or otherwise does not necessarily constitute or imply its endorsement, recommendation, or favoring by the United States Government or any agency thereof. The views and opinions of authors expressed herein do not necessarily state or reflect those of the United States Government or any agency thereof.

PATENT STATUS

This technical report is being transmitted in advance of DOE patent clearance, and no further dissemination of publication will be made of the report without prior approval of the DOE Patent Counsel.

TECHNICAL STATUS

This technical report is being transmitted in advance of DOE review, and no further dissemination or publication will be made of the report without prior approval of the DOE Project/Program Manager.

REFERENCES

1. Galegher, S. J., M. D. Mann, and M. D. Johnson, "Evaluation of Treated Gasification Wastewater as Cooling Tower Makeup," Presented at 13th Bienniel Lignite Symposium, Bismarck, ND, DOE/FE/60181-120 (1985).
2. ANG Coal Gasification Company, "A Wastewater Treatment System for a Lurgi Coal Gasification Plant," *DOE/RA/20225-T3* (1982).
3. Gallagher, J. R., and G. G. Mayer, "Process Performance of Pilot-Scale Activated Sludge Treatment of Pretreated Coal Gasification Wastewater," *Proc. 40th Annual Purdue Indust. Waste Conf.*, Purdue, Univ., West Lafayette, IN (1985).
4. Blaszczyk, M., E. Galka, E. Sakowicz, and R. Mycielski, "Denitrification of High Concentrations of Nitrites and Nitrates in Synthetic Medium with Different Sources of Organic Carbon. III, Methanol," *Acta Microbiologica Polonica*, 34:195-206 (1985).
5. Focht, D. D., and A. C. Chang, "Nitrification and Denitrification Processes Related to Waste Water Treatment," *Adv. Appl. Microbiol.*, 19:153-186 (1975).
6. U.S. EPA, *Process Design Manual for Nitrogen Control* (1975).
7. American Public Health Association, *Standard Methods for the Examination of Water and Wastewater*, 15th ed., APHA, Washington, D.C. (1985).
8. Wiltsee, G. A., "Low Rank Coal Research," *UNDERC Quarterly Technical Progress Report* (April-June 1986).
9. Grady, C. P. L., and H. C. Lim, *Biological Wastewater Treatment: Theory and Applications*, Marcel Dekker, New York (1980).
10. Stratton, F. F., and P. L. McCarty, "Prediction of Nitrification Effects on the Dissolved Oxygen Balance of Streams," *Environ. Sci. Technol.*, 1:405-410 (1967).
11. Adams, C. E., Jr., "Treatment of a High-Strength Phenolic and Ammonia Wastestream by Single and Multi-Stage Activated Sludge Processes," *Purdue Indust. Waste Conf.*, 29:617-630 (1974).
12. Metcalf and Eddy, Inc., *Wastewater Engineering: Treatment, Disposal, Reuse*, 2nd ed, McGraw-Hill, New York (1979).
13. Reynolds, T. D., *Unit Operations and Processes in Environmental Engineering*, Brooks/Cole Engineering, Monterey, CA (1982).
14. Rozich, A. F., and D. J. Castens, "Inhibition Kinetics of Nitrification in Continuous Flow Reactors," *J. Water Poll. Control Fed.*, 58:220-226 (1986).
15. Downing, A. L., *et al.*, "Nitrification in the Activated Sludge Process," *J. Instit. Sewage Purif.*, 130-158 (1964).
16. Poduska, R. A. and J. F. Andrews, "Dynamics of Nitrification in the Activated Sludge Process," *Purdue Indust. Waste Conf.*, 29:1005-1025 (1975).

68 USING ROTATING BIOLOGICAL CONTACTORS FOR THE TREATMENT OF COAL GASIFICATION WASTEWATERS

Charles D. Turner, Associate Professor
Department of Civil Engineering
University of North Dakota
Grand Forks, North Dakota 58202

Kevin Wernberg, Project Engineer
Widseth, Smith & Nolting Consulting Engineers
Brainerd, Minnesota 56401

INTRODUCTION

The University of North Dakota Energy Research Center (UNDERC) is involved in an extensive low-rank coal research program. This research includes basic coal science and topics relating to coal combustion and conversion processes. UNDERC is conducting studies on the treatment of coal-derived wastewaters as part of the coal conversion research program. Wastewaters for these tests were generated by two sources: UNDERC's slagging, fixed-bed, pilot scale gasifier operated on Indian Head lignite from North Dakota and the Lurgi Mark IV gasifiers at the Great Plains (GP) gasification plant located north of Beulah, North Dakota. Raw gas condensate from these processes is contaminated with dissolved organics and ammonia. This liquor is not suitable for either discharge or direct reuse within a coal conversion system and must be pretreated prior to biological processing.

OBJECTIVE

The objective of this research was to determine the treatability of UNDERC's and Great Plains' coal gasification wastewaters using a bench scale four-stage rotating biological contactor (RBC). The treatability testing included an evaluation of organic removal rates in the first stage and the overall rates in the last three stages using the Stover-Kincannon model. Nitrification was evaluated at various loading rates.

BACKGROUND

The feed wastewaters used in this research were first pretreated by a series of physical/chemical processes. Wastewater condensate from the UNDERC gasifier was treated using oil/water separation followed by filtration to remove tars, solvent extraction to remove dissolved organics, and steam stripping to reduce levels of ammonia and acid gases. The Great Plains raw gas liquor was treated by tar oil water separation followed by the Phenosolvan process to remove phenolic compounds and then by the Phosam process reduce ammonia concentrations. Following these treatment schemes, the wastewater is termed stripped gas liquor (SGL).

The SGL was then frozen and stored for later use. Tables I and II present the characteristics of each SGL.

A comparison of the two wastewaters shows that analytical differences in organic compounds exerting biochemical oxygen demand (BOD) are due to the 99 mg/L of phenol and 148 mg/L of thiocyanate in the UNDERC wastewater while the GP SGL contained 14 mg/L and 8 mg/L of the same compounds. The UNDERC wastewater contained 27 mg/L of cyanide as compared to 5 mg/L for the GP wastewater. The mean BOD_5 of the UNDERC wastewater was, however, only 690 mg/L while the mean BOD_5 for GP SGL was 1300 mg/L. All BOD_5 analyses were run on filtered samples.

EQUIPMENT

The four-stage RBC system was constructed of plexiglass and PVC pipe. Each stage contained 39 plexiglass discs 4.83 inches in diameter and 0.125 inches thick. Discs in the first stage were separated

577

Table I. UNDERC SGL Wastewater Characteristics

Parameter/ Constituent	Mean mg/L	Range mg/L	Std. Deviation mg/L
BOD$_5$	690	360 to 999	145
COD	3,570	2,720 to 4,580	358
TOC	1,149	950 to 1,300	82
Ammonia	552	360 to 640	57
Phenol	99	60 to 130	17
Alcohols	216	170 to 250	24
Fatty Acids	914	–	–
Total Cyanide	27	20 to 50	9
Thiocyanate	148	100 to 200	20
Phosphate[a]	25	10 to 40	8
Sulfate	55	20 to 70	15
Alkalinity	1,338	1,020 to 1,550	161
pH	8.9	8.5 to 9.3	0.2
5-5-Dimethyl Hydantoin	825	680 to 950	87
5-Ethyl-5-Methyl Hydantoin	277	230 to 310	26

[a]Added as phosphoric acid because UNDERC SGL is phosphorus deficient.

by 0.375 inches while those in the remaining stages were separated by 0.25 inches. The total disc surface area per stage was 10.5 square feet. The discs were mounted on a 0.375 inch diameter threaded rod. The shafts were rotated using a modified variable speed pump drive. Each stage was covered with a plexiglass lid to minimize evaporation losses.

A schematic of the RBC bench scale model is shown in Figure 1. Each stage was divided into three compartments with each compartment being fed individually. Flow between stages was by gravity. Rotation speed was maintained at 6.5 revolutions per minute. Inlet and outlet ports were offset to minimize shortcircuiting. The outlet ports were set to maintain a 40% submergence of the discs in each stage. Effluent from the outlet ports on the fourth stage was pumped to a clarifier. Effluent from the clarifier was collected in a polyethylene container for analyses.

A dye test was run on the system prior to operation with wastewater to determine mixing efficiency. An 11.6 mg/L solution of methylene blue was placed into one cell of stage 1 while the remaining cells and stages were filled with distilled water. Distilled water was pumped into the test cell of stage one at a rate of 2.4 gpd (9.08 L/d). The hydraulic detention times for stages 1, 2, 3, and 4 were calculated as 2.3, 1.9, 1.9, and 2.9 hours, respectively. The dye test results are shown in Figure 2. The curves represent a decay pattern characteristic of a series of complete mix units operated in series.

Table II. Great Plains SGL Wastewater Characteristics

Parameter/ Constituent	Mean mg/L	Range mg/L	Std. Deviation mg/L
BOD$_5$	1,300	960 to 1,600	208
COD	2,785	2,300 to 3,280	365
TOC	853	608 to 1,280	160
Ammonia	757	580 to 864	70
Phenol	14	5 to 26	6
Alcohols	297	216 to 385	45
Fatty Acids	707	–	–
Total Cyanide	5	1 to 10	5
Thiocyanate	8	4 to 10	2
Hydantoins	<10	–	–
Phosphate	26	0 to 93	30
Sulfate	35	20 to 50	9
pH	9.2	8.9 to 10.0	0.3

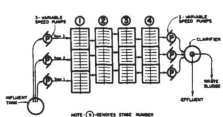

Figure 1. Schematic of four-stage, bench-scale RBC.

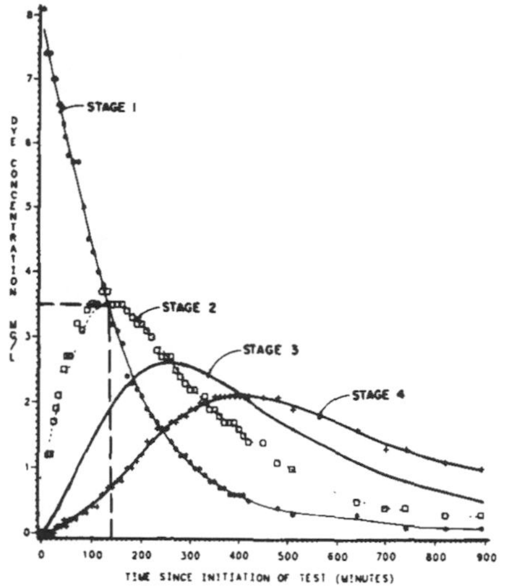

Figure 2. Dye test results from four-stage, bench-scale RBC.

PROCEDURE

Each wastewater was applied to the RBC at three separate loading rates. The UNDERC SGL was the first wastewater tested. The hydraulic and organic loading rates are given for both wastewaters in Table III. Earlier research using UNDERC SGL had shown that loading rates in the 2 lbs/d/1000 ft² rage were at a level where removals were near a maximum [1]. Therefore, loadings were started at a BOD₅ loading rate of 0.58 lbs/d/1000 ft² which corresponded to 1.0 gallon per day feed rate to the RBC. The first loading rate for the UNDERC SGL was continued for 120 days while the second and third BOD₅ loading rates of 0.79 and 1.09 lbs/d/1000 ft² were continued for 18 and 16 days, respectively.

Upon completion of testing with the UNDERC SGL, the RBC was disassembled and cleaned. All biomass was removed from the discs and weighed. Startup with the GP SGL as feed was initiated using seed biomass from the UNDERC SGL RBC operation.

The range of the GP SGL loading rates was increased in response to the results from the UNDERC SGL loading rates. The GP SGL rates were 1.0, 2.0, and 3.0 gallons per day which corresponds to BOD₅ loading rates of 1.12, 1.88, and 2.61 lbs/d/1000 ft². These rates were maintained for 70, 118, and 39 days, respectively.

Table III. RBC Loading Rates

Feed Rate gal/d	Hydraulic Load gal/d/1000 ft²	BOD₅ lbs/d/1000 ft²	COD lbs/d/1000 ft²	TOC lbs/d/1000 ft²
		UNDER SGL		
1.0	95	0.58	2.77	0.89
1.5	143	0.79	4.66	1.43
2.0	191	1.09	5.75	1.86
		GP SGL		
1.0	98	1.12	2.51	0.78
2.0	196	1.88	3.94	1.09
3.0	293	2.61a	5.81	1.54

aValue based on TOC/BOD ratio.

Table IV.   Primary Constituents and Methods of Analysis

Chemical Constituent or Physical Parameter	Analytical Method Used
Soluble 5-day Biochemical (BOD$_5$)	Standard Methods, Sect. 507 Oxygen Demand
Filtered Chemical Oxygen Demand (COD)	Standard Methods, Sect. 508C
Total Organic Carbon (TOC)	Standard Methods, Sect. 505A
Total Ammonia (NH$_3$-N)	Standard Methods, Sect. 417 A & D
Nitrate (NO$_3$)	Standard Methods, Sect. 429
Nitrite (NO$_2$)	Standard Methods, Sect. 429
Phosphate (PO$_4$)	Standard Methods, Sect. 429
Sulfate (SO$_4$)	Standard Methods, Sect. 429
Alkalinity (CaCO$_3$ to pH 4.5)	Standard Methods, Sect. 403
Cyanide	Standard Methods, Sect. 412
Thiocyanate	Standard Methods, Sect. 429
Fatty Acids	Ion Chromatography
Phenols	Capillary Gas Chromatography
Hydantoins	Capillary Gas Chromatography
Alcohols	Capillary Gas Chromatography

Note: *Standard Methods* [3].

Daily maintenance and operational procedures consisted of replenishing the feed, effluent disposal, pH and dissolved oxygen measurements, and sample collection. Grab samples from the individual stages were directly pipetted from the basins at the effluent ports. One third of the total sample was obtained from each of the three cells comprising a single stage. TOC samples were examined five days a week for stages one and four. All other feed and effluent parameters were examined twice weekly during the course of the study. Table IV is a listing of the main constituents analyzed and methods used in quantifying them.

TOTAL ORGANIC LOADING METHOD

The total organic loading approach, as presented by Stover and Kincannon, was applied to this research [2]. The assumption of the first order removal kinetics in multiple-stage RBCs requires two removal rate constants. One constant is developed for the first stage of the system where the majority of the carbonaceous substrate is removed, and the other constant is developed for the remaining stages. Assumptions of second order kinetics, while requiring only one constant, cannot predict substrate removals at or beyond the point of substrate saturation and/or oxygen limitation where zero order removal occurs.

The total organic loading approach is capable of predicting substrate removals and treatment efficiencies over a wide range of loading conditions regardless of the kinetics displayed by the system. Using the total organic loading approach, graphs are plotted relating the total amount of organics applied to the system to the total amount of organics removed by the system in terms of BOD$_5$, COD, or TOC per day per 1000 square feet of disc surface area.

As the applied load is increased, a point is reached where the amount removed is at a maximum due to substrate saturation and/or oxygen limitation. At this point the system is functioning at its maximum removal rate, and no further increase will be achieved by increasing the applied loadings. Stover and Kincannon then fit a hyperbolic function, similar to the "Monod equation," to approximate the relationship.

$$L_R = L_R max \, Lo/(Ks + Lo) \tag{1}$$

where:

Lo = Applied loading in lb/d/1000 ft^2
L_R = Amount removed in lb/d/1000 ft^2
$L_R max$ = Maximum amount removed in lb/d/1000 ft^2
Ks = Applied loading rate at which the removal rate is one-half the maximum rate, or the saturation constant (in lb/d/1000 ft^2)

A relationship of the form y = mx + b may be developed by rearrangement of the above equation as follows:

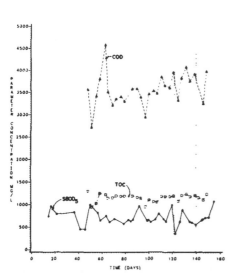

Figure 3. BOD, COD, and TOC variations in UNDERC SGL feed.

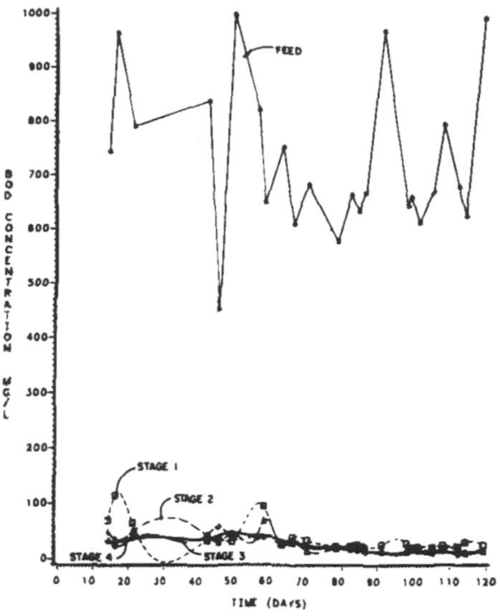

Figure 4. BOD concentrations for feed rate of 95 gal/d/1000 ft².

$$\frac{1}{L_R} = \frac{Ks}{L_Rmax} \times \frac{1}{Lo} + \frac{1}{L_Rmax} \tag{2}$$

The kinetic constants L_Rmax and Ks are determined by plotting the reciprocal of the applied loading against the reciprocal of the amount removed using the slope and y-intercept values. The y-intercept equals $1/L_Rmax$ as $1/Lo$ approaches zero, and the slope is equal to Ks/L_Rmax.

UNDERC SGL RESULTS

The UNDERC SGL feed variation in terms of BOD_5, COD, and TOC is shown in Figure 3. COD had the largest variation with a range of 2720 mg/L to 4580 mg/L and a standard deviation of 358 mg/L. The BOD_5 varied between 360 mg/L and 1000 mg/L with a mean of 690 mg/L and a standard deviation of 145 mg/L. The mean TOC was 1150 mg/L with a standard deviation of 82 mg/L.

Growth on the first stage was dark brown in color with a weight range varying from 1050 grams at the 95 gal/d/1000 ft² to 1510 grams wet weight at the 191 gal/d/1000 ft² loading. By contrast the second stage weights varied from 40 to 70 grams for the same loading rates. The weight of the biomass on stage three and four was 10 grams or less. Bridging by biomass on the discs was not encountered on any of the stages.

The performance of the system at the 0.58 and 1.09 lbs/d/1000 ft² BOD_5 loading rates is shown in Figures 4 and 5. Effluent BOD_5 concentrations from the first stage were 24, 34, and 48 mg/L for the 0.58, 0.79, and 1.09 lbs/d/1000 ft² loading rates. This amounts to BOD_5 reductions of 98%, 95%, and 93% for the first stage. All BOD_5 concentrations for the fourth stage effluent were less than 20 mg/L.

COD concentrations for the system are shown in Figures 6 and 7 for the 2.77 and 4.66 lbs/d/1000 ft² COD loading rates. Stage 1 removed 0.92 and 1.78 lbs COD/d/1000 ft² at these loading rates while at the highest loading rate (See Table V) 1.66 lbs/d/1000 ft² were removed. The COD not removed was due principally to the hydantoins in the wastewater which are very poorly biodegradable [4]. Stage 4 effluent COD concentrations were often slightly higher than the stage 1 effluent concentrations. Virtually all COD removal was accomplished in stage 1.

Nitrification began in stage 4 after 100 days of operation. Figure 8 shows the ammonia concentrations in the feed and the ammonia, nitrite, and nitrate concentrations for stage 4 at the highest loading rate of 191 gal/d/1000 ft². Thiocyanate was degraded to ammonia in stage 1 which increased the ammonia effluent concentration entering stage 2 to nearly 700 mg/L. Effluent ammonia from stage 4

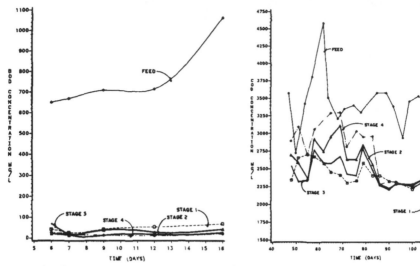

Figure 5.   BOD concentrations for feed rate of   Figure 6.   COD concentrations for feed rate of
191 gal/d/1000/ft².   95 gal/d/1000/ft².

contained less than 400 mg/L near the end of the test period. The nitrate concentration rose to over
700 mg/L while the conversion of nitrites to nitrate had not begun after 155 days of operation.

GREAT PLAINS SGL RESULT

The Great Plains SGL feed concentrations are shown in Table V. Mean BOD_5 concentrations varied
between 1070 and 1375 mg/L.

A heavy buildup of dark brown/black biomass accumulated on the discs during Great Plains SGL
testing. The 1.88 and 2.61 lb BOD_5/d/1000 ft² loading rates generated large quantities of sloughed
material from stage 1 and smaller amounts from stage 2. Stage 3 and 4 had very light biomass

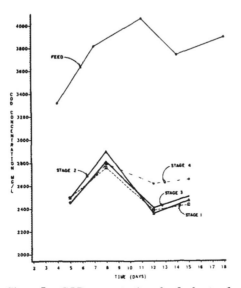

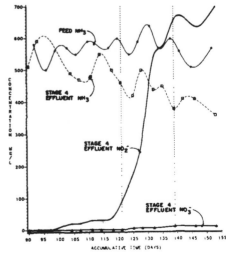

Figure 7.   COD concentrations for feed rate of   Figure 8.   Concentrations of NH_3, NO_2^-, and
143 gal/d/1000 ft². NO_3^- in feed and stage 4 effluent.

Table V. Great Plains SGL Feed Characteristics

Hydraulic Loading gal/d/1000 ft²	BOD₅ mg/L	SDᵇ mg/L	COD mg/L	SDᵇ mg/L	TOC mg/L	SDᵇ mg/L
98	1,375	176	3,064	155	949	123
196	1,150	174	2,410	43	669	28
293	1,070ᵃ	—	2,353	78	622	17

ᵃBOD₅ value is based on TOC/BOD₅ ratio.
ᵇSD—standard deviation.

accumulation similar to that encountered when treating the UNDERC SGL. The dissolved oxygen concentration in stage 1 was 1.0 mg/L or less when loading at the 1.88 lb BOD₅/d/1000 ft² rate while in stage 2 the dissolved oxygen was always greater than 4.0 mg/L. The 2.61 lbs BOD/d/1000 ft² loading rate resulted in dissolved oxygen concentrations less than 0.5 mg/L in stage 1 with odors characteristic of anaerobic conditions.

Figure 9 illustrates the mean BOD₅ concentrations for the feed and effluent from each stage operating under the three loading rates. BOD₅ removals were 93% or better through the first stage for each of the three loading rates. The mean effluent BOD₅ values from stage 1 were 80, 83, and 52 mg/L for the lowest to highest loading rate. Mean effluent BOD₅ concentrations from stage 4 were 13, 16, and 40 mg/L for each successively higher loading rate.

COD concentrations for the feed GP SGL and effluent COD values for each of the four stages at the three loading rates are shown in Figure 10. COD removal rates in stage 1 were 66, 69, and 73% for the lowest to highest loading rates, respectively. Stage 4 effluents were similar in that increasing loading rates gave increasing removals of 69, 73, and 76%. Respective stage 1 mean COD effluent concentrations were 1040, 756, and 630 mg/L.

Nitrification occurred at all three loading rates using the Great Plains SGL. Figure 11 shows the ammonia nitrogen and nitrate nitrogen concentrations for the influent wastewater and the effluent concentrations for each of the four stages at the intermediate loading rate. The ammonia nitrogen concentration was reduced from 654 mg/L in the feed to 235 mg/L in the effluent from stage 4. The stage 4 effluent ammonia nitrogen concentrations at the low and high loading rates were 85 and 363 mg/L, respectively.

KINETICS

The total organic loading method of Stover and Kincannon was used to generate the kinetic constants shown in Table VI [2] The method fit the results of this study very well. For BOD₅, the saturation constant, K_s, and maximum removal rate constant, $L_R max$, were both significantly higher for the Great Plains SGL than for UNDERC SGL. This result is consistent with the makeup of the wastewaters since the UNDERC SGL has much higher concentrations of cyanide, thiocyanate, phenol and hydantoins which are virtually absent in the Great Plains SGL.

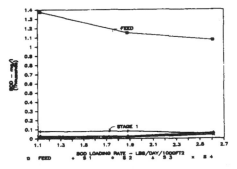

Figure 9. BOD loading and removal for RBC treating GP SGL.

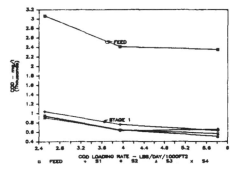

Figure 10. COD loading and removal for RBC treating GP SGL.

Table VI. Summary of Stage 1 Kinetic Constants for Bench — Scale Treatment of SGL

Feed SGL	BOD$_5$		COD		TOC	
	Ks	L$_R$Max	Ks	L$_R$Max	Ks	L$_R$Max
UNDERC	12.5	12.3	20.0	56.0	3.7	11.7
GP	30.0	30.8	−20.0	−32.8	−40.0	−52.6

A comparative plot of the graphs for determining the COD constants is shown in Figure 12. Note that the slope of the UNDERC SGL plot is steeper than that for the Great Plains SGL. As the slope approaches 45°, the amount of COD removed approaches the amount of COD applied. Since the loading rates and removal rates are plotted as their reciprocals, the plot with the least slope has a higher removal rate. The slope (Ks/L$_R$max) for the UNDERC SGL is 2.80 and for the GP SGL it is 1.64.

Negative kinetic constants for COD and TOC were caused by the relatively steep slope of the fitted line for the loading rates tested. If higher loading rates had been used, the plots would have become flatter and possibly curved upwards as they neared the vertical axis indicating that removal rates had reached a maximum and might decrease due to toxicity and/or oxygen transfer limitation.

Kinetic constants were also evaluated for UNDERC SGL by treating stages 2, 3, and 4 as one unit. The L$_R$max was -0.07 lb BOD/d/1000 ft^2, and Ks was -0.19 lb BOD/d/1000 ft^2. These values are based on the removal of an additional 5% of the incoming BOD$_5$ in the last three stages. Very little COD and TOC were removed after the first stage, and removal was erratic at times with apparent slight increases as the wastewater passed through successive stages.

CONCLUSIONS

Stage 1 accounted for most of the removal of alcohols, fatty acids, phenol, and thiocyanate from both UNDERC SGL and for alcohols and fatty acid removal from the GP SGL. The 2, 3, and 4 stages accomplished very little additional organic removal in either system. Biodegradable organic removals remained high in the first stage of the GP SGL test run despite anaerobic conditions in the first stage.

Stage 1 BOD$_5$ removals were over 93% for all loading rates for both UNDERC SGL and GP SGL. These high removal rates would likely be maintained during scaleup at loading rates less than 1.5-2.0 lbs BOD/d/1000 ft^2 based on work done by Brenner [5] and Stover and Kincannon [2]. At loading rates of 2.0 lbs BOD/d/1000 ft^2 or higher, removals have been lower for prototypes than for bench and pilot scale models treating other wastewaters. Both UNDERC SGL and GP SGL tend to polymerize organics and precipitate these organics as solids. This polymerization phenomena appeared to occur with the bench scale RBC testing at higher loading rates thereby increasing removal rates. Further investigation into the degree of polymerization occurring in the RBC may be merited.

Nitrification occurred in stages 2, 3, and 4 during the treatment of both the UNDERC SGL and the GP SGL. Nitrification did not begin until after 100 days of operation when treating the UNDERC SGL and removals never reached 50%. Nitrification was 88% complete under the low loading rate for

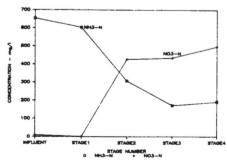

Figure 11. RBC system nitrification for the treatment of GP SGL at a loading rate of 196 gal/d/1000 ft^2.

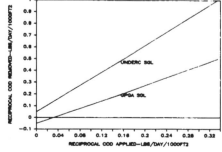

Figure 12. COD kinetic development for bench-scale RBC treating UNDERC SGL and GP SGL.

GP SGL with 70 mg/L of ammonia nitrogen remaining from the 735 mg/L in the influent. Nitrification decreased rapidly, however, with increasing loading rates. Only 70 and 51% of the ammonia was removed at the next two higher loading rates, respectively. Given the large surface area required and the sensitivity to loading, RBCs do not appear to be well suited for nitrification of these wastewaters.

Earlier work with two-stage activated sludge by Turner and Diepolder had shown that hydantoins could be biologically degraded during nitrification [6]. Hydantoins were not degraded during nitrification of UNDERC SGL by the bench scale RBC. Since the nitrification process was sensitive to loading, hydantoin removal should not be expected when using the RBC process for treating UNDERC SGL.

ACKNOWLEDGMENT

This research was funded by U.S. Department of Energy Cooperative Agreement DE-FC21-83FE60181.

REFERENCES

1. Strain, John, and Charles Turner, "Use of Rotating Biological Contactors for Treating Coal Gasification Wastewater," *Proceedings of the North Dakota Academy of Science*, Vol. 38 (April 1984).
2. Stover, Enos L., and Don F. Kincannon, "Rotating Biological Scale-Up and Design," *Water/ Engineering and Management Reference Handbook* (1982).
3. *Standard Methods for the Examination of Water and Wastewater*, 16th edition, American Public Health Association, Washington, D.C. (1985).
4. Aaslestad, H. G., and A. D. Larson, "Bacterial Metabolism of 2-Methylalanine," *Journal of Bacteriology*, Vol. 88, No. 5 (Nov. 1964).
5. Brenner, Richard C., et al., "Design Information on Rotating Biological Contactors," Municipal Environmental Laboratory, Cincinnati, OH., U.S. Environmental Protection Agency, *600/52-84-106* (July 1984).
6. Turner, Charles D., Paula Diepolder, and John Strain, "Biological Nitrification and Hydantoin Removal in Coal Gasification Wastewater," *Proceedings of the 40th Industrial Waste Conference*, Purdue University, Butterworth Pub., Stoneham, MA (1985).

69 PILOT STUDY OF UPGRADING OF EXISTING COKE OVEN WASTE TREATMENT FACILITY WITH TRICKLING FILTER

John P. Zurchin, Project Engineer

Meint Olthof, General Manager

John J. Schubert, Project Manager
Duncan, Lagnese and Associates, Inc.
Pittsburgh, Pennsylvania 15237

David Peck, Project Manager
Eichleay Engineers, Inc.
Pittsburgh, Pennsylvania 15206

Jerry Penrose, Senior Development Engineer
LTV Steel Company
Cleveland, Ohio 44101

INTRODUCTION

The coke plant wastewaters at LTV's Pittsburgh Works are treated by a dephenolization plant and a free and fixed leg ammonia still. This combination of physical/chemical treatment results in relatively low ammonium and phenol concentrations in the final effluent. However, the concentrations are not low enough to be in compliance with the Best Available Technology (BAT) limits.

Rather than applying EPA's standard technology for upgrading coke oven wastewater treatment plants, which is second stage biological treatment by activated sludge, LTV proposed to apply a fixed film biological treatment system. This technology was expected to require less space, result in lower capital and operating costs and also in a more reliable operation. As a backup to the trickling filter, a parallel study was also conducted on a bench-scale activated sludge unit. This paper presents the findings of the pilot study.

EXISTING FACILITIES

The present generalized arrangement for treating all wastewaters from the byproduct plant and the coke oven batteries at Pittsburgh Works is shown in Figure 1. The mixed and equalized waste is first fed through the free leg of the ammonia still. In this operation, the free ammonia and free cyanide is removed by steam stripping. The waste is then pumped through the dephenolization plant. This unit operation uses solvent extraction to remove phenol and other organics from the wastestream. The phenols are recovered and mixed with tar for sale as a byproduct. The solvent is recycled, and a small amount of makeup solvent is added to the process. Following the dephenolizer, the waste is discharged to the fixed leg of the ammonia still. Caustic is added to raise the pH so that the ammonium salt will be converted to the free form. By using steam stripping, ammonia is then driven off in the gas stream.

PROPOSED TREATMENT TECHNOLOGY

In order to meet the new BAT guidelines, it was proposed to add a fixed film biological growth reactor to remove phenol and ammonia. The proposed treatment scheme was a tall trickling filter with plastic medium. The trickling filter is a widely used unit operation for a variety of industrial waste streams. However, trickling filters have not been used in the steel industry. Generally, advantages of this treatment technology are:

1. Because of the low organic load and therefore the low sludge production, it would be difficult to maintain a healthy biomass in a suspended growth system.

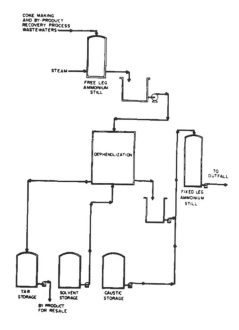

Figure 1.   Schematic of wastewater flow at the
Pittsburgh Works by-product department.

2. In fixed film biological reactors there is a better chance for bacteria to develop which degrade specific priority pollutants.
3. Because of the low sludge growth, no clarifier would be required.
4. The space limitations at the plants are severe. With trickling filters it is possible to provide tall towers while with activated sludge the maximum depth of a basin is limited.
5. The operation of the trickling filter is relatively easy while operating an activated sludge plant can be quite delicate.

Trickling filters have been used extensively and are successfully operated for a variety of waste streams. However, in the steel industry the activated sludge process is predominant. A large number of studies have been conducted on applying activated sludge treatment to this waste stream and there are also numerous installations in operation. References [1-19] present results on activated sludge plants treating coke plant wastewaters. In general, the articles support the site specific nature of the treatability of coke plant wastewaters due to the variability of wastewater constituents resulting from the variations of coke and byproduct plant operations.

Phenol removal in all of the articles reviewed was in the range of 90 to 100%. Nitrification was not as consistent. References 2, 4, 12, 13, 16, 17, and 19 discuss activated sludge systems that nitrified. References 1, 6, and 10 discuss activated sludge systems that did not nitrify, and references 7 and 18 discuss activated sludge systems where the effluent ammonia concentration was greater than the influent concentration. The stability of the process in each of the studies that achieved nitrification is difficult to ascertain since many were batch tests and short-term bench or pilot plants. Reference 12 discussed some of the factors that may be possible causes of biological nitrification process instabilities observed in coke plant wastewater treatment facilities.

Table I presents the typical influent concentrations to existing coke plant wastewater biological treatment facilities. The LTV Pittsburgh Works Coke Plant wastewaters fall within the range of typical plant influents with the exception of the phenol concentration, which is considerably lower due to the operation of a solvent extraction dephenolization plant. The phenol concentration at the Pittsburgh Works generally falls between 0.5 and 5.0 mg/L.

Generally, coke oven wastewaters are biologically treatable. The activated sludge process depends to a very large extent for successful operation on a final clarifier with attendant sludge settling and recycle. Clarifier upsets are regular, resulting in treatment plant upsets. In case of biomass washout, the performance of the treatment operation will be greatly reduced. For example, in LTV's particular

Table I. Typical Coke Plant Biological Treatment Facility Influent Wastewater Characterization[a]

Parameters	Typical Mean[b]	LTV Pittsburgh
pH	8.5	11 to 12
Phenol mg/L	330	4
Cyanide, T mg/L	5	9
Ammonia-N, mg/L	50	45

[a]From Coke Plant Wastewater Biological Treatment Facility—Cost Comparisons of Alternate Components, Association of Iron and Steel Engineers (January 1984).
Concentrations are after dilution to a mean value of 54% of the initial strength.
[b]Undiluted and prior to neutralization.

application, it is estimated that the amount of excess biological sludge formed during the biological degradation would amount to only 30 to 50 mg/L due to the very low organic content of the wastewater. Therefore, if the overflow from the clarifier contained more than that amount of solids, it would be impossible to sustain an active biomass. Therefore, only by using a sand filter after the final clarifier could the biomass be retained for an activated sludge system. In a trickling filter, however, the biomass would be retained on the medium and only excess cells would be washed away from the system. Therefore, a trickling filter would not depend on a clarifier to retain the biomass. In this particular case, because of the low solids growth, it would even be possible to operate without a final clarifier since the solids concentration would be well below the BAT limit.

Another major advantage of a trickling filter would be the ease of operation. The only mechanical parts of the process are recycle pumps. Therefore, the full scale system would be on line for a very high percentage of the time. Since the process would not depend on clarifier performance, the likelihood of process upsets would be also greatly reduced.

DESCRIPTION OF PILOT PLANT

A flow schematic of the pilot trickling filter is shown in Figure 2. The feed to the trickling filter was a bleed stream of the fixed ammonia still effluent. The wastewater has a temperature of approximately 100°C and a pH greater than 12 when it exits the ammonia still. A heat exchanger was required to maintain the wastewater temperature in the trickling filter at temperatures compatible with biological treatment (20–30°C), and a pH controller was installed to feed sulfuric acid to neutralize the

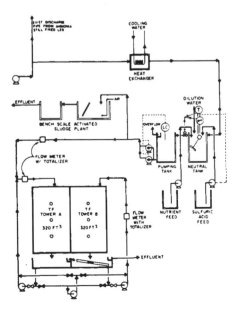

Figure 2. Flow schematic of trickling filter pilot study.

wastewater to the desired pH. The wastewater is deficient in phosphorus; therefore, a sodium ortho-phosphate solution was also fed into the neutralization tank.

The conditioned feed to the trickling filter was pumped to a spray header on top of Tower A. A recycle system was installed to pump a portion of the treated effluent from Tower A into a common pipe with the conditioned feed to Tower A. The effluent from Tower A flowed by gravity to a sump under Tower B. The feed to Tower B was taken from this sump. This arrangement provided a convenient method to keep the trickling filters wetted. In order to promote more of a "series" flow pattern, the Tower A effluent was directed towards the Tower B feed pump. Tower A, the first stage, was packed with 320 ft³ of Munter's BIODEK media model 27060. This packing is an oriented, self-supporting, PVC modular media with a specific surface area of 30 ft²/ft³ and a void ratio greater than 95%. Tower B, the second stage, was packed with 320 ft³ of Munter's BIODEK media model 19060. The media in Tower B is essentially the same construction as in Tower A but with a specific surface of 42 ft²/ft³. The selection of the media type was based on the potential biological growth thickness, air requirements, and as a comparative performance evaluation. A media with larger openings was used in the first stage to provide the air required by the biomass to oxidize the more readily degradable organics. Also, since the loading of organics are higher in the first stage, the larger openings are desirable to minimize plugging problems associated with a heavier growth. Less biomass is generated in the second stage due to the removal of organics in the first stage resulting in a lesser likelihood of plugging using a more dense media. The media depth on each tower is 20 ft. The media distributes the water laterally; therefore, to direct water that would channel along the walls of each tower into the media, plastic collars were installed at six foot intervals from the top of the tower.

Inspection/sampling ports were installed at 5 ft. intervals along the height of each tower. The media was cut into removable keys at each inspection/sample port to allow inspection of the biological growth inside the tower. The sample ports were 1 inch pipes which were cut in half to form a trough which extended across the width of the trickling filter. The pipes could be turned to collect water only when sampling.

In addition to the trickling filter, a bench scale activated sludge unit was operated as a control treatment system. This system consisted of a 75 liter reactor with a 52 liter aeration volume and an integral 9 liter clarifier, a second clarifier, and an air blower.

CHRONOLOGY

Operation of the pilot trickling filter and activated sludge systems began in late June 1984 and continued through April 1985. A discussion of the chronology is presented below.

Phase I

The feed to the trickling filter was on a batch basis for the month of July, following an initial sludge seed, with the addition of 30 gallons of wastewater per day. Continuous flow began in late July at 1 gpm undiluted waste.

During the month of August the trickling filter and activated sludge units were seeded weekly with biological sludge from a municipal treatment plant and a coke plant wastewater treatment system. The municipal sludge was selected to obtain nitrifying bacteria while the coke plant sludge was selected to provide the phenol degrading portion of the population.

Phenol removal reached 90 to 100% in early August in both the trickling filter and activated sludge units. By late August it became apparent that nitrification was being inhibited since there was no ammonia removal and little or no nitrate and nitrite production which are products of ammonia oxidation. The existence of some toxic substance in the wastewater was suspected. To reduce the concentration of the toxic substance, the feed was diluted by 50% with service water. This practice was continued until it became apparent that consistent nitrification was still not occurring.

Seeding with municipal sludge continued weekly through September and then only following a major upset. The quantity of sludge used in each instance to seed the trickling filter was 50 gallons while 1 gallon of sludge was now typically added to the activated sludge unit. The seed sludge improved the performance throughout both systems as evidenced by nitrate and nitrite productions which climbed from less than 1 mg/L in late August to 30 mg/L by the end of September.

A second clarifier was added to the activated sludge unit in mid-September to increase the sludge age. The second clarifier was decanted daily with the solids being recycled to the aeration tank. This amounted to about 1.5 liters of return sludge per day and stabilized the phenol removal at 97 to 100%. Nitrification, however, remained very unstable.

Inspection of internal pieces of the trickling filter media in late September revealed a loose spotty growth. At this time the wetting rate was reduced from the media manufacturer recommended wetting rate of 1.0 gpm/ft² to the minimum manufacturer recommended wetting rate of 0.5 gpm/ft² by adjusting the recycle flow rate. Immediately following the wetting rate reduction, the frequency and size of the growth increased. The wetting rate was not further reduced because flow distribution problems would result.

The operation of the trickling filter and the activated sludge units remained unchanged from late September through mid-November when sodium bicarbonate addition was initiated. The sodium bicarbonate was added to both raise the pH in the activated sludge unit and to ensure a sufficient inorganic carbon source required by the nitrifying bacteria.

Continuous erratic and poor nitrification suggested the possibility of a toxic organic that inhibits nitrification. To remove the toxic organic, powered activated carbon (PAC) was added on a batch basis to the activated sludge aeration tank. PAC is frequently used to adsorb toxins in the activated sludge process and also serves as an attachment medium for the bacteria, improving the settleability.

An initial dose of 500 mg/L of PAC was added to the aeration tank and maintenance dosage was added at a rate of 50 mg/L of influent wastewater. The PAC addition gradually raised the MLVSS from about 2500 to 4500 mg/L, but with no apparent performance change.

Having had sporadic success with sustaining a nitrifying bacteria population by conventional means, a mutant bacteria program was initiated on November 29. The bacteria were added daily at three times the recommended dosage to Tower B, the second stage of the trickling filter. The mutant bacteria were added to the activated sludge system from December 6 to 11. The mutant bacteria had no effect on improving ammonia removals.

Phase II

In late November it was decided to abandon the trickling filter for nitrification due to the history of upsets. The activated sludge system performance was superior to the trickling filter as far as nitrification was concerned, and operation was continued for both ammonia and phenol removal. Although better than the trickling filter, the activated sludge unit also had poor and erratic nitrification. To further reduce the load, the hydraulic retention time was increased. The trickling filter was continued to be operated to optimize phenol removal without dilution water. Therefore, during December the dilution water flow rate was gradually reduced to zero.

During Phase II, the load was gradually increased to determine the kinetics for removal of phenol in the trickling filter.

Phase III

Throughout the study, good breakdown of thiocyanate was achieved. This results in an increase in ammonium concentration in the effluent. It was noticed, however, that the thiocyanate breakdown was less at higher loadings. Therefore, a few tests were conducted to determine the relationship between organic load and the degradation of thiocyanate.

PERFORMANCE

Nitrification

The initial basis for loading the trickling filter was derived from work done by the Munters Company as presented in their book entitled *Plastic Packed Trickling Filters*. The limiting factor for loading is ammonia removal. For nitrifying secondary effluent from a municipal treatment plant with a BOD of 30 mg/L and a NH_3-N concentration of 20 mg/L, the loading could be up to 10 lb NH_3-N/ 1000 ft³/day. The actual loadings on the trickling filter and the activated sludge unit are summarized in Table II for the Phase I period. Also shown in Table II are the loadings based on the total degradable nitrogen load which is the sum of the ammonia plus thiocyanate. These values indicate the ammonia loading based on the influent ammonia plus the ammonia that could be produced through the conversion of thiocyanate. Even when considering the total possible ammonia loadings, the trickling filter was fed at well below the recommended value. The overall trickling filter loading is equal to 1/2 the first stage loading since the 1st and 2nd stages are of equal volume.

The principal parameters of concern to meet BAT limits are ammonia and phenol. The pilot plant influent and effluent wastewater quality is summarized for ammonia and phenol in Phase I in Table

Table II. Pilot Plant Loadings Summary-Phase I

	1st Stage Tower A	2nd Stage Tower B	Activated Sludge
NH₃-N			
Avg. Loading, lb NH_3-N/1000 Ft³/Day	2.1	4.8	2.9
Standard Deviation	2.3	3.8	2.6
Maximum	9.8	18.5	8.3
NH₃-N + SCN-N			
Avg. Loading Lb/1000 Ft³/Day	3.6	5.8	4.4
Standard Deviation	2.7	4.6	2.7
Maximum	11.9	23.3	13.4
SOC			
Avg. Loading Lb/1000 Ft³/Day	9.1	4.3	11.4
Standard Deviation	5.8	3.3	7.2
Maximum	19.6	15.8	26.9

III. There was a considerable increase in the ammonia concentration due to the breakdown of thiocyanate, but there was good phenol removal.

The overall performance of the trickling filter and the activated sludge unit was good for phenol removal, but nitrification was sporadic. A chronological plot of the parameters of concern for nitrification is given in Figures 3 and 4. The products of the biological degradation of ammonia are nitrate and nitrite with the sum of the nitrate and nitrite nitrogen being equal to the amount of ammonia nitrogen that was degraded. Inspection of the chronological plot for the sum of the nitrate and nitrite nitrogen shows that the performance of both the trickling filter and the activated sludge units was erratic. As shown in Figure 3, even under the more controlled laboratory conditions (12/11 to 1/31), nitrification was not achieved consistently in the activated sludge unit. The abrupt decrease in the sum of the nitrate and nitrite occurred when a fresh drum of feed stock was started. This supports the suspicion of some toxic constituent.

The loadings on the trickling filter and activated sludge pilot plants were based on removing both phenol and ammonia with ammonia being the limiting factor. Phenol is more readily degradable than ammonia and was consistently removed to an average of 0.03 and 0.05 mg/L in the activated sludge and Tower B effluents, respectively. The loadings used in Phase I of the study were low for the possible application of this technology specifically for phenol oxidation only (i.e., no nitrification).

Nitrification Batch Tests

Batch tests on the trickling filter and the activated sludge unit were conducted to determine the ammonia removal rate under controlled conditions.

The first batch test was conducted in the field to determine the ammonia removal rate in both stages of the trickling filter and the activated sludge unit. The trickling filter batch test procedure was to turn off the feed, add a spike of ammonium sulfate to each tower sump and allow the trickling filter to recycle. For the activated sludge unit, the feed was simply turned off and the reactor spiked with ammonium sulfate. Samples were collected periodically, and the ammonia concentration was measured in each. The results of the test are given in Table IV (Test No. 1), which shows that the activated

Table III. Ammonia and Phenol Concentration Summary

	Feed	Stage 1 Effluent	Stage 2 Effluent	Activated Sludge Effluent
NH₃-Phase I				
Avg. Concentration, mg/L	31	48	47	62
Standard Deviation	32	37	32	79
Phenol-Phase I				
Avg. Concentration	1.7	0.2	0.05	0.03
Standard Deviation	1.5	0.3		

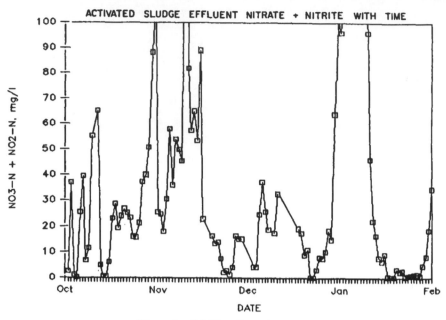

Figure 3.   Trickling filter pilot study.

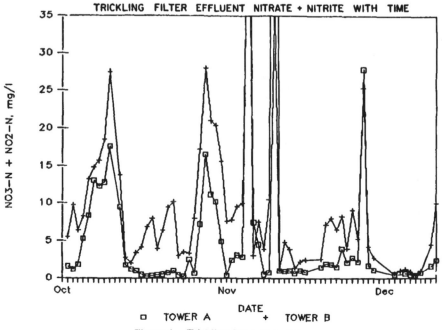

Figure 4.   Trickling filter pilot study.

Table IV. Batch Test Summary

	lb NH$_3$-N Removed/ 1000 ft^3/day	g NH$_3$-N Removed/ g VSS/day	Addition To Feed
Test No. 1			
T.F. Tower A	0.05		100 mg/L NH$_3$-N
T.F. Tower B	0.30		100 mg/L NH$_3$-N
Activated Sludge	5.4	0.03	100 mg/L NH$_3$-N
Test No. 2			
Lab-A.S.	6.3	0.04	100 mg/L NH$_3$-N
	7.5	0.05	200 mg/L NH$_3$-N
Test No. 3			
Lab-A.S.	2.3	0.015	90 mg/L NH$_3$-N
	2.5	0.016	90 mg/L NH$_3$-N + 33 mg/L Solvent
From Literature	10	0.02	

Note: VSS = 2500 mg/L

   T.F. volume of water estimated at 180 gallons

sludge ammonia removal performance was superior to the trickling filter. Although the activated sludge removal rate was better than the trickling filter, it was still a very low removal rate much like that seen by others on coke plant wastewater [8, 13, 16].

Concurrent with the above field test, a similar test was conducted in the laboratory on a sample of the activated sludge mixed liquor spiked with ammonium chloride and sodium thiocyanide. The results are given in Table IV (Test No. 2). The removal rates were in agreement with the field test on the activated sludge unit.

The poor ammonia removal performance in the activated sludge unit brings out the possibility of some toxic constituent of the wastewater causing an inhibition. The dephenolization system at the Pittsburgh Works is atypical to plants that have successfully supported nitrification in coke plant wastewaters. The dephenolization plant utilizes an extraction solvent and typically about 1 mg/L of solvent remains in the fixed ammonia still effluent. Batch tests were run with an ammonia spiked sample of the existing wastewater and a similar sample with an additional amount of solvent. The results are given in Table IV (Test No. 3).

The ammonia removal rate appears to be unaffected by the solvent addition. The batch tests showed that during periods of nitrification, the removal rate was comparable to literature data. The problem in this study was to sustain nitrification for an extended period of time.

Phenol Removal

Phenol removal was consistently good throughout the course of the study. During the first phase of the pilot study (i.e., nitrification), the pilot plant was underloaded (if operated solely for phenol removal) and the potential phenol removal performance could not be adequately evaluated. After nitrification was abandoned, the hydraulic and phenol loading was increased. When the loading rates were increased, on several occasions, samples were collected at five 5-foot intervals over the height of the trickling filters. From the analysis of these samples the relationship between flow rate, phenol concentrations, depth in the trickling filter, and phenol removal rate was determined. The kinetic data collected during this study are shown in Figure 5. These data are for a wetting rate of 0.5 gpm/ft^2 by using the following equation:

$$r = \frac{dc}{dD} = kc^n \tag{1}$$

where:

   r = rate of change of phenol concentration (mg/L/ft)
   c = concentration of phenol
   D = depth of filter

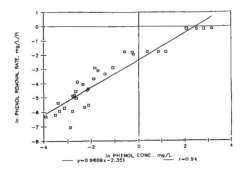

Figure 5.   Kinetics of phenol removal.

K = rate constant
n = reaction order

K and n can be obtained from Figure 5.

For any particular situation the depth (D) can be calculated from

$$\frac{C_o}{C_e} \left| \frac{C^{1-n}}{k(1-n)} \right. = D \tag{2}$$

where C_o and C_e are influent and effluent phenol concentration.

The calculations require trial and error to balance desired effluent quality, hydraulics, and practical limits to D.

Thiocyanate Conversion to Ammonia

Thiocyanate conversion to ammonia during the first phase of the study was essentially complete as shown in Table V.

The thiocyanate removal rate versus organic loading is shown in Figure 6. This data supports the hypothesis that thiocyanate breakdown can be inhibited by organic loading. It was shown that for thiocyanate concentrations below 200 mg/L (typical for most coke plants), the conversion of thiocyanate was inhibited by SOC concentrations greater than 70 lb SOC/1000 ft³/day. Below 70 lb SOC/1000 ft³/day and with an initial thiocyanate concentration less than 200 mg/L, the thiocyanate removal follows first order kinetics. At thiocyanate concentrations greater than 200 mg/L, the thiocyanate is degraded to ammonia at SOC loadings compatible with adequate phenol removal. The relationship between SOC loading and thiocyanate degradation inhibition was not further pursued.

DISCUSSION

The purpose of the biological systems described in this study was primarily to remove ammonia and phenol to concentrations that will allow LTV to meet BAT limitations. Comparing the actual average effluent concentrations from the trickling filter and the activated sludge units with the required effluent concentration of 25–50 gpm (depending on flow rate) shows that neither of the biological treatment schemes provided sufficient ammonia removal capabilities. It must further be noted that the majority of the operating data was collected at 50% dilution of the feed and therefore the normal

Table V. Phase I-Typical Performance Summary

	%NH₃-N Removed[a]	% Phenol Removed	% SCN Removed
Tower A	–300 to 0	90 to 100	50 to 90
Tower B	–5 to 5	50 to 80	30 to 90
Activated Sludge	–300 to 0	90 to 100	60 to 100

[a]Negative NH₃-N removals are due to the breakdown of organic nitrogen and thiocyanate.

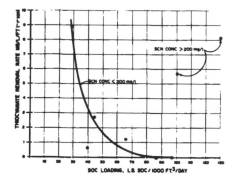

SCN CONC > 200 mg/l

SCN CONC < 200 mg/l

Figure 6.   Thiocyanate removal rate versus SOC loading.

flow range would be doubled. At that dilution even the phenol removal requirements also become difficult to achieve, although phenol removal was good with no dilution.

The effluent ammonia from Tower B and the activated sludge unit was generally about equal to the TKN, which indicates that little nitrification of ammonia took place. In general, this conversion of thiocyanate and organic nitrogen doubled the concentration of ammonia in the effluent as compared to the influent in both the trickling filter and activated sludge systems.

Nitrification was very erratic and periodically subject to complete cessation as can be seen by the sum of the nitrate and nitrite production in Figures 3 and 4. Inspection of the data indicated no change in any of the operating parameters to induce the decreased performance. The most probable cause of the upsets is some toxic constituent in the wastewater that appears periodically. The standard adjustments to the systems to remove or decrease the effect of the toxins, were employed with no performance improvement. Dilution of the wastewater should reduce the inhibition caused by any toxic substances. The addition of powdered activated carbon is the standard practice to reduce the effects of an organic toxic substance. Neither of the above steps had any noticeable effect on ammonia removal.

The biological removal of phenol remains a viable alternative. Phenol removals were acceptable and stable throughout the course of the study. The required effluent phenol concentration is between 0.05 and 0.100 mg/L, depending on the flow, and typical effluent concentrations in the trickling filter and activated sludge pilot plants were below 0.05 mg/L. The variation of the influent concentration did not affect the phenol removal performance. A variation of 3 or 4 mg/L at the normal low concentrations is a change of as much as 800% (0.5 to 4.0 mg/L) and was handled successfully.

The production of ammonia due to thiocyanate degradation is a function of organic loadings and thiocyanate concentrations. In terms of sizing, it places a constraint on the maximum size of the trickling filter, i.e., oversizing can cause unnecessary production of ammonia.

REFERENCES

1. Adams, C. A., R. M. Stein, and W. W. Eckenfelder, "Treatment of Two Coke Plant Waste-waters to Meet Guideline Criteria," *Proceedings 29th Purdue Industrial Waste Conference*, 864 (1974).
2. Adams, C. A., "Treatment of High Strength Phenolic and Ammonia Wastestream by Single and Multi-stage Activated Sludge Processes," *Proceedings 29th Purdue Industrial Waste Conference*, 617 (1974).
3. Association of Iron and Steel Engineers, *Coke Plant Wastewater Biological Treatment Facility— Cost Comparison of Alternate Components* (January 1984).
4. Bridle, T. R., W. K. Bedford, and B. E. Jank, "Biological Treatment of Coke Plant Wastewaters for Control of Nitrogen and Trace Organics," *53rd Annual Water Pollution Control Federstion*, (September 1980).
5. Cooper, R. L., and J. R. Catchpole, "Biological Treatment of Phenolic Wastes," *Iron and Steel Institute Publication No. 128* (1970).
6. Cusins, W. G., and A. B. Mindler, "Tertiary Treatment of Weak Ammonia Liquor," *Journal Water Pollution Control Federation*, Vol. 44, No. 4 (April 1972).
7. Ganczarczyk, J., and D. Elion, "Extended Aeration of Coke Plant Effluents," *33rd Purdue Industrial Waste Conference*, 895 (1978).

9. Jones, L., "Experimental Investigations of Biological Treatment of Coke Plant Ammonia Still Effluent," *M.S. Thesis*, Carnegie-Mellon University (1978).
10. Kostenbader, P. D., and J. W. Flecksteiner, "Biological Oxidation of Coke Plant Weak Ammonia Liquor," *Journal Water Pollution Control Federation*, Vol. 41, No. 2 (February 1969).
11. Luthy, R. G., and J. T. Tallon, "Experimental Analysis of Biological Oxidation Characteristics of Hygas Coal Gasification Wastewater," *Department of Energy, FE-2496-27.*
12. Neufeld, R. D., and Greenfield, J. H., "Nitrification Kinetics as Influenced by Coke Plant Wastewater," *Proc. Symp. on Iron and Steel Pollut. Abatement Technol. for 1982, EPA-600/9-83-016*, U.S. EPA Washington, D.C., NTIS PB83-258665 (1983).
13. Olthof, M., E. Pearson, N. Mancuso, and I. Wittmann, "Biological Treatment of Coke Oven Wastewater Including Provisions for Nitrification," *Iron and Steel Engineer* (June 1980).
14. Olthof, M., and Oleszkiewicz, J., "Benzol Plant Wastewater in a Packed Bed Reactor," *37th Purdue Industrial Wastewater Conference* (May 1982).
15. Osantowski, R., et al., "Two Stage Biological Treatment of Coke Plant Wastewater," *EPA 600/2-81-052* (April 1981).
16. Wong-Chong, G., and S. Caruso, "Advanced Biological Oxidation of Coke Plant Wastewaters for the Removal of Nitrogen Compounds," Carnegie-Mellon Institute (April 1977).
17. Wong Chong, G., "Design and Operation of Biological Treatment of Coke Plant Wastewaters," Carnegie-Mellon Institute (1978).
18. Neufeld, R. D., "Thiocyanate Bio-oxidation Kinetics," *ASCE, Journal Environment Engineering Division*, Vol. 107, No. 5 (October 1981).
19. Nutt, S. G., H. Melcer, and J. H. Pries, "Two Stage Biological Fluidized Bed Treatment of Coke Plant Wastewater for Nitrogen Control," *54th WPCF Conference*, Detroit (1981).

70 PERFORMANCE EVALUATION AND ECONOMIC ANALYSIS OF PRE-LIMING AMMONIA STEAM STRIPPING FROM COKE-PLANT EFFLUENTS

J. J. Ganczarczyk, Professor
Department of Civil Engineering
University of Toronto
Toronto, Ontario M5S 1A4

R. M. Chin, Engineer
Ontario Ministry of the Environment
Toronto, Ontario

M. J. Kelleher, Engineer
MacLaren Plansearch
Toronto, Ontario

INTRODUCTION

Coke-plant effluents contain free and combined (fixed) ammonia. The free ammonia may be stripped out by application of steam, but to strip combined ammonia, an alkalization of the wastewater, usually by the addition of lime, is required. The fate of ammonia nitrogen in the treatment of coke-plant effluents is presented in Table I. Poor performance of stripping facilities is caused by several factors such as fluctuation of wastewater flow to the still, insufficient alkalization, insufficient steam flows, and others. To obtain satisfactory performance of ammonia stripping, proper design and operation of stripping facilities is required. To achieve an advanced performance, in addition to the above requirements, an improved process control is needed.

TECHNOLOGY OF AMMONIA STEAM STRIPPING

Ammonia steam stripping from coke-plant effluents is based on Henry's law and is performed in stripping towers (stills). The effectiveness of the operation depends on the applied hydraulic loading rate (mass flux), the steam-to-liquid ratio, and the treatment time (stripping tower heights and number of trays in the tower).

The following general types of stripping systems can be considered.

- application of a stripping still with a free leg only,
- application of a conventional free-and-fixed leg still (conventional liming system),
- application of a pre-liming system, and
- application of a modified pre-liming system.

Table I. Fate of Ammonia Nitrogen in Treatment of Coke-Plant Effluents [1-3]

	Concentration (mg/L)
Raw Wastewater	3,000
After Stripping of Free Ammonia	1,500
After Stripping of Free and Fixed Ammonia	
Poor Performance	300
Satisfactory Performance	100
Advanced Performance	30
After Biological Dephenolization	70–150
After Biological Nitrification	>1

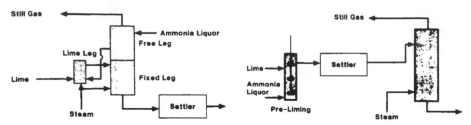

Figure 1. Conventional liming in ammonia still.

Figure 2. Pre-liming for ammonia stripping.

The concept of the conventional liming system is presented in Figure 1. The coke-plant effluent (ammonia liquor) is introduced to the upper part of an ammonia still and flows down through several bubble-cap trays where it is brought into contact with the counter current flow of steam. After losing its contents of free ammonia in the part of the still called "free leg," it is directed to the "lime leg" where lime is introduced for alkalization and the mixture is then returned back to the still to allow the residual ammonia, now made free, to be stripped by the steam flow in the part of the still called "fixed leg."

The pre-liming system for ammonia steam stripping was developed by Bethlehem Steel [4]. The concept of this system is presented in Figure 2. Ammonia liquor is alkalized with lime in a separate contact chamber preceding a settler and a steam stripping still. Such "pre-liming" essentially converts all ammonia to the free form. This sequence helps to avoid the serious operating problems associated with the conventional liming process such as the still plugging and scaling of the trays. It is also claimed that it helps to design and operate the stripping still more effectively.

A possible adaptation of the pre-liming concept to the existing conventional liming system is shown in Figure 3 as a modified pre- liming system. This concept has been implemented at the DOFASCO plant in Hamilton, Ontario, to upgrade the original conventional liming system for ammonia stripping from coke-plant effluents.

CONVENTIONAL AMMONIA STRIPPING AT DOFASCO

The conventional ammonia stripping system that existed prior to 1978 at DOFASCO is presented in Figure 4. It included a 425,000 US gallon equalization tank that received raw excess ammonia flushing liquor from the coke oven gas flushing liquor recirculation system. The liquor was pumped to the 17th tray of the ammonia still using two 165 US gpm liquor pumps. The ammonia still, which was 6 ft. in diameter and approximately 53 ft. high, was manufactured by Koppers Co. Inc. of Pittsburgh, Pennsylvania, and was equipped with 17 bubble cap plate trays (4 in the free leg and 13 in the fixed leg), a dephlegmator on top of the still to cool the ammonia vapours, and a 4 ft. diameter and 9.5 ft. high cylindrical lime leg contact tank. The lime was stored in a 10 ft. diameter lime storage bin with a 40 ton capacity. A lime slaker and feeder, with a capacity of 1,000 pounds of lime per hour, produced lime water that was discharged to a 500 US gallon lime water storage/stirring tank. Two lime pumps, each with the capacity of 10 US gpm, were used to lift the lime water from the stirring tank to the lime leg. Steam generation facilities with a capacity of 213,000 pounds per day and two 170 US gpm waste liquor sump pumps were also part of this installation.

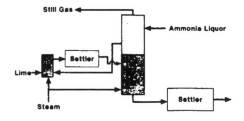

Figure 3. Modified pre-liming for ammonia stripping.

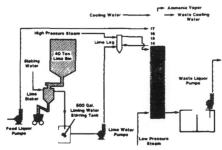 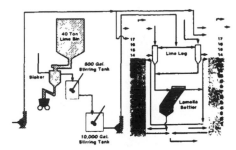

Figure 4.   Dofasco conventional liming system.   Figure 5.   Dofasco modified pre-liming system.

MODIFIED PRE-LIMING SYSTEM AT DOFASCO

The above ammonia stripping system was modified (Figure 5) by the addition of the following units:

- a 10,000 US gallon lime storage /stirring tank to supplement the existing 500 US gallon tank,
- a second (identical) ammonia still with a lime leg,
- a clarifier on the return line to the ammonia still from the lime leg (Axel-Johnson 1500/ 55° Lamella settler/thickener).

The designed operating conditions for the modified system are for liquid feed rates of approximately 20 m³/hr per still with a capacity of 27 m³/hr to each still. When only one still is in operation, the flow rate may be adjusted to 35 m³/hr. The actual flow rates recorded during the period of data collection varied between 14 to 17 m³/hr per still. When both stills are in operation, the normal steam flow rate is approximately 4,000 Kg/hr/still. This rate is increased to 7,000 Kg/hr when only one still is in operation. Since the actual steam flowrate depends on a number of factors including production rates, liquid flow rates and coal moisture, the steam flow rate is adjustable. The plant operator tries to maintain the steam/liquid ratio at 200 Kg/m³. The water/lime ratio of the lime slurry is maintained between 29.1 and 33.3 L/Kg of lime. The normal capacity of the lime slurry feed system is 1.2 m³/hr/ still when both stills are in operation but is increased to 2 m³/hr when only one still is in operation. The lime slurry feed rate is also adjustable depending on the pH of the return feed to the stills.

SCOPE OF THE STUDY

The DOFASCO coke-plant in Hamilton, Ontario, used the previously described conventional liming ammonia stripping system until 1978 when a modified pre-liming system was introduced. The operational data from both of these systems (operation in 1975 and 1982) were collected and statistically processed. Wastewater quality sampling was conducted on a routine basis by the DOFASCO personnel for all critical parameters of the systems including pH, suspended solids, and ammonia. This sampling was done on both influent and effluent waste streams of the ammonia stills and, in the case of the modified pre-liming system, also on the overflow from the Lamella settler. The sampling procedure used consisted of obtaining a single grab sample every hour at each of the sampling locations. These samples were then composited to obtain a 24 hour composite sample. Statistical processing of the data included distribution analysis of the influent and effluent ammonia concentration (Figure 6 and 7) and descriptive statistics of the effluent ammonia concentrations (Table II). Some scattergrams, the T-test analyses, as well as some correlation tests, were also performed.

RESULTS

The influent to both systems differed in terms of quality (Figure 6); the 1982 influent was more concentrated (by about 1,000 mg/L) than the influent in 1975. This may perhaps be a result of a lower dilution of the wastewater by lime water in 1982 due to a more efficient preparation of the latter.

The new system significantly improved the quality of the effluent both in terms of the achieved levels of the residual ammonia as well as their variability (Figure 7, Tables II and III).

Specifically, the new modified pre-liming system was capable of achieving 100 mg/L or less of ammonia in the effluent for up to 85% of the time compared to only 40% of the time for the

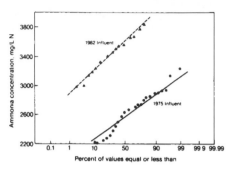

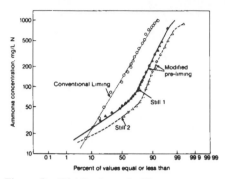

Figure 6. Distribution of influent concentration.

Figure 7. Distribution of effluent concentration.

conventional liming system. Similarly, the modified pre-liming system had a greater ability to eliminate the higher ammonia concentrations than the conventional liming system. Therefore, the modified pre-liming system is a more appropriate pre-treatment operation than the conventional liming system if effluent nitrification is to be considered.

In the study of the 1975 data, it was observed that the normal distribution of the ammonia concentration in the influent changed to the log-normal distribution of the ammonia concentration in the effluent. Such a change in the type of data distribution is common in the performance study of various technological facilities and indicates their instability (skewness towards somewhat higher number of cases of poorer performance). However, the 1982 data cannot be explained in a similar way. The distribution of the ammonia concentrations in the influent was normal, but the distribution of the ammonia concentration in the effluent was apparently bi-modal and could be presented on the log-normal probability graph as a composition of two straight lines of two different slopes (Figure 7). The left side branch has a lower slope than the line for the conventional liming, indicating a more stable system, but it was followed by a line at the level of 70–80% of values equal to or less than practically of the same steepness as the line for the 1978 data, indicating an occurrence of some poor performance situation. Surprisingly, a presentation of the 1982 data as log-values produced a straight line on the log- normal distribution graph. This might imply the existence of a log-log-normal distribution but most likely was a consequence of some set of coincidences. A scattergram analysis of the 1982 effluent data has showed that during that year only two serious upsets occurred in the treatment system, which otherwise consistently produced an effluent of well below 100 mg/L of residual ammonia.

As expected, no statistical correlation was found between the pre-limed effluent clarification and treatment effectiveness. The beneficial role of the clarifier in this system, in addition to the increased longevity of the still without cleaning, may be in an increased equalization capacity of the system.

Table II. Descriptive Statistics of Effluent Ammonia Concentration

Stripping System	Mean, mg/L	Standard Deviation	Variance
Conventional liming	287.1	476.7	227,278
Modified pre-liming (still #1)	91.8	150.2	22,564
Modified pre-liming (still #2)	76.3	130.0	16,866
Modified pre-liming (single still operation)	70.8	112.0	12,458

Table III. Distribution Analysis of Effluent Ammonia Concentration

System	Skewness	Kurtosis
Conventional liming	5.12	39.4
Modified pre-liming (still #1)	4.4	23.5
Modified pre-liming (still #2)	3.7	15.8
Modified pre-liming (single still operation)	3.7	15.3

The differences in the performance of the stills, specifically in the one still operation mode, indicate that further improvement of the treatment is possible through some optimization of operational strategy.

ECONOMICAL ANALYSIS OF STEAM AMMONIA STRIPPING

Information collected from various North American plants applying steam stripping of ammonia from coke-plant effluents indicates that the capital and annual operating costs of such treatment can be presented in the form of two general equations:

$$\text{Capital costs} = 580,000 + 5,700 \times Q$$
$$\text{Operation costs} = 176,000 + 1,580 \times Q$$

where costs are expressed in January 1986 Canadian dollars and Q is the wastewater flow rate in L/min [5].

The operating costs of steam ammonia stripping can generally be divided into the costs of chemicals, steam, and labour. In the case of lime application, the costs of chemicals amount to 7% of total operating costs for lime and 5% for chemicals used in the still cleaning. The steam costs are in the range of 48% of total operating costs while the costs of labour amount to 40% and can be divided into the operational labour (30%) and the maintenance labour (10%).

Although a substitution of caustic soda for lime increases the cost of alkalization chemicals by a factor of 8.5, it leads to substantial savings in operational and maintenance labour and in the costs of chemicals used for the still cleaning. According to some estimates these costs and savings balance each other.

The application of pre-liming ammonia stripping, and perhaps also of modified pre-liming such as the system at DOFASCO, should lead to savings on steam consumption, cleaning chemicals, and maintenance labour [4]. The relevant DOFASCO data were not made available to the authors of this paper.

CONCLUSIONS

The modifications of the ammonia stripping system have resulted in a substantial improvement in the effectiveness of the wastewater treatment. The residual concentrations of ammonia nitrogen were below 100 mg/L for 85% of the time as compared with only 40% of the time for the previous system.

The modifications also caused a change in the effluent quality data distribution. Previously, this distribution was log-normal, and for the modified system it appears to be bi-modal.

It is possible that in addition to the increased longevity of the still without cleaning, the clarifier in a modified pre-liming system also provides an increased internal equalization to the system.

It appears that some changes in operational strategy of the modified pre-liming system may result in a further improvement of the treatment results.

The steam ammonia stripping is an expensive operation. Therefore, all possible ways to make it less costly should be thoroughly investigated.

ACKNOWLEDGMENTS

The authors are indebted to the DOFASCO personnel for the information about the ammonia stripping systems used at the DOFASCO plant. Without this help this work would not have been possible.

REFERENCES

1. Ganczarczyk, J., "Fate of Basic Pollutants in Treatments of Coke-Plant Effluents," *Proc. 35th Industrial Waste Conference*, Purdue University, 325–331 (1980).
2. Ganczarczyk, J., "Pre-Treatment of Coke-Plant Effluents," *Waste Treatment and Utilization*, Vol. 1, 119–226, Pergamon Press Oxford (1980).
3. Ganczarczyk, J., "State of the Art in Coke-Plant Effluent Treatment," *CRC Critical Reviews in Environmental Control*, 13, 2, 103–115 (1983).
4. Rudzki, E. M., Burcaw, K. R., and Horst, R. J. "An Improved Process for the Removal of

Ammonia from Coke-Plant Weak Ammonia Liquor," *Ironmaking Proceedings*, 36, 525–533 (1977).
5. Ganczarczyk, J., and Kelleher, M., "Costs of Pre-treatment of Coke-Plant Effluents," *Proc. of the EPA Symposium on Iron and Steel Pollution Abatement Technology for 1982, EPA-600/9-83-016, NTIS PB83-258665 (1983).*

Section Ten
DYE WASTES

71 FATE OF AZO DYES IN THE ACTIVATED SLUDGE PROCESS

Glenn M. Shaul, General Engineer

Clyde R. Dempsey, Environmental Engineer

Kenneth A. Dostal, Environmental Engineer
Industrial Wastes and Toxics Technology Division
Water Engineering Research Laboratory
U.S. Environmental Protection Agency
Cincinnati, Ohio 45268

Richard J. Lieberman, Graduate Research Assistant
Department of Civil and Environmental Engineering
University of Cincinnati
Cincinnati, Ohio 45221

INTRODUCTION

The U.S. Environmental Protection Agency's (EPA) Office of Toxic Substances (OTS) evaluates submissions to the Premanufacture Notification (PMN) process under section 5 of the Toxic Substances Control Act (TSCA). Azo dyes constitute a significant portion of these submissions. OTS is concerned because some of the dyes, dye precursors or their degradation products such as aromatic amines, which are also dye precursors, have been shown to be, or are suspected to be, carcinogenic [1]. Therefore, EPA's Water Engineering Research Laboratory, Office of Research and Development has an ongoing study to determine the fate of specific azo dye compounds in the activated sludge process (ASP).

The objective of this study is to dose pilot plant ASP systems with various azo dyes and monitor each dye compound through the system, analyzing both liquid and sludge samples. The fate of the parent dye compound was assessed via mass balance calculations. From these data, it was possible to determine if the compound was removed by adsorption, apparent biodegradation or not removed at all. This report presents results to date (5/86) from this study which is being conducted at EPA's Test and Evaluation Facility in Cincinnati, Ohio.

EXPERIMENTAL PROGRAM

A previous paper [2] described in detail the background, objectives and analytical methods employed for this study. This paper describes work that used wastewater from the Greater Cincinnati Mill Creek Sewage Treatment Plant as the influent to three pilot-scale ASP systems (two experimental and one operational control). Each dye was dosed as commercial product to the two experimental systems operated in parallel at 1 and 5 mg/L of influent flow (henceforth referred to as the low and high spike systems, respectively). The principal focus of this work is on the ASP and as such the primary sludge was not sampled. These selected dosages were based on expected concentrations in publicly owned treatment works [2]. Table I presents a summary of the operating conditions of the pilot plant systems. Each system consisted of a primary clarifier, complete-mix aeration basin and secondary clarifier.

Before each intensive data collection phase, dye recovery studies were conducted in organic free water, influent wastewater, and mixed liquor. These studies were run in duplicate and each recovery study was repeated at least once in order to ensure that the compound could be extracted from these samples. Purified dye standards were analytically prepared from the commercial dye product as per Tincher [3] since standards are not available from conventional sources. All samples were analyzed by

Table I. Summary of Operating Conditions

Parameter	Value
Influent Flow Rate, L/d	720
Primary Sludge Flow Rate, L/d	6
Primary Effluent Flow Rate, L/d	714
Mixed-Liquor Wastage Flow Rate, L/d	67
Secondary Effluent Flow Rate, L/d	647
Solids Retention Time, days	2.7
Hydraulic Retention Time, days	0.28
Dissolved Oxygen, mg/L	2.0–4.0
Influent Spike Dosages, mg/L	
Low	1
High	5
Influent pH, pH units	7.0–8.0
Aeration Basin Temperature, °C	21–25

high performance liquid chromatography (HPLC) with an ultraviolet-visible detector. The targeted range of recovery from all samples was 80 to 120%. If recovery analyses were within this range, then it was assumed that little, if any, chemical transformation was occurring. Previous work had found that recovery of the dye from sludge was less than the goal of 80 to 120%. However, through a change in the solvents used in the published methods [3,4] acceptable recoveries were achieved for sludge samples, and as such, all data reported herein are as measured and have not been corrected for recoveries. All samples were analyzed by the University of Cincinnati's Kettering Laboratory of the College of Medicine.

All systems were operated for at least three times the solids retention time (SRT) to ensure acclimation prior to initiation of data collection. During the intensive data collection period the following analytical tests were performed:

- Total suspended solids (SS) analyses on the influent (INF), primary effluent (PE), mixed liquor (ML) and activated sludge effluent (ASE) samples. All samples were analyzed using EPA approved methods [5].
- Specific dye analyses on the PE, ML, and ASE samples. The ML samples were filtered through 0.45 micron pore size filters and the soluble dye in the filtrate (SOL) and suspended residue (RES) were analyzed as separate samples. On occasion, the INF samples were analyzed to ensure that targeted dosages were maintained. No standard method exists for these analyses; however, Tincher [3] and Dickson [4] have developed an analytical method.

All samples were 24-hour composites made up of six grab samples collected every four hours and stored at 4°C. Each test run was sufficiently long to allow at least six sets of samples collected over three to five weeks (after acclimation) for the specific dye analyses. If acclimation appeared to be continuing during the data collection phase, the intensive data collection period was usually extended accordingly.

In addition to the pilot plant studies, adsorption isotherms were developed for each of the dyes using activated sludge solids as the adsorbent. Purified dyes prepared from the commercial product were used in these studies in order to minimize interferences from the impurities of the commercial formulations. These studies were conducted with sludge solids obtained from the control pilot plant system which did not receive a dye spike. In order to prevent potential biodegradation from influencing the adsorption measurements, the biological solids were inactivated by freeze-drying of the solids with subsequent desiccation of the dry biomass in an oven for a minimum of three hours at 105°C. This procedure was reported [6] to be the simplest method for inactivation of the biomass while generally retaining the structural integrity of the biomass. Flocculation and settling properties of this freeze-dried, desiccated biomass resembled viable activated sludge solids upon rehydration. In addition, photomicrographs of this material revealed the presence of structurally-intact organisms.

Freeze-dried, desiccated biomass was utilized in kinetic studies to determine the contact time required for the test dyes and sludge solids to reach equilibrium. Previous studies have reported contact times of 40 to 120 minutes [7], 30 minutes [8] and 20 minutes [9] for some azo dyes. However, 24 hours was chosen as the standard contact time for all dyes investigated in this study since the kinetic experiments revealed that although for some dyes equilibrium was reached within one or two hours, other dyes require six or more hours of contact between the test dye and biomass for adsorption equilibrium to be attained.

Adsorption isotherm experiments were performed once the equilibrium contact time was established. A typical experimental setup consisted of a series of stoppered glass vessels which contained increasingly greater concentrations (typical range was 0.5 to 8.0 g/L solids concentrations) of the inactivated sludge solids which were rehydrated in distilled water. A known concentration of dye was spiked into each vessel and the contents of each were well mixed for 24 hours via magnetic stirring apparatus. A quality control sample included in this setup consisted of a dye blank (no sludge solids) to assess potential changes in soluble dye concentration due to dye molecules adhering to the inside walls of the container or photodegradation. Another quality control sample was obtained from a vessel containing sludge solids in water which did not receive the dye spike. This was then used to assess potential analytical interferences originating from the biomass. The contents of all vessels were sampled at the end of the test. These samples were centrifuged at 10,000 revolutions per minute for fifteen minutes, and the centrate was subsequently filtered through a 0.45 micron pore size filter.

High performance liquid chromatography (HPLC) was used to analyze for the residual dye in the sample filtrates. The amount of residual dye was then utilized with the amount of biomass in each vessel to calculate an adsorption isotherm according to the Freundlich relationship. This is an empirically derived [10] relationship with the general form:

$$X/M = K \, C_e^{1/n} \tag{1}$$

where:

X = amount of solute (dye) adsorbed, mg/L
M = concentration of adsorbent (sludge solids), g/L
X/M = amount of dye adsorbed per unit mass of sludge solids, mg/g
C_e = equilibrium concentration of dye in solution, mg/L
K,n = empirical constants

The logarithmic form of the Freundlich equation is employed to obtain a straight line relationship:

$$\log (X/M) = \log K + (1/n)(\log C_e) \tag{2}$$

This linear form of the Freundlich model may be illustrated graphically with a plot of X/M versus C_e on log-log paper. The line of best fit was calculated using least squares regression analysis. The slope of this line $(1/n)$ indicates adsorption intensity whereas the constant K indicates adsorption capacity.

Other adsorption models could have been evaluated. For example, Langmuir's equation and the Brunauer-Emmett-Teller (BET) equation are commonly utilized in adsorption studies [11]. However, Dohanyos, et al [7] applied all three models to data for 22 textile dyes and found the Freundlich relationship to be the most suitable for 18 of these dyes with activated sludge solids as the adsorbent.

RESULTS AND DISCUSSION

Seven acid azo dyes have been studied and are listed below by the Colour Index Name and Number [12] with Figure 1 presenting the chemical structure for each:

NAME	NUMBER
C.I. Acid Blue 113	26360
C.I. Acid Orange 7	15510
C.I. Acid Red 1	18050
C.I. Acid Red 88	15620
C.I. Acid Red 151	26900
C.I. Acid Red 337	–
C.I. Acid Yellow 151	13906

The possible removal mechanisms for a dye compound in the ASP system include adsorption, biodegradation, chemical transformation, photodegradation, and air stripping. During the spike recovery studies for each dye, several wastewater and sludge samples from the control ASP were spiked and recovery was assessed. Recovery for all dyes was within the targeted range of 80 to 120%; thus, it appeared that little or no chemical transformation occurred for any dye due to contact with the variable wastewater and/or sludge matrix. In addition, photodegradation potential was investigated with each dye added to distilled/deionized water and no measurable decrease in concentration

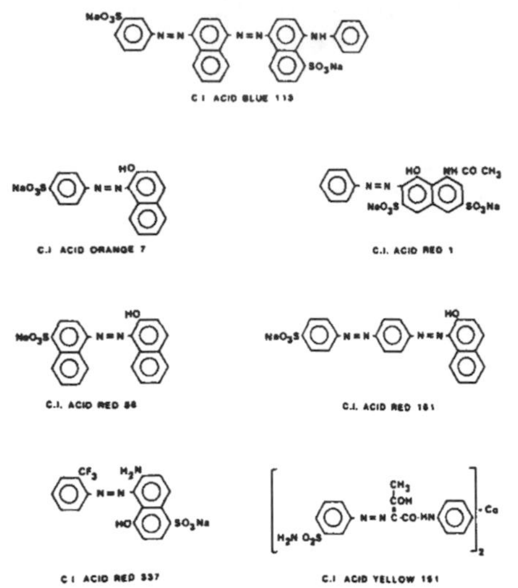

Figure 1.   Dye structures.

was observed for samples held in light for more than two weeks. Moreover, the estimated Henry's law constant for each dye tested is less than 10^{-15} atm-m^3/mol and as such, air stripping is very unlikely [13]. Therefore, adsorption and/or biodegradation appeared to be the possible removal mechanisms.

In order to determine the fate of the parent dye compound in the ASP systems and to allow a comparison between the adsorption isotherm data and the ASP data, mass balance calculations were made. The mass balance calculations are presented below:

DYE IN   = (PE FLOW RATE, L/d) (PE DYE CONC., mg/L) = mg/d
DYE OUT   = (DYE OUT 1) + (DYE OUT 2)
DYE OUT 1 = (DYE OUT VIA SECONDARY EFFLUENT)
DYE OUT 2 = (DYE OUT VIA WASTING MIXED LIQUOR, WAS)
DYE OUT 1 = (ASE FLOW RATE, L/d) (ASE DYE CONC., mg/L) = mg/d
DYE OUT 2 = (WAS FLOW RATE, L/d) (WAS SOL DYE CONC., mg/L)
            + (WAS FLOW RATE, L/d) (WAS ADSORBED DYE CONC.,
            mg/L) = mg/d

with

(WAS ADSORBED DYE CONC.) = (WAS SS CONC., g/l (RES DYE CONC., mg/g)
        DYE Recovered,% = (DYE OUT/DYE IN) x 100

Table II presents the mean SS and dye concentrations and standard deviations calculated from the ASP raw data while Table III presents the mass balance closure data calculated from the values in Table II. Upon inspection of the data from Table II, it can be seen that wide fluctuations in dye removal (PE versus ASE) were measured. (Note that two runs were conducted for C.I. Acid Orange 7).

From Table III, it can be seen that three of the seven dyes studied had total mass balance closures within the 80 to 120% range. This indicates that there was little or no biodegradation of these three dyes. The three dyes were C.I. Acid Red 1, C.I. Acid Red 337 and C.I. Acid Yellow 151. C.I. Acid Red 1 had very little adsorption onto the sludge with the majority of the dye compound exiting the ASP in the secondary effluent stream. Both C.I. Acid Red 337 and C.I. Acid Yellow 151 were found to be partially adsorbed (9 to 17%) onto the sludge; however, the majority of these dyes was found in the secondary effluent.

The data in Tables II and III show that only 4 to 27% and < 1 to 7% of the influent dye could be accounted for in the ASE and WAS for C.I. Acid Orange 7 and C.I. Acid Red 88, respectively. Thus, it is apparent that both of these dyes underwent significant biological degradation.

Both C.I. Acid Blue 113 and C.I. Acid Red 151 were strongly adsorbed (see WAS RES Dye in Table II). Between 22 and 33% of the amount of these dyes entering the aeration basin could not be

Table II. Data Summary[a]

		Primary Effluent (PE)		Waste Activated Sludge (WAS)			Activated Sludge Effluent (ASE)		Percent Removal[b]
		SS mg/L	Total Dye mg/L	SS mg/L	SOL Dye mg/L	RES Dye mg/g	SS mg/L	Total Dye mg/L	%
C.I. Acid Blue 113									
Low Spike	Mean	190	0.84	1490	0.04	4.10	7	0.07	92
	Std. Dev.	61	0.20	330	0.02	2.43	3	0.03	
	Number	5	5	5	5	4	5	4	
High Spike	Mean	220	4.55	1150	0.44	20.65	13	0.84	82
	Std. Dev.	105	0.33	252	0.10	7.74	2	0.17	
	Number	5	4	5	5	4	5		
C.I. Acid Orange 7 Run 1									
Low Spike	Mean	140	0.95	950	0.15	<0.02	17	0/18	81
	Std. Dev.	20	0.06	240	0.20	–	20	0.19	
	Number	6	6	6	6	6	6	6	
High Spike	Mean	90	5.34	910	0.14	<0.02	21	0.24	96
	Std.Dev.	41	0.32	74	0.15	–	39	0.17	
	Number	6	6	6	6	6	6	6	
C.I. Acid Orange 7 Run 2									
Low Spike	Mean	160	1.07	1010	0.22	<0.03	21	0.29	73
	Std. Dev.	53	0.23	236	0.19	–	30	0.21	
	Number	10	10	10	10	10	10	10	
High Spike	Mean	220	5.57	970	0.53	<0.02	41	0.73	87
	Std. Dev.	69	1.03	230	0.61	–	43	0.67	
	Number	10	10	10	10	10	10		
C.I. Acid Red 1									
Low Spike	Mean	190	0.90	1860	0.83	<0.01	25	0.88	2
	Std. Dev.	40	0.17	280	0.21	–	8	0.12	
	Number	8	8	8	8	8	8	8	
High Spike	Mean	260	4.43	1690	4.42	0.05	12	4.48	0
	Std. Dev.	58	0.75	66	0.98	0.02	6	0.91	
	Number	8	8	8	8	7	7	8	
C.I. Acid Red 88									
Low Spike	Mean	140	0.68	910	<0.02	0.10	14	0.04	94
	Std. Dev.	53	0.06	362	–	0.08	11	–	
	Number	6	5	6	6	6	6	6	
High Spike	Mean	150	3.96	1300	<0.01	0.08	22	<0.01	<99
	Std. Dev.	70	0.61	234	–	0.05	19	–	
	Number	6	6	6	6	6	6	6	
C.I. Acid Red 151									
Low Spike	Mean	140	0.56	910	0.08	2.89	14	0.17	70
	Std. Dev.	53	0.13	362	0.02	1.25	11	0.05	
	Number	6	5	6	6	6	6	6	
High Spike	Mean	150	3.64	1300	0.17	19.80	22	0.44	88
	Std. Dev.	70	0.76	234	0.03	5.30	19	0.10	
	Number	6	6	6	6	6	6	6	
C.I. Acid Red 337									
Low Spike	Mean	190	1.13	1490	0.68	1.15	7	0.93	18
	Std. Dev.	61	0.07	330	0.11	0.14	3	0.07	
	Number	5	5	5	5	5	5	4	
High Spike	Mean	220	5.06	1150	3.61	4.08	13	4.40	13
	Std. Dev.	105	0.39	252	0.48	0.87	2	0.51	
	Number	5	4	5	5	5	5	5	

Table II Continued

| | | Primary Effluent (PE) | | Waste Activated Sludge (WAS) | | | Activated Sludge Effluent (ASE) | | Percent |
		SS mg/L	Total Dye mg/L	SS mg/L	SOL Dye mg/L	RES Dye mg/g	SS mg/L	Total Dye mg/L	Removalb %
C.I. Acid Yellow 151									
Low Spike	Mean	190	0.67	1860	0.26	0.46	25	0.48	28
	Std. Dev.	40	0.20	280	0.05	0.17	8	0.10	
	Number	8	8	8	8	8	8	8	
High Spike	Mean	260	4.05	1690	2.66	4.31	12	4.08	0
	Std. Dev.	58	0.60	66	0.58	1.23	6	0.44	
	Number	8	7	8	8	8	7	8	

aUnits: SOL Dye = Concentration of soluble dye in filtrate.
RES Dye = Mass of dye measured per mass of SS in the suspended residue.
bPercent Removal = (PE - ASE)/PE × 100.

accounted for in the ASE and WAS. Thus, there may have been some limited degradation of C.I. Acid Blue 113 and C.I. Acid Red 151.

Table IV presents a summary of the adsorption isotherm data showing the adsorption constants (1/n and K) and the linear regression correlation coefficient (r), and Figures 2 and 3 present these data graphically. The ASP data can be graphically presented on the isotherm plot as well. The value for C_e is directly related to the SOL values from the ASP data, each representing the equilibrium concentra-

Table III. Mass Balance Summary

| | Percent of Primary Effluent Dye Feed in: | | | |
| | Activated Sludge Effluent | Waste Activated Sludge | | ASE + WAS |
	ASE %	SOL %	RES %	Total %
C.I. Acid Blue 113				
Low Spike	8	<1	68	76
High Spike	17	1	49	67
C.I. Acid Orange 7				
Run 1 Low Spike	17	1	<1	19
High Spike	4	<1	<1	4
Run 2 Low Spike	25	2	<1	27
High Spike	12	1	<1	13
C.I. Acid Red 1				
Low Spike	88	9	<1	97
High Spike	92	9	<1	101
C.I. Acid Red 88				
Low Spike	5	<1	1	7
High Spike	<1	<1	<1	<1
C.I. Acid Red 151				
Low Spike	28	1	44	73
High Spike	11	<1	66	78
C.I. Acid Red 337				
Low Spike	75	6	14	94
High Spike	79	7	9	94
C.I. Acid Yellow 151				
Low Spike	65	4	12	80
High Spike	91	6	17	114

Table IV. Adsorption Isotherm Results

	Slope 1/n	Intercept[a] K	Correlation Coefficient r
C.I. Acid Blue 113	1.4	7.3	0.959
C.I. Acid Orange 7	1.2	0.11	0.975
C.I. Acid Red 1	24	1.0×10^{-17}	0.879
C.I. Acid Red 88	0.82	2.5	0.969
C.I. Acid Red 151	0.98	9.1	0.981
C.I. Acid Red 337	1.3	0.67	0.952
C.I. Acid Yellow 151	1.5	1.3	0.957

[a]When $C_e = 1$ mg/L.

tion of dye in solution (mg/L). Similarly, the value of X/M is directly related to the RES values from the ASP data, each representing the amount of dye adsorbed per unit weight of sludge solids (mg/g).

If the ASP data fall on or near the calculated line of best fit from the isotherm, then one could conclude that the isotherm provided a good estimate of the dye removal due to adsorption in the ASP studies. The adsorption isotherms provided reasonably good estimates of adsorption in the ASP studies for three of seven compounds. These were C.I. Acid Red 1, C.I. Acid Red 337, and C.I. Acid Yellow 151, the same three dyes for which no biodegradation occurred.

The ASP data for C.I. Acid Orange 7 and C.I. Acid Red 88 were not plotted on Figures 2 and 3, respectively, since many of the ASE values were reported as less than the detection limit (generally

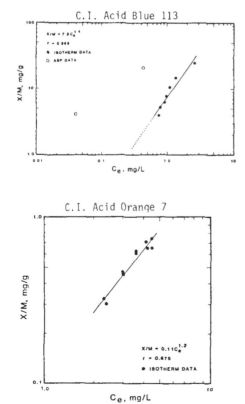

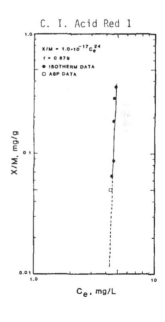

Figure 2. Adsorption isotherms.

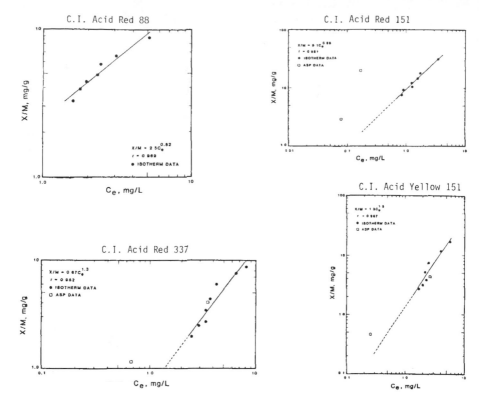

Figure 3.   Adsorption isotherms.

0.01 mg/L), and thus, a finite ASE concentration could not be calculated. The ASP data from C.I. Acid Blue 113 and C.I. Acid Red 151 do not fall near the isotherm line. More apparent adsorption was calculated for each of these dyes in the ASP systems than was predicted by the isotherms. This significantly higher apparent adsorption cannot be explained at this time.

STATUS

No overall prediction about the fate of azo dyes in ASP systems can be made at this time as several additional compounds need to be tested before generalities can be attempted. Once these additional data are available, relationships between physical and chemical properties and the fate of the compounds will be investigated.

CONCLUSIONS

C.I. Acid Red 1 appeared to be very nominally adsorbed but not biodegradable.

C.I. Acid Red 337 and C.I. Acid Yellow 151 appear to be moderately adsorbed but not biodegraded.

C.I. Acid Blue 113 and C.I. Acid Red 151 appear to be strongly adsorbed and possibly partially biodegraded.

C.I. Acid Orange 7 and C.I. Acid Red 88 appear to be moderately adsorbed and significantly biodegraded.

REFERENCES

1. Helemes, C. T., et al., "A Study of Azo and Nitro Dyes for the Selection of Candidates for Carcinogen Bioassay," *Journal of Environmental Science and Health*, Volume A19, No. 2, pp. 97–231 (1984).
2. Shaul, G. M., Dempsey, C. R., and Dostal, K. A., "Removal of Azo Dyes by the Activated Sludge Process," *Proceedings of the 40th Purdue Industrial Waste Conference*, Purdue University, West Lafayette, Indiana, Butterworth Press (1985).
3. Tincher, W. C., "Survey of the Coosa Basin for Organic Contaminants from Carpet Processing," *Contract No. 3-27-630*, Environmental Protection Divison, Department of Natural Resources, State of Georgia (1978).
4. Dickson, D. N., "The Effectiveness of Biodegradation in the Removal of Acid Dyes and Toxicity from Carpet Dyeing Wastewater," *M. S. Thesis*, Georgia Institute of Technology (1981).
5. "Methods for Chemical Analysis of Water and Wastes," *EPA-600/4-79-020*, Environmental Monitoring and Support Laboratory, U.S. EPA (1979).
6. Blackburn, J. W., et al., "Prediction of the Fates of Organic Chemicals in Activated Sludge Wastewater Treatment Processes, *EPA-600/2-85/102*, U.S. EPA (1985).
7. Dohanyos, M., Madera, V., and Sedlacek, M., "Removal of Organic Dyes by Activated Sludge," *Progress in Water Technology*, Vol. 10, Nos. 5/6, pp. 559–575, Pergamon Press (1978).
8. Hitz, H. R., Huber, W., and Reed, R. H., "The Adsorption of Dyes on Activated Sludge," *Journal of the Society of Dyers and Colorists*, 94, pp. 71–76 (February 1978).
9. Nakaoka, M., Tamura, S., Maeda, Y., and Azumi, T., "Adsorption of Synthetic Dyes on Activated Sludge," *Sen i Gakkaishi*, 39 (2) pp. 64–74 (1983).
10. Freundlich, H., "Kapillarchemie," *Akademische Verlagsgesellschaft M.B.H.*, Leipzig, East Germany (1922).
11. Morris, J. C., and Weber, W. J., Jr., "Adsorption of Biochemically Resistant Materials from Solution, No. 1," *Report No. 999-WP-11*, U.S. Department of Health, Education and Welfare, Public Health Service, Division of Water Supply and Pollution Control (1964).
12. *Colour Index*, Revised 3rd Edition, Seven Volumes, published by The Society of Dyers and Colourists (1982).
13. Lyman, W., Reehl, W., and Rosenblatt, D., *Handbook of Chemical Property Estimation Methods*, McGraw Hill Book Company, New York (1982).

NOTICE

72 ECONOMICAL PRETREATMENT FOR COLOR REMOVAL FROM TEXTILE DYE WASTES

George R. Brower, Associate
Barge, Waggoner, Sumner and Cannon
Knoxville, Tennesse 37929

Gregory D. Reed, Professor
Civil Engineering Department
University of Tennessee
Knoxville, Tennessee 37996

INTRODUCTION

Small quantities of dye in a waste often contribute to the presence of color in very large quantities of wastewater and often the dye is affected very little in domestic wastewater treatment processes. The dye passes through the biological treatment process essentially unaltered producing a colored effluent. It is not practical for the municipality to remove the dye at their treatment facility since the quantity of waste is so large. It is more economical for the industry to remove the dye at its facility. However, if the dye is removed by precipitation at the industry, this usually results in sludge handling which is quite expensive.

Chemical coagulation is the chief method for color removal in dye wastes [1–6]. Typical coagulants used are alum, ferrous sulfate, ferric sulfate and ferric chloride with lime or sulfuric acid for pH control. Each coagulant has an optimum isoelectric point for a given dye which must be determined experimentally. Some dyes are not effectively coagulated with any known economical coagulant [1].

Nemerow [3] found that alum completely removed the apparent color from a sewage-dye waste at a dosage of 200 mg/L at pH 8.3 and 140 mg/L at pH 7.0. Parsons [4] reported alum doses of 2000 to 2500 mg/L of alum to remove the color of a different dye waste while doses of only 500 to 700 mg/L of ferrous or ferric sulfate were required. Lime was also added in each case with the best results produced by the addition of lime prior to the addition of ferric sulfate and after the addition of ferrous sulfate. Crowe et al. [7] discovered that cationic polymers enhanced the effectiveness of alum while nonionic and anionic polymers were ineffective. Polymer dose requirements were reported to have a stoichiometric relationship with the concentration of dispersing agent while the alum dose requirement increased in a non-linear fashion with increasing concentration of dispersing agent. This cationic polymer enhancement was also reported by Stahr et al. [8] who had found the alum to be ineffective without the polymer. Dosages were 100 to 300 mg/L of alum and 50 to 300 mg/L of polymer.

While alum has been a common coagulant, iron coagulants have also been used with success [9,10]. Hydrolysis effects had been proposed to explain the removal of the dye micelles; however, Weber [11] demonstrated that the effect is negligible and that the treatment is solely the result of coagulation and precipitation. Iron coagulant doses generally are in the hundreds of mg/L for either ferrous or ferric. These reactions produce a quick settling, stable floc which can easily be removed by sedimentation at the treatment plant [12].

Oxidizing agents have been tried with variable success [13]. Chlorine or hydrogen peroxide doses of 100 to 300 mg/L have been able to bleach the color out of some dye wastes while in others this technique was ineffective.

Treatability studies were conducted to determine if the colors of two textile wastes could be destroyed by oxidation or removed by coagulation since the literature indicated that either might be effective. The studies were conducted to determine if the dye, once removed by coagulation, would remain insoluble even after prolonged exposure to sewage so that the solids could be removed by the domestic treatment plant and relieve the industry from sludge handling operations.

LABORATORY ANALYSIS

Wastewaters tested contained a mixture of dyes with the dominant color being either a dark blue or a dark red. Color removal was measured as the percent change in light transmission at 445 nm (zeroed against new sewage). The pH of these wastes ranged between 7 and 9. Each sample was tested with two oxidizing agents, sodium hypochlorite and hydrogen peroxide. Sample No. 53 was tested first and the test was stopped when a dosage of 2.1% NaOCl was required to achieve only 70% reduction. This dose was considered too high for economical use, and the testing on the other samples was stopped at the doses shown in Table I when no color reduction was detected. The same relationship was true for the hydrogen peroxide as well.

The wastes were then treated with ferric chloride as a coagulating agent and sodium hydroxide to maintain the pH between 8 and 9. Sodium hydroxide was chosen so that additional sludge would not be produced.

The results are shown in Table II.

The sludge formed was stable when mixed with raw municipal sewage and there was no measurable color change to the sewage.

A second set of samples from a different source was analyzed in the same manner. The dominant color in these samples was a dark red. The dosages of ferric chloride required to remove the color and of sodium hydroxide to maintain a pH of 7 are shown in Table III.

In this case it was found that the sequence of the addition of ferric chloride and sodium hydroxide was important. Adding the sodium hydroxide first produced a weak floc with little color removed while adding the sodium hydroxide after coagulation produced the results in Table III. As with the previous samples, mixing the waste with sewage did not release any of the color and the combined stream did not show any increase in color. A slow mix period of five minutes was necessary to produce a large floc which settled rapidly.

APPLICATION

One of the industries studied dyed spools of yarn by introducing the dye in a solution under pressure and high temperature. After equilibrium was reached, the spools were removed and the solution was discharged to the municipal wastewater plant. Since this created problems for the municipality, they required pretreatment for color removal. The municipality established a policy that if 20:1 dilution of the dye waste with sewage essentially produced a colorless to the eye waste, then it was acceptable for discharge. The industry produced approximately 150 gallons per minute of wastewater flow and had an equalization tank to dampen out fluctuations in flow.

This industry's wastewater required the addition of alkalinity after the ferric chloride was added to achieve a good flocculation of the color. Based on the laboratory results, approximately 500 pounds of technical grade, 50% sodium hydroxide was required to be added per day or 40 gallons per day of sodium hydroxide technical grade. Also, ferric chloride was required to be added at 1154 pounds technical grade-per-day or approximately 100 gallons per day technical grade 35% ferric chloride. At

Table I. Oxidizing Agents

| Sample | NaOCl | | H₂O₂ | |
	Dose, mg/L	% Color Reduction	Dose, mg/L	% Color Reduction
51	1050	0	450	0
52	682	0	240	0
53	21000	70	765	10

Table II. Ferric Chloride

Sample	Dose, mg/L	% Color Reduction
51	328	70
52	328	90
53	281	75

Table III. Ferric Chloride/Sodium Hydroxide

Sample No.	23	24	25
FeCl$_3$·6 H$_2$O, mg/L	355	224	374
NaOH, mg/L	140	92	120
Final pH	6.9	6.6	6.9
Final Transmittance	89	95	90
% Color Reduction	85	92	86

local prices this amounted to approximately $300 per day for ferric chloride and $50 per day for sodium hydroxide.

Since the color was only present at times when the dye tanks were being discharged and flow from the plant was continuous, the owner wished to operate the treatment system only when the color was present to save money. To accomplish this task, a light sensitive photo-electric cell was used to detect when dye is present in the effluent. To insure that sufficient dye is present, a time delay was added to the circuit so that the chemical feed pumps would not cycle unnecessarily. Once the time delay is complete the chemical feed pumps and the propeller mixer in the reaction mix tank start and continue to run as long as dye is present. When the effluent is void of any further dye, a second time delay is used to provide a further flow of chemicals to the reaction mix tank for adequate precipitation of the dye and to monitor the effluent for another dye discharge. When the second delay is complete the pumps are de-energized until the effluent again contains enough dye to complete another cycle. To

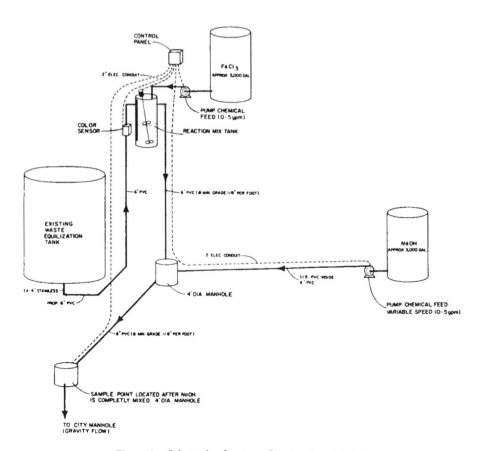

Figure 1.   Schematic of proposed pretreatment system.

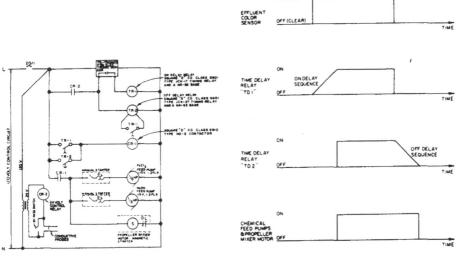

Figure 2. Schematic of wastewater pretreatment control system.

Figure 3. Graphs of sequence of events.

insure that the pretreatment system does not operate when effluent flows are low, a level sensor was provided to isolate the controls. A schematic diagram is shown in Figure 1, and the control system is shown in Figure 2. Figure 3 shows how the time delay system works. Since the system has been installed, it has functioned reasonably well except when flows fluctuate considerably from the design amount and the pumps are not appropriately adjusted. If this continues, a feed back control mechanism may have to be installed. However, most of the time, this does not present a problem.

The second industry studied had a much smaller flow of approximately 25 gallons per minute. Their system has worked superior to the above industry due to the reduced flow and fewer fluctuations from design levels. They have not had to add caustic since their waste has a higher alkalinity. The second industry's estimated cost was approximately $700 per month, total, for ferric chloride.

The second industry deposited dye on paper which was then transferred to fabric by a heat process similar to iron-on decals for tee shirts except the paper was the width of the fabric which could be up to 72-inches wide. Each time the colors were changed, the equipment had to be cleaned up and this resulted in the dye waste. The dyes from each of the manufacturers reacted quite similarly to the addition of ferric chloride, and the dye remained in a precipitated form that would settle out at the wastewater treatment plant.

The schematic design for the second industry was the same as the first except much smaller tanks were used and smaller mixers. The second industry has not had any problems with controlling their system except in normal maintenance operations.

REFERENCES

1. Nemerow, N. C., *Industrial Water Pollution*, Addison-Wesley Publishing, Reading, MA (1978).
2. Nemerow, N. C., "Textile Dye Wastes," Proceedings of 7th Purdue Industrial Waste Conference, Purdue University (May 1952).
3. Nemerow, N. C., "Color and Methods for Color Removal," Proceedings of 11th Purdue Industrial Waste Conference, Purdue University (May 1956).
4. Parsons, W. A., *Chemical Treatment of Sewage and Industrial Wastes*, National Lime Association, Washington, D.C. (1965).
5. Kreyer, W. C., et al., "Polymer Aided Coagulations of Textile Dyeing and Finishing Wastes," Proceedings of 27th Purdue Industrial Waste Conference (May 1972).
6. Porges, R., et al., "Chemical Precipitation of Textile Wastes and Studies of pH Control," Sewage Water Journal, 11:823 (1959).
7. Crowe, T., et al., "The Coagulation of Disperse Dyes," Proceedings of 32nd Purdue Industrial Waste Conference (May 1977).

8. Stahr, R. W., et al., "Textile Waste Treatment: Color Removal and Solids Handling Characteristics," Proceedings of 35th Purdue Industrial Waste Conference (May 1980).
9. Kace, J. S., and Linford, H. B., "Reduced Cost Flocculation of a Textile Dyeing Wastewater," Journal WPCF, 7:1971 (1975).
10. Souther, R. H., and Alspaugh, T. A., "Textile Waste Treatment Studies," Proceedings of 23rd Purdue Industrial Waste Conference (May 1958).
11. Weber, W. J., Discussion of paper by Zuckerman and Molof, Journal WPCF, 42:456 (1970).
12. Shelly, M. C., et al., "Evaluation of Chemical-Biological and Chemical-Physical Treatment for Textile Dyeing and Finishing Wastes," Journal WPCF 48:4:753 (1976).
13. Chamberlain, N. S., "Application of Chlorine and Treatment of Textile Wastes," Proceedings of 3rd Southern Municipal and Industrial Waste Conference (March 1954).

Section Eleven
LANDFILL LEACHATE

73 LEACHATE GENERATION IN A LINED LANDFILL. A CASE STUDY OF THE CENTRAL SOLID WASTE FACILITY AT SANDTOWN, DELAWARE

N. C. Vasuki, General Manager
Delaware Solid Waste Authority
Dover, Delaware 19903

INTRODUCTION

Solid waste disposal is a major environmental problem throughout the United States and most industrialized countries. Contamination of ground and surface water from previously abandoned disposal sites and some currently operating sites have brought extraordinary media attention to the disposal problem. Although the population is running scared of landfills, the nation will continue to rely on landfills as the major repository for unprocessed or unburned solid waste for at least the next ten to fifteen years. In the meantime, many communities are moving towards beneficial combustion of solid waste as fuel for steam and electricity production.

Even with such plants, a landfill is needed for ash disposal, emergency by-pass, and non-combustible wastes. Consequently, the nation will continue to rely on landfills as the ultimate repository of residues and wastes. Surprisingly, despite this extraordinary dependence, the design of landfills and leachate collection systems is still based upon empirical procedures.

Landfills generate methane gas as a major constituent of the decomposition gases. Migration of landfill gas and the emission of volatile organic compounds are also perceived as threats to the environment. It has now become practical to capture the landfill gas for beneficial use although the mechanisms of gas production, collection, and movement within the landfill are not fully understood and the design depends on empirical procedures.

The State of Delaware relies on groundwater as the major source of drinking water. Prior experience with leachate contamination of aquifers has made the state very sensitive to methods of disposal of solid wastes and sewage sludge. In order to forestall an impending crisis in solid waste disposal, the Delaware General Assembly authorized, in 1975, the creation of the Delaware Solid Waste Authority.

Three new landfills have been developed by the Delaware Solid Waste Authority to serve each of the three counties of the state, and a major complex for materials recovery and energy recovery has been constructed in the most populous northern county.

CENTRAL SOLID WASTE FACILITY–SANDTOWN, DELAWARE

The first new landfill constructed by the Authority is in Kent County, Delaware. The County's population of 100,000 (1985) resides in an area of 594 square miles. Dover, the capital of the State, is located in Kent County. The major industries in Kent County include 1) General Foods Corporation—a manufacturer of chocolate, coconut, rice, and other prepared food products; 2) International Playtex Corporation—manufacturer of rubberized goods for consumer and industrial applications; 3) Scott Paper Company—manufacturer of paper towels and industrial wipes; 4) Reichhold Chemicals, Inc.—manufacturer of synthetic latex materials for paint and other chemical products; 5) PPG Industries, Inc.—manufacturer of latex paints; 6) Country Pride Foods—poultry processing; 7) Dover Air Force Base—the largest Military Air Transport Command Base on the eastern seaboard; 8) ILC Industries—manufacturer of space suits; and 9) State of Delaware—State Government capital complex.

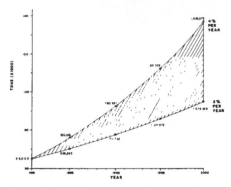

Figure 1. Solid waste generation projections.   **Figure 2. Central solid waste facility.**

WASTE GENERATION

The actual generation data from select communities in Kent County, Delaware, indicate a rate of 2.74 lbs/capita/day (1000 lbs/person/year) . During calendar year 1985, the landfill received 100,000 tons. Of this amount, approximately 50,000 tons are from households in the county. The remaining is a combination of commercial and industrial wastes, including fly ash from a coal-fired co-generation plant at General Foods. In Kent County, the mean household population is 2.6, and the amount of solid waste generated is 1.3 tons/year.

In order to design a solid waste disposal program for a twenty year period, the Authority has made estimates of total waste generation over a twenty year period between 1981 and the year 2001. The growth rate is based on several factors including population increases, changes in per capita genera-tion of solid waste, and higher industrial activity in Kent County. The Authority's projections include an envelope ranging from 2% to 4% compounded growth rates as shown in Figure 1. This is very important in a long range disposal program because the site acquisition process must result in acquiring space for the estimated maximum amount of solid waste expected. The actual growth rate will determine the rate of usage of the remaining landfill capacity and migration of wastes across county and state lines can materially reduce space available for landfills. However, through exercise of "Flow Control" (an important issue in the solid waste industry) landfill projects can be managed. Flow control means the ability to direct solid wastes generated in a geographical area to a specific disposal site.

In Kent County, the goal was to design, construct, and operate a new landfill system which could accommodate the solid waste generated during a 20 year period and prevent damage to the environ-ment.

THE SITE

The Authority's Central Solid Waste Facility is comprised of 547 acres. The Authority owns approximately 460 acres and leases the remaining 87 acres. The site (Figure 2) is planned to accommo-date a series of seven cells which will cover 218 acres for landfilling.

Assuming a refuse to cover ratio 4:1, "in place" density of 1000 lbs/cu yd and 40 feet of refuse depth, the total capacity of the landfill site is over five million tons. Therefore, the site is more than ample to dispose the estimated two million tons expected between 1981 and the year 2001.

Prior to start of construction, the Authority conducted detailed hydrogeologic [1] and baseline studies of the site [2]. The hydrogeologic studies used data from six deep wells and eight soil profile borings in addition to the available information from the Delaware Geologic Survey. Four major aquifers were identified at the site. They are shown as Figure 3. The water table and the Cheswold Aquifers are the most important source of water in western Kent County.

DESIGN AND CONSTRUCTION FEATURES

In 1980, there was no rational design model for constructing and operating a landfill on a predomi-nantly sandy soil strata. No local clay deposits were nearby and, therefore, clay could not be used as a liner material. Imported bentonite clay was considered as an option. However lacking sufficient

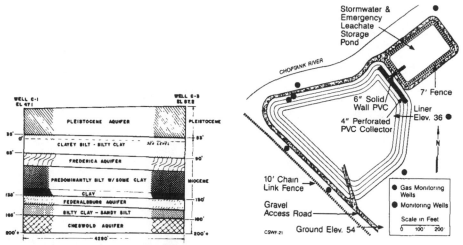

Figure 3. Cross-section of central solid waste facility.

Figure 4. Leachate collection system of cell "A".

information on effectiveness of such clay liners for landfill application, the decision was made to go with synthetic liners. The studies conducted by Haxo et al. [3] were useful in making the synthetic liner selection. Some limited information on usage of such liners at a landfill in Long Island induced more confidence in selecting synthetic liners.

The first landfill Cell "A" was constructed in a nine acre abandoned sand pit previously used by the Delaware State Highway Department. Inspection of the site showed the presence of water at a depth of 18 feet. There was no visible groundwater flow in the side walls of the pit. The nine acre pit was cleared of debris and several deep holes were backfilled with locally available soil. The subbase was compacted to accept the liner. Special care was taken to ensure a smooth surface devoid of protrusions. The pit bottom was graded to yield a slope of 0.5% to induce the leachate to collect at the northwestern end of the cell. The liner was overlaid with two feet of clean, sandy soil devoid of large stones, twigs, or other debris. This layer protects the liner from materials which could puncture the membrane and also acts as a collector of leachate. Inspection of the tracks left by the compactor on the first day of operations showed only indentations of approximately eight inches in the soil cover, even when the compactor started from rest. The pit depth averaged six feet with the northwestern end at elevation 36 feet.

The 30 mil PVC fabric (in 300 ft. long−70 ft. wide strips) was placed on the side and bottom of the pit. A transverse leachate collection line was constructed with an outlet into a pond (Figure 4). The pond bottom and sides were also graded and compacted prior to placement of a 36 mil Hypalon liner. A special boot was prepared to allow the penetration of the leachate line through both liners. The liner was anchored as shown in Figure 5. Initially, the pond was designed to hold the anticipated estimated flow of 3300 gallons/day of leachate from Cell "A". A series of groundwater monitoring wells were constructed surrounding the cell in order to measure the water levels in the aquifers as well as the water quality. A typical well cross section is shown in Figure 6.

Cell "B" was constructed in 1982, and its design was an improvement over Cell "A". Its size is 17.5 acres. The leachate collection system for Cell "B" was designed with a herringbone pattern as shown in Figure 7. As before, the subgrade was carefully graded and compacted to allow the leachate to flow to the northern end. A 30 mil PVC liner was selected and the entire cell, including the sides, was covered with the synthetic fabric. Two feet of clean, sandy soil was placed to cushion the liner. Figure 8 shows the cross section of the leachate collection trench. Unlike Cell "A", there is no penetration of the liner. Instead, a sump is provided and the leachate collected is pumped from the sump into fiberglass leachate storage tanks. The liners of adjacent cells were connected as shown in Figure 9 thereby maximizing the use of lined areas.

During construction of Cell "B", two fiberglass storage tanks (each 20,000 gallons capacity) were installed to hold the leachate generated. The line carrying leachate from Cell "A" into the pond was connected to the new fiberglass storage tanks by a three way valve. The leachate from Area "A" is no

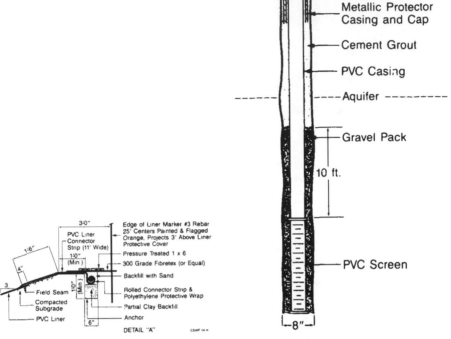

Figure 5. Liner anchor.

Figure 6. Typical monitoring well.

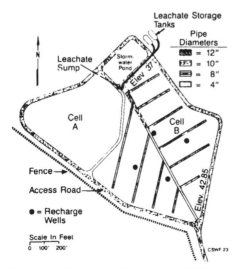

Figure 7. Plan view of leachate collection system for cell "B".

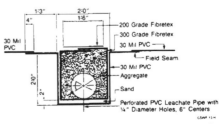

Figure 8. Typical leachate collection trench for cell "B".

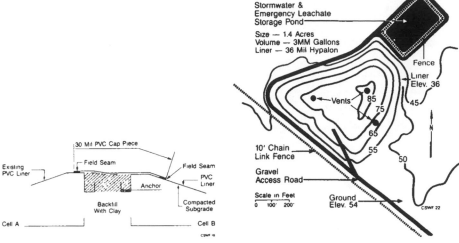

Figure 9.   Liner connection.   Figure 10.   Completed fill contours for cell "A".

longer allowed to go directly into the pond. The pond is now used for stormwater storage. It is also available for storage of leachate in the event of an emergency.

A heavy downpour prior to start of solid waste disposal in Cell "B" showed the efficient movement of rainwater to the northern end while the southern end remained dry. There is further room for improvement in the design of leachate collection systems. Our newest landfill near Wilmington, Delaware, uses a grid system and no point within the landfill is more than 200 feet from a leachate collection line.

After Cell "A" was placed into operation (October 1980), the limited quantity of leachate generated became diluted with rainwater accumulating in Cell "A" and the leachate holding pond. After testing and with the concurrence of the State Regulatory Agency, the accumulated water in the pond was discharged into the Choptank River whenever the BOD was less than 30 mg/L. Cell "A" was filled to an average height of 40 feet. Figure 10 shows the finished contours of Cell "A". In accordance with the permit conditions, the cell was capped with two feet of compacted soil and vegetated with grass.

Table I and Figure 11 shows the leachate generation pattern for Cell "A" ever since it was filled and capped. Over a period of three years, the leachate production has dropped from a 600 gpd (peak flow) to approximately 60 gpd. Leachate generation rate, using the EPA water balance method, was estimated at 3300 gpd. In actual practice, leachate generation rate can be reduced substantially below those estimates if the landfill cells are designed and operated to induce maximum runoff. The reasons for the discrepancy are not known at this time.

A break in the liner and consequent escape of the leachate has been ruled out because the ground-water monitoring system and visual inspection along the Choptank River (including vegetation) do not show any leakage. Figure 12 shows the potentiometric head in the groundwater table aquifer. Over the years there has been no increase or anomalies in the potentiometric head. Figures 13 and 14 show potentiometric heads in the Frederica and Cheswold Aquifers, respectively.

As can be seen, both are artesian aquifers. The probability of the leachate contaminating these deeper aquifers, especially the Cheswold Aquifer, is low as long as the system is properly managed. Three aquifers are monitored on a routine basis because there may be leakage elsewhere between the aquifers.

Perhaps the reasons why the leachate generation rates vary significantly from the predicted rates are as follows: 1) The infiltration rate may be much lower than assumed for a water balance calculation; 2) The mound system of landfilling induces more efficient runoff, thereby reducing the potential for infiltration; 3) The high compaction rate achieved prevents movement of the leachate through the mound; and/or 4) The waste within the fill may not have reached full field capacity.

Dr. John Mather of the University of Delaware, who developed the water balance method [4] for estimating leachate generation rates, is currently investigating the cells. Attempts will be made to

Table I. Leachate Generation Pattern
Area of Cell "A"–9 Acres
Cell Start – October 23, 1980
Cell Finish – October 9, 1982
Solid Waste in Cell – 140,000 Tons

Month/Yr.	Average Rate Gallons/Day[a]
1. Dec. '81	Trickle
2. Jan. '83	325
3. March	425
4. April	600
5. June	206
6. Sept.	248
7. Dec.	250
8. March '84	280
9. May	400
10. Sept.	118
11. Nov.	80
12. March '85	132
13. June	124
14. Sept.	69
15. Dec.	54
16. March '86	74

[a]Preliminary Data not corrected for recycling.

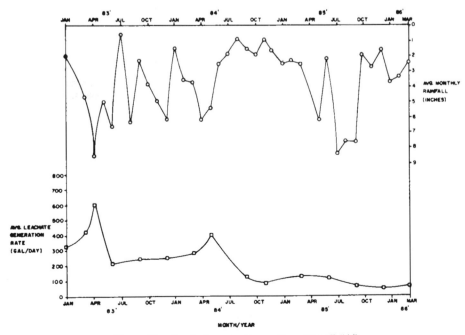

Figure 11.   Leachate generation pattern for cell "A".

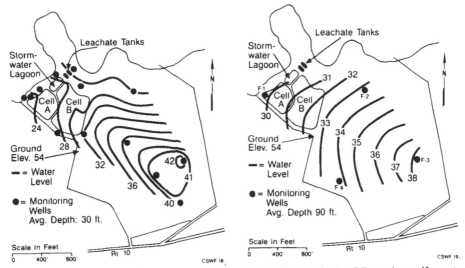

Figure 12.   Groundwater table contours.   **Figure 13.   Frederica (Miocene) acquifer artesian head contours.**

capture runoff from known areas of the landfill surface to determine actual rates of infiltration. In addition, piezometers are being installed in Cell "A" to determine the potentiometric head of the leachate within the fill.

LEACHATE TREATMENT

Based on Pohland's [5] work on leachate recirculation in controlled laboratory cells, large scale leachate recirculation has been attempted. Special wells to allow percolation of leachate have been constructed within the landfill cells as shown in Figure 15. Periodically, leachate accumulating in the storage tanks is pumped into the leachate recharge wells.

This process has its own problems. The first well became saturated after attempts to recirculate. This showed that modifications were needed to spread the flow of leachate. The two remaining wells in Cell "A" were constructed with coarse gravel trenches (radiating from the wells and between the lifts of solid waste). This system seems to work much better than the first attempt although rest periods are needed between pumping cycles. The reason for difficulties in recharging through wells may be the non-homogenous nature of the solid waste, and the presence of plastics, trash bags, plastic sheets, carpet backing, etc. prevent uniform spread of leachate. Disposal of dewatered sludge and fly ash also affect dispersal of leachate.

A traveling spray irrigation system is used to spray leachate (when conditions permit) on unused portions of Cell "B". Leachate cannot be sprayed on Cell "A" because of the high iron content which seems to plug the soil and kill the grass cover. These attempts to recirculate leachate have shown that: 1) leachate recirculation systems must be designed similar to leachate collection systems in order to get the reverse effect of spreading the leachate rather than collecting it; 2) landfill operator must assure that there exists a pathway for leachate between lifts; 3) the number of leachate recharge wells may have to be increased, perhaps to one per acre of landfill; and 4) if landfill gas collection systems are planned, then the leachate recirculation systems could double as gas collection devices.

In Cell "A" the leachate recharge wells double as gas vents. From periodic tests performed, one out of three wells show anticipated gas production, i.e., 50% methane and 45% CO_2. The other two wells do not indicate significant gas production. Perhaps the cell is quite dry in those areas. Once the piezometers are installed, it may be possible to determine why gas production is low in some portions of the cell. While leachate recirculation is useful to moderate the characteristics of the leachate, a separate leachate treatment plant becomes necessary, particularly in winter months and during periods of heavy precipitation.

A treatability study of the leachate generated at Sandtown was conducted by Gaudy and Rozich of the University of Delaware. They will report their findings in the next paper.

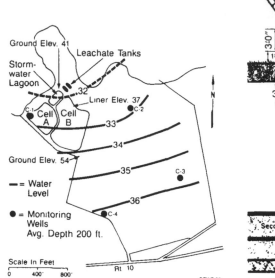

Figure 14.   Cheswold (Miocene) aquifer arte-   Figure 15.   Gas vent/recharge well details.
sian head contours.

COSTS

The cost of land acquisition, engineering and construction of Cell "A" are shown in Table II. The cost of engineering and construction of Area B is given in Table III. In Table IV, the costs of construction of Cells "A" and "B" are compared and it shows that Cell "B" costs/acre were lower than that of Cell "A". This is mainly due to experience gained by the contractor, the consulting engineer, and the Authority staff. The construction cost of the land, subgrade preparation, liner, and leachate collection system to dispose 675,000 tons turns out to be $1.43/ton—a real low cost!

The operating costs of the entire facility for FY 87 are shown in Figure 16. The disposal fee charged is $18.65/ton although the actual cost is $21.63/ton. The difference of $2.98 is a rebate provided by the Authority from income surpluses from previous years. The environmental monitoring costs (i.e., water quality and landfill gas tests) are only about $.42/ton of solid waste or less than a nickel per family per month. The monthly disposal cost per dwelling unit in select communities of Kent County vary between $1.62 and $2.02 as shown in Table V.

Table II. Central Solid Waste Facility
Cell "A"—9 Acres
Construction Cost

1. Land	$ 5,067
2. Subgrade Preparation	$ 153,395
3. Liner and Leachate Collection System	$ 182,322
4. Access Roads and Parking	$ 186,445
5. Stormwater Pond	$ 143,090
6. Maintenance Building and Office	$ 133,950
7. Scales Plus Computer	$ 80,000
8. Security Fence	$ 105,070
9. All Other	$ 42,647
Total	$ 1,031,986

Table III. Central Solid Waste Facility
Cell "B" — 17.5 Acres
Construction Cost

1. Land	$ 9,900
2. Subgrade Preparation	$ 163,000
3. Liner and Leachate Collection System	$ 457,000
4. Leachate Storage Tanks	$ 66,000
5. All Other	$ 57,935
Total	$ 753,835

Table IV. Central Solid Waste Facility
Construction Cost

Item	Cell A	Cell B
1. Land	$ 5,100	$ 9,900
2. Subgrade Prep	$ 153,400	$ 163,000
3. Liner and Leachate Collection System	$ 182,300	$ 457,000
Total	$ 340,800	$ 629,900
4. Cell Area (Acres)	9	17.5
5. $ Cost/Acre	37,900	36,000
6. Tonnage Disposed	140,000	535,000
7. $/Ton	2.43	1.18

Table V. Delaware Solid Waste Authority
Central Solid Waste Facility

Item	Dover	Smyrna	Milford
Population (1985)	27,000	4,712	2,660
Service Units (1985)	8,150	2,124	1,299
Annual Solid Waste Generation for 1985 (Tons)	10,587	2,200	1,507
Annual Solid Waste Generation per Service Unit (Tons)	1.30	1.04	1.16
Monthly Disposal Cost @ $18.65/Ton	$2.02	$1.62	$1.80

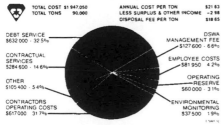

Figure 16.   Operating costs for FY 87.

SUMMARY

Landfills are an essential part of any solid waste disposal system.

Despite the malignment, they will continue to be the mainstay disposal system for the nation during the next decade.

Landfills are the lowest cost option for solid waste disposal outside urban centers.

Leachate production in landfills can be controlled by constructing mounded cells and by maximizing runoff.

Leachate recycling on a large scale basis is difficult through the use of only recharge wells. Spray irrigation systems are also needed.

Leachate cannot be directly applied to vegetated areas of landfill because it will kill the vegetation.

The treated leachate of a landfill could be beneficially used to maintain grass cover on side slopes and top of the landfill mounds.

During wet weather and freezing conditions, alternate treatment and disposal systems for leachate will be needed.

REFERENCES

1. Stanislawczyk, D., "Geotechnical Evaluation of the Kalish Farm and Adjacent 66 Acre Tract," Gilbert-Commonwealth, Reading, Pennsylvania (1980).
2. Klemas, V., Ackleson, S., Sprague, L., and Wethe, C., "Environmental Brochure Study of the Central Solid Waste Facility," *Center for Remote Sensing Bulletin*, College of Marine Studies, University of Delaware (1980).
3. Haxo, H. E., Jr., "Liner Materials Exposed to MSW Landfill Leachate," *Proceedings of the Fifth Annual Research Symposium — Municipal Solid Waste: Land Disposal, EPA — 600/9-79-023A*, U.S. Environmental Protection Agency (1979).
4. Mather, John, "The Use of Water Budget in Evaluating Leaching Through Solid Waste Landfills," *Water Resources Center Bulletin, University of Delaware (1978)*.
5. Pohland, F. G., Shank, D. E., Benson, R. E., and Timmerman H. H., "Pilot Scale Investigations of Accelerated Landfill Stabilization with Leachate Recycle," *Proceedings of the Fifth Annual Research Symposium — Municipal Solid Waste: Land Disposal, EPA — 600/9-79-023A*, U.S. Environmental Protection Agency (1979).

74 TREATABILITY STUDY OF HIGH STRENGTH LANDFILL LEACHATE

Anthony F. Gaudy, Jr., Professor

Alan F. Rozich, Associate Scientist

Stephanie Garniewski, Graduate Student
Department of Civil Engineering
University of Delaware
Newark, Delaware 19716

INTRODUCTION

The sanitary landfill was devised over four decades ago as a relatively inexpensive, hygienic way to bury solid refuse rather than continue the then prevalent practice of open dumping. At the time little consideration was given to the potential for groundwater pollution. The landfill is an attractive technology because, in addition to its simplicity and relatively low cost, it is an ultimate method of disposal (as, for example, compared to incineration of municipal refuse). As the use of this technique grew, there was increasing concern over pollution of groundwater by organic materials leached from the municipal refuse and, since many municipalities accepted industrial wastes as well as municipal refuse, there was increasing concern over the possibilities for entry of hazardous waste components into the groundwater resource. Today, mounting evidence for groundwater contamination mandates that modern landfill design provide means for containing all seepage, or leachate, from the fill.

The sanitary landfill operated at Sandtown, Delaware, by the Delaware Solid Waste Authority is a state-of-the-art facility in which leachate is collected and held on the site, but as yet there is no provision for its treatment. The facility accepts solid industrial wastes as well as municipal refuse. The operational and managerial strategies for the facility are presented elsewhere [1]. Approximately 25,000–30,000 gallons of leachate are produced daily. Its BOD_5 is approximately 100 times higher than that of municipal sewage and it contains significant amounts of heavy metals. Since the amenability of municipal landfill leachate to aerobic biological treatment has been rather widely reported and, since some of the work recently reported from our bio-environmental engineering laboratories has dealt with biological treatment of waste containing heavy metals [2,3] and inhibitory and toxic compounds [4,5,6,7] as well as conventional pollutants, the Delaware Solid Waste Authority approached us to discuss the possibilities for investigating the biological treatment of the Sandtown site leachate. Our task was essentially three-fold.

There was need for additional information on the chemical characteristics of the leachate. The Authority, from time to time, had obtained BOD, COD, and metals analyses on the waste, as well as analyses for volatile acids concentrations, but a more systematic program of sampling and analysis was needed to gain some idea of the variations in composition which might be expected, as well as the average values.

Since it is widely known that leachate composition varies from site to site, it is equally well-recognized that the amenability of the leachate to biological treatment can also vary. Thus, it was important to perform a treatability investigation on the leachate over a period of time sufficient to allow assessment of the biochemical response to variations which might occur in the waste composition. One of the best approaches to treatability studies which we have found is to determine values of the biokinetic parameters which govern the process. These are also the kinetic characteristics needed for insertion into kinetic models for prediction of performance and of the effects of variation in engineering control parameters on system performance. That is, the study should include the determination of the following biokinetic "constants": μ_{max}, the maximum specific growth rate; K_s, the saturation constant; K_i, the inhibition constant; k_d, the specific decay rate; Y_t, the true cell yield; and k_a, the autodigestion constant. In our view, such studies require the running of both a continuous-flow pilot plant and batch experiments.

It was intended that the knowledge gained in the conduct of studies relating to items 1 and 2 above should lead to the design of a treatment scheme, i.e., selection of the unit process chain for treatment and the essential control strategies.

MATERIALS AND METHODS

General Constraints and Conditions

In making the investigation, the time factor was of urgent concern. Leachate was accumulating and there was need for speed in making a decision regarding treatment. The amount of money available for the investigation was also rather limited. After considering the problem, it was felt that a 6-month period would probably suffice, and, because the project represented a somewhat experimental approach to research sponsorship and industrial partnership, the university shared some costs usually borne by research sponsors. It should also be pointed out that, in approaching the problem, we knew that the Delaware Solid Waste Authority had land available which could be used for spray irrigation of treated leachate, and it was anticipated that such a polishing treatment would be employed.

Experimental Approach and Methods

Samples of leachate were obtained from a leachate collection sump between the landfill and the leachate storage facility. The volume of liquid in the sump was approximately 2,000 gallons, or about $1/15$ of the daily leachate flow. Periodically, leachate was delivered to the Bioenvironmental Engineering Laboratories. Samples were taken for analysis and the remainder was frozen in volumes needed for the daily experimental routine. In order to obtain the data to determine k_d and Y_t, i.e., to prepare a maintenance plot, a once-through chemostat was operated at various dilution rates, D. The pH was maintained at 7.0 by automatic addition of sodium hydroxide. The impeller speeds and the airflow rates were such as to maintain complete mixing and the dissolved oxygen level was always above 4 mg/L. The system was run at 25° ± 1°C. Analyses for NH_3-N [8], total Kjeldahl nitrogen [9], and phosphorus [9] were performed, and the leachate was found to be deficient in both nitrogen and phosphorus. These nutrients were added in slight excess of the usual ratios of BOD: N = 20:1 and BOD: P = 100:1 to assure that carbon source was the growth-limiting nutrient. Soluble COD's [10] were run on the feed and the effluent. BOD's [10] were run, but less often. Mixed-liquor suspended solids concentrations were determined by filtration through a 0.45 μm poresize filter, drying at 105°C and weighing. Mixed-liquor volatile suspended solids were also determined as prescribed by *Standard Methods* [9]. Samples were taken for analyses by atomic absorption [9] for soluble iron, manganese and zinc, which were known to be in the leachate.

During each run at a specific hydraulic detention time, mixed liquor from the chemostat was used as seed material for batch growth studies on various dilutions of the leachate. Growth was measured as absorbance (optical density) at 540 nm. These results were used to obtain numerical values for μ_{max}, K_s, and K_i. Numerical values were obtained by fitting the μ versus S data to either the Monod (non-inhibitory) or Haldane (inhibitory) equations using nonlinear least squares fitting techniques [6]. A batch reactor was set up as an aerobic digester to obtain some insight into values of the aerobic digestion constant, k_a, for activated sludge developed on the leachate. An additional batch experiment was run in which sludge was fed leachate at a fairly low initial feed to biomass ratio in order to obtain additional data regarding growth rate and also to obtain an estimate of the residual COD which might be routinely expected.

RESULTS AND DISCUSSION

Leachate Characterization

Table I shows the general characteristics of the leachate over the sampling period of approximately six months. It is apparent that the leachate was slightly acid and that it is a rather strong wastewater. It can be seen that the BOD_5 and COD are due almost solely to soluble organic materials with a rather high soluble BOD/COD ratio of 0.8. The waste is seen to be deficient in nitrogen and phosphorus, and it contains significant amounts of iron, manganese and zinc. The BOD/COD ratio suggests that the waste should be highly amenable to aerobic biological treatment.

Table I. Characteristics of the Leachate from the Sandtown, Delaware Landfill over the Period 5/14/85–11/21/85

	Average[a]	High[a]	Low[a]
pH	5.4	5.5	5.4
SS	750	1,630	180
VSS	560	1,400	270
COD_{TOT}	30,500	36,000	14,000
BOD_{TOT}	24,700	31,500	19,000
COD_{SOL}	29,800	34,500	13,000
BOD_{SOL}	23,600	31,000	18,300
TKN	370	580	160
NH_3-N	700	780	650
PO_4^{-3}	0.1	—	<0.1
Fe_{TOT}	1,550	1,700	1,350
Mn_{TOT}	41	50	30
Zn_{TOT}	120	210	65

[a]All concentrations are in units of mg/L, except pH.

Chemostat Studies

In all, six chemostat runs were made. Four were made at the constant temperature of 25°C, and two runs were included in which the temperature was allowed to increase autothermally. Data from the runs at 25°C were used to estimate k_d and Y_t from a maintenance plot. It would have been desirable to have six, rather than four, plotting points, but recent reports on studies of autothermal aerobic digestion of sludge [11] prompted us to reason that, because of the high strength of the leachate, autothermal operation of the activated sludge reactor might also be possible, and it was important to look into this aspect while the study was in progress. Figures 1 and 2 show the chemostat results for typical runs at 25°C. Figure 1 shows data taken when a 1:10 dilution of leachate was fed at a dilution rate of 0.5 day^{-1} ($\theta_c = \bar{t} = 2$ days), whereas the results shown in Figure 2 were obtained when undiluted leachate comprised the feed and the dilution rate was 0.17 day^{-1} ($\theta_c = \bar{t} = 6$ days). Figure 2 shows that a rather high residual soluble COD concentration can be expected, approximately 2800 mg/L for a feed COD averaging 29,000 mg/L. Soluble BOD_5 was not routinely run, but several analyses made during each steady state run indicated that, when feeding the 1:10 dilution of leachate, residual BOD_5 ranged between 20 and 30 mg/L and, when feeding undiluted leachate, soluble BOD_5 was approximately 200 to 300 mg/L. Also, it was found that the mixed-liquor suspended solids concentration contained a rather high ash content of approximately 50% (e.g., see Figure 2).

The average values ("steady state") for all six runs are shown in Table II. The values for detention times of 10, 6, and 2 hours at 25°C were employed to determine k_d and Y_t. When the temperature was allowed to rise autothermally, it leveled off at approximately 38°C under the insulating conditions employed, i.e., the sides of the glass cylindrical reactor were simply wrapped with insulating tape. The results of the autothermal runs are presented in Table II primarily to show that considerable metabolic heat energy is released and that it may be rather easily retained in the system to elevate the operating temperature and consequently exert an effect on the observed cell yield; i.e., it may reduce sludge production. Only two runs could be made, and the three-week running time for each was not really sufficient to develop a steady condition, but the decrease in Y_o when θ_c increased from 2 to 6 days would appear to be greater for the higher temperature. These preliminary results indicate that further consideration should be given to the possibilities of insulating the aeration tank to take advantage of the heat energy in the high-strength leachate.

A maintenance plot for the data obtained at 25°C is shown in Figure 3. A decay constant, k_d, of 0.10 and a true yield Y_t of 0.5 mg volatile suspended solids/mg COD removed were estimated.

Metal analyses were performed periodically during runs with the full-strength leachate. In general, it was found that only very low amounts of metals remained in solution after biological treatment. Thus, there was some reason to expect that in the prototype plant the metals could be swept down with the biomass during quiescent settling. This high degree of removal of metals is a fortuitous benefit of the biological treatment process which has been found in studies of synthetic wastes and on mixtures of municipal and industrial wastes [2,3,12].

It should be noted that, due to the high concentration of mixed liquor suspended solids (resulting from the high strength of the waste and the high ash content of the MLSS), settling characteristics

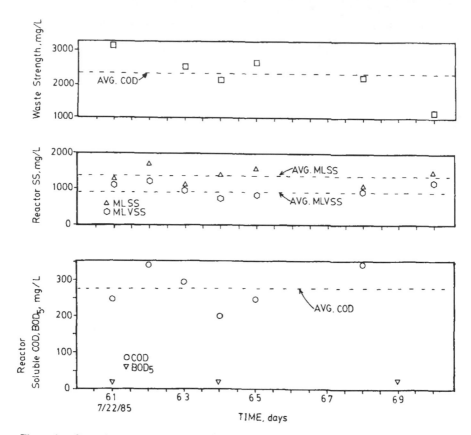

Figure 1.   Operational results for a once-through continuous-flow growth reactor fed a 1:10 dilution of landfill leachate. θ_c = 2 days.

were poor throughout the course of the pilot plant studies. Thus, it appeared that some sort of chemical or mechanical assistance to settling would be necessary in a full scale operation. It was not possible to investigate fully the possibilities for centrifugation or chemically-aided coagulation. However, several studies were made to assess the effects of addition of polymer to the mixed liquor, and two important observations were made. The results showed that for a mixed liquor suspended solids concentration of 14,000 mg/L, a supernatant of approximately 250 mg/L suspended solids could be readily attained with a one-hour settling time. Also, it was found that initial total heavy metal concentrations in the mixed liquor of 1650, 68, and 140 mg/L of iron, manganese and zinc, respectively, were reduced to 22, 6, and 9 mg/L, respectively, in the settled supernatant. From the results of the biological reactor studies and the brief study of effects of polymer addition, it would appear that biological treatment will not only provide removal of organic matter but will also achieve removal of significant amounts of heavy metal as well, and we concluded that removal of metals from the leachate prior to biological treatment was not necessary.

Batch Growth Studies

During steady state operation at 25°C, six sets of batch growth studies were made, each including six dilutions of leachate (i.e., initial COD levels of leachate) using cells from the chemostat as seed. These studies were made using cells taken over the entire operating range of growth rates, i.e., θc = 2 to 10 days, in an effort to estimate the range of biokinetic constants which might be expected. A typical growth curve is shown in Figure 4. The specific growth rate was measured as the slope of the straight line portion of the curve. The first set of growth studies gave cause for concern because they yielded a plot typical of growth on an inhibitory or toxic carbon source. Figure 5 shows a plot of the

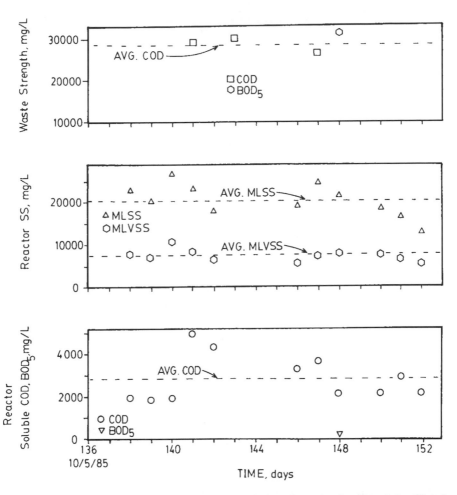

Figure 2. Operational results for a once-through continuous-flow growth reactor fed undiluted landfill leachate. θ_c = 6 days.

values of μ versus initial substrate concentration measured as soluble leachate COD. It is clear that, as initial COD concentration increased, μ decreased, as is typical for an inhibitory carbon source. These data were fitted to the Haldane equation, which has previously been shown to be suitable for fitting data for growth on inhibitory substrates [5,6]. The fitting procedure is described elsewhere [6]. The plotted points are the observed data points while the curve was calculated using constants determined from the data. All other runs yielded data which more closely fit the Monod relationship, i.e., the non-toxic relationship (e.g., Figure 6). Values of the biokinetic constants for all six runs are given in Table III. It is seen from Figures 5 and 6 that the curves plotted using the biokinetic constants determined from the data provide a reasonably good description of the relationship between μ and S.

Since Run No. 1 was the only one of the six runs which gave an indication of an inhibitory effect, it appears safe to proceed, although cautiously, on the premise that the leachate can be considered as a noninhibitory carbon source for growth of aerobic microorganisms. The values for the biokinetic constants, μ_{max} and K_s, obtained from the growth studies, do not differ greatly from those for other readily metabolized wastes. The range for K_s values was 29 to 437 mg/L and that for μ_{max} was 0.096 to 0.622 hours^{-1}. From larger samples of values obtained on readily available nontoxic carbon sources, statistically significant ranges for these constants have been found to be approximately 50 to 200 mg/L for K_s and 0.4 to 0.6 hr^{-1} for μ_{max} [13]. It is not unexpected to find values outside these ranges

Table II. Average Biomass Production and Substrate Removal for Six "Steady State" Runs in the Chemostat

Θ_c days	T °C	X mg/L	S_i mg/L	S mg/L	Y_o mg/mg
10	25	700	2,900	130	0.26
6	25	7,570	28,900	2,850	0.29
2	25	990	2,370	280	0.47
2	25	9,520	27,600	2,450	0.38
2	~38	13,210	29,000	2,020	0.49
6	~38	4,790	26,270	3,840	0.20

Θ_c = mean cell age; the reciprocal of μ_n, the net specific growth rate, days.
T = temperature of reaction mixture, °C.
X = concentration of biological solids in the reactor, mg/L.
S_i = concentration of carbon source in the influent (feed), mg/L (may be measured as COD and/or BOD).
S = concentration of carbon source in the reactor effluent, mg/L (measured as COD or BOD).
Y_o = observed cell yield, mg biomass produced per mg substrate (COD) utilized, mg/mg.

for small numbers of samples such as were obtained in this short study on the landfill leachate. However, average values for both constants are within the ranges of values usually observed.

It is not possible to combine these data with those from other studies on leachate to obtain more statistically meaningful values. Studies of this type are not common, and biokinetic data of any kind that are pertinent to aerobic treatment of landfill leachate are rather scant. Several recent studies have included determination of some of the biokinetic constants while others have simply determined BOD and/or COD removal efficiencies and requirements for treatment.

Robinson and Maris [14], studying a leachate with an initial BOD_5 of 3000 mg/L, reported a yield value of 0.44 mg solids per mg COD removed at a Θ_c of 10 days. Uloth and Mavinic [15] studied a leachate with an initial BOD of 35,000 mg/L and determined the following values for biokinetic constants: Y = 0.332 lb solids per lb BOD_5; k_d = 0.0025 day^{-1}; μ_{max} = 0.09 hr^{-1}; K_s = 21,375 mg/L. (The K_s value reported for this waste is approximately 100-fold higher than the normal range for this constant.) Cooke and Foree [16] found values somewhat similar to those found in the present study. Their constants were Y_t = 0.4, k_d = 0.05 and K_s = 175. Other constants were not evaluated. The values of Y_t and k_d were determined using only two points and are therefore subject to doubt. Variations in the strength of the wastes studied and differences in other characteristics such as BOD/COD ratio, toxicity, and metals content make it impossible to compare data for different landfills. It is apparent from even a brief survey of the literature that treatment of each landfill leachate must be considered on an individual basis using experimental data for that leachate. Other investigators have reached the same conclusion [17].

In addition to the batch growth studies, which necessarily employed very low initial biomass concentrations, an experiment was run to assess the rate of COD removal in a larger batch system with

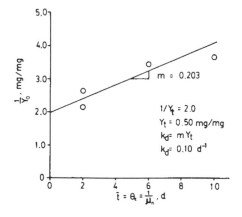

Figure 3.   Maintenance plot for determination of the true cell yield, Y_t, and the decay constant, k_d. Results plotted are from continuous-flow runs at 25°C (see Table II).

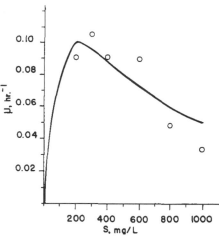

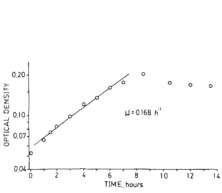

Figure 4.   Semi-logarithmic plot of optical density readings (as a measurement of biomass) versus time for a typical batch growth system. The initial leachate concentration was 300 mg/L COD.

Figure 5.   Observed specific growth rate μ versus S (plotted points) for the first set of growth studies, showing an inhibitory effect of the leachate. The curve was calculated using the Haldane equation and the biokinetic constants given in Table III: μ_{max} = 0.264 hr^{-1}, K$_s$ = 164 mg/L, K$_i$ = 251 mg/L.

a higher initial solids concentration. This experiment also served as a check on the residual COD observed in the continuous flow studies. A batch reactor was seeded with 1200 mg/L of sludge from the continuous flow reactor and brought to an initial concentration of 3600 mg/L leachate COD. Nitrogen and phosphorus were added in slight excess and the system was aerated. The resulting COD removal and biomass accumulation curves are shown in Figure 7. It is seen that the rate of COD removal was slow, even though a rather high initial suspended solids concentration was employed. After 24 hours 82% of the COD had been removed whereas with rapidly metabolized wastes at the same ratio of solids to COD, most of the COD would be removed in less than half that time. The remaining COD was removed very slowly. These results compare well with those from the batch growth studies; that is, since growth in the system used to evaluate the constants depends on substrate removal, slow substrate removal and slow growth rate are synonymous. The growth rates measured in this study were generally in the lower part of the range usually observed with other wastes such as

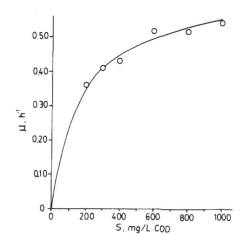

Figure 6.   Observed specific growth rate μ versus S (plotted points) for growth on landfill leachate, showing a non-inhibitory effect which was more typical of the growth studies other than those shown in Figure 5. The curve was calculated using the Monod equation and the biokinetic constants given in Table III: μ_{max} = 0.622 hr^{-1}, K$_s$ = 152 mg/L.

Table III. Values of the Biokinetic Constants, μ_{max}, K_s and K_i, Determined for the Landfill Leachate at Sandtown, Delaware

Expt. No.	μ_{max} hr^{-1}	K_s mg/L	K_i mg/L
1	0.264	164	251
2	0.182	169	—
3	0.096	437	—
4	0.183	29	—
5	0.622	152	—
6	0.342	126	—
Average	0.282	180	

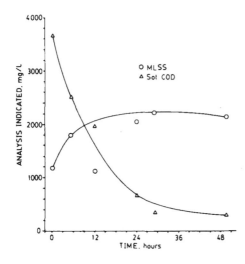

Figure 7. COD removal and MLSS accumulation in a batch system with high initial biomass concentration. Initial leachate COD = 3600 mg/L; initial MLSS = 1200 mg/L.

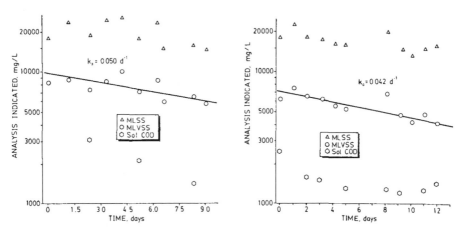

Figure 8. Batch aerobic digestion experiments for determination of the aerobic digestion constant, k_a. Sludge was taken from the reator treating leachate.

Table IV. Average Values of Biokinetic Constants for Activated Sludge Treating Landfill Leachate from Sandtown, Delaware

Parameter	Value
μ_{max}	$0.282\ hr^{-1}$
K_s	180 mg/L
Y_t	0.5 mg VSS/mg COD removed
k_d	$0.10\ day^{-1}$
k_a	$0.46\ day^{-1}$

municipal sewage and many industrial wastes. This is in agreement with the rather slow rate of COD removal observed in the larger batch study shown in Figure 7. Both the study shown in this figure and the continuous flow studies indicate that approximately 5 to 10% of the influent COD is present as residual after a reasonable aeration period; i.e., this portion of the organic material in the leachate is degraded only at a very slow rate and may contain some portion of non-biodegradable material.

Aerobic Digestion Studies

Two aerobic digestion studies were run. One-liter batches of mixed liquor were removed from the chemostat and aerated in the absence of leachate feed for an extended period of time. Figure 8 shows semi-logarithmic plots of total and volatile solids as these experiments proceeded. Also shown are the values of the autodigestion constant, k_a, calculated from the data, 0.042 and 0.050 day^{-1}, respectively. Soluble COD removal was measured in both experiments. Soluble BOD samples were taken during one of the experiments ($k_a = 0.05\ day^{-1}$), but for clarity these results are not plotted in the figure. However, it is important to note that the initial soluble BOD_5 of 200 mg/L (i.e., the chemostat residual BOD) was reduced to 20 mg/L after 15 days of aeration. This finding provides some evidence that the BOD exiting the treatment reactor can be subjected to further removal, but the rate of exertion of this residual BOD can be expected to be rather slow.

Average values for all biokinetic constants determined in this study are given in Table IV.

Design Recommendations

It is recalled that an aim of the study was to arrive at a suggested design strategy for treating the leachate, keeping in mind that it would be possible to use spray irrigation as a polishing treatment.

The leachate represented a rather high strength waste consisting largely of soluble organic matter, most of which exerted a biochemical oxygen demand. There was some evidence for periodic presence of toxicants or inhibitory substrates, and the waste was found to contain high concentrations of several heavy metals. The acid pH and the rather low nitrogen and especially low phosphorus concentration indicated that these deficiencies should be corrected for optimum biological treatment.

The numerical values of the biokinetic constants were in the general range of values expected for biodegradable wastes. These studies and the pilot plant studies on the waste showed that it was highly amenable to biological treatment which concurrently removed both the organic material and the heavy metals. There was no need to pretreat waste for removal of heavy metals.

There was ample indication throughout the study at mean sludge retention times from 2 to 10 days that a turbid effluent would be obtained; preliminary studies indicated that addition of polymer could be used to reduce effluent suspended solids.

The high residual COD made it difficult to make direct use of various design models which are available to predict effluent COD. For wastes of extremely high strength, the design models predict much lower COD values than are actually observed, and such was the case when the biokinetic constants were inserted into such design and analytical equations used in these studies. Under such conditions, the models predict more accurately if ΔCOD, rather than initial COD, is used as a measure of the strength of the waste and if the plant performance is measured against an experimentally-determined base-line residual COD. In the current studies sufficient time was not available to perform such studies. However, the batch study indicated that the ΔCOD for an average feed of 30,000 mg/L would be approximately 27,000–27,500 mg/L, i.e., residual, COD_e, would be expected to amount from 2,500 to 3,000 mg/L. This was generally found to be the residual COD range when high strength waste was fed to the pilot plant. However, much more work is needed to establish a firm base-line COD_e. It is also important to point out that, when appropriate predictive model equations were employed, it became apparent that for such a high strength waste one would

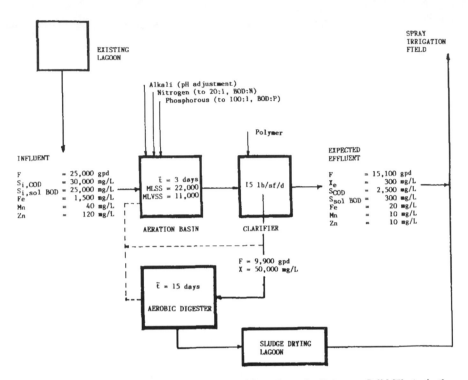

Figure 9. Recommended treatment scheme for leachate from the Delaware Solid Waste Authority's landfill at Sandtown, Delaware.

gain little by the use of sludge recycle. The biomass concentration one calculates for the once-through reactor is rather high and, when one is dealing with a biomass of high ash content, there is very little to be gained by incurring the extra expense of suspending a large portion of non-working suspended solids. Hence our analytical studies led us to the conclusion that once-through reactors would suffice. Using the equations for a once-through reactor, one can estimate that, for a mean hydraulic retention time of 3 days, there would be approximately 22,000 mg/L mixed liquor suspended solids, i.e., 11,000 mg/L mixed liquor volatile suspended solids in the reactor. It was also felt that because of the expected effluent turbidity problem, a conservative clarifier surface loading should be chosen. The proposed treatment scheme, along with expected influent and effluent characteristics, is shown in Figure 9.

It is proposed that all clarifier underflow be channelled to an aerobic digester. Based on the limited number of polymer studies which were performed, it would appear that a 5% sludge can be readily obtained. Using the k_a values obtained in the study, the reduction in volatile suspended solids can be calculated from the following equations:

$$X_t = X_o [1/(1 + k_a\bar{t})] \tag{1}$$

$$\% \text{ reduction in VSS} = (X_o - X_t)(1/X_o)(100) \tag{2}$$

Thus, the reductions in VSS for detention times of 10 and 15 days are 31 and 41%, respectively. The longer detention time is recommended since the greater degree of sludge stabilization will enhance retention of the heavy metals upon drainage and storage at the site. Even so, it may be desirable to lime the digested sludge to prevent a drop in pH and release of heavy metals.

Both clarified effluent and liquid drained from the digested sludge can be channelled to the spray irrigation field. It is also proposed that part of the clarifier effluent be recirculated to the landfill to enhance and control aerobic digestion of the solids in the fill.

The presently recommended strategy should suffice to provide rather good total treatment of the leachate. More work on the effect of autothermal operation on the rates of purification and decay of biomass is needed to determine conclusively whether insulation of both the reactor and the aerobic digester would provide a real economic advantage. However, there is little doubt that insulation of these tanks could buffer the system against the seasonal temperature drop and thus provide further stabilization of operational conditions. Also shown on the flow diagram (dashed lines in Figure 9) are sludge recycle lines to the aeration tank from the clarifier and/or the aerobic digester. Installation of these recycle facilities would provide considerable flexibility, and in case the strength of the leachate decreases, as usually occurs in landfills, recycle sludge could then be effectively used to produce a more purified effluent.

In conclusion, while there is some feeling among environmental engineers that the sanitary landfill is marked for extinction as a disposal method, its most usually mentioned replacement, materials recovery and incineration of combustibles, has not as yet proven to be a generally workable alternative for the landfill. Such facilities have operational problems and under the best conditions there is a solid waste which is subject to final disposal. Sanitary landfills, on the other hand, can continue to solve solid waste disposal problems, but their operation certainly requires more than compaction and covering of the daily load to form hygienic disposal cells. It would seem that the sanitary landfill itself can, with continued studies, be operated as a controlled anaerobic bioreactor and that this reactor can be coupled with a second reactor system to treat the leachate. In the present case, it is proposed to couple the landfill reactor with an aerobic leachate treatment system. By extending the scope and, to some extent, the technological complexity of the total landfill operation, its continued use as a mainstay for solid waste disposal for municipal and industrial wastes can be assured at least for medium-sized facilities.

REFERENCES

1. Vasuki, N. C., "Leachate Generation in a Lined Landfill. A Case Study of the Central Solid Waste Facility at Sandtown, Delaware," *Proceedings of the 41st Annual Industrial Waste Conference*, Purdue University, Lafayette, Indiana (May 1986).
2. Gaudy, A. F., Jr., Manickam, T. S., Lowe, W. L., and Button, M. P., "A Biological Process for Removal of Heavy Metals and Regeneration of Sorptive Capacity," Presented at the 54th Annual Conference, Water Pollution Control Federation, Detroit, Michigan (October 1981).
3. Lowe, W. L., and Gaudy, A. F., Jr., "Removal of Cadmium at High and Low Dosages by an Extended Aeration Process," *Proceedings of the 40th Annual Industrial Waste Conference*, Purdue University, Lafayette, Indiana, pp. 431–442 (1985).
4. Gaudy, A. F., Jr., Gaudy, E. T., Feng, Y. J., and Brueggemann, G., "Compatibility of Organic Waste and Cyanide during Treatment by the Extended Aeration Process," *J. Water Pollution Control Fed.*, 54, 153–164 (1982).
5. Rozich, A. F., Gaudy, A. F., Jr., and D'Adamo, P. C., "Predictive Model for Treatment of Phenolic Wastes by Activated Sludge," *Water Research*, 17, 1453–1466 (1983).
6. D'Adamo, P. C., Rozich, A. F., and Gaudy, A. F., Jr., "Analysis of Growth Data with Inhibitory Carbon Sources," *Biotechnol. Bioeng.*, 26, 397–402 (1984).
7. Rozich, A. F., and Gaudy, A. F., Jr., "Response of Phenol-Acclimated Activated Sludge Process to Quantitative Shock Loading," *J. Water Pollution Control Fed.*, 57, 795–804 (1985).
8. Ecker, R. E., and Lockhart, W. R., "Specific Effect of Limiting Nutrient on Physiological Events During Culture Growth," *J. Bacteriol.*, 82, 511–516 (1961).
9. American Public Health Association. *Standard Methods for the Examination of Water and Wastewater*, 15th Edition (1980).
   Total Kjeldahl nitrogen (TKN), Method 420A
   Phosphorus, Method 424
   Volatile suspended solids (VSS), Method 209E
   Metals, atomic absorption methods, 315A, 319A and 328A for iron, manganese and zinc, respectively.
10. Hach Chemical Company, *Hach Water Analysis Handbook*, Loveland, CO, 2-144-146 (1980).
11. Deeny, K. Heidman, J., and Smith, J., "Autothermal Thermophilic Aerobic Digestion in the Federal Republic of Germany," *Proceedings of the 40th Annual Industrial Waste Conference*, Purdue University, Lafayette, Indiana, pp. 959–968 (1985).
12. Elenbogen, G., Sawyer, B., Lue-Hing, C., Rao, K. C., and Zenz, D. R., "Studies of the Uptake of Heavy Metals by Activated Sludge," *Proceedings of the 40th Annual Industrial Waste Conference*, Purdue University, Lafayette, Indiana, pp. 493–505 (1985).
13. Gaudy, A. F., Jr., and Gaudy, E. T., *Microbiology for Environmental Scientists and Engineers*, McGraw-Hill Book Co. (1980).
14. Robinson, H. D., and Maris, P. J., "The Treatment of Leachate from Domestic Wastes in

Landfills. I. Aerobic Biological Treatment of Medium-Strength Leachate," *Water Research*, 17, 1537–1548 (1983).

15. Uloth, V. C., and Mavinic, D. S., "Aerobic Bio-Treatment of a High-Strength Leachate," *J. Environ. Engin. Div., ASCE*, EE4, 647–661 (1977).

16. Cook, E. W., and Foree, E. G., "Aerobic Biostabilization of Sanitary Landfill Leachate," *J. Water Pollution Control Fed.*, 46, 380–392 (1974).

17. Keenan, J. D., Steiner, R. L., and Fungaroli, A. A., "Landfill Leachate Treatment," *J. Water Pollution Control Fed.*, 56, 27–33 (1984).

75 REVERSE OSMOSIS REMOVALS OF CHLORINATED HYDROCARBONS

Riley N. Kinman, Professor

Janet Rickabaugh, Research Associate

Janette Martin, Graduate Research Assistant
Department of Civil and Environmental Engineering
University of Cincinnati
Cincinnati, Ohio 45221

Ronald F. Lewis, Microbiologist
U.S. EPA-ORD-HWERL
Cincinnati, Ohio 45268

INTRODUCTION

Background

In the past few years, the contamination of water and groundwater supplies with toxic organic compounds has challenged scientists and engineers with the task of finding innovative treatment techniques to remove these hazardous constituents. The ideal solution would be to treat the contaminated water in such a way that the organics were completely destroyed. This could be accomplished by biological, physical or chemical methods. Prior to treatment, the concentration of the contaminants into as small a volume as possible would be highly desirable. Reverse osmosis (RO) is a unit process that is used to separate solute from solvent. RO concentrates and isolates the contaminants in the solution, reducing the volume which must be treated.

Purpose and Scope

The purpose of this study was to determine the effectiveness of reverse osmosis treatment in removing chlorinated hydrocarbons and pesticides from a hazardous waste leachate. The leachate was produced by passing water through columns packed with 25 kg of contaminated soil collected from the Chem-dyne site in Hamilton, Ohio. These soil columns generated a leachate which contained significant (although low) concentrations of various chlorinated hydrocarbons and pesticides. In order to manage the waste produced from the column study, reduction of the volume of contaminated leachate was necessary. Spiral wound cellulose-acetate (CA) and polyamide thin film composite (PA) membranes were used to treat the leachate in a reverse osmosis unit. Their contaminant removal capabilities were compared, but the mechanism for the removal was not investigated.

METHODS AND MATERIALS

Equipment

The reverse osmosis unit used in this study was a pilot scale Osmonics OSMO 1919-SB equipped with a 100 mesh prefilter screen and operated at a pressure of 190 psi. Two different membranes were used. The first was a cellulose acetate spiral wound membrane model SEPA-97E manufactured by Osmonics, with approximately 19 sq.ft. of surface area. The manufacturer's specifications for this membrane predicts a 94 to 97% salt rejection and a lower molecular weight cut off of 200 for organics. This type of membrane is typically used for desalination of water, salt recovery, organic removal, BOD and COD reduction [1]. The system could operate between a pH range of 6 and 8, at approximately 22% permeate recovery per pass when using the CA membrane. The second membrane was a spiral wound polyamide thin film composite membrane. It also had approximately 19 square feet of surface area and was manufactured by Osmonics, Inc. The PA membrane is used primarily for desalination of water with a salt rejection of greater than 99%. It is also used for removal of organic

639

compounds from waste water but is not designed to remove compounds with a molecular weight of less than 100 [2]. The PA membrane was a high rate membrane, operating at approximately 44% permeate recovery per pass with a pH range between 5 and 12. This membrane is sensitive to chlorine and cannot withstand a solution containing greater than 0.5 mg/L chlorine [1].

Stainless steel fifty-five gallon drums were used as feed tanks for the leachate and also for the concentrate stream if the volume was sufficient. Five gallon glass jars were used to hold smaller volumes of permeate and concentrate.

Experimental Design

A total of nine experiments were run during the course of this study, the first four using the CA membrane as shown in Figure 1. The last five used the PA membrane as shown in Figure 2. The feed stream of the first experiment (CA membrane) was column leachate. The feed stream of the second CA experiment was the concentrate from Experiment 1, modeling a double pass system. The third CA experiment used the concentrate from Experiment 2 as the feed stream, modeling a triple pass system. The fourth CA experiment was an attempt to "polish" the permeate from Experiments 1 and 2.

The PA membrane was used for Experiments 5 and 6, which were conducted in the same manner as Experiments 1 and 2. Experiments 7 and 8 modeled a double pass run for the PA membrane, using column leachate as the initial feed to duplicate the results of Experiments 5 and 6. Experiment 9 used the concentrate from Experiments 6 and 8, modeling a triple pass system for the PA membrane.

Experimental Procedure

Leachate Production. The primary feed solution (column leachate) used in the experimentation was obtained by passing water through six soil columns packed with contaminated soil collected from the Chem-dyne site located in Hamilton, Ohio. The soil columns were approximately 5 ft. tall and 6 in. in diameter and were each packed with approximately 25 liters of soil (30 kg). The soil columns were used in a different study investigating the degradation of chlorinated hydrocarbons and pesticides by various microbial populations. A solution used to feed the column (1-2 liters per day) contained a high population of microorganisms, nutrients, and 1 mg/L Tween 80, a surfactant to promote wetting. As this solution passed through the columns, it became contaminated with various hydrocarbons and pesticides, producing approximately 1 liter of leachate per day per column. The concentration of compounds in the leachate was significant (ppb range), simulating a dilute hazardous leachate.

The column leachate was stored in capped five gallon glass jars in a hood until sufficient volume had accumulated for an experiment. The primary feed solution used in the reverse osmosis experiments consisted of approximately 50 gallons of composited column leachate.

Operation of Reverse Osmosis System. For each experiment, the pump was primed with distilled water, and the system was brought up to full operating pressure (190 psi). The intake line was then

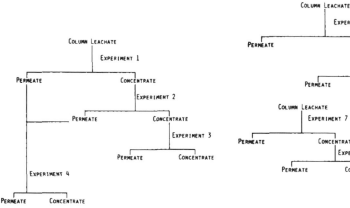

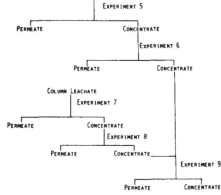

Figure 1. Experimental design flowchart for cellulose acetate membrane.

Figure 2. Experimental design flowchart for polyamide membrane.

Table I. Leachate Contaminants

Compound	MDL[a] (μg/L)
1,3-DCB	1.3
1,2-DCB	0.6
1,3,5-TCB	0.3
1,2,3-TCB	0.2
Hexachlorobutadiene	0.6
Hexachlorobenzene	0.1
Heptachlor	0.1
Heptachlor Epoxide	0.2
Dieldrin	0.1
Endrin	0.1

[a]MDL = Minimum Detection Limit.

switched to the feed tank being used for that experiment. Approximately the first two liters of permeate and concentrate were collected separately and not included in the final sample to prevent dilution by the distilled water remaining in the system after priming. At the end of each run, the membrane was backwashed by reducing the operating pressure to 50 psig and flushing with approximately 5 gallons of distilled water. The permeate and concentrate streams from the flushing procedure were collected separately and not included in the final composite sample.

Analysis

The feed streams were sampled approximately half way through each run while the permeate and concentrate streams were sampled at the end of the run, resulting in composite samples.

Three hundred mL samples of each stream were acidified and analyzed for COD and TOC content. A Beckman Tocamaster model 915 Total Organic Carbon Analyzer was used to determine the total organic carbon (TOC) content of the samples. The COD of the samples was determined according to *Standard Methods* [3]. All streams were analyzed for the chlorinated hydrocarbons and pesticides listed in Table I.

One liter samples were extracted according to EPA Method 3520 [4] over an 8 hour extraction period followed by concentration to 5 mL. The concentrated sample was analyzed by EPA Method 8120 [4] using a Varian gas chromatograph Model 3700 equipped with an electron capture detector. The column used was a Supelcoport 100/120 mesh coated with 1.5% SP-2250/1.95% SP-2401 packed in a 183 cm long by 4 mm I.D. glass column. A temperature program of 75 to 201°C at 5/min was used, with an injection temperature of 150°C.

RESULTS

Cellulose Acetate Membrane

The rejection (% concentration reduction) of the chlorinated hydrocarbons and pesticides was determined using the following equation:

$$\% \text{ Conc. Red.} = [1 - (P/F)]*100 \qquad (1)$$

where

P—concentration of compound in permeate (μg/L)
F—concentration of compound in feed (μg/L)

The removal from Experiments 1 through 4 is shown in Table II. Table III contains concentration data for Experiment 2, typical of these experiments. The rejection of individual compounds for single pass through the CA membrane (Exp. 1) ranged from 16% to greater than 93%. The rejection of the chlorinated hydrocarbons with a molecular weight greater than 200 is generally greater than those with a molecular weight of less than 200. Pesticides were removed to the greatest extent, ranging from 82% to greater than 89% rejection. Endrin was the only compound to be completely rejected (the detectable limit concentration was used to calculate rejection). 1,3-DCB showed a negative concentration reduction, indicating that the permeate concentration was greater than the feed concentration. This did not occur in any of the other experiments and is probably due to experimental error. All other

Table II. CA Percent Removals
Experiments 1–4

	Exp 1 %	Exp 2 %	Exp 3 %	Exp 4 %
1,3-DCB	−24.6	17.8	57.6	69.1
1,2-DCB	74.7	68.3	75.6	57.8
1,3,5-TCB	16.6	22.8	59.7	60.1
1,2,3-TCB	42.0	36.5	62.8	52.5
Hexachlorobutadiene	64.9	58.7	73.3	64.5
Hexachlorobenzene	83.5	78.5	95.3	47.9
Heptachlor	84.2	89.6	92.7	32.8
Heptachlor Epoxide	82.3	85.6	93.5	33.1
Dieldrin	89.6	89.6	94.8	4.4
Endrin	>93.6	>93.1	>93.7	*
COD	95.2	94.9	96.1	10.1
TOC	100.	100.	93.8	−21.2

compounds were present in significant concentrations, and the permeate could not be discharged. The COD and TOC content of the feed was reduced by 95.2 and 100%, respectively. These gross parameters indicate that the CA membrane was effective in removing compounds that were not analyzed for in this study.

Chromatograms of CA membrane Experiment 2 are shown in Figures 3 and 4. The peaks on the permeate chromatogram are somewhat smaller than those of the feed. These chromatograms illustrate the fact that a significant amount of contaminants remained in the permeate solution even though the data indicates fairly large reductions in the concentrations of the measured compounds.

Experiments 2 and 3 modeled a double and triple pass system, using the concentrate from the previous experiment as its feed stream. The rejection data (see Table II) show the same tendencies as Experiment 1. The lower weight hydrocarbons (less than 200 MW) are rejected to a lesser extent than hexachlorobenzene or hexachlorobutadiene. In most cases, the pesticides again show the highest concentration reduction. For these two experiments, the concentrations of the compounds remaining in the permeate were generally large enough that it could not be discharged.

Experiment 4 used the combined permeate streams from Experiments 1 and 2 as the feed stream. This was done to see if the permeate could be "polished". This proved to be unsuccessful because the permeate of this run still contained a significant concentration of the compounds in question. The chromatograms of the feed and permeate solutions from Experiments 2, 3, and 4 were all similar in appearance to those from Experiment 1. This result illustrates the fact that a large fraction of the contaminants remained in the permeate solution.

Table III. Concentration Data Experiment 2

	Concentration (μg/L)			Con. Red.
Compound	F	C	P	(%)
1,3-DCB	36.6	40.0	30.1	17.9
1,2-DCB	50.0	16.3	15.8	68.3
1,3,5-TCB	11.3	8.18	8.72	22.9
1,2,3-TCB	12.8	6.36	8.13	36.5
Hexachlorobutadiene	227.	59.7	93.7	58.7
Hexachlorobenzene	11.6	9.01	2.50	78.4
Heptachlor	9.77	4.75	1.02	89.6
Heptachlor Epoxide	41.6	32.5	5.98	85.6
Dieldrin	2.80	3.76	0.29	89.6
Endrin	1.45	1.75	BDL	93.2
COD (mg/L)	259.	342.	13.2	94.2
TOC (mg/L)	134.	98.0	0.0	100.
Volume (L)	112.2	86.2	26.0	

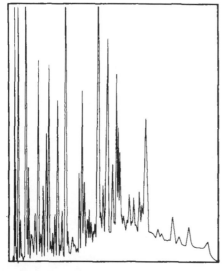

Figure 3.   Chromatogram of feed for CA membrane.

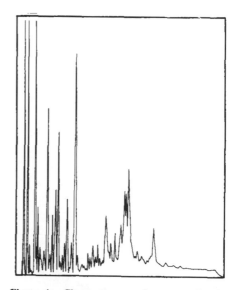

Figure 4.   Chromatogram of permeate for CA membrane.

It is interesting to note that in Exp. 4, the percent removal of the chlorinated hydrocarbons was higher than that of the pesticides. Chian and Fang [5] suggested that higher removals of pesticides in a natural water system could be expected than in an artificial environment due to complexation of the pesticides with humic and fulvic acids. It is possible that the initial feed solution for this study contained these substances. Since the humics are quite large and would have been removed during the first two passes through the system, the pesticides remaining would have been uncomplexed resulting in lower rejection.

Adsorption. A mass balance was calculated for Experiments 1–4 and expressed as a percent loss. The results are shown in Table IV. It appears that a portion of the chlorinated hydrocarbons and pesticides was adsorbed by the system during Experiments 2, 3, and 4, as indicated by positive percent loss values in Table IV. The compounds may have been adsorbed onto the tubing within the R.O. unit or onto the membrane itself. It is probable that a large portion was adsorbed onto the membrane, a phenomena reported by other researchers [5–8]. At this time, the membrane has not been analyzed for the presence of contaminants so this cannot be confirmed. In some cases, negative adsorption occurred. The negative adsorption values are probably due to experimental error.

Fouling. During the course of Experiments 1–4, the flux remained fairly constant, and the membrane exhibited no signs of fouling.

Table IV. Percent Loss Experiments 1–4

Compound	Exp 1	Exp 2	Exp 3	Exp 4
1,3-DCB	−48.0	−3.0	67.0	5.5
1,2-DCB	5.8	67.6	76.3	91.4
1,3,5-TCB	−11.2	26.5	45.5	39.3
1,2,3-TCB	8.0	70.2	78.4	73.1
HCBD	−1.2	47.1	59.6	52.4
HCBZ	1.1	35.3	85.3	42.3
Heptachlor	−53.5	60.3	79.7	86.4
Hept Epox	−22.9	36.7	83.1	50.3
Dieldrin	−9.5	−5.5	76.1	80.6
Endrin	28.1	7.5	100.0	*

Table V. Experiment 5 Data

Compound	Concentration (μg/L)			Con. Red. (%)
	F	C	P	
1,3-DCB	19.2	49.9	BDL	>93.2
1,2-DCB	8.84	12.6	BDL	>93.2
1,3,5-TCB	5.68	10.2	BDL	>94.7
1,2,3-TCB	5.72	10.7	BDL	>96.5
Hexachlorobutadiene	97.3	161.	BDL	>99.4
Hexachlorobenzene	5.61	15.6	BDL	>98.2
Heptachlor	2.98	19.6	BDL	>96.6
Heptachlor Epoxide	16.2	92.0	BDL	>98.8
Dieldrin	6.32	3.16	BDL	>98.4
Endrin	0.62	0.88	BDL	>83.9
COD (mg/L)	320.	475.	95.6	70.2
TOC (mg/L)	137.0	191.0	37.6	72.6
Volume (L)	176.9	99.2	77.7	

Polyamide Membrane

The data from Experiment 5 is shown in Table V. The concentrations of all compounds in the permeate stream were below detectable limits. The rejection of the various compounds was essentially 100%. However, the percentages reported are conservative and use the detectable limit concentration in the calculation. Figures 5 and 6 show the chromatograms of the feed (980 mL sample volume, 1.2 μL injection volume) and permeate (975 mL sample volume, 1.4 μL injection volume) streams for Experiment 5. A comparison of these two figures illustrates the fact that the PA membrane not only removes the compounds being measured but other contaminants as well.

Experiment 6 used the concentrate from Experiment 5 as its feed stream, modeling a double pass system. As seen in the earlier PA runs, the concentration of chlorinated hydrocarbons and pesticides in the permeate of this run was below the detectable limit.

Experiments 7 and 8 modeled a double pass run using column leachate as the initial feed solution. The results duplicated those of Experiments 5 and 6 with the concentrations of the target organic compounds below detectable limits in the permeate.

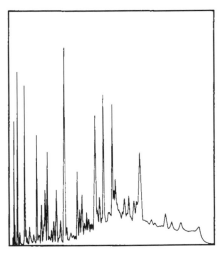

Figure 5.   Chromatogram of feed for PA membrane.

Figure 6.   Chromatogram of permeate for CA membrane.

The concentrate from Experiments 6 and 8 was used as the feed solution for Experiment 9, modeling a triple pass system. The concentrations of contaminants in the permeate were once again below the detectable limits. The permeate chromatograms from Experiments 6, 7, 8, and 9 were all similar in appearance to the permeate chromatogram from Experiment 5 emphasizing that the membrane was efficient at removing all of the compounds in the solution.

Adsorption. The solutions from the flushing and cleaning of the PA membrane showed no detectable concentration of contaminants with the exception of those contaminants also found in the blank. For this reason, it was felt that probably no adsorption of these compounds onto the PA membrane had occurred.

Fouling. The flux remained constant during Experiments 5, 6 and 7. During Experiment 8, the permeate flux decreased from 1050 mL/min to 930 mL/min, indicating membrane fouling. Immediately after Experiment 9 began, the permeate flux declined to 380 mL/min. At this point, the experiment was halted, and the membrane was cleaned according to the manufacturer's directions. After cleaning, Experiment 9 was completed with no other apparent problems. The permeate flux during this run decreased from 1000 mL/min to 860 mL/min over the 30 min run. The fouling of the membrane could have been due to a build up of a gel layer on the membrane surface caused by organics such as humic substances and biological slimes or a porous layer caused by precipitation of calcium and magnesium salts and hydroxides [9]. Fouling did not occur during Experiments 5, 6, 7, and 8 because the concentration of the foulants in the feed was not yet high enough to cause problems. Since the feed for Experiment 9 had already been passed through the R.O. unit twice, the concentration of the foulants was high.

SUMMARY

The use of a low pressure reverse osmosis system combined with a polyamide thin film composite membrane was shown to be successful for the removal of chlorinated hydrocarbons and pesticides from a hazardous waste leachate. The polyamide membrane was found to be superior to the cellulose acetate membrane in both removal of the contaminants and reduction of volume of the contaminated leachate. For each pass, using the PA membrane, approximately 44% of the original volume of leachate was recovered as "clean" water and could be discharged. After three passes, this amounted to an 83% volume reduction of the leachate, leaving only 17% retained for disposal.

DISCLAIMER

Although the research described in this paper has been funded wholly or in part by the United States Environmental Protection Agency, it has not been subjected to Agency review and, therefore, does not necessarily reflect the views of the Agency, and no official endorsement should be inferred. This research was funded under EPA contract No. 68-03-3210-01.

REFERENCES

1. Osmonics Inc., "Sepa Membrane," *Technical Bulletin No. 109*, Osmonics Inc., Minnetonka MN. (1978).
2. Spatz, D. D., and R. H. Friedlander, "Rating the Stability of RO/UF Membrane Materials," *Water and Sewage Works* (Feb. 1978).
3. Franson, M., *Standard Methods for Examination of Water and Wastewater*, 15th Edition, American Public Health Association, Washington D.C. (1980).
4. U.S.E.P.A., *Test Methods For Evaluating Solid Waste, Physical/Chemical Methods*, SW-846 (1982).
5. Chian, E. S., H. P. P. Fang, and W. N. Bruce, "Removal of Pesticides by Reverse Osmosis," *Environmental Science and Technology*, 9, 52–59 (1975).
6. Deinzer, M., R. Melton, and D. Mitchell, "Trace Organic Contaminants in Drinking Water; Their Concentration by Reverse Osmosis," *Water Research*, 9, 799–805 (1975).
7. Edwards, V. H., and P. F. Schuberty, "Removal of 2,4-D and Other Persistent Organic Molecules from Water Supplies by Reverse Osmosis," *Journal American Water Works Association*, 610–616 (Oct. 1974).
8. Klein, E., J. Eichelberger, C. Eyer, and J. Smith, "Evaluation of Semipermeable Membranes for Determination of Organic Contaminants in Drinking Water," *Water Research*, 9, 807–811 (1975).
9. Binovi, R. D., *A Replacement for the Silt Density Index; Permanganate Demand to Predict Reverse Osmosis Membrane Fouling*, Ph.D dissertation, University of Cincinnati, p. 11 (1983).

Section Twelve
METAL WASTES

76 WASTE METAL RECOVERY CASE HISTORIES

C. S. Brooks, President
Recycle Metals
Glastonbury, Connecticut 06033

INTRODUCTION

Sequential solvent extractions at low pH have been conducted with actual industrial wastes to provide selective recovery of metals such as chromium, copper, nickel, and zinc in the presence of iron, the principal contaminant. The evaluated extraction agents were primarily commercial materials of diversified chemical functional types such as alkyl phosphates, oximes, alkyl sulfonates, a β-diketone and a carboxylic acid. The multi-component metal waste systems were a mineral acid and acid solubilized sludges from electroplating hardware and electrochemical machining (ECM) industrial operations. Highly efficient separations were consistently obtained for copper with efficiencies ranging from poor to good for nickel, zinc, and chromium, depending on the specific waste system.

SIGNIFICANCE OF THE PROBLEM

After the precious metals, strategic non-ferrous metals are of most interest for recycling [1,2,3]. The non-ferrous component of municipal solid waste is estimated to be about 2%, with aluminum the predominant metal. The amounts of the non-ferrous metals, although minor components of municipal waste, are an impressive 2,000,000 tons annually with about 500,000 tons of metals other than aluminum. Metal wastes generated by mining, mineral processing and metallurgical industries in the United States are much more voluminous, estimated to amount to 1,800,000,000 tons annually [3,4].

The recycling of waste metal sludges is essentially non-existent because of the lack of economical technology, lack of appropriate technology, lack of appropriate markets, and the absence of compelling incentives in the form of government regulations. However, disposal costs are high and are consistently increasing. It is anticipated that high disposal costs and increased regulatory pressures to minimize landfill disposal options will increase the incentives for development of improved technology and economics for recycling many of the non-ferrous metals.

The most promising potential for recycling appears to be with the strategic non-ferrous metal wastes of the metal finishing and electronics industries.

The present study consists of a bench scale evaluation of solvent extraction separation options for recovery of metals such as chromium, copper, nickel, and zinc from multi-component metal wastes: pickling acids and acid solubilized sludges from the plating, hardware and electrochemical machining (ECM) metal finishing industries. Separation processes evaluated in bench scale experimentation [5,6] include precipitation, partition, and ion exchange. Separations attainable by solvent extraction at low pH with actual industrial wastes are reported here.

EXPERIMENTAL PROCEDURES

Waste Systems

A waste mineral acid and three acid solubilized waste metal sludges were used in this study. The acid was obtained from a Connecticut treatment company and the sludges were provided from brass, plating, and electrochemical machining (ECM) industry operations in Connecticut.

The waste acid was used as received without pretreatment. The sludges were solubilized with 10 grams ($\approx$ 30 wt% solids) in 100 mL of 10 wt% concentrated H_2SO_4 and the insolubles separated by filtration.

Extraction Agents and Solvents

The evaluated extraction agents were reagent grade acetyl acetonate (Aldrich P 775–4), lauric acid (MacAllister Bicknell LX0060–3) and 8-Hydroxyquinoline (MacAllister Bicknell HX0885–3) and commercial materials for DI-2-Ethylhexyl phosphoric acid (Albright and Wilson DEHPA 95 wt%), 2 Hydroxy 5 nonyl benzophenone oxime (Henkel HS LIX 64N), an oxime (Henkel LIX 622) and Dinonyl naphalene sulfonic acid (King Industries SYNEXDN-052).

The solvents were technical grade naptha, toluene, and xylene and reagent grade isoamyl alcohol (Aldrich M3265–8). A mixed solvent of isoamyl alcohol (20 vol.%) and toluene (80 vol%) was used for the acetyl acetonate and the 8-hydroxyquinoline extractants (Table I).

Extraction Procedure

The solvent extractions were conducted in a 250 mL separatory funnel with 15 to 45 mL of the acid waste solutions using 30 to 50 mL of extraction solution in one or two stages with 5 minute agitation. The extraction agent concentration ranged from 30 to 20 wt% providing ratios from 5 to 25 of the extraction agent to metal present in the aqueous phase. Transfer efficiency of the metal to the solvent was determined from analysis of the metal composition in the aqueous equilibrium phase following the shake tests. Metal analyses were based on ASTM colorimetric procedures or atomic absorption, the latter being conducted by Griswold & Fuss Experimental Laboratories, Manchester, Connecticut.

The solvent extractions were conducted sequentially with progressive pH adjustment with NaOH as indicated in Figure 1.

Waste Compositions

The metal wastes had significant amounts (in excess of several hundred mg/L) of one or more of the four metals selected for this study (chromium, copper, nickel, or zinc). The waste acid had copper, nickel, or zinc in amounts large enough to be of interest for recovery, along with minor amounts (< 100 mg/L) of metals such as chromium, manganese, lead, and tin. The three waste sludges solubilized in 10 wt% H_2SO_4 were filtered to remove the acid insolubles which ranged from 4.7 wt% for the brass industry sludge to 1.14 wt% for the ECM sludge and 0.28 wt% for the plating sludge. The acid

Table I. Extraction Agents

Extraction Agent	Wt%	Solvent	Metals	pH	Extraction Stage
DI2EHPA	10–20	Naphtha	Fe, Zn	1–1.5	1
LIX64, LIX622	10–20	Naphtha	Cu	1.5–2.5	2
Lauric Acid or	3–10	Xylene	Ni	2.5–4.5	3
SYNEX-NO52 + LIX63 (50/50)	10–20	Naphtha	Ni	2.5	3
Acetyl Acetonate	10–20	Isoamyl Alcohol + Toluene (20/28)	Cr	3.0–4.5	4

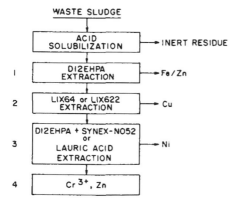

Figure 1.   Solvent extraction separations.

Table II. Solvent Extraction of Waste Mineral Acid

Agent	Wt % Extracted				pH	Stage
	Fe	Cu	Ni	Zn		
DI2EHPA	46–94			20–65	~1	1
LIX64		56–85			~1	2
Lauric Acid			11		4.5	3

Table III. Solvent Extraction of Acid Solubilized Plating Sludge

Agent	Wt % Extracted				pH	Stage
	Fe	Cr	Cu	Ni		
DI2EHPA	37–99				1–1.5	1
LIX64			99		2.5	2
Lauric Acid				1.6	2.5	3
Acetyl Acetonate		17			3.0	4

solubilized brass industry waste had significant amounts of chromium, copper, nickel, and zinc. The acid solubilized plating sludge contained chromium and nickel, and the acid solubilized ECM sludge contained principally nickel in amounts sufficient to be of interest for recovery. All the wastes contained appreciable iron as contaminant.

Solvent Extraction Hierarchy

Early efforts to achieve separation of the contaminant iron from acid waste systems by hydroxide precipitation with NaOH and NH$_4$OH led to undesirable losses of metals of interest, notably copper and nickel [5]. The sequential solvent extraction hierarchy used for recovery of the metals of interest (chromium, copper, nickel, and zinc) was established in a prior investigation [6]. Subsequent study established that there was considerable advantage to conducting sequential solvent extraction with progressive pH adjustment in the following order of removal as indicated in Figure 1: 1) iron; 2) copper; 3) nickel; and 4) finally chromium. In part, zinc separation occurs along with the iron and in part ends up in the residual solution from the successive extractions. When DI2EHPA is used as the extraction agent, separation from iron can be obtained by using H$_2$SO$_4$ acid to strip the zinc from the organic solvent phase. In the acid wastes studied thus far, the zinc is very resistant to solvent extraction by the agents used and accumulates in the residual solution of the fourth stage of extraction.

The extraction agents are predominantly commercial materials with good stability and demonstrated extraction efficiency at low pH. In the present evaluation the reported separations correspond to 5 minute mixing of the aqueous phase and the extraction agent in appropriate solvent following the procedure described above.

Case Histories

The results of these solvent extraction separations with the four acid metal waste systems are given in Tables II, III, and IV.

In the case of the waste mineral acid (Table II), the iron is separated efficiently (46 to 94%) in the first extraction stage, and copper is separated efficiently (56 to 85%) in the second extraction stage. The separation efficiency for nickel in the third extraction stage is low, attaining only 11%. Variable

Table IV. Solvent Extraction of Acid Solubilized ECM Sludge

Agent	Wt % Extracted			pH	Stage
	Fe	Cr	Ni		
DI2EHPA	28–99			1–2	1
SYNEX-NO52 + LIX63			61	2.5	2
Acetyl Acetonate		81		3.5	3

Table V. Solvent Extraction of Acid Solubilized Brass Pickling Sludge

Agent	Wt % Extracted					pH	Stage
	Fe	Zn	Cu	Ni	Cr		
DI2EHPA	91	23				1–1.5	1
LIX622			75			2	2
Lauric Acid				28		3.5	3
Acetyl Acetonate					99	3.5	4

separation efficiencies in the range of 20 to 65% were obtained for zinc along with iron in the first stage.

High separation efficiencies to 99% were obtained for copper and iron in the first and second extraction stages for the acid solubilized plating waste sludge (Table III). A quite low separation efficiency, less than 2%, was obtained for nickel in the third extraction stage. A separation efficiency of 17% was obtained for chromium in the fourth extraction stage.

The acid solubilized ECM waste sludge (Table IV) provided relatively high separation for the three metals of interest (chromium, iron, nickel) with 99% for iron in the first stage, 61% for nickel in the second stage and 81% for chromium in the third extraction stage.

The acid solubilized brass industry waste sludge (Table V) containing all five metals provided a high efficiency at 91% for iron removal in the first extraction stage. A somewhat lower separation efficiency of 75% was obtained for copper in the second stage. Nickel provided the lowest separation efficiency of 28% in the third stage. A chromium separation efficiency to 99% was obtained in the fourth stage. Zinc was separated only to the extent of 23%, along with iron, in the first extraction stage.

DISCUSSION

This scheme of sequential solvent extractions at low pH with progressive pH adjustment has provided some promising high separation efficiencies, notably for iron extraction with DI2EHPA and for copper extraction with LIX64 and LIX622. The results for iron and copper were consistently high for all four waste systems in spite of the diversity of their compositions.

Zinc recovery was variable in efficiency, occurring predominantly along with iron in the first stage extraction with DI2EHPA. Ultimate separation of the zinc from the iron can be obtained by using selective stripping with 10 to 40 wt% aqueous H_2SO_4 acid of the DI2EHPA-naphtha solvent phase. Iron can be selectively stripped from the solvent phase with 10 to 40 wt% aqueous HCl acid.

Quite variable recovery efficiencies were obtained with nickel, depending on the system. Nickel separation efficiencies in the range of 1.6 to 28% were obtained for two of the acid solubilized sludges (brass and plating industries) and for the waste mineral acid. In an earlier study [5] using lauric acid extractions, quite high separations of 83% were obtained for nickel from this same brass industry acid solubilized waste and 90% from a waste chromic acid, whereas negligible separation of less than 2% was obtained for the same waste mineral acid. In this earlier study the iron had been removed by pH adjustment to 2 with NaOH and separation of the hydroxide precipitates by filtration, rather than by solvent extraction. As has been observed above, the earlier procedure [5] leads to substantial coprecipitation loss of metals, such as copper and nickel, present in the system.

The highest separation efficiency of 61% was observed for nickel with a mixed extraction system consisting of Synex-NO52 (50 wt%) and LIX63 (50 wt%) in naphtha.

The chromium also provided quite variable results ranging from 17% separation efficiency for the acid solubilized plating waste to 99% for the acid solubilized brass industry sludge. These promising high separation efficiencies obtained with acetyl acetonate for trivalent chromium are considered particularly noteworthy since solvent extraction of trivalent chromium is seldom undertaken. This system could be improved by using solvents, such as chloroform, providing a more favorable solubility distribution ratio for the metal acetyl acetonate between the solvent and the aqueous phase. In the present study the solvents were selected to provide the lowest toxicity and volatility compatible with favorable solubility properties.

CONCLUSIONS

Sequential solvent extraction at low pH with extraction agents effective at low pH is considered a promising approach to achieving efficient recoveries of such metals as chromium, copper, nickel, and zinc from multicomponent acid waste systems. Efficient separations for copper following iron extraction for four diverse waste systems had been demonstrated. Efficient separation of nickel following iron and copper extractions is quite variable, dependent upon the system. Stripping iron and copper by solvent extraction in a mixed waste system, combined with alternative nickel separations by cation exchange or electrowinning, can provide useful separation options.

The promising high solvent extraction separations obtained with trivalent chromium using acetyl acetonate for at least two of the acid solubilized sludge systems (brass industry waste and ECM filter cake) encourage further optimization and evaluation of this solvent extraction system.

REFERENCES

1. Spendlove, M. J., "Recycling Trends in the United States; A Review," Bureau of Mines Information Circular 8711 (1976).
2. U.S. Dept. of Commerce "Base Line Forecasts of Resource Recovery 1972–1990," *NTIS, PB245924* (1975).
3. Hanna, H. S., and C. Rampacek, "Resources Potential of Mineral and Metallurgical Wastes," *Fine Particles Processing*, Vol. 2, pp. 1709–1730, P. Somasundaran, Ed., A.I.M.E., New York, NY (1980).
4. Hill, R. D., and J. L. Auerbach, "Solid Waste Disposal in the Mining Industry," Op. Cit., pp. 1731–1753.
5. Brooks, C. S., "Metal Recovery from Waste Acids," *Proc. 40th Industrial Waste Conference*, Purdue University, p. 551, Butterworths, Ann Arbor Science Books (1985).
6. Brooks, C. S., "Recovery of Waste Metals by Solvent Extraction," *First Int. Conf. on Separations Science and Technology, Div. Ind. & Eng. Chem.*, 191st National American Chemical Society Meeting, New York, NY, Paper No. 256 (1986).

77 WASTE TREATMENT FOR SILICON VALLEY COMPUTER AND ELECTRONIC SUPPORTING INDUSTRIES

Ellen Lee, President

William Strangio, Vice President

Boon Lim, Sanitary Engineer
Lee, Strangio and Associates, Inc.
San Jose, California 95126

INTRODUCTION

During the last ten years, through the economic fluctuations of the computer and electronic industries, Lee, Strangio and Associates, Inc., has designed and installed more than eighty industrial waste treatment systems. Five of these installations have won the "Best Plant of the Year" and "Honorary Mention" awards of the California Water Pollution Control Association. Because of our project planning, designs, construction supervision, turnkey installations, and maintenance services we provide, we have accumulated an abundance of knowledge, information and experience which could be useful to engineers and industrial entrepreneurs in starting and operating similar industries.

The Silicon Valley industries which discharge hazardous/toxic waste can be divided into the following four categories:

- Metal finishing of parts and computer chassis
- Semiconductor/wafer production
- Manufacture of printed circuit boards
- Manufacture of computer rigid discs

The metal finishing technologies involved in these categories have existed for many years. Elements in the waste can exhibit wide variations in type of metals as well as in concentration. In most cases, the waste contains cyanide, hexavalent and trivalent chromium, and metals such as copper, nickel, lead, zinc and iron in most cases, and silver and cadmium in some cases. However, this wastewater is relatively easy to treat with the proper stream separation, special provision for handling of concentrated chemical dumps, and by the use of properly designed equipment. The treatment process is relatively straight forward and does not vary significantly even if the process chemicals are changed. The treatment processes have been subjected to many research efforts over the years, and an abundance of information is available in the literature.

Although the semiconductor and wafer design is relatively intricate and complicated, the production process does not produce highly complex waste. In most cases, only neutralization is adequate.

The discussion presented in this paper will focus on the industries manufacturing printed circuit boards and computer rigid discs where evolution of production technology is still taking place.

SPECIAL CONSIDERATIONS

To plan and design a successful treatment facility for printed circuit board or rigid disc wastewater, the following information is needed at the preliminary phase of the project.

- Type of concentrated chemical dumps
- Frequency and volume of each concentrated chemical to be dumped
- Review of chemical process tanks and rinse tanks
- Flow rate of rinse waste, average flow and peak flow
- Identify chemicals or rinse waste which needs to be separated and handled independently
- Identify the chemicals or waste which are "untreatable" and which shall be hauled away to a hazardous waste landfill
- Perform bench scale tests to develop treatment processes and design criteria

652

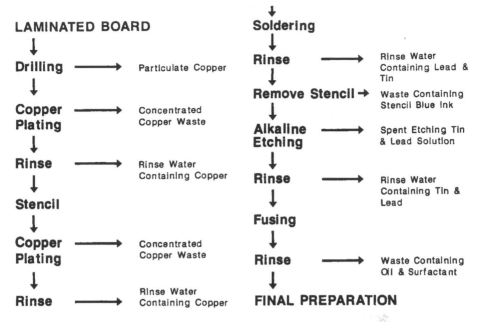

Figure 1.   Printed circuit board manufacturing process.

- Determine the type of chemicals to be used for treatment processes and the required dosages
- Identify any problem waste which may require special handling and equipment
- Evaluation of alternatives for metal recovery and reuse of concentrated wastes
- Cost effective analysis of treatment and disposal alternatives
- Identify hazardous conditions and design for reliability and safety

Thorough considerations of all this information and detailed planning can reduce the construction and operation costs and can increase the reliability and efficiency of the treatment system.

PRINTED CIRCUIT BOARD MANUFACTURING

The simplified printed circuit board manufacturing process is shown on Figure 1. In most cases, the wastewater contains copper, lead, and tin. A few production processes utilize chromic acid instead of sulfuric acid for etching and, therefore, discharge chromium into the waste stream.

The chemical forms of the metals are usually as follows:

Copper sulfate
Ammonia copper persulfate
Lead in fluoboric acids
Stannous tin
Chromic acid
Concentrated caustic solution

These heavy metals by themselves normally are relatively easy to treat. However, for the printed circuit board manufacture, these chemicals are mixed with other chemicals, in most cases, with chelating agents to keep heavy metals from settling in the manufacturing process. These additives of chemicals vary in quality and quantity depending on the suppliers' formula. The formula of the additives is often proprietary information and is not readily available from the suppliers.

To design a successful treatment system, whether the detailed information on chemical ingredients is available or not, a series of bench scale tests are imperative. Samples of combined wastewater and each major chemical utilized in the production process which will eventually become part of the wastewater are collected and tested. For a new industry where actual waste samples are not available,

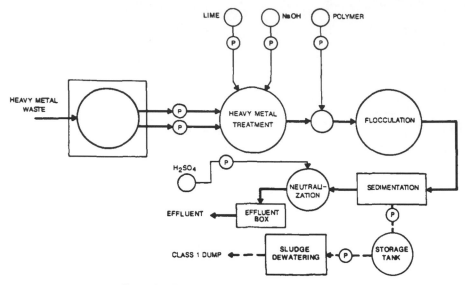

Figure 2.   Schematic of waste treatment process.

the chemicals planned to be used in the future are collected and tested individually and in combination to assimilate the actual wastewater.

Over 90% of the printed circuit waste can be treated with sodium hydroxide. The process schematic diagram is shown in Figure 2. Most of the printed circuit wastes are discharged into the treatment system at pH 4 to pH 6. By increasing the pH of the wastewater to 11.0–12.0, most of the heavy metals such as copper, lead, and tin are converted to hydroxide forms of metals. Copper, lead, and tin hydroxides normally are relatively insoluble in water. The optimum pH for treatment is higher than the theoretical optimum pH due to the presence of various chelating agents in these chemical solutions. After the pH adjustment it is necessary to provide for 20 to 30 minutes of reaction time. An anionic polymer is added to coagulate the heavy metal particles. The polymer utilized by our treatment system is Dow Chemical PuriFloc A-23. The dosage ranges from 3 to 5 mg/L. The wastewater flows through a lamella media type of clarifier. The sludge is settled and collected in the hopper section of the clarifier. The wastewater from the clarifier still has a pH of 10 to 11. The effluent pH is lowered with H_2SO_4 to meet the discharge requirement. Most cities have effluent pH requirement of 5 to 10 or 6 to 9.

The metal hydroxide sludge is withdrawn from the hopper shape bottom of the clarifier into a storage tank for gravity thickening and/or to a sludge dewatering device. Filter presses are most popular with this type of small industrial wastewater treatment system due to their simplicity and reliability in application.

In some systems, it is necessary to use lime in replacement or in combination with sodium hydroxide. Some other wastes also require treating with phosphoric acid, ferrous sulfate, or calcium chloride. The use of these chemicals is shown in Figure 3.

Most of the treatment systems for printed circuit board production fall in one of the above mentioned alternatives. The process may change when the chemicals used in the production or supplier of the chemicals is altered. Therefore, it is imperative to perform a treatability test prior to design and prior to the altering chemical or the chemical supplier.

A successful treatment system is capable of reducing the copper content to less than 1 mg/L and lead and tin down to trace level.

RIGID DISC MANUFACTURING

During the last five years, the rigid computer memory discs of 3.5 to 5 inch diameter have been developed and created a highly competitive market. Rigid discs have become an essential element of

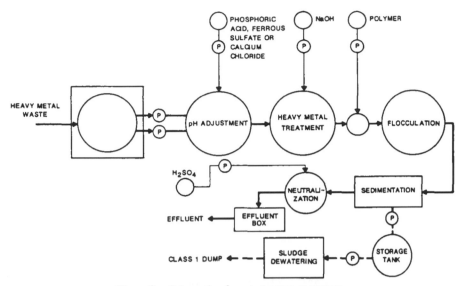

Figure 3.   Schematic of waste treatment process.

personal computers and its smaller size and higher capacity is highly desirable for use in small computer systems.

Rigid disc manufacturing is an extremely precise process where a single 3 millimicrons nodule on the disc is considered an unacceptable flaw. The chemical solution, the plating time, and handling of the disc has to be very precise. The technology is called " High Tech" because of the high degree of sensitivity and precision involved in the production technology. Since the plating solution is the fundamental element toward successful production of rigid discs, it is kept in check at all times. The slightly deteriorated or contaminated plating solutions have to be discarded.

The simplified production process shown in Figure 4 indicates the sources of wastewater to be treated and/or disposed of. The chemical solutions particularly electroless nickel plating solution, are highly chelated. In most cases, special considerations and handling would be needed.

TREATMENT OF RIGID DISC WASTE

From a typical rigid disc manufacturing process, normally 5 categories of wastes are generated. The characteristics of these wastes are so diversified, they often have to be treated differently and separately to be cost effective.

Combined Rinse Water Waste

This waste includes all the rinse water from the production which contains nickel, zinc, and aluminum. The concentration of each metal is approximately 10 to 30 mg/L and is usually slightly acidic. Since the concentration of metal is low and the volume from each rinse process is approximately equal, it is impractical and unnecessary to separate each rinse water stream.

Aluminum Oxide Slurry Waste

The aluminum oxide slurry is generated from the polishing process using aluminum oxide to polish nickel plated disc. The waste contains approximately 2% of aluminum oxide and fine nickel particles of lower concentration. The aluminum oxide particles are as small as one micron. In the past, the slurry was combined with the rinse water waste and treated together. The upcoming technology is to remove the aluminum oxide particles through use of a precoated filter press, or by using a cyclone which is capable of removing two micron particles. However, the waste still needs to be treated with the rinse water stream because those methods are incapable of lowering the nickel concentration to an acceptable level.

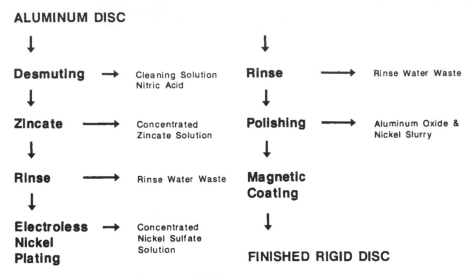

Figure 4.   Rigid disc manufacturing process.

Concentrated Zinc Waste

Depending on the production work load, the zinc bath is replaced once a day. When discarded, the solution still contains 5% zinc oxide and 15% caustic soda. Since the spent solution is dumped in batches, it has to be contained and stored. The treatment includes electroless plating or discharging in small quantity into the combined waste stream for treatment. Electroplating for zinc recovery has proved so far to be unsuccessful. Zincate solution is relatively compatible with the rinse water waste and no significant quantity of chelating chemicals are present in the chemical to interfere with the treatment process.

Concentrated Spent Nickel Solution

Electroless nickel bath is probably the most critical chemical solution in the production process in terms of producing a good disc. During a normal production the solution is replaced daily. The chemical solution when discarded still contains 6 grams/l of nickel. In terms of treatment, the electroless nickel bath is the most difficult chemical due to the presence of strong organic chelating agents. However, the spent nickel solution can be successfully recovered by electro or electroless process. With the bulk of nickel removed, the remaining solution can be discharged in small quantities to the rinse water waste stream.

Concentrated Acid Waste

Nitric acid is used to desmut the disc as a preliminary step of production. The solution when discarded contains 50% of nitric acid and 500 to 1,000 mg/L of nickel and zinc. The acid waste does not present special difficulty for treatment. It is normally discharged at a regulated rate into the combined waste stream with rinse water as the largest in volume and with nickel and zinc mostly from the concentrated waste with or without pretreatment. The key to successful waste treatment appears to depend on the proper regulation of the concentrated discharge especially the electroless nickel solution into the system. Even with pretreatment of the nickel solution, the quantity of chelating agent allowed in the system should be regulated and controlled. The characteristics of the combined waste varies depending on what chemical is being discharged. Normally, the pH of the wastewater is 6 or higher when zincate solution is being discharged.

Several treatment processes have been used. The typical process is shown in Figure 5. The processes and their effectiveness in removing heavy metals are described in the following sections.

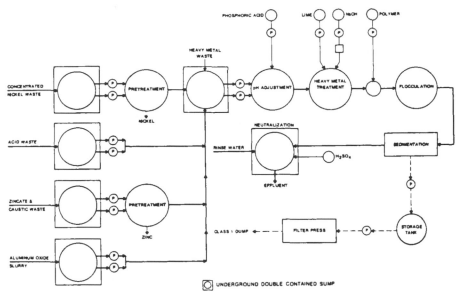

Figure 5.   Schematic of rigid disc waste treatment process.

Method A. Addition of calcium chloride when the pH of the waste is above 10. Addition of 3 to 5 mg/L anionic polymer. The process produces acceptable effluent only when incoming waste has a relatively high pH, preferably above pH 10.5.

Method B. Adjustment of pH with phosphoric acid between pH 4 to 5. Addition of lime or sodium hydroxide or both to adjust pH between 10 to 11. Anionic polymer 3 to 5 mg/L. The process can reduce the nickel, zinc, and aluminum content to less than 0.1 mg/L on a consistent basis. Lime provides more reliable treatment and better effluent, however, it produces a larger quantity of sludge.

Method C. Addition of phosphoric acid to pH 6-8. Addition of 40 to 80 mg/L of alum or ferrous sulfate. Addition of lime to raise pH between 10 to 11. Anionic polymer 3.5 mg/L. The process produces effluent with heavy metal all less than 0.1 mg/L. However, the use of alum and ferrous sulfate proves to be unnecessary in most cases.

Method D. Adjust pH with phosphoric acid or sulfuric acid to 6.0. Addition of metabisulfite based on metal content. Addition of sodium borohydrite based on metal content with combination of NaOH. Addition of polymer. The process produces acceptable effluent, however, gas formation rules out the gravity sedimentation.

Judging from the criteria such as ease of operation, chemical cost, reliability of operation, and effluent quality, Method B provides the best treatment and flexibility.

REGULATORY CONSIDERATIONS

Since the waste produced by printed circuit boards and rigid disc production is classified as toxic and hazardous waste and is closely monitored by the regulatory agencies, all the systems installed have to comply with the following requirements:

- The piping for collection of waste is to be double contained. Double pipe and pipe in the coated trench are both acceptable.
- Containment of the treatment facility is required. The containment should be able to hold 10 to 15% more than the largest tank in the containment area.
- The production area where hazardous chemicals are being used needs to be contained.
- The floor at the production area where the chemical may drip or spill needs to be coated with protective and chemical resistant coating.
- Some local regulatory agencies require separation of contaminated stream from non contaminated stream.

- Some local regulatory agencies prohibit clean rain water from entering the sanitary sewer system.
- This means the treatment facility has to have a roof over it.
- Waste hazardous chemicals and solids are not to be stored for more than 90 days.
- All hazardous chemical storage areas are to be contained.
- Special fire rated wall is to be installed between the building and hazardous chemical storage.
- Some local agencies require a minimum distance between chemicals and the property line.
- A minimum solid content is required for hazardous sludge to be hauled away to a disposal site.
- All hazardous chemicals and sludge are to be disposed of to a licensed hazardous waste dump site.

CONCLUSION

The density of hazardous waste producing industry is relatively high in the Silicon Valley. An intensive enforcement program has been developed and carried out. In addition to producing acceptable effluent, a successful project has to meet local, State, and Federal regulations.

78 DESIGN AND CONSTRUCTION OF A 200-GALLON PER MINUTE ULTRAHIGH PURITY WATER SYSTEM

James D. Edwards, Associate

John W. Cammarn, Senior Chemical Engineer
Burgess & Niple, Limited
Columbus, Ohio 43220

INTRODUCTION

When a large Midwestern manufacturing company decided to integrate forward to produce finished electronic computer components, it needed a new treatment system for ultrahigh purity, sterile water. The minimum water quality standard was 18 megohm which contains less than 30 μg/L of total dissolved solids. Due to the extremely short deadline, the company retained Burgess & Niple Service Corp. (B&NSC) to act as Construction Manager. Burgess & Niple, Limited was retained to complete the technical design. The project included design and construction of both the ultrahigh purity water system and an associated heavy metal wastewater treatment system. Two major equipment suppliers were used, along with general/civil, mechanical, and electrical subcontractors. Design specifications, fabrication, and installation of the systems were completed in less than 6 months with a total cost below the original project estimate.

The process engineering evaluation of the main line production system determined that approximately 200 gpm of finished water would be required. The recommended system involved three stages: treating incoming municipal water, treating recycle effluent, and polishing both streams to produce finished water of the required volume and purity. The available city water was a combination of surface and well water which is treated using the soda/lime softening process.

PROCESS DESIGN

The process design began in July of 1984 for both the water and wastewater treatment systems. Both systems required a great deal of background development. The basic design parameters for the system were developed and discussed with a number of equipment suppliers and some electronic components manufacturers.

One item was universally expressed — when there is a problem in product quality, the probable cause is water quality. Early electronics applications specified 4 megohm, then 9 megohm, and now the standard appears to be 18 megohm quality water. As an example, a certified U.S. Environmental Protection Agency (U.S. EPA) or Department of Health laboratory does not require even 1 megohm water. There is limited information on the corrosion properties of water purer than 1 megohm. Above 1 megohm water becomes very aggressive and can lose its "purity" quickly.

The primary key to obtaining of high purity water is the reverse osmosis (RO) membrane. There are three basic types: the original cellulose acetate, tubular membrane; the Arimid hollow fiber membrane; and the spiral wound composite membrane. The cellulose acetate membranes require the most pressure and are seldom used in the new larger RO systems. Hollow fiber membrane units allow the maximum passage per square foot of floor space, but since the fiber is only about 42 microns (0.0016 inches) in diameter, they require the most careful pretreatment. A new, spiral wound composite membrane manufactured by Dow Chemical was selected.

An RO membrane typically has a 0.1 to 1.0 micron (0.000004 to 0.000039 inches) thick "skin" on a porous substructure. In a spiral wound cartridge, the membrane is formed into an envelope over a support backing. The open end of the envelope is attached to the center product tube. The membrane envelopes are alternated with plastic feed-channel spacers and wrapped in a spiral 8 inches OD by 40 inches long. The feed water stream passes between the spirals parallel to the product tube. The product water (permeate) which passes through the membrane passes to the center product tube and flows out countercurrent to the feed flow. The membrane itself is selectively permeable to water and the feed side is maintained at a pressure above the osmotic pressure of water in the system. Less than

659

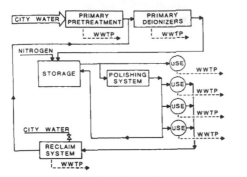

Figure 1.   Overall flowsheet of water system.

Figure 2.   Flowsheet of primary city water system.

3.5% of the total dissolved solids, no filterable solids, and essentially none of the organic molecules, e.g., oils, virus, or bacteria pass through the membrane.

Through a typical RO unit nominally 50% of the water can pass through the membrane in a single stage. In this case, there were six cartridges in series per stage. The units used were two stage units so that 75% of the water was passed through the membrane. Because the municipal water supply used soda/lime softening, pH control and a sequestering agent were added to prevent supersaturation and premature fouling of the membrane.

At the time of the process design, there was no record of a significantly sized high purity water system which was reclaiming plating rinse waters. The rinse waters were lower in dissolved solids than the municipal water supply, but are subject to in-plant upsets, carry trace quantities of chelating agents or other undefined organics, and may be subject to pH swings. Several contacts indicated that as long as 25 to 50% of the rinses were not recycled, operating problems were minimized. Reclaiming nominally one-third of the higher purity rinse water in this case saved $200,000 per year in city water and sewer charges.

PROCESS DESCRIPTION

The overall process flow is shown in Figure 1 and in greater detail in Figures 2, 3, and 4. Approximately 170 gpm of city water enters the Primary Water System. After the Primary RO, 125 gpm of treated water is combined with 70 gpm from the Reclaim System before the Primary Deionizers. The combined flow, which is now 9 megohm water, is stored in a 40,000 gallon DI storage tank. A small quantity of water is used as 9 megohm DI water. The majority is further deionized and sterilized before joining to the various process uses. To maintain quality and pressure, some water is always recirculated to the DI storage tank.

City water is first treated with a nuclear grade of activated carbon to remove chlorine, which could be harmful to the RO membrane and organics. There are some facilities, particularly on the East

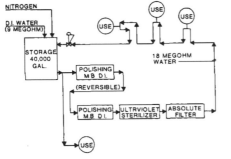

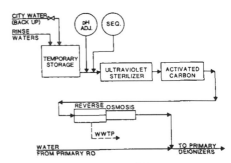

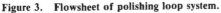

Figure 3.   Flowsheet of polishing loop system.

Figure 4.   Flowsheet of reclaim water system.

Coast, who do not use activated carbon ahead of the RO because of the possibility that the carbon can become a breeding area for bacteria. The colonies may eventually tear loose, migrate to the RO membrane, and can grow and foul the RO membrane. Where silicates are present, most agree that the activated carbon is required. The RO membranes have only been cleaned once for organics in 1 1/2 years, and there has been no sign of organics having caused any fouling. The activated carbon has been backwashed regularly but not sterilized or replaced. The 10 micron filter upstream of the RO unit provides some fouling protection but is mainly for erosion protection.

Normally, RO membranes are rated 77°F. The membrane material selected is suitable for use from 40° to 95°F and a pH range of 4 to 11. RO membranes are temperature sensitive and can lose around 1.5% capacity for every 1°F drop in temperature. The water temperature was expected to be at about 55°F and the system was suitably deaerated. When the system was started up, the city water temperature was 40° to 45°F. Some waste heat has now been made available which should prevent the temperature from falling below 60°F. The RO unit is a 3 x 2 two stage system operating at near 400 psig.

After the primary RO unit, the pretreated city water is combined with the treated reclaim waters and enters one of two mixed bed deionizers. Even with the very low inlet dissolved solids, these units can require regeneration up to once per week. The waters leaving the Primary Deionizers are at least 9 megohm quality water. This water passes through an absolute filter to prevent any growth of bacteria which might somehow have entered the system from contaminating the DI Storage Tank. The DI Storage Tank is also blanketed with nitrogen. As a backup, if the DI Storage Tank ever goes to a negative pressure, there is a HEPA filter on the air inlet.

The DI Storage Tank is a 40,000-gallon, field erected, fiberglass reinforced polyester (FRP) tank. During design applications, experts for the major resin manufacturers were contacted. None would recommend their material for above 1 megohm water, and each could cite at least one failure with above 1 megohm water. However, three different FRP tanks were found in 4 to 9 megohm water service. All were shop built and used the same resin and curing agent. The DI Storage Tank has been in use for 1 1/2 years with no deterioration of the FRP, and the resins have not contaminated the water.

The water from the DI storage tank is pumped through two mixed bed deionizers. The units are piped to allow for alternating the lead unit. When the lead unit must be regenerated, about every 3 months, the second deionizer becomes the lead unit. After regeneration, the fresh bed becomes the lag unit. Before final distribution, the polished water is sterilized with ultraviolet light and is passed through an absolute filter.

Waters leaving the second polishing unit have typically been 19 to 21.5 megohm water. By the time they pass through the passivated stainless steel in the booster pumps, sterilizer piping and filter housing, they have picked up sufficient dissolved solids to drop to 18 to 18.5 megohm water. The distribution piping is designed to maintain continuous flow. The Owner has piped the continuously flowing mainline close to the use points to minimize dead areas. The use of absolute filters or an ultrafilter at critical use points was considered, but this has not been necessary.

The Reclaim System is a miniaturization of the city water system, except that a sterilizer was included. The reclaim water is exposed to the air and could pick up airborne contaminants. The reclaim waters pass through the sterilizer, nuclear activated carbon, and a 2 x 1 two-stage RO system operating at about 250 psig. Antisealant and pH control are included.

DESIGN AND CONSTRUCTION

The original flowsheet and bid packages were completed and sent out for bid on September 4, 1984. Award of contract was scheduled for September 19, 1984. At the last minute, the wastewater treatment process was changed which greatly reduced the waters which could be reclaimed. The flowsheet for both the water and ultrahigh purity water system had to be revised.

The Owner had originally allowed an area 40' x 40' for both the DI water and wastewater treatment systems. The total area was expanded to 60' x 70' and the equipment was located on two levels. In order to determine the best equipment layout and pipe routing, a simple model was constructed. This model was an invaluable tool. In the final layout, equipment was first located by physical height and then by logical flow pattern. One of the last pieces set was the primary RO which had to be jacked up within 6 inches of the roof truss and the structural beams then installed.

The project was fortunate to have a good local contractor who was retained as a general and civil contractor to Burgess & Niple Service Corp. As equipment was laid out and loads determined, the local contractor designed the structural steel. The mechanical and electrical subcontractor were also familiar with the Owner's plant. The contractors were instrumental in maintaining a smooth and tight field schedule.

It was necessary to erect, cure, and sterilize the DI storage tank over a very cold Thanksgiving holiday. This was done in an unheated warehouse while concrete was being poured some 20 feet away. Because silicates in concrete can be harmful to high purity water, a temporary plastic enclosure was built around the DI Storage Tank during construction, curing, and sterilizing. All openings were closed and plastic bags were sealed over the nozzles until the piping was installed.

The original schedule called for the system to be on line and operating on January 20, 1985. The last major piece of equipment arrived on January 26, 1985 and ultrapure water was delivered about February 26, 1985. It was only necessary for the Owner to use a temporary water source for his pilot facility for about 3 weeks. The ultrapure water system was on line in less than 8 months from the initial contact and within 6 months from the time that an order was received for the water/wastewater systems.

SYSTEM PERFORMANCE

The ultrahigh purity water system has never failed to deliver 18 megohm water. At times, the capacity has been marginal. This has been due to a combination of low temperature, a tight sizing by the system supplier, and RO membranes which were out of specifications.

There was deep concern early in the operation; after a few weeks of operation, a distinct organic odor was detected in static laboratory lines and later in the nitrogen in the DI storage tank. Samples were sent to two different firms with Mass Spectrophotometers capable of detecting volatile organics in the 0.2 to 10 μg/L range. Both found methyl ethyl ketone (MEK), tetrahydrofuran (THF), toluene, and xylene. One laboratory determined 42 μg/L, 78 μg/L, 40 μg/L, and 117 μg/L, respectively. The other found between 100 to 200 μg/L MEK and THF and 10 to 20 μg/L toluene and xylene. One lab found no other detectable organics. The other found a few halomethyl and haloethyl compounds in the 0.2 to 3.0 μg/L range, but the occurrences were inconsistent.

The entire water system was drained and flushed three times over one weekend. The MEK and THF dropped drastically to less than 10 μg/L and eventually disappeared. The toluene and xylene did not disappear, but persisted at a level of around 10 μg/L which would slowly increase to 20 μg/L unless the entire system was drained. A testing program was instituted to determine the source. The toluene and xylene were both continuously discharged from the four mixed bed deionizers. The manufacturer claims there is not even any toluene or xylene used or stored within his facility. After 6 months, the levels of toluene and xylene gradually decreased to the 3 to 10 μg/L range and they have apparently not affected the quality of the Owner's product.

PVC pipe has been known to bleed MEK, THF, and sometime phthalate esters into the material being pumped for periods up to 6 weeks. An alternate would have been to use Kynar pipe. Kynar pipe would have added almost $60,000 to the price of the prepackaged, skid mounted equipment for the ultrahigh purity water system and delayed this equipment up to a month. The costs and delays would have carried through the entire project, so it was decided to use all PVC.

SUMMARY

An ultrapure water system was designed, constructed, and brought on line in less than 6 months. The systems were installed for less than the originally estimated price. The system as described has consistently delivered 18 megohm quality water with no discernible organics other than those which can be accounted for as coming from the pipe cement and from the DI resins. The levels of organics has been only on the parts per billion range, and has caused no problem to the computer components which are being produced.

The system is capable of producing 200 gpm of ultrahigh purity water, which we believe is the second largest single ultrahigh purity water system now in operation. This system and Control Data's system are the only ones of significant size to incorporate reclaiming of high quality rinse waters.

ACKNOWLEDGEMENTS

Burgess & Niple Service Corp. and Burgess & Niple, Limited wishes to acknowledge the cooperation and faith that a trusted client showed in awarding this contract. This project could not have been completed without their trust and the assistance of three fine field contractors and two recognized equipment suppliers. There were numerous equipment representatives, water experts, college professors, resin experts, and many others whose comments, both positive and negative, were considered and in some way may have been reflected in the final system. In addition there are many at Burgess & Niple, Limited who worked many hours making this system work who should be acknowledged.

REFERENCES

1. Applegate, L. E., "Membrane Separation Processes," *Chemical Engineering* (June 11, 1984).
2. Cartwright, P. S., "Reclaiming Ultrapure Rinse Waters," *Water World News* (September/October 1985).
3. Sinisgalli, P. D., and McNutt, J. L., "Industrial Use of Reverse Osmosis," *Journal AWWA* (May 1986).

79 A UNIQUE METHOD FOR NEUTRALIZING METAL CHLORIDE WASTE PRODUCED IN THE PURIFICATION OF POLYSILICON

William C. Breneman, Chief Design and Process Engineer
Electronics Division
Union Carbide Corporation
Washougal, Washington 98671

David M. Reeser, Environmental Process Manager
RUST International Corporation
Portland, Oregon 97225

INTRODUCTION

Union Carbide Corporation has constructed a polycrystalline silicon purification facility on a site near Moses Lake, Washington. The plant, covering 40 acres, contains two facilities. The first, utilizing proprietary technology, converts metallurgical-grade silicon to ultra-high-purity silane. Using another proprietary process, the other facility converts the silane to semiconductor-grade polycrystalline silicon.

Process design for the facilities was the work of a combined team of engineers from Union Carbide Corporation and RUST International Corporation. The silane area process was developed from original Union Carbide work done for the Department of Energy. The polysilicon facility process was based on a proprietary license of the Komatsu Corporation of Japan. Process design of the waste treatment systems was the work of the process team based on original concepts of the authors.

Design of the waste treatment facilities was accomplished by first developing a preliminary flow diagram. Because of uncertainties in part of the process, a research laboratory in Salt Lake City, Utah was rented and product reaction tests were made. The process flow diagrams were revised in accordance with the findings of these tests. After finalization of the process flow diagram, a conceptual styrofoam model of the treatment facility was constructed, followed by a $^3/_4''$ to $1'$ scale construction model. Construction began in July 1982, with startup of the plant occurring in the fall of 1984.

The facility was awarded both the Washington State and Pacific Northwest regional industrial pollution control awards for 1985 by the Pacific Northwest Pollution Control Association.

PROCESS DESCRIPTION

Polycrystalline silicon is a basic element of the electronics industry. Prior to this facility, it had most commonly been made by chlorosilane vapor deposition using a power-intensive technology developed when electricity and fuel were inexpensive. Union Carbide Corporation's process, however, is based on recently developed, low-energy technology and builds on the firm's manufacturing experience with chlorosilane intermediates.

Metallurgical silicon is first fed to a pressurized, heated vessel where it is reacted with a mixture of hydrogen and chlorosilanes to yield a higher silicon content chlorosilane liquid. This liquid is alternately distilled and catalytically rearranged to yield silane and unconverted chlorosilanes. Reclaimed chlorosilanes are recycled back to the initial reactor to react with more metallurgical silicon. The silane is decomposed in a heated vessel to yield both a high- purity silicon product and hydrogen, which is also totally recycled back to the initial reactor. After cooling, the product silicon is removed in rod form, crushed, cleaned, and packaged for shipment.

The impurities extracted from the silicon raw material are concentrated internally in this closed loop process and are periodically purged either to the flame hydrolyzer or a batch hydrolyzer for neutralization in a lime slurry. The research and development of a controllable hydrolyzation process are subjects of major import in this paper and are also the subjects of a process patent application now pending in the U.S. Patent Office. The products of hydrolysis, SiO_2, and HCl, are scrubbed before being discharged to the atmosphere, and the scrubber liquid is neutralized with calcium hydroxide to recover calcium chloride.

A portion of the purified polysilicon product rods is etched with nitric and hydrofluoric acid prior to shipment.

PROCESS WASTEWATER TREATMENT

Sources of Wastewater

Wastewater is generated in both silane and polysilicon production areas as well as in support facilities such as the cooling towers, water treatment system, and sanitary system. Waste generated is separated into four categories for treatment: high-chloride process wastes, low-chloride process wastes, nonprocess cooling wastes, and sanitary wastes. High-chloride process waste streams include water that has been in contact with chlorosilanes and water containing hydrochloric acid.

Low-chloride process streams include those containing hydrochloric acid rinse water and non-chloride acids and waste from the reactor wash stations and product handling area and the regeneration waste from the demineralizer units. Fluorides in the spent acid are precipitated with lime and removed as a calcium fluoride precipitate, which is dried for proper disposal. Nitric acid is neutralized with lime to produce soluble $CaNO_3$, which is sent to the Moses Lake Publicly Owned Treatment Works (MLPOTW).

Nonprocess cooling waste streams are primarily the blowdown from cooling towers.

Sanitary wastes are those associated with the employees' restrooms, lunch room, and shower facilities.

Table I summarizes the waste characteristics, flow rates, and plant treatment of each stream. Process flow diagrams are shown in Figures 1 and 2.

Table I. Wastewater Production

	Flow (gpm)
Sanitary Wastewater	
Administration and Maintenance Areas	2.5
Reactor Building	2.5
Silane Area Control Building	2.5
Product Handling	2.5
Low Chloride Wastewater	
Reactor Building	0.5
Filament Processing (Product Handling)	0.5
Area 01 Expansion – Miscellaneous	5.5
Effluent Area Hose Station, Vacuum Pump Seals	3.5
Area 03, 04 (Miscellaneous Hoses, Wet Vacuum System) + Area 07 Vacuum Headers	15.5
Area 07 – Pump Seals, Wet Vacuum System	15.0
Area 03, 04 Reactor Rinse (Existing)	2.8
Area 07 Reactor Rinse (New)	1.4
Acid Neutralization	0.3
Filament Cleaning	1.5
Product Washing	0
Demineralizer System 5 MOHM	13.7
Filter System	0.2
High Chloride Wastewater	
Hydrolyzer Scrubber	0.3
McGill Scrubber MakeUp and Wash Down	1.4
Area 02 Expansion – Pump Seals, Misc.	0.08
STC Receiver	0.7
Area 02 Chemical Storage Pad Drainage	0.7

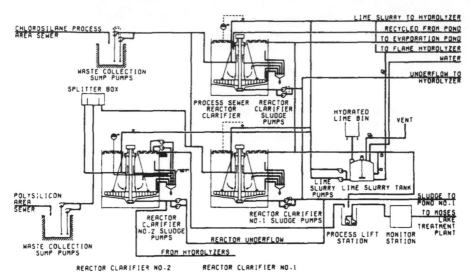

Figure 1.   Schematic of reactor clarifier system.

Design Basis

The facility is designed to operate 24 hours per day, 7 days per week, 50 weeks per year. The wastewater treatment system is designed to accommodate the wastes generated during all operating hours. The quantity and quality of the waste streams were predicted based on previous developmental work for these processes.

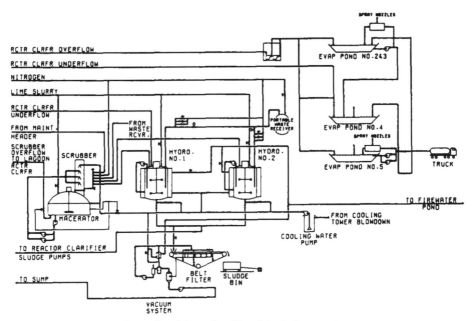

Figure 2.   Schematic of batch hydrolyzer system.

Process Effluent Systems

Aqueous process waste streams are collected at seven waste collection sumps throughout the facility. The waste streams are separated into those having high levels of chloride and hydrochloric acid and those with relatively low levels of chloride. Waste streams are neutralized with lime slurry in three parallel reactor clarifiers. The high-chloride level wastes are then transferred to one of three evaporation ponds with spray towers and are evaporated onsite. The low-salt concentration wastes are neutralized and released to the Moses Lake treatment plant via the city sewer.

A recirculating calcium hydroxide loop is provided to facilitate immediate response to calcium hydroxide demands in the various neutralizing units. Two progressive-cavity positive displacement pumps provide flow in the circulating systems. Rated capacity for these pumps is 75 gpm. An automatic plug valve at the reactor clarifiers and the metal chloride slurry hydrolyzer meters the lime slurry into the reaction vessels. The valves on the reactor clarifiers are controlled by pH meters, and the valve into the hydrolyzer is controlled by a batch flowmeter. A nuclear density meter is provided in each loop to maintain a consistent slurry density.

Sanitary wastes are discharged by a separate sewer to the City of Moses Lake sanitary sewer system and, therefore, are not a portion of the process.

Evaporation of High-Chloride Effluent

A solar evaporation system capable of concentrating high chloride waste to a level at which the brine becomes a saleable chloride source is included in the system. This process was selected to take advantage of the low rainfall (8" average annual) and high net evaporation (60" a year) that occur in the Moses Lake area. The basis for design of these ponds was the work of John Ball, University of Alabama, and G.O.G. Lof, et al., Colorado State University.

Effluent from the high-chloride neutralization system, along with supernatant or filtrate from the hydrolyzation process, is collected in one of three spray evaporation ponds. Reactor clarifier underflow is sent to one of two sludge collection ponds for drying. All ponds utilize natural evaporation. The spray ponds utilize spray nozzles to increase their evaporation capacity. Of the 5.5 gpm total to evaporate, 4.0 gpm is by natural evaporation and 1.5 gpm is by the spray towers. Existing Ponds 2 and 3 and new Pond No. 5 have the capacity to handle the waste produced by the facility for a year. Ponds 2, 3, and 5 are equipped with recirculating pumps to supply the spray evaporator system. One hundred percent redundant capacity is provided in accordance with accepted Washington's DOE guidelines for designing this type of system. The spray pond recirculation pumps have the capacity to increase flow or pressure if greater spray evaporation is required.

Provisions have been made to allow transferring wastes between ponds. All ponds are lined with 100-mil polyethylene sheets to provide a leak-tight barrier and all include a detector system.

Cooling Tower Blowdown

The manufacturing processes produce streams of noncontact cooling water that are suitable for irrigation. The facility operates with three cooling towers in the polysilicon area. Blowdown rate from the three units at two cycles of concentration is approximately 333 gpm.

The cooling towers operate between 2 and 3.5 cycles of concentration. The sodium absorption ratio (SAR) at 2 cycles is 1.6, and the EC ratio (salinity expressed as conductivity) is 0.26 millimhos/cm (mmho/cm). For 3.5 cycles, the SAR and EC values are 2.1 and 0.46 mmho/cm, respectively. These figures indicate a superior water quality to that presently found in Moses Lake and a water that is suitable for unrestricted irrigation.

A portion of the cooling water is used for landscape irrigation onsite. The remainder of this water is discharged into the farming unit drainage ditch, which is a portion of the Columbia Basin Irrigation District Collection System, eventually finding its way into Pot Holes Reservoir. Because this water is superior in quality to the water currently in Moses Lake reaching Pot Holes Reservoir, it is more beneficial to use it in this manner than to discharge it to MLPOTW. Moreover, the hydraulic capacity of the Moses Lake sewage treatment plant is not interfered with if the water is disposed of in this manner.

Hydrolyzation of Metal Chloride Slurries

Upon development of the initial process flow sheet for the hydrolyzation of the metal chloride waste, an inspection trip was made to facilities in West Virginia handling similar waste. In all of these cases the waste slurries were aqueous and very dilute. Standard treatment was to add large quantities of lime to these solutions, which neutralized them but resulted in violent exothermic reactions and the release of hydrogen and hydrogen chloride gases, often producing both explosive and toxic hazards from the process.

In order to develop a new process, the research laboratories of Eimco Corporation in Salt Lake City were rented and the authors spent a week developing the basis for the process. Initially, it was postulated on the old high school conundrum, "May his rest be long and placid for he put water in his acid," that it would be sensible to titrate the waste into a lime slurry. Using the new Union Carbide process, the waste could be obtained as a nonaqueous slurry, thereby greatly reducing the quantity of waste to be treated.

Initial tests aimed at hydrolyzing the proposed waste in a reactor clarifier by adding it to lime slurries revealed that this technique worked, but that the hydrolyzed slurry was not settleable. Further tests indicated that the slurry was easily filterable on a standard belt filter. Other characteristics of the titration determined by these tests were that the amount of gases escaping from the titration was a function of temperature, which was a function of feed rate and pressure. It was determined that, if operated under a slight pressure, controlled hydrolysis could take place with a minimal production of free hydrogen chloride gas.

The chemical function of the hydrolysis process is to titrate the accumulated, concentrated waste into a 15% calcium hydroxide slurry until the pH of that slurry is reduced to a level between 6.5 and 7. The primary end products of this titration are silicon dioxide hydrolyzate and calcium chloride. Small amounts of ferric and aluminum hydroxide are also produced, as well as the oxides and/or hydroxides of any and all metallic chlorides present in the concentrated slurry.

This reaction takes place in one of the 2,000-gallon, glass-lined, water-cooled hydrolyzer equipped with a two-blade mixer. The waste receiver and hydrolyzer operate under a nitrogen pad, and all exhaust gases are passed through a trace scrubber. One-thousand-five-hundred gallons of calcium hydroxide are transferred from the lime slurry system into the hydrolyzer. The waste metal chlorides are forced by nitrogen pressure into the hydrolyzer until the maximum operating temperature of the resultant slurry is reached. At this temperature, the feed valve is closed, and the temperature of the slurry is reduced. This cycle is repeated, feeding small amounts of the waste until either all the waste is hydrolyzed to the pH of the slurry in the hydrolyzer, or until it drops to the 6.5 to 7 range.

The hydrolyzer is equipped with the necessary pH meters and temperature control devices to allow semi-automatic operation. Because of the highly reactive exothermic hydrolyzation that takes place in the vessel, a small amount of hydrogen chloride gas is released in the process. For this reason, all gases exiting the hydrolyzer pass through a spray type lime slurry cascade scrubber, where hydrogen chloride content of the exit stream is reduced to levels suitable for atmospheric discharge as established by the Grant County Clean Air Authority.

A cross-section of a typical hydrolyzer is shown in Figure 3. Significant parts are identified in Table II. Components of the hydrolyzer critical to this operation are the water jacket for temperature control and the dual-bladed mixer. The center blade of the mixer is designed for shear mixing. The metal slurry wastes are introduced immediately below this mixture which disperses them into a broad band of lime slurry. The bottom blade of the mixer is a recirculation mixer which keeps fresh slurry passing the shear mixer during the times of injection of waste. The unit is equipped with instrumentation to control pH and temperature, and the wastes are fed intermittently within the control range until the lime slurry is spent.

Upon completion of the hydrolysis process, the neutralized silicon hydrolyzates are washed and dewatered on a vacuum belt filter. Dewatered solids are loaded into a "dumpster" box for disposal in a nonhazardous landfill. The calcium-chloride-rich filtrate is pumped to the high-chloride evaporation pond system.

IMPACTS OF THE EFFLUENT

Subsequent to the issuance of the permit for the original Union Carbide plant, the City of Moses Lake switched its effluent disposal from the Pelican Horn of Moses Lake to a series of infiltration basins in the Sand Dunes, approximately six miles south of town. Several constraints regarding effluent volume, salinity and sodium adsorption ratio imposed on the plant at the time of initial

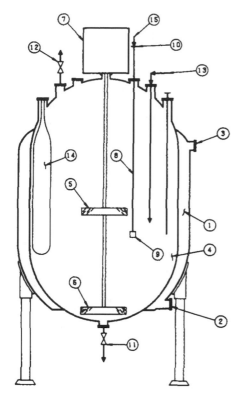

Figure 3. Hydrolyzer cross-section.

design have been made less important because of this change. The polysilicon facility no longer has any direct impact on the waters in Moses Lake.

Two areas of concern have been the sodium and chloride compounds which, if in sufficient quantity, can turn water brackish and render it useless for consumption by humans and animals or for irrigation. In addition, these compounds may have an adverse impact on aquatic water in the Potholes Reservoir.

Table II. Hydrolyzer Components

1. Cooling Water Chamber
2. Cooling Water Inlet
3. Cooling Water Outlet
4. Reaction Chamber
5. Shearing Mixer Blades
6. Recirculation Mixer Blades
7. Mixer Drive
8. Waste Injection Tube
9. Check Valve
10. Waste Supply Line
11. Drain Valve and Line
12. Pressure Valve and Vent
13. Lime Slurry Supply
14. Mixing Baffle
15. Pressurization Nitrogen Supply
16. pH Probe

After diffusion in the Potholes Reservoir, however, the impact of the chlorides is not significantly different than current conditions when the equivalent level of chlorides enters the reservoir via Moses Lake.

CO-PRODUCTS

Several co-products are produced incidental to the main polycrystalline silicon product. These co-products (fume silica from flame hydrolysis of chlorosilanes, precipitated silica from lime slurry hydrolysis of chlorosilanes, and 35% calcium chloride solution) meet the criteria for commercial materials because their chemical, physical and toxicological properties meet published quality standards for materials that are commonly used by a variety of industrial customers in concrete, building materials, oil well drilling mud, etc. It is intended that these saleable co-products be marketed.

PROCESS SOLID WASTES

The plant produces three solid and three liquid process wastes. The solid wastes are silicon hydrolyzate, fumed silica and calcium fluoride. The liquid wastes are primarily calcium chloride solutions.

An attempt was made to make a book classification of the designation of the wastes in accordance with WAC 173-303-9902, but the results were inconclusive. It was then decided to classify the materials by the oral rat toxicity and bio-assay methods. In addition, chemical analysis of the samples was performed along with EPA leachate tests. These tests corroborate the fact that the process wastes are non- dangerous and are suitable for disposal in a municipal sanitary landfill.

SPILL PREVENTION, CONTROL, AND COUNTERMEASURE PLAN

As a part of the effluent system design, an SPCC (Spill Prevention, Control, and Countermeasure Plan) was developed to meet the requirements of Washington's DOE. The SPCC plan is used for emergency response and operator training and is kept on file in the facility control rooms.

Principal parts of the plan consist of a deluge and hazardous waste spill control system which has been incorporated in the expanded silane area. All areas containing potentially hazardous material are diked. Flammable material areas are diked and equipped with a deluge system. A lagoon sized to contain the entire contents of the fire water supply pond is provided to collect major spills. Any acid wastes spilled to this area are neutralized by batch treatment with calcium hydroxide.

Storm water ditches were designed to separate contaminated and uncontaminated storm water. Contaminated storm water is processed with the high-chloride wastes. Uncontaminated storm water is diverted to the USBR wasteway drain.

CONCLUSION

Operation of the plant over the last 12 months has proven the viability of the system design.

REFERENCES

1. Ball, John, "Evaporation Lagoon for Humid Regions," *Proceedings of the Industrial Waste Conference*, Purdue University, 386–393 (1979).
2. Breneman, W. C., "Feasibility of the Silane Process of Producing Semiconductor Grade Silicon," Jet Propulsion Laboratory/U.S. Department of Energy, *Final Report* (June 1985).
3. Hohmann, P. Eric, and Brian Day, *Vacuum Filtration Testing of a Chlorosilane Waste Stream for Union Carbide*, Eimco Process Equipment Company, Salt Lake City, Utah (November 1982).
4. Reeser, David M., et al., *Engineering Report on Effluents*, submitted to Washington State Department of Ecology (April 1985).
5. U.S. Department of the Interior, "Concentration of Brines by Spray Evaporation," *Research and Development Progress Report*, No. 764.

80 DEVELOPMENT AND IMPLEMENTATION OF WATER QUALITY BASED NPDES PERMITS FOR THE METAL FINISHING INDUSTRY IN NEW ENGLAND

William R. Beckwith, Environmental Engineer

David O. Lederer, Environmental Engineer
U.S. Environmental Protection Agency
Boston, Massachusetts 02203

INTRODUCTION

Nature of the Toxics Problem in New England

The industrial profile of New England is diverse with a wide variety of raw materials, manufacturing processes, and finished products. One common ground shared by many facilities is that their wastewater is discharged to one of the region's numerous small streams. Many of these small watercourses do not provide sufficient dilution during low flow periods to adequately assimilate the toxic effects that can result from industrial discharges.

Facilities that perform metal finishing operations constitute a significant portion of the industrial base in New England. The wastewater resulting from such activities is generally concentrated with toxic metals of particular concern because of their potential toxicity to aquatic organisms.

REQUIREMENTS OF PERMITS UNDER THE CLEAN WATER ACT

The Clean Water Act (CWA) [1] requires that the effluent from point source discharges to waters of the United States satisfy minimum technology and water quality requirements.

Technology Requirements of the CWA

The initial or "first round" National Pollutant Discharge Elimination System (NPDES) permits issued by the United States Environmental Protection Agency (EPA) concentrated primarily on the regulation of conventional pollutants (BOD, TSS, pH, oil and grease, and fecal coliform) and other selected pollutants (ammonia, metals, and phenols) at levels that met the technology requirements applicable at the time, Best Practicable Control Technology Currently Available (BPT).

National Effluent Guidelines for the Metal Finishing Industry were promulgated on July 15, 1983 [2]. In EPA Region I, which covers the New England states, most of the permit limitations established in the initial permits to satisfy BPT were stringent enough to satisfy the BAT (Best Available Technology Economically Achievable) requirements that were to be achieved by July 1, 1984.

Water Quality Requirements of the CWA

Water quality standards are required by the CWA to be adopted by each of the states. These standards must designate uses applicable to each stream segment within the state (fishing, swimming, for example). The standards must also designate minimum instream criteria sufficient to protect each stream use (minimum dissolved oxygen concentrations, for example).

Section 301 (b)(1)(C) of the CWA requires that all discharges to waters of the United States meet limitations necessary to ensure compliance with applicable state water quality standards. All state water quality standards forbid the discharge of pollutants in amounts that would interfere with the attainment or maintenance of the most sensitive receiving water uses. In addition, all state standards include a narrative statement specifically forbidding the discharge of toxic compounds in amounts or combinations that may cause toxicity in the receiving waters.

The main focus of water quality concerns during first round permit issuance was the control of conventional pollutants to insure adequate dissolved oxygen levels in the receiving waters and to avoid

671

odor and other aesthetic problems. Toxic metals were regulated where known to be present (i.e., metal finishers) but on a minimum technology basis. During the recent reissuance of NPDES permits, EPA Region I has emphasized the regulation of toxic metals, organics, and other toxic substances to ensure compliance with the state water quality standards concerning toxicity.

An increased emphasis on toxics control has been made possible in recent years by the availability of the EPA National Water Quality Criteria Documents, promulgated on November 28, 1980 [3] (with subsequent additions and updates), and the increased commercial availability of standardized, reproducible, short term acute toxicity testing methods.

REGION I PERMITTING STRATEGY FOR TOXICS CONTROL

A toxics control permitting strategy was developed and implemented by the Region in 1984. A flow chart of the strategy is presented in Figure 1.

Screening

Permit applications are reviewed, toxic pollutants in the discharges are identified, and available instream dilution is determined. Calculated pollutant levels after dilution instream are compared to EPA national instream criteria, state water quality criteria, or other toxicological information available for specific pollutants. Pollutant levels instream resulting from maximum plant discharge and 7Q10 low flow receiving water conditions are compared against acute instream criteria. Pollutant levels instream resulting from average plant discharge and 30Q2 low flow receiving water conditions are compared against chronic instream criteria. Historical receiving water flow data from U.S.G.S. gage stations are used whenever available. In the absence of historical data, 7Q10 and 30Q2 low flows are estimated by regression analysis of flows for gaged streams with drainage areas that are similar in size, geography, and topography.

The toxics control strategy is invoked when this screening process indicates that pollutant levels more stringent than BPT/BAT technology requirements are necessary to achieve the criteria levels after the facility discharge is diluted by the receiving water.

The BAT levels required by the Metal Finishing Guidelines, the EPA national water quality criteria for the pollutants regulated by those guidelines, and the dilution required to meet the criteria are presented in Table I. Inspection of the column describing the "dilution required" to meet instream water quality criteria yields information regarding the expected relative toxicity of the metals listed. Note that the silver, cadmium, copper, lead, and cyanide levels meeting BAT limits require over 100:1

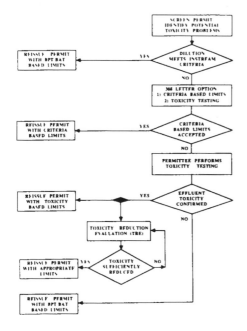

Figure 1. Region I toxics permitting strategy.

Table I. Dilution Required for Metal Finishing Guidelines

Pollutant	Water Quality Criteria[a]		Final Metal Finishing Guidelines[b]		Dilution Required to Meet Criteria[c]	
	Chronic μg/L	Acute μg/L	Mo. Ave mg/L	Daily Max mg/L	Average	Maximum
Cadmium	0.66[e]	1.8[e]	0.26	0.69	394	383
Copper	6.5[e]	9.2[e]	2.07	3.38	318	367
Lead	1.3[e]	34[e]	0.43	0.69	331	20
Cyanide	5.2	22	0.65	1.20	125	55
Zinc	47[e]	159[e]	1.48	2.61	31	16
Nickel	56[e]	641[e]	2.38	3.98	42	6
Chromium III	120[e]	980[e]	1.71[g]	2.77[g]	14	3
Silver	[0.12]	0.44[e]	0.24	0.43	2000	977
Chromium VI	11	16	—	—	—	—
Mercury[d]	0.012	2.4	—	—	—	—
TTO	f	f	—	2.13	—	—
Oil & Grease	—	—	26	52	—	—
TSS	—	—	31	60	—	—

[a]Freshwater aquatic life protection criteria from 45 FR 79318, November 28, 1980 or from 50 FR 30784, July 29, 1985. Numbers in brackets are levels cited for chronic effects but not suggested criteria.
[b]From 48 FR 32486, July 15, 1983.
[c]More recent draft criteria used for calculation, when available.
[d]Mercury criteria is for divalent inorganic Hg, the criteria for methyl mercury would be lower.
[e]At hardness of 50 mg/L CaCO₃.
[f]Listed by individual compound.
[g]Total Chromium.

dilution to meet the national criteria to protect against chronic and acute effects. These toxics are, therefore, the ones of primary concern when screening for the potential impacts of a given discharge on aquatic life.

When the screening procedure indicates that sufficient dilution is available to meet the water quality criteria at permitted facility flow levels, and when other site specific information does not indicate that more stringent limits are needed to protect water quality, the permit is reissued with BPT/BAT based limitations.

Section 308 Letter

If the discharger is identified as having the potential to cause toxic conditions instream if allowed to discharge at BPT/BAT levels, a letter is sent, under authority of Section 308 of the CWA, that proposes effluent limitations suggested by the dilution calculations involving the criteria. The letter offers the permittee the option of either accepting the criteria based limitations or performing a series of four static acute toxicity tests in an effort to demonstrate that limits less stringent than those based on national criteria are justifiable. The second option is offered because it is recognized that the chemical form of the pollutants in a particular discharge, antagonistic interactions between pollutants, and site specific receiving water factors (such as hardness) can influence the behavior of pollutants and possibly render an effluent less toxic than predicted by an evaluation using national criteria. While the toxicity testing provides the opportunity to develop site specific criteria less limiting than the national criteria, there is also the possibility that the testing will confirm the national criteria based limitations.

Toxicity Testing

If the permittee elects to perform the toxicity testing, a 60 day period is allowed for the completion of four 48 hour static acute toxicity tests. An analysis of effluent chemistry for the pollutants of concern in each sample undergoing toxicity testing is included in the report. In addition to metals and organics, parameters such as chlorine residual, ammonia, and hardness are included in the analysis of effluent chemistry.

Each toxicity test is performed using an invertebrate species, *Daphnia pulex* (Daphnid-water flea), and a vertebrate species, *Pimephales promelas* (Fathead minnow) [4]. Whole effluent composite samples are used with dilution water generally collected from the receiving stream at a point upstream of the discharge.

The result of the tests is the determination of the No Observed Acute Effect Level (NOAEL) and an LC_{50}. The LC_{50} is defined as the effluent concentration that is lethal to 50% of the test organisms within the prescribed time. The NOAEL is defined as the effluent concentration at which 90% or more of the test organisms survive.

If the results of the tests indicate that limits based on BPT/BAT are adequate to protect water quality, the permit is reissued with such limits.

Toxicity Reduction Evaluation

If the results confirm the need for toxicity test or criteria based limits, EPA will issue the permit with these water quality based limitations unless the permittee elects to pursue a Toxicity Reduction Evaluation (TRE). The goal of the TRE is to identify the sources of toxicity in the effluent and identify options which will allow the discharge to meet the applicable water quality standards in a reissued permit.

A TRE may include one or more of the activities mentioned below:

1. Isolation of internal wastestreams that may have caused the bulk of the toxicity measured in the earlier tests. These discharges may be targeted for further treatment or elimination.
2. Drastic reduction in effluent flow which increases the effective dilution available in the receiving waters.
3. Relocation of the discharge to a point offering greater dilutional capacity (larger stream, POTW).
4. Changes in manufacturing practices as a method of reducing the quantities of particularly toxic substances discharged.
5. Installation of additional or advanced treatment technologies such as filters, ultrafiltration, sulfide precipitation, and ion exchange.

Following the completion of the TRE, a permit is issued for the discharge based on all data submitted to EPA. Limitations in the reissued permit may be expressed in terms of a minimum NOAEL (whole effluent limitations), individual chemical limits for each suspected toxicant, or both.

A case study example of a permit that has been issued using the toxics permitting strategy is presented in the next section.

CASE STUDY-VERY LARGE INCORPORATED

Introduction

Very Large, Inc. operates a large industrial facility which employs some 6000 persons. A wide variety of metal finishing activities take place at the plant.

Drainage Plan of the Site

Metals and other contaminants are discharged from the present treatment system into an open trench that carries the waste 1/4 mile to an unnamed brook. The brook flows approximately 1/2 mile to a 10 acre shallow impoundment which empties under a railroad embankment forming the southern boundary of the plant site.

Existing Wastewater Treatment at the Facility

Cyanide rinses and spills are collected, raised to a pH of approximately 11.5, and chlorinated to achieve cyanide destruction. Final neutralization is then used to adjust the pH to approximately 9.0. Segregated chromium wastes are collected, reduced with sodium bisulfite to trivalent chromium, and discharged with other segregated metal wastewaters to a sedimentation system. A flocculation tank is employed with alum and a polymer before tube clarification. Sludge is decanted and hauled for reclamation and disposal.

Screening

A comparison of the maximum and average plant flow conditions with the low stream flow of the unnamed brook yielded an estimate that the process discharge contributes some 60% of the stream at periods of 7Q10 stream flow and maximum plant flow (potential acute toxicity conditions). Similarly the monthly average plant flow and the 30Q2 stream flow showed the process discharge to be contributing 40% of the total flow of the brook during potential chronic toxicity conditions.

Employing the above dilutional capacity calculations, an allocation of toxic metals and cyanide for Very Large, Inc. was generated using the suggested EPA criteria. For most of the contaminants, the predicted allowable levels were far below the limits contained in the expired permit and the levels shown in the discharge monitoring reports the company had submitted in fulfillment of the requirements of the expired permit. Therefore, a Section 308 letter offering the option of either accepting the criteria based limits or performing toxicity testing was sent to the permittee.

Effluent Toxicity Testing

The company opted to perform toxicity testing on the effluent in an effort to show that higher concentrations of pollutants were allowable. The results showed NOAELs for daphnia pulex of 5.6% effluent and 14% effluent for two tests. The tests for fathead minnows yielded NOAELs of 56% effluent and 100% effluent. A comparison of the above NOAELs with the estimated contribution of process wastewater to the total stream flow at 7Q10 conditions (approximately 60%) confirmed EPA's contention that the potential for a toxicity problem existed at the facility. In addition, the company analyzed water column samples taken at the pond outfall. These analyses showed metals concentrations greatly in excess of the national water quality criteria and further confirmed that a problem existed.

Based on these results and following a meeting between the company, EPA, and the state pollution control agency, a second round of toxicity testing and instream evaluation was agreed to by Very Large, Inc.

Daphnia pulex was selected as the test species because the results of the first two tests showed daphnia to be significantly more sensitive to this particular wastewater.

Toxicity Reduction Evaluation

Acute Toxicity Tests. The permittee performed several more toxicity tests on samples collected from the discharge. Tests were also run for samples collected from the receiving water at the point of the discharge and from the pond influent. The results of these tests are presented below.

Sample Location	Range of LC_{50}'s (Percent Effluent)	Range of NOAEL (Percent Effluent)
Outfall	0.1%–2.5%	0.1%–1%
Unnamed Brook at Outfall	1.69%–5%	0.1%–1%
Unnamed Brook at upstream end of Pond	1.91%	<1%

As in the previous two tests, the discharge was shown to be very toxic to daphnia. The results also demonstrated the potential of an acute toxicity problem at the instream sites upstream of the pond. The NOAELs for both instream sites were far below the percent contribution of process discharge to the total flow of the unnamed stream at low flow conditions.

Benthic Organism Study. The study concluded that there were significantly reduced populations of infauna and insect larvae in the pond as compared to the control station to which no discharges are made. In addition, a state water quality survey team found only pollution tolerant genera downstream of the pond.

Indigenous Fish Study. This study showed the presence of brown bullheads, golden shiners, and pumpkin seeds in the pond, all known to be pollution tolerant species. A state conducted bioaccumulation study of fish captured at the pond showed significant elevation of cadmium, copper, nickel, and silver in fish tissues as compared to tissue in fish captured at the control site. These metals are discharged from the facility.

Pond Sediment Study. Sediment composition and chemistry data showed that the Pond is acting as a retention basin for the unnamed brook's flow, trapping suspended solids, as well as dissolved and

Table II. Case Study, Sulfide Precipitation Pilot Test Results

Parameter	Test 1	Test 2	Test 3	Test 4
LC_{50} (%)	100	100	100	100
NOAEL (%)	100	50	100	100
7-day NOEL (%)	12.5	20	60	80
Cadmium (mg/L)[a]	0.004	0.006	0.002	0.002
Chromium VI (mg/L)[a]	0.0005	0.035	0.0005	0.0005
Copper (mg/L)[a]	0.009	0.041	0.014	0.005
Lead (mg/L)[a]	0.001	0.043	0.0015	0.023
Silver (mg/L)[a]	0.005	0.025	0.0008	0.0005
Total Cyanide (mg/L)	0.06	0.21	0.09	0.06

[a]total metals.

suspended metals. Chemical analysis conducted on the pond sediment showed highly elevated concentrations of heavy metals such as cadmium, chromium, copper, lead, nickel, platinum, silver, tin and zinc. These contaminants were discharged to the unnamed brook under the terms of the expired permit.

Based on the study data EPA Region I concluded that more stringent effluent limitations than those in the expired permit were required should Very Large wish to continue discharging process water to the unnamed brook and the pond.

Pilot Testing

Having firmly established the toxicity of the discharge, pilot testing was performed by the permittee using various advanced wastewater treatment systems. The most promising was a sulfide precipitation process followed by multimedia filtration. Sulfide precipitation theoretically is more efficient than standard hydroxide precipitation due to the lower solubilities of the complexes formed with the sulfide ion than with hydroxide.

Table II shows the results of four pilot tests run on the wastewater flows. Effluents from the sulfide precipitation were analyzed for the significant pollutants, and toxicity tests were run. The results demonstrated that an effluent not causing acute toxicity could be produced and that there was the potential to produce an effluent that would not cause chronic toxicity.

The Reissued Permit

Following the completion of the pilot study, EPA proceeded to reissue the permit. NPDES regulations at 40 CFR 122.44 require that reissued permits contain limitations based on the more stringent of the following: 1) technology based effluent limitations; 2) limitations at least as stringent as those in the expired permit (with certain exceptions); or 3) Section 301 (b)(1)(C) of the CWA, which requires that permit limitations be stringent enough to comply with state water quality standards.

In the case of this permit, the water quality based limitations proved to be the most stringent of the three requirements for the toxics limited.

The permit was drafted employing the "whole effluent" approach of monitoring and limiting toxicity on a regular basis. The NOAEL of the effluent is required to be at least 60% in each of the tests performed. Testing with *Daphnia pulex* is required twice per month for enforcement purposes.

Metals limitations were based on the no effect levels determined in the various toxicity testing submissions. A comparison of the effluent limitations in the expired permit with those in the reissued permit is shown in Table III.

The reissued permit also contains a requirement that the flow exiting the pond be sampled for toxics and other water quality parameters. The monitoring of the pond effluent will indicate the extent to which metal laden sediment is leaving the pond and whether water quality objectives are being achieved in the water bodies downstream of the discharge. Sampling is required once per month.

CASE STUDY SOLUTIONS AND SUMMARY

Of the 170 major industrial permits in Region I that were screened by the procedure described above, 70 were found to have the potential to cause toxicity instream. Many of these dischargers agreed to criteria based limitations early in permit development. Section 308 letters offering the

Table III. Case Study, Expired and Reissued Metals Limitations

	Monthly Average Limitations	
Contaminant (total metals)	Expired Permit (mg/L)	Reissued Permit (mg/L)
Cadmium	1.2	0.025
Chromium VI	1.05	0.090
Copper	1.5	0.150
Lead	No Limit	0.030
Silver	1.05	0.007
Total Cyanide	0.5	0.130

	Daily Maximum Limitations	
Contaminant (total metals)	Expired Permit (mg/L)	Reissued Permit (mg/L)
Cadmium	1.2	0.025
Chromium VI	1.1	0.090
Copper	2.0	0.150
Lead	No Limit	0.150
Silver	1.1	0.050
Total Cyanide	1.0	0.185

Table IV. Solutions to Toxicity Problems in Region I

Company	Maximum Permitted Flow, gpd	Approximate Dilution Available At 7Q10	Proposed or Existing Solution
Company I	100,000	zero	Proposed: a) Partial tie-in. b) Isolation of most toxic effluent for utlrafiltration.
Company II	43,000	zero	Move process to new location.
Company III	25,000	zero	Proposed: Divert to a larger river via a city storm drain.
Company IV	475,000	20:1	a) Sliding scale permit. b) Constructed force main to carry waste to larger river. c) Treatment consists of conventional system with rapid sand filtration and equalization of flow.
Company V	10,000	12:1	Complete recycle.
Company VI	16,000	30:1	Complete recycle of rinses using evaporative system.

option of either criteria based limits or toxicity testing were mailed to 25 companies. Of these, 11 accepted limitations based on the national criteria, while 14 elected to perform toxicity testing. Another 16 companies in a multiple discharge situation requiring a toxics load allocation were mailed Section 308 letters requiring that toxicity testing be performed.

A number of permittees that have elected to perform toxicity testing have justified limitations less stringent than those based on the national criteria. In many cases, however, the magnitude of the relief has not been sufficient to allow compliance without additional treatment or other changes at the facility. Some examples of routes that have been taken to solve toxicity related problems are shown in Table IV.

The strategy has worked fairly well. Toxicity testing results graphically demonstrate toxicity problems to the dischargers and, as a result, most have been cooperative in searching for solutions. EPA Region I intends to continue use of the strategy. Also, the Region plans to conduct studies to determine how successful the program has actually been in reducing toxic effects in the receiving waters. Information from these studies will identify areas were adjustments to the strategy might be needed and will allow for the fine tuning of permits.

REFERENCES

1. *The Clean Water Act*, as amended through December 1981, Public Law 97-117, 33 USC 466, et seq., U.S. Government Printing Office (1982).
2. *Effluent Limitations for the Metal Finishing Point Source Category*, 48 FR 32462, 40 CFR Part 433, (July 15, 1983).
3. *Water Quality Criteria Documents*, 45 FR 79319-97378 (November 28, 1980).
4. Peltier, W., *Methods for Measuring the Acute Toxicity of Effluents to Aquatic Organisms*, 3rd edition, EPA-600/4-83-00, Cincinnati, Ohio (1978).

81 METAL FINISHING WASTEWATER TREATMENT INCORPORATING MATERIALS RECOVERY AND LOW TEMPERATURE EVAPORATION

Ernest R. Ramirez, Director of Environmental Engineering

Richard C. Ropp, Vice President of Technology
Ramirez Associates, Inc.
Far Hills, New Jersey 07931

INTRODUCTION

The electroplating and metal finishing industries traditionally are operations that have low profit margins. This is especially true in the case of job shop operations where competitive bidding leads to minimized earnings. In order to keep capital costs down, many plants still operate electroplating equipment that is ten and even thirty years old. In general, innovation in this industry is not extensive because this involves capital funds which electroplaters usually do not have.

While Federal EPA Effluent Limitations Guidelines for (non-integrated) Electroplating Point Source Category has mandated that this industry be in compliance by April 27, 1984, there is still a large number (perhaps 40%) of job shop operations which do not meet the USEPA Effluent Limitations Guidelines for Electroplating Point Source Category.

The conventional heavy metal precipitation and cyanide destruction will cost (capital) a metal plater no less than $250,000 (installed), will occupy no less than 800 ft² of area, and last, but not least, will generate a sludge that must be manifested (expensive). Additionally, operating costs for the above wastewater treatment facility will vary annually from $25,000 to $50,000. There is no surprise that in view of the above, many metal platers are not in compliance with the effluent discharge limitations.

This paper is a case study involving the design and engineering of a plating operation discharging 14,000 gpd. Metals involved are: chromium, nickel, copper and the cyanide ion.

NEW WASTEWATER TREATMENT CONCEPT WHICH LOWERS PLANT OPERATING COSTS

The new concept of wastewater treatment involves the operations of material recovery and low temperature evaporation. In reality these methodologies have been used and are still used by precious metal platers where materials recovery is paramount to their staying in business. This paper refers to the technology as new simply because platers have been unaware of its effectiveness, low capital costs and its overall contribution to reduced operating costs when chemical dragout volumes are large. This is typical in the barrel plating operations. Also, plating parts with blind holes will markedly lower operating costs due to the value of the recovered chemicals. In these two cases wastewater pretreatment would actually lower overall costs and raise profits, contrary to the metal finisher's normal perception.

CONCEPT—MATERIALS RECOVERY WITH LOW TEMPERATURE EVAPORATION

Materials Recovery

The guiding principles in materials recovery are straightforward and simply entail the following concepts:

- Contains all (99%) chemicals that leave the electroplating tank on the plated part.
- Use still (non-flowing) dragout tanks directly in series with the electroplating tank.
- The size and the number of dragout tanks will be determined by an engineering material balance encompassing the factors of: 1) parts plating load; and 2) the desired percentage of recovery (often set at 99%).
- Each working day, the dragout tanks are pumped out once, as follows: 1) 1st dragout

679

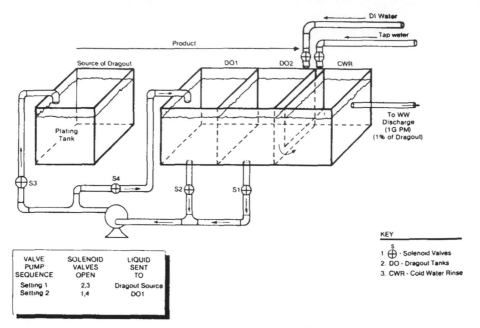

Figure 1. Schematic of wastewater treatment concept. Two dragout tanks with or without evaporator. 99% recovered per 8-hour day.

tank is pumped to plating tank; 2) 2nd dragout tank is pumped into 1st dragout tank; and 3) last dragout tank is pumped into its lesser number, and it in turn is filled with deionized water (Figure 1).

- It is absolutely mandatory that no floor dripping be generated in the transfer of the plated parts from the plating tank through the last dragout tank. Free passages between the tanks must have floor drip shields to catch all dripping liquids.

Evaporation Rate

Since each day the 1st dragout tank is pumped into the plating tank, it follows that a volume of water equal to the volume of the dragout tank must be evaporated daily. In many instances, where the plating tank temperature is high (130°F), operating evaporation losses may be equal to the volume required. In this case no further evaporation equipment is needed. If, however, the daily operating evaporative losses are less than the volume of a dragout tank, then low temperature forced air evaporation via a packed tower arrangement will be required (Figure 2).

Each plating tank will need "in series" dragout tanks (Figure 1). Supplemental evaporation will be needed if operating evaporative losses are inadequate. In order to keep the required evaporation to a minimum the volume of the dragout tanks must be kept as reduced as practical. This often leads to no need for a supplemental evaporation source.

WASTEWATER TREATMENT PLANT AUDIT

This paper is designing and engineering a wastewater treatment facility for a northern New Jersey captive market electroplating shop. A minimum information needed in the audit includes:

- Hydraulic volume flow of the plant.
- Layout of the plating workplace.
- Volume of parts plated per day—alternatively the square feet of product processed per day in each plating tank.
- List of metals plated and documentation on pounds of each metal discharged per day.

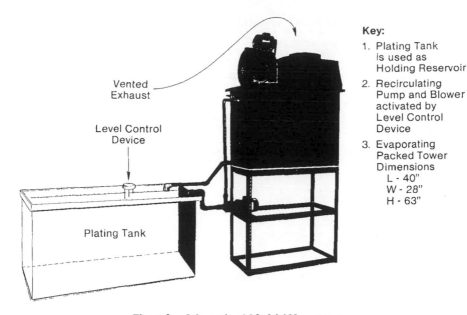

Key:

1. Plating Tank
   is used as
   Holding Reservoir

2. Recirculating
   Pump and Blower
   activated by
   Level Control
   Device

3. Evaporating
   Packed Tower
   Dimensions
   L - 40"
   W - 28"
   H - 63"

Figure 2.   Schematic of Model 233 evaporator.

- Dimensions of each plating tank and operating temperature of each tank. This will establish the operating evaporating volume of each plating tank.
- An evaluation of bright dipping operations because they are a strong contributing source to heavy metal dragout.

The above wastewater audit for the New Jersey plating facility is given in Table I.

DESIGN AND ENGINEERING OF MATERIALS RECOVERY AND EVAPORATION METHODOLOGIES

The final design of the plating wastewater treatment facility is given in Table II. The design and engineering parameters used are based on the following criteria:

- 99.3% of all plating dragout solutions shall be recovered from each plating tank.
- Drip shields will recover better than 95% of the normal floor losses.
- The brass bright dipping operation will not have dragout tanks. This operation will be serviced by a flow-through reticular cathode electrodeposition cell using platinized (or equivalent) titanium anodes. This cell shall recover 92% of the copper and approximately 60% of the zinc generated in the bright dipping operation.

Table I. Wastewater Treatment Plant Audit

1. Wastewater generated per day 16 hours of operation	14,000 gpd
2. Metals Plated	Product Area[a] Ft[2]
2.1 Chromium decorative (rack)	480
2.2 Chromium (barrel)	480
2.3 Nickel (rack)	500
2.4 Nickel (barrel)	460
2.5 Copper (cyanide) rack	500
3. Metals Bright-Dipped	
3.1 Brass	300

[a]This includes inside and outside surfaces.

Table II. Wastewater Treatment — Design and Engineering Data

Plating Tank	Dimensions L × W × H (inches)	Dragout Volume from Plating Tank, Gal/Day	Pounds Metal in Dragout Per Day	Temperature of Plating Solution	Oper. Evaporational Losses, Gal/Day	Volume of Dragout Tank, Gal	Supplemental Evaporation Needed, Gal/Day
1a	80 × 41 × 39	2.1	0.52	110	12	50	38
2b	67 × 39 × 33	2.0	0.50	95	5	50	45
3c	144 × 60 × 39	2.0	0.41	130	55	50	0
4d	81 × 42 × 33	2.0	0.42	90	5	50	45
5e	56 × 24 × 39	1.5	0.10	130	11	50	39
6f	15 × 15 × 15	—	0.1–Cu 0.05–Zn	Ambient	—	—	—

a1 = Chromium (rack)
b2 = Chromium (barrel)
c3 = Nickel (rack)
d4 = Nickel (barrel)
e5 = Copper-Cyanide (rack)
f6 = Bright-Dipping (Brass-not a plating process)

CAPITAL EQUIPMENT COSTS AND OPERATING COSTS

Notable advantages inherent in materials recovery and evaporation concept are given below:

- Additional space requirements are small and often equal to zero.
- Simplicity of operation – no instrumentation or controls.
- Extremely low labor component.
- The wastewater treatment facility generates no sludges.
- Low cost of capital equipment – simplicity of installation.
- Low operating costs.

For the captive electroplating shop under study, no additional area was needed to incorporate the wastewater system. An equipment index and capital costs required for the plant are given in Table III. Daily operating costs for the wastewater pretreatment system are shown in Table IV.

The operating costs given in Table IV are simply expenses required above and beyond the cost experienced prior to the installation of the wastewater pretreatment system.

CONCLUSIONS

Material recovery coupled with low temperature evaporation is an inexpensive option applicable to wastewater pretreatment.

The concept is especially fitting to small platers (less than 20,000 gpd).

Table III. Equipment Index[a] and Capital Costs

1. Ten, 50 gallon dragout tanks (for five plating tanks)	$ 6,150
2. Four, Forced Air Evaporators (capacity 20 gal per hour @ 120 °F and 50% R.H.)	$ 16,450
3. Five, Automatic Pumping Systems (15 gal/min)	$ 7,500
4. Reticular Cathode Electrodeposition Cell (600 amperes at 6 VDC)	$ 13,100
Total Equipment Cost	$ 43,200
5. Installation	$ 16,200
Grand Total	$ 59,400

aThe plant already had pH control and wastewater monitoring; therefore, this equipment was not needed.

Table IV. Daily Operating Costs[a]

	Costs Per Day
1. Added Steam to evaporate 167 gallons of water per day use figure of $0.25 per gallon	$ 41.75
2. Electrical Power	
2.1 Four evaporators with 1.3 kw each @ $0.10 per kw hour	$ 4.20
2.2 Electrodeposition Cell — 600 amperes D.C. @ 4 volts	$ 1.92
3. Labor — 1 manhour per day @ $20/hour	$ 20.00
4. Chemicals Recovered	
4.1 Chromium trioxide — two pounds	$ (3.00)
4.2 Nickel Sulfate hexahydrate — 3.6 pounds	$ (3.60)
4.3 Copper Cyanide — 0.14 pounds	$ (0.40)
Total	$ 60.87
Annual Cost — 250 Days	$ 15,275

[a]Operating Cost use previous costs as a zero data base.

Platers may find it expedient to use forced air evaporation over the plating tanks in order to increase the evaporated volume per day. In some cases this will replace the need for a low temperature evaporator. This will reduce capital costs and space needs.

This new wastewater pretreatment concept is labor-effective.

No sludges or manifestable materials are generated in the pretreatment operation.

There is no reliance on instrumentation (which can malfunction) in this methodology.

Since each day the recovery system is calibrated there is no possibility of an accumulation time factor effecting the operation.

It is conceivable that certain large automatic plating lines may find this concept inappropriate due to space, time or design incompatibility. In this case the conventional cyanide destruction-heavy metal precipitation technology would have to be used.

REFERENCES

1. *Federal Register*, Vol. 48, No. 137 (July 15, 1983).
2. Ramirez, E. R., "Innovative Design, and Engineering of a Wastewater Pretreatment Facility for a Metal Finishing Operation," *Proceedings of the 39th Annual Purdue Industrial Waste Conference* (May 9, 1984).
3. Graham, A. K., *Electroplating Engineering Handbook*, Reinhold Publishing Corporation, Second Edition (1962).
4. Ramirez, E. R., "Engineering and Design of A Wastewater Treatment System for a Vibrobot Plating," *Proceedings of the 40th Industrial Waste Conference*, Purdue University, Page 483 (1985).
5. *EPA 625/10-80-001*, "Environmental Regulation and Technology," The Electroplating Industry (August 1980).
6. *EPA 625/8-80-003*, "Control Treatment Technology for Metal Finishing Industry — Sulfide Precipitation," (April 1980).
7. Cherry, K. F., *Plating Waste Treatment*, Ann Arbor Science (1982).
8. *EPA 625/5-79-016*, "Environmental Pollution Control Alternatives," (June 1979).
9. Steward, F. A., and Mclay, W. J., Series on "Waste Minimization," *Metal Finishing*, Page 63 (October 1985).

82   A CASE STUDY IN MEETING PRESENT AND FUTURE WATER QUALITY STANDARDS

Edward L. Von Stein, Vice President
The Center for Engineering, Inc.
Wallingford, Connecticut 06492

Elizabeth F. Wadge, Engineer
Nathan L. Jacobsen and Associates, Inc.
Chester, Connecticut 06412

INTRODUCTION

Using analytic techniques, the authors discuss the affects of changing water quality regulations on technical design characteristics and on project economics.

This paper discusses a 100 GPM wastewater treatment system recently designed to meet current regulations. The Connecticut Department of Environmental Protection (ConnDEP) has recently completed the public hearing process in anticipation of promulgating revised water regulations.This process represents the initial stage in the development of a State Pollution Discharge elimination system, expected to be completed in phases over the next few years. A review of trends in future regulations forms the basis for our analyses of additional treatment requirements and costs.

Our case study discusses an electroplating facility which is planning to relocate in Connecticut in December 1986. Engineering designs and regulatory reviews were completed in early 1986. Included in this case study is an electroplating facility already confronted by more stringent effluent limitations. Engineering staff at this facility recently completed pilot studies of three treatment methods.

REGULATORY REVIEW

Current Regulatory Framework

Federal statutes are increasingly important with respect to enforcement; "non-statutory" private actions are not available due to court decisions. Actions at the state level, including those based on common law and state statutes, are seen as being appropriate for localized situations. Through the Federal Water Pollution Control Act (FWPCA) and other Acts, the role of the Environmental Protection Agency (EPA) is increasingly important [1].

Due to the conscious effort of the EPA to develop state authority for monitoring and regulating discharge into state waters, Connecticut, as many other states, has drafted new State Pollution Discharge Elimination System (SPDES) permit requirements. We expect that initially these changes will include procedural requirements with steeper fines and clearer documentation of policies.

Once these amendments have been adopted, ConnDEP personnel indicate that actual discharge limits will emphasize receiving water quality. Existing NPDES and SPDES permits are issued based on the levels of concentration of various parameters within the mixing zone of a stream. Minimal background water quality considerations are included at present.

Based on Draft 1985 Revisions to Connecticut Water Quality Standards [2], it is expected that future permits limits will be calculated from the LC_{50} toxicity of priority pollutants to indicator organisms for specific dilution rates. LC_{50} represents the concentration of any parameter which will cause death to 50% of a population. Applicants will require individual toxicity studies or Discharge Toxicity Evaluations (DTEs).

Toxicity-based regulations are designed to "ensure that a healthy environment is maintained for the aquatic biological community . . . and . . . to support designated uses . . ." [3] of receiving waters. The Connecticut regulations will have the effect of making enforceable the water quality standards prohibiting concentrations which accumulate in aquatic biota. "Where bioconcentration of toxics and human health issues are a major concern in making NPDES permitting decisions, DEP will use best

Table I. Estimated Effluent Limitations

	Concentrations, mg/L	
	Most Stringent[a]	Most Probable[a]
Copper	0.0056	0.07
Nickel	0.056	0.10
Lead	0.0008	0.005
Silver	0.0012	0.004
Cyanide (Total)	0.0035	0.008
Source	b	c

[a]Hardness at 50 mg/L as $CaCO_3$, except cyanide.
[b]Reference [6]
[c]Authors' estimate.

professional judgment in cooperation with the State Department of Health officials to set appropriate effluent limits" [3].

To ascertain the effect of water quality based regulations in neighboring states, we discussed permitting throughout New England with EPA staff. We understand that, of about 170 permit renewals requested in EPA Region I in 1985, 70 industrial applicants were required to make system changes requiring substantial capital investment. Other industries elected to move or to meet the new standards by using flow reduction techniques, additional treatments steps, and/or recirculation of flows.

Since exact discharge levels are unclear presently, this case study considers estimated effluent limitations and three treatment technologies which offer ways to meet these estimated limits.

Anticipated Effluent Requirements

To estimate probable effluent requirements, we obtained comments from regulatory personnel and from others in the metal finishing industry. Consensus values could not be obtained. The expected discharge requirement levels are as shown in Table I; most stringent and most probable values are shown.

EPA "white book" values [4], as shown in Table I, and our estimate of "most probable" values are lower than those established in a 1985 permit issued to a metal finisher in Massachusetts. We understand that "white book" values would be used as a starting position by regulators, and that levels above those would be established based on applicant's DTE study. The "most probable" values are approximately median values and as such are conservative.

BACKGROUND

Introduction

Our client is headquartered in central Connecticut. The firm is planning a new plating facility, replacing and expanding the present plating operation. The firm manufactures relays, timers, and switches. Only its metal finishing wastewaters are considered herein [5].

Wastewater Characteristics and Flows

For the purposes of wastewater treatment design, baths and rinsewaters were segregated by wastewater characteristic; thus, the contaminant of concern was determined for each plating room tank and tanks-both rinse and concentrated bath-with like contaminants were collected separately. The concentrated baths could be bled from storage tanks into the appropriate dilute rinsewater flow for treatment. Wastewater segregation categories and constituents are summarized below [5] and in Table II [5]:

1. *Cyanide Wastes*—Rinsewaters from copper strip, cyanide, reverse current clean, copper plating, silver plating all contain cyanide. Most plating baths are dumped only when contaminated.
2. *Nuvat Wastes*—The rinses following the Nuvat alkaline soak contain chelates which interfere with the precipitation of nickel and other metals.

Table II. Wastewater Characteristics [5]

Waste	Principal Contaminant	Concentration	Flow Rates Max, gpm	Avg, gpd	Frequency
Cyanide	CN = 400 mg/L pH = 10+	Rinses Baths	38.5	45	≤ 1/mo
Nuvat	pH = 11 Some Metals	Rinses Baths	17.5	60	≤ 1/wk
Electroless Nickel	pH = 5 (e)Ni = 50 mg/L	Rinses Baths	1.5	4	≤ 1/2 mo
Nickel	Ni = 1000 mg/L Chelating Agents	Rinses Baths	25.5	40	Varies
Gold	–	Rinses	5.0	–	–
Acids	pH	Rinses & Neutralization Baths	95.5	175	Varies
Floor Spills	–	–		50	Varies
Total				46000 gpd	

3. *Electroless Nickel Wastes* — Electroless nickel plating solution is also chelated. There are two (2) electroless nickel plating tanks and one rinse tank. The two plating tanks are dumped when contaminated.
4. *Nickel Wastes* — The rinses following the Watts nickel plating, electroless nickel, and nickel strike contain high concentrations of nickel which must be precipitated at a high pH. Nickel rinsewater flows include the pretreated Nuvat and electroless nickel rinses. Watts nickel baths and nickel strike baths are dumped when contaminated.
5. *Gold Rinsewaters* — Rinsewater from gold plating and stripping is pumped through resin column for gold reclaim and then discharged to the cyanide rinse line.
6. *Acid Wastes* — All rinsewater from acid baths, as well as the pretreated chelated nickel, and cyanide wastes, are collected for pH adjustment.
7. *Floor Spills* — The drag-out minimization techniques are expected to reduce floor spillage of baths and rinsewater.

By way of comparison, untreated cyanide concentration is in excess of published values for common plating streams while nickel is well within the ranged of published values.

Effluent Limitations

The effluent from the treatment facility was designed for discharge to a stream or a Publicly Owned Treatment Works (POTW). While there are EPA Pretreatment Standards for the metal finishing industry, certain Connecticut effluent limitations for metal finishing wastes are more stringent. In designing the facility, we acknowledged that the ConnDEP standards will prevail.

The current limits for effluent quality for metal finishing wastes are given in Table III.

Bench-Scale Treatment Studies

To determine the treatability of the various waste streams, we performed bench-scale studies using wastewater from the existing facility. Using expected concentrations and waste stream volumes, we modelled waste influent and prepared samples which represented the composition of wastewater to be treated in the new facility. The results of the treatment studies are as follows [2] and as presented in Table IV [5]:

1. *Treatment of Cyanide Wastes* — A two stage alkaline chlorination treatment system was planned for the cyanide wastes. Treatment of cyanide wastes with a 10% sodium hypochlorite solution reached the reaction endpoint in the first stage of cyanide destruction. In the second stage of cyanide destruction, 5N concentrated sulfuric acid was added.
2. *Treatment of Nuvat Wastes* — The Nuvat alkaline baths contain chelates which interfere

Table III. Current Limits

Parameter	Maximum Average Daily Concentration (mg/L)
Copper	1.0
Gold	0.1
Lead	0.1
Nickel	1.0
Silver	0.1
Tin	2.0
Cyanide (amenable)	<0.1

Table IV. Outline of Conventional Treatment System as Designed [5]

Waste	Treatment	pH	Detention Time Minutes	Predicted Results[a]
Cyanide	Alkaline Chlorination			
	• CN I		<20	
	• CN II	8.5	60	0.04 mg/L ND
Nuvat	Calcium Chloride @ 2 lb/12 gal; Nickel	6	—	Requires Nickel Treatment
Nickel	Sodium Hydroxide	12	20	0.1 mg/L
Electrodes	Proprietary Chemicals; Nickel	—	20	Requires Nickel Treatment
Acids	Neutralization • Sodium Hydroxide • Sulfuric acid	6.5-8	30	pH 6.5-8
Gold	Ion Exchange Recovery's CN I & II	—	—	0.1
Floor Spills	Batch, as required			—

[a]Authors' analyses.

with the precipitation of metals as hydroxides. Nuvat wastes were pretreated with calcium chloride at a rate of 2 pounds per 12 gallons of wastewater.

3. *Treatment of Electroless Nickel Wastes* — Electroless nickel baths and rinses also contain chelates which interfere with the precipitation of nickel. However because of their high treatment costs and low volume, they were treated separately from the Nuvat wastes.

4. *Treatment of Nickel Wastes* — After treatment of the Nuvat and electroless nickel, these wastes are mixed with the Watts nickel and nickel strike wastes to precipitate nickel at a high pH. The initial nickel content is 1,000 mg/L. Adjustment to pH 12 with caustic soda under mixed conditions lowers the final nickel concentration to 2-3 mg/L.

Proposed Treatment System

The system included 1) waste collection in sumps; 2) Nickel and Nuvat waste treatment; 3) two stage cyanide destruction; 4) pH adjustment, settling, metering and discharge; 5) various batch treatment steps include those addressing floor spills and sludge; and 6) chemical feeds.

Historic and Predicted Success

The existing plant was permitted in 1981 by DEP for discharge of 25,000 gpd. During 1983, 1984, and 1985 the average monthly concentrations and the frequency with which the effluent quality was violated were as reported in Table V. In general, limits were exceeded over 30% of the time during the three year period.

The system was designed to control the parameters being monitored for compliance reporting. The operator's success in meeting guidelines will not deteriorate with the new system. The conventional system as designed is expected to meet current permit limits on a consistent basis.

ALTERNATIVES

Introduction

Three alternative treatment schemes were analyzed. Consideration of probable effluent limitations (Table I) in light of current literature and site visits to similar facilities indicated that conventional treatment could be followed by one of three alternatives polishing steps. These are: 1) Ultrafiltration; 2) Sulfide precipitation; and 3) Ion exchange.

The flowsheet for the study of alternatives is similar for all: each process was considered as an effluent polishing step to follow the pretreatment system as originally designed.

Pilot studies [6] indicated that polishing technology was available and that toxicity-based regulations could be met in systems consistently.

Ultrafiltration

Ultrafiltration is a low pressure (50 psig) membrane filtration process. Various proprietary membranes and membrane support systems are available; a typical membrane offers pore openings of about 0.005 microns (0.005 μ). Systems engineering should consider useful life of the membrane and its ability to resist plugging, high operating temperature, fluctuations in pH and solvent exposure [7].

Representatives of the industry [7] report that ultrafiltration looks attractive as a post treatment filtration step to obtain lower and chelated nickel and copper. While we considered bench scale treatability tests by system vendors to be essential, we expect that after waste segregation, nickel could be treated by means of reducing chemistry, blended with copper which had undergone alkaline reduction, and introduced to a recirculating ultrafiltration system.

We understand that removals of metallic copper on the order of 99.9% have been reported in treatment systems employing sodium borohydrate and ultrafiltration.

Sulfide Precipitation

Sulfide precipitation relies on the low solubility of metal sulfides relative to metal hydroxides to form a highly insoluble metal sludge. Sulfide ion is made available in either a soluble form (sodium sulfide, Na_2S, or sodium hydrosulfide, NaHS) or a relatively insoluble form (ferrous sulfide, FeS). Table VI reports solubilities of metals as both hydroxides and sulfides and the ratios of their solubilities.

Design criteria for sulfide precipitation systems should consider: 1) limitation of excess sulfide remaining in effluent; and 2) control of H_2S, toxic hydrogen sulfide gas.One proprietary process [8] deals with this aspect through the addition of iron sulfide.

The case study alternative consisted of an upflow clarifier and dual media filter. Sulfide precipitation is essentially a polishing system; as such substantial sludge volumes are created. To verify the suitability of such a system, treatability or pilot testing would be recommended.

The system would leave cyanide untouched.

We considered also the results obtained by a similar facility in central Massachusetts. This integrated manufacturer had recently completed a pilot study of the three methods: sulfide precipitation was the method of choice, based on that system.

Table V. Historic Probability of Meeting Limits[a]

	mg/L Limit	1983 x[b]	1983 f[c]	1984 x[b]	1984 f[c]	1985 x[b]	1985 f[c]
Copper	1.0	4.5	1.0	2.2	0.8	1.4	0.7
Gold	0.1	0.0	0.0	0.1	0.3	0.1	1.0
Lead	0.1	1.0	1.0	0.2	0.6	0.2	0.9
Nickel	1.0	3.0	1.0	3.0	1.0	1.7	0.8
Silver	0.1	0.0	0.0	0.1	0.3	0.0	0.0
Tin	2.0	2.5	0.3	1.0	0.0	0.9	0.0
ACN	<0.1	0.0	1.0	0.2	0.7	0.1	0.3
TSS	20.0	21.3	0.5	10.4	0.2	4.0	0.0

[a]Author's analysis.
[b]"x" is average of monthly grab samples.
[c]"f" is frequency of exceedance, i.e. = no. of months above limit/12.

Ion Exchange

Positively or negatively charged ions are removed from solution by the surface exchange of resin particles of like charge. Ion exchange is a metal recovery technique and is well suited to dilate solutions; as such it is acceptable as a polishing step in our analysis.

"The operating cost of ion exchange can be significantly reduced if the system is operated as a polishing process rather than a primary treatment process" [8]. Literature indicates that the application of ion exchange would be successful, but specific resins would require selection based on laboratory tests to optimize removals of copper, cyanide, gold, silver, lead, tin and nickel. Duplicate parallel systems would permit continuous operation during regeneration of the resin. Sources indicate 99% removals of cyanide, tin, silver, copper and nickel for electroplating facilities studies [9].

An ion exchange unit with a fixed bed and counter current operation was considered for this case study due to availability and success in similar applications. A cylindrical ion exchange bed with storage tanks for solutions and wastewater are required.

COST EVALUATION

Guidance in preparing a cost effective conventional waste treatment design was provided by the client's experience, that of ConnDEP staff and our own, and by published sources [10]. Our system, for example, segregates cyanide solutions from other streams, eliminating a source of hard-to-treat wastewater.

Table VII presents gross comparisons of cost for the conventional system and alternatives.

In September of 1985, we estimated the capital cost of the system as $372,000.00.

It is of value to note that this capital cost estimate represents the following: $3700 per gpm of treatment capacity.

Table VI. Metal Solubilities [8]

	Hydroxide	Sulfide Concentrations, mg/L	Column 1 / Column 2
Copper	2×10^{-2}	2×10^{-13}	1×10^{11}
Gold	—	—	—
Lead	2×10^{0}	6×10^{-9}	3×10^{8}
Nickel	7×10^{-1}	6×10^{-8}	1×10^{7}
Silver	2×10^{-1}	2×10^{-12}	5×10^{12}
Tin	—	—	—

Table VII. Cost Comparisons [4,7,8,9,10]

	Conventional System	Ion Exchange	Sulfide Precipitation	Ultrafiltration
Capital Cost				
Unit Capital Cost[a]				
$/gpm	3,700	33,500	3,000	27,000
$/gpd	3	23	2	19
Annual Operating Cost[a]				
Unit Annual Cost				
$/gpy	0.02	0.08	0.02	0.03

[a]Unit cost basis: 100 gpm; 144000 gpd; 12000000 gpy.

REFERENCES

1. Yates, Edward E., "Federal Water Pollution Laws: a Critical lack of Enforcement by the Environmental Protection Agency," *San Diego Law Review*, Vol. 20: 945 (1983).
2. Water Compliance Unit, Connecticut Department of Environmental Protection, *Connecticut Water Quality Standards*, 1985 Revisions, Draft.
3. Water Compliance Unit, Connecticut Department of Environmental Protection, *Water Quality-based Permitting Strategy for Protection of Aquatic Life from Acute and Chronic Toxicity* (January 1986).
4. United States EPA, *1985 Technical Support Document for Water Quality Based Toxics Control*, USGPO (1985).
5. Revised Engineering Report for New Plating Facility and Associated Wastewater Treatment System, The Corporation, prepared by TRC Environmental Consultants, Inc.
6. Telephone discussions and field visits (Spring 1986).
7. Sales literature, EPS Ultrafiltration Systems, undated.
8. Sales literature, Sulfex System, The Permutit Company, Inc.
9. United States EPA, IERL, *Summary Report, Control and Treatment Technology for the Metal Finishing Industry, Sulfide Precipitation*, Cincinnati, OH (April 1980).
10. United States EPA, ITD, *Environmental Pollution Control Alternatives Reducing Water Pollution Control Costs in the Electroplating Industry*, Cincinnati, OH (September 1985).

83 TREATMENT OF DIE CASTING WASTEWATER USING ULTRAFILTRATION

Terrence J. Allen, President

David L. Clayton, Plant Engineer
PEMCO Die Casting Corporation
Bridgman, Michigan 49106

Charles D. Fifield, Project Manager

Jeffrey K. Pugh, Project Engineer
ISD/McNamee, Porter and Seeley
Ann Arbor, Michigan 48104

INTRODUCTION

On October 30, 1985, the U.S. EPA published its final regulations on the discharge of process wastewater from foundries. These categorical standards define the allowable limits of pollutants discharged by foundries into navigable waters and to publicly owned treatment works (POTW's). PEMCO Die Casting Corporation falls under these regulations as an aluminum and zinc die caster.

PEMCO began supplying die castings as well as finished components and assemblies in 1946. The Company operates a plant of 135,000 square feet in the southwestern corner of Michigan and employs approximately 150 people. Products produced in both aluminum and zinc-based alloys go to diverse markets, including automotive, office furniture, office equipment, power tools, recreational equipment, and casket hardware.

Casting, machining, and finishing operations at PEMCO require the use of a variety of cleaning solutions, coating compounds, coolants, cutting fluids, and lubrication products. A substantial degree of process water recycle is practiced which results in a minimal wastewater discharge to the local POTW. The recycle rate is well in excess of the 96% recommended in the foundry categorical standards, resulting in a buildup of dissolved solids, metals, and oxygen consuming substances in PEMCO's process water between biannual system blowdowns.

To explore the options available for pretreating their wastewater, PEMCO retained ISD/ McNamee, Porter and Seeley. This paper presents the results of the pretreatment facilities study completed at PEMCO and describes how ultrafiltration was found to be the best available technology in terms of ease of operation, reliability, degree of treatment attainable, economics and adaptability.

BASELINE MONITORING

Based on a preliminary screening of PEMCO's wastewater completed in November of 1984 and anticipated categorical standard and local POTW ordinance limits, a baseline monitoring program was structured. The program was completed in June of 1985 from which it was determined that the process wastewater pollutants of major concern at PEMCO are BOD_5, COD, oil, grease, iron, and zinc.

Six 24-hour composite samples were collected from PEMCO's wastewater holding/recycle tank over a one week period of typical operation. Table I summarizes the data obtained for the parameters of major concern during the baseline monitoring program and lists the respective treatment goals developed based on the anticipated limits.

ALTERNATIVES

Of the many alternatives available for treating PEMCO's process waste stream, three were chosen for further investigation based on ease of operation, reliability, degree of treatment attainable, economics, and adaptability to changes in plant operations or effluent requirements. The three

Table I. Data Summary for Major Pollutants

Parameter	Analytical Data, mg/L			
	High	Low	Average	Goal
BOD$_5$	770	600	682	300
COD	8500	6500	7383	600
FOG[a]	133	84	109	50
Iron	10.8	9.24	10.1	3.0
Zinc	5.78	3.30	4.58	1.0

[a]Freon extractable oil and grease.

alternatives were: 1) dissolved air flotation (DAF) followed by extended aeration; 2) the PACT® process; and 3) a membrane filtration process.

The *first alternative*, DAF/extended aeration, would include metals removal either within the DAF unit or within an intermediate metals removal package system. It was hoped that metals could be adequately removed along with the oil and grease by elevating the pH of the influent to the DAF unit. Assuming that the COD was largely biodegradable, the extended aeration process would bring the COD down to the treatment goal.

The *second alternative*, PACT®, is a proprietary process which combines powdered activated carbon with the mixed liquor in the activated sludge process. The activated carbon helps to destabilize the oily emulsions, removes nonbiodegradable COD and, to some extent, removes metals. It was assumed that additional metals removal would be accomplished with the flotation of oily floc and the wasting of mixed liquor organisms.

The *third alternative*, a membrane filtration process, would include either one or two stages. The first stage would operate in the ultrafiltration range while the second stage would operate in the reverse osmosis range and only treat a fraction of the permeate generated in the first stage. Due to the small size of metal ions and most organic molecules, it was assumed that the second stage would be required to meet the treatment goals.

The membrane filtration process requires the least amount of operator attention of the three alternatives while being the most reliable in terms of meeting treatment goals. The DAF process requires daily bench testing to maintain the proper chemical dosages and the PACT® system requires routine testing to minimize carbon usage. The membrane process only requires routine inspection of the mechanical components to ensure proper operating conditions along with minimal testing or possibly on-line turbidity measurement. Since the membrane process relies on a mechanical barrier rather than gravity settling for the removal of contaminants from the waste stream, it is more reliable than the other two alternatives [1].

While there was some question as to whether DAF followed by extended aeration could bring the COD down to the required level, for the preliminary evaluation it was assumed that all three alternatives could meet the required effluent limitations.

Table II was developed to provide a cursory economic analysis of the three alternatives. Since the table was developed prior to any pilot testing the costs, particularly for operation and maintenance, were used only to give a rough idea of the magnitude of the differences among alternatives.

The cost estimates presented here are based on a wastewater flow rate of 5 gpm and do not include the labor requirement or the cost of residuals hauling and disposal.

Both the PACT® and the DAF/extended aeration alternatives require close control of operational parameters to ensure meeting effluent quality criteria while the membrane process requires only

Table II. Economic Summary of Treatment Alternatives

	Capital Cost	Equivalent Annual Cost of Capital[a]	Annual O&M Cost	Total Equivalent Annual Cost
DAF/Extended Aeration	$197,000	$32,000	$18,000	$ 50,000
PACT (TM)	$185,000	$30,000	$61,000	$ 91,000
Membrane Filtration	$255,000	$41,500	$76,000	$117,500

[a]based on 10 years at 10%.

Table III. Chemical Dosages and Annual Cost

	Dosage	Annual Cost
Ferric Chloride (FeCl₃)	300 mg/L	$3861
Sodium Hydroxide (NaOH)	450 mg/L	$2266
Percol E-10	2 mg/L	$ 100

minimal attention. Based on the cost estimates presented in Table II and the degree of operator attention required, the PACT® system was screened out in the preliminary evaluation of alternatives. The cost estimates placed on the membrane filtration alternative were discouraging, but the simplicity of the process was an incentive to complete testing with a unit operating in the ultrafiltration range. To determine the degree of treatment attainable and to firm up the economics, it was decided to conduct a pilot study on both ultrafiltration (UF) and DAF including respirometry work to determine the biodegradability of the wastewater.

RESULTS AND DISCUSSION

Dissolved Air Flotation

A series of bench tests revealed that the suspended solids, oil and grease, iron, zinc and a fraction of the COD could be removed in a single step DAF process with initial pH elevation to 9.5. The effluent from the process still contained high concentrations of COD and dissolved solids. The laboratory procedure for flotation by Eckenfelder [2] was followed in this testing.

A 30-second rapid mix time followed by a 2-minute flocculation time and flotation with 33% recycle resulted in a rise rate of 2 feet per minute and a float volume of 50 gallons per 1000 gallons treated.

Table III lists the chemical dosages required to achieve the desired treatment goals along with their annual cost based on an average wastewater flow rate of 5 gpm. The results of the DAF testing gave support to the total equivalent annual cost figure of $50,000 presented in Table II.

In summary, the DAF alternative with integral pH adjustment offered a cost effective solution to PEMCO's wastewater treatment requirements, but it was unknown whether the residual COD would be amenable to biological treatment.

Ultrafiltration

A pilot scale ultrafiltration unit was installed on-site and produced an effluent of extremely high quality. All of the parameters monitored were significantly reduced, and only the BOD_5 and COD remained at concentrations above their target levels.

The UF pilot plant operated at a pressure of approximately 40 psi and a permeate flux rate of approximately 20 gfd (gallons per square foot per day). The unit utilized 4 tubular membranes with a combined area of 8.8 square feet and a nominal molecular weight cutoff of 35,000. The pilot system was operated over a one month period in August of 1985, and four sets of composite samples were taken over a one week period for laboratory analysis.

Table IV lists the data obtained during the UF testing. Chemical conditioning of the process wastewater was not required to attain the permeate values listed below. The pH of the process wastewater varied from 7.1 to 7.4 with an average value of 7.2.

It was unexpected that a membrane process with a molecular weight cutoff of 35,000 could retain metals at neutral pH and also reject such a large fraction of the COD. One possible explanation is that the metal ions and lower molecular weight organic compounds become tied up within relatively large micelles which are retained by the membrane. Micelles are large colloidal particles typically formed when soap molecules surround a small amount of oil or grease in aqueous solution. This phenomenon occurs due to the soap molecule structure which is basically a hydrocarbon with a negatively charged carboxylate group at one end. The hydrocarbon ends of the soap molecules dissolve into the oil or grease droplet and the charged carboxylate groups point outward. Another mechanism which could account for the observed membrane rejection capabilities is the formation of a gel layer. de Filippi [3] describes the precipitation or gelation of rejected solute at the ultrafiltration membrane surface which results in the formation of a gel layer. In theory, the thickness of the gel layer remains constant once steady state is reached due to mass transfer of solute back into

Table IV. Ultrafiltration Pilot Testing Data

		Parameter, mg/L						
		BOD_5	COD	SS	DS	FOG	Fe	Zn
Aug. 7	Raw	700	6200	512	2380	118	13.4	2.1
	Permeate	600	620	2	1230	1	1.51	0.61
Aug. 9	Raw	820	6350	423	2750	303	12.4	2.2
	Permeate	560	610	4	1270	6	2.79	0.70
Aug. 12	Raw	660	6350	417	3210	162	12.9	2.2
	Permeate	530	680	5	1750	7	4.92	0.64
Aug. 14	Raw	900	7400	503	3160	151	15.3	2.0
	Permeate	740	710	2	1600	7	4.1	0.49

the bulk fluid stream. The presence of the gel layer enhances the solute rejecting capability of the membrane, but the actual mechanism of enhancement is not fully understood.

With the data obtained in the UF pilot testing a refined cost estimate could be developed for the membrane process alternative. The capital cost of the system was estimated to be approximately $150,000 which yields an equivalent annual cost of capital of $24,412 assuming 10 years at 10%. The annual operation and maintenance cost of the UF system becomes $36,933 assuming an energy cost of $560 per horsepower per year and a membrane replacement interval of 3 years. So, the total equivalent annual cost of the ultrafiltration system adds up to $61,345 which is approximately one-half the value listed in Table II for the membrane process alternative.

Figure 1 compares the results observed during the UF testing with those obtained in the DAF bench testing. The removal efficiencies are similar for most of the parameters but the UF alternative does a much better job of reducing the COD and dissolved solids concentrations. Eliminating the need for the second, reverse osmosis stage improved the economics of the membrane process significantly.

Respirometry

An electrolytic respirometer was utilized to determine the biodegradability of the COD present in the raw process wastewater as well as that contained in the permeate of the UF process. The electrolytic respirometer is operated similarly to a BOD bottle in the standard BOD test but allows continuous monitoring of the BOD reaction with a larger sample volume and a reduced dilution requirement. Unlike the BOD bottle in the standard BOD test, the reaction vessel of the electrolytic respirometer is mixed with a magnetic stir bar, contains a mechanism for continuous removal of metabolically produced carbon dioxide and oxygen is electrolytically produced at the rate of utilization by the organisms within the reaction vessel. The methods used in the respirometry work were those outlined by Young and Baumann [4].

It was assumed that the BOD reactions observed followed first order kinetics and the constants in the BOD equations were determined using the Thomas [5] method. Table V summarizes the data obtained in the respirometry testing. The values obtained for the rate constant, k, in the BOD equation using the electrolytic respirometer are approximately double those obtained in the standard, long term BOD procedure [6]. This phenomenon has been attributed to the absence of mixing in the standard BOD procedure.

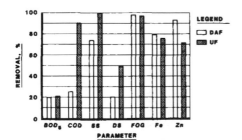

Figure 1.   UF testing results compared to DAF bench testing.

Table V. Respirometry Data

	Rate Constant k, sec^{-1}	Ultimate BOD, mg/L	Initial COD, mg/L	Final COD, mg/L
Raw	0.1804	1019	2088	992
Permeate	0.1052	1374	2073	693

The rate constant, k, for the raw process wastewater has a value typical of that expected for domestic settled raw sewage while the rate constant for the permeate is approximately 60% of the raw value. The ultimate BOD of the raw waste amounted to 49% of the initial COD concentration within the reaction vessel while that of the permeate totalled 66%. At 21 days the BOD reaction is essentially complete in each of the vessels as determined by theoretical calculations. At 21 days filtered samples were tested for residual COD concentration, and the values obtained appear in the last column of Table V. The filtered samples measured slightly lower than the calculated residual COD concentrations but were within 7% of the calculated values in each case.

The results of the respirometry testing indicate that the ultrafiltration process is preferentially retaining a portion of the non-biodegradable COD. At the same time it appears that the process passes a large fraction of the more slowly biodegradable material which makes up a lesser fraction of the raw COD concentration.

SUMMARY AND CONCLUSIONS

The respirometry testing has shown that the COD in the raw process wastewater is biodegradable to a large degree but DAF followed by extended aeration would result in an unacceptable COD concentration as compared with the goal of 600 mg/L. With a COD removal of 25% in the DAF step and up to 60% removal with extended aeration, alternative 1 has an overall COD removal capability of only 70% and falls short of that obtained using ultrafiltration alone. The ultrafiltration alternative provides more complete pretreatment of the process wastewater in a single step without any chemical preconditioning. It is expected that some improvement in metals removal, without adverse effects on the removal of other parameters, may be possible by elevating the pH of the feed to the UF system. The COD concentration within the permeate will be close to the goal of 600 mg/L and, if required, can be effectively polished by a biological process as shown in the respirometry testing.

The refined economic analysis for the ultrafiltration system brings the total equivalent annual cost of this alternative back in line with that anticipated for the DAF/extended aeration alternative. With equivalent anual cost estimates of approximately $61,000 for the UF alternative and $50,000 for the DAF/extended aeration alternative, ultrafiltration was chosen as the superior technology due to ease of operation, reliability, effluent attainability and adaptability to changes in plant operations or effluent requirements.

Since the composition of wastewater generated at an industrial facility will depend upon such variables as materials used, housekeeping practices, makeup water quality, degree of water reuse, and other site specific variables, caution should be exercised when applying the results obtained at one location to other similar cases. It is expected that the results presented in this paper will provide a basis for investigation at other locations and subsequent confirmation by technically qualified personnel.

REFERENCES

1. Tran, Tam V., "Advanced Membrane Filtration Process Treats Industrial Wastewater Efficiently," *Chemical Engineering Progress*, 29–33 (March 1985).
2. Eckenfelder, W. Wesley Jr., *Industrial Water Pollution Control*, 52–61, McGraw Hill Book Company, New York (1966).
3. *Filtration Principles and Practices Part I*, 475–518, Clyde Orr (Ed.), Marcel Dekker, Inc., New York (1977).
4. Young, J. C., and E. R. Baumann, "Demonstration of the Electrolysis Method for Measuring BOD," *Engineering Research Institute Report* (ERI 72153), 25–74, Iowa State University, Ames, Iowa (1972).
5. Thomas, Harold A., Jr., "Analysis of the Biochemical Oxygen Demand Curve," *Sewage Works Journal*, Volume 12, 504–512 (May 1944).
6. Young, J. C., and E. R. Baumann, "The Electrolytic Respirometer—II," *Water Research*, Vol. 10, pp. 1141–1149 (1976).

84 EVALUATION OF A SCREW PRESS FOR DEWATERING STEEL INDUSTRY SLUDGES

Michael T. Unger, Associate Scientist
Laboratories for Environmental Research
Indiana University Northwest
Gary, Indiana 46408

BACKGROUND

The dewatering study was part of a research effort entitled, "Feasibility of Recycling Zinc Bearing Dusts and Sludges Generated by the Northwest Indiana Steel Industry."

Results of that research suggest that more than 515,000 tons (2,000 pounds, dry weight basis) of wet sludges are produced in northwest Indiana from blast furnace (BF), basic oxygen furnace (BOF), electric furnace (EAF), and terminal treatment plant (TTP) sources with weighted average water concentrations of 32.5, 32.1, 17.4, and 47.0%, respectively. This implies a total wet weight (for sludges from northwest Indiana) of nearly 735,000 tons annually. These steel industry sludges must be effectively dewatered before they can be easily or economically handled through reuse or disposal.

Most existing dewatering systems in the steel industry utilize conventional vacuum filters or centrifuges. However, limited data exist for newer dewatering technologies such as the screw press.

DESCRIPTION OF SCREW PRESS

Pilot work was done on various sludges generated by the northwest Indiana steel industry to determine the dewatering capabilities of the Korean Hoilim Iron Works (HIW) screw press. The pilot unit's screw measured 300 mm in diameter, was 2500 mm in length, and typically produced three dry tons of sludge per day. A schematic of the screw press is shown in Figure 1.

The screw press has four main components: the driving gear, screw, strainer drum and effluent basin. The wetted parts are made of stainless steel to minimize abrasion. A low screw rotation speed (0.5 to 6 rpm) accounts for the small energy requirements of this dewatering system. This particular unit was driven by a two horsepower (4.7 amp, 1.5 kilowatt) electric motor and the entire system was mounted in a tractor trailer.

The positive displacement sludge pump was a variable speed Gorman Rupp pump with adjustments from 0 to 100% stroke. The drive was a 0.5 horsepower (5.5 amp) electric motor and at 100% stroke would pump approximately 20 gallons per minute.

The polymer pump was positive displacement with variable settings of 0 to 10. Maximum stroke was equivalent to 20 gallons per hour.

The sludge tank, polymer tank, and flocculation tank had volumes of 350, 160, and 39 gallons, respectively.

Adjustable operating parameters of the screw press include: strainer size (openings), screw rpm, pressure plate position, and the type and dosage of conditioning polymers.

METHODOLOGY

Prior to the pilot testing, Calgon Corporation and/or Nalco Chemical did bench testing on the sludges to be dewatered and determined the polymer type and approximate dosage to be used.

The first day in the field at a particular location was utilized to install the screw press. Services were connected (electrical, water and steam), access to the sludge source was made, and the entire system was checked out.

On the second day, trials began and operational parameters were recorded. The detention time through the screw was typically 12 to 17 minutes; therefore, a trial was 30 to 60 minutes in duration (defined as a minimum of 3 detention times). It was assumed that the process had equilibrated after 3 detention times and samples were then taken of the influent sludge, cake, and filtrate.

RESULTS

Due to limited time, these trials emphasized the feasibility of dewatering rather than the development of optimum operating conditions. The results of the pilot work are presented in Tables I and II.

Cake Solids

The data suggest that the screw press produces cakes which typically range from 40 to 77% solids. At less than 35% moisture, all of the steelmaking sludges tested had a rigid cake-like texture and consistency and did not yield "free water" upon standing. Blast furnace sludges dewatered very well and had an overall average of 72% solids. These cakes were quite "firm" and crumbled when broken.

Basic oxygen furnace sludge is much finer in particle size than blast furnace sludge which may account for the lower overall average of 57% solids.

Biological sludge and oily sludges (those found in terminal treatment plants) produced cakes which contained the most moisture, ranging from 20 to 70% solids.

Blending

The ability to dewater difficult sludges (biological, oily, or very fine grained) can be enhanced by the blending of other sludges which dewater well such as blast furnace sludge or by the addition of materials such as coke breeze or cellulose fiber.

The addition of coke breeze to BOF sludge increased the cake moisture but markedly decreased filtrate solids concentration. The blended cake, though higher in moisture, offers the benefit of having a carbon source (coke fines) as a constituent, thereby being in a potentially more valuable form for sintering or recycling.

The addition of 2:1 (v/v) blast furnace sludge to biological sludge increased the cake solids dramatically—from 20 to 63% solids.

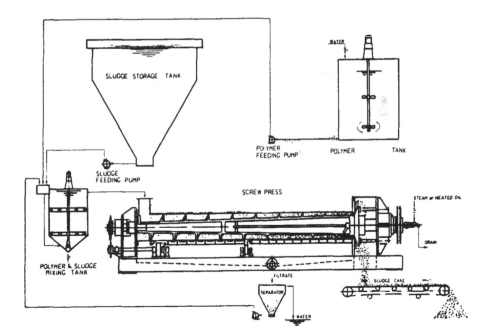

Figure 1.   Schematic of screw press.

Table I. Percent Solids

Sludge Type	Trial I.D.	Feed	% Solids Cake	Filtrate
Blast Furnace	A-1	24.16	72.39	2.83
	A-2	22.10	72.45	1.16
	A-3	16.78	71.24	6.31
	A-4	20.17	73.31	1.92
	B-1	5.95	72.10	0.20
	B-2	3.86	67.01	0.22
	B-3	6.85	71.32	0.26
	B-4	5.89	72.68	0.41
	C-1	9.91	74.25	0.54
	C-2	7.79	72.54	0.52
	C-3	9.93	73.94	0.39
	C-4	3.52	74.48	0.36
Basic Oxygen Furnace	D-1	21.17	59.11	0.15
	D-2	21.53	62.54	0.24
	D-3	21.54	56.84	0.21
	D-4a	13.08	60.09	1.54
	D-5a	20.12	48.46	0.18
	D-6a	22.77	52.76	0.21
	E-1	38.67	53.84	—
	E-2	17.95	53.09	0.47
	E-3	17.95	53.40	2.12
	E-4b	24.76	43.73	0.37
	F-1	11.92	51.80	0.31
	F-2c	14.24	59.73	0.45
	F-3	16.23	50.97	0.44
	G-1	53.21	63.12	—
	G-2	32.36	64.87	—
Biological	H-1	3.38	20.36	—
	H-2	2.87	20.40	1.40
	H-3	3.41	17.59	1.77
	HA-1	11.00	65.65	2.56
	HA-2	10.21	61.31	1.53
Terminal Treatment	J-1	6.37	56.00	—
	J-2d	10.20	60.80	—
	J-3d	10.42	52.13	—
	J-4d	—	—	—
	J-5	7.01	55.80	—
	J-6d	11.20	53.70	—
	J-7	16.30	55.50	—
	J-8	15.20	56.00	—
	K-1	13.26	43.93	1.21
	K-2	12.97	43.58	0.62
	K-3	12.64	35.98	1.67
	K-4	—	38.31	0.62
	L-1	6.41	65.44	—
	L-2	6.50	74.96	—
	L-3	—	77.42	—
	M-1	—	67.92	—
	M-2	32.19	70.17	—
	M-3	32.30	71.92	1.34
	M-4	32.42	70.77	1.44
	M-5d	31.64	74.14	—
	M-6d	30.59	71.05	—
	M-7	25.16	69.16	—

awith steam applied to screw (75°C).
bmixed with approximately 6% coke breeze (–3mm).
cmixed with approximately 6% cellulose fiber.
dwith steam applied to screw (80°C).

Table II. Characteristics of Operation

Sludge Type	Trial I.D.	Capacity (dry ton per day)	Capture %	Poly Use (lb. per dry ton)	Poly Dose PPM (v/v)
Blast Furnace	A-1	3.39	87.7	07.5	721
	A-2	3.18	95.5	03.3	182
	A-3	2.87	69.4	03.2	198
	A-4	3.48	92.0	01.4	091
	B-1	3.22	97.9	02.3	083
	B-2	2.29	95.9	02.4	047
	B-3	3.29	96.8	02.2	117
	B-4	2.88	94.3	01.0	032
	C-1	2.62	95.2	04.9	147
	C-2	3.55	93.1	01.5	045
	C-3	2.40	97.5	03.8	112
	C-4	3.62	93.0	03.0	099
Basic Oxygen Furnace	D-1	3.32	99.5	06.6	494
	D-2	3.02	99.0	07.3	634
	D-3	3.31	99.4	06.6	618
	D-4a	3.11	93.9	05.3	309
	D-5a	2.65	99.3	05.9	473
	D-6a	2.55	99.5	06.5	488
	E-1	—	—	—	—
	E-2	1.76	97.7	09.7	418
	E-3	1.88	94.6	07.8	383
	E-4b	1.57	98.8	06.2	333
	F-1	1.73	97.9	11.3	379
	F-2c	3.98	97.5	04.0	372
	F-3	2.59	98.0	03.6	303
	G-1	—	—	—	—
	G-2	2.38	—	10.3	1144
Biological	H-1	—	—	—	—
	H-2	0.20	67.6	153	855
	H-3	0.26	61.4	96	775
	HA-1	2.36	76.3	06.2	249
	HA-2	2.69	87.6	03.6	220
Terminal Treatment	J-1	1.76	98.3	04.3	134
	J-2d	1.09	—	06.7	330
	J-3d	0.33	—	26.7	142
	J-4d	—	—	—	325
	J-5	1.45	—	07.6	280
	J-6d	1.62	—	06.8	268
	J-7	1.40	—	07.9	385
	J-8	1.45	—	08.3	394
	K-1	1.55	95.0	09.6	366
	K-2	2.29	96.9	04.9	294
	K-3	1.73	92.0	05.1	261
	K-4	—	—	—	465
	L-1	0.99	—	14.3	519
	L-2	1.19	—	10.9	449
	L-3	0.89	—	21.8	900
	M-1	1.76	—	14.5	601
	M-2	1.52	—	15.6	608
	M-3	4.14	90.7	04.2	515
	M-4	3.91	89.1	03.2	441
	M-5d	2.03	—	10.8	515
	M-6d	2.64	—	12.5	608
	M-7d	3.24	96.8	06.3	441

awith steam applied to screw (75°C).
bmixed with approximately 6% coke breeze (–3mm).
cmixed with approximately 6% cellulose fiber.
dwith steam applied to screw (80°C).

Table III. Mass Balance on Oil

	Percent[a]		Recovered Oil
Trial	Feed	Cake	in Filtrate
M-5	13.86	10.99	278 g oil/min
M-6	17.24	11.02	262 g oil/min

[a]weight oil/weight of dry sludge.

Capacity

The data show that the pilot unit produced an average of 2.6 dry tons of cake per day (dtpd) and typically ranged from 2 to 3.5 dtpd. Under optimized conditions, all materials (BF, BOF, and TTP) could be processed at a rate in excess of 3 dry tons per day.

Capture Rate

The capture rate (cake solids divided by influent solids) was generally greater than 90% for all materials tested, except for biological sludge which had capture rates of 61 to 88%. The lower capture for biological sludge was partially due to the very small concentration of sludge (2–11%) being fed into the screw press for these materials. The corresponding filtrate concentrations were larger (1.4 to 2.6%) than the typical filtrate concentrations (0.2 to 1%) for other materials tested.

Polymer Usage

Polymer usage varied from 1.0 to 11.3 pounds per dry ton of cake produced, except for biological sludge which required larger quantities for effective treatment (96 to 153 lbs/dry ton). This was largely a result of the very small influent sludge concentration and the large openings in the strainer drum of the screw press.

Generally, the polymers were added in an emulsion form and had the generic structure of a polyacrylamide. Polymers were largely anionic for blast furnace and basic oxygen furnace sludges and were cationic for biological and terminal treatment sludges. No trends were observed for polymer molecular weights as these ranged from low to very high.

The emulsion was diluted to provide a feed concentration of 0.5% (v/v). Polymer usage was reported in pounds of wet product (the emulsion) per dry ton of cake produced. Actual polymer usage would be somewhat less if expressed on a dry weight basis (the percent active component in the emulsion would have to be known).

Oil Recovery

The ability to recover oil from oily sludges was tested by applying steam (1.39 pounds/minute) through the hollow shaft of the screw and allowing the sludge to be heated to 75 to 80°C. The results are shown in Table III.

The results suggest that 29% (by weight) of the oil was removed from the sludge. When the mean values for oil in the feed and cake are compared by a one-sided t-test, a significant difference was indicated at the 95% confidence level.

Trial M-6 suggests that 262 grams of oil per minute can be recovered from the filtrate which is equivalent to 256 pounds of oil per dry ton of sludge. The concentration of oil in the filtrate suggests a 74% (by weight) recovery. The mass balance for oil was 29% removal as measured in the sludge and 74% recovered as measured in the filtrate. This difference is probably due to the small number of samples and the difficulty encountered in obtaining a representative filtrate sample.

Even though the results suggest that some oil was recovered, further testing is required to more accurately assess the amount of recovery and economics of the process.

Bulk Density

Bulk density was studied to evaluate the reduction in volume of the sludge as the percent moisture decreased. Bulk density was measured by placing a known volume of sludge in a container and weighing it. Using percent moisture, the wet weight was converted to dry weight and the bulk density expressed in milliliters per gram of dry sludge. These data were plotted against the percent solids as

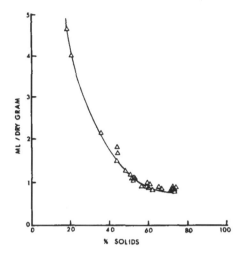

Figure 2.   Variation of bulk density with percent solids in sludge.

shown in Figure 2. The data in this figure suggest that: 1) bulk density decreases exponentially; and 2) the dewatering of steel industry sludges to approximately 65% solids achieves the maximum benefit as far as volume reduction is concerned. If the sludge is dried to a greater percentage solids, no decrease in the number of truckloads transported can be expected. This does not mean, however, that further economic benefits could not be achieved from additional dewatering if the sludges were to be recycled in a process that required heat. Moisture content in the sludges, in this case, would have a direct relationship to the energy costs associated with heating the sludges to the elevated temperatures required in a recovery process.

SUMMARY

This research indicates that the screw press provides a viable dewatering alternative for specific applications within the steel industry. Two steel mills in northwest Indiana are currently pursuing the application of this technology. The work further suggests that oil can be recovered by the application of steam and should be investigated further to quantify this recovery and establish the economics of oil recovery by this means.

85 RESIDUAL CHLORINE BASED AUTOMATIC PROCESS CONTROL FOR THE ALKALINE CHLORINATION PROCESS TREATING CYANIDE WASTE

Abbas Zaidi, Process Development Engineer

Larry Whittle, Process Technologist
Wastewater Technology Centre
Environment Canada
Burlington, Ontario L7R 4A6

Mike Jovanovic, Project Manager
McClaren Engineering, Inc.
Toronto, Ontario M5E 1E7

INTRODUCTION

The traditional treatment approach for destruction of cyanide contained in metal finishing and gold milling wastewaters has been the alkaline chlorination process and automatic chlorine addition based on the measurement of effluent redox (reduction-oxidation) potential has been widely used for controlling the process [1]. Environment Canada's Wastewater Technology Center (WTC) has conducted extensive evaluations of full scale alkaline chlorination systems treating cyanide bearing wastewaters from gold mining and milling operations [2]. These evaluations and in-house bench scale studies proved that good chlorine dosage control was difficult to achieve using redox potential feedback because of the complex relationship which exists between chlorine dosage and redox potential. The bench scale tests also showed that the use of chlorine residual measurement would provide a superior feedback signal for process control and could result in a reduction in chlorine consumption.

To demonstrate this residual chlorine control strategy, a mobile alkaline chlorination pilot plant incorporating the necessary instrumentation and control equipment was constructed and tested at a metal finishing plant where cyanide was used in the process.

This paper presents the results of the pilot plant investigations.

BACKGROUND

Conventional Redox Potential Control

Redox potential is a measure of the ability of a solution to reduce or oxidize another chemical that may be added to it. It is measured in millivolts (mV) and is usually referenced to a common base such as the potential of a Standard Hydrogen Electrode (S.H.E.) [3].

During alkaline chlorination (a typical system is shown in Figure 1) of a cyanide solution, the addition of chlorine results in the oxidation of cyanide to cyanate (Equations 1 & 2). When all of the cyanide has been oxidized to cyanate, the redox potential of the solution rises rapidly to a new level known as the cyanate potential as shown in Figure 2. If the pH of the solution is adjusted at this point and chlorination is continued, the redox level of the solution rises again when all the cyanate has been destroyed (Equation 3).

$$CN^- + H_2O + ClO^- \rightarrow CNCl + 2OH^- \tag{1}$$

$$CNCl + 2OH^- \rightarrow CNO^- + Cl^- + H_2O \tag{2}$$

$$6Cl^- + 6OH^- + 2CNO^- \rightarrow 2HCO_3^- + N_2 + 6Cl^- + 2H_2O \tag{3}$$

In ideal situations involving chlorination of pure cyanide solutions, the straightforward relationship between redox potential and chlorine dosage results in a simple control strategy. However, in wastewater treatment situations, the relationship between redox potential and chlorine dosage is affected by variations in temperature and pH and by the presence of complex metal cyanides typically present in gold milling and metal finishing wastewaters.

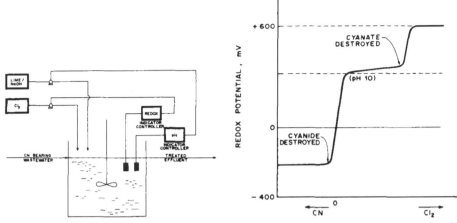

Figure 1. Typical redox control system.

Figure 2. Typical redox response during chlorination of simple cyanide solutions.

The effect of solution temperature on redox potential is minor, representing a change of less than 1 mV for each degree centigrade change in temperature. The effect of pH on redox potential, however, is more significant. For every one unit change in pH, the redox potential can change by 60 mV. Therefore, fluctuations in pH tend to confound the changes in redox potential associated with the oxidation of cyanide by chlorination, and this creates difficulty in maintaining the optimum dosage based on a measured redox potential.

The presence of metals in solution, as found in metal finishing wastewaters or in gold milling effluents, causes the formation of complex metal cyanides and can result in a change in the relationship between redox potential and chlorine dosage. A redox potential/chlorine dosage relationship observed for a solution containing 200 mg/L cyanide is illustrated in Figure 3. A rapid rise in redox occurred after the addition of approximately 0.5 g/L of chlorine. Figure 4 illustrates the changes in the redox response observed for solutions containing 200 mg/L to which copper, zinc, nickel and iron were added separately. Although the addition of copper and zinc caused a slight change in the shape of the redox potential/chlorine dosage relationship, a major change occurred when nickel and iron were added. Figure 5 illustrates the changes to the redox response curve when various metal cyanide complexes were all present simultaneously in the solution. The relationship changed radically so that although two dosages of chlorine existed for a given redox potential, only one of them represented complete cyanide oxidation. It is this phenomenon that creates difficulty when redox potential is used to control the alkaline chlorination process treating wastes from either metal finishing or gold milling operations.

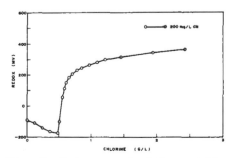

Figure 3. Redox curve of solution containing cyanide.

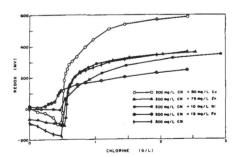

Figure 4. Redox curve of solutions containing cyanide and individual metals.

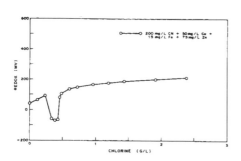

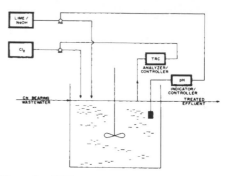

Figure 5. Redox curve of solution containing cyanide, copper, iron, and zinc simultaneously.

Figure 6. TRC control system concept.

Residual Chlorine Control Concept

Cyanide in metal finishing effluents and cyanide and thiocyanate in gold mill effluents react rapidly with chlorine. The addition of chlorine to these effluents results in the immediate reduction of chlorine to chloride until the cyanides have been oxidized. At this point, further addition of chlorine to the wastewater results in the appearance of residual chlorine which may be used as a basis for the residual chlorine control strategy. Figure 6 illustrates the residual chlorine control system concept. As the residual chlorine/chlorine dosage relationship is not subject to the same problems as the redox potential/chlorine dosage relationship, this control strategy can be expected to provide more stable operation with respect to cyanide oxidation.

EFFLUENT TREATMENT SYSTEM AT THE METAL FINISHING PLANT

The metal finishing plant where the pilot tests were conducted had a number of still and barrel lines to perform brass, bronze, copper, nickel, and zinc plating. The plant had an alkaline chlorination system for treating cyanide wastes.

Figure 7 schematically illustrates the alkaline chlorination system at the metal finishing plant. The system had a volume of approximately 3400 L providing a theoretical hydraulic retention time of 34 minutes at the maximum design flow. The treatment system had three separate stages, namely a mixing chamber, a baffled reaction chamber and an effluent sump.

The monitoring instruments (pH and redox potential electrodes) for the system were located in the 1420 L mixing chamber. Sodium hydroxide was added on demand using set-point control at pH 11.5. Gaseous chlorine was supplied to the chamber by recirculating the reactor contents through a chlorine eductor. The chlorinator operated at a fixed rate, and dosage control was accomplished by on-off switching of the feed pump when the redox potential deviated from the set-point of $+450$ mV.

Samples of the full scale system effluent were taken at the outfall of the baffled section prior to entering the effluent collection sump.

EXPERIMENTAL PROGRAM

Pilot Plant Description

An alkaline chlorination pilot plant together with the process control system was assembled at the WTC in a mobile trailer and later moved to the metal finishing plant located 100 km away.

Figure 8 schematically illustrates the pilot plant. It had two completely mixed reactors in series. Each reactor provided 30 minutes hydraulic retention time at the design flowrate of 15 L/min. The first reactor received the oxidant and was equipped with the required instruments to provide feedback signals for process control. The second reactor provided additional retention time but did not form part of the process control loop. Redox potential and pH, however, were monitored in the second reactor. A side stream of the plating wastewater was brought into the pilot plant and the overflow from the second reactor was returned to the metal finishing plant's effluent treatment system.

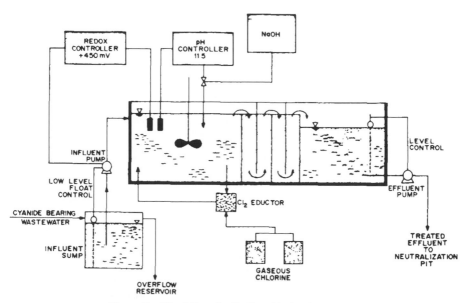

Figure 7.   The full scale alkaline chlorination system.

The pH was adjusted by off-on control using a solution of NaOH. Calcium hypochlorite solution, the source of oxidant, was prepared in a pre-mix tank and decanted into a stock tank. Calcium hypochlorite addition to the reactor could be controlled based on the residual chlorine analyzer feedback signal or the reactor redox potential, depending on the mode of operation. The residual chlorine measurements were accomplished using two on-line analyzers. During the testing of the residual chlorine control concept, the control set point was 3.5 mg/L.

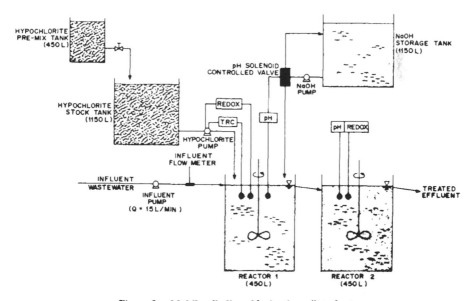

Figure 8.   Mobile alkaline chlorination pilot plant.

Two commercial residual chlorine analyzers, the Unimonitor CS221, manufactured by Rossell Fluid Control Ltd., and the Anachlor series 17B4200, manufactured by Fischer & Porter Co., were used in the pilot plant. The Rossell unimonitor was capable of measuring free residual chlorine (FRC), and the Fisher & Porter Anachlor was capable of measuring both total residual chlorine (TRC) and FRC.

The redox potential and pH measurements were made using instruments manufactured by Analytical Measurements Ltd.

A Commodore Business Machine Computer Model 8032 was used in conjunction with a Control Micro Systems CMS 2000 process control package to control pilot plant operations. The computer was interfaced to the process instrumentation by means of a Dynage MPC-8 input/output processor. The Fischer & Porter analyzer was initially set up and calibrated to measure TRC. However, towards the end of the experimental program, the analyzer reagents were changed to measure FRC.

Pilot Plant Operation

The two chlorine analyzers were set up in a manner which allowed a sidestream of the contents of reactor No. 1 to be directed to each unit. The Rossell unit received a pumped flow of 1.5 L/min, while the Fischer & Porter unit received a gravity flow adjusted to 0.15 L/min. During plant operation, one of the two analyzers provided the feedback control signal to the process while the second analyzer was used for simultaneous, comparative residual chlorine concentration measurements.

The experimental testing of the residual chlorine control concept was performed five days per week. The duration of the residual chlorine control runs ranged from 1 hour to 13 hours. While the pilot plant was operating, reagent consumption (NaOH and calcium hypochlorite) was monitored, and comparative samples of the pilot plant effluent and the metal finishing plant's treatment system effluent were collected and analyzed for total and weak acid dissociable cyanide and total and dissolved metals.

Calibration of both analyzers was performed daily, and the residual chlorine was spot checked frequently during each day. If poor analyzer response was noted, the units were dismantled, cleaned, recalibrated, and placed back on line.

Calibration of the instruments for analyzing TRC was accomplished by adjusting the output signals (4–20 mA) generated by the analyzers to correspond to the residual chlorine concentration in the sample stream as measured by the PAO Back Titration method [4]. For FRC, the DPD method was used [4]. Linear calibration equations were developed (i.e., slope and intercept) to convert the output signals to units of mg/L residual chlorine.

Tuning of the controllers was performed using the Ziegler-Nichols method [5].

A period of time during the experimental program was designated in which the pilot plant alkaline chlorination system was operated using redox control to define the treatment system performance using this control strategy. Chlorine utilization efficiency was monitored by measurement of the residual chlorine concentration in the effluent from the process.

The redox set-point during the redox control phase was +325 mV. This control set-point was chosen since it approximated the redox potential at which 3.5 mg/L of residual chlorine was present in the treated effluent. The redox set-point being used at that time in the full-scale alkaline chlorination system was +450 mV. The higher set-point in the full scale system was to ensure complete destruction of cyanide even though it resulted in excessive consumption of chlorine.

RESULTS AND DISCUSSION

Effectiveness of Residual Chlorine Control

The pilot plant was operated for a period of four months using the residual chlorine control strategy. During this period, fourteen sets of grab samples of the treated effluent from both the pilot plant and the full scale alkaline chlorination system were collected to compare the effectiveness of the two plants. The grab samples were collected simultaneously and preserved for cyanide and metal analyses. Table I summarizes the results of the analyses and indicates the variability that was measured in the two streams.

The mean total cyanide (CN_T) concentration in the effluent from the full-scale treatment system, which was operating under redox control, was 1.50 mg/L. The pilot plant system, operating under residual chlorine control, produced an effluent containing 1.30 mg/L total cyanide.

Table I. Comparative Effluent Quality from Full-Scale Redox Control System and Pilot Plant Residual Chlorine Control System

Parameter[a]	Full-Scale System (Redox Control)			Pilot Plant (Residual Chlorine Control)		
	Mean	Std. Dev.	% Variability	Mean	Std. Dev.	% Variability
CN_T	1.50	1.20	80.0	1.30	0.96	73.8
CN_W	0.31	0.13	41.9	0.26	0.13	50.0
Cu_T	2.80	2.03	72.5	2.88	2.27	78.8
Cu_D	0.22	0.59	268.2	0.08	0.05	62.5
Fe_T	0.82	0.41	50.0	0.86	0.55	64.0
Fe_D	0.59	0.34	57.6	0.44	0.19	43.2
Ni_T	0.74	2.49	336.5	0.78	2.63	337.2
Ni_D	0.18	0.55	305.6	0.18	0.55	305.6
Zn_T	2.36	1.64	69.5	2.43	1.47	60.5
Zn_D	0.04	0.02	50.0	0.05	0.03	60.0

[a]T = total; W = weak acid dissociable; D = dissolved.

Comparison of the two means indicated that no statistically significant difference existed at the 95% confidence level. This was also found to be true for all the other parameters measured in the effluents.

The major difference between the two systems was the quantity of residual chlorine being discharged in the effluent and the variability in the effluent redox potential. Table II presents the measured variability of the redox potential and residual chlorine in the two effluent streams during the days when comparative effluent sampling occurred. The average residual chlorine concentration from the full-scale plant was 46 mg/L compared to the controlled concentration of 3.5 mg/L in pilot plant. The redox potential in the full-scale plant was controlled at + 450 mV, while the redox potential in the residual chlorine controlled pilot plant was highly variable, averaging + 300 mV.

Chlorine usage at the metal finishing plant where this study was conducted used approximately 1370 kg/year. It is estimated that the use of residual chlorine based control could reduce the chlorine consumption by approximately 685 kg/year, or 50% of the consumption observed with redox based control.

Residual Chlorine Analyzer Operation & Maintenance

During the pilot plant investigation, the Rossell analyzer required cleaning after an accumulated operating time of 103 hours. The Fischer & Porter analyzer required cleaning after 80 hours of cumulative operating time. The maintenance time required to dismantle and clean the residual chlorine analyzers was approximately 15 minutes for each unit.

Operational problems were experienced with both analyzers. Problems with the Rossell unit were primarily associated with the pH control system. The Rossell unit was very sensitive to pH, and poor control of pH caused erratic behavior of the analyzer. The internal pH control system that was an integral part of the Rossell unit was ultimately replaced with an external pH control system.

The Fischer & Porter analyzer experienced other types of operating problems. After operating the pilot plant for two months, the analyzer cell became insensitive to changes in chlorine concentration and could not be recalibrated. In order to overcome the calibration problems, the analyzer had to be respanned by changing an internal load resistor. This modification returned the performance of the Fischer & Porter analyzer to previous levels. On several occasions, the tubing to the Fischer & Porter analyzer cell plugged with solids and accumulated scale. Regular manual cleaning of the tubing prevented this problem from recurring.

A ground fault in the electrical system at the metal finishing plant caused erratic behavior in both of the analyzers. After installing a separate ground in the pilot plant, this problem was eliminated.

During the investigation, problems in the form of erroneous measurements were observed with the analyzers. High concentrations of bronze (copper and tin alloy) in the wastewater produced a positive interference in the Rossell analyzer, although the Fischer & Porter analyzer appeared to be unaffected. When the interference was observed, both of the analyzer cells were dismantled. Examination of the Rossell analyzer showed a copper-colored deposit on the platinum electrode. Cleaning the electrode with a solution of 10% nitric acid caused the deposit to turn a green color, potentially indicating copper. A new platinum electrode was installed and the Rossell cell was reassembled. The Fischer &

Table II. Comparative Effluent Redox and TRC from Full Scale and Pilot Plant Treatment Systems

Sampling Day	Full-Scale System (Redox Control)		Pilot Plant (TRC Control)	
	Effluent Redox (mV) Set Point = 450 mV	Effluent TRC (mg/L)	Effluent Redox (mV)	Effluent TRC (mg/L)
84/10/16	430	30	234–327	3.5
84/11/12	460	45– 63	185–324	3.5
84/11/13	450–480	28– 38	88–474	3.5
84/11/14	430	30– 38	135–445	3.5
84/11/15	420–520	64– 80	119–573	3.5
84/11/16	400–460	50– 90	84–501	3.5
84/11/21	460	13– 35	145–612	3.5
85/01/03	450	100–120	78–359	3.5
85/01/04	475	30– 32	214–395	3.5
85/01/07	480	30	225–348	3.5
85/01/09	420	43	234–321	3.5
85/01/10	450	38	224–452	3.5
85/01/11	450	30	244–380	3.5
Mean	450 mV	46 mg/L	300 mV	3.5 mg/L

Porter cell had a slight visible deposit on the gold electrode and the copper electrode had a very dull finish. The electrodes were cleaned, using soap and a damp cloth and the cell was reassembled. After these repairs, the analyzer operated satisfactorily.

On one occasion, it was observed that the Fischer & Porter unit was indicating significantly higher levels of residual chlorine than the Rossell analyzer. The Fischer & Porter analyzer at the time was measuring TRC while the Rossell was measuring FRC. Investigation showed that a corrosion inhibitor, containing cyclohexylamine and morpholine, was being added to the boiler system in the plant. Since steam condensate was discharged to the alkaline chlorination system, addition of chlorine may have caused the formation of organic chloramines which were measured as TRC by the Fischer & Porter analyzer. The Rossell unit, which was measuring FRC, did not measure the chloramine component. When the Fischer & Porter analyzer was set up to measure free chlorine residual, the two analyzers indicated the same concentrations of residual chlorine.

Discrepancies were found to exist in the measured concentrations of residual chlorine by the two analyzers, although the analyzers were calibrated using the same method; differences in concentration were normally ±1 mg/L of residual chlorine. This discrepancy is not unexpected as the analyzers are based on different measurement principles.

Difficulties with Redox Control

Grab samples of the pilot plant influent were collected daily and titrated to determine the redox potential/chlorine dosage relationship. Figure 9 represents the redox potential/chlorine dosage relationship observed on some of the days during the study. This redox/dosage relationship illustrates the difficulties which can be encountered in operating the treatment system using redox control. A redox

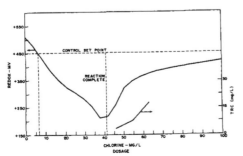

Figure 9. An example of wastewater redox curve.

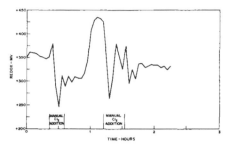

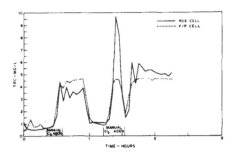

Figure 10.   An example of redox levels during redox based automatic control of the pilot plant.

Figure 11.   Pilot plant effluent TRC during an attempt at automatic redox control at a set point of 325 mV.

potential of + 450 mV could be achieved at two chlorine dosages; 6 mg/L and approximately 130 mg/L. At the lower chlorine dosage (6 mg/L), incomplete cyanide destruction would result and no chlorine residual would be present in the effluent. At the higher dosage, complete cyanide oxidation would be achieved but an excessive amount of residual chlorine would be present in the effluent.

The pilot plant was operated on a limited basis using redox control to determine if the control set-point could be lowered from the + 450 mV to + 325 mV to minimize chlorine consumption. Figures 10 and 11 show the results from an attempt at using redox feedback as the control parameter in the pilot plant. Figure 10 illustrates the redox potential during a 2.5 hour test run. Control was achieved 1.5 hours after switching the plant to automatic control. Twice during this time period, manual chlorine addition was required to regain process control. Figure 11 shows the TRC during this experimental run. At the time when manual addition was required, residual chlorine in the effluent was not detectable. Under these circumstances, cyanide oxidation would be incomplete. Other attempts during this study to achieve simultaneous optimization of cyanide oxidation efficiency and chlorine utilization showed that it was not possible to do so using the conventional redox control strategy. With redox based control, stable process performance was achieved only at chlorine dosages considerably higher than those required for the residual chlorine based control.

PRACTICALITY OF RESIDUAL CHLORINE CONTROL

Operation of the pilot plant using residual chlorine control produced an effluent of equal quality to the effluent produced by the redox-controlled full scale chlorination plant. This effluent was produced at significantly lower chlorine dosages in the pilot plant compared to the dosage requirements of the full-scale system. Furthermore, the pilot plant testing demonstrated that simultaneous optimization of cyanide oxidation and chlorine dosage by operating the alkaline chlorination process with a redox based control resulted in unstable process operation.

Alkaline chlorination systems treating waste from small metal finishing operations may not be able to justify the cost of residual chlorine based control. However, the use of residual chlorine control at alkaline chlorination plants which use large quantities of chlorine would appear to be very attractive based on the potential savings in chlorine cost. For example, some of the full-scale alkaline chlorination systems operating at Canadian gold mills use several hundred to several thousand kilograms of chlorine per day [2]. Use of residual chlorine control at these plants would result in sufficient savings in chlorine to justify the additional cost associated with this control strategy. Furthermore, in situations where dechlorination of the treated effluent is required prior to discharge, the residual chlorine control approach would have an additional economic benefit of reducing the chemical requirements associated with dechlorination.

CONCLUSIONS

Based on concurrent sampling of the redox controlled full scale treatment plant and the residual chlorine controlled pilot plant treating a sidestream of the same wastewater, no significant difference in treated effluent quality was observed.

Based on the measured residual chlorine concentrations in the effluents from the redox controlled full scale plant and the residual chlorine controlled pilot plant, the application of residual chlorine control in the full-scale system would have reduced the chlorine usage by approximately 50% while maintaining the same treated effluent quality.

Difficulties observed when the pilot plant was operated on redox feedback control mainly involved loss of control. Manual chlorine addition was required to regain control. Operating with redox control at a higher millivolt set-point resulted in more dependable operation, but excess chlorine consumption.

The economic feasibility of residual chlorine control will depend on the reduction in chlorine usage achievable, and whether there is a requirement to dechlorinate the wastewater prior to discharge. It appears that large users of chlorine could justify the cost of residual chlorine control on the basis of savings in chlorine consumption.

ACKNOWLEDGEMENTS

The pilot tests would not have been possible without the cooperation and assistance of the management at the metal finishing plant. Mr. Stephen Nutt of Canviro Consultants had the overall responsibility for coordination and data interpretation for the pilot tests.

REFERENCES

1. White, G. C., *Handbook of Chlorination*, Van Nostrand Reinhold Company, New York, New York (1972).
2. Zaidi, S. A., and J. B. Brodie, "Performance Evaluation of Full-Scale Alkaline Chlorination Systems Operating at Canadian Gold Mills," presented at the 13th Annual Hydrometallurgical Meeting of the Canadian Institute of Mining and Metallurgy, Edmonton, Alberta (August 1983).
3. Shinsky, F. G., *pH and pIon Control in Process & Waste Streams*, John Wiley & Sons Inc., Toronto, Ontario (1973).
4. APHA-AWWA-WPCF, *Standard Methods for the Examination of Water and Wastewater*, McGraw-Hill Book Company, Toronto, Ontario (1965).
5. Coughanowr, D. R., and L. B. Koppel, *Process Systems Analysis and Control*, McGraw-Hill Book Company, Toronto, Ontario (1965).

Section Thirteen
LAWS, REGULATIONS, AND TRAINING

86 A MUNICIPAL PROGRAM TO PREVENT HAZARDOUS MATERIAL SPILL PROBLEMS

Richard T. Shogren, Senior Engineer
Camp Dresser & McKee Inc.
Louisville, Kentucky 40223

John B. Weil, Industrial Waste Manager
Metropolitan Sewer District
Louisville, Kentucky 40202

INTRODUCTION

On Friday, February 13, 1981, a major sewer explosion occurred in Louisville, Kentucky, caused by the negligent discharge of hexane into the City's sewers. A local food processing plant caused the explosion when an existing containment basin was inadequate to contain the hexane. Neither fire nor emergency response agencies in the City were contacted about the spill. More than 2.5 miles of sewer were destroyed in an area encompassing 40 city blocks. To date, the industry responsible for the spill has paid damage claims exceeding $33 million to local government agencies, residents and businesses in the affected area [1].

With a county population of 690,000, Louisville and Jefferson County have experienced other major spills. In March, 1977, an intentional and illegal spill of hexachlorocyclopentadiene into City sewers caused a new 105 mgd plant to be contaminated for 2 1/2 months. Raw sewage was directly discharged into the Ohio River during that period. In late 1982, the Kentucky Division of Water (DOW) fined a local industry $20,000 for a major phenol spill into the Ohio River. Phenol discharges into the Ohio River reached a maximum of approximately 1700 pounds per day, over 600 times the National Pollutant Discharge Elimination System (NPDES) permit standard of 2.76 pounds per day. Between 1980 and 1983, reported hazardous material spills exceeding 25 gallons averaged 43 per year. In 1984 and 1985, reported hazardous material spills exceeding 25 gallons have averaged 84 per year.

LOCAL SPILL PREVENTION LAW

In the fall of 1984, Louisville's Mayor appointed a task force to make greater progress in protecting the community from hazardous material spills. This followed additional major spills which caused serious problems to the Metropolitan Sewer District (MSD) and the Louisville and Jefferson County Health Department due to late notification. The task force consisted of eleven government agency representatives, three elected officials, and eight representatives from major local industries. The task force drafted an ordinance to improve spill reporting by industries and to implement an industry spill prevention program. The ordinance was adopted by both Louisville and Jefferson County in August, 1985.

The new law requires that businesses with hazardous materials must submit a plan for each business site as to how they will respond in the event of an accidental discharge of that hazardous material. A Hazardous Material Spill Prevention Control (HMPC) plan must be submitted by any business which manufactures, uses, or stores hazardous materials in minimum designated quantities at their business location. The source for defining a hazardous material is a master list established by 40 CFR Part 302 as part of the Resource Conservation and Recovery Act (RCRA). Hazardous materials defined in the Federal Water Pollution Control Act, The Comprehensive Environmental Response Compensation and Liability Act (CERCLA) and the Toxic Substances Control Act (TOSCA) are also included in the definition.

MSD was designated administrating agency for both the City and County. Under the ordinance, businesses which have a minimum "reportable quantity" (RQ) of a hazardous material on their business site must submit an HMPC plan to MSD. Businesses which must submit plans are generally defined by a Standard Industrial Classification (SIC) code in the ordinance. The ordinance exempts residents who have hazardous materials on site for personal use, consumer product and food stuff manufacturers who are covered under the Food and Drug Act, agricultural operations which are handling and not selling chemicals for application on a farm, and instances where hazardous materials do not exceed reportable quantities and the administrating agency sees no danger to public health.

A major emphasis of the ordinance is that hazardous material releases must be promptly reported to a local "911" emergency number. Businesses and industries which fail to provide prompt notification are subject to a fine from MSD of up to $5,000. Penalties of up to $1,000 may be levied if a release occurs due to negligence or because of previous spill occurrences by the same business. In all cases, businesses are responsible for promptly cleaning up hazardous material discharges or paying for cleanup.

The ordinance provides for creation of an Appeals Board to mediate disagreements over acceptability of an HMPC plan submission or fines which have been levied. The nine-member Appeals Board is made up of four representatives from industry, three regulatory appointees and two general public appointees. The Appeals Board hears cases and makes recommendations to the MSD Board of Directors for final action.

HAZARDOUS MATERIAL SPILL PREVENTION CONTROL (HMPC) PLAN PROCESS

The requirement that HMPC plans be submitted is crucial to reducing spills. Four questions were considered in developing the HMPC program: 1) who must submit a plan; 2) what was a reasonable fee structure to recover inspection and review costs; 3) how should businesses be contacted; and 4) how could business participation and response be guaranteed.

Criteria for Requiring an HMPC Plan

Two issues are most important in determining whether a business must submit a plan: the definition of a hazardous material and its reportable quantity (RQ).

The ordinance identifies hazardous materials as those listed by EPA in 40 CFR Part 302 [2] which defines federal spill notification responsibilities for specific hazardous materials. A copy of this list of materials was provided to all businesses. However, other materials may have hazardous characteristics that are not specifically listed. MSD has used the *1984 Emergency Response Guidebook* [3] published by the U.S. Department of Transportation as an additional source for potential hazardous materials. This guidebook is designed for citizen use and includes a list of materials that are much more extensive than the source officially recognized by Congress. Between these two sources, about 3100 hazardous materials (including petroleum base products) have been identified.

The second issue is the definition of reportable quantity (RQ), the minimum amount of a hazardous material which must exist on a business site to require a HMPC plan. The list from 40 CFR Part 302 defines RQs, but the more extensive *Emergency Response Guidebook* does not. The ordinance addresses the issue as follows: "unlisted hazardous waste designated as hazardous material have a reportable quantity of 100 lbs . . ." Oil and oil base products are designated in the ordinance as having a reportable quantity of 56 gallons. The limited information on relative hazard of materials is likely to be debated for many materials. Ultimately, as new hazardous materials are developed and identified, MSD and its staff will add specific materials and define appropriate reportable quantities.

Businesses to be Contacted

Industry representatives on the task force defined the businesses to be contacted according to SIC classification as shown in Figure 1. Landscape and horticultural services were selected because of their storage of pesticides and agricultural fertilizers. Manufacturing establishments were entirely included in the list because industry can have substantial quantities of hazardous materials due to their manufacturing characteristics. Transportation, communication and public utility establishments were selected because of the likely storage of gasoline for vehicle fleets. The wholesale establishment SIC categories (5043 through 5199) were selected because they are likely to store chemicals, petroleum products or distilled beverages in substantial quantities. Gasoline service stations are listed because of their fire and explosion potential. Industrial and commercial services (SIC 701 through 721) include industrial and commercial launderers and services to motels and hotels. Disinfection and exterminating services and photofinishing laboratories are an obvious potential source of hazardous pollutants. Passenger car rental and leasing facilities are listed because of possible gasoline storage concerns.

MANUFACTURING, TRANSPORTATION & OUTDOOR SERVICES		
SIC GROUP	DESCRIPTION	NO. ESTB.
078	LANDSCAPE AND HORTICULTURAL SERVICES	56
2011-3999	MANUFACTURING	841
4011-4953	TRANPORTATION, COMMUNICATION & PUBLIC UTILITIES	461
WHOLESALING & GASOLINE SERVICE STATIONS		
SIC GROUP	DESCRIPTION	NO. ESTB.
5043	PHOTOGRAPHIC EQUIPMENT AND SUPPLIES	4
5085	INDUSTRIAL SUPPLIES	66
5161-5199	WHOLESALE TRADE GROUP FOR CHEMICALS, PETROLEUM PRODUCTS, DISTILLED BEVERAGES & MISCELLANEOUS	205
554	GASOLINE SERVICE STATIONS	314
RETAIL AND OTHER SERVICES		
SIC GROUP	DESCRIPTION	NO. ESTB.
7011-7218	INDUSTRIAL AND COMMERCIAL SERVICE	206
7342	DISINFECTING AND EXTERMINATING SERVICES	19
7395	PHOTOFINISHING LABORATORIES	16
7512-7513	CAR & TRUCK RENTAL AND LEASING	34
7538-7549	AUTOMOTIVE REPAIR SHOPS & AUTO SERVICES	200
8062-8069	MEDICAL HOSPITALS	14

SOURCE: U. S. Bureau of Census,

1982 County Business Patterns for Kentucky.

1984

Figure 1.   Establishment of hazardous materials ordinance for Jefferson County, Kentucky by SIC group.

Finally, hospitals are listed because of the possibility of unusual chemical storage in substantial amounts.

1982 County Business Patterns for Kentucky was used as a source for number of businesses and employees according to SIC classification. From this reference, about 2400 businesses were expected to be contacted for possible submission of plans.

To insure adequate coverage and response to the ordinance, a personal letter was sent to each business which might need to submit a plan. To establish a business contact list, a cross index of SIC codes against telephone index listings was utilized. Using selected headings in the telephone directory, a master listing of business name, address and phone number with probable SIC code was directly entered to a mini-computer. The ultimate mailing list numbered 4500 businesses.

Exemption from HMPC Plan

Once it was determined that all businesses would have to be contacted, two forms were developed for possible response by that business. One of these forms is a request for exemption from the requirement for an HMPC plan. The other form is an HMPC plan application form. To guarantee a response from a business, one or the other form *must* be completed and returned to MSD with all questions answered.

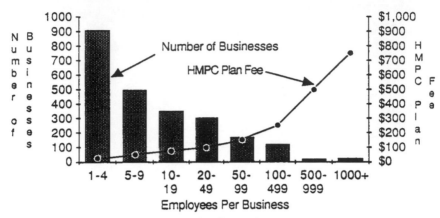

Figure 2.   Hazardous material ordinance fee structure for plan review.

The request for exemption form was designed to be no longer than two pages and requires no attachments. The request for exemption form requires facility information (business name and address), the type of business at the site including number of employees and hours of operation, names and phone numbers of local owners and managers, a description of hazardous materials (if any) which are stored on site and their quantities, and a certification by the manager that he does "not store, handle, or process at this facility hazardous materials in reportable quantities prescribed by the ordinance." The owner or manager must sign and date the exemption form and also must submit the form within a time frame defined in the letter.

Fees for Plan Review

The ordinance provides that a fee may be levied that would pay for MSD's cost to review the plans. To develop a fee plan, the number of employees in businesses identified in the County Business Pattern report were reviewed and likely possible fees were estimated. In addition, the time for technicians and engineers to review a plan was estimated and a preliminary budget of $120 to $140 thousand was defined for the entire process. This budget excludes the cost of any inspections of sites that may be necessary. It assumes an average review time of four hours for each HMPC plan and an average agency cost of $80 per plan. No fee is required for submitting a request for exemption form.

A progressive fee structure which increases the plan application fee as number of business employees increases was selected for recovering costs. This fee structure was selected due to potential employee and employer liabilities at a business. An alternate method of defining fees according to relative hazard was considered too difficult to objectively define. The fee structure shown in Figure 2 ranges from a $25 cost for each HMPC plan submitted by employers having 1 to 4 employees to a maximum cost of $750 for employers having in excess of 1,000 employees. If 70% of businesses submit plans, an expected revenue of $78 per plan will be provided by this fee structure.

HMPC APPLICATION FORM

The six-page HMPC plan application form is divided into six major sections. Section A, Business Information, requests the business name, nature of business, local ownership and key persons to contact in event of an emergency. The information requested in section A provides basic identification information about the character of the business and the ownership.

Section B requests a description of hazardous materials and their location. This section consists of four attachments, three of which are information maps. Attachment A is a summary form at the back of the application which describes the names of hazardous materials, key identification numbers tied to the Department of Transportation (DOT) identification series and the Chemical Abstract Series (CAS) for identifying materials, and storage characteristics. A general location map (Attachment B) is required to show the location of the facility relative to the surrounding area and major streets. A site facility map (Attachment C) will provide a schematic layout of the entire facility and key characteristics of the site. The materials map (Attachment D) is a map showing the detailed location and design

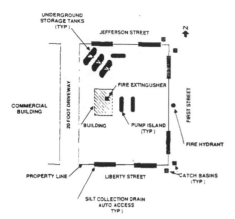

Figure 3. Typical facilities/materials map.

of areas where hazardous materials are stored. The materials map should show critical spill control devices such as valves, berms or other structures designed to contain a spill, and drainage contours from work areas. An example combined facilities/materials map is shown in Figure 3.

Section C of the application requests a description of the emergency response training procedures that have been or will be provided to employees. Knowledge of emergency response procedures will help protect employees or residents from hazardous material releases that may threaten life or damage to individuals. Section D describes how each business is prepared to control a hazardous material release that occurs. Where is emergency equipment located? If the spill occurs, how will it be cleaned up, and where will disposal of waste occur?

Section E describes the existing prevention, detection, and containment measures that are present at the facility. In addition, proposed improvements of the existing program must also be described and a schedule provided of when these improvements will occur.

Section F at the end of each application requires that the owner certify the validity of information provided in the application. The owner has major responsibility for implementing the HMPC plan, and he is to provide a signature that appears under the certification. Existing MSD regulations provide that the HMPC plan application does not need to be prepared by a professional engineer but may be prepared by the manager of the facility in question.

The application has been organized with extensive instructions that are designed to provide assistance to the person completing the application. In addition, applicants are encouraged to contact MSD to obtain answers.

IMPLEMENTING THE ORDINANCE

Staffing Requirements

Employee staffing requirements have been greater than expected due to the number of businesses contacted. The initial receipt of the submitted forms and fees is logged by the Customer Service Department of MSD and approximately two employees are dedicated to the tracking of customers through the review and approval process. Tracking is maintained on an IBM 370 computer utilizing the TIF software package.

Plan and exemption request reviews and inspections are handled by MSD's Industrial Waste Department. A staff engineer and either one or two cooperative interns conduct the reviews. A check list approach to the review process was found to be the most effective means of evaluation of these forms. Two temporary and one permanent inspectors conduct inspections. Inspections are conducted for a sample of businesses submitting plans and for all exemption requests.

In the Industrial Waste Department, tracking of information is maintained on an IBM PC/XT using the dBase III software package. A secretary commits about 6 hours/day to coordinate the paperwork of this process.

Interagency Cooperation

Historically, the agencies with the most experience in the hazardous materials field in Jefferson County have been the Louisville Fire Department and the Louisville and Jefferson County Health Department. Although MSD has the lead administrative and enforcement role under the Ordinance, a cooperative agreement involving the three agencies has developed. This has been formalized in Memoranda of Understanding and put into effect through a free exchange of information and a sharing of the review and inspection duties by all three agencies. Copies of all forms and exemptions are provided by MSD to the other two agencies. In cases where there are businesses that have identifiable hazardous material problems (repeat spills, poor housekeeping, etc.), joint inspections by two or all agencies are conducted. However, in most instances, inspections by only one agency is necessary. These site visits are primarily for the purpose of confirming materials inventories and proper storage and handling practices.

The three agencies usually respond to chemical spill incidents together. The combined expertise that is afforded in joint agency response has led to a more effective execution of containment and cleanup results. This greater level of information sharing and joint response to emergencies has worked well to date.

Progress in Reviews

To manage the large volume of business contacts, MSD divided the over 4500 businesses to be contacted into five priority groups which have different submission dates spread over 12 months. With the sequential phase-in of Priority Groups, MSD is approximately halfway through the implementation process. Exemption Forms constitute about sixty percent of all application forms with 1200 forms submitted to date. Almost 800 Plan Application Forms were processed between December 1, 1985 and May 1, 1986.

Progress in Inspections

After administrative review of the forms, inspections are conducted on all businesses that have submitted Requests for Exemptions and on all businesses that have submitted Application Forms that have significant deficiencies. Site visits for businesses that have submitted exemption forms are conducted to confirm the correctness of the request. As a result of these inspections, about half of all Requests for Exemption are denied and a Plan Application is subsequently required.

In the site visits to the businesses that have submitted Plan Applications with significant deficiencies, inspectors provide assistance to the business owner to successfully complete the Plan Application. The inspector will instruct, educate, and advise the business owner where necessary to arrive at an approvable plan. In this way, continued correspondence and the attendant paperwork and delays are avoided.

Spill Response Experience

Since the Ordinance became effective on December 1, 1985, MSD has averaged between thirty and forty-five chemical spill incidents per month. Of these, five have resulted in fines, all of them due to either non-notification or delayed notification. Fines have ranged from $500 to $5,000 per incident. In only one case has a business appealed for a reduction in the fine amount.

The three major participating agencies have continued the spirit of cooperation in the field during chemical spill incidents. Each contributes their own particular expertise to control, contain and mitigate the impact of chemical spills.

Future Implementation

With the progress made thus far, maintenance of the HMPC plan program in the future is being planned. On-going contact with those businesses that have been identified as users of hazardous materials is to be the keystone of the program in years to come. A likely method for carrying this out is an annual mailout to these businesses, with a required response on a postcard indicating changes of ownership, materials usage and inventory, and spill control containment procedures. More significant changes may call for complete plan revision by the business owner.

As new businesses are established, they will be contacted by MSD to determine anticipated materials usage and planned spill control and containment measures. Several avenues are being investigated for information on new business startups. The local Chamber of Commerce and the Louisville Water Company may provide listings of new business activity which can be used for this purpose.

MSD has provided support for the Jefferson County HAZMAT Mutual Aid Group, an association of agencies and companies that are concerned about the safe handling and hazard mitigation of hazardous materials. Exercises and regular training sessions among concerned members are the chief activities that are used to foster awareness of chemical hazards and to develop a spirit of mutual cooperation of the people involved in emergency incidents.

Ultimate benefits to businesses are expected to be lower insurance rates, better understanding of hazardous materials, a greater sensitivity to problems that may exist at the business site due to hazardous materials and quicker notification to government agencies when spills occur. For local government, the benefits are expected to be quicker response to spills, avoidance of sewer contamination and corrosion and a knowledge of where hazardous materials exist if a spill or a fire occurs at a business.

Hazardous chemical incidents can range in magnitude from very minor discharges causing no adverse health effects to major incidents with the potential to affect hundreds of people. The Louisville and Jefferson County HMPC program emphasizes the responsibilities of businesses and industries which handle hazardous material to protect residents and the environment from adverse health effects. The legislation is a continuing effort by local government to improve the environment of the Louisville community.

REFERENCES

1. *Louisville Courier Journal & Times* (Dec. 6, 1984).
2. 40 CFR Part 302, published in *Federal Register*, Vol. 50, No. 65, effective date of July 1, 1985 (April 4, 1985).
3. U.S. Department of Transportation, *1984 Emergency Response Guidebook* (A Guidebook for Hazardous Materials Incidents), Washington, D.C.: U.S. Department of Transportation, Materials Transportation Bureau, edition DOT p. 5800.3 (December 1983).

87 INDUSTRIAL WASTEWATER TREATMENT PLANT OPERATIONS

James C. O'Shaughnessy, Associate Professor

Frederic C. Blanc, Professor
Department of Civil Engineering
Northeastern University
Boston, Massachusetts 02115

David Bassett, Project Coordinator

Susan Moulton, Executive Director
Bay State Skills Corporation
Boston, Massachusetts 02110

INTRODUCTION

A large number of industrial facilities in the Commonwealth of Massachusetts are presently being required to install wastewater treatment or pretreatment systems operated by licensed operators. In Massachusetts a program is presently being implemented to provide separate examinations and licensing for industrial wastewater operators and municipal wastewater operators. Typically the municipal operators deal with biological secondary treatment plants while the industrial operators are involved in physical-chemical treatment of wastewaters from metal finishing, electroplating, electronics, and other high-tech industries. An existing training program for municipal wastewater operator training did not cover physical chemical treatment areas in any depth. The Commonwealth therefore decided to establish training courses for the more than 2000 individuals likely to apply for licensing.

BACKGROUND

In February, 1984, a suit which had been brought by the Attorney General's Office on behalf of the MDC, against an electronics company was settled. Part of the settlement agreement specified actions that the Bay State Skills Corporation should take, i.e.:

"Within 30 days of entry of Judgment, and again on February 1, 1985, on February 1, 1986, on February 1, 1987, and on February 1, 1988, the Company shall grant $100,000 (for a total of $500,000) to the Bay State Skills Corporation ("Bay State"), established by G.L. C. 401. The Grants shall be for the purpose of establishing courses of instruction concerning the operation and maintenance of industrial wastewater treatment facilities that discharge to metropolitan sewer systems. The courses of instruction shall be developed by Bay State in consultation with technical personnel from the Metropolitan District Commission ("MDC"), the Department of Environmental Quality Engineering ("DEQE") and other entities that are familiar with the operation and regulation of wastewater treatment facilities."

Bay State Skills Corporation is a quasi-public corporation established by state law to encourage and facilitate cooperative relationships between government and industry and develop or expand programs of skills training that are consistent with employment needs. Accordingly, Bay State Skills Corporation awarded a contract to develop and test industrial wastewater operations course materials for use in an industrial operator training program to Northeastern University. A total of five courses were developed. The courses were one introductory course, two intermediate courses, one advanced course, and one course based on a series of lecture modules for various types of industrial wastewaters generated in Massachusetts.

Initial planning meetings were held between Bay State Skills Corporation, the Massachusetts Department of Environmental Quality Engineering (DEQE), and the Metropolitan District Commission (MDC) which has since been reorganized as the Massachusetts Water Resources Authority (MWRA). Following the planning meetings, an Industrial Advisory Board was formed. The Industrial Advisory Board was comprised of members of Massachusetts Industry, DEQE and MWRA personnel. During the period in which the courses were developed and taught, the Industrial Advisory Board met bimonthly to provide valuable input to the course development. Areas of assistance included the development of profiles of the typical background of the operators for which each course level would be targeted.

In determining the kind of training that was lacking, it was desirable to evaluate how waste treatment operators are used within the industry. Massachusetts industry, even larger companies (such as those represented on the Industry Advisory Committee), operate small establishments, whose waste treatment plants typically only require between one and three operators during first shift and usually one (or maybe two) on second shift. This differs from municipal waste treatment plants where there may be up to 20 or 30 personnel on a given shift due to the large size of the treatment plants. An even larger share of Massachusetts industry is made up of small job-shop operations which may have a maintenance person running the treatment plant (which may only be a part of their job).

The difference between industry and municipalities was a key factor. Municipalities do not have (or need to have) a majority of their personnel familiar with how the entire plant operates. Therefore, the skill level of entry-level municipal waste treatment personnel is not advanced. Personnel who operate industrial waste treatment plants must know how the entire treatment plant operates because of its small size and the fact only a few (or maybe one) person(s) are responsible for the operation of the plant. In larger industry, and even in a small job shop, the operator must know a great deal more about treatment plant operation than the entry-level municipal operator.

It was determined that there were three types of training needs for wastewater treatment plant operators. First, there is a need to provide training to senior-level waste treatment personnel. Persons taking this course would have worked in an industrial waste treatment plant for several years and taken other courses (i.e., chemistry, math, municipal wastewater treatment courses). The advanced course was targeted for senior waste treatment operators and would fulfill this need. Second, another course would provide sufficient training that someone could come in and actually be a junior operator (or run a small, simple waste treatment plant). Persons taking this course may have maintenance background, have taken some related courses, possibly either have worked in the past at a municipal treatment plant or are now working at a municipal treatment plant but wish to become qualified as an industrial wastewater operator. Finally, there is a need to introduce personnel interested in becoming industrial waste treatment operators to the field. This would not qualify personnel to be a junior operator at an industrial waste treatment plant but would provide them with an introduction to let them figure out whether they wish to pursue further education and experience in the field.

TRAINING MATERIALS

One of the objectives of this training program was the development of training materials which could be used for industrial wastewater operator training. There were some wastewater operations training manuals in existence [1,2,3,4]; however, such manuals were designed for municipal wastewater treatment operations. Usually such materials emphasized biological treatment operations for the secondary treatment of wastewater. One excellent 3-volume manual [3] was designed for a self study program on municipal biological treatment operations. Such self study materials are not easily adapted for use in classroom teaching. Areas such as chemical oxidation and reduction, neutralization, electrochemical methods, and membrane technologies are not well covered in such training manuals. Consequently, it was decided to prepare a series of training course lectures which would serve as a reference for the students. The lecture notes contained all the figures, tables, and charts which would be used in each course. Students received a copy of the lecture notes at the beginning of the lecture. Space was provided throughout the printed documents which permitted the students to take additional notes directly on the document.

After the completion of each course, the lecture notes were edited and combined with course outlines and copies of examinations into a complete training document. The documents were then given to other colleges located strategically throughout the Commonwealth of Massachusetts to which Bay State Skills Corporation awarded development monies to initiate industrial wastewater operations training programs.

PRETESTING

In order to properly tailor the lecture materials and lectures to the students involved in each course, pre-examinations were administered at the first meetings of each course. The pre-examinations were designed to determine the level of proficiency of the students in chemistry, wastewater treatment

Table I. Potential Educational and Training Response to Industry Needs for Industrial Wastewater Treatment

Level of Training	Skill Requirements & Work Experience	Institutions	Workforce Needs
Advanced Level	Bachelors and/or Masters and Significant Field Experience	• Universities & Colleges	Professional/Technical • Supervisory/Managerial Positions in the Operation of Large Facilities
Intermediate Level (Upgrading)	• Certificate or Other Bonafide Skills Training • Associate Degree • Some Industrial or Municipal Experience • Related Coursework • High School Diploma or GED • Related B.A. Degree (Biology, etc.)	• Community Colleges • Universities & Colleges	Technical/Para-Professional • Operators of Small to Medium Wastewater Treatment Plants (Junior Operators) • Line Workers in Large Company Wastewater Treatment Facilities
Introductory Level (Entry-Level)	• Related B.A. Degree (Biology, etc.) • High School Diploma or GED • High School Level Math	• Community Colleges • Voc.-Tech. Schools • Skills Training Centers	Technical • Entry Level Line Workers in Large Facilities • Preparation for Further Training

fundamentals, fluid mechanics/hydraulics, and mathematics. For the introductory course, a total of three pre-examinations were administered, one in elementary mathematics, one in fluid mechanics principles, and one in chemistry. It was felt that mathematics, fluid mechanics, and chemistry are the three most important subject areas for wastewater treatment operations training.

Pre-examinations for the two intermediate courses and the advanced course involved some questions related to fundamental principles which were developed in the introductory course. The pretesting for the introductory course indicated that most students found it extremely difficult to quantitatively set up and solve problems which required the most elementary calculations. This indicates that an elementary applied mathematics course would be helpful as a prerequisite in many cases.

TYPE OF TRAINING

When designing the training courses, there was a great deal of deliberation regarding the type of training which could be effectively performed in a four-hour lecture classroom session at a junior college or college. "Hands-on-training" is always deemed desirable in operations training programs. This type of training requires equipment, facilities, often at an on-site location, instructors familiar with the on-site facilities, small instructional groups and large blocks of lecture time. Because of the four hour lecture format, the only type of hands on training which could be provided consisted of sessions in environmental analysis, unit operations, and hydraulics laboratories. Techniques of flow measurements, composite sampling, chemical bench scale testing, and routine wastewater analysis were covered in such laboratory sessions.

Field trips to industrial biological, metal finishing, and other hi-tech wastewater treatment facilities were incorporated into the courses.

COURSES

The five courses were designed to develop operational skills at three levels of operations. A summary of training levels, the operators skill requirements, work experience, workforce needs, and appropriate institutional level of instruction are presented in Table I.

The introductory level training course is intended for personnel just entering the wastewater treatment operations field. Materials developed include: 1) review of mathematics, chemistry, and fluid mechanics; 2) the characterization of industrial wastewaters, industrial pretreatment, and basic waste-

Table II. Industrial Wastewater Treatment Operations Introduction Course

Week	Topic	Hours
1	Introduction to Industrial Wastewater Treatment *Pre-Exam: Mathematics*	4
2	Mathematics Review *Pre-Exam: Fluid Mechanics*	4
3	Fluid Mechanics and Hydraulics *Pre-Exam: Chemistry*	4
4	Chemistry Review	4
5	Industrial Waste Characteristics	4
6&7	Sampling, Flow Measurements, and Pumps	4
8	Sampling, Flow Measurements, and Pumps Laboratory Session	4
9	Wastewater Analysis Laboratory	4
10	Pretreatment of Inorganic Industrial Wastes	4
11	Chlorination and Chemical Handling	4
12	*Final Examination*	4

water treatment principles; and 3) principles of sampling, flow measurements, pumps and wastewater analyses. Lectures on chemical handling regulations, and O & M principles were also developed.

At the intermediate level both a physical chemical and biological course were designed to cover the theory and daily operational problems associated with wastewater treatment systems. The instructional level is for operators who are currently working with a system and have had some previous training or experience in related subject areas.

In the Physical Chemical Processes course, the major areas covered in this course dealt with physical chemical applications of industrial wastewater treatment. Topics included neutralization, water conservation, chemical feeding and chemical calculations, sludge handling and disposal, instrumentation and control systems, activated carbon systems, solid and hazardous waste handling.

The course Biological Applications in Industrial Wastewater Treatment covered topics which included theory of biological treatment, operational controls lab, activated sludge systems, aerobic lagoons, aerobic fixed film systems, anaerobic systems, sludge handling, land application, instrumentation, and data handling.

The advanced level course is for operators, engineers, and other management personnel who have had some involvement with industrial waste treatment operations. Professionals enrolling in this course should have a basic understanding of industrial wastewater, its chemistry, sampling, and basic wastewater treatment principles. The topics presented in this course included applied fluid mechanics and hydraulics, environmental chemistry and analysis, microbiology, biological treatment, and industrial waste biological treatment. Also, unit operations and processes (with lab), process calculations and trouble shooting, regulatory requirements, surcharge considerations and rationale, and EPA development documents were presented. Additional lectures on source reduction, hazardous waste considerations, records operation, and industrial safety were also included.

In addition, an Overview of Industrial Wastewaters was prepared to give operations personnel a feeling for the different types of industrial wastewaters. This course presented an overview of a number of industrial wastewaters. Industries reviewed included metal finishing and plating, pulp and paper, food processing, inorganic chemicals, tanneries and fabrics, and steam electrical generation. A review of the processes, wastewater characterization and generation, treatment methods, and solids handling was covered.

COURSE CONTENTS

Table II presents the topics covered in the introductory course. The course was presented in a 12 week format with four hours of instruction each week. Three pretest and two regular examinations were used as a method of testing. This course did not provide for specific field trips, although portions of lectures 6, 7, and 8 could well be conducted at an actual industrial wastewater treatment facility.

Table III. Intermediate Industrial Wastewater Treatment Operations Biological Operations

Week	Topic	Hours
1	Pretest Treatment Rational, Key Parameter Review	4
2	Theory of Biological Wastewater Treatment Systems	4
3	Operations Control Laboratory	4
4	Operation of Activated Sludge Systems	4
5	Operation of Aerobic Lagoon Systems	4
6	Operation of Aerobic Fixed Film Systems	4
7	Operation of Biological Sludge Removal Systems	4
8	Operation of Anaerobic Treatment Systems	4
9	Instrument - Applications & Operation	4
10	Land Application and Subsurface Disposal Operations	4
11	Field Trip	4
12	Data Collection and Presentation	4

Tables III and IV present the topics included in the intermediate level biological and physical chemical treatment courses. The same pretest was used for each course, and three regular examinations were given during the course. Two separate field trips were conducted in conjunction with each course.

Table V presents the topics presented in the advanced level course. A pretest was used during the development phase but was not considered to be a normal requirement at this level of training. Field trips to existing locations were not conducted.

Table VI summarizes the types of industrial wastes covered in the industry overview modules. These materials are designed to give more breath in the training of industrial wastewater treatment plant operators. Each module followed the following format:

1. Industry Description
2. Manufacturing Operations
3. Wastewater Generation
4. Important Wastewater Characteristics

Table IV. Intermediate Industrial Wastewater Treatment Operations Physical Chemical Operations

Week	Topic	Hours
1	Pretest Review of Key Treatment Parameters	4
2	pH Control/Neutralization	4
3	Chemical Feed Systems	4
4	Chemical Handling Calculations	4
5	Water Conservation, Dragout, and Rinse Control Systems	4
6	Sludge Handling and Disposal from Pretreatment Systems	4
7	Oxidation/Reduction Reactions & Controls	4
8	Evaporation Distillation and Steam Stripping Systems for Wastewater Treatment	4
9	Field Trip/Laboratory	4
10	Activated Carbon Systems	4
11	Electrochemical Heavy Metal Removal	4
12	Hazardous Waste Operations	4

Table V. Industrial Wastewater Treatment Operation Advanced Course

Week	Topic	Hours
1	Applied Fluid Mechanics and Hydraulics	4
2	Environmental Chemistry and Analysis	4
3	Microbiology, Disinfection, and Biological Treatment Principles	4
4	Biological Treatment Applications, and Flow Equalization	4
5	Physical Chemical Treatment Applications/Screening, Neutralization, Filtration, and Ion Exchange	4
6	Physical Chemical Treatment Applications/Membrane Processes and Sedimentation	4
7	Physical Chemical Treatment Applications/Flotation, Aeration, Chemical Oxidation and Reduction	4
8	Unit Operations Laboratory Session	4
9	Chemical Precipitation, Mass Balances, and Discharge of Wastewater	4
10	Regulatory Requirements, Discharge and Sewer Use Regulations, U.S.E.P.A. Development Document	4
11	Hazardous Wastes, Source Reduction, and the "Right to Know Law"	4
12	Safety	4

Table VI. Industrial Wastewater Treatment Operations — Industry Overview Modules

Topics
Introduction
Metal Finishing and Metal Plating
Textile and Tannery Industries
Food-Processing Industries
Pulp and Paper Industries
Steam-Power Generation Industries
Inorganic Chemical Industries

5. Wastewater Treatment Requirements
6. Typical Treatment Technologies
7. Operational Considerations and Strategies
8. Residuals Source Reduction and Hazardous Waste Considerations

The use of numerous field trips to typical industrial wastewater treatment operations will prove beneficial to students during this level of instruction. A student should have completed the intermediate level courses prior to taking the industry overview modules.

During 1986 ten different universities and community colleges throughout Massachusetts will conduct courses related to industrial wastewater treatment operations. Most of these schools have received funding from Bay State Skills Corporation for the purpose of establishing industrial wastewater operator training programs. In addition a number of these schools have included the developed or similar courses as part of their regular academic curriculum.

OPERATOR CERTIFICATION

The Commonwealth of Massachusetts has seven grades of certified operators. Until 1986 all grades of operator certification were based on a municipal type examination with some industrial questions. Grade 1 represents the lowest level of responsibility while Grade 7 represents the Chief Operator responsible for larger wastewater treatment facilities. As of July 1986 the Massachusetts Board of Certification will administer examinations relating to municipal and industrial certification by means of a separate testing mechanism in each category.

The Board of Certification of Wastewater Treatment Plant Operators began its examination program in May of 1973 utilizing a combined method of testing with both municipal and industrial wastewater treatment plant operators being certified under this examination process. This method was presented as the best means of obtaining certification for all concerned based on the Board's financial structure. Considerable comments were received by the Board from the industrial segment of the industry as to this method of testing. Industry sought a separate examination for its operators. The Board sought to accomplish this task, but budgetary, administrative and financial restraints prevented the Board from establishing a separate examination for industrial wastewater treatment plant operators.

Using funding from the same settlement, the Board of Certification was able to establish a separate industrial wastewater treatment plant operators examination. Concurrent with the training program development, the Board of Certification initiated a separate industrial wastewater operator examination process. Thus the Board, BSSC, and the Associated Boards of Certification of Operating Personnel developed this separate industrial testing process. Separate examination for industrial wastewater treatment plant operators will be administered during the July 1986 testing in Grades 1,2,3, and 4. This will allow operators passing these examinations to be certified as Grades 1,2,3, and 4 industrial operators. The examination material contained in each testing will be strictly of an industrial nature and category utilizing testing material gained from BSSC's industrial training programs and ABC's testing material based on Need-To-Know and job specific testing matrixes. This is a unique procedure and will be a "first in the nation" testing program based on such training and testing pattern.

Examinations will also be presented by the Board in separate testings for Grade 1,2,3, and 4 municipal operators utilizing a separate municipal testing process. The Board will continue to give Grade 5 and 6 examinations utilizing the combined industrial and municipal testing approach; however, operators gaining certification in these grades during and after the July 1986 testing will be allowed to crossover downward in the industrial category.

Operators who have gained certification prior to the July 1986 testing will not lose any certification privileges obtained by either method of certification-grandfathering, examination or reciprocity.

The Board further intends to grant certification to pretreatment industrial wastewater treatment facility operators. These facilities are those which pretreat industrial flows and then discharge to municipal facilities. Previous administrative, personnel, budgetary, and financial restraints prevented the Board from developing a program of certification for pretreatment facilities. The requirement of certification for these operators was always present in the legislation establishing the Board. Increased funding, more staffing and better overall administration of the program through passage of Chapter 606 of 1983 now allow the Board to develop new programs in the pretreatment area. To develop this aspect of industrial certification, the Board has developed a pretreatment category in the certification program.

Highlights of this program include the following:

1. Development of a procedure allowing the Board to rate pretreatment facilities which previously had not been rated.
2. Establishment of a grandfathering concept for pretreatment industrial wastewater treatment plant operators.
3. Finalization of an examination program for pretreatment industrial wastewater treatment plant operators for the July 1987 testing.

SUMMARY

The use of funding resulting from an industrial wastewater treatment violation has lead to the establishment of a comprehensive training and certification program which will greatly help improve the treatment of industrial wastewaters. New course materials at all levels of operations are being used by approximately ten universities and colleges to enhance wastewater treatment operation. Additionally, the Commonwealth of Massachusetts has a separate examination process for the certification of operators in the areas of industrial and municipal wastewater treatment.

REFERENCES

1. *Manual of Instruction for Wastewater Treatment Plant Operators*, New York State Department of Environmental Conservation, distributed by Health Education Service, Albany, N.Y., Vol. I, II (1978).
2. "Industrial Wastewater Control Program for Municipal Agencies," *MOP OM-4, Operations and Maintenance*, Water Pollution Control Federation (1982).
3. Kerri, K. D., et al., *Operation of Wastewater Treatment Plants*, U.S.E.P.A., Office of Water Program Operations, 2nd ed., Vol. I, II, III (1980).
4. "Operation of Wastewater Treatment Plants," *Manual of Practice No. 11*, Water Pollution Control Federation, Washington, D.C. (1976).

88 CASE HISTORY: DEVELOPMENT OF A DEDICATED SLUDGE APPLICATION SITE BY AMOCO CHEMICALS CO. AND THE ASSOCIATED PERMIT TO OPERATE BY SOUTH CAROLINA DEPARTMENT OF HEALTH AND ENVIRONMENTAL CONTROL

Marie H. Bushway, Environmental Engineer Associate III
Division of Industrial and Agricultural Wastewater
South Carolina Bureau of Health and Environmental Control
Columbia, South Carolina 29201

Geoffrey L. Gilman, Senior Chemical Engineer
Amoco Chemicals Company
Cooper River Plant
Mt. Pleasant, South Carolina 29464

INTRODUCTION

Completed in 1978, the Cooper River Plant, located northeast of Charleston, SC, is the newest of the five major domestic manufacturing facilities owned by Amoco Chemicals Company. With annual shipments in excess of 1.3 billion pounds, the Cooper River Plant is the world's largest producer of purified terephthalic acid, a chemical intermediate used primarily in the manufacture of polyester fibers, films and molded plastics. Crude terephthalic acid is produced by the catalyzed air oxidation of paraxylene and then purified by a proprietary hydrogenation process.

BACKGROUND

Amoco Waste Treatment Plant

The manufacturing processes, along with the Utilities Unit, produce about 1.5 million gallons per day of wastewater normally containing 2500 mg/L of total organic carbon or TOC. The wastewater is processed on-site through a conventional waste activated sludge treatment plant. The Waste Treatment Unit consists of two parallel biological oxidation (biox) trains each having three aeration basins in series. Wastewater feed and recycled mixed liquor are introduced into the first and second stage basins only. Each first stage basin contains eight low speed mechanical aerators, each second stage basin, eight high speed aerators, and each third stage basin, five high speed aerators.

The biox trains are preceded by a primary basin with a six day retention time. In addition to providing equalization, the primary basin acts as a clarifier to allow suspended solids to settle from the wastewater reducing the waste feed concentration to less than 50 mg/L suspended solids. The settled solids are sold as boiler fuel.

Each biox train is followed by its own secondary clarifier. The overflow is passed through a polishing pond with a ten day retention time and a multi-media filter prior to discharge to the Cooper River. The underflow is recycled at a one-to-one ratio with the feed. A side stream of the underflow is wasted to one of two anaerobic lagoons to maintain the proper sludge age.

The wasted sludge digests in the anaerobic lagoons and is then pumped to a holding basin prior to land application. The supernate is returned to the biox trains as a nutrient source. Anaerobic digestion retention time is about four months with an additional two month storage capacity in the holding basin. The total digestion process reduces total suspended solids by 30% and produces a sludge with an average concentration of two percent total solids.

726

Sludge Farm Design

Amoco Corporation is proud of its record and reputation as a good steward of the ecology and considers itself a leader in the implementation of sound environmental practices. This is evidenced by the careful consideration given by the plant design team to the problem of wastewater sludge disposal. Charged with the task of developing an environmentally sound, ultimate disposal process at a reasonable cost and wishing to keep the sludge on the plant site, the corporate environmentalists settled upon the concept of land application or sludge farming. This approach eliminates the costs of mechanical dewatering and auxiliary fuel and makes use of the 2000 acres of land on the plant site.

The Amoco sludge farm is a hybrid of a traditional dedicated land disposal site and an agricultural farm. The sludge farm was designed to receive sludge year after year at nitrogen loadings somewhat higher than normal agronomic practices might dictate, but within the published nutrient uptake rates of the cover crops grown. As a safeguard, groundwater monitoring wells were included in the design.

HISTORY OF THE PERMIT PROCESS

Amoco contacted the Solid Waste Management Division of SCDHEC in May, 1975, to discuss options for sludge disposal. In April, 1976, Amoco received a permit to construct the industrial and sanitary wastewater treatment systems which allowed for separate approval of the dedicated sludge application site. Site inspections by the Groundwater Quality Control Division were conducted July and September, 1978, to collect preliminary data for the proposed sludge application areas at Amoco. This survey revealed a potential for leachate generation at the site due to high annual rainfall and very permeable surficial sands, the probability of leachate attenuation below a depth of one meter due to the clayey nature of the subsoil and marginal potential for surface water contamination. This resulted in a recommendation for site approval with precautions for run-off control from the disposal area.

In February, 1979, Amoco submitted a Preliminary Engineering Report (PER) which outlined the concepts and proposals for site development. The quantity of sludge was estimated based on the analysis of clarifier underflow at a similar plant in Decatur, Alabama. Sludge quality was extrapolated from analysis of this sludge and two analyses of the Cooper River Plant sludge. A total of 138 acres of the plant site was targeted for sludge application. The estimated design loads for the various parameters are shown in Table I.

A formula to calculate plant available nitrogen of:

$$N \text{ (available)} = NH_3\text{-}N + 0.3 \, N \text{ (organic)} \tag{1}$$

was proposed based on the premise that organic nitrogen is initially found in the sludge and is not immediately released in the soil. During the first three years, the organic matter and nitrogen will cycle through the biodegradation process with the result that 70% of the nitrogen will be lost to the atmosphere. An ultimate nitrogen loading rate of twice the crop uptake capacity was proposed. This loading rate could result in leaching of the excess nitrates. In this case, any nitrate and ammonia leachate would move to the lowland periphery of the area in groundwater flow. The estimated nitrogen removal capacity of the marsh and lowland exceeded the total nitrogen load, eliminating the discharge of nitrates to surface waters.

The PER also detailed the techniques of sludge application, site development, sludge spreading operations, and proposed a monitoring program which included monitoring of sludge quantity and quality, soil quality and groundwater quality.

Table I. Estimated Design Load for AMOCO Chemicals Company Industrial Wastewater Sludge (1979)

Parameter	Design Load	Unit Loads on Application Area
Volume	14.6 mg/year	3.9 in/year
Sludge Solids	2,920,000 lbs/year	21.2 lb/acre/year
Chromium	190.0 lbs/year	1.37 lb/acre/year
Cobalt	22,000 lbs/year	159 lb/acre/year
Zinc	7,240 lbs/year	52.5 lb/acre/year
Total Nitrogen	63,000 lbs/year	456 lb/acre/year
Total Phosphorus	5,140 lbs/year	37.2 lb/acre/year
Total Potassium	7,120 lbs/year	51.6 lb/acre/year

Upon receipt, the PER was reviewed and evaluated by SCDHEC. It was found that the concept of using groundwater to transport the ammonia and nitrates to the lowland swamps for uptake was in violation of the State's Water Classification Standards System (Regulation 68). Additional information on soil background conditions, groundwater flow data, and plans to control run-off and ponding water was requested. The PER was also reviewed and approved by the SC Coastal Council with the provision that all parts of the sludge disposal site be adequately diked and maintained to prevent discharge seepage or spill into the Cooper River and adjacent wetlands.

Upon receipt of the requested information, the PER was approved. A permit was issued in February, 1980, which contained the following conditions:

1. The land application sites shall be operated in accordance with the engineering plans and reports submitted to this office.
2. Lime shall be added as necessary to maintain a soil pH of 6.5 or above.
3. Groundwater monitoring program data shall be submitted to this office on a quarterly basis.
4. The sludge, soil, and plant monitoring program data shall be collected on an annual basis. The plant monitoring program shall follow the procedure for the particular plant growth as outlined in Jones and Steyn [1].
5. This permit is valid for one year from date of issue. During this time period, inspections will be made by SCDHEC.
6. Sludge sites are restricted to the sludge that comes from Cooper River Plant main process wastewater treatment system.

SLUDGE FARM DESCRIPTION

Soil Characteristics

The Cooper River Plant lies within the South Carolina Coastal Plain Physiographic-geologic province. The area soil consists of marine sediments laid down during successive periods of high sea levels. The surficial soils of the plant site belong to the Craven-Duplin-Dunbar-Coxville Association. The original sludge application areas consist of borrow areas from plant construction, powerline easements, and croplands existing from the former plantation. The sites were chosen to avoid clearing any timberland. Pedologically, the sites are classified as poorly to moderately well drained with sandy loam surface soils and sand clay subsoils. The soils are described by Clemson University as being moderately suited to farming.

Prior to site development, 30 soil borings were performed to define the subsurface stratigraphy and hydrology of the area and to serve as the original groundwater monitoring well network. Several soil strata were identified by the borings.

The topsoil is generally silty clayey fine sand ranging in depth from 6 to 8 inches. Permeation is moderate at an infiltration rate of 0.6 inches per hour. The second stratum found only in the higher elevations consists of silty fine sand. A third stratum, found in all locations, contains sandy clay or sandy clayey silt and extends to depths of 4 to 5.5 feet. With a permeability of only .015 in/hr, this soil is believed to partially confine the movement of groundwater. Underlying this layer is a stratum of silty sand extending beyond sea level and interest of the test boring study.

The first groundwater aquifer is found at a depth of 1 to 12 feet below grade and at elevations of 2 to 25 feet above mean sea level. The groundwater flows laterally west to the Cooper River and north and south to two tributary tidal creeks (Figure 1).

Sludge Characteristics

Amoco's sludge can best be described as almost composted. Having digested for 4–6 months, the sludge is extremely well stabilized when land applied. The Kjeldahl nitrogen content averages 8.5% on a dry weight basis or 1600 mg/L as applied. The average ammonia nitrogen concentration is 500 mg/L as applied while the nitrate nitrogen concentration is insignificant. The suspended solids concentration ranges up to three percent but normally runs between 1.5 and 2.0%. The wet sludge contains an average 350 mg/L total phosphorus of which about 30% is soluble. The only heavy metals found in the sludge in excess of naturally occurring levels are cobalt and manganese. They are present in wet concentrations of 150 and 500 mg/L, respectively. A typical sludge analysis appears as Table II.

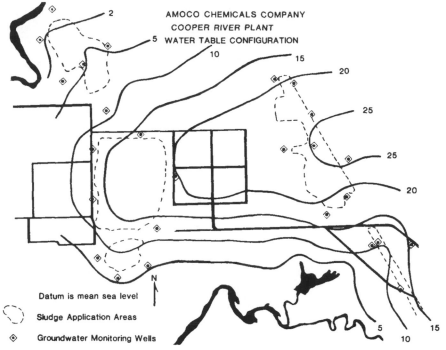

Figure 1.   Watertable configuration.

Sludge Application

The sludge is applied to the fields exclusively by surface spraying and is not subsequently incorporated into the soil. This practice maximizes ammonia volatilization. The sludge is pumped from the sludge holding basin into an 8000- gallon tank truck and hauled to one of the sludge fields. The sludge is then pumped from the truck with a portable diesel pump through a retractable reel irrigation system and discharges from a reciprocating sprinkler mounted on wheels. The irrigation reel contains 1000 feet of 4" polypropylene hose. After each complete retraction of the reel, the complete apparatus is indexed about 200 feet to the right or left. In this manner an entire field can be sprayed at a uniform sludge application rate. Amoco uses a contract service for all sludge application and field maintenance.

Amoco maintains field cover throughout the year. Coastal bermudagrass is grown wherever possible during the warm weather months. Bahiagrass is substituted in areas where bermuda won't grow well. The fields are oversowed with ryegrass in the winter months. This arrangement allows for maximum nitrogen removal during the year. Sludge is applied year round, weather permitting. However, the greatest quantities are applied in the spring and late fall between growing seasons. The fields are mowed when necessary. The clippings are baled and sold for landscaping mulch.

PERMIT REISSUANCE

1981

Prior to reissuance of the permit, responsibility for land application of industrial sludges was transferred to the Industrial and Agricultural Wastewater Division. In June, 1981, Amoco submitted a request to review the permit along with a PER detailing a program to land apply sludge to an 80–100 acre onsite wooded area. This additional land was deemed necessary due to a year's accumulation of sludge prior to the initial permit issuance. Nitrogen was the limiting factor in the sludge and an application rate of 275 pounds per acre per year was requested. A monitoring program, similar to the one approved for the grass land sites, was proposed.

Table II. Sludge Monitoring Summary

	As Applied	Dry Basis
pH	7.6 SU	N/A
Total Solids	18,839 mg/L	100%
Total Suspended Solids	17,203 mg/L	91.3%
Volatile Suspended Solids	11,935 mg/L	63.4%
Ash Suspended Solids	5,268 mg/L	28.0%
Total Dissolved Solids	1,636 mg/L	8.7%
Volatile Dissolved Solids	532 mg/L	2.8%
Ash Dissolved Solids	1,104 mg/L	5.9%
Total Organic Carbon	125 mg/L	6,640 mg/kg
Total Kjeldahl Nitrogen	1,600 mg/L	8.5%
Ammonia Nitrogen	500 mg/L	2.7%
Nitrate Nitrogen	2 mg/L	110 mg/kg
Soluble Phosphorus	52 mg/L	2,760 mg/kg
Potassium	85 mg/L	4,510 mg/kg
Sodium	361 mg/L	19,100 mg/kg
Calcium	294 mg/L	15,600 mg/kg
Iron	184 mg/L	9,760 mg/kg
Manganese	378 mg/L	20,000 mg/kg
Magnesium	24 mg/L	1,280 mg/kg
Aluminum	269 mg/L	14,300 mg/kg
Zinc	15 mg/L	770 mg/kg
Copper	1 mg/L	58 mg/kg
Nickel	5 mg/L	281 mg/kg
Chromium	3 mg/L	138 mg/kg
Cobalt	163 mg/L	8,640 mg/kg
Cadmium	N/D	N/D
Lead	N/D	N/D

In March, 1982, the Industrial and Agricultural Wastewater Division issued a permit with no expiration date for the dedicated land application site with average sludge loading rates based on the estimated design loading as follows:

Hydraulic Rate—3.8 inches/year
Suspended Solids—20,860 lbs/acre/year
Nitrogen—450 lbs/acre/year
Phosphorus—36.7 lbs/acre/year
Potassium—50.9 lbs/acre/year
Cobalt—157 lbs/acre/year.

The special conditions were similar to those detailed in the original permit. In addition, the following pertained to the forested area:

The present sludge spreading program shall be augmented for one year from the date of the permit for the onsite wooded area. All monitoring requirements on the original land disposal facility shall apply. The following application rates shall apply for the wooded area.
1. Hydraulic rate—2.4 inches/year.
2. Available Nitrogen load—275 lbs/acre/year.
3. Total Sludge applied—10 million gallons.

In January, 1983, approval to continue the forest farming of sludge was granted for 15 million gallons of sludge at a rate of 1–1.25 million gallons per month provided the established monitoring program was maintained.

1984

In September, 1984, Amoco requested a meeting with SCDHEC to discuss the possibility of upgrading the groundwater monitoring well system and to review the permit limits. An informal internal environmental audit conducted by the Company revealed that several procedural changes over the previous two years had been based on verbal approval from SCDHEC and interpretation of general statements made in the PER. Amoco wished to confirm these approvals in writing. At the same time, it was realized that the 138 acre permitted sludge field site had not been surveyed correctly and, in actuality, consisted of only 85 acres. This resulted in a loading rate of 750–850 lbs/acre plant available nitrogen. Amoco requested that the permitted application of plant available nitrogen, phosphorus, and potassium be increased to twice the crop uptake rate, as detailed in Table III. This request was based on EPA recommendations for dedicated sludge application sites [2], US Dept. of Agriculture fertilizer recommendations [3], a study describing the nitrate removal mechanisms, which exist in soils similar to those at Amoco [4], and Amoco's interpretation of the original permit and PER.

PERMIT PARAMETERS

It should be noted that at that time, SCDHEC had no established guidance procedures for permitting an industrial dedicated sludge application site. The EPA Process Design Manual "Land Application of Municipal Sludge" [5] was available, but didn't address industrial sludges. The permit review and revision provided an educational experience in sludge application systems for both parties.

The first step was a review of the existing groundwater monitoring system and quarterly monitoring data. Several wells showed nitrate contamination slightly above the drinking water standard of 10 mg/L. Due to inappropriate placement and poor well construction, it was felt that some of the groundwater samples obtained were not representative. Amoco agreed to conduct a comprehensive groundwater monitoring study and proposed a program to include plans to upgrade existing monitoring wells and/or locate additional monitoring wells. The study was reviewed and approved by the SCDHEC Groundwater Protection Division prior to implementation. Amoco was then asked to proceed with plans to expand the sludge farming operations on cleared fields from 85 acres to the permitted 138 acres. This was completed in early 1985.

Hydraulics

The year-to-date hydraulic loading information was reviewed. It was noted that at a limit of 3.8 inches/year, permitted hydraulics become the site limiting factor. This condition was revised to eliminate a hydraulic limit and restrict loading to management practices which would prevent puddling and sludge run-off to receiving waters.

Nitrogen Loading

The EPA Process Design Manual recommends a nitrogen loading of twice the crop uptake rate for dedicated sites. This is based on ideal site conditions in which the depth of groundwater is 10–50 feet below the surface. The depth to groundwater at the Amoco site is 1–12 feet below the surface, so a more conservative approach was called for. The application of nitrogen, phosphorus, and potassium is limited to the uptake rate of the seasonal crop until data from soil analysis or a test field indicated an increase or decrease is in order. Since Amoco planted coastal bermudagrass or bahiagrass in the warm weather and winter rye in the cool season, a cumulative total of the two was allowed. This resulted in a limit of 775 lbs/acre plant available nitrogen per year, as detailed in Table IV.

Cobalt

Cobalt was the only heavy metal of concern present in the sludge. Although there are no guidelines for cobalt disposal, some guidelines have been published for plant and animal intakes. Since Amoco does not introduce any crops into the food chain the animal intake rate is not of major concern. The EPA [7,8] allows up to 5 mg/L cobalt in irrigation waters if proper soil pH is maintained. Amoco applies approximately 20 million gallons of sludge containing an average 150 mg/L cobalt to 138 acres of land. Assuming the cobalt is 10% leachable and an eight-to-one rainfall to sludge ratio, the cobalt loading corresponds to 1.6 mg/L in irrigation water. A limit of 25,000 lbs/year cobalt was thus set. Quarterly groundwater monitoring, monthly sludge analysis and annual plant tissue analysis for this metal were required. Application of organic suspended solids was limited to 45 tons/acre/year, a rate at which soil clogging would not become a problem.

Calculation of Available Nitrogen

The final remaining issue to resolve was the method of calculating plant available nitrogen applied with the sludge. The EPA Process Design Manual, which takes into account the residual nitrogen over a 20 year period uses the formula:

Table III. Requested Limiting Rates for AMOCO's Dedicated Sludge Application Site (1984)

AMOCO Chemicals Company
Cooper River Plant

Cover Crop	Available N (lbs/acre/year)	Phosphorus (lbs/acre/year)	Potassium (lbs/acre/year)
Coastal Bermuda	1140	290	800
Bahia	1140	290	800
Winter Rye	210	150	270
Annual Forages	240	60	120
Forest	275	100	100

$$N \text{ Available} = 0.5 (NH_3\text{-}N) + 0.2 (N \text{ Organic}) + (N \text{ Residual}) \tag{2}$$

This is based on a site design of surface application of sludge in which 50% of the ammonia nitrogen is lost through volatilization, and stable, anaerobically digested sludge contains no nitrate nitrogen. Calculations showed that a 190 acre dedicated site would be needed to satisfy ultimate disposal requirements using this approach. This was rejected by the company because they felt it was economically unwise to develop a site that was not needed for years. Both SCDHEC and Amoco decided to abandon the yearly residual nitrogen calculations because the procedure is both time consuming and could generate confusion in the future. SCDHEC then proposed the following conservative formula:

$$N \text{ Available} = 0.5 (NH_3\text{-}N) + 0.5 (N \text{ Organic}) \tag{3}$$

which was rejected by the company. Amoco felt their sludge was extremely stable based on the treatment processes. First, it is aerobically processed for 20 days, followed by six months of anaerobic digestion. Thus, mineralization is well along prior to land application. Neither EPA nor SCDHEC had any data on land application of this type of sludge and there was no data in the literature. It was agreed that the Amoco process of sludge treatment was so different from municipal sludge used by EPA to develop its equations that they were inappropriate. Amoco recommended the following formula:

$$N \text{ Available} = 0.8 (.5 NH_3\text{-}N + .2 \text{ Organic } N) \tag{4}$$

based on 50% ammonia volatilization, 20% mineralization of organic nitrogen, and a 20% overall loss by denitrification. They felt that this formula was sufficiently conservative to prevent groundwater contamination until a soil test was conducted to define a true residual mineralization rate.

By way of a compromise, and based on the fact that the fields had been loaded at a rate of 750–850 lbs/acre plant available nitrogen, using a similar formula, for four years without nitrate contamination of the groundwater, the following formula was agreed upon:

$$N \text{ Available} = 0.5 (NH_3\text{-}N) + .25 (\text{Organic } N) \tag{5}$$

Table IV. Nutrient Uptake by Specific Agricultural Crops

Crop	(lbs/acre/year)		
	Nitrogen	Phosphorus	Potassium
Corn	170	70	48
Tobacco	85	15	155
Soybeans	252	49	87
Peanuts	140	22	35
Johnson Grass	890	150	630
Fescue	135	65	105
Coastal Bermuda Grass	570	145	400
Bahiagrass	570	145	400
Winter Rye Overseed	205	75	135
Forest Area	250	100	100

Table V. Sludge Field Soil Summary

	1979		1985					
	0"–18"		0"–6"		6"–12"		12"–18"	
	Avg	Range	Avg	Range	Avg	Range	Avg	Range
CEC (meq/100gm)	6.2	3.7–9.6	1.6	0.2–3.4	4.1	0.2–8.0	1.6	0.3–6.0
pH (SU)	5.0	4.5–5.9	6.5	5.1–7.7	6.3	5.3–7.8	6.4	5.6–7.5
TOC[a] (mg/L)	3500	2300–4700	31	15–53	30	12–52	29	21–37
TKN (mg/L)	694	500–1200	1249	107–3159	553	151–890	992	214–3904
NH$_3$-N (mg/L)	4.4	2.4–10	0.2	0–1.0	0.1	0–1.1	0.1	0–1.0
NO$_3$-N (mg/L)	1.1	0.1–1.5	1.6	0.3–6.6	2.0	0.3–4.4	1.8	0.5–2.8
P[b] (mg/L)	122	0–930	8.9	3–19	6.3	1.0–18	11	2.0–22
Na (mg/L)	11	3–37	72	20–210	137	25–325	149	25–305
K (mg/L)	56	24–120	59	10–195	34	10–75	56	5.0–200
Zn (mg/L)	8.7	2.5–53	20	6.9–33	19	5.2–31	16	2.2–29
Co (mg/L)	5.0	1.0–29	129	60–270	166	30–325	125	60–250
Mn (mg/L)	32	1.8–210	310	110–605	444	50–835	328	70–685

[a]1979 TOC results are total. 1985 TOC results are soluble only.
[b]1979 P results are total. 1985 P results are soluble only.

Soil Analysis

Amoco agreed to amend its soil sampling program to include a multi-sample composite from each field at three separate depth ranges; 0"-6", 6"-12", and 12"-18". This approach will allow the plant to track any migration of nitrates downward through the soil and buildup of organic nitrogen in the upper layer. The 1979 background and 1985 soil analyses are shown in Table V. At this time no significant nitrate levels are seen at any soil level.

Other Conditions

The company requested a permit condition allowing them to develop a test field of up to 40 acres to determine the effects of accelerated rates of sludge application and evaluate alternate cover crops. Rates in excess of these described in Table IV were limited to twice the crop uptake rate. The plant was required to submit a best management practices plan detailing the operations and maintenance of the sludge farm. The plan also includes a proposed upgrading of the forest application program discontinued since mid-1984. A 10 year Permit to operate the dedicated sludge application site was issued in May, 1985.

GROUNDWATER MONITORING PROGRAM

Concurrent with the development of the new permit regulating sludge application rates, Amoco worked closely with both the Industrial Wastewater and Groundwater Protection Divisions of SCDHEC to upgrade the plant's groundwater monitoring program. This involved the relocation of several well sites, the physical improvement of some existing wells, and the development of an approved sampling protocol.

Although generally spaced around the periphery of the various sludge application sites, the original monitoring wells were located primarily to develop soil profiles and a piezometric groundwater contour map. Over the years since installation, a number of the wells were damaged by logging vehicles, lost in the understory and replaced without following proper procedures. By 1984, it was felt that the original system was no longer reliable. This was evidenced by sporadic high nitrate nitrogen analyses in some of the wells. It was not uncommon for a sample from a given well to show no nitrates one quarter, over 50 mg/L the next quarter, and then no nitrates again the following quarter. Inspection of the suspect wells revealed that the specified cement grout seals had never been placed around any of the wells. Subsequently, several of the wells were pulled and were found to also be missing one or both bentonite clay seals (Figure 2). Since all of the suspect wells were located in areas of direct sludge application or low lying areas receiving occasional sludge run-off, it was theorized that sludge was flowing down the exterior wall of the well casing and directly entering the well.

Following initial discussions with SCDHEC, the plant called upon the Amoco Corporate Hydrogeology Group to prepare the upgrade plan. After meeting again with SCDHEC and incorporating some

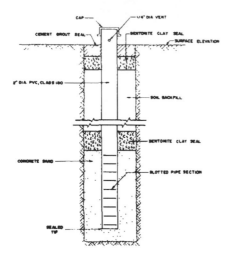

Figure 2.   Typical monitoring well installation.

minor changes, a plan was agreed upon whereby the plant would maintain 26 active groundwater monitoring wells. These were located so that each application area would have at least one upgradient and three downgradient wells. In addition, four wells were located in direct application areas. The new system consists of 18 newly constructed wells and 8 existing wells to which were added an upper bentonite seal and a cement grout cap. Fourteen of the 33 pre-existing wells were pulled and the cavities sealed. The remaining 11 were abandoned in place and may be used for water level determination. Of the current 26 monitoring wells, only one exhibits a nitrate nitrogen concentration in excess of 10 mg/L. That well is upgradient to present sludge application sites and is in a low area which previously received an inordinate amount of sludge run-off from a less than successful forest application program. The average analytical values and ranges of values for various groundwater parameters are presented in Table VI.

The surficial confined groundwater aquifer below the Cooper River Plant and entire surrounding area can be characterized as relatively poor to unusable for human consumption. The pH is generally acidic and can range as low as 4.3. It is not uncommon for samples to contain over 3000 μg/L iron which causes a sample to look almost bright orange. The total dissolved solids level ranges up to 600 mg/L and the natural sodium levels range up to 50 mg/L.

THE AUTHORS' POINTS OF VIEW

DHEC

At this time, there are no Federal regulations or guidelines which deal specifically with land application of industrial sludge. Also, little information on this subject has been published. Since the issuance of the permit for Amoco's dedicated sludge application site, SCDHEC formed a Sludge Task Force which has developed a draft Land Application of Sludge Guidance Manual and is currently

Table VI. Groundwater Summary

	1979		1985	
	Avg	Range	Avg	Range
pH (SU)	7.0	5.9–7.7	6.5	4.3–7.7
TDS (mg/L)	262	35–600	186	30–551
TOC (mg/L)	13	0–40	7.4	4–14
NO_3-N (mg/L)	0.3	0–5	1.0	0–7
P (mg/L)	0.03	0–.18	0.03	0–.25
Zn (μg/L)	6.8	0–130	29	3–38
Co (μg/L)	0	0	7.4	0–51
Mn (μg/L)	31	0–180	105	0–431

working on development of the State Regulations. Unlike many states, our Guidance Manual considers industrial sludges. The information collected and reviewed during this permitting process has resulted in a better understanding of the fate of nutrients and metals in industrial sludge application systems. Working in a cooperative atmosphere benefited both Amoco and the State of South Carolina. The company's attitude is commendable.

Amoco

Amoco found the process of renegotiating an existing permit to be frustrating, educational, and ultimately rewarding. What began as a request to clarify some discrepancies and nebulous statements in the permit and engineering report turned into the development of an entirely new permit, a process which lasted nine months. During this period, Amoco engaged the services of two consultants, devoted well over 100 man-hours in researching and developing the company's point of view, and met on numerous occasions with SCDHEC to both negotiate and educate them on the subject of industrial sludge land application. Amoco now has a concise permit with which it can live. More importantly, the company has a better understanding of and a better working relationship with SCDHEC and has had the opportunity to help formulate the guidelines for future industrial sludge land application projects in South Carolina.

A bit of counsel to others embarking upon the permitting process:

> Research your position thoroughly and be ready and willing to educate your regulator. Treat your regulator as a professional remembering that you are your company in the eyes of the regulator. Recognize the goals and interests of the regulator; they probably are not too far from your own. Above all, GET IT IN WRITING, but don't forget to maintain face-to-face relationships.

Adapted from the thoughts of
R. J. Ganim, Attorney,
Amoco Corporation

REFERENCES

1. Jones, T. B., Jr. and W. J. A. Steyn, *Sampling, Handling, and Analyzing Plant Tissue Samples*, pp. 249–270 (1973).
2. "Municipal Sludge Management: Environmental Factors," *EPA 430/9-77-004*, MCD-28 (October 1977).
3. "Fertilizer Recommendations for South Carolina," *Circular 476*, Clemson University Cooperating with United States Department of Agriculture (November 1977).
4. Engler & Patrick, "Nitrate Removal From Flood Water Overlying Flooded Soils and Sediments," *Journal of Environmental Quality*: 3(4): 409–413 (1974).
5. "Process Design Manual: Land Application of Municipal Sludge," *EPA-625/1-83-016*, (October 1983).
6. U.S. Department of Agriculture, *USDA Waste Manual*, p. 26 (1975).
7. *Water Quality Criteria*, USEPA. (1972) (excerpts re livestock water and irrigation water).
8. Hodgson, J. F., "Cobalt Reactions with Montmorillonite," *Soil Science Society American Proc.*, 24(3): 165–168 (1960).

Section Fourteen
MISCELLANEOUS

89 ACID RAIN AND THE SPACESHIP EARTH

Nolan A. Curry, Consultant
Troy, New York 12180

INTRODUCTION

The planet earth is a spaceship, orbiting endlessly around the sun. Its orbit is controlled by its speed and by the mutual attractions of the sun and other orbiting bodies. At present, no adjustment of speed or of direction is required.

However, food supply for an indefinite voyage requires recycling of essential atoms. An essential atom is one whose absence causes low or zero growth, abnormal growth or plant death. No other element can be substituted for it. Atoms vary in combinations from raw materials to finished plants and back to raw materials again.

In nature, only plants can take simple inorganic molecules containing atoms essential to plants and reform them into the large complex molecules. These form the plant structure.

The next part of the cycle involves the use of the plant as food. The complex molecules are broken up, to form the simple inorganic molecules usable as nutrients by plants. This is done by animal digestion, by biodegradation, by oxidation, or by a combination of two or more processes. In the process, energy is released.

An understanding of the final products is important. Four of the essential nutrients (carbon, hydrogen, nitrogen and sulfur) form gases when combined with the fifth (oxygen). These gases escape into the atmosphere. The other ten essential nutrients (calcium, magnesium, potassium, phosphorus, iron, manganese, copper, zinc, boron, and molybdenum) are also united with oxygen. These form solids in the residual ashes. These fifteen essential nutrients must be re-assembled in the required proportions to generate plant growth.

FORMATION OF ACID RAIN

Acid rain is the name given to the condensing water vapor and its absorbed oxides of nitrogen, of carbon, and of sulfur. The condensing is necessary to keep from excessive buildup in the air. It is also necessary to return these nutrients to the earth's surface as raw materials for plants. These are in a usable form. The very acidity is a part of nature's plan to release by ion exchange the solid nutrients. These nutrients may occur as solid carbonate, as ions absorbed by natural zeolites or held by decaying vegetation.

The naturally gaseous fertilizers have an excellent means of distribution. Wind circulation and diffusion scatter the gases over the surface of the earth. The return to earth is difficult and sparse in areas of little or no rainfall.

For the main components of acid rain to be converted into neutral growth, the other essential nutrients must be supplied in the proper proportions. The following equation represents the overall reaction:

$$\text{Acid Rain} + \text{Nitrogen Fixation} + \text{Carbon Dioxide} + \text{energy (sun)}$$
$$+ \text{Plant Residuals gives Plant Growth} + \text{Oxygen} \tag{1}$$

Unfortunately, the mode of transport is such that there is generally an excess or deficit of one or more of the materials required in the equation. Any imbalance tends to reduce the potential growth.

The ratio of nitrogen to sulfur in acid rain is about one to two. In most plants the ratio is between eight and fifteen to one. In the wilderness, the difference is provided by nitrogen fixation from the air. The enzyme nitrogenase has been isolated from many biological nitrogen fixing systems. M. N.

Table I. Products from Incineration of Dried Leaf Tissue

Found in Gases		Obtained in Incineration Residues			
Element	Weight	Element	ppm	Element	ppm
Oxygen	44.3%	Calcium	15,000	Manganese	40
Carbon	43.6%	Potassium	15,000	Boron	40
Hydrogen	6.2%	Phosphorus	2,500	Zinc	40
Nitrogen	2.0%	Magnesium	3,000	Copper	25
Sulfur	0.2%	Iron	100	Molybdenum	1

Hughes reports that an iron protein and an iron molybdenum protein are active ingredients. Both contain sulfur, metal and protein.

Carbon dioxide has limited solubility in acid rain. The concentration of carbon dioxide in the air is normally .03% by volume at a level slightly above the earth's surface. Fortunately, plants can absorb carbon dioxide as a gas. If the plant contains 45% of carbon by weight, 1.65 pounds of carbon dioxide are needed in its synthesis.

EFFECTS OF ACID RAIN

The plant residual on incineration will contain the ten essential solid nutrients as oxides plus a varying amount of non-essential elements. Some of these may promote the synthesis of the organic compounds. Others may retard the reactions. It is generally best to reduce the concentrations of any one nonessential element to as low a value as possible. To replenish the depleted essential nutrients, it is advisable to avoid residuals with excessive amounts of any element.

The ten essential nutrients found in plant residues pose a considerable problem in recycling to the higher areas. Their flow is generally by gravity; however, in the older geological areas, there has been a steady movement of these solids away from the higher elevations. The fertile valleys, the flood plains, the deltas and finally the ocean floor are being enriched. The mountain tops and ridges are left depleted. Their tops contain sparse soil with little or no essential nutrients left. Granite is the underlying rock in many cases. It is a mixture of quartz (silicon dioxide) and of feldspar (mixed silicates). At their slow rate of weathering, they contribute little of the needed solid nutrients.

The first evidence of depletion occurs as a reduction in yearly growth. The growth rings of trees show a reduction in widths. As the ten essential solid nutrients are depleted, one by one, the excess of the acidic nutrients is not converted to growth. They are neutralized by the excess nutrients and by non-essential cations in the soil. This increases the salinity. The final phase will be the increase in acidity.

In this phase, the trees show a reduction in sap flows and in resin. Deficiency in essential proteins will reduce their ability to fight disease and to resist insect penetration. The result is a dying environment.

Most research effort seems to be directed to proving that the acidity causes the dying vegetation. Some try to prove that the vegetation in dying causes the acidity. The more logical approach seems to be to convert the excess acidity into neutral plant growth. This would solve both problems.

In nature, the tropical rain forest is perhaps the best natural example of nearly complete recycling. Demands of plant growth are supplied by decay and decomposition of an equivalent amount of previous growth. The reservoir of essential atoms is sufficient to absorb the time delay between growth and recovery of nutrients used in that growth. The result is a lush, self-perpetuating environment. It can support an elaborate system of life (as long as all used wastes and dead material are contained and recycled). When man harvests products for use elsewhere, the rain forest dies. At the edge of the rain forest, there is a gradual loss of nutrient and a gradual loss of the forest.

In hydroponics (water or sand culture of plants) scientists have worked out the desirable mixture of the fifteen nutrients essential for plants. In these solutions, growth can be perpetuated by replacing the fifteen essential nutrients as they are used. This method demonstrates the similar needs.

These two examples maintain an acceptable balance between the acid nutrient stream and the alkaline mix of solid material. Table I gives the relative amounts of the various essential elements found in dried leaf tissue. Variation in the ratio between carbohydrates and amino acid containing compounds will change the proportion of the carbon, hydrogen, oxygen, nitrogen, and sulfur atoms. The other ten nutrients can also vary within the working range for each. Table II lists the twenty

Table II. Amino Acid Data [1]

Name	Formula	Iso-Electric Point	Percent by Weight of Element				
			Carbon	Nitrogen	Sulfur	Hydrogen	Oxygen
Alanine	$C_3H_7O_2N$	6.1	40.4	15.7		7.9	35.9
*Arginine	$C_6H_{14}O_2N$	10.8	41.3	32.1		8.0	18.4
Aspartic Acid	$C_4H_7O_4N$	3.0	36.1	10.5		5.3	48.1
Cysteine	$C_3H_6O_2NS$	5.0	30.0	11.6	26.7	5.0	26.7
Cystine	$C_6H_{10}O_4N_2S_2$	5.0	30.2	11.8	26.7	4.2	26.7
Glutamic Acid	$C_5H_9O_4N$	3.1	40.8	9.5		6.1	43.5
Glycine	$C_2H_5O_2N$	6.1	32.0	18.6		6.7	42.6
*Histidine	$C_6H_9O_2N_3$	7.6	46.4	27.1		5.8	20.6
Hydroxyproline	$C_5H_9O_3N$	5.8	45.8	10.7		6.9	36.6
*Isoleucine	$C_6H_{13}O_2N$	6.0	54.9	10.7		9.9	24.4
*Leucine	$C_6H_{13}O_2N$	6.0	54.9	10.7		9.9	24.4
*Lysine	$C_6H_{14}O_2N_2$	9.5	49.2	19.2		9.6	21.9
*Methionine	$C_5H_{11}O_2NS$	5.7	40.2	9.4	21.4	7.4	21.4
*Phenolalanine	$C_9H_4O_2N$	5.9	65.4	8.5		6.7	19.4
Proline	$C_5H_9O_2N$	6.3	52.1	12.2		7.8	27.8
Serine	$C_3H_7O_3N$	5.7	34.3	13.3		6.7	45.7
*Threonine	$C_4H_9O_3N$	5.6	40.3	11.8		7.6	40.3
*Tryptophan	$C_{11}H_{12}O_2N_2$	5.9	64.7	13.7		5.9	15.7
Tyrosine	$C_9H_{11}O_3N$	5.6	59.6	7.7		6.1	26.5
*Valine	$C_5H_{11}O_2N$	6.0	51.3	12.0		9.4	27.3

*Essential

amino acids found in plants. Their chemical formulae and the percent of the five basic elements are given.

Farmers correct deficiencies of elements in acid rain and in soil by the use of fertilizers. Lack of water is generally covered by irrigation. As this is not a problem in areas of acid buildup, it will not be discussed here.

Reduction of emissions of sulfur and of nitrogen oxides will reduce the potential for plant growth where there is an adequate supply of the ten solid nutrients. In cultivated areas it will mean either reduced yields or increased cost of fertilizer.

In the areas deficient in one or more of the residual nutrients, the acid rain will remain unconverted. All that is received will either run off to increase the demands of the lower slopes or contribute to the acid buildup. It will not neutralize the acidity already present. It will not permit resumption of plant growth without replacing the lost solid nutrients.

Data from Germany [2] reports a decrease in sulfur dioxide in the last fifteen years. Sulfur dioxide was lowest where the trees were sickest. It also reports the lichens are thriving on trees dying of Waldesterben (forest death). Lichens are known to be susceptible to sulfur dioxide in the air. Trees were also stated to be starving for magnesium.

In Germany the entire tree, branches, and foliage are removed for use elsewhere. The atoms of carbon, hydrogen, oxygen, nitrogen and sulfur are replaced by the acid rain or from the air. The ten residual nutrients are not replaced. The result eventually is a buildup of acidity when the ten solids are depleted.

Appalachia, the Adirondacks, Eastern Canada, Sweden, Czechoslovakia, Switzerland, and France are now showing similar patterns. The reports of reduced growth, thin soil over a granite base rock are common.

SOLUTIONS TO THE PROBLEM

It is time that we realize that a material balance must be maintained in areas of critical acidity. If we want growing trees on the high slopes, we must balance the input of acid rain with proper proportions of the ten essential solid nutrients. Initial additions must have sufficient quantities of the essential residual nutrients to convert the acidity present before treatment as well as the expected acidity to be added by rain.

Table III. Nutrient Requirements

Element	as Oxide	as Carbonate	
Calcium	10.5	18.8	
Potassium	9.0	13.3	
Magnesium	2.5	5.2	
Phosphorus	2.9	2.9	both as oxide
Total	24.9	40.2	
Plus 5%	26.1	42.2	

Material to use should be cheap, available, easily handled and contain the same proportion of nutrients required by the plants. The principle has worked well in the rain forest. There the material balance is maintained by natural recycling. The necessary nutrients are contained within the confines of a specific area.

At least for initial experiments, the actual ash residue from healthy trees could serve as the fertilizer. Take all of the waste from harvesting a healthy tree farm (bark, faggots, thinnings, litter and sawdust). Burn them in an incinerator. Collect the fly ash and bottom ash. Analyze for the ten essential residual elements. Fresh ash should have a pH of about 11. It may be desirable to allow ash to absorb carbon dioxide from the air to reduce the alkalinity. Conversion of calcium, potassium, magnesium, and zinc to carbonate should lessen the solubility and stabilize the residue.

Most of the nitrogen is provided by fixation from the air. The carbon dioxide is provided from the air. Therefore, the amount of ash needed is best estimated based on the sulfur in the acid rain. Table I shows that the sulfur in the dried leaf is .2%. The total of the four major nutrients (calcium, potassium, magnesium, and phosphorus) is 35,500 ppm. This equals 3.55% of the weight of the dried leaf. The ratio of these four major components to the sulfur is 17.75. For each pound of sulfur deposited per acre per year, Table III gives the required amount of the four major nutrients and ash per acre per year.

The concentration of the other six nutrients is sufficiently small so that an arbitrary 5% of ash can cover the needs of all six. Phosphorus does not form a carbonate. In the actual incineration, a glassy fusion of the various elements is formed. This may interfere with complete carbonate formation. If there is excessive leaching of nutrients down the slope, extra ash may be required. This formula is only a starting point. Further checks will have to be made at specific locations.

If the ash comes from trees suffering from a deficiency, it may need enriching. Ash from any vegetable waste or sewage may be usable. It will depend on the analysis.

Many sewage plants accept metal wastes. While it can remove the metals from the water, it accumulates them in the sludge. This may make the incinerated ash unsuitable for land disposal. The problem was addressed in a paper [3] given at the Eleventh Mid-Atlantic Industrial Waste Conference at Penn State in July 1979. Tables IV, V, VI, and VII summarize these data. Incinerated ash from such sludges should not be used without experimental determination of its effect.

The use of a layer of plant ash as a means of improving plant growth is not new. Nature has used fires set by lightning or by spontaneous combustion. By this method, old vegetation and weeds can be incinerated. Five of the elements escape as gases. They form acid rain to act as one part of the fertilizer. The ash produced is spread by the winds for the second part of the fertilizer. With energy from the sun, carbon dioxide from the air, and nitrogen fixation from the air, nature has provided the necessary raw materials. The result is new, vigorous, healthy plant growth.

The Indians of the plains set fire to the dead grass and weeds at the end of the season. They were careful to move their portable camp to safety behind a water barrier.

Man has banned this practice. His dwellings are fixed. They cannot be moved temporarily to a safe place. While the practice of burning vegetation in place is no longer acceptable, the same effect can be obtained by other means. Waste can be incinerated in a safe manner. The ash can then be applied by air to the areas showing the effects of acidity. The feasibility of acid rain being applied to a layer of ash has proven effective.

Our present method of handling wastes ignores the basic need of a spaceship — that of recycling the essential nutrients. A municipality of 100,000 people generates approximately 200 tons of refuse and 20 tons of sludge per day. Our only aim seems to be to find a safe place to put it so that it will not generate pollution. This represents a tremendous volume of the fifteen essential nutrients that should

Table IV. Heavy Metal Concentrations

	1975 Average Analysis Wastewater Sludge (ppm) Phase II Report June 1976 Interstate Sanitation Commission					New Jersey-New York Metropolitan Area Camp, Dresser & McKee and Alexander Potter Associates			
#	Cd	Cr	Cu	Hg	#	Ni	Pb	Zn	Plant
a	68.4	530.	1,020	4.3	a	102.9	590	1,435	#24
b	74.	700.	1,510	15.3	b	230.	780	5,500	Linden Roselle
c	58.	3,070.	1,900	5.0	c	370.	830	3,160	Bergen
d	11.	53.	740	4.8	d	40.8	240	1,690	Middleton
e	59.	545.	530	1.2	e	150.	840	2,800	Passaic V.
f	310.	460.	880	3.1	f	160.	880	2,960	Middlesex
g	11.3	60.	620	3.6	g	35.8	560	2,010	W. Long Beach
h	30.2	610.	1,140	8.0	h	220.	530	1,210	Bay Park
i	6.0	38.8	435	7.2	i	17.1	280	720	Long Beach
j	63.	1,240.	2,110	7.	j	225.	1,030	2,820	Bowery Bay
k	11.	470.	1,830	12.	k	126.	1,530	1,760	Coney Island
l	30.3	1,035	2,619	25.9	l	1,129	1,037	2,675	Hunt's Point
m	25.	580.	1,930	9.	m	240.	1,290	2,130	Jamaica
n	100.	2,230	2,020	8.	n	260.	6,370	1,920	Newton Creek
o	19.	660.	2,180	9.	o	370.	1,960	2,000	Owls Head
p	5.6	130.	310	14.6	p	100.	800.	1,490	Pt. Richmond
q	14.	380.	1,630	10.	q	90.	880.	1,240	Rockaway
r	31.	730.	1,270	12.	r	400.	730	4,390	Tallman's Is.
s	189.	997.	2,150	13.	s	307.	1,090	2,746	26th Ward
t	11.	500.	2,290	4.9	t	80.	1,240	1,520	Ward's Island
u	30.	180.	1,020	5.2	u	80.	600	1,315	Yonker's

be recycled. Our latest solution is to send this waste to sea, incinerate it, and dispose of the residual ash in the ocean. The gases will be discharged to the air to form acid rain. The ash covering the ocean will be lost for thousands of years. This practice overlooks the need to reunite the two nutrient streams to grow food. Unfortunately, nature also tends to move these ten nutrients towards the ocean to rest on its floor.

Nature's method of lifting old ocean floors up to mountain heights is slow. The erosion and depletion of nutrients exceeds the amount of nutrients stored on old slopes. The result is acidification of the higher elevations. If the environment is to be saved from creeping acidity, man must intercept the flow of the ten nutrients and recycle them to the higher, acidic areas. Nature's method is too slow.

Table V. Toxic Organics in Sewage Sludges
New Jersey – New York Metropolitan Area
Concentrations on a Dry Solids Basis

Compound	mg/kg
Aldrin	0.35
Dieldrin	0.02
Endrin	0.30
Heptachlor	0.15
Heptachlor Epoxide	0.03
DDT	0.20
pp-DDE	0.05
pp-TDE	0.05
Methychlor	0.24
Chlorane	2.50
Lindane	0.20
Mirex	5.6
Polychlorinated Biphenyls (PCBs)	2.2

Table VI. Ratio of Calcium to Sodium Uptake

Ca/Na	By Barley Roots		By Pea Roots	
of Clay	Kaolin	Bentonite	Kaolin	Bentonite
9/1	0.72	0.45	8.38	5.66*
1/1	0.46	0.37	–	3.23
1/9	0.34	0.24	–	1.7

*6.70 for 50–50 mix of Kaolin to Bentonite.
CEC of Roots Pea   71. (8 days old)
              Barley 22.7 (3 weeks old)
The greater the acidoid content of the roots, the higher is the Ca/Na ratio of the cations taken up by the roots.
Total takeup of Na and of Ca is greater from Bentonite than from Kaolin.

In addition to the acidity, a common complaint is the damage to man's treasured structures. His favorite building materials seem to be those that nature requires for nutrients. Limestone, marble, concrete, cement, mortar iron are prime examples. The choice between food and life of a structure is easy. Nature's needs must come first. In case of conflict, man must choose materials resistant to the elements, shelter them, protect them with an acid rain resistant coating or protect them from the weather. Some of the early structures were protected by building at the edge of the desert.

Table VII. Elements Found in Plants

Abundant Essential	Abundant Beneficial
N Nitrogen	Na Sodium
P Phosphorus	Cl Chlorine
K Potassium	MICRONUTRIENTS
Ca Calcium	Fe Iron
Mg Magnesium	Mn Manganese
S Sulfur	Cu Copper
O Oxygen	Zn Zinc
C Carbon	B Boron
H Hydrogen	Mo Molybdenum

Micro Quantities (Non-essential)

Li Lithium	I Iodine
Be Beryllium	Cs Cesium
F Fluorine	Ba Barium
Al Aluminum	La Lanthanum
Si Silicon	Pr Praseodymium
Ti Titanium	Nd Neodymium
V Vanadium	Sm Samarium
Cr Chromium	Eu Europium
Co Cobalt	Gd Gadolinium
Ni Nickel	Tb Terbium
Ga Gallium	Dy Dysprosium
As Arsenic	Yb Ytterbium
Ge Germanium	Er Erbium
Se Selenium	Au Gold
Rb Rubidium	Hg Mercury
Sr Strontium	Tl Thallium
Y Yttrium	Pb Lead
Ag Silver	Ra Radium
Sn Tin	Th Thorium
U Uranium	

SUMMARY

The planet earth is a spaceship. Recycling and reuse of nutrients is a necessity.

Acid rain is an important part of the recycling of the nutrients that form gasses in their cycling. We must learn to live with it and to use it efficiently.

To prevent acid buildup, the acids must be converted to plant growth (amino acids and proteins).

Ten essential nutrients (calcium, potassium, magnesium, phosphorus, iron, manganese, zinc, copper, boron and molybdenum) are required to catalyze and to promote the reactions.

Man must make adjustments to insure that an adequate supply of the ten nutrients is available to meet the needs of acid rain.

Nature does not supply enough of these nutrients on higher elevations to meet the needs of plants and to provide for the leaching and erosion.

Use of ash from the incineration of dead decaying waste matter from plant growth offers the best possible source for the solid nutrients.

REFERENCES

1. Frear, D. E. H., *Agricultural Chemistry I Principles*, Vol. 1, D. Van Nostrand, pg. 102.
2. Kiester, E., Jr., "Death in the Black Forest," *Smithsonian*, 16, 211 (Nov. 1985).
3. Curry, N. A., "Problems in Toxic Element Disposal on Land," *Proc. 11th Mid-Atlantic Ind. Waste Conf.*, Penn State (1979).
4. Thompson, L. M., *Soils and Soil Chemistry*, 2nd Edition, McGraw-Hill (1957).

90 THE EFFECTS OF CARBONATE ON CALCIUM FLUORIDE PRECIPITATION

Joseph G. Rabosky, Private Consultant
Pittsburgh, Pennsylvania 15202

James P. Miller, Jr., Professor Emeritus
University of Pittsburgh
Pittsburgh, Pennsylvania 15202

INTRODUCTION

The adverse health impacts of fluorides in our environment are primarily linked to excessive concentrations in drinking water supplies. However, in order to provide a viable source of potable water, contamination must be prevented. In the United States the procedure chosen to safeguard water supplies is effluent control, i.e., contamination flowing into natural waters is regulated by means of standards which impose maximum permissible concentrations. Regulation of these discharges requires treatment to reduce contaminants to within maximum allowable limits; consequently, wastewater treatment is widely practiced in the United States. By providing proper wastewater treatment, our drinking water supplies will be protected.

Past research describes several defluoridation techniques in which the sole objectives were to reduce fluoride concentrations to levels which either approached solubility or exhibited few adverse health effects. The reporting authors had varying degrees of success with their respective methods [1,2,3,4,5,6]; however, little understanding was gained as to why some research was more successful than others when identical removal processes were employed.

The primary objective of this study was to determine the effects of pH and carbonate on defluoridation using calcium salts. An ion that competes with fluoride for calcium, namely carbonate, was introduced into the calcium fluoride equilibrium system to study its effect on precipitation at various pH levels. Because calcium carbonate is slightly less soluble than calcium fluoride, carbonate effectively competes for available calcium. When the solubility product of calcium carbonate is exceeded, calcium is removed from solution in the carbonate form. Establishment of this competitive scheme in a pH range of 5 to 11 permitted the effects of the carbonate equilibrium on calcium defluoridation to be demonstrated.

The results of the study revealed several significant observations. The first suggests that pH influences defluoridation efficiency by affecting species form. Essentially, if ions are in the proper species form to combine with calcium, the resulting competition between these species and fluoride for available calcium reduces defluoridation efficiency. Secondly, particle size analysis indicated that a pore diameter of 0.45 microns may be too large to filter calcium fluoride whose particle size ranged from 0.25–0.75 microns for samples collected during this study. Lastly, with the exception of only dilute solutions, incorporation of activity coefficients into solubility calculations are necessary to accurately predict theoretical fluoride concentrations. Wastewaters may not fall into the category of dilute solutions; therefore, consideration must be given to ionic strength and inclusion of activity coefficients for calculation of theoretical predictions.

PRELIMINARY DISTILLATION

During the initial stages of the study, the effect of interferences on fluoride determinations was checked. Samples were collected and divided for comparison purposes. Analytical results indicated only very slight differences between distilled and undistilled samples; therefore, the study was continued without distillation. All reported results were based on this premise.

ANALYTICAL TECHNIQUES

Throughout this study, deionized and distilled water were used, and all solid reagents were dried for a minimum 12 hour period at temperatures ranging between 101° and 108°C. Solution temperatures of 26 ± 1°C were maintained in a constant temperature water bath.

All fluoride determinations were made using a specific ion electrode as described in *Standard Methods*. TISAB II was employed as the buffering agent for these analyses.

Calcium concentrations were determined by atomic absorption spectrophotometry using the same instrument and analyst.

However, for the carbonate and bicarbonate determinations, wet chemistry was employed, i.e., the concentrations of both carbonate and bicarbonate in individual samples were determined by titrating for alkalinity and mathematically converting to carbonate or bicarbonate.

CALCIUM OPTIMIZATION STUDY

Initially, determination of the optimal calcium concentration was necessitated. Tests were conducted in the presence of fluoride which was added as the sodium salt. Nominal concentrations of 200 mg/L fluoride were selected to insure that initial concentrations exceeded solubility limits. No carbonate was added, and the pH was varied between 5 and 11. The basic chemical equation representing this portion of the testing is:

$$Ca^{++} + 2F^- \rightleftarrows CaF_{2(S)} \tag{1}$$

The data for this phase of the study is presented in Figure 1 and Tables I, II, III, IV, and V and is summarized in Table VI.

Calcium concentrations during this phase approximated five different levels as follows:

1. 311 ± 26 mg/L   4. 647 ± 34 mg/L
2. 399 ± 10 mg/L   5. 979 ± 30 mg/L
3. 510 ± 7 mg/L

Table I illustrates fluoride data for calcium concentrations approximating 310 mg/L. As observed, fluoride residuals vary from 3-6 mg/L after calcium precipitation with defluoridation efficiencies in the 97-98% range. In Table II, which presents results for calcium concentrations near 400 mg/L, fluorides range from 2.5-4.0 mg/L after treatment with fluoride removal percentages approximating 98-99%. For calcium ion concentrations approaching 510 mg/L, the data presented in Table III show that fluorides seldom exceed 3.0 mg/L, and only three samples indicate fluoride removals below 99%. The data in Table IV, which is illustrative of the 650 mg/L calcium test run, are essentially the same as the test results obtained using 510 mg/L calcium. At approximately 980 mg/L calcium ion concentration (see Table V), all defluoridation efficiencies are above 99%, and several samples indicate fluoride residuals below 2.0 mg/L.

The laboratory results presented in Tables I through V, and which are projected in Figure 1, illustrate that very little incremental benefit is achieved using calcium concentrations greater than 400 mg/L. Only a 0.5% removal difference is detected between the 400 mg/L calcium and 980 mg/L calcium additions; however, defluoridation efficiencies are noticeably different between the 311 mg/L concentration and the higher levels of calcium addition. The regression equations for the straight lines represented in Figure 1 are presented in Table VI.

Generally, the regression equations predict relatively flat curves for defluoridation in a pH range of 5-11, thus, pH does not significantly influence fluoride removal in this pH range if only calcium fluoride solubilities are considered. This study, therefore, confirms that defluoridation using calcium salts varies very little with pH but is primarily governed by the solubility product relationship.

In natural water systems and in wastewaters, other ions are present which may affect the calcium fluoride solubility. The following discussion introduces the carbonate anion which competes with fluoride for calcium and which exhibits both soluble and insoluble forms depending upon pH. Carbonate is common to almost all water systems.

INTRODUCTION OF CARBONATE

During this phase, different concentrations of the $CO_3^=$ anion were introduced into 200 mg/L fluoride solutions which contained approximately 400 mg/L of calcium. Nominal $CO_3^=$ concentrations utilized were 50, 125, 250, and 395 mg/L. Chemical equations expressing these equilibria are:

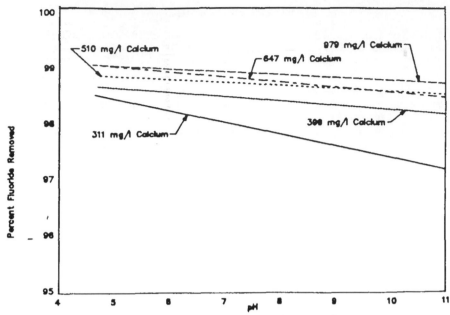

Figure 1.   Percent fluoride removed versus pH.

Table I. Results Obtained Using 311 ± 26 mg/L Ca⁺⁺

Sample Number	pH	mg/L F⁻ Initial	mg/L F⁻ Final	mg/L F⁻ Removed	% F⁻ Removed	mg/L Ca⁺⁺ Initial	mg/L Ca⁺⁺ Final
1	4.35	205	3.65	201	98	286	70
2	4.90	200	2.94	197	99	337	115
3	4.95	225	3.93	221	98	331	90
4	5.35	178	3.05	175	98	293	100
5	5.52	224	3.70	220	98	334	125
6	5.91	210	3.85	206	98	320	110
7	6.58	202	3.35	199	99	330	145
8	6.59	212	3.50	208	98	325	120
9	6.78	164	3.05	161	98	333	195
10	6.88	195	3.45	192	98	337	165
11	7.04	231	5.65	225	97	327	105
12	7.64	208	4.45	204	98	324	140
13	8.05	209	5.34	204	98	299	70
14	8.40	219	6.18	213	97	289	55
15	8.75	229	4.22	225	98	320	105
16	8.85	196	4.90	191	97	302	95
17	8.94	202	3.74	198	98	298	120
18	9.10	194	5.05	189	97	292	70
19	9.62	224	5.55	218	97	314	85
20	9.74	202	4.70	197	98	325	130
21	10.30	205	5.58	199	97	301	70
22	10.40	201	6.18	195	97	293	70
23	11.00	198	4.65	193	97	322	125
24	11.05	203	5.70	197	97	327	120

Table II. Results Obtained Using 399 ± 10 mg/L Ca⁺⁺

Sample Number	pH	mg/L F⁻ Initial	mg/L F⁻ Final	mg/L F⁻ Removed	% F⁻ Removed	mg/L Ca⁺⁺ Initial	mg/L Ca⁺⁺ Final
1	4.70	209	2.55	206	99	398	160
2	5.15	208	2.70	205	99	400	175
3	5.40	196	2.43	194	99	403	180
4	5.55	216	3.18	213	99	406	185
5	5.70	192	2.58	189	98	406	185
6	6.15	185	2.52	182	98	407	190
7	6.60	218	3.08	215	99	408	175
8	6.65	212	2.91	209	99	400	170
9	6.75	203	2.86	200	99	397	170
10	6.75	210	2.70	207	99	406	175
11	6.90	198	2.91	195	98	401	195
12	7.20	199	2.79	196	98	390	170
13	7.40	211	2.90	208	99	395	165
14	7.60	207	2.95	204	99	401	165
15	7.70	200	3.18	197	99	407	195
16	7.85	210	2.96	207	99	396	155
17	7.90	205	2.80	202	99	398	165
18	8.20	207	2.91	204	99	401	175
19	8.80	226	3.48	223	99	393	140
20	8.80	206	3.00	203	99	391	160
21	9.10	201	3.15	198	99	398	170
22	9.20	202	3.35	199	99	406	175
23	9.65	206	3.97	202	98	395	175
24	9.85	211	2.80	208	99	400	165
25	9.90	228	3.48	225	99	389	135
26	10.05	209	3.50	205	98	404	165
27	10.35	201	3.45	198	99	390	170
28	10.50	198	3.25	195	98	388	160
29	10.60	187	3.42	184	98	406	185
30	11.00	213	3.85	209	98	394	155

$$Ca^{++} + 2F^- \leftrightarrows CaF_{2(S)} \text{ and } Ca^{++} + CO_3^= \rightleftarrows CaCO_{3(S)} \qquad (2)$$

The data for this phase of the experimentation is presented in Tables VII, VIII, IX, and X, Figure 2, and is summarized in Table XI. In Table VII, which illustrates results for normal carbonate concentrations of 50 mg/L, fluoride residuals increased significantly above a pH of 7.3. Carbonate ion was detected at pH values greater than 8.65, the minimum pH where fluoride removal efficiencies decreased. An increase in initial carbonate concentrations to 125 mg/L, as depicted by the data in Table VIII, resulted in a significant residual fluoride increase for pH values greater than 7.5-8.0. Defluoridation efficiency decreased at these minimum pH values also. Carbonate was detected at the lowest pH value of 9.55 which is two standard pH units higher than the pH where fluoride residuals consistently increased. Identical observations of higher fluoride residuals and minimum pH values for carbonate detection occurred at a pH of 8.25 and 8.80, respectively, for 250 mg/L carbonate (see Table IX).

Percent fluoride removals significantly decreased above pH values of 8.25 as is also indicated in Table IX. Table X, which presents data for the 395 mg/L carbonate testing, indicates reduced defluoridation efficiency, initial residual carbonate, and increased fluoride residuals at pH values ranging from 7.95-8.30. Thus, fluoride residuals consistently increased at pH values approximating 8.0 resulting in consequential reductions in the efficiency of fluoride removal. Residual carbonate ion appeared at pH values higher than those minimum pH readings where fluoride residuals increased.

Residual calcium concentrations decreased with increasing pH indicating that calcium combined with both carbonate and fluoride to precipitate calcium carbonate and calcium fluoride in the higher pH range where the carbonate species predominated. Therefore, low calcium residuals in high pH regions reflect the combination of calcium with carbonate, and the soluble nature of bicarbonate is suggested by high calcium residuals in intermediate pH ranges. Fluoride removal percentages indicate

Table III. Results Obtained Using 510 ± 7 mg/L Ca^{++}

Sample Number	pH	mg/L F$^-$ Initial	mg/L F$^-$ Final	mg/L F$^-$ Removed	% F$^-$ Removed	mg/L Ca^{++} Initial	mg/L Ca^{++} Final
1	4.85	196	2.22	194	99	517	300
2	5.15	204	2.25	202	99	505	280
3	5.40	215	2.20	213	99	514	280
4	5.75	218	2.35	216	99	513	260
5	5.80	206	2.88	203	99	507	280
6	6.10	205	2.25	203	99	508	280
7	6.20	221	2.25	219	99	511	270
8	6.30	195	2.18	193	99	509	290
9	6.45	194	2.34	192	99	513	290
10	6.60	207	2.88	204	99	505	280
11	7.15	200	2.28	198	99	517	290
12	7.20	200	2.21	198	99	513	280
13	7.20	195	2.42	193	99	508	290
14	7.25	201	2.42	199	99	509	280
15	7.60	203	2.75	200	99	507	280
16	7.65	204	2.40	202	99	513	280
17	7.85	200	2.52	197	99	507	290
18	8.30	207	2.36	205	99	516	280
19	8.65	209	2.63	206	99	511	270
20	8.75	217	2.59	214	99	505	260
21	8.90	193	2.43	191	99	516	290
22	9.35	220	2.58	217	99	516	270
23	9.50	204	3.17	201	99	513	270
24	9.80	205	2.88	202	99	503	270
25	9.80	217	2.73	214	99	512	280
26	10.15	196	2.98	193	98	508	290
27	10.30	185	2.55	182	98	516	290
28	10.65	212	2.90	209	99	512	270
29	10.90	223	3.42	220	99	511	250
30	11.05	212	3.56	208	98	506	280

that free carbonate successfully competed with fluoride for precipitation in a pH range of 7.5-8.0 to approximately 11.0, the high end of the study.

As the data show, the presence of carbonate began to compete with fluoride for available calcium at a pH near 7.5. As both carbonate concentration and pH increased, defluoridation efficiency generally decreased. The exception to this trend was realized in the lower carbonate series of 54 and 127 mg/L. These two runs showed relatively constant competitive levels of removal with no discernible difference in defluoridation efficiency between these two concentration levels.

These data indicate that defluoridation does not vary directly with pH as the literature has suggested but is a function of the formation of the carbonate species from bicarbonate which is pH dependent. The inverse relationship between fluoride removal and pH is caused by the presence of carbonate and the competition it exerts for the available calcium cation in solution to precipitate as calcium carbonate. This reduces the availability of calcium to combine with dissolved fluoride to form the calcium fluoride precipitate thereby reducing defluoridation efficiency. Also, the reduction of solution calcium concentration by precipitation with carbonate causes a dissolution of the calcium fluoride precipitate to maintain equilibrium conditions. Therefore, residual fluoride concentrations increase because all equilibrium solubility relationships must be satisfied.

In summation, the effects of introducing a competitive ion into the calcium fluoride equilibrium system are observable. Laboratory data indicate that defluoridation efficiency in the high pH range is inversely proportional to carbonate concentration. Although the data suggest this conclusion, a confirmation of laboratory results is suggested. The following section involving solids analysis serves this purpose.

Table IV. Results Obtained Using 647 $\pm$ 34 mg/L Ca^{++}

Sample Number	pH	mg/L F⁻ Initial	mg/L F⁻ Final	mg/L F⁻ Removed	% F⁻ Removed	mg/L Ca^{++} Initial	mg/L Ca^{++} Final
1	4.70	229	2.15	227	99	613	350
2	4.70	227	2.00	225	99	613	350
3	4.80	225	2.09	223	99	649	350
4	4.95	208	2.03	206	99	657	410
5	5.05	213	1.96	211	99	650	400
6	5.20	201	2.03	199	99	650	450
7	5.25	206	2.24	204	99	647	430
8	5.45	212	2.32	210	99	648	380
9	5.61	216	2.05	214	99	646	420
10	5.85	215	2.70	212	99	633	370
11	6.26	203	2.33	201	99	646	460
12	6.35	214	2.18	212	99	666	460
13	6.40	226	2.18	224	99	671	470
14	6.58	218	3.00	215	99	681	500
15	6.60	211	1.87	209	99	651	400
16	6.69	195	2.37	193	99	673	500
17	6.74	240	2.55	237	99	659	430
18	7.14	230	2.65	227	99	642	450
19	7.23	198	2.80	195	98	644	470
20	7.35	217	2.23	215	99	665	380
21	7.70	207	2.45	205	99	645	370
22	7.82	203	2.75	200	99	659	460
23	8.12	195	2.70	192	98	674	500
24	8.55	202	2.35	200	99	639	390
25	8.85	228	2.67	225	99	634	350
26	8.90	219	2.81	216	99	651	400
27	8.90	206	3.08	203	99	663	400
28	9.25	212	2.30	210	99	665	400
29	9.25	218	2.50	215	99	642	410
30	9.35	207	2.55	204	99	632	370
31	9.35	201	2.70	198	99	632	370
32	9.90	212	2.95	209	99	658	390
33	10.00	199	2.15	197	99	644	420
34	10.20	197	3.14	194	98	634	390
35	10.40	193	3.27	190	98	643	390
36	10.55	220	3.60	216	98	653	390
37	11.05	205	2.83	202	99	660	440
38	11.10	207	2.95	204	99	639	400
39	11.10	210	3.37	207	99	651	430
40	11.10	188	2.80	185	98	631	430

Table V. Results Obtained Using 979 $\pm$ 30 mg/L Ca^{++}

Sample Number	pH	mg/L F⁻ Initial	mg/L F⁻ Final	mg/L F⁻ Removed	% F⁻ Removed	mg/L Ca^{++} Initial	mg/L Ca^{++} Final
1	4.60	204	1.84	202	99	962	700
2	4.70	236	1.86	234	99	953	700
3	4.80	210	1.88	208	99	977	650
4	4.90	215	1.81	213	99	982	700
5	5.40	219	1.75	217	99	990	750
6	5.55	212	2.05	210	99	993	750
7	5.69	199	1.82	197	99	982	750
8	5.71	209	1.83	207	99	976	750
9	5.80	205	1.87	203	99	988	750
10	6.03	217	1.90	215	99	990	850
11	6.12	191	1.87	189	99	991	850
12	6.20	220	1.75	218	99	991	700
13	6.48	211	1.92	209	99	956	750
14	6.53	206	2.15	204	99	949	800
15	6.75	221	2.70	218	99	977	700
16	6.80	221	2.78	218	99	990	700
17	6.82	214	2.70	211	99	962	750
18	7.28	214	2.62	211	99	959	750
19	8.10	212	2.50	209	99	969	700
20	8.25	195	2.38	193	99	974	600
21	8.52	236	2.36	234	99	1004	850
22	8.55	207	2.45	205	99	985	650
23	8.65	254	2.52	251	99	994	800
24	8.66	216	2.15	214	99	972	750
25	8.75	223	2.09	221	99	996	700
26	8.90	216	2.61	213	99	987	700
27	8.90	226	2.95	223	99	1009	750
28	8.95	204	2.63	201	99	999	750
29	9.60	208	2.80	205	99	974	650
30	10.00	212	2.65	209	99	996	700
31	10.40	205	2.16	203	99	987	750
32	10.47	204	2.82	201	99	966	800
33	10.50	213	2.47	211	99	987	750
34	10.51	203	2.57	200	99	978	800
35	10.80	209	2.10	207	99	984	750
36	10.85	217	2.27	215	99	990	750
37	11.00	221	2.42	219	99	986	750
38	11.10	200	2.50	197	99	983	750

Table VI. Summary of Calcium Variation Study

Item	Calcium Concentration, mg/L				
	311 ± 26	399 ± 10	510 ± 7	647 ± 34	979 ± 30
Fluoride Concentration, Initial mg/L					
Low	163	185	185	188	191
High	231	228	223	240	254
Fluoride Concentration, Final mg/L					
Low	2.94	2.43	2.18	1.87	1.75
High	6.18	3.97	3.56	3.60	2.95
Fluoride Removed, mg/L					
Low	161	182	182	185	189
High	225	225	220	237	251
Calcium Remaining, mg/L					
Low	55	135	250	350	600
High	195	195	300	500	850
mg/L F^- Removed =	$-0.47pH + 205$	$+0.38pH + 200$	$+0.47pH + 199$	$-2.31pH + 226$	$-0.24pH + 213$

Table VII. Results Obtained with 54 ± 14 mg/L $CO_3^=$, 400 ± 8 mg/L Calcium

Sample Number	pH	mg/L F^- Initial	mg/L F^- Final	mg/L F^- Removed	%mF^- Removed	mg/L $CO_3^=$ Initial	mg/L $CO_3^=$ Final
1	4.90	206	2.75	203	99	59	0
2	5.05	196	2.66	193	98	42	0
3	5.20	210	2.95	207	99	52	0
4	5.30	203	2.50	200	99	51	0
5	5.75	212	2.70	209	99	51	0
6	5.90	214	2.90	211	99	57	0
7	6.10	221	2.85	218	99	50	0
8	6.65	220	3.15	217	99	54	0
9	6.75	220	2.90	217	99	51	0
10	6.90	218	3.15	215	99	45	0
11	7.05	203	2.90	200	99	49	0
12	7.10	221	2.95	218	99	47	0
13	7.30	205	5.95	199	97	42	0
14	7.45	223	5.75	217	97	47	0
15	7.95	210	4.95	205	98	44	0
16	8.10	217	5.60	211	97	49	0
17	8.30	197	6.25	191	97	40	0
18	8.45	197	5.35	192	97	46	0
19	8.65	195	7.60	187	96	55	2
20	8.80	198	8.00	190	96	50	18
21	9.15	200	7.50	192	96	51	8
22	9.55	208	6.35	202	97	63	0
23	9.65	198	6.40	192	97	68	8
24	9.90	196	7.30	189	96	52	22
25	10.20	200	7.95	192	96	52	8
26	10.20	214	6.50	207	97	47	7
27	10.45	205	5.50	199	97	46	4
28	10.50	218	6.70	211	97	57	1
29	10.85	215	6.30	209	97	46	0
30	11.05	218	6.25	212	97	47	4

Table VIII. Results Obtained with 127 ± 8 mg/L $CO_3^=$, 396 ± 14 mg/L Calcium

Sample Number	pH	mg/L F⁻ Initial	mg/L F⁻ Final	mg/L F⁻ Removed	%mF⁻ Removed	mg/L $CO_3^=$ Initial	mg/L $CO_3^=$ Final
1	4.75	196	2.40	194	99	119	0
2	4.80	203	3.75	199	98	133	0
3	4.80	210	2.58	207	99	122	0
4	5.40	204	2.40	202	99	127	0
5	5.80	207	2.55	204	99	130	0
6	5.85	202	2.60	199	99	130	0
7	6.45	201	2.55	198	99	125	0
8	6.50	211	2.79	208	99	126	0
9	6.50	214	2.64	211	99	127	0
10	6.70	206	2.95	203	99	130	0
11	6.80	210	2.86	207	99	135	0
12	7.15	218	3.00	215	99	124	0
13	7.30	209	2.95	206	99	130	0
14	7.40	202	2.80	199	99	135	0
15	7.55	215	3.65	211	98	126	0
16	8.25	206	5.00	201	98	132	0
17	8.40	197	3.65	194	98	121	0
18	8.70	189	5.85	183	97	134	0
19	8.70	204	5.90	198	97	129	0
20	9.50	209	6.70	202	97	135	0
21	9.55	204	6.65	197	97	121	4
22	9.75	198	6.95	191	96	128	5
23	10.00	209	8.30	201	96	135	8
24	10.25	202	6.65	195	97	132	6
25	10.35	204	7.30	197	97	130	7
26	10.35	192	6.10	186	97	128	8
27	10.75	210	6.55	203	97	126	14
28	10.85	206	6.95	199	97	131	8
29	11.00	215	7.60	207	96	130	18

SOLIDS ANALYSIS

The solubilities of calcium fluoride and calcium carbonate are very similar. At 26°C, the solubility product of calcium fluoride is 3.95×10^{-11}, and the solubility product is 0.87×10^{-8} for calcium carbonate at 25°C [7]. These values convert to solubilities of 2.15×10^{-4} moles per liter and 0.93×10^{-4} moles per liter for calcium fluoride and calcium carbonate, respectively. In essence, the solubility of calcium fluoride is just slightly more than double that of calcium carbonate.

With a solubility ratio slightly greater than 2, measuring residual concentrations is inadequate if system conditions result in very low residuals. Sensitivity to operator and experimental error significantly influences results for these types of experimental conditions. Consequently, analysis of the solid precipitates is considered essential to thoroughly analyze the calcium fluoride-calcium carbonate equilibria system.

A summary of analytical results is presented in Table XII and illustrated in Figures 3, 4, 5, and 6. Table XII illustrates pertinent data from both SEM studies and X-ray diffraction analysis. Figures 3 through 6 illustrate typical scanning electron microscope (SEM) micrographs for each sample. In these micrographs the calcium fluoride is identified by the cubic crystals while the calcium carbonate crystals are rhombohedral.

The micrographs illustrate some interesting results. Where pH was controlled at approximately 5, generally no rhombohedra exist indicating the absence of calcium carbonate. This is understandable because bicarbonate is the primary species at this pH. The general absence of calcium carbonate is verified by the results of X-ray diffraction. Only the 395 mg/L nominal carbonate sample shows detectable calcium carbonate with one percent having been reported. This is well within the operator error of ± 10%. Also, sufficient calcium and carbonate were added to this sample to exceed the

Table IX. Results Obtained with 251 ± 14 mg/L CO$_3$$^=$, 398 ± 11 mg/L Calcium

Sample Number	pH	mg/L F$^-$ Initial	mg/L F$^-$ Final	mg/L F$^-$ Removed	% F$^-$ Removed	mg/L CO$_3$$^=$ Initial	mg/L CO$_3$$^=$ Final
1	4.70	204	3.30	201	99	238	0
2	4.75	209	2.95	206	99	252	0
3	5.15	202	2.63	199	99	245	0
4	5.25	192	2.55	189	98	245	0
5	5.40	207	3.10	204	99	251	0
6	5.80	201	2.85	198	99	251	0
7	5.85	199	2.87	196	98	246	0
8	6.00	215	2.87	212	99	249	0
9	6.40	202	4.75	197	98	252	0
10	7.00	209	3.98	205	98	258	0
11	7.15	229	3.80	225	98	254	0
12	7.25	210	3.55	206	98	238	0
13	7.25	196	4.70	191	97	249	0
14	7.30	203	2.75	200	99	251	0
15	7.30	214	3.90	210	98	258	0
16	7.65	206	3.38	203	99	251	0
17	7.80	207	3.60	203	98	241	0
18	8.05	212	4.15	208	98	252	0
19	8.15	201	4.80	196	98	260	0
20	8.15	212	4.60	207	98	253	0
21	8.25	203	6.50	196	97	244	0
22	8.40	204	13.9	190	93	250	0
23	8.40	196	8.40	188	96	265	0
24	8.80	202	15.5	186	92	255	1
25	8.85	195	10.8	184	94	238	0
26	9.40	223	17.8	205	92	251	5
27	9.50	205	14.0	191	93	255	2
28	9.65	210	18.5	191	91	263	6
29	9.90	207	19.3	188	91	264	6
30	10.25	199	13.4	186	93	252	10
31	10.50	212	16.5	195	92	246	18
32	10.60	205	19.0	186	91	257	27
33	11.00	205	20.4	185	90	256	70
34	11.05	221	21.3	200	90	250	40

calcium carbonate solubility product by 1000 times even assuming that sufficient calcium combined to completely precipitate all fluorides. A possible explanation is that perhaps a slight amount of calcium carbonate formed which did not dissolve in the acidic conditions and remained suspended in the sample of solution which was filtered for analysis.

Samples where pH was controlled between 6.8 and 7.0 indicate the presence of calcium carbonate because one or two larger rhombohedra were detected. The results from X-ray diffraction verify the presence of calcium carbonate with 5%. This larger percentage probably arose because this pH range is very close to the transition zone where the carbonate ion began to compete for calcium to form the calcium carbonate precipitate.

Samples collected at a pH approximating 9.5 indicate that substantial quantities of both calcium fluoride and calcium carbonate are present. Calcium carbonate rhombohedra and fluorite, calcium fluoride, are readily distinguishable in the micrographs at both magnifications. X-ray analysis verifies both are present in substantial quantities.

Samples approximating a pH of 11 clearly show the presence of both crystalline forms which are expected in this pH range, and X-ray analysis verifies that both calcium fluoride and calcium carbonate are present in substantial percentages.

The particle size in all samples is important to note. The small cubes of fluorite, calcium fluoride, are consistently smaller than one micron when compared to the bar graph at the bottom of each

Table X. Results Obtained with 395 $\pm$ 18 mg/L $CO_3^=$, 400 $\pm$ 10 mg/L Calcium

Sample number	pH	mg/L F⁻ Initial	mg/L F⁻ Final	mg/L F⁻ Removed	% F⁻ Removed	mg/L $CO_3^=$ initial	mg/L $CO_3^=$ Final
1	4.65	220	3.15	217	99	398	0
2	4.85	215	3.05	212	99	395	0
3	5.35	210	3.50	206	98	412	0
4	5.50	207	3.85	203	98	397	0
5	6.05	200	3.15	197	98	403	0
6	6.35	213	3.45	210	98	388	0
7	6.45	195	3.10	192	98	401	0
8	6.50	214	3.25	211	99	400	0
9	6.60	209	3.70	205	98	401	0
10	6.80	213	3.50	209	98	397	0
11	7.50	200	5.70	194	97	397	0
12	7.70	213	4.40	209	98	398	0
13	7.80	206	3.55	202	98	397	0
14	7.95	205	6.70	198	97	396	0
15	8.25	220	10.2	210	95	408	0
16	8.25	191	14.8	176	92	392	0
17	8.30	205	13.3	192	94	399	2
18	8.40	211	19.2	192	91	396	2
19	8.75	210	25.8	184	88	389	8
20	8.95	215	30.5	184	86	390	9
21	9.03	212	47.0	165	78	393	8
22	9.20	213	38.0	175	82	377	13
23	9.45	205	43.0	162	79	378	14
24	9.90	197	34.0	163	83	394	32
25	10.10	206	39.2	167	81	402	38
26	10.10	204	55.0	149	73	398	40
27	10.55	211	40.0	171	81	391	64
28	10.60	197	48.5	148	75	408	66
29	10.85	199	49.5	149	75	401	74
30	11.05	218	44.5	173	79	402	117

micrograph. In fact, the size of the cubes appears to be in the 0.25–0.75 micron range. The calcium carbonate rhombohedra approximately approach 1.0–1.25 microns. Past research has suggested that samples should be filtered through a 0.45 micron filter; however, particle size analysis suggests that a smaller size filter membrane, e.g., 0.1 micron, is necessary to capture the calcium fluoride particles.

In summary, solids analyses of the precipitates resulting from calcium fluoride – calcium carbonate equilibrium studies reveal the following: 1) Confirmed is the transformation of bicarbonate to carbonate at a pH between 7 and 9.5 with a consequential change in precipitate constituents from calcium fluoride only to a combination of calcium fluoride and calcium carbonate; and 2) The crystalline particle size of the precipitate constituents consistently range from 0.25 to 0.75 microns for calcium fluoride and from approximately 1.0 to 1.25 microns for calcium carbonate.

The above revelations should be considered prior to conducting filtration studies involving calcium fluoride precipitates because they are sufficiently important to affect final results.

INCORPORATION OF ACTIVITY COEFFICIENTS INTO SOLUBILITY CALCULATIONS

In a number of technical publications, the theoretical solubility of fluoride has been suggested as the target for residual fluoride concentrations. Most authors indicate that theoretical concentrations are difficult, if not impossible, to achieve by calcium precipitation methodology.

The results of this study show that theoretical predictions are extremely difficult to achieve when activity coefficients are not incorporated into the solubility calculations. Inclusion of these coefficients substantially modifies theoretical fluoride residual predictions to provide more acceptable agreement with laboratory observations; exclusion of activity coefficients in fluoride calculations results in theoretical predictions much lower than both laboratory observations and theoretical values

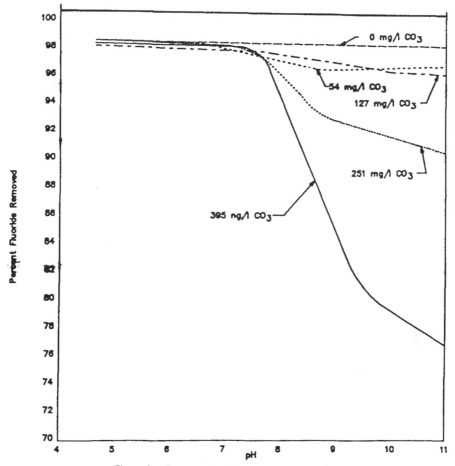

Figure 2.   Percent fluoride removed versus pH.

calculated including activity coefficients. Essentially, this research agrees with literature suggestions in that neglect of activity coefficients in solubility calculations yields unattainable values for residual fluorides.

Therefore, in order to predict more realistic target residual fluoride concentrations, efforts must be directed toward approximating activity coefficients. In very dilute solutions activities usually are ignored; however, in highly contaminated wastewater solutions ignoring activity coefficients may lead to erroneous conclusions. Incorporating activity coefficients into solubility product calculations for wastewaters will project more realistic fluoride residual concentrations.

Figure 7 presents a typical comparison between the electrode or experimental results and the solubility product, Ksp, with and without the consideration of activity coefficients (γ). In this figure approximately 400 mg/L of calcium were used. Data agreement between electrode results (actual) and solubility data with γ are better than the results between experimental and theoretical data without inclusion of activity coefficients.

SUMMARY AND CONCLUSIONS

This research achieved more efficient fluoride removal at lower pH values (5–7) than in the high pH range (8–11). This suggests that defluoridation efficiency is higher in the lower pH range than at high pH values which contradicts other research. The conversion of bicarbonate to carbonate and the

Table XI. Summary of Results of Carbonate Study

	Carbonate Concentration, mg/L			
	54 ± 14	127 ± 8	251 ± 14	395 ± 18
Fluoride Concentration, Initial mg/L:				
Low	195	189	192	191
High	223	218	229	220
Fluoride Concentration, Final mg/L:				
Low	2.50	2.40	2.55	3.05
High	8.00	8.30	21.3	55.0
Fluoride Removed, mg/L:				
Low	187	183	184	148
High	217	215	225	217
Calcium Concentration, Initial mg/L:				
Low	393	382	387	390
High	408	413	408	409
Calcium Concentration, Final mg/L:				
Low	115	60	10	1
High	180	180	320	200
Carbonate Concentration, Initial mg/L:				
Low	40	119	238	377
High	68	135	265	412
Fluoride Removed, %				
Low	96	96	90	73
High	99	99	99	99

subsequent competition between fluoride and carbonate for the calcium ion provides the most logical explanation for the inverse relationship between fluoride removal and pH.

To obtain the most precise fluoride predictions using theoretical solubility product relationships, activity coefficients must be considered. For this study inclusion of activities into calculations provided much better agreement between theoretical and observed data. Without consideration of activity coefficients, theoretical predictions did not closely agree with laboratory observations.

The results indicate that calcium fluoride precipitation is only slightly affected by pH and is primarily a function of calcium fluoride solubility. However, pH affects the presence of ion species which in turn influences defluoridation because other ions in solution compete for available agents necessary for precipitation. In this research carbonate successfully competed with fluoride for available calcium in the pH region of carbonate ion predomination when the solubility product of calcium carbonate was exceeded. As predicted by the solubility product for calcium fluoride, lower calcium concentrations due to precipitation of calcium carbonate result in higher fluoride residuals. Thus, the presence of other ions in solution directly influences defluoridation and the pH where maximum defluoridation occurs.

Lastly, solids analysis revealed that the calcium fluoride crystal size could be less than 0.45 microns—the accepted standard differentiating suspended and dissolved solids; however, the calcite (calcium carbonate) particle approached 1–1.25 microns. As theory suggests, calcium fluoride formed throughout the entire pH range studied; however, calcium carbonate only formed above pH values of 7.5–8.0 which supports the theory of competitive precipitation.

In summation, a more thorough understanding of the calcium fluoride defluoridation process has gained by this research. The calcium fluoride-calcium carbonate system was examined to study the effects of extraneous variables on their simultaneous equilibria. Data generated during this work revealed that misconceptions in prior studies involving calcium fluoride precipitation resulted in generation of erroneous conclusions. The ideas presented in this study attempted to clarify concepts involving calcium defluoridation for the benefit of future research.

Table XII. Solids Analysis

Figure Number	Magnification	pH	Nominal Solution $CO_3^=$ Concentration mg/L	% CaF_2	% $CaCO_3$
3	21,000X	4.80	127	100	<0.01[a]
4	21,000X	6.80	395	100	<0.01[a]
5	21,000X	9.40	251	35	65
6	21,000X	10.85	395	28	72

[a]Not detected.

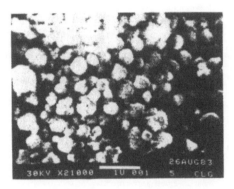

Magnification: 21,000X   pH = 4.8

Figure 3. Scanning electron micrograph of 127 mg/L carbonate sample.

Magnification: 21,000X   pH = 6.8

Figure 4. Scanning electron micrograph of 395 mg/L carbonate sample.

Magnification: 21,000X   pH = 9.4

Figure 5. Scanning electron micrograph of 251 mg/L carbonate sample.

Magnification: 21,000X   pH = 10.85

Figure 6. Scanning electron micrograph of 395 mg/L carbonate sample.

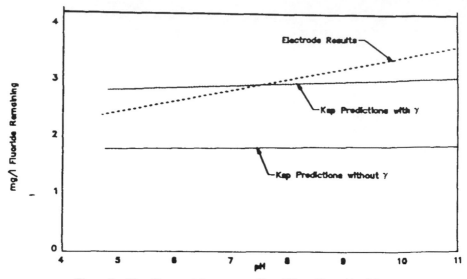

Figure 7.   Fluoride remaining versus pH at 300 ± 10 mg/L calcium.

REFERENCES

1. Zabban, W., and H. Jewett, "The Treatment of Fluoride Wastes," *Proceedings of the 23rd Purdue Industrial Waste Conference*, Lafayette, Lafayette: Purdue University Press, 706–716 (May 2–4, 1967).
2. Rohrer, K., "An Integrated Facility for the Treatment of Lead and Fluoride Wastes," *Industrial Wastes*, Vol. 118, 36–39 (September/October 1971).
3. Anon, "Treating Lead and Fluoride Wastes," *Environmental Science and Technology*, Vol. 6, No. 4, 321–322 (April 1972).
4. Rabosky, J. G., "Fluoride Removal by Lime Precipitation and Alum and Polyelectrolyte Coagulation, *M. S. Thesis*, Department of Civil Engineering, School of Engineering, University of Pittsburgh (1973).
5. Rabosky, J. G., and J. P. Miller, Jr., "Fluoride Removal by Lime Precipitation and Alum and Polyelectrolyte Coagulation," *Proceedings of 29th Purdue Industrial Waste Conference—Part 2*, Lafayette, Lafayette: Purdue University Press, 669–676 (May 7–9, 1974).
6. Parker, C. L., and C. C. Fong, "Fluoride Removal: Technology and Cost Estimates," *Industrial Wastes*, Vol. 122, 23–27 (November/December 1975).
7. *Handbook of Chemistry and Physics*, 63rd ed., Boca Raton: CRC Press, Incorporated, B-242 (1982–1983).

91 ENVIRONMENTAL ENGINEERING INFRASTRUCTURE: PROBLEMS AND NEEDS

Marcia H. Bates, Associate Professor
School of Civil Engineering
Oklahoma State University
Stillwater, Oklahoma 74078

Stephen P. Shelton, Professor
Civil Engineering Department
University of New Mexico
Albuquerque, New Mexico 87131

INTRODUCTION

Infrastructure represents the basic life support systems of society and their relationships to one another, and, as such, it is one of the nation's most important assets. Environmental infrastructure includes a wide variety of public services such as water supply, water and wastewater treatment, air pollution control, and solid and hazardous waste treatment and disposal. The failure of any environmental infrastructure is potentially harmful to a wide segment of society and may have far reaching ramifications. A contributing factor to the failure of many ancient societies was their inability to deal with the byproducts of their growth; for example, often disease occurred because of improper use of the land, improper use of the water, or both. Generally, societies have risen and failed in direct relationship to the intelligent use of their resources.

In the 1980's, the United States is faced with a massive need to plan and construct new infrastructure and to repair and maintain existing structures. The increasing needs of a growing society coupled with a previous "out of sight, out of mind" attitude have created an exorbitant demand for efficient, workable environmental facilities. Engineers of the 1980's and 1990's will truly be challenged by the task of repairing problems posed by today's deteriorating infrastructure.

TECHNICAL ISSUES

The technical infrastructure issues facing environmental engineers encompass several facets of the environment, including water, air, and solid waste. Identifying and characterizing these issues is prerequisite to developing appropriate, effective solutions.

Water supply needs are both quantitative and qualitative with water shortages resulting from depleted and/or contaminated sources. Sixty percent of the United States receives only 25% of the country's annual rainfall. This uneven distribution of precipitation, coupled with a growing population and increased emphasis on enhanced agricultural productivity, has prompted overpumping of aquifers in the west and southwest. The Ogallalah aquifer, which extends from South Dakota through Texas, is being reduced by 14 million acre-feet per year with recharge approximating only 3 million acre-feet annually. Part of the aquifer is already depleted, and experts warn of even greater depletion in the next century. Houston, Texas is sinking because of overpumping, and the rapid use of groundwater in Florida is causing massive sinkholes. The limited availability of water is obviously a problem, and increasing water supplies is the prime objective of the south central and western regions of the United States [1].

In other areas of the country, water supplies are plentiful but contaminated so their use is limited. Trichloroethylene (TCE) has been detected in ground waters from the east coast to the west coast. In Florida, New Jersey, and Louisiana, salt water intrusion is limiting the beneficial use of ground water that was formerly salt-free. Both surface and ground waters across the United States are characterized by the presence of one or more priority pollutants; also, acid rainfall, coupled with the atmospheric deposition of heavy metals, polycyclic aromatic hydrocarbons, and other pollutants, is rendering water supplies potentially dangerous. Protecting water supplies is the major objective of the northeast and midwest regions of the United States [1].

The key to effective use of polluted waters is adequate treatment. The United States is moving into an era where the best water treatment technology will be required to provide potable water. However, in addition to the need for advanced technology, there is also the problem of old, deteriorating facilities. A lack of preventive maintenance and the "out of sight, out of mind" attitude has resulted in the progressive deterioration of many facilities; the magnitude of the decay is just now being realized with the onset of many serious problems. According to Boland [2], some utilities are already confronted with reduced quality and reliability, and many others will face these problems over the next 10–20 years. The achievements of the 19th and 20th century engineers and recent progress in water treatment technology have led most people in the United States to believe that a safe, inexpensive supply of water will always be available. The general public has forgotten that in the 19th century the death rate from water borne diseases was very high. The American public assumes that such events will never occur again. However, even in the 20th century, we are still plagued by water borne disease outbreaks which have affected more than 150,000 people [3]. In 1983, 40 water borne disease outbreaks that afflicted 20,905 people were reported to the Center for Disease Control and EPA. The state of Pennsylvania has more water borne disease outbreaks than any other state in the nation, and inadequate water treatment facilities are to blame.

Several professional and technical engineering societies have assessed U.S. water treatment needs. ASCE, WPCF, and NSPE have all recognized a need to enlarge and/or upgrade 5,832 existing water treatment facilities by the year 2000. In addition, 6,627 new plants are needed.

Water distribution systems, some more than 100 years old, are also one of the nation's major infrastructure needs. Leaks in water distribution systems are causing water shortages in areas where raw water is plentiful. It is estimated that more than 1 million gallons of tap water are lost daily through leaks beneath the streets of Berwyn, Illinois [4]. Another one million gallons are lost daily in New York City. In 1981, there were 547 watermain breaks in New York City and in January of 1982, there were 170 water main breaks in Milwaukee, Wisconsin. Thirty nine percent of New York's watermains and 8% of its valves need replacing during the 1980's, and in Chicago, replacements are needed for 92 miles of watermains. In New Jersey, a watermain break completely shut down the distribution of treated water to 300,000 people.

Many of the nation's wastewater treatment facilities do not produce effluents that meet Federal clean water standards. The discharge of these effluents into receiving streams reduces surface water quality and endangers the water sources of downstream towns. This problem does not exist only in small communities, as one might expect, but in major cities with large discharges such as Houston, Texas.

Wastewater collection systems across the nation are experiencing similar infrastructure problems. Infiltration into old, decaying pipes is overloading already stressed sewage treatment facilities and reducing the efficiency of wastewater treatment. During intense rainfall, infiltration also results in overflowing sewers, which create a health hazard. In Chicago, sewer systems overflow regularly, causing raw sewage to enter homes, lakes, and rivers.

The disposal of sewage sludges, municipal refuse, and hazardous waste by incineration, landfilling, or land spreading can cause significant and potentially harmful air pollution problems. The development of new disposal alternatives or the upgrading of existing facilities can control emissions of heavy metals, hydrocarbons, and other airborne pollutants to the atmosphere. For instance, a more effective approach to emission control may be the improvement of incinerator design and operating conditions. Similarly, upgrading or properly operating existing landfills can reduce the production of gases from the site. Pretreating the waste and installing a gas collection and control system can control air pollution problems at or near hazardous waste land disposal sites.

During the last four decades, municipal and industrial production of solid and hazardous wastes has been increasing substantially, both in the United States and around the world. In 1977, it was estimated [5] that 17–20 million disposal facilities in the U.S. were handling about 7 billion cubic meters of waste each year. It is estimated that more than 200 million tons of solid wastes will be generated in 1985 [6]. Technology to contain and treat these wastes has been lagging. This is due partly to society's failure to recognize the significance of the potential problems associated with these wastes, but is primarily the result of reliance on older technologies. In some cases, these technologies merely transferred the materials from one environmental sector to another; however, in all cases, the "out-of-sight, out-of-mind" approach was apparent.

Landfills, surface impoundments, and underground storage tanks are the major disposal alternatives for solid and hazardous wastes. Conventional landfill techniques, which rely on clay and/or synthetic liners to prevent pollutant transport, are often no longer adequate. The EPA has, in effect, banned land disposal of certain hazardous wastes because of the land's inability to contain them.

Deterioration of the underground storage tanks used to contain wastes and prevent their migration into the environment is also a major infrastructure area of concern. It is estimated that of the 1.4 million underground storage tanks in use around the country, 75,000 to 100,000 are leaking and contributing substantially to groundwater pollution.

DEVELOPMENT OF SOLUTIONS

Four key interrelated components are required to yield an effective strategy for solving the infrastructure problem: prioritization, funding, management, and education.

Societal need, or perceived need, will always exceed available resources. Thus, stewardship of existing resources and prioritizing future requirements is essential for long term success in dealing with important environmental issues. Jasper C. Davis, chairman and chief executive officer of Davis Water and Waste Industries, Inc. in Thomasville, Georgia, addressed the setting of priorities in an 1984 article published in *Water Engineering and Management* [7]. He suggests that, in the case of water systems, rebuilding should be started at the point nearest the end use. The water consumed should be metered to the user so that the facility providing the water can assess the actual demand. If the user is immediately charged for the water, capital will then be available for operation and depreciation. For wastewater treatment systems, the priorities should be reversed and the starting point should be upgrading the wastewater treatment facility. Once the plant has been upgraded and a suitable effluent has been obtained, the collection system can be repaired.

This is but one example of how system priorities could be developed. The key to successful prioritization is taking each need back to the most basic level and noting how it evolves. Then all needs and perceived needs can be compared equally to formulate priorities.

During the past decade, the nation has been committed to funding to improve the environmental infrastructure. However, much of the legislation committed to this funding has already or will soon expire. The costs of meeting environmental infrastructure needs are phenomenally high. In 1981, it was estimated that $31 billion would be needed before 1986 to construct and upgrade sewer systems to meet existing water pollution control standards. Total treatment needs after 1985 have been assessed at $118 billion [8].

The American Association of General Contractors suggests that $138.6 billion will be needed to repair potable water treatment facilities and that wastewater treatment needs will cost $507.8 billion [9]. EPA estimates that by the year 2000, $106 billion will be needed to upgrade existing wastewater collection systems to meet standards.

Various funding options are available. According to Grigg [10], more has been written about the financial aspects of infrastructure than has been written about any other aspect of the problem. So far, the capital outlay has been low, because the public avoids appropriating funds for maintenance, replacement, and construction of infrastructure. The infrastructure problem is national in its scope and will require both a Federal commitment and the assumption of management and funding responsibilities by state and local governments [1]. A decline in spending at all government levels will inevitably lead to infrastructure deterioration.

It is possible that the Federal government could provide necessary help for state and local governments by permitting them to issue tax free bonds and allowing states to generate new tax sources. It is also suggested that manpower development should be a part of capital development, and that private sector groups should assume responsibility for services now provided by other agencies. According to a 1982 article in *Newsweek* [4], the private sector is already assuming a greater share of the infrastructure burden. New home owners in a Danville, California subdivision are providing funds for water tanks, a fire engine, and an elementary school through a $15,000 increase in the price of new homes [4]. The private sector is also providing funds for the restoration of the Statue of Liberty and the cable cars in San Francisco.

The American Society of Civil Engineers has produced a pamphlet outlining recommendations of the 1982 National Water Symposium. Suggested potential sources of public works financing include local debt financing through bonds, state assistance in local debt financing, fee systems, leasing, direct financial assistance, and privatization. A 1985 article published in *Water Engineering and Management* [11] discusses privatization as a potential solution to infrastructure needs. This process, in which a private entity becomes the legal owner of a facility, is one approach for aiding municipal projects. According to Boland [2], the two basic types of privatization are lease-purchase and contract operation. Privatization is attractive to municipalities because of their inability to raise capital. Henry L. Longest II, the EPA's Acting Deputy Assistant Administrator for Water, states that privatization is thought to be advantageous for rapidly growing communities. The company builds, operates, and maintains the treatment facility so the community does not have to draw up a capital budget or issue bonds. Privatization also provides a transfer of risk, tax incentive advantages, and greater private-sector efficiency. The biggest disadvantage of privatization is loss of control over the facility, which is inherent in this process. Thus, the municipality should spell out in the contractual agreement the areas of control for construction, monitoring, and operation in which they want to participate.

EPA suggests a state revolving fund (SRF) structure to meet wastewater infrastructure needs and to provide a mechanism for communities to establish a structure for long-term financing of future wastewater projects. Other funding alternatives include Infrastructure Banks, which have been introduced in New Jersey and in Massachusetts, and increasing the role of subsidies in providing and maintaining environmental facilities.

The key to dealing with prioritization and funding—knowing what to do and when to do it—is management. The deterioration of existing infrastructure shows that this is perhaps the least developed area of the problem. Development of expertise in management areas such as program planning and capital budgeting, financial management, executing needs surveys and condition assessments, decision making, development of maintenance information systems and rehabilitation, and repair and replacement programs is essential in providing a solid effective approach to solve infrastructure problems [10]. Beyond these, the ability to present convincing arguments for maintenance and reconstruction of existing infrastructure is essential to guarantee success.

Once priorities are established, funds obtained, and management strategies developed, the programs must be implemented. Successful implementation requires people educated in all aspects of infrastructure maintenance, renovation, and construction. As early as 1972 [12], engineers were warning that there were manpower shortages and that they would get worse. In addition, a number of studies have predicted that there will not be enough environmental engineers to handle infrastructure problems. Consequently, there is a need for significant, short-term programs, for refresher courses, and for long-term commitment at all levels to satisfy the need for water pollution control engineers.

On July 18-19, 1985, the National Academy of Engineers Education Advisory Board and a group of the Deans of Engineering met to discuss engineering education and practice in the United States, particularly the problems that engineering education faces relative to America's technological and economic future. Both groups agree that engineering education is facing a critical problem. They further agree that the political and social structure of the United States is ignoring this problem and that the problem is fast reaching crisis proportions. To achieve the kind of technological and social future we want for our society, we must improve our educational systems to produce both the *quantity* and the *quality* of engineers that are needed. Currently, engineering enrollment is stagnating, not only because young people are not entering the field, but because many of the universities have limited their enrollment due to inadequate amounts of equipment, space, and faculty. The trend toward starting new engineering schools seems to be a guarantee for *mediocrity* since the new schools face exactly the same problems of faculty shortage, equipment, money, etc., and will have to develop solutions from the same pool that existing schools are using.

This document wants to address not only the general problems of engineering education but also the more specific one of how civil engineering education can meet the challenge of environmental infrastructure needs. It appears that engineering schools are currently not meeting these needs very well and that the situation is getting worse.

The truly critical problem is how to convince educational institutions and society as a whole of the need for planned maintenance and revitalization of the environmental infrastructure. It appears likely that catastrophes will have to strike several times in several cities before this issue becomes vital to the public. An occasional collapse of a bridge or failure of a plane seems to keep society aware of potential hazards in these areas. However, the public has not recently seen the catastrophic failure of a water distribution or sewage treatment system. The environmental engineers have done our work too well; as a result, it has been invisible. It has produced a safe environment, and the body politic has not awakened to the impending crisis.

University administrations have little concern for future infrastructure problems. This lack of concern is related to the way in which universities are organized and the ways various academic elements within a university are viewed by senior administrative officials. At engineering colleges, all university administrators have been forced to meet progressively higher teaching loads with fewer facilities, older equipment, and less per-student funding. Furthermore, in most universities, central administrators are rarely engineers, and they strongly resent the cost of engineering education. The statement is often made that engineers have more space than other faculties on campus, ignoring the extremely high needs for laboratories.

Within the current framework of higher education, both deans of engineering and chairpersons of civil engineering departments must first be concerned with maintaining and funding a base that assures continuance of a quality undergraduate program within Accrediting Board for Engineering and Technology (ABET) criteria. ABET must find some way to guarantee some sort of a generalized undergraduate engineering education that includes the environment. It is clear that the four-year undergraduate program can no longer produce a civil engineer who can handle all the problems that he/she will face. We must either consider having a five-year program or having a general undergraduate civil engineering degree with an MS being the first professional degree and, perhaps finding some way to include such requirements in the licensing process of each state.

It appears that the infrastructure of civil engineering education, as well as the entire engineering educational system, is in trouble all over the country. With declining resources, we can no longer continue to add to already overloaded curricula. It is apparent that we must soon consider either eliminating civil engineering as a discipline and offer "specialty" engineering education much earlier than we have ever wanted to; or, we must consider offering a broad civil engineering degree, that would be fairly uniform among all schools, with the first specialty level at the Master's Degree level. Under such a system, universities would have to start funding and treating engineering schools as they do schools of law and medicine.

CURRENT EFFORTS TOWARDS A SOLUTION

Current efforts toward solving infrastructure problems can be categorized as congressional, state, local, and individual. The National Infrastructure Act, which was introduced in June 1984 but not enacted, was a Congressional attempt to deal with the infrastructure problem. The bill provided for the creation of a $30 billion infrastructure fund with allocations of $3 billion per year for 10 years. The money would be made available, interest-free, to states and local governments. Repayments would be placed in interest-bearing accounts with the interest creating a permanent fund for infrastructure needs. The Act would have provided a combination of subsidy and long-term debt [2].

Public Law 98-501, the Public Works Improvement Act of 1984, and the Water Research Foundation Act of 1985 also address infrastructure problems. Public Law 98501 defines the resources needed to upgrade and maintain the infrastructure at a level needed to sustain economic growth.

The Water Research Foundation Act of 1985 (Senate Bill 904), proposed by a group of bipartisan senators, would create a foundation for research planning and for dissemination of information. It would also have the authority to make grants for nonFederal research in the public and private sectors.

Several states have sought solutions to infrastructure problems. During 1984, Michigan spent $1.3 billion on 1700 infrastructure projects [9]. To comply with Clean Water Act requirements, the state's total wastewater needs are estimated to be $3.7 billion with $1.8 billion needed for their wastewater treatment facilities. The water and sewer department of Miami, Florida is working to maintain and rehabilitate its treatment systems using a budget renewal, replacement, and rehabilitation fund [9]. New York has established a $1.25 billion "Rebuild New York" program [13]. Pittsburgh has assessed its infrastructure needs and California has formed an Infrastructure Review Task Force which is making recommendations to the state [13]. Florida has introduced two constitutional amendments addressing the financing of infrastructure needs, and Montana has established a Reconstruction Trust Fund to rehabilitate public works [13].

SUMMARY

Over the past four decades, much progress has been made in environmental infra-structure development, and there have been substantial gains in efforts to protect both human and natural environments. Public water supplies have been expanded to provide most of the population with clean drinking water. Treatment of public and industrial waste sources has yielded an improved air, land, and water environment. However, a number of circumstances are combining to compromise the future of these successes, and it is these issues that are of major concern.

With the building of the environmental infrastructure, little thought was given to systematic preventive maintenance, major renovation, and ultimate replacement. Furthermore, few institutions established sinking funds or other financial instruments to provide capital funding for replacing a system at the end of its useful life. Such approaches have resulted in a large project budgetary shortfall over the next four decades for maintaining and enhancing the environmental infrastructure.

Institutional pressures have been applied to further enhance many aspects of environmental quality. Unfortunately, many of these pressures have been misdirected due to the lack of technical expertise among those who write and enforce environmental regulations. These approaches have misallocated resources by funding programs with exotic solutions to problems with poorly documented impacts at the expense of maintaining and renovating viable existing systems with better known merits.

Many of the recent environmental laws have generated institutional disincentives for progress. The laws typically require a public agency to promulgate regulations as a framework for compliance. In so doing, regulations have become increasingly more technology-specific, thereby forcing use of existing technologies rather than encouraging development of new, more innovative solutions.

Perhaps the largest problem to be addressed in overcoming infrastructure problems is not money, manpower, or institutions, but the adjustment of public opinion. The American public is voting down many issues that will require any additional expenditure from them, no matter how small. Citizens must be educated about the seriousness of the problem and its far-reaching ramifications. In general, the American public is unaware and often apathetic or even antipathetic to the problems. It is hard for citizens who have enjoyed safe, clean, adequate water supplies to believe that both the quality and quantity could decrease. It is also difficult for the general public to be concerned about treating or disposing of wastes, since most people do not even think about their wastes once they leave their homes.

Public recognition and understanding of infrastructure problems and the whole-hearted support of technical and financial endeavors to solve these problems are of vital importance. Public support and backing will help pass the bond issues and legislation needed to raise funds. However, developing citizen support for infrastructure needs is a problem. It should not take a major disaster, such as a dam failure or a water borne disease outbreak to capture the public's attention. It is the responsibility of the professional and technical societies and of community leaders to educate the public and provide them with examples of the serious ramifications of infrastructure failure.

Without public consent and understanding, the difficulties encountered in maintaining, upgrading, and expanding the nation's environmental infrastructure will be monumental. Only with the public's help can the nation supply the expertise and training, the manpower, and the financial backing needed to sustain the quantity and quality of services prerequisite to a safe and productive environment.

If the status quo is maintained, the state of the environment and the associated environmental infrastructure will decline. It is suggested that environmental quality is now declining after four decades of improvement. Furthermore, maintenance of the existing status quo will only accelerate this decline over the next decade. Unless we are able to redirect the status quo relative to funding, public policy, and institutional direction, the environmental infrastructure and the associated well-being of society will diminish.

ACKNOWLEDGEMENTS

This paper was prepared by the ASCE Environmental Engineering Research Council. Many members of the council contributed ideas, time, and effort. Special thanks go to the following people for their work.

1. Dr. John Bandy, U.S. Army Construction Engineering Research Laboratory.
2. Dr. Steve Field, Louisiana State University.
3. Dr. Charles Jennett, Dean of Engineering, Clemson University.
4. Dr. Joe Middlebrooks, Provost and Vice President, Tennessee Tech. University.
5. Dr. Clifford Randall, VPI University.
6. Dr. Ed D. Smith, U.S. Army Construction Engineering Research Lab.

REFERENCES

1. National Infrastructure Study, "Hard Choices," Prepared for the Joint Economic Committee of the United States Congress (1984).
2. Boland, J. J., "Alternative Solutions," *Water Engineering and Management*, Vol. 132, No. 5, 33 (May 1985).
3. Lippy, E. C., et al., "Waterborne Disease Outbreaks—1946–1980—A 35 Year Perspective," *J. American Water Works Association*, Vol. 76, No. 2, 60 (1984).
4. Anon., "The Decaying of America," *Newsweek* (1982).
5. U.S. EPA, "Report to Congress: Waste Disposal Practices and Their Effects on Groundwater," Office of Water Supply, Office of Solid Waste Management Programs (1977).

6. Vesiland, P. A., and A. E. Rimer, *Unit Operations in Resource Recovery Engineering*, Prentice Hall, Englewood Cliffs, New Jersey, 452 (1981).
7. Davis, J. C., "Priorities, Opportunities and the Infrastructure Task," *Water Engineering and Management*, Vol. 131, No. 1, 12 (January 1984).
8. Anon., "ASCE Official Warns of Overworked Water Infrastructure," *AWWA Mainstream* (1984).
9. Langworthy, V. W., "How Big is the Problem," *Water Engineering and Management*, Vol. 132, No. 5, 13 (1985).
10. Grigg, N. S., "Interdisciplinary Research Needs Related to Physical Infrastructure," Paper for Office of Interdisciplinary Research, National Science Foundation (1984).
11. Westenhoff, G. P., "Privatization Needs the Professional Approach," *Water Engineering and Management*, Vol. 132, No. 2, p. 25 (1985).
12. Middlebrooks, E. J., et al., "Manpower Needs of Consulting Engineering Firms," *JWPCF*, Vol. 44, 1865 (1972).
13. Anon., "Infrastructure Money and Methods," *Civil Engineering ASCE*, 62 (September 1984).
14. Crook, J., "Water Reuse in California," *Water Reuse Symposium III*, San Diego, California (1984).
15. Lauer, W. C., "Denver's Potable Water Reuse Project—Current Status," *Water Reuse Symposium III*, San Diego, California (August 1984).
16. Maddeus, W. O., et al., "Reducing Water Demand and Wastewater Flow," *J. American Water Works Association*, Vol. 75, No. 7, 330 (1983).
17. Wright, J. D., et al., "A Stepwise Approach to Environmental Systems Maintenance Management," ASCE Environmental Engineering Specialty Conference, Boston, Massachusetts (July 1985).

INDEX

Printed and bound by CPI Group (UK) Ltd, Croydon, CR0 4YY
22/10/2024
01777622-0017